PROJECT MANPOWER MANAGEMENT

PROJECT MANPOWER MANAGEMENT

DECISION-MAKING PROCESSES IN CONSTRUCTION PRACTICE

S. D. Anderson
Fluor Technology, Inc.

R. W. Woodhead
University of New South Wales

A Wiley-Interscience Publication
JOHN WILEY & SONS
New York • Chichester • Brisbane • Toronto • Singapore

Library of Congress Cataloging in Publication Data:

Anderson, S. D. (Stuart D.)
Project manpower management.

"A Wiley-Interscience publication."
Bibliography: p.
Includes indexes.
1. Construction industry—Personnel management.
2. Industrial project management. I. Woodhead, Ronald W. II. Title.

HD9715.A2W647 1987 624′.068 86-26728
ISBN: 0-471-86189-8

Printed in the United States of America

10 9 8 7 6 5 4 3 2 1

Preface

This book is concerned with the construction industry's approach to project manpower management. The material covered includes the labor component of a project and the decision processes that relate to its determination and management. Unlike most books on project and construction management, this does not include such typical topics as equipment and materials management, except insofar as these directly impact the labor component management area. Our reasons for this apparently limited scope are many and are directly related to the importance of the labor component and the level of detail that we feel is necessary for its proper structuring and understanding on an industry-wide basis. Thus, we have covered the area in a logically linked and interactive manner from project initiation to completion. We have, for example, included discussion of crew planning and crew management, which emerges naturally in our treatment as field activity converts plans into construction in place. In order to accomplish our objectives, we have limited our efforts to the consideration of the management of small and large work groups and exclude details on industrial relations and the behavioral science aspects of individuals and personal interactions.

Our general approach to the topic reflects our assessment of the management role and challenge. In our view management is concerned with the detection, evaluation, and resolution of problems. Managers make decisions and implement solutions and strive to achieve cost, schedule, and quality goals. This role especially challenges the manager when new problems emerge that are not adequately addressed by traditional procedures and techniques. In these cases managers must develop different approaches to decision making that often require creative and lateral thinking. Thus, the manager must be constantly aware of the dangers associated with being confined to traditional modes of operation and yet fully understand the significant effort required to change existing procedures and attitudes.

The writing of this book has been a considerable challenge to us personally. We have striven to provide a reference framework and methodology from which to view the entire spectrum of management effort relevant to the labor component area in the U.S. construction industry. At the same time, we have provided details on management and supervisory roles on a technical and professional scale ranging from the foreman to top management. In accomplishing this, we have collected in this text the most complete set of manpower-related management functions, policies, documents, and management processes that exist. Consequently, the text has a number of unusual features that distinguish it from other books on construction and project management.

The entire management process is captured in twelve independent chapters which are logically linked and integrated. Each stand-alone chapter corresponds to an area of professional and supervisory responsibility. The structure of these chapters is unique in the manner of developing a problem scope statement that leads to procedures, management functions, and to concepts and techniques relevant to decision making. The techniques emerge when concepts are applied to the description of current management approaches in practice. Each chapter enables the reader to look at one specific decision area from a variety of viewpoints and to ascertain how and why different management approaches emerge. Although the book relates concepts to techniques the scope of the book prohibits a full description of current techniques used in practice. A bibliography, organized by chapter, is provided to encourage the reader to follow up on details relevant to specific techniques.

The stand-alone chapters are also effectively linked by four continuing presentations that characterize different management approaches. Thus, the reader can concentrate on the small firm management approach by selecting the third section in each decision process chapter. Similarly, section four in these chapters relates to the modus operandi of the medium-sized firm, and sections five and six to the management approach of the large firm. It should be emphasized that the contractor is not necessarily confined to the management approach that corresponds to its size. Depending on the project type and its characteristics, the contractor may select an approach that is different from the contractor's normal practice. Furthermore, the contractor may operate within a given management approach while, at the same time, using some of those techniques more commonly associated with another management approach.

Each stand-alone chapter can be, and is, linked in a variety of ways that reflect the flow of information and documents through the different management phases of a project. The manner in which these chapters are integrated is highlighted in Chapter 16, Project Management Overview. This chapter also stresses schedule and cost control linkages. Additionally, an extensive set of management policies is provided for each management approach. These policies provide management directives and pressures for the ways in which management functions and decision processes are to be performed and effectively linked and integrated.

A significant portion of the text is devoted to document and information flows, detailed decision process networks, and Management Information System formulations of decision-making processes. For example, four complete sets of labor-related documents are presented together with 55 decision process networks and over 300 figures that represent the structures and details of the decision processes involved in project manpower management. We feel that this material gives the book a unique feature and provides a comprehensive and rigorous description of project and construction management.

The book can be used in a variety of ways. Those interested in a specific disciplinary area or decision process should first understand the focus, language, and presentation format by reading the first three chapters, and then the decision process chapter of interest. A reading pattern of Chapters 1, 16, and 17 should enable the reader to grasp our concept of the management challenge. Some readers may prefer to concentrate on information flows and related documents or on the Management Information System formulations, especially since much of the material incorporated within each decision process network is not specifically handled in the text. For such readers we recommend a reading pattern of Chapters 3 and 16, and then specific portions of the text selected using the document listing and pagination structure of the Appendix. The problem-focused design and nature of the index can aid the reader in following through a selective reading on a management problem.

In a book such as this that has been long in preparation, it is difficult to adequately acknowledge the contributions and assistance that we have received. Various portions of the book have been read by management personnel in small, medium-sized, and large contracting firms and client organizations that have in-house construction divisions. A number of academics and several professional researchers have contributed useful suggestions and often pointed criticisms. Finally, we wish to thank the various secretaries and typists who have provided back-up support.

It remains to acknowledge our great debt and appreciation to our wives, Deborah France Anderson and Kathleen Anne Woodhead who have patiently and faithfully supported our endeavors over so many years.

STUART D. ANDERSON
RONALD W. WOODHEAD

Irvine, California
Kensington, Australia
October, 1986

Contents

PROJECT MANPOWER MANAGEMENT

CHAPTER 1

Manpower Management; A Decision Process Approach

Construction is the creative activity that transforms physical resources into useful structures. The basic resources involved are *materials, machines, manpower,* and *money*. The effective and coherent development of any construction effort, however, also requires the authority and direction of a management team. Construction objectives are not achieved without directed effort, but must be planned and striven for over a considerable period of time. Consequently, the effectiveness of field activity depends upon the quality of management's leadership and ability to motivate the work force. It also depends on the skill and effort of the management team in the procurement and use of resources. Thus, *management* must be considered the essential integrating resource in construction.

The construction process itself consists of the fabrication and field erection of specially designed facilities at specific locations defined by the customer-owner. In realizing this goal a *project management* relationship develops between the customer-owner and the contractor involved in the construction process. Construction is, therefore, both field and *project* oriented.

In construction, as in any other enterprise, people are an indispensable resource. Construction is an inherently satisfying activity for those involved, because they can readily see the results of their work effort and can easily identify with the physical product. People are very complex and often unpredictable, however. It is not surprising, then, that the *management of people* consumes a large portion of the management effort on any construction project. This interest in the management of manpower is necessarily broadened to incorporate all aspects associated with the management of the *labor component* of the construction project. *Project manpower management* is, therefore, a complex process impinging on all phases of a project as they are relevant to the labor component. This holistic approach is reflected in the many different *decision problems* that management must address and resolve, and in the variety of ways that decisions can be reached and implemented.

This chapter introduces basic concepts of project manpower management, and suggests a framework for viewing the project manpower management decision-making processes that are currently used in construction practice.

1.1 THE SCOPE OF MANPOWER MANAGEMENT

The management of the labor component of a project, referred to hereafter as *manpower management,* spans such diverse concerns and activities as (1) the management and behavior of individuals, (2) the establishment and management of temporary work groups (such as crews and design or project team sections), (3) the size, acquisition, and effective use of the entire project work force, and (4) the relationships that exist and develop between management and organized labor. These manpower management concerns can be readily divided into the following four major hierarchical areas:

1. Human relations
2. Personal management of labor
3. Impersonal management of labor
4. Industrial relations

The hierarchical nature of these four management areas reflects a progressive growth in the organizational ways in which people are managed while at the same time portraying a progressive loss of individual identity. These areas are described as follows as they relate to construction management:

1. *Human relations*. This most fundamental human management area is concerned, first, with those aspects of individual human behavior that relate personality and skill level to the requirements of the work environment, and second, with the detection and resolution of difficulties that arise between individuals.
2. *Personal management of labor*. This management area is concerned with face-to-face contact with and management of small work groups. It is here called *personal management of labor* because the relevant field managers and supervisors often know the individual members of the work group by name, skill level, and reliability, and therefore strive to develop a close rapport with the group as a whole.
3. *Impersonal management of labor*. This management area involves the decision processes and management effort associated with the planning, scheduling, enumerating, and performance monitoring of large groups of personnel at the field work force, design group, and project management team levels. It can be considered as an *impersonal* management effort to the extent that manpower totals and overall performance indicators are considered in decision processes.
4. *Industrial relations*. This last management area concerns itself with the contractual and jurisdictional aspects of dealing with craft and organized labor groups, the availability of skilled labor, and the terms under which individuals are hired, paid, and under which they work. In the majority of construction projects industrial relations guidelines are negotiated and/or established before construction work begins. These guidelines thus become constraints within which all other management areas operate.

Each of these four manpower management areas has been described, at least in broad outline form, in terms of areas of management concern. Each area has also been focused on representative construction management problems at the work face, crew, and project levels. Using these management areas and descriptions as a frame of reference, the scope of manpower management can be outlined in terms of the decision processes addressed, and by the techniques, documents, and information flows that comprise these decision processes.

The project manpower management decision processes considered in this book primarily address the personal and impersonal manpower management areas described above. While they are naturally affected by both human relations and industrial relations, the last two manpower management areas will not be considered in any detail beyond the ways in which they affect personal or impersonal manpower management efforts. Using this approach, one can define manpower management in terms of all the decision processes relating to:

1. The determination of the size of the project work force (whether field work force, field management team, or head office design groups)
2. The acquisition of labor and the growth cycle of the project work force over the life of the project
3. The structure and breakdown of the project work force into work groups such as crews and management sections
4. The sizing and composition of specific crews in relation to the work content of project activities and the work assignments given to individual crew members
5. The planning, scheduling, directing, monitoring, and control of manpower activity at the work face, crew, and project levels.

Finally, by analyzing the decision processes associated with various aspects of project manpower management for a particular project, a company's mode of operation can be identified and described.

1.2 MANPOWER MANAGEMENT APPROACHES IN CONSTRUCTION PRACTICE

A variety of *management approaches* to project management in general, and to project manpower management in particular, exists in practice. These different management approaches have evolved as a result of the diversity of needs and pressures placed on project management by the construction and management environment of the client, the project, and the contractor.

Different management approaches emerge as a result of a construction environment that can be characterized, for example, in terms of projects of vastly different size, type, and complexity. These characteristics dominantly influence the nature of the labor component of any given project, and hence, the requirements for manpower management.

The manner in which a contractor solves the problems created by the separation between its head office and its various project work sites significantly influences the management approach it will adopt for a particular project. A common and practical solution to the problem of separation is for the contractor to locate all general business functional groups in a head office or a conveniently located regional office together with relevant professional project-oriented staff support groups, and to locate all field construction and field management functional groups in the field at the project sites. Individual projects are then integrated into the contractor's operations through the performance of project manpower management functions. These project management functions can be performed by head office or field-based personnel, or by an employee who is both head office and field based. Each alternative exists in practice and creates its own unique project management environment. The selection of a particular management form is a critical decision significantly influencing the management environment, the project team structure, the professional and decision-making roles of the project team members involved, and the functions performed by field management personnel.

Finally, the extent of the client's participation in the management of a project, especially when coupled with contractual obligations, may necessitate manpower management techniques that normally exceed the standard management practices employed by the contractor.

Each project, from the contractor's point of view, has a life cycle with recognized birth, growth, decay, and death phases. Each life phase has its own management, organizational, and decision problems, and involves different professional and functional groupings of company employees. In this management environment the company must continuously spawn project management organizations to support field effort during project start-up and field construction. It must also reabsorb key personnel back into the head office (for later reassignment) while punch lists, wrap-up, and hand-over aspects of the project are being finalized. These management and organizational problems are magnified for remote or inconveniently located projects.

The different types of manpower management approaches that exist in construction practice can be described in terms of the relationships between three specific decision phases of a project, namely, project definition, field construction, and project management. The first phase, project definition, consists of the determination and evaluation of the anticipated labor content of a project. This phase establishes labor budgets and translates these budgets into manpower utilization plans over the life of the project. The field construction phase focuses on the planning, direction, and control of the field work force effort during the period of construction. This phase provides the field management format for manpower management, and establishes the professional and work motivation environment that is essential for any effective coordinated work effort. The final phase, project management, concerns itself initially with monitoring the manpower involvement in the project, and the effectiveness of labor performance. Using the resulting information and knowledge of current project status, project management can focus on the implications that follow from project projections and on the determination and implementation of corrective and remedial management action. These last activities, analysis and corrective action, are the main purpose of the project management phase.

When these three major components of management effort are analyzed in terms of when (timing) and where (field or head office) they are performed during the life cycle of a project in relation to the characteristics of the construction environment, distinct manpower management approaches emerge. For example, small projects constructed

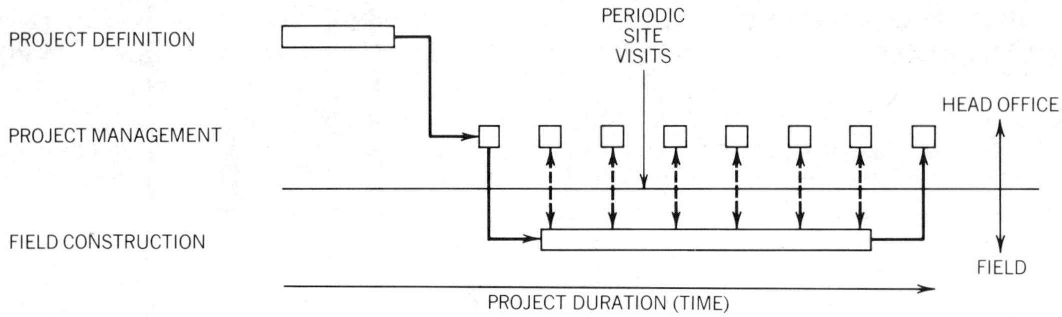

Figure 1.1 Project Team Component approach.

locally would generally result in the sequential execution of project definition, field construction, and project management, because the problem of the separation of the head office from the field site is virtually a nonissue. Alternatively, projects located in remote areas will ultimately require at least a partial shift of all three phases to a field location in order to strengthen the field management effort.

Four different management approaches can be readily identified. They are listed as follows:

1. The Project Team Component (PTC) management approach
2. The Traveling Project Manager (TPM) management approach
3. The Field-Based Project Management management approach
4. The Total Field Autonomy management approach

A schematic representation of the Project Team Component management approach is presented in terms of the three project phases (here represented in bar form on a project life time scale) in Figure 1.1. The major sequential flow of management linkage between these project phases is illustrated by single lines, as is the periodic project management input (site visits by the project manager whose major manpower management role is in the head office). Whenever the project manager faces a management overload situation, in terms of the number of projects to be managed and other business and management involvement, the management linkage between the three phases shown in Figure 1.1 tends towards the purely sequential interaction stressed in the PTC management approach formulation.*

In the Traveling Project Manager management approach, an attempt is made to strengthen the management linkages between the three management phases. This is achieved by introducing a new type of project manager who is head office based, but whose major manpower management role is exercised in the field. He or she accomplishes this by continuously cycling out to the project field site to meld together the project definition and field construction management phases. As shown schematically in Figure 1.2, the project management phase is both head office and field located and is shown as a continuous bar, much larger than that used in Figure 1.1 to indicate the PTC approach. The result of this TPM management approach is a much greater management capacity in the field, and an effective linkage between head office and project site that enables larger and more complex projects to be handled effectively by the construction company.†

*The small construction firm operates in a local area in a highly competitive bidding environment under typically self-generated business and management constraints. This tends to limit total construction dollar volumes. Management effort is inherently limited by the small number of permanent employees and management personnel. These staffing and management capacity constraints can be partly overcome if field staff are encouraged to complement the limited management potential of the firm by delegating to them almost full responsibility for field construction activities. This management compromise encourages the emergence of the PTC management approach.

†Medium-sized construction firms secure some of their projects through owner negotiation and most others through competitive bidding. The projects in this environment generally require higher levels of technical competence, capitalization, and management effort than those normally associated with small firms. General building and heavy or highway construction are the most common types of construction projects performed by medium-sized firms. The number of permanent key salaried personnel is substantially larger than that of the small firm. A significant percentage of these permanent personnel are either superintendents, project managers (TPMs), or head office support staff. This increase in management capacity is related to the increasing size and complexity of the company's projects and leads to the emergence of the TPM management approach in medium-sized firms.

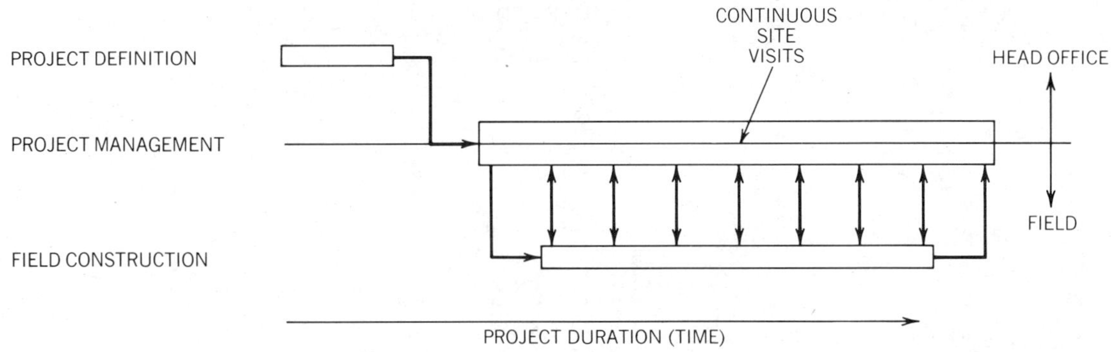

Figure 1.2 Traveling Project Manager approach.

At a certain level of project size and complexity, it becomes desirable, and for remote sites essential, to both strengthen field management effort and broaden project team responsibilities. This management need stems from the requirement to cope with an increased level of detailed project information, and the desirability of increasing the effectiveness of field management by relieving key field personnel of management function overloads. A characteristic management solution is to assign relevant project-oriented head office manpower management functions for field based execution by incorporating into an enlarged project team structure personnel capable of performing the transferred head office functions. In this way, a complete project team structure emerges. It is characterized by dual lines of communication; internally between the field based project management team headed by a full time on-site project manager, and externally to each relevant head office functional area. Since more manpower management functions are shifted and are staffed at the field level, a more balanced and capable field organization emerges. This management approach, called the Field-Based Project Management management approach, is shown schematically in Figure 1.3.

In certain project environments it becomes desirable to strengthen the position of the project team by creating a team that is fully integrated and autonomous in terms of its project management approach. A common management technique for achieving this goal is to assign complete responsibility to the project team and provide it with the freedom to manage the project with total autonomy. This means that the field is capable of executing all manpower management functions that were previously based in the head office, including engineering and design functions as required. The project team

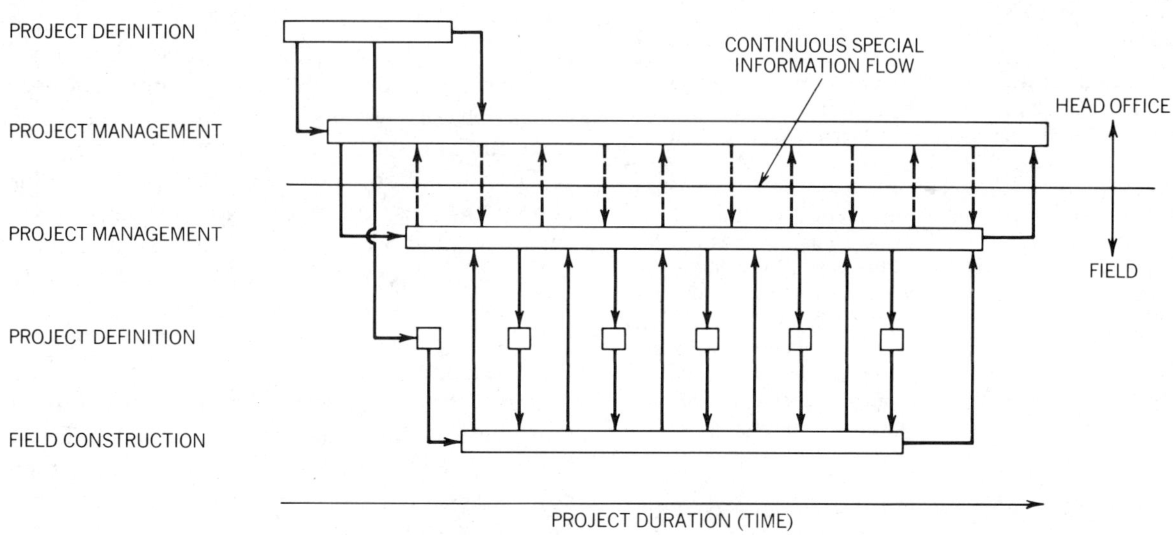

Figure 1.3 Field-Based Project Management.

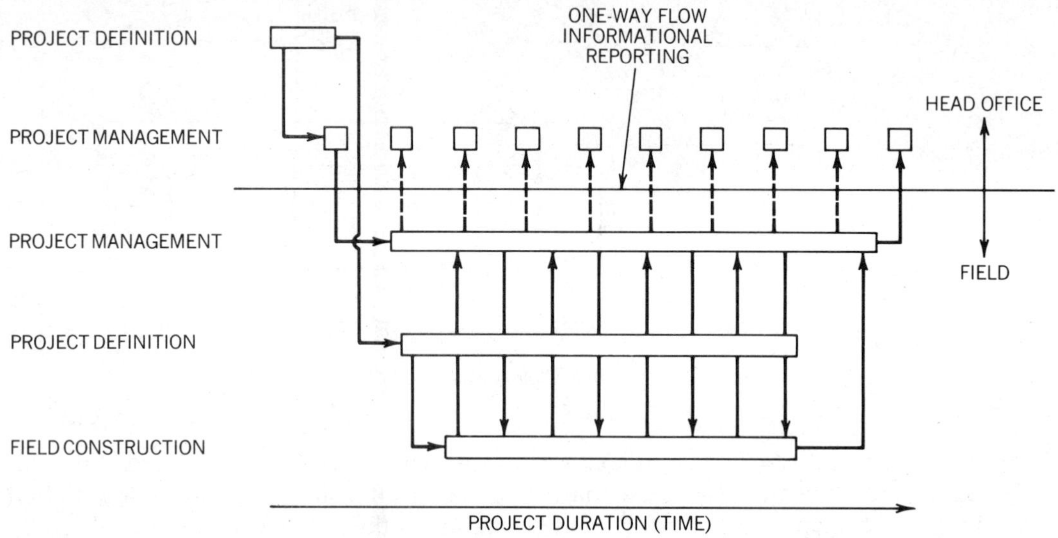

Figure 1.4 Total Field Autonomy.

is enlarged at both the head office and field levels, with the head office staff transferring to the field as the field project organization increases in size and as field project execution intensifies. Without exception, lines of communication between the head office and the field must remain open in order to keep executive company management apprised of project status and progress. The overall project manager is often an executive, however. This gives him increased authority to act, especially when he is quartered at the project site. In this way the project management approach moves closer to the Total Field Autonomy concept. This management approach is shown schematically in Figure 1.4.*

*The large firm construction environment can be characterized by: large and complex projects that are technical in nature, such as a chemical processing plant; projects that are often located in isolated areas and/or near client offices and operational areas; large project labor forces that are often required, including many highly specialized crafts; and an engineering, design, and construction life cycle of up to five years or possibly greater. Projects of the magnitude and complexity associated with the large firm environment dramatically change the organizational structure and project approach, making it quite different from that found at the medium-sized firm level. The number of decision makers increases sharply and each becomes specialized within a given functional area. The client is often extensively involved in the project management process with a project organization structure of its own. This can place considerable management pressure on the project team approach of the large firm. The large firm environment generally fosters the Field-Based Project Management and Total Field Autonomy management approaches.

The four project manpower management approaches described herein form the basis in this book for viewing the different decision-making processes currently found in construction practice at the small, medium-sized, and large firm levels.

1.3 DECISION PROCESS METHODOLOGY FOR PROJECT MANPOWER MANAGEMENT

Managers often make decisions and must subsequently ensure that their decisions are implemented. Management decisions become necessary whenever problems are identified or anticipated, and a solution is required. This act of making decisions can be defined as a process. Thus, *decision processes* can describe management action and a collection of decision processes, when problem focused, can portray a management area.

The fundamental nature of the problem identification and resolution decision process is shown schematically in Figure 1.5. A first step is to formulate the problem in a *problem statement* form. Then, one gathers the relevant data and information for the appropriate decision process technique so that the solution can be found. As is often seen in practice, there are a variety of possible solutions for any given problem, and this is reflected in the different techniques that can be used in the decision process. These techniques in turn have their attendant different requirements for input data and information. Thus, a general decision process can take dif-

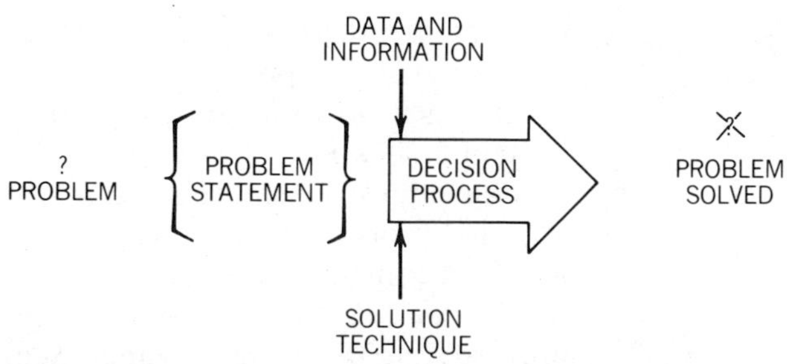

Figure 1.5 Problem Identification and Resolution decision process.

ferent technical forms that also often reflect different management approaches to problem solution.

The scope of the project manpower management area can be indicated and described in terms of a collection of specific problem statements and the decision processes that ensue. These decision processes are commonly met in construction practice. As shown in Figure 1.6 and developed in detail in Chapter 2, the project manpower management area is described in terms of over 300 component problem statements and equivalently in sixty-one problem area statements and twelve basic decision process. For such a framework to be useful for the portrayal of management practice it is necessary to focus on management roles, policies, and styles. Thus, each decision process can be illustrated by describing the manner in which managers implement their decisions through management techniques that are commonly used in practice. The manner in which this is done establishes a methodological approach to decision processes in general. The elements of the proposed methodology are developed in Chapters 2 and 3. The methodology is illustrated in the remainder of the text in terms of twelve decision processes that can be readily identified in practice and which collectively span the project manpower management area.

In developing such a methodology, there is a need:

1. To identify the scope of each decision process that project manpower managers must address
2. To identify and structure the decision processes in terms of manpower management functions that must be performed (Once this is accomplished, the specific selection of management techniques for the performance of a particular function provides a means for describing a specific decision process in construction practice.)
3. To portray the different roles that managers and supervisors adopt in the making of decisions and in their implementation.

Each decision process can thus be formulated in terms of the management functions that need to be performed during its execution. These component management functions generate, manipulate, and present data and information relevant to the decision. Thus, generic and structured functional formulations can be developed for any given decision process.

A comprehensive set of management functions and major problem decision areas suitable for the description of project manpower management decision processes is given in Chapter 3. Each decision process can be formulated as a specific management information system by introducing document and information flows and enumerating policies. Thus, once specific management techniques and documents are prescribed for the manner in which each decision component function is to be performed, each decision process takes on the character of a completely defined and problem focused information system. The manner in which this can be accomplished is also developed in Chapter 3 and is

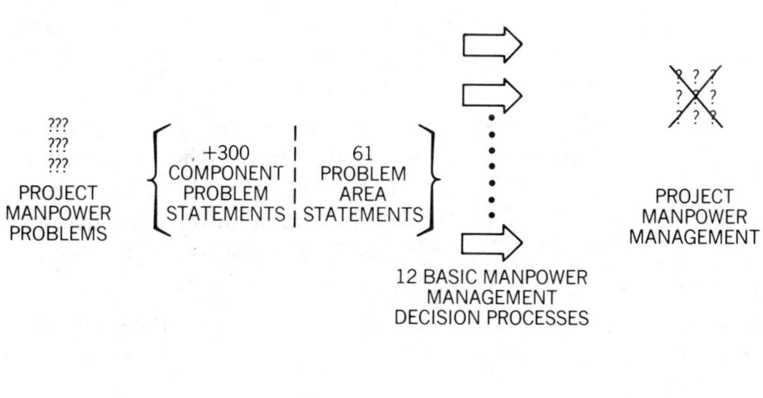

Figure 1.6 Project Manpower Management decision process description.

used later throughout the text for each decision process, where it is brought to bear on the description of the various management approaches that exist in practice.

The third and final stage in the methodological formulation of the project manpower management area is to describe the role of each manager or supervisor in each decision process in which he or she is involved. In this way, the roles of individuals and groups of individuals in an organization can be established. This is addressed throughout the text and more specifically focused in the final overview, Chapter 16.

The overall scope of a decision process can be defined in terms of the major manpower management problem areas it addresses. Consequentially, each of the twelve basic decision processes is introduced by a statement of the problems and basic management issues to be solved.

A common format has been adopted for each decision process chapter (Chapters 4 through 15). This highlights cross references and facilitates the comparative reading and study of the formulation of the scope of different management areas and the emergence of different management approaches to the identification and resolution of problems. Finally, it provides an opportunity for the listing and the evaluation of the adequacy of management concepts that lead to different functional performance techniques, and changing management roles. Each decision process chapter has the following format:

Section 1. *The Basic Decision Process:* This is a broad introductory statement of the scope of the decision process to be considered. It includes a table of major decision areas and their relevant component problem statements, a description of the general decision process procedure in terms of management tasks and duties, a functional statement of the management functions involved in the decision process, and a description of the major linkage relationships with other manpower management decision processes.

Section 2. *Functional Performance Techniques:* This section introduces a variety of management concepts which form the basis for specific management techniques that have emerged in practice as problem-solving methods for each decision process.

The focus on concepts rather than on techniques is both necessary and fundamental to our purpose. First of all, the proper presentation of the many techniques used in practice would be too lengthy to be compressed into a single volume. Secondly, and more importantly, the specific formulation and details of techniques are constraints to the reader in the sense that they require him or her to understand, if not to accept, particular points of view on how to do something. That is, they are inward looking, unlike conceptual formulations that are outward looking and thus imply freedom as to how a concept can be applied. This is especially important in a presentation that highlights *different* management solutions and which encourages the reader to question why and how a special development of a concept has occurred in practice.

In construction practice a number of approaches are used to generate feasible management solutions to manpower management problems. Different solutions emerge because of the widely varying management environments that characterize project management. Each management solution will have its own techniques, with each based on its own set of concepts. The four approaches identified in Section 1.2 have been selected to illustrate different problem-solving decision processes as follows:

Section 3. *The Project Team Component Approach:* This section describes the manner in which the decision process is performed by the small firm using the PTC Approach to project manpower management.

Section 4. *The Traveling Project Manager:* This section describes the manner in which the decision process is performed by the medium-sized firm using the TPM Approach to project manpower management.

Section 5. *Field-Based Project Management:*

This section describes the manner in which the decision process is performed by the large firm using the Field-Based Project Management Approach to project manpower management.

Section 6. Total Field Autonomy: This section describes the manner in which the decision process is performed by the large firm using the Total Field Autonomy Approach to project manpower management.

While four different management approaches have been selected to illustrate each decision process, other management solutions are possible. Thus, it may be of interest to note differences between the illustrative examples and to examine the nature and causes of these differences. The reader should keep his or her mind open and alert for other approaches, or even new management approaches to project manpower management.

1.4 ENTRY VIEWPOINTS

The layout of this book, together with the format of the decision process chapters, enables the reader to read selectively. The book has also been designed to be self-contained, yet the concepts developed have different levels of focus and different potential applications for the reader. This approach provides the reader with the option to select those subjects within the text which most closely correspond to his or her areas of interest, background, and professional experience.

Although the material can simply be read, it has a certain rigor. There are chapters and sections that require careful study if a full and comprehensive understanding of the concepts and subject matter they present is to be gained. The reader should carefully peruse Chapter 1 in order to gain an understanding of the scope of project manpower management. Chapters 2 and 3 are crucial in that they set forth the methodology used to describe project manpower management decision processes. A thorough grasp of these concepts will be of assistance in selecting the appropriate entry point. The multiple entry facet of the book broadens its applications for both the professional and the student reader.

The reader who is interested in quickly grasping the methodological approach and/or in the assessing of its usefulness for application in a management area other than project manpower management could read the first three chapters, then select and peruse any one decision process chapter and conclude with the Project Management Overview and the Project Management Challenge presented in Chapters 16 and 17 respectively. Similarly, the reader who wishes to focus quickly on a specific professional area in project management could also follow the above prescription.

The student or professional interested in an in-depth reading of the decision environment surrounding a project team member could initially focus on specific chapters of interest and then, with the aid of the index, focus on the decision processes that most directly impact his or her own areas of professional responsibility. Such readers may find Figures 1 through 3 in each decision process chapter useful for this purpose. Thus, for example, the project manager may be initially interested in Chapter 4, Initiate Project Action; Chapter 5, Project Planning; Chapter 8, Field Staffing; and Chapter 14, Project Management Control. The scheduling engineer could read Chapter 5, Project Planning; Chapter 7, Project Scheduling; Chapter 12, Field Monitoring; and Chapter 13, Project Status, Analysis, and Reporting; and maybe in Chapter 9, Crew Planning. The superintendent would certainly be interested in Chapter 9, Crew Planning; Chapter 10, Crew Management; Chapter 11, Work Execution; and Chapter 12, Field Monitoring.

Since each decision process chapter presents four different practical management approaches to problem solving which are currently found in construction practice, the point of entry into a decision process chapter may be selective. In other words, the reader may wish to concentrate on the particular approach of the small firm, the medium-sized firm, or the large firm, or may be interested in evaluating the professional differences between the management approaches of these types of firms. In this respect the reader may wish to examine critically the adequacy of the problem statement scope description of a decision process area, and the extent to which each management approach addresses that area. In this way, the reader may be induced to venture into a creative formulation of possibly more specifically focused problem statements relevant to a decision process.

Taken together, the material in each decision process chapter relating to one particular management approach provides a comprehensive description of that firm level. Hence, the reader who is interested in the small firm could focus on the first and third sections of Chapters 4 through 15, whereas for the medium-sized firm, the first and fourth sections of these chapters would be appropriate.

Those readers interested in the design of, and documentation used in, project management information systems will find specific case studies at the small, medium-sized, large, and total field autonomy levels. In this respect, the text contains complete sets of documents and functional network and information flow diagrams for the project manpower management area in each of these firm categories. Because each of these management information systems is focused on meeting the practical requirements of the relevant management approach, they exhibit different degrees of systematic formulation and integration of field labor costing, cost and schedule control systems, and so on, as well as different degrees of manpower control systems focused on the integration of estimating, scheduling, and crew planning. Thus, the reader may wish to study the documents also, and will find the cross referencing of documents with figure and page numbers in Appendix A useful for this. Other readers may be interested in the evaluation of the effectiveness of an information system. They should refer to Chapter 16 for a method that the authors have found helpful in this regard.

Finally, those readers interested in specific manpower management procedures may find it instructive to follow the sequence of management effort through the various decision processes that are relevant, and which are brought into play in the resolution of specific problem areas. Thus, project productivity programs have their genesis in management's initial failure to achieve planned productivities. Program needs are confirmed in Field Monitoring and Project Status, Analysis, and Reporting, and are initiated via Project Management Control. Program implementation is focused on the response of craftsmen and laborers. Therefore, several specifically focused decision processes are involved in management's approach to solving productivity problems. The reader who wishes to follow this or similar types of manpower management related problems may find the problem focused nature of the index useful in guiding him or her through the relevant chapters.

The twelve decision process chapters have been designed to provide many possible entry points. The interested and careful reader will readily locate a point of entry into the text that will address his or her specific personal requirements.

CHAPTER 2
Manpower Management Decision Processes

A construction company's mode of behavior is indicated by the manner in which it acquires, organizes, and handles its portfolio of projects. At any one time these projects are being sought, bid for, added to the company's portfolio, and removed when completed. From the company's point of view, the management of a particular project within its current portfolio of projects may only be distinguishable in terms of its numerical aspects such as dollar size, complexity, or incidence of problems requiring specific top management attention. Nevertheless, a whole range of management functions and decision processes must be focused on each project in its turn.

Considerable insights into the behavior of a construction company can be gained by considering these management functions and decision processes, and by analyzing the way it addresses, evaluates, and implements major decisions relating to policies and its portfolio of projects. By defining the problems addressed by these major decisions, a set of interacting decision processes can be formulated.

This chapter introduces the basic project manpower management decision processes that are considered relevant to the planning and management of construction projects. Collectively they address a wide spectrum of problems ranging from those relating to the determination of estimated project manpower requirements, through those concerning the establishment and effective use of the project work force, to those arising from the correlation of manpower management efforts with the orderly achievement of project goals.

2.1 PROJECT MANAGEMENT DECISION PHASES

In contract construction a number of legal milestones are established for the contractor by the terms of the contract, and by normal contract administration. These include the notification of bid award, the notice to proceed, and the completion date established by the contract time. These last two dates effectively bracket the field construction and general management phases of a project. Other dates of significance to the contractor are those associated with the calling and submitting of bids. These dates limit the project definition phase for most contractors. Thus, these legal milestone dates envelop three major components of management effort, namely project definition, field construction, and project management.

The project definition phase initiated by the *decision to bid* either terminates on *bid submission* in the event the contractor is unsuccessful, or is continued until the project is *issued to the field* in the event the contractor is awarded the project, with additional as-needed effort continuing during field construction. After company management decides to bid on a project, the specific management functions initially encountered in the project definition phase are planning and scheduling. Planning investigates labor availability, evaluates labor productivities, and actuates the estimating function, which is the principal function for determining the project's labor content and total project cost. The scheduling function commits a project to a specific time frame by defining a logical construction sequence and determining the duration of each activity in rough or approximate times compatible with available labor. The result is a schedule that will ensure that the project can be completed within the specified contract time. As part of pre-job project definition, the head office must establish an initial project team and allocate the project to a suitable project manager.

The field construction phase is visibly bracketed by the *issuing of the project to the field* and by the milestone dates corresponding to *substantial completion* and/or the cessation of field construction operations. Specifically, from the point of view of manpower management, field management will define and detail construction methods, schedule and supervise the project's actual work force, allocate manpower to each work task, monitor crew performance in relation to activity progress, make manpower management decisions, and handle labor relations problems such as hiring and firing and the resolution of jurisdictional disputes.

The project management phase, as mentioned earlier, is dependent upon the manner in which the contractor organizes around the problems created by the physical separation of the head office and the project field site. At the small firm level, for example, the primary communicative link between the head office and the field during construction is project information flow (payroll data, construction progress reports, and information concerning quantities in place). This field input is the ingredient required by the project management component to generate project labor cost data and such related information as may be needed by accounting and payroll. The basic output provided by this recording function is then translated into project labor cost reports and company payroll documents, and is also used for updating project schedules. These labor cost reports and revised construction schedules serve as the basis for head office monitoring and overall project management decisions. The basic field reports, for example, either convince management that field progress is satisfactory, or raise questions concerning labor performance that may result in site inspections and field conferences.

A number of manpower-oriented problems must be addressed by management in these three major management decision phases. These problems are encountered on every construction project, and the consequential need to evaluate and solve them leads to the initiation and performance of a variety of decision processes. Many of these decision processes, for example, those associated with estimating and crew management, are the responsibility of specific construction management personnel, namely the estimator and foreman-superintendent. These processes have taken on a universal character. Others are more or less formalized, however, depending on the magnitude of the problems that exists on a given project, and in the management approach of the project team.

2.2 DISSECTION FRAMEWORK FOR MANPOWER MANAGEMENT DECISION PROCESSES

As indicated in Section 2.1, management must address a wide range of manpower focused problems during the various life stages and management phases of a project. In addition, the recurring nature

TABLE 2.1 DECISION PROBLEM AREAS IN PROJECT MANPOWER MANAGEMENT

Major Decision Problems	Major Decision Process
1. What are project management requirements:	DP1 INITIATE PROJECT ACTION
2. What project team:	
3. Who is available:	
4. When available:	
5. How is project team to be managed:	
6. What is project composition:	DP2 PROJECT PLANNING
7. What is project strategy:	
8. What are project manpower requirements:	
9. What is manpower availability:	
10. What manpower policies:	
11. What is level of anticipated labor productivity:	
12. What will be subcontracted:	

TABLE 2.1 *(Continued)*

Major Decision Problems	Major Decision Process
13. What are estimate requirements: 14. What labor productivity: 15. What is man-hour content: 16. What is estimated labor cost:	DP3 ESTIMATE PREPARATION
17. What are schedule requirements: 18. What is best schedule: 19. What are manpower requirements: 20. What is best use of manpower:	DP4 PROJECT SCHEDULING
21. What field team: 22. Who is available: 23. When available: 24. How is field team to be managed: 25. What are work force requirements:	DP5 FIELD STAFFING
26. What is to be done: 27. Who can do it: 28. How will it be done: 29. When will it be done:	DP6 CREW PLANNING
30. How many crews: 31. What are hiring and firing practices: 32. What crew manipulation: 33. What crew direction: 34. What personal management:	DP7 CREW MANAGEMENT
35. What is work environment: 36. What is management environment: 37. What do I have to do: 38. What is the work motivation:	DP8 WORK EXECUTION
39. What is being done: 40. What has been done: 41. What field data: 42. What documentation is required: 43. What workforce is available: 44. What work site control is required:	DP9 FIELD MONITORING
45. How is status and progress defined: 46. What is labor productivity: 47. What is current status and progress: 48. What is planned status and progress: 49. What deviations can be identified: 50. What are implications of deviations: 51. What is impact of work scope changes:	DP10 PROJECT STATUS, ANALYSIS, REPORTING
52. What is project management information system: 53. What is current position: 54. What should be done: 55. What are procedures for implementing decisions: 56. How effective is project management information:	DP11 PROJECT MANAGEMENT CONTROL
57. What data should be kept: 58. How will data be kept: 59. How will source data be gathered: 60. How will data be analyzed: 61. How will data be released:	DP12 HISTORICAL DATA RETRIEVAL

TABLE 2.2 PARTIAL SCOPE OF FIELD MONITORING PROBLEM AREA

Major Decision Problem	Component Problem Statements
39. What is being done:	Who is doing what? Is it being done correctly? Is sufficient skill evident? Are materials, supplies, construction equipment, and tools sufficient to keep the crews busy?
40. What has been done:	Just how much has been done? Has it been done correctly and at an acceptable quality? Who has been doing this work? Is it sufficient? How can the desired output and quality of work be achieved? How will I assess the work that has been done? Can I remember the details, or will I write it down?

of many of these problems, and their significant impact on the contractor's mode of operation has led to the formulation of specifically focused decision processes, and in some cases to the emergence of a number of key professional areas. It is convenient then to focus on identifying the various manpower oriented problems commonly met in construction practice.

In order to establish a framework for the identification of basic manpower management oriented decision processes, it is helpful to list the major decision problems that arise in the project manpower management area. In order to accomplish this, the authors have identified sixty-one decision problem areas that are considered relevant to project manpower management. They are listed in numerical sequence in Table 2.1. Those numbered from 1 through 20 are commonly associated with the project definition management phase, those numbered from 21 through 38 concern field construction, and those numbered from 39 through 61 can readily be identified with the project management phase of project manpower management.

Using these sixty-one major decision areas as a framework, a covering set of twelve basic manpower management decision processes (i.e., four each from the three management phases of project definition, field construction, and project management), is identified (also shown in Table 2.1). These basic manpower management decision processes can readily be observed in construction management. In Table 2.1 they are labeled and numbered from DP1 through DP12 and form the structure for Chapters 4 through 15. Thus, for example, the manpower focused (DP3) Estimate Preparation decision process addresses the four relevant major decision problems listed in Table 2.1, indicating the need to determine estimate requirements (No. 13), labor productivity (No. 14), man-hour content (No. 15), and the estimated labor cost (No. 16). The Estimate Preparation decision process is fully presented in Chapter 6. In a similar fashion, each of the twelve basic manpower management decision processes can be described in terms of the major decision problems it addresses.

Each decision process can be further described by the fine dissection of each of its component major problem areas. Thus, for example, Table 2.2 describes more minutely two of the six major decision problem areas associated with the Field Monitoring decision process (DP9). As shown in Table 2.2, each problem is divided into a number of component problem statements. The major decision problem area, "What is being done" (No. 39), is divided into four component problem statements, and "What has been done" (No. 40) into seven component problem statements.

In like manner, each of the twelve basic project manpower management decision processes has its component major decision problem areas, which are in turn divided into the relevant component problem statements. Such dissections considerably enlarge the scope and increase the detail of each of the decision processes. Thus, the set of twelve ba-

TABLE 2.3 HIERARCHICAL FLOW OF DECISION-MAKING PROCESS DEVELOPMENT

Decision Phases	(3)
Decision Processes	(12)
Decision Problem Areas	(61)
Component Problem Statements	(300+)

TABLE 2.4 GENERAL PROCEDURE FOR FIELD MONITORING

Procedural Step	Management Duties and Tasks
1.	Visual inspection of individual crew member activity, work progress, and quality of finished product by foreman.
2.	Visual inspection of crew performance and resource use by field supervision staff.
3.	Troubleshooting detection of problem.
4.	Preparation of time cards, plant hire reports, quantities in place, percent progress.
5.	Control of field work force movements, equipment movement, and materials requisitions by work force.
6.	Consideration of effectiveness of location of material dumps, tool boxes, and construction equipment, and other factors affecting smooth work progression.
7.	Observation of crew activity and attitude to work, and the effectiveness of the foreman's control of the crew.

sic project manpower management decision processes is obtained from, and can be described in terms of, sixty-one major decision problem areas, and more than 300 component problem statements. Table 2.3 is a graphic representation of this hierarchy.

Finally, it is possible to describe the manner in which to address each decision process by establishing a set of general management duties and tasks. Thus, a general procedure is developed for each decision process. Table 2.4, for example, presents a general procedure for the Field Monitoring decision process by listing a number of procedural steps and their corresponding general management duties and tasks. Similar procedural descriptions are developed for each basic decision process in the following chapters.

2.3 THE MANPOWER MANAGEMENT DECISION PROCESSES IN CONSTRUCTION PRACTICE

As indicated in Section 1.2, a variety of management approaches to project manpower management exist in practice. These different management approaches emerge because of the different construction and management environments in which they operate. They differ especially in the manner in which they solve the problems raised by the physical separation of the field construction site from the head office. In addition, however, these approaches represent solutions to different manpower focused problems in so far as the mix and magnitude of these major decision problems varies with the variations in project size and complexity that generally result from these different construction and management environments. Consequentially, there are significant differences in the execution of the manpower management decision processes for construction firms operating at different levels.

The Project Team Component management approach addresses several basic manpower management problems at the small firm level. Those problems addressed by the head office are the determination of the project labor content and the establishment of work force guidelines for field management. At the field level the foreman-superintendent must establish crew sizes, schedule crews, and solve crew member work relations problems in order to meet the required work load of the project. Finally, head office management strives to monitor work progress effectively, as well as evaluate and revise project scope, schedule, and cost efficiency. The head office must simultaneously actively pursue new projects and perform project definition and business operation service functions.

These manpower management problems can be referenced against the basic decision processes involved in project definition, field construction, and overall project management. These decision processes are in turn referenced against seven basic project milestones common to the small construction project: (1) decision to bid; (2) bid submission; (3) bid award; (4) issue to field; (5) substantial competition; (6) change orders; and (7) contractual completion.

The integration of these decision processes with their problem focus and project milestones produces the general structure of the decision process, which is illustrated in Figure 2.1. The sequential nature of the Project Team Component approach can be clearly identified from the natural progression of the decision processes, beginning with project initiation and continuing through to project management control. Notice that at the field level both the crew planning and crew management decision processes receive revised scope and schedule

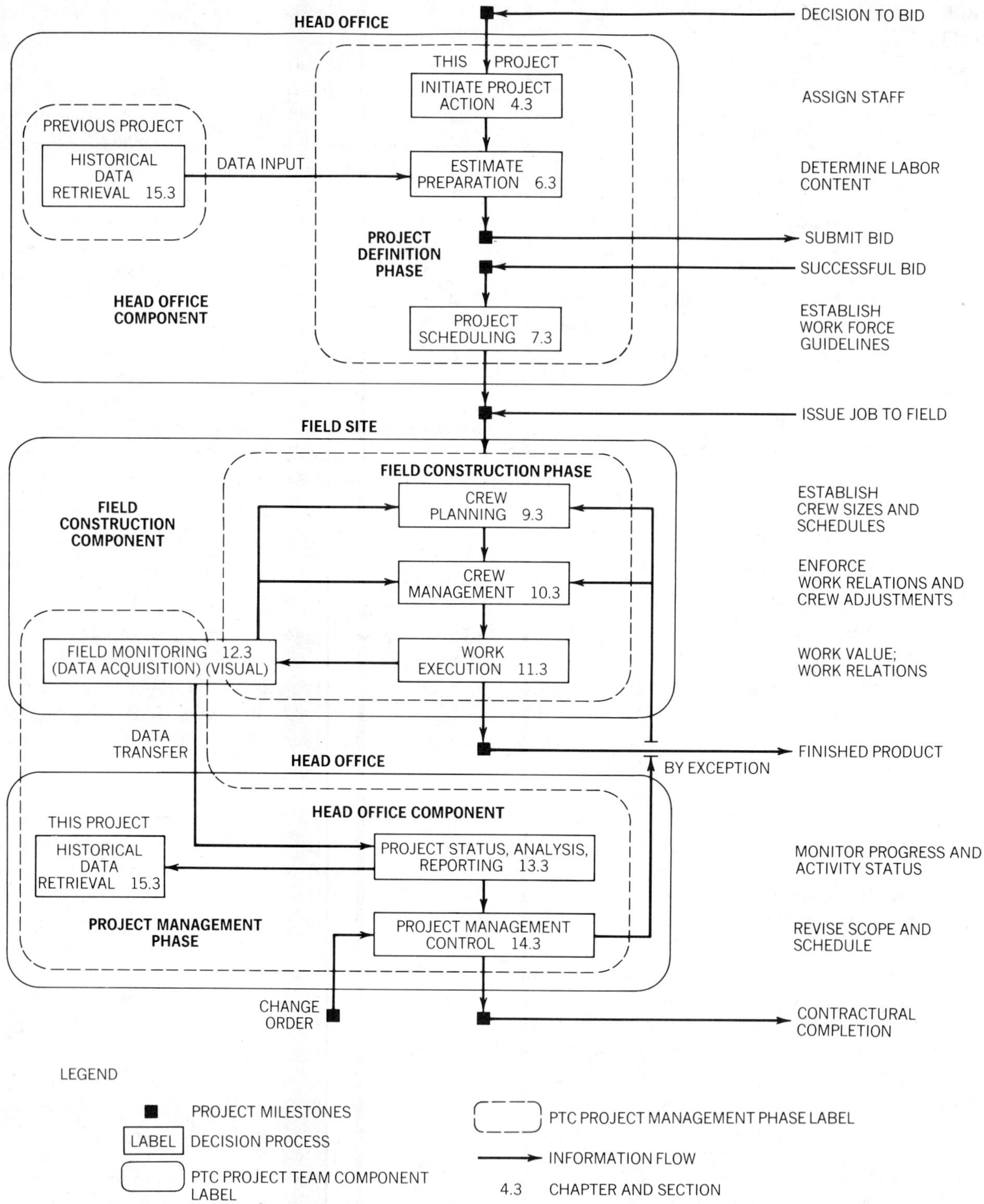

Figure 2.1 Project Team Component (PTC) manpower management decision process.

changes from the head office project management phase.

At the PTC level, only ten of the twelve basic decision processes are required, as shown in Figure 2.1. The two decision processes that are not required (i.e., DP2 PROJECT PLANNING, and DP5 FIELD STAFFING), primarily address manpower management problems associated with projects that have both a large manpower content and significant project team organizations, such as for projects in the large firm environment. Thus, both of these decision processes are essentially nonissues at the small firm level, and are assimilated into other related decision processes. For example, project planning is incorporated into estimate preparation, and field staffing into project initiation.

The Traveling Project Manager management approach found primarily in the medium-sized firm environment focuses on manpower management problems similar to those of the small firm, but significant differences arise in the manner in which they are solved. For instance, those problems initially addressed by the head office remain the determination of the project labor content and the establishment of work force guidelines for field management. To ensure that these guidelines are both acceptable to field management and that they are subsequently carried to the field, the key project superintendent participates in project scheduling. As a result, field staffing emerges prior to project scheduling as a salient manpower management problem in the project definition phase.

At the field level the superintendent and foreman must establish crew size and solve crew work relations problems in order to meet the cost and scheduling objectives for the project. The project manager and possibly other field staff will often assist the superintendent and foreman in crew planning and management. The project manager coordinates the monitoring of work progress to ensure that proper field data are recorded, processed, and evaluated in a timely manner. In this capacity the project manager maintains a degree of involvement in field monitoring such that critical field problems demanding immediate action can be resolved without top-level management concurrence. Finally, the project manager informs top-level management of project status and the effective use of project resources, and in turn provides field management with directives when cost efficiencies, revised scope, and schedules alter the construction plan.

The general structure of the Traveling Project Manager decision process can be examined by comparing the pertinent decision processes with the same basic project milestones introduced in Figure 2.1. A schematic illustration of the resulting overall decision process of the medium-sized firm is shown in Figure 2.2. Notice, however, that only eleven of the twelve basic decision processes are required. The decision process which is not required (i.e., DP2 PROJECT PLANNING), addresses manpower management problems associated with projects that have a large manpower content, such as projects in the large firm environment. This decision process is absorbed into the estimate preparation and project scheduling decision processes. In contrast to the small firm, the medium-sized firm does require the field staffing decision process (DP5 FIELD STAFFING), but only in terms of the need for early staffing of key job superintendents. While the field staffing problem becomes more critical at this level, it does not have the significance of field staffing at the large firm level, where large field management teams emerge.

The two critical changes in the decision process structure of the medium-sized firm, when compared with the small firm, are field staffing in the head office, and head office and field interface during the project management decision phase. The former permits field participation in overall project planning and scheduling, while the latter allows the Traveling Project Manager to interact continuously with the field.

The decision problem focus changes significantly under the Field-Based Project Management approach which is found in the large firm environment. In the head office, the project team is initially designed and developed in detail to reflect the desired level of management control by ensuring that all functions are properly staffed with decision makers. The manning of the project team is carefully considered and mapped for the life of the project. As part of the overall project team development, the field management team is also designed and key positions are staffed early in the project definition phase, so that the personnel who will hold these positions can assist in project scheduling.

The basic determination of project manpower content and work force guidelines must be integrated in order to evaluate the field work force magnitude in terms of total project manpower and craft manpower. Depending on the nature of the contract (i.e., design-construct on a cost plus fee basis), various stages in the development of estimates and schedules will often occur. At each stage, previous estimates and schedules will be revised to reflect the better information available as engineering and design progresses.

Because of the magnitude of projects in this environment, the traditional field management team can no longer effectively plan and schedule crew manpower to the level of detail required. This problem is solved by transferring two project definition decision processes, project planning and project scheduling, to field-based execution. In this way, detailed definition of the work force guidelines and work force content can be established as the basis for planning, scheduling, and allocating crew manpower. These detailed assessments of the work content will be consistent with the overall project plans and schedules by virtue of project planning and project scheduling being a component of project definition.

At the field level, field management, primarily the superintendents and foremen, must establish crew sizes for the work areas, solve work relations problems, and make crew adjustments to meet the cost and scheduling objectives of the project. These field managers and supervisors are assisted in crew planning and management by many project team personnel who continually supply cost and scheduling data from their area of expertise. Many personnel are also involved in monitoring field work progress and evaluating and reporting on status to provide field management with timely information. Because timely reporting is desirable, the majority of the project statusing decision process is performed in the field. Furthermore, the field management team, through its project manager, has the authority to act on field problems. This makes the project management control decision a vital field management responsibility. Critical field problems can be solved and actions taken without delay. Project scope, costs, and schedules can be revised in the field, thus ensuring that the impact is immediately known to all concerned.

The many manpower management problems referred to above form the basis for the decision process focus of the large firm. These problems, together with their relevant decision segments and the basic project milestones, provide the structure of the Field-Based Project Management decision process. Because of increasing complexity, the decision process is illustrated in three parts which correspond to the project definition, field construction, and project management phases in the construction process (see Figure 2.3). The example in Figure 2.3 is based on the large design–construct project in order to delineate the impact of engineering and design on the decision process. The decision process for the pure construction project could easily be illustrated by removing the engineering/design decision segment and those flow lines associated with it.

The decision processes that comprise project definition are shown in Figure 2.3*a* and include Initiate Project Action, Project Planning, Project Scheduling, and Estimate Preparation. There are several important aspects of the project definition decision area which emerge at the large firm level. First is the simultaneous performance of project planning, project scheduling, and estimate preparation. The interface of project scheduling and estimate preparation is particularly important because estimating data are required for the evaluation of project manpower profiles. Second, because of the heavy field commitment, project planning and project scheduling also have a field focus (see Figure 2.3*a*). Third, the head office interfaces with field staff early on in the project in order to identify and establish key field management personnel and release them to the project to support the other project definition decision processes. Finally, an early link with the project management phase burgeons as a result of the desire to control the growth and monitor the management effectiveness of the project team, especially of the engineering and design groups.

In terms of project milestones, two critical milestones of project definition are the client approval of control base estimates and control base schedules. In addition, as design develops client approved change orders enter the system and are incorporated into the planning, scheduling, and estimating decision processes.

The field construction phase in the project and construction management process has the same basic structure as that used in small and medium-sized firms (Figure 2.3*b*). The significant change occurs with the flow of data from the Field-Based Project Definition phase through the project scheduling and project management directives provided by project management. This type of input improves both the field planning and scheduling of crews, as well as crew management.

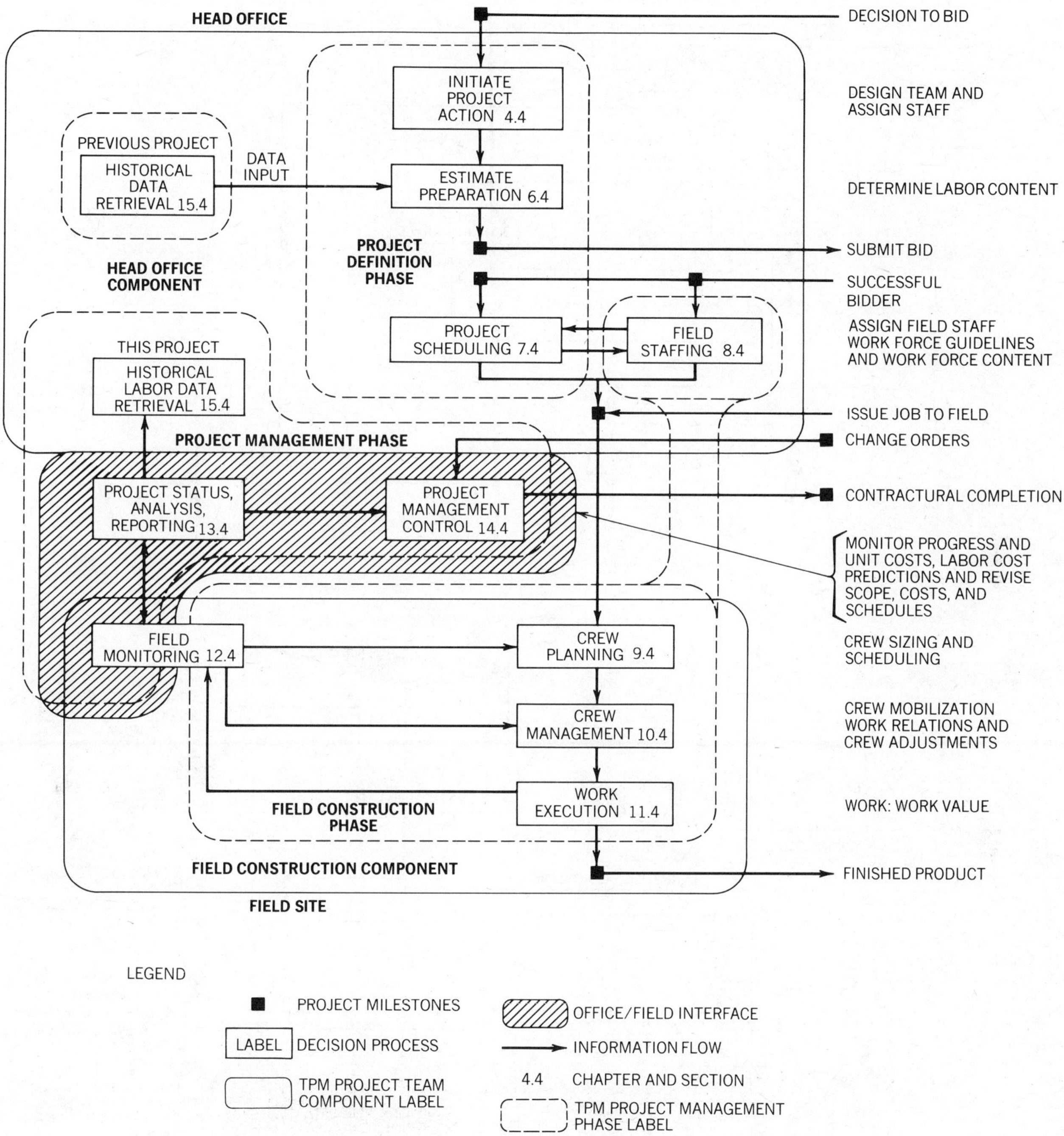

Figure 2.2 Traveling Project Manager (TPM) manpower management decision process.

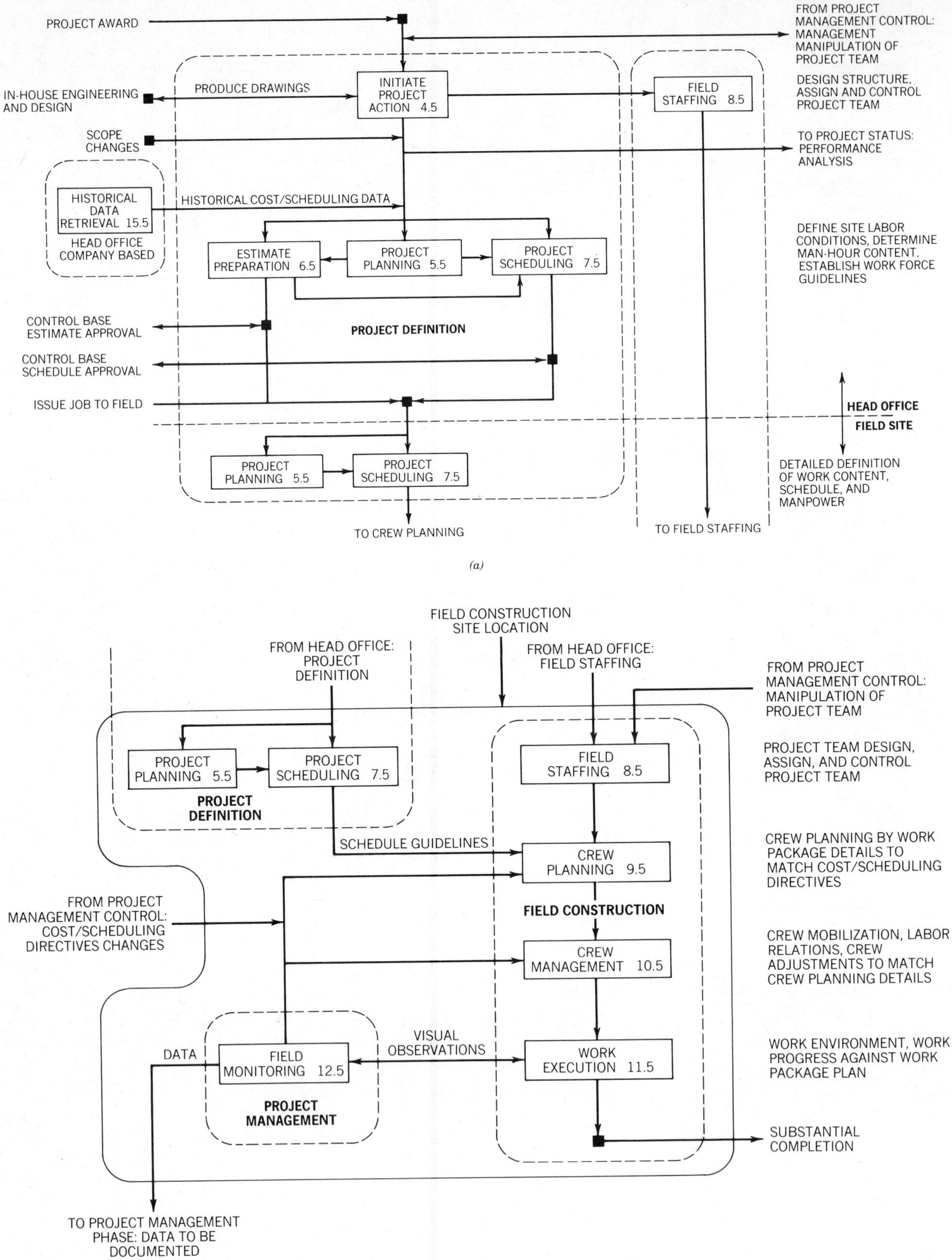

Figure 2.3 Field-Based Project Management manpower management decision process. (*a*) Project definition phase. (*b*) Field construction phase.

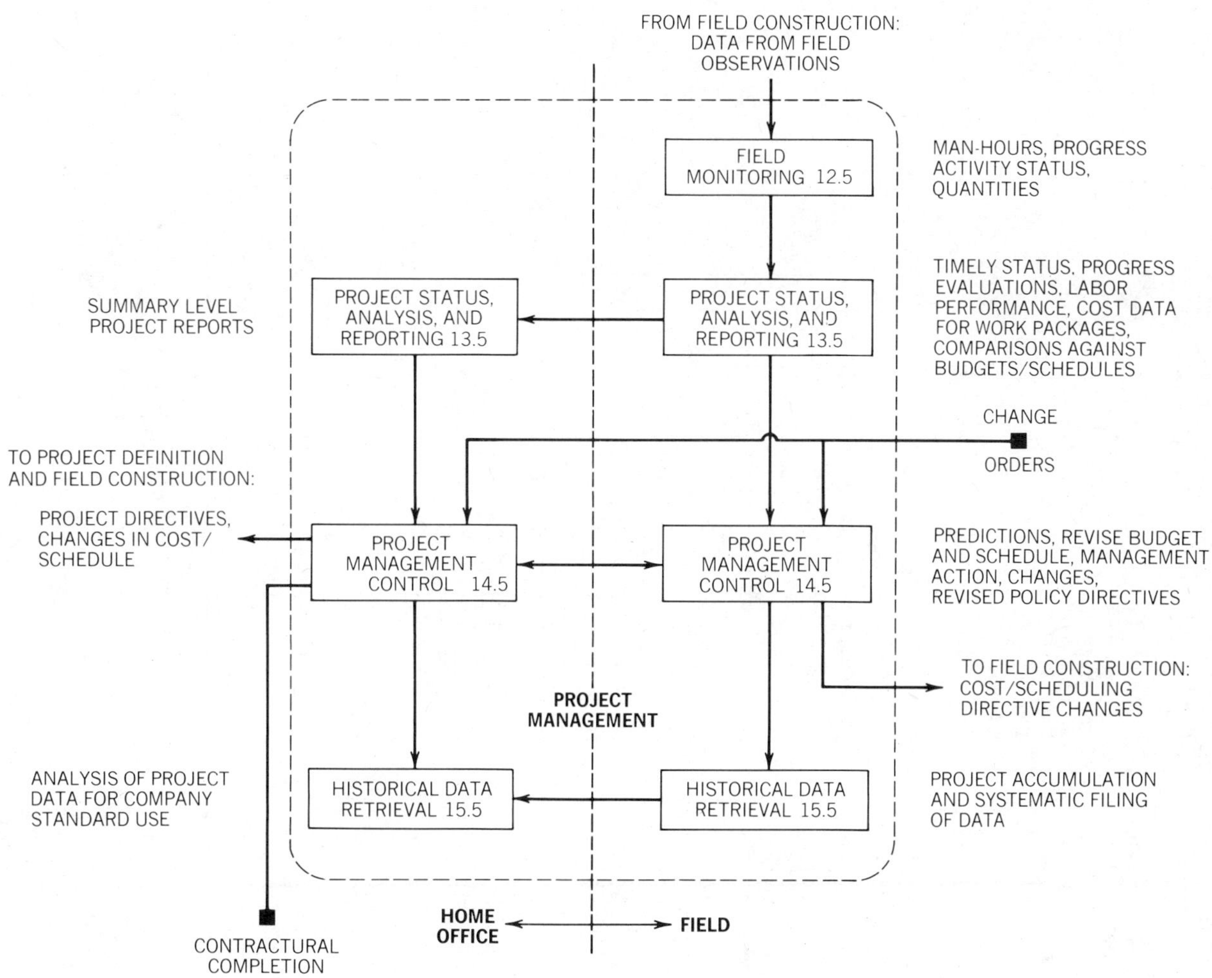

Figure 2.3 (*Continued*) (*c*) Project management phase.

The project management phase is comprised of the Field Monitoring; Project Status, Analysis, and Reporting; Project Management Control; and Historical Data Retrieval decision processes. As depicted in Figure 2.3*c*, the decision processes of the project management phase are performed at both the head office and field levels. In the head office the focus is twofold: First, the head office monitors and controls the project team effort, including engineering/design manpower and man-hour expenditures, and engineering/design progress and performance. Second, it supports construction by providing summary level reporting, scope revisions, and cost and scheduling assistance. If engineering and design is not applicable, then the role of the head office in this decision process is greatly reduced. Thus, the critical change in the decision approach of the large firm over that of the medium-sized firm is the heavy field commitment toward project management. This expands the decision-making capacity of the field project team and begins to provide an autonomous field operation through the field-based performance of all four project management decision processes, as shown in Figure 2.3*c*.

The decision problems, the general form of the decision process, and the functions involved in each decision segment of the Total Field Autonomy approach are similar in most respects to those of the Field-Based Project Management approach. The major obvious change in the overall decision process is the potential location of all decision pro-

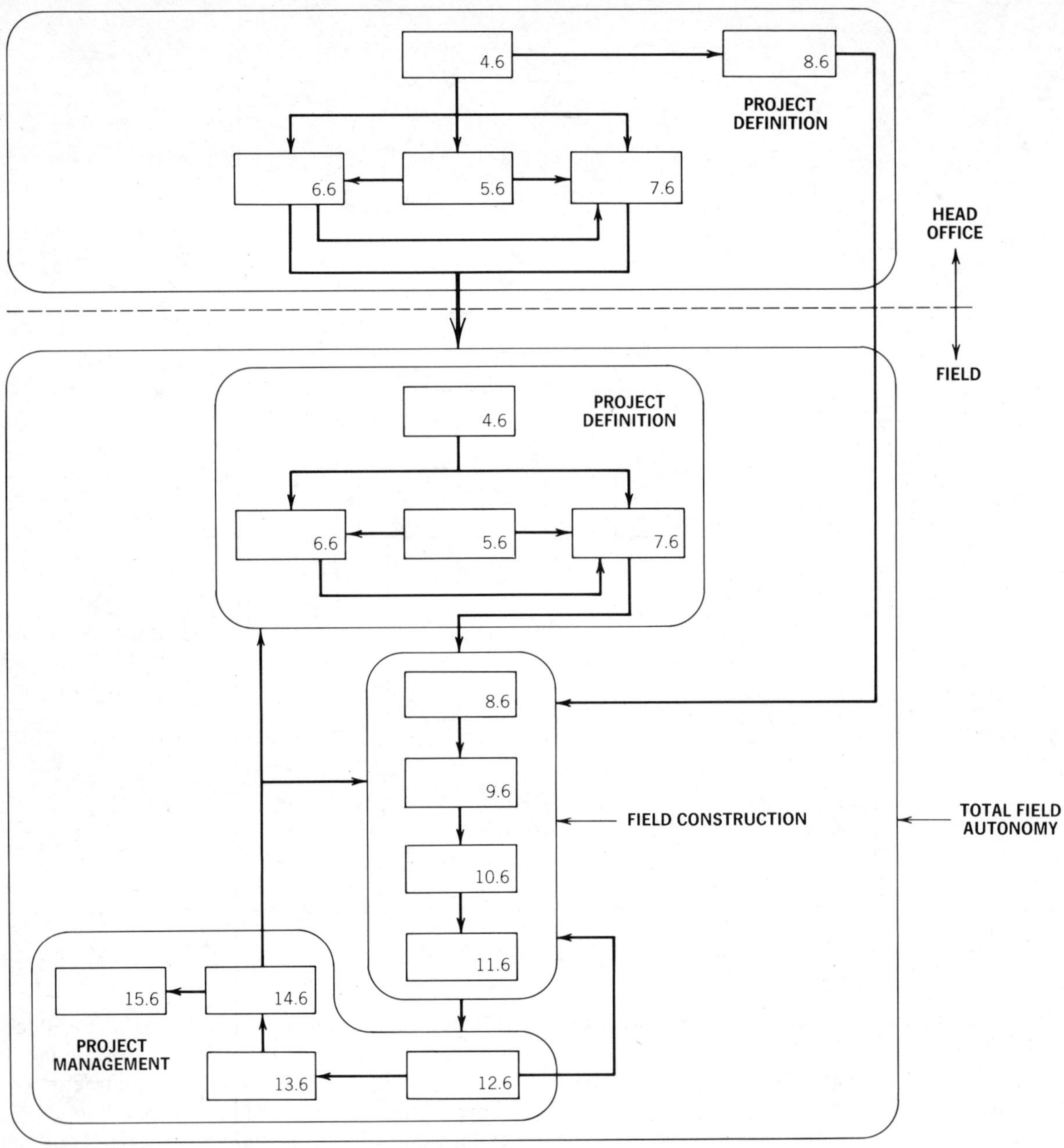

Figure 2.4 Total Field Autonomy manpower management decision process.

cesses in the field. A schematic illustration of the Total Field Autonomy decision process approach is provided in Figure 2.4, but with only limited detail. As shown in this example, a total project definition phase is located in the field. The degree of project involvement in terms of project initiation, planning, estimating, and engineering/design in the field, would depend heavily on client requirements and on the location of the project. In almost every project, the head office will support the field with a project definition focus. The head office commitment will, of course, depend on the degree of field project definition.

CHAPTER 3

Portrayal of a Decision Process

Managers are always making decisions. Some decisions are easily made, especially if the manager can equate the decision problem to an appropriate policy or to a previous situation in which he or she was involved. Some decisions, however, can be made only after careful consideration and discussions with key personnel. These tougher decisions are based on the acquisition and processing of relevant project data and on the study of the possible implications of a decision to both the particular project and company operations in general. In these more complex situations decision making requires the prior completion of a number of management functions. Thus, management functions when given a problem focus become components of the decision process.

A decision process can therefore be defined in terms of the specific management functions that need to be performed during the decision-making process. Many of these component management functions will be performed sequentially, whereas others will interact with each other so that in some cases the iterative or repeated performance of management functions will become necessary. Hence, the structured relationships between the component management functions must be established in the description or definition of the decision process. Finally, given the sequence of the component functions together with the data, documents, and policies required by each component function, a complete information system description of the decision process can be established.

Generally speaking, the policies and documents that guide the performance of each function and the decision process as a whole will be unique to the company, so that it is often possible to describe how a specific company solves a particular problem. In addition, if management roles are identified with and incorporated into a statement or description of a decision process, the management approach of specific groups to solving decision problems can be revealed and examined.

This chapter introduces a functional approach to the structuring and description of decision processes, and to the portrayal of Management Information Systems. These concepts form the basis for the descriptions and illustrations of the various project manpower management decision processes, at the small, medium-sized, and large firm levels, that appear in later chapters.

3.1 FUNCTIONAL APPROACH

Management tasks have traditionally been divided into the broad functional areas of organizing, staffing, planning, directing, and controlling. This traditional broad dissection and sequencing of management functions corresponds naturally to project management requirements. Thus, once management has been empowered to act, its initial efforts must be directed to organizing itself, deciding upon its scope of action, and formulating the policies under which it will act. When this has been accomplished, project management effort can be directed toward the completion of the project. Figure 3.1

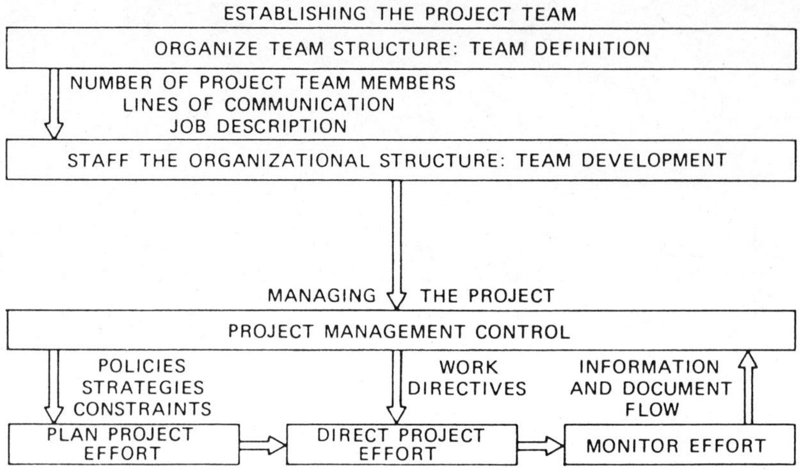

Figure 3.1 Relationship between general management functions.

indicates schematically the relationship between these management functions and their functional groupings.

The particular breakdown of these functions depends upon the nature of the project. The breakdown shown in Figure 3.2 dissects planning into two components, one of which is associated with *planning* what to do, the other with *scheduling* when to do it. Similarly, directing is divided into the active *allocation* of staff and labor resources to a job and the subsequent directing of them in their *work* activities. Finally, the control function is associated with decision making based upon information gathered in the *monitoring* and *recording* of current project status and work progress, and with the consequential impact of decision making on the planning, scheduling, allocation, and work functions of the project effort.

The six general project management functions (*planning, scheduling, allocating, working, monitoring,* and *recording*), show a progressive movement of management concern from the planning and enumeration of required resources to the commitment, use, and evaluation of the effective use of these project resources. The natural sequence of these functions corresponds to the general movement and relocation of management concern and effort from the head office environment (concerned with planning and scheduling), to the field for work execution, and then back to the head office environment (for final resource use assessment and related decision management efforts). This sequence of the general project management functions therefore provides an ideal dissection scheme for both resource use management and construction management purposes.

A schematic illustration of these general project management functions in the context of a typical contractor head office and construction site environment is given in Figure 3.3. Each project management function is indicated in terms of simple project situations, and the relationships between these functions are shown in terms of basic project management information flows.

Using the six general management functions of Figure 3.2 as a frame of reference, a more specifically focused set of project management functions can be obtained by further dissection. Consider, for example, a dissection rationale for each general project management function, based upon various management interests, which would span the construction hierarchy from the foreman (with a crew and work face focus on construction management), through the project management staff, to a top-level executive manager (with a project staffing and overall project management focus).

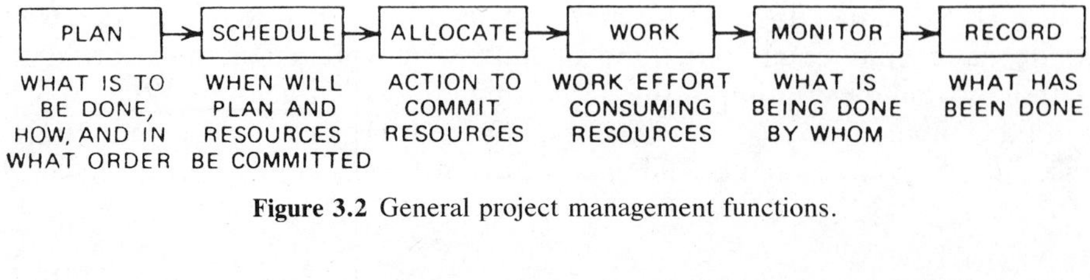

Figure 3.2 General project management functions.

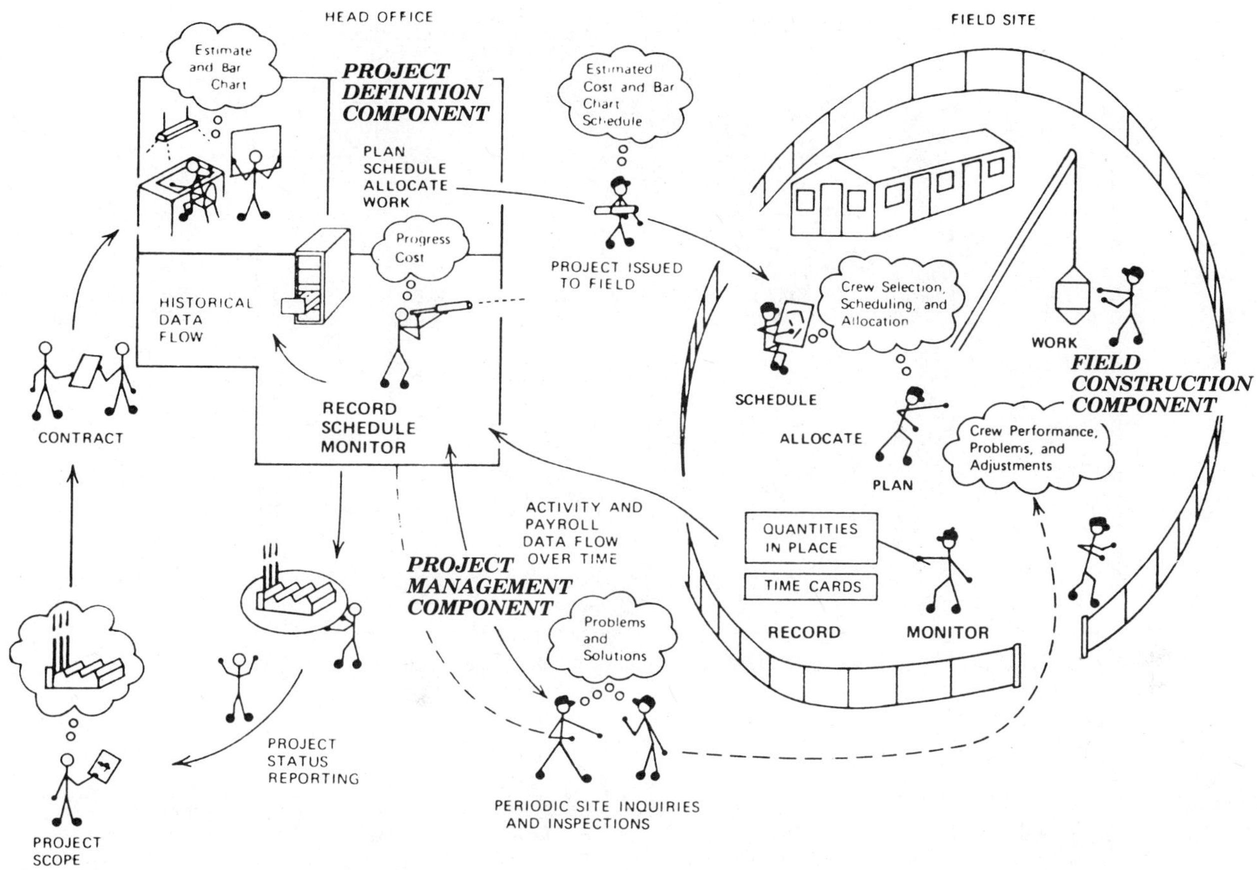

Figure 3.3 Project Team Component (PTC) concept.

This dissection rationale exposes the different levels of management problems that each decision maker encounters. For instance, the superintendent-foreman is interested in determining the size of the crew and in allocating a crew to an activity, whereas the project manager may be interested only in total project manpower. Figure 3.4 illustrates an initial step in such a dissection.

Using the two-dimensional reference framework of the management process functions and the management interests of staff in the construction management hierarchy, forty-six project manpower management functions are identified and described in Figure 3.5.* The forty-six management functions are not unique. They represent, in the authors' view, a balance between, on the one hand, the need for an adequate description of the various decision processes involved in construction management and the roles that individual decision makers play in these decision processes and, on the other hand, the problems associated with the codification and understanding of a large number of specialized management functions obtained from too fine a dissection of each general project manpower management function.

The forty-six basic project manpower management functions as collected together in Figure 3.5 are identified by an alphanumeric label. Thus the CONSTRUCTION RECONNAISSANCE management function is labeled P1 because it is the first of the planning functions. Similarly, the DETERMINATION OF THE NUMBER OF FIELD CREWS is labeled A4.†

*The reader is referred to Chapter 2 and Appendix A of *Project Manpower Management: Management Processes in Construction Practice* by S. D. Anderson and R. W. Woodhead, John Wiley & Sons, Inc., New York, 1981 for a full treatment of this development.

†Notice that whenever specific mention is made in the text to one of the basic management functions listed in Figure 3.5, the function name is capitalized. Notice further that the function names are often altered slightly to fit the syntax of the textual sentence in which they appear. In this case function A4 appears as DETERMINE NUMBER OF FIELD CREWS in the figure, but is referenced above as the DETERMINATION OF THE NUMBER OF FIELD CREWS, because this descriptive title is more appropriate to the sentence structure.

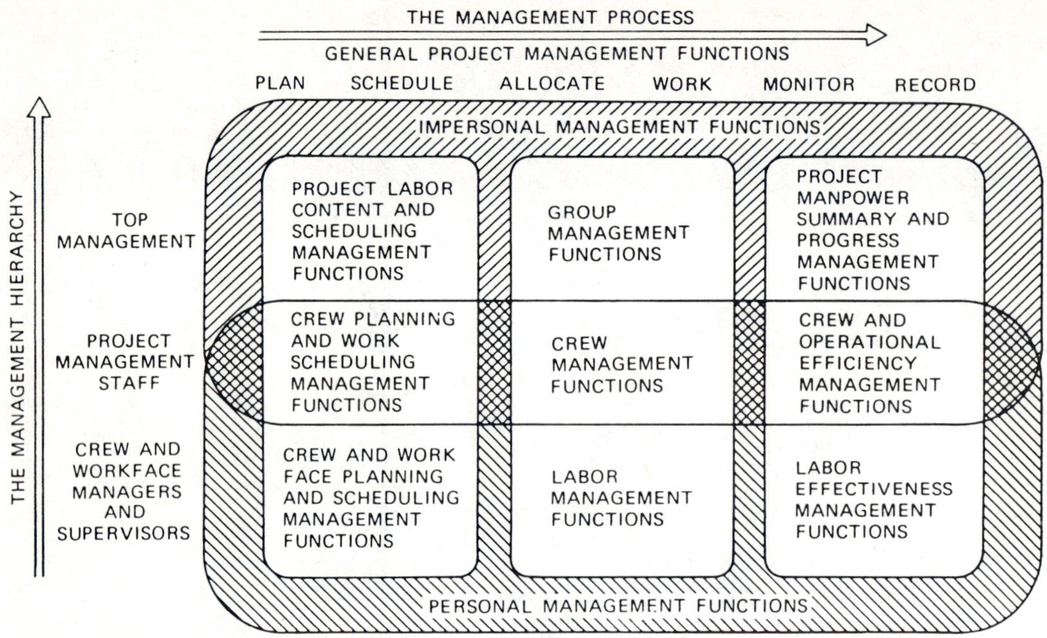

Figure 3.4 Two-dimensional reference framework for project manpower management functions.

	PLAN P	SCHEDULE S	ALLOCATE A	WORK W	MONITOR M	RECORD R
1	P1 CONSTRUCTION RECONNAISSANCE	S1 CONSTRUCTION SEQUENCE	A1 ALLOCATE OFFICE, ENGINEERING, AND DESIGN STAFF	W1 RELEASE STAFF TO PROJECT	M1 SUMMARY PROJECT REPORTS	R1 HISTORICAL DATA BASE
2	P2 ACTIVITY DEFINITION	S2 ESTABLISH CONSTRUCTION SCHEDULE	A2 ALLOCATE FIELD STAFF	W2 PRE-JOB LABOR CONFERENCE	M2 CURRENT PROJECT STATUS	R2 GENERATE LABOR MANAGEMENT DATA
3	P3 QUANTITY TAKE-OFF	S3 DETERMINE PROJECT AND TRADE MAN-HOUR PROFILES	A3 DETERMINE PROJECT AND TRADE MANPOWER PROFILES	W3 WORKFORCE MANAGEMENT	M3 PROJECT FORECAST	R3 PROJECT COST ACCOUNTING
4	P4 ESTIMATE LABOR CONTENT	S4 DEVELOP WORK PACKAGE DETAILED LOGIC	A4 DETERMINE NUMBER OF FIELD CREWS	W4 ENFORCE JURISDICTIONAL AND SKILL LEVEL WORK ASSIGNMENTS	M4 ACTIVITY ANALYSIS	R4 UPDATE AND REVISE WORK PACKAGE SCOPE
5	P5 DEFINE WORK PACKAGE	S5 DEFINE CURRENT PROJECT STATUS	A5 ADJUST CREWS	W5 ESTABLISH FIELD CREWS	M5 LABOR ANALYSIS	R5 LABOR PRODUCTIVITY
6	P6 DETAILED FIELD QUANTITY TAKE-OFF	S6 REVISE CONSTRUCTION SCHEDULE	A6 ASSIGN INDIVIDUAL WORK TASKS	W6 SOLVE HUMAN RELATIONS PROBLEMS	M6 WORK PROGRESS	R6 QUANTITIES IN PLACE
7	P7 ESTABLISH CREW REQUIREMENTS	S7 SCHEDULE PROJECT CREWS		W7 INITIATE INDIVIDUAL WORK TASKS	M7 RESOURCE USE	R7 LABOR HOURS
8	P8 PLAN WORKFACE LAYOUT	S8 SCHEDULE CREW MEMBER WORK SEQUENCES		W8 EXECUTE WORK	M8 FIELD INSPECTION	R8 FIELD AUTHORITY DOCKETS

Figure 3.5 Project manpower management functions.

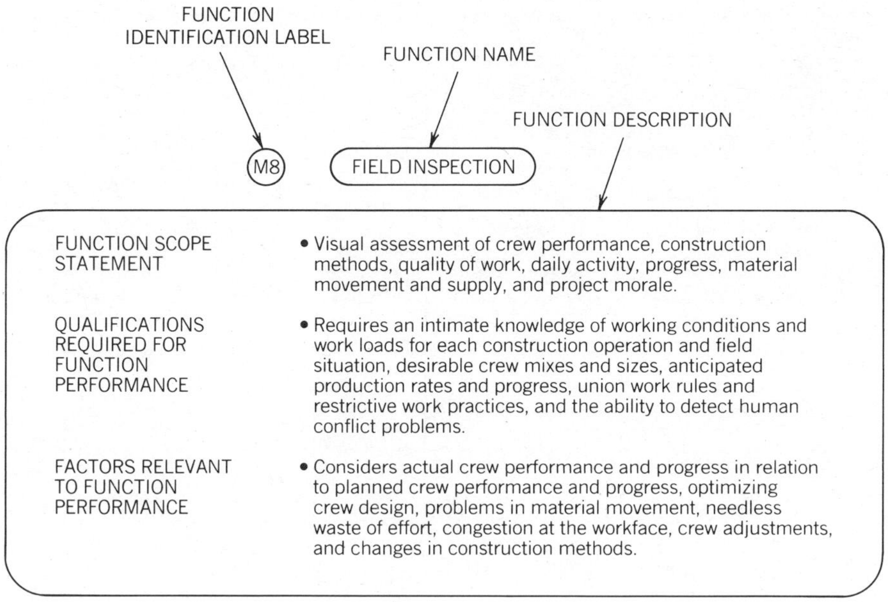

Figure 3.6 A management function.

Each individual basic management function can be further described by a function scope statement, a job description statement relevant to the performance of the function, and a statement of factors that may be considered relevant to the performance of the function. Figure 3.6, for example, gives such a description for the FIELD INSPECTION (M8) management function. Detailed descriptions of each of the forty-six project manpower management functions are left for later chapters, when they become relevant to establish the scope of each basic decision process.

The functions relevant to each basic project management decision process are indicated schematically in Figure 3.7, using simple functional plots. Thus, the Project Planning decision process is shown as involving three project management functions, namely: CONSTRUCTION RECONNAISSANCE (P1), CONSTRUCTION SEQUENCE (S1), and ACTIVITY DEFINITION (P2). An initial step toward a fuller description of the Project Planning decision process thus involves an analysis of these three basic management functions and their formulation, along the lines indicated for the management function illustrated in Figure 3.6. This approach, among others, is illustrated in the succeeding chapters of this book.

The functional plots for the Estimate Preparation and Project Scheduling decision processes, as shown in Figure 3.7, have common function interfaces with the Project Planning decision process in the ACTIVITY DEFINITION (P2) and CONSTRUCTION SEQUENCE (S1) project manpower management functions, respectively. The general nature of the interaction between the basic decision processes is of course more complex than the interfacing of one or more management functions. Some aspects of this interaction relate to the need for decisions and the related requirement for transfer of information and documents. Others relate to the passage of time and the changing management role of individuals in the project team that results. It is convenient at this stage, however, to simply indicate schematically the manner in which such interfaces will be portrayed (see Figure 3.8), and leave for later chapters the full description and portrayal of the relationships between the basic project manpower management decision processes. The next Section 3.2, however, develops a schematic approach to the description of the internal structure of a decision process.

3.2 FUNCTIONAL PERFORMANCE VIEW OF A DECISION PROCESS

Management functions can be defined and described in a variety of ways. In some cases a descriptive title and a written description of the scope of a function may be sufficient. An example is that shown in Figure 3.6. A more detailed method would be to refer to the documents and procedures used in

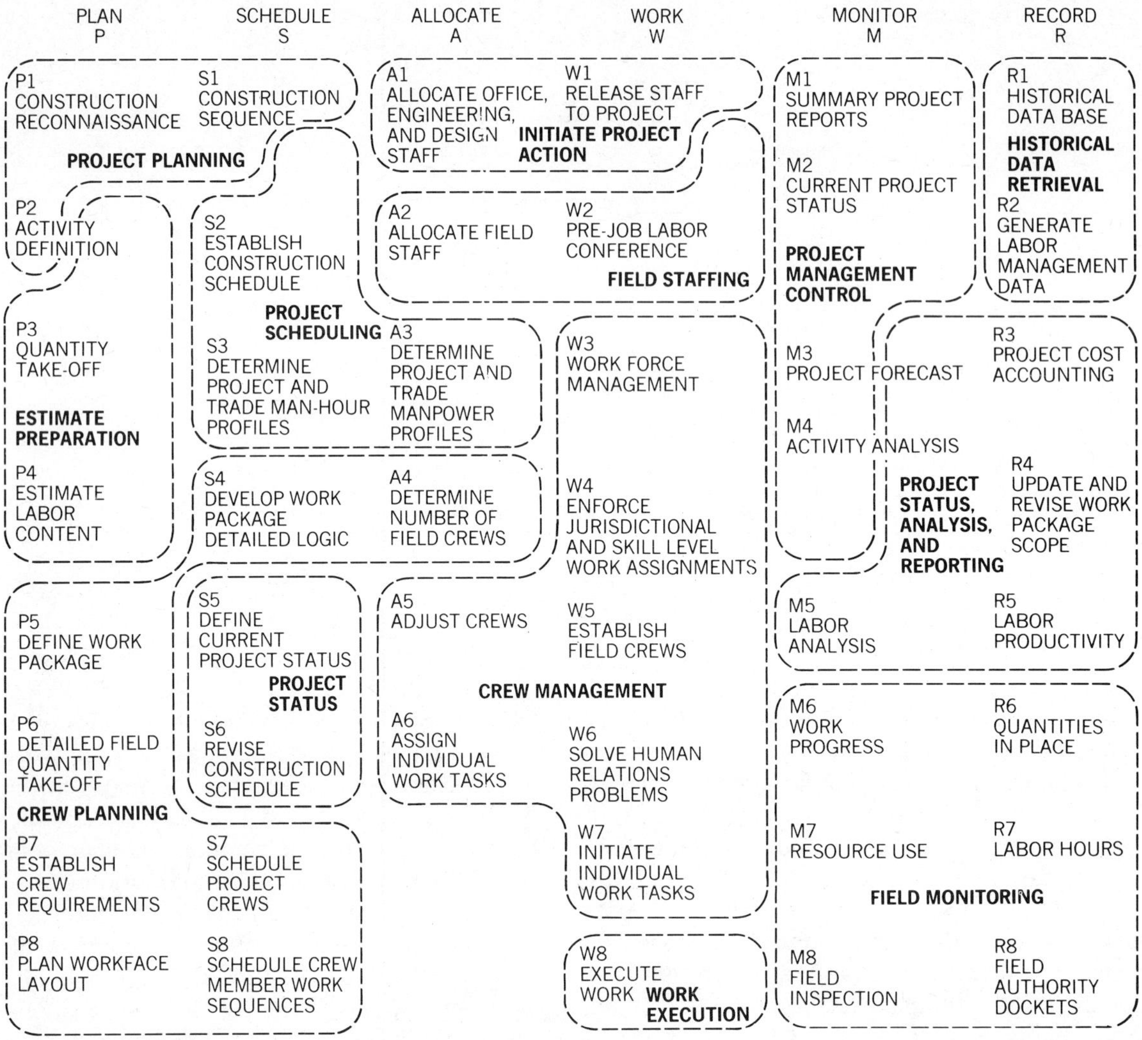

Figure 3.7 Functional break-out of the twelve project manpower management decision processes.

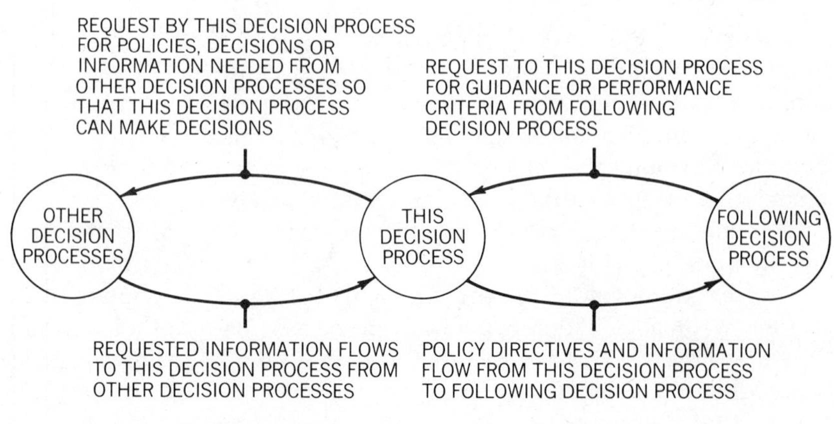

Figure 3.8 Schematic illustration of the possible nature of decision process interfaces.

the performance of the function, and prepare procedural performance manuals for major or critical functions. The approach adopted will depend upon the magnitude and importance of the function, the expertise and experience of the employee involved, and the level of sophistication required by management in the performance of the function.

In most cases the basic elements of performing a management function are the same regardless of whether the approach is ad hoc or formalized. In addition, these elementary steps have a close correlation with information processing concepts, especially when associated with problem solving via a decision process. For instance, the employee must collect and process data. The resulting information often needs to be highlighted and manipulated into a form that can be readily absorbed by others. Finally, steps may be required in order to relay the final information to other personnel so that further management functions may be performed. When collecting data the employee must determine what data are required and from what sources the data can be gathered. Similarly, the employee must determine how the data should be processed and how the final information should be presented. Finally, the employee must consider who requires the information and when and how the information should be transmitted. Whether the employee carries out the function formally or mentally, a data-to-information processing effort is involved. Thus, management functions can be thought of as information flow processes. In many cases, of course, the information flow is by word of mouth, observation, or via the handling of documents, and decisions are made based upon past experience. For some functions, however, management performance can be directly correlated with the various steps involved in using information processing systems (whether automated or not).

A schematic illustration of a decision process should incorporate symbols that represent information flows such as the generation and transmittal of data on documents, and the impact of policy directives and the like on the performance of the component management functions. This requirement is adhered to throughout this text by the use of a simple network technique which uses flowchart symbols to represent functions, policies, documents, and information flow. The technique used in this text for portraying a decision process is shown in Figure 3.9, in terms of the elements involved in the performance of a management function.

In Figure 3.9*a*, F1 represents the function to be performed, and a circle is used to symbolize the requirement to perform the function. The information required in the performance of the function and that generated during its performance is shown in braces above and below the function symbol (see Figure 3.9*b*). The line symbol entering at the top of the function circle in Figure 3.9*b* represents data from another source, such as information input which can either be transmitted verbally by direct contact between employees or in the form of a written document. In the former case, the flow line would simply represent verbal information transfer. If the information is transferred on a document, a third symbol (flowchart symbol for documents) is used, as shown in Figure 3.9*c*, and a flow line becomes a documented information transfer. Further, Figure 3.9*c* shows the three types of documents that could emerge in performing a management function. These are (1) a policy document (P-DOC), (2) an internally generated document (I-DOC), and (3) an external document (E-DOC).

A *policy document* (P-DOC) is a direct expression of a policy statement. It may be a specially printed form or a specially prepared reference list. A form for calculating and summarizing quantities is an example of a policy document. Many companies utilize their own standard cost coding system manual. This manual serves as a reference list for generating a cost code system designed specifically for a project, in which case the cost code reference list would be considered a policy document.

An *internal document* (I-DOC) contains the output of a function and may or may not be a derivative of a policy document. A project scope list developed from a company standard cost code manual (policy document) would be considered an internal document and a derivative of a policy document. Another example of an internal document is the formulation and graphic display of a bar chart schedule. The internal document would result from a scheduling function for which a policy statement would direct only the use of the bar chart method. In many management situations, a standard form is not available, and therefore the document is generated internally by the person performing the function.

The *external document* (E-DOC) is a document that enters the decision process from sources outside the company. Drawings and specifications could be considered external documents when provided by an architectural firm who is not affiliated

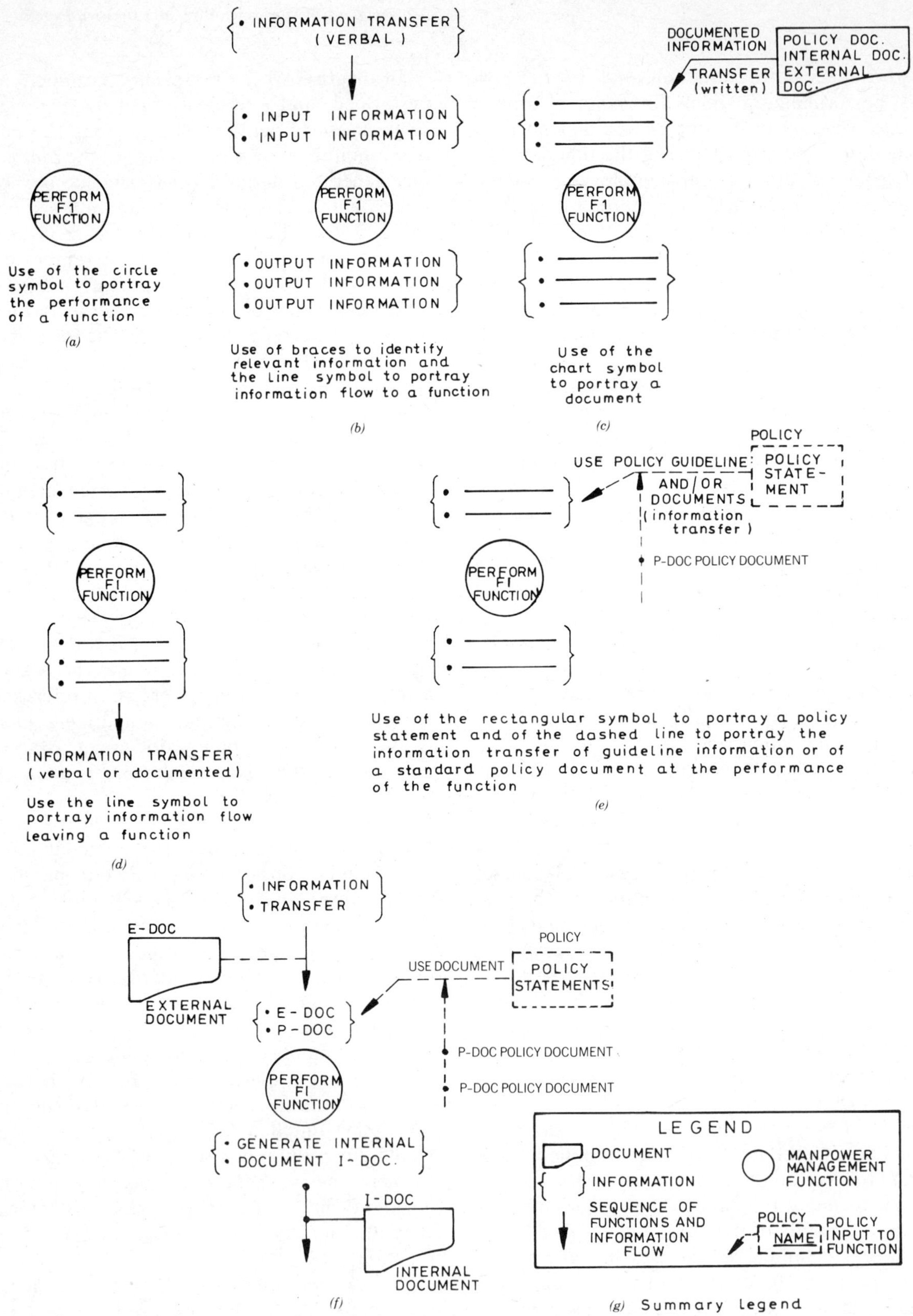

Figure 3.9 Elements of the network technique used to describe a management function information system.

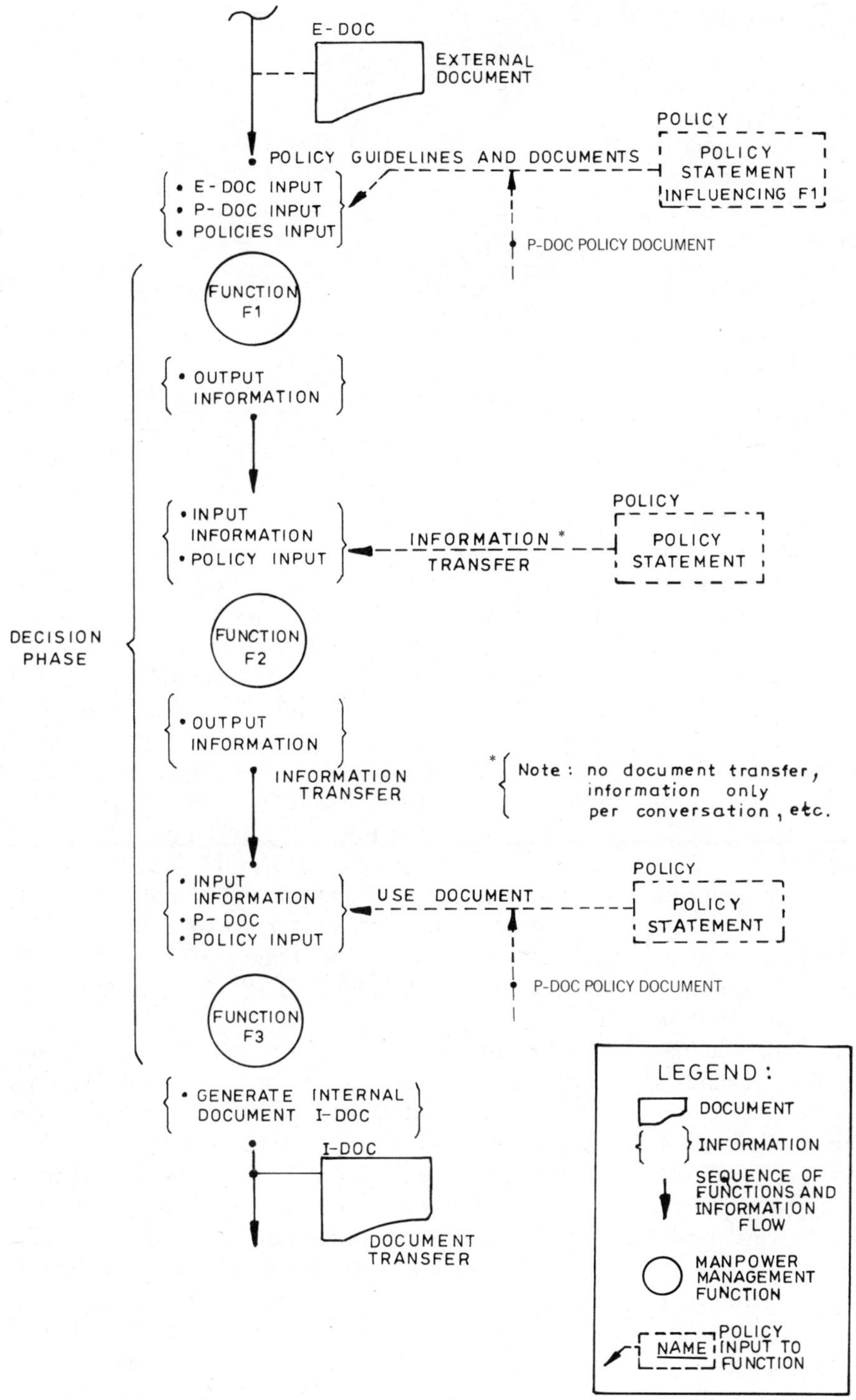

Figure 3.10 Management function decision network.

with the contractor. A union agreement is a second example of an external document.

All documents enter a management function on flow lines, as information input, and after processing are forwarded to subsequent functions as information input for those functions. The flow line leaving the bottom of the function circle in Figure 3.9*d* represents this transfer of information. Again, this could be in the form of verbal data and instructions or a written document. In the latter case the existing line is further elaborated to become the chart symbol in a similar manner to that used in Figure 3.9*c* to indicate an entering document.

The performance of a function is often further influenced by a policy, either in the form of a suggested guideline or a directed method for performing a function. The policy statement is identified by a rectangle symbol, as shown in Figure 3.9*e*. The broken line that flows from the policy statement to the function in Figure 3.9*e* represents information transfer regarding the policy statement, which is in the form of guidelines. The guidelines could be either written or verbal instructions, a document to be used, or an inherent knowledge of function performance requirements. Thus, a policy statement is either a set of formal written procedures and directives involving usable documents, or simply a statement of the method required to perform the function that does not include formal instructions or procedures.

The total of each component element in the management function process is illustrated in Figure 3.9*f*. In this illustration all the possible information flows, documents, and policy inputs are summarized. Finally, an abbreviated summary legend is presented in Figure 3.9*g*.

A decision process normally focuses on a specific problem which requires the performance of one or more functions. Thus, several functions must be performed in succession to solve the decision problem. By connecting the appropriate functions, a decision network can be generated which portrays and defines the decision process for solving the problem. The technique developed in Figure 3.9 is the basis for illustrating a decision process as is done in Figure 3.10.

Each decision process can therefore be described in terms of the functions involved, policy statements, documents, and information flow. Using the techniques introduced above, one can gain considerable insights into the structure of a decision process. In addition, the careful description of the relevant management function, its associated policies, and document design can reveal the manner in which the management function and the overall decision process is performed. As an example, the large firm policies influencing the performance of the ESTIMATION OF LABOR CONTENT (P4) project manpower management function are shown in schematic form in Figure 3.11. A more complete tabular format description of the management function and the relevant policies is given in Figure 3.12. Finally, when the relevant information flows and documents required to perform the function are incorporated with the function and its policies, an information processing system focus emerges. This concept is illustrated in Figure 3.13 for LABOR CONTENT ESTIMATING (P4) at the large firm level.

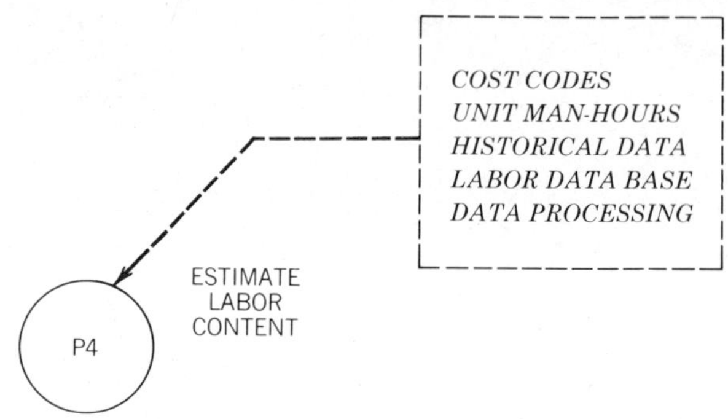

Figure 3.11 A schematic illustration of policies influencing the performance of the ESTIMATION OF THE LABOR CONTENT (P4) Project Manpower Management function.

In Figure 3.13, the subscript "L" is used to describe both the policies and the policy documents uniquely associated with the large firm. This notation system will be used throughout the text when describing the unique policies and documents relevant to small and medium-sized firms and to the Total Field Autonomy management approach. Table 3.1 summarizes the policy and document notation system used in Chapters 4 through 15.

3.3 THE DECISION PROCESS: A MANAGEMENT INFORMATION SYSTEM

Management functions can be thought of as information flow processes, because their performance requires the collecting and processing of data and subsequent formulation of processed data into a readily usable format for decision making. For

P4—ESTIMATE LABOR CONTENT

- Rationalizes the labor content for each project activity.
- Requires an intimate knowledge of crew sizes and capacities for each proposed construction method and intended level of equipment use, as well as an extensive knowledge of labor productivities, site conditions, and historical labor data base.
- Considers the impact of site conditions on labor productivity and the magnitude of activity work content in estimating activity duration and labor content. Extends direct labor content to include the labor burden as well as considering the labor content involved in the field management of labor and in the engineering and general office functions.

POLICY STATEMENTS

COST CODES	Systematic classification of work categories segregated by unique cost codes is the basis for structuring the project scope into work tasks, categorizing estimated costs, reporting and controlling costs and manpower, and developing a historical data base.
UNIT MAN-HOUR	Realistic evaluation of the labor content for a work task is essential for successful project planning and cost control. Therefore, labor costs must be based on work task man-hour content in relation to work volume, construction methods, productivities, wage rates, and labor availability.
HISTORICAL DATA	Historical labor data are basic to a realistic determination of the labor content for a work task.
LABOR DATA BASE	Historical labor data must be analyzed and documented in a meaningful way in order to be useful for estimating and planning manpower for future projects.
DATA PROCESSING	Maximum use of electronic data processing is often the most economical means of processing data into useable information.

Figure 3.12 Detailed description of the manner in which the large firm performs the ESTIMATE LABOR CONTENT (P4) project manpower management function

some management functions, the structure of processed data is predetermined by standard policies and processed data are reflected on policy documents. These documents are often uniquely associated with a specific function and thereby provide a guideline for performing that management function. In this way, a single management function can be considered a documented information system. An example of a documented information system for ESTIMATNG THE LABOR CONTENT (Function P4) at the large firm level is illustrated in Figure 3.13 using the network technique shown in Figure 3.9.

When management solves a specific problem, normally one or more functions must be performed and incorporated into the decision process (see Figure 3.10). Since each function of the decision process may have its own unique documented information system, the succession of functions which describes the decision process must be performed in a manner that integrates the information generated through one function with that of another function. Consequently, the information on a document for one function must be readily incorporated into a document required for the performance of another related function within the decision process. Thus, when the documents are designed to integrate the information flow through a decision process, the decision process becomes a Management Information System.

When viewed from an overall project management perspective, the information system concept has a broader application. For instance, an information system can link together several more specifically focused decision processes in response to generic project-oriented management problems such as the contractor's need for cost control during the field construction of a project. Consequently, the design and implementation of a project's cost control system must consider the requirement that information be exchanged between the various decision processes that are involved with project cost control. Furthermore, the information processed in one decision process should be structured in a manner which permits it to be readily incorporated into other decision processes. The degree to which this is accomplished will determine the extent to which the information system for cost control is integrated.

When describing a project focused Management Information System (MIS), consideration must be given to certain basic criteria. First, the management problem area must be identified and its scope

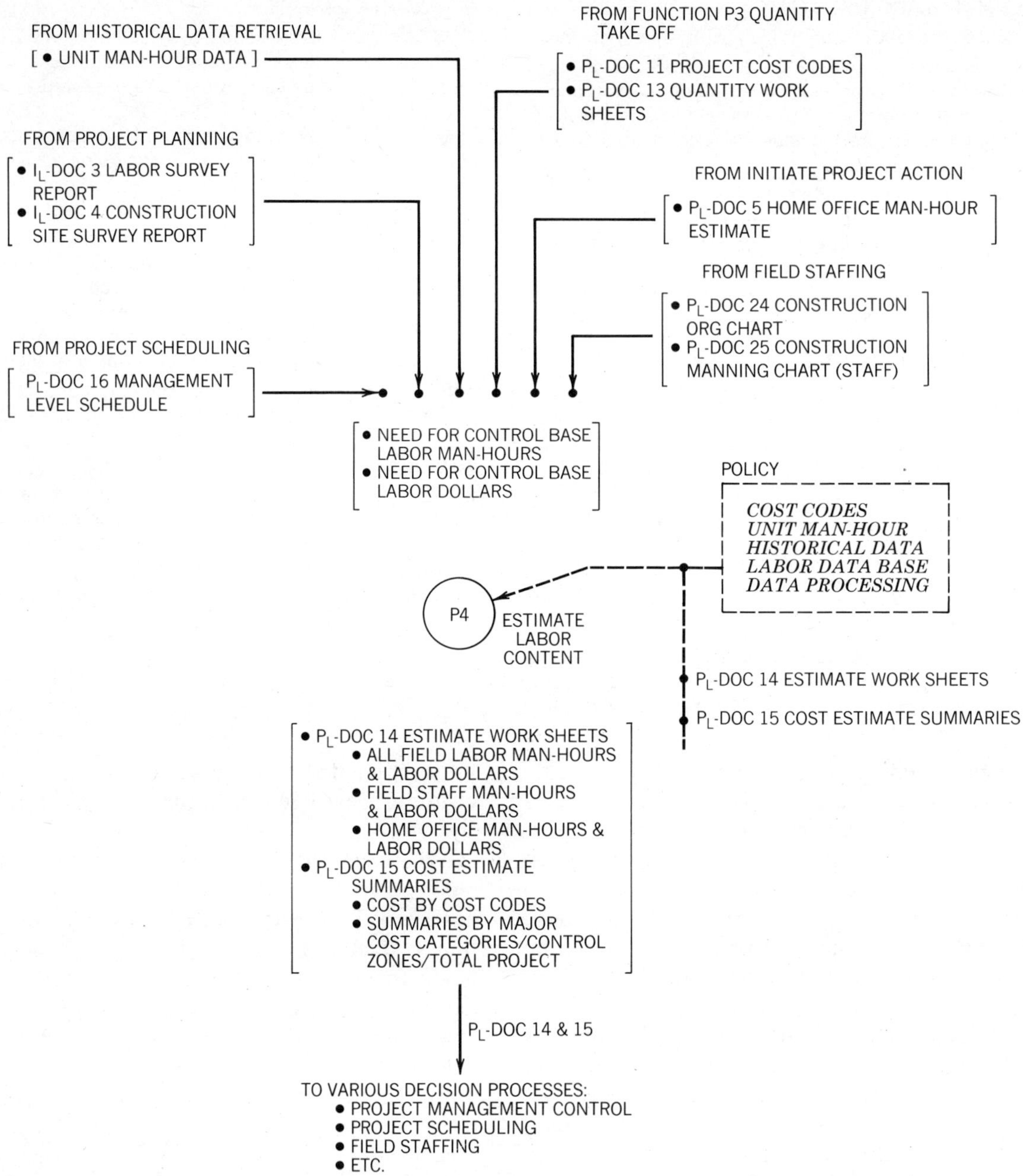

Figure 3.13 Information processing system for ESTIMATING THE LABOR CONTENT (P4) at the large firm level.

TABLE 3.1 POLICY AND DOCUMENT NOTATION SYSTEM

Manpower Management Approach	Classification	Policies	Documents
Project Team Component	Small firm	P_S No.	P_S-DOC No./I_S-DOC No.
Traveling Project Manager	Medium-sized firm	P_M No.	P_M-DOC No./I_M-DOC No.
Field-Based Project Management	Large firm	P_L No.	P_L-DOC No./I_L-DOC No.
Total Field Autonomy	Large firm	P_T No.	P_T-DOC No./I_T-DOC No.

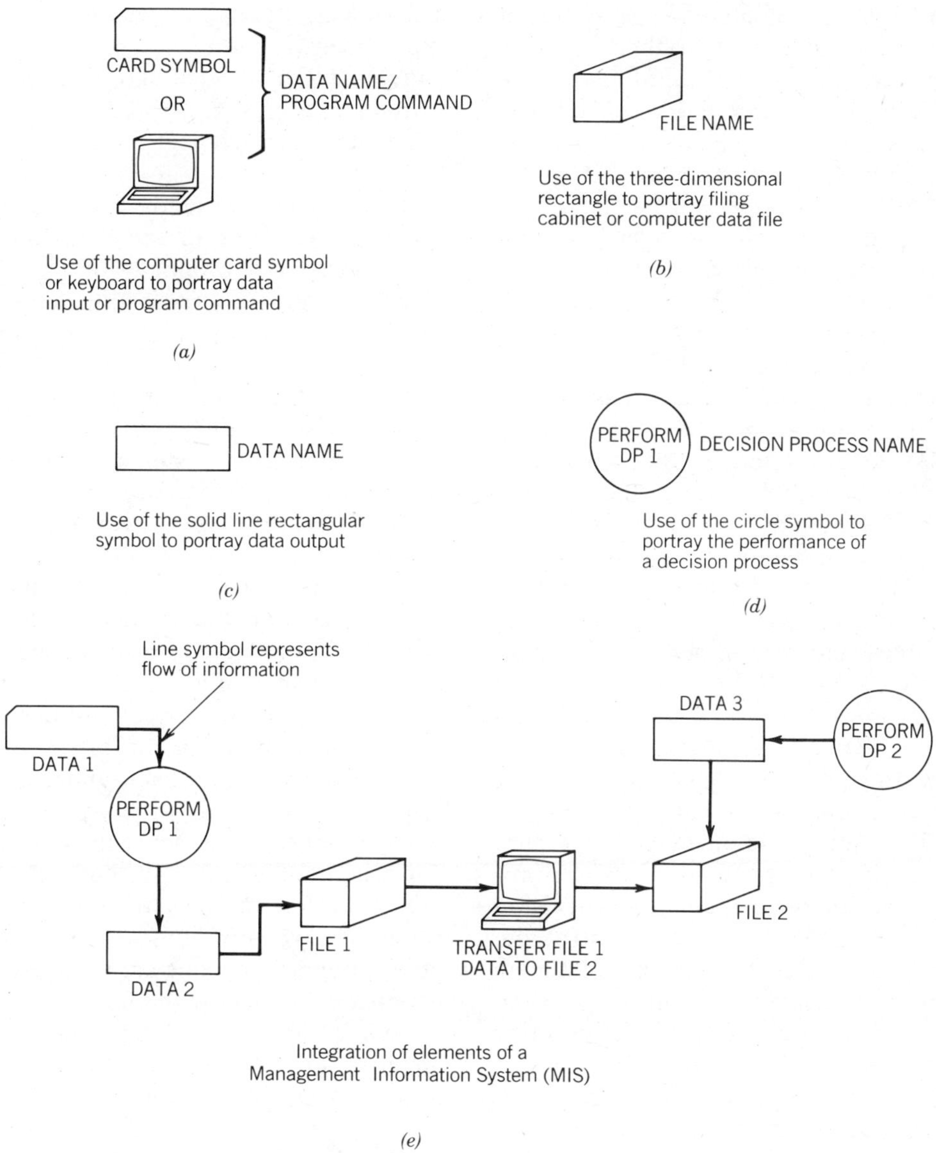

Figure 3.14 Elements of the network technique used to describe a project Management Information System (MIS).

defined (e.g., the project cost control area). The information required to solve the problem must then be identified in order to determine the specific types of data to be processed and the appropriate format of this processed data (i.e., document design and format). In addition, the method of processing and storing data must be selected (e.g., manual or computer) and the structure of the database determined. Finally, the flow of processed information between decision processes must be evaluated to ensure that the proper data is channeled to the appropriate locations for further processing. A simple schematic format, different from that used for decision process networks, will be used in this text to characterize the relevant project MISs used in construction practice to serve manpower management decision processes. A technique for describing an MIS in terms of the elements involved in defining its structure is shown in Figure 3.14.

In Figure 3.14*a*, a flowchart symbol for a computer card or a keyboard schematic is used to symbolize either data input or a user selected program initiation or modification command. In terms of computer applications, data could either enter the system through card input or keyboard input. In a manual system, only the card symbol would be used and would in this case represent data on a *document*. The name of the data appears next to the card. Figure 3.14*b* uses a three-dimensional parallelepiped to symbolize a file such as a filing cabinet or computer data file (e.g., disk). The file name is used to describe the contents of the data file. A solid line rectangular symbol is used in Figure 3.14*c* to represent data output and normally will reflect a formalized document generated either manually or by computer. In Figure 3.14*d*, DP represents the decision process to be performed and a circle is used to symbolize the performance of the decision process. Finally, the integration of each component element of the project management information system is illustrated in Figure 3.14*e*. This integration is expressed by a line symbol which represents the flow of documented information between the component elements of the management information system being described.

A practical illustration of a management information system describing the initial implementation of a project labor cost control system is shown in Figure 3.15, using the technique developed in Figure 3.14. In Figure 3.15, estimate documents are processed through the Estimate Preparation decision process to establish the project budget control file. Labor cost analysis documents and budget estimate data from the project budget control file become the input to the Project Status, Analysis and Reporting decision process. During the initial performance of this decision process the requisite labor budget estimate data is transferred to the appropriate labor cost analysis documents, such as a labor distribution report (budget versus actual labor unit costs). This transfer of the initial estimate data establishes the project labor cost control file. This file would be updated frequently as actual labor costs and actual quantity information are generated during field construction.

The structure of the project management information system as represented in Figure 3.15 is determined by the company's approach to solving the problems related to cost estimating and cost control. In this case, for example, company management must determine whether or not labor unit costs are sufficient for estimating and controlling the labor component of a project, as opposed to using a man-hour estimating and control approach. These decisions will depend to some extent on the type of construction performed, the size of the company, whether automated or manual control system techniques are used, and the level of detail and documentation deemed necessary for management control. These are just a few of the pertinent factors that must be evaluated. Many contractors often select a given system and maintain their management systems as originally designed, while others frequently update and modernize their management systems as a means of improving their competitive position.* Whatever the case, the contractor must understand the management information system he or she uses to manage construction projects, and must see that these systems operate most efficiently when they are integrated, regardless of the level of sophistication of the system and its related technique.

The formal aspects of a management information system are presented in Figure 3.15. In this illustration, data is reflected on documents derived from the performance of a manpower management decision process. However, not all decision processes produce data in document form and, depending on the management approach and construction environment, a decision process may be performed more formally at one level than at another level in construction practice. Consequently, informal management information systems emerge in which data is received by word of mouth or as a result of visual observations, decisions are made based on past experiences coupled with the characteristics of the particular decision-making circumstance, and the resulting decisions and data are passed on to others directly through verbal instructions.

The Crew Planning decision process is an example of a typical decision-making area where informal management information systems are typically found in construction practice. In this decision area, external documents such as drawings and specifications are analyzed and translated into activities for which crew requirements, construction methods, and the like are planned and scheduled in relation to other current on-site activity. The field managers and supervisors responsible for this effort consider and evaluate the physical aspects of the job, its location, access, working environments, material requirements, and the technical details of

*This is especially apparent with the increased capabilities of microcomputers and software such as spreadsheets and data base management systems.

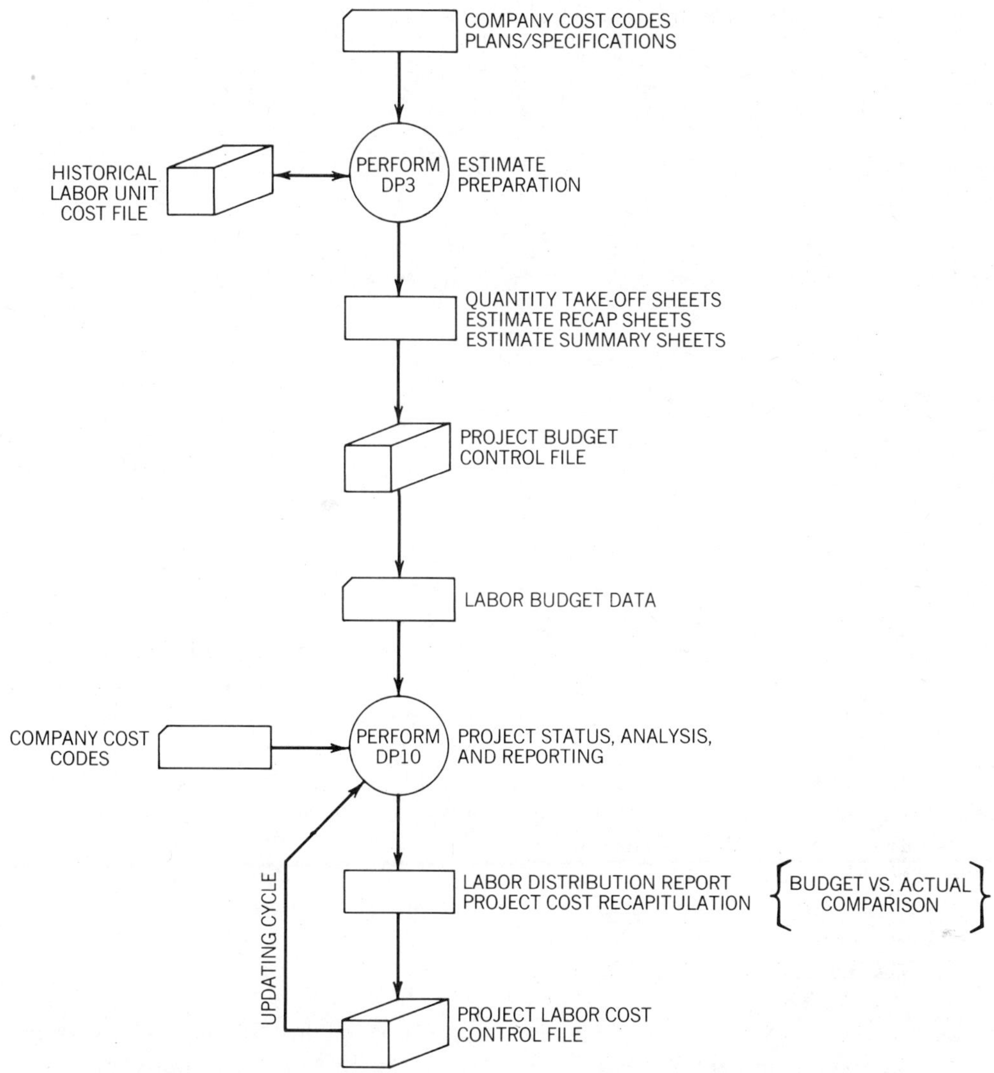

Figure 3.15 A Management Information System to implement a project labor cost control system at the small firm level.

the construction method to be implemented. They relate individual crew members to the various basic and cooperative skills involved in the performance of the construction method. This process occurs during all stages of the crew's planned and scheduled activity. Crew Planning decisions are made from experience and "gut" feelings, and information and instructions are disseminated verbally to other decision-making areas. This constitutes the informal management information system of the Crew Planning decision process.

The chapters to follow will expose typical informal and formal management information systems currently found in practice in the construction industry in relation to the twelve project manpower management decision process areas.

CHAPTER 4
Initiate Project Action

The decision to commit staff resources to a speculative, feasibility, or other type of prebid investigation of a potential project represents the first step in project manpower management in the event the firm is successful in winning the project, or the necessary cost to stay in business in the event the effort is unsuccessful. In either case, this initial decision involves the selection of key technical, engineering, and/or management staff. It further involves the setting up of a task group that must expand its scope of action into the organization, development, and maintenance of a project team charged with the detailed planning and management of whatever project is being investigated. Thus, this decision area considers the manpower impact of the contractor's own staff involvement in a project for whatever time period becomes necessary.

In some cases, and especially at the small firm level, the contractor's image and reputation rests on the efforts of a small "cast of characters." In this situation, the decision to initiate action on a prospective project simply adds a work burden to that small group. For large projects, however, the selection of a suitably qualified project team for this initial project effort is considerably more difficult. This is especially true when a highly specialized team, capable of working with a sophisticated client, must be quickly established by a contractor who has access to a large pool of technical and professional people. In the case of large projects the decision maker cannot simply rely on personal knowledge of a limited number of staff, as he or she could in a small firm. Instead, staff selection becomes a multi-staged, progressively defined process. It first requires the selection of key people, and then relies upon the meaningful use of personnel files to build up the project team progressively. The decision process must be responsive to the project needs of the client, the perceptive evaluation of staff, and must match staff with the needs of a project in a professional manner that will enhance the reputation of the firm.

The decision area is critical to the contractor for a number of reasons relating to:

1. The need to establish the level of risk and profitability potential of the project.
2. The need of the company to maintain a portfolio of projects and construction volume which will, it is hoped, contribute to the profit of the company.
3. The opportunity the project provides for the company to satisfy a client, improve its reputation, and gain favorable treatment with future work.

The successful achievement of all these desirable objectives is heavily dependent upon the ability of top management to sense the requirements of a project and to match those needs against the selection, availability, and skill of suitable key personnel and support staff. Thus, this decision process should always be the responsibility of top management.

4.1 THE BASIC DECISION PROCESS

Initiate Project Action is a decision area that initially commits staff to a project either in a preliminary project investigation before the decision to bid, or in a full scale project proposal effort. Ignoring for the time being the technical and professional work that these staff will perform, the decision area focuses on the management aspects of the contractor's involvement in a construction project. It is a decision area that is very critical and significant for large construction companies, but one that may be automatic or almost irrelevant to smaller companies who simply schedule, and impose, an additional work load on a very limited "cast of characters." At the large firm level, the "cast of characters" for the project team itself is often larger than the entire staff of many construction companies. Consequently, the focus of the project initiation effort changes drastically from what is basically a nonissue for small projects, to the very critical and sensitive problem on large projects of designing, implementing, and managing a project team which may comprise hundreds of people over the life of the project.

Initiate Project Action commences with either a decision to bid or a formal contract award. Top company management personnel review current policies and standard operating procedures in relation to specific contract and client requirements, and then establish basic management control philosophies for the project. This determines the nature, type, and expertise required of key project team positions such as the project manager, lead engineers, estimators and other relevant support personnel. The selection of the personnel to staff each key position can then be made and the design of the project team structure begins.

As the project effort gains momentum and the complexity and scope of the project becomes more definite, the work load and team size requirements and planned growth for the initial project team begins to emerge. On large projects, the rate of growth of the project team necessitates the development of a viable method of evaluating its performance. This requires a continuous matching of the size, structure, and management objectives of the project team effort with clarification of the project scope, technical requirements, and project management control systems to be implemented. Thus, the project team becomes a dynamic and live organization which must respond to the management demands imposed by the needs of the project, including those of the client/owner.

The Initiate Project Action decision area is therefore concerned with the following manpower management problems:

1. The development of a management control approach that forms the basis for determining the anticipated management work load for the project effort.
2. The selection and releasing of key personnel to the project.
3. The staged development and growth of this initial group to a fully operating project team, whether for a short-term initial effort only, or for a full project life effort if the project is won.
4. The subsequent control by management manipulation of the project team positions and staff to meet both management and project objectives and the resulting work loads.

When a construction company begins to address these major problems it must be prepared to focus management attention on a range of interacting decision problems such as those listed in Table 4.1. This table provides a scope description of the complex considerations that can be involved in Project Initiation in terms of a number of major decision problems and their component problem statements. The attention and detail with which these problems are considered by decision makers will depend on the construction environment in which the contractor operates, current staffing profiles and commitments to other ongoing projects, and the short- and long-term needs of the new project effort.

The general procedure for performing Initiate Project Action is given in Table 4.2. The procedure is described in terms of a number of management activities referred to as Management Duties and Tasks, which are arranged in a sequence that generally corresponds to the order of the major decision problems in Table 4.1. The performance of these management duties and tasks is the management action of the decision process and becomes the basis for identifying the relevant Manpower Management Functions included in the functional plot for the Initiate Project Action decision process (see Figure 4.1). Thus, top management must first AL-

TABLE 4.1 SCOPE OF INITIATE PROJECT ACTION PROBLEM AREA

Major Decision Problem	Component Problem Statements
1. What are project management requirements:	What is the nature, scope, and magnitude of the project? What special project contract requirements impact management effort? What are project management objectives and goals? What management approach should be adopted? What are relevant management control systems? What is best project team structure?
2. What project team:	What are key staffing requirements? What is engineering, design, management, and construction breakdown of required project effort? What magnitude of staff effort is required? What is best project team size and staffing distribution?
3. Who is available:	Who are suitable staff? What is location and availability of suitable staff?
4. When available:	When should key staff be released to the project? When can suitable support staff be released/assigned to the project? When are staff training courses required?
5. How is project team to be managed:	How is the effectiveness of the project team to be measured? What staff have been released? How is organizational growth of the project team proceeding? How is the size and growth of the project team being controlled? What management areas need additional support?

LOCATE OFFICE, ENGINEERING, AND DESIGN STAFF (A1) before RELEASING THESE STAFF TO THE PROJECT (W1). In terms of these two basic Manpower Management Functions, the functional statement of the Initiate Project Action decision process is given in Figure 4.2. Note that both the functional scope statements and listing of factors relevant to function performance correspond generally to the component problem statements and related management duties and tasks of Tables 4.1 and 4.2, respectively.

The relationships between the Initiate Project Action decision process and the other manpower management decision processes with which it may interact over the life of a project are shown in Figure 4.3.

4.2 FUNCTIONAL PERFORMANCE TECHNIQUES

A contractor's initial efforts on most projects are those the firm expends in maintaining a search for suitable projects, gathering and considering of relevant information and documents on prospective projects, and ascertaining bid closure dates and related bidding procedures.* Once the contractor has received the plans and specifications and has become aware of a project's basic and special requirements, the identification of the types of expertise required for the project's evaluation and planning

*A brief description of the contractor's initial efforts in this regard is left to the next Section 4.3.

TABLE 4.2 GENERAL PROCEDURE FOR INITIATE PROJECT ACTION

Procedural Step	Management Duties and Tasks
1.	Identify client, contract and/or top management requirements of project management.
2.	Evaluate desirable project team structure and management control systems in relation to project scope.
3.	Establish basic project management guidelines, policies, and procedures.
4.	Identify key project team positions and assess availability of personnel for those positions.
5.	Select key staff and release them to the project.
6.	Design project team structure and identify desirable growth pattern of project team.
7.	Assess availability of other potential key and support staff, select personnel, and release to project.
8.	Monitor management effectiveness of project team, levels of performance, and quality of effort.
9.	Adjust project team size and structure as project conditions, scope, and client/owner requirements change.

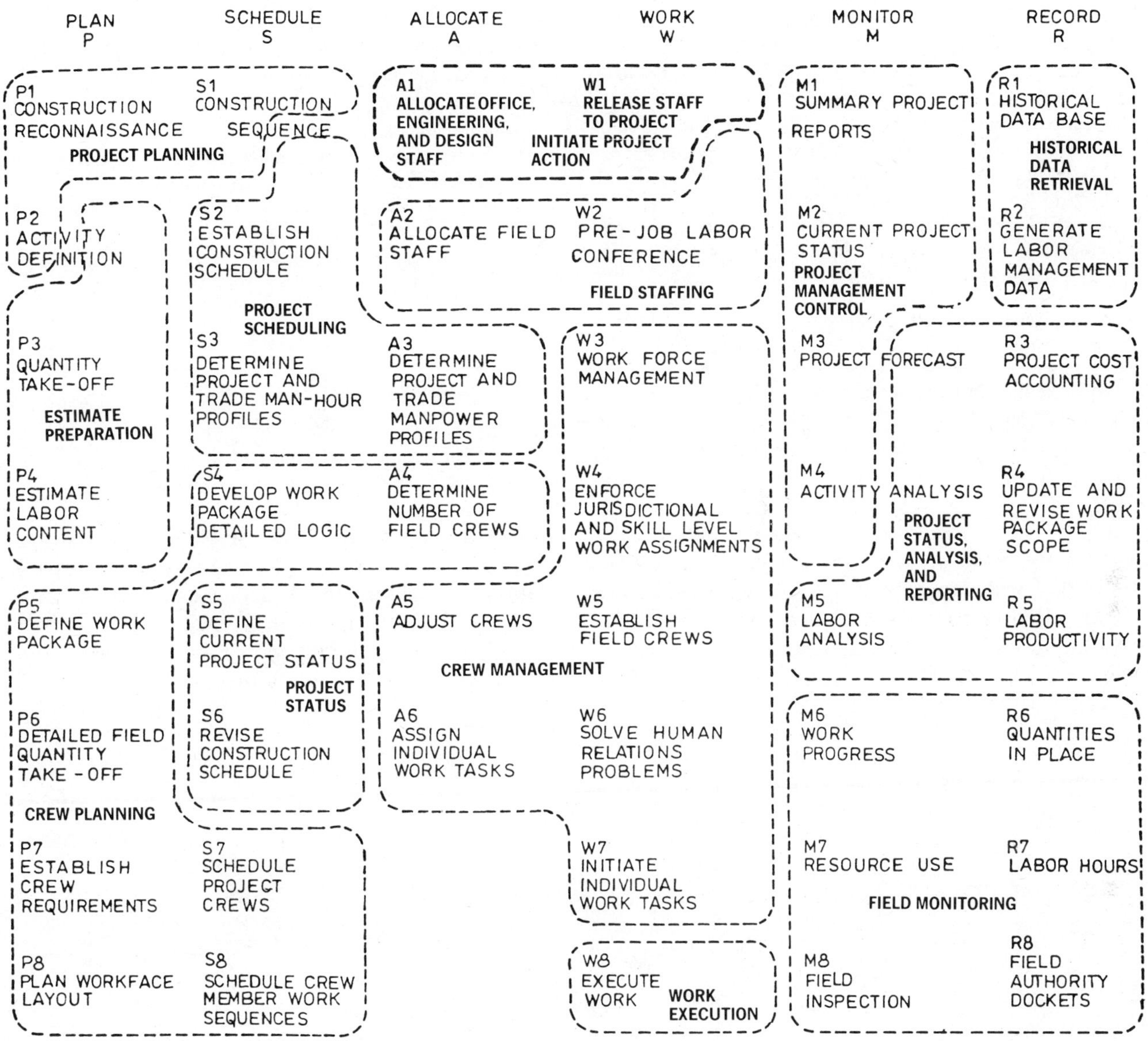

Figure 4.1 Functional plot of the Initiate Project Action decision process.

efforts can be initiated. Hence, except in so far as manpower is used to search for new projects and to collect the pertinent project documents, this initial assessment of staff requirements is the first manpower-focused contact with the project, leading to the Initiation of Project Action.

A similar approach is involved with design-construct situations in which the client, in his project initiation efforts, must indicate the nature and scope of the project to the contractor, and formulate user requirements for the functions that the completed facility must perform. In this environment the contractor must begin to identify project engineering, design, and construction requirements and to nominate the types of people with the particular expertise required by the project. Thus the design-construct contractor must also address a manpower-focused decision problem.

In the construction industry the contractor is usually under considerable time constraints and work volume pressures to mobilize the necessary staff expertise quickly, and to release these staff to

A1 ALLOCATE OFFICE, ENGINEERING, AND DESIGN STAFF

- The selection, allocation, and commitment of best suited staff to the project.
- Requires a knowledge of project scope and magnitude of effort required in project planning, engineering, and scheduling, and the qualifications and skill level required to perform the project definition phase.
- Considers time available; type of project team structure required for project and time frame relationship between engineering, design, planning, estimating, and scheduling groups; work load on and availability of existing staff; need to establish a well-balanced staff; the possibility of supplementing initial effort by transferring staff from other projects or by new hires; and methods of managing the project team's growth, redeployment, and effectiveness as a managing body.

W1 RELEASE STAFF TO PROJECT

- Directs office, engineering, design, and field staff to begin project work.
- Requires authorization from top management and release from previous assignments.
- Considers availability of project decision makers, relevant contract documents, office space, and desirable rate of buildup of the head office and field project teams.

Figure 4.2 Functional statement of the Initiate Project Action decision process.

initial project efforts. Concurrently, the contractor must keep close control of the project team size since this will impact overhead costs. If staff are tardily released to the project, however, critical decisions may either be delayed or be based on inadequately prepared data. Conversely, if staff are released and made available to the project too quickly, they often must wait for properly prepared data before making major decisions and thus cannot be immediately effective. Hence, the contractor must balance conflicting views in the determination of the timely release of staff to the project team. In

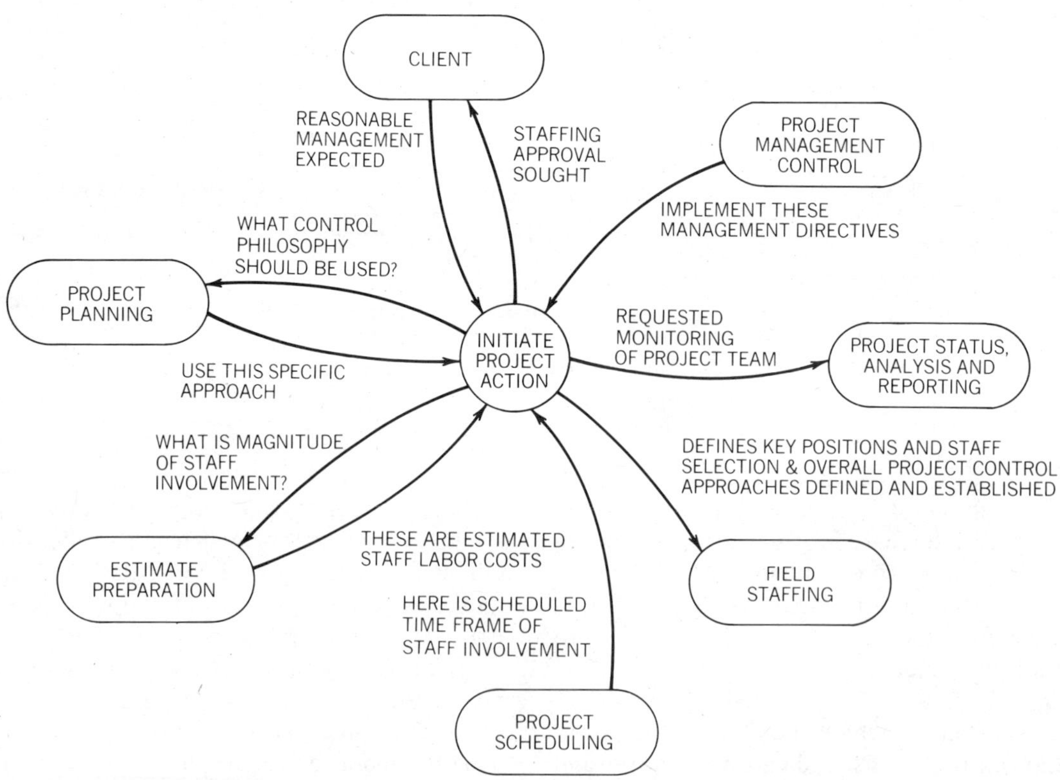

Figure 4.3 Schematic illustration of Initiate Project Action decision process interfaces with other manpower management decision processes.

many cases the contractor relies on past experience when making these staffing decisions. On larger projects it may become necessary to have a more viable and quantitative methodology that can aid the decision maker when making critical decisions on staffing such as planned manpower profiles.

An obvious and very practical approach to the determination of staff requirements for the initial effort on large projects is to refer to previous project experience and to relate this experience in some way to the perceived magnitude and/or complexity of a new project. One method is to define historical factors that are relevant to project team size, such as the number of staff involved, their periods of involvement, and their cost in relation to similar technical areas on previous projects. These factors could be related to simple project measures such as the type of contract, volume of preplanning work required, and the project's estimated capital cost. Other factors that are relevant to the description of the organizational form of the initial project team are the size and distribution within the company of the specialized groups involved, their rate of involvement in response to work loads generated by critical stages in the decision process, and the requirements for information flow and interaction within the decision process.

Most of the factors mentioned above can be readily quantified with little effort, but others need to be carefully documented and processed before they can be made useful as aids in decision making. If more meaningful data is required, management must focus on parameters that reflect the nature of the project involved. Thus, work loads on specific projects can be developed in terms of rough project parameters such as linear feet of piping, cubic yards of concrete, tons of fabricated steel, and so on. This approach can be extended to critical technical and management areas such as, for example, the number of technical drawings required per component facility or the number of specialized management reports normally or specifically requested by the client. Similarly, the size of the contractor's field management team is often related to the number of on-site control areas that will need to be established, or to the magnitude of the total field craft manpower to be managed, and possibly to the number of methods engineers or work study teams needed to achieve planned field productivities. The extent to which historical information on these factors is useful to a contractor will depend upon the ease and rapidity with which such information can be marshaled and focused directly on an impending decision, the critical significance of the impact each factor makes on a project evaluation, and the needs and sophistication of the client.

Simple concepts relating to project team size, and line and staff organizational development are shown in Figures 4.4, 4.5 and 4.6. They present different practical approaches that are based on significantly different management focuses regarding the design and management of project team groups. Figure 4.4 illustrates the management adoption of a standard project team in which the organizational structure and size are fixed for most projects. Figure 4.4*a* illustrates the situation where project team members adopt standard roles so that the only project staff records required are those focusing on staff involvement in a project (see Figure 4.4*b*). On larger projects, work load parameters for the magnitude of specialized group involvement become a more desirable ingredient in project team design. This approach is illustrated in Figure 4.5, wherein a standard departmental structure is pro-

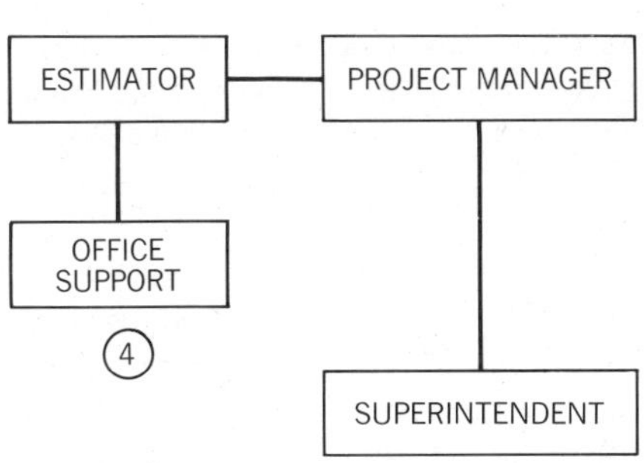

(*a*)

ESTIMATOR	1 MONTH
PROJECT MANAGER	3 MONTHS FULL-TIME, 6 MONTHS PART-TIME
SUPERINTENDENT	8 MONTHS
OFFICE SUPPORT	2 DAYS/WEEK FOR 9 MONTHS

(*b*)

Figure 4.4 Standard project team approach. (*a*) Standard project team. (*b*) Staff involvement estimate.

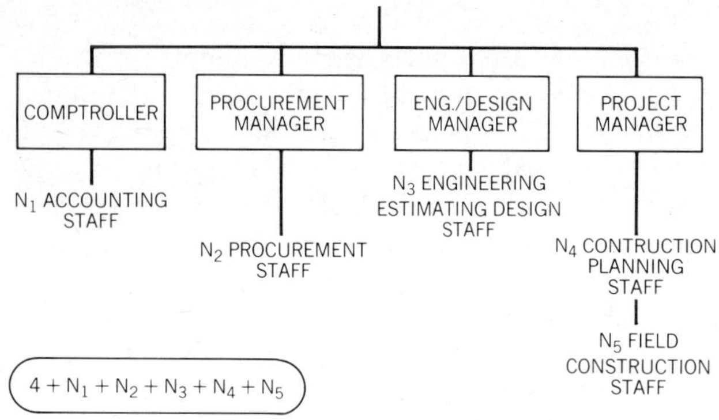

(a)

STAFF REQUIREMENTS

ACCOUNTING	C_1	$\$ \div C_1 = N_1$
PROCUREMENT	C_2	$\$ \div C_2 = N_2$
ENG./DESIGN	C_3	
CONSTN. PLANNING	C_4	
FIELD CONSTRUCTION	C_5	

\$ = PROJECT COST (ORDER OF MAGNITUDE)

(b)

Figure 4.5 Historical factors approach. (*a*) Standard departmental project team approach. (*b*) Historical factored estimate.

posed, but with departmental staff involvement related through historical factors to the estimated capital cost of the new project.

To represent the other end of the spectrum, Figure 4.6 illustrates the management situation where the project team is designed when the project scope has been sufficiently defined and can be quantified in some manner. Thus Figure 4.6*a* develops total man-hour estimates for project team effort in terms of a physical breakdown of the project into individual facilities for which parameters reflecting professional group involvements have been determined. Figure 4.6*b* depicts typical manning profiles which describe the time commitment and magnitude of the professional groups involved. Figure 4.6*c* relates the estimated man-hours and peak manpower profiles of a professional group to a specific project team structure designed to meet the management objectives of the project. While Figure 4.6 illustrates the engineering/design component of a project, the approach is also relevant to the planning and construction functions. Whether a rule-of-thumb approach is used for project team design and sizing instead of considerations based on a set of specially prepared case study reports for similar past projects, depends upon the time frame in which

FACILITY: CRACKING UNIT
COMPONENT: HYDROTREATER
DESIGN:

CATEGORY	NO. DRAWINGS	M-HRS/ DRAWING	M/HR ESTIMATE
FOUNDATIONS	15	100	1500
. . .	. . .	. . .	. . .
TOTAL			N_1

ENGINEERING:

DISCIPLINE	MAN-MOS.	HRS/ MO.	M-HR ESTIMATE
COST ENGR.	11	173	N_2
PROCUREMENT	10	173	N_3
. . .	. . .	. . .	. . .

(a)

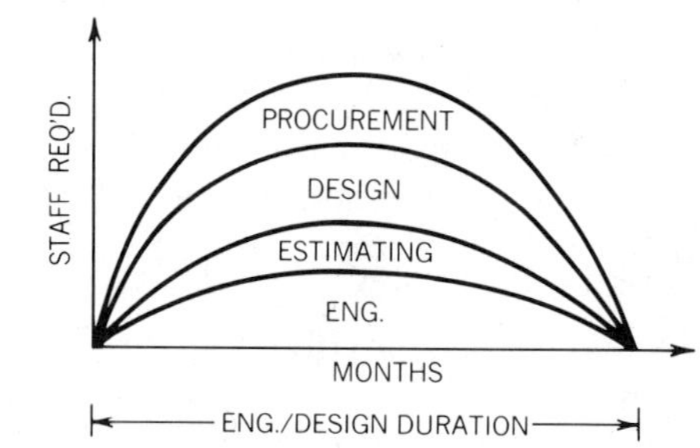

(b)

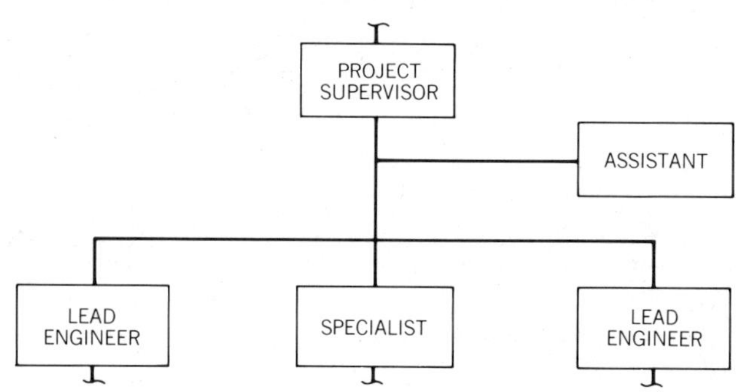

(c)

Figure 4.6 Scope estimate approach. (*a*) Physical component drawings of task man-hour estimate approach. (*b*) Staff involvement profile. (*c*) Organization chart for staff involvement profiles.

decisions must be made, on scope definition, and on the attitudes of top management and the client.

In very large firms it is essential to maintain staff location, work experience, and availability files. These files must be current and include data relevant to an employee's past experience and measures of his or her personal strengths and weaknesses. Thus the staffing needs of a project must be

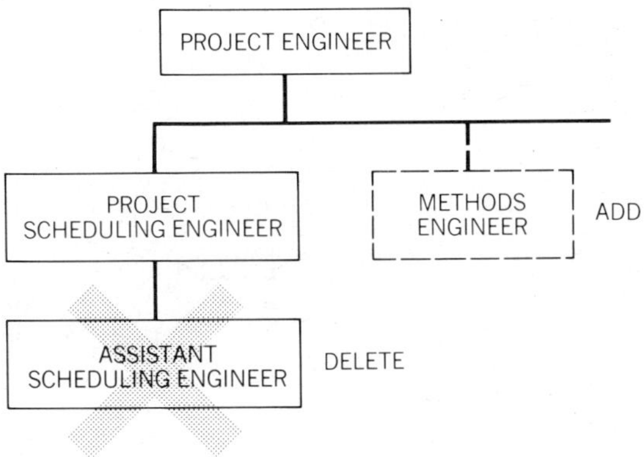

Figure 4.7 Modified project team structure.

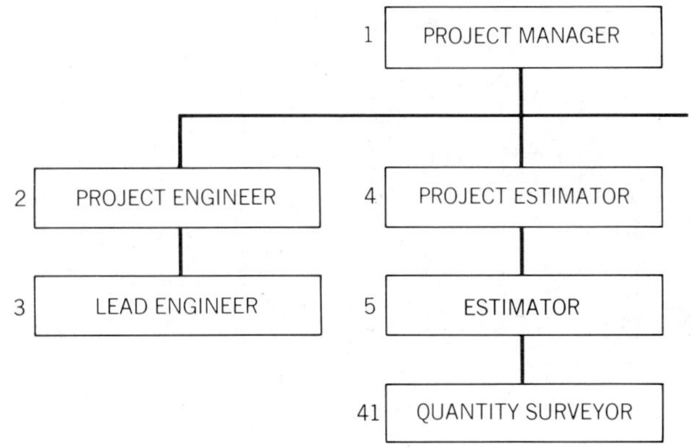

(a)

POSITION	TITLE	AVAILABILITY
1	PROJECT MANAGER	6-86 THRU PROJECT LIFE
2	PROJECT ENGINEER	6-86 THRU PROJECT LIFE
3	LEAD ENGINEER	7-86 THRU 12-87
4	PROJECT ESTIMATOR	8-86 THRU 10-86

(b)

Figure 4.8 Authorized project team structure and manning chart. (*a*) Authorized project team positions. (*b*) Project manning chart.

matched against the firm's inventory of skilled professional staff. While this can often be accomplished by personal knowledge of suitable staff once key staff have been selected, it is very useful to have listings of all possible qualified staff that can be quickly made available during the staffing decision process.

It is essential for project management to be constantly aware of the effectiveness of the project team and to respond to indications of management ineffectiveness such as those characterized by low project team morale or the failure to make timely decisions. Measuring management effectiveness is frequently subjective and therefore difficult. However, by reviewing the frequency with which documents are used and establishing productivity targets, management has available at least some measures for judging project team effectiveness. These measures are more relevant to project control concepts and the role of project management in general. However, changes in project team structure and size, and the releasing and replacing of staff are normally functions of the work load and project life phase unless significant client-initiated scope changes emerge or the contractor needs to exercise more control in specific technical or management areas.

On large projects and especially on cost plus contracts, the project staff is the subject of close client attention and approval. In some cases a contractor will have a standard project team structure and team, and will argue for its adoption. Figure 4.7, for example, represents a portion of such a standard structure, in which the client has requested a modification—replacement of an Assistant Scheduling Engineer with a Methods Engineer. This type of *body transfer* modification to the contractor's proposed project team is often the only means a client can possibly influence a contractor's management approach to a project. A more flexible approach to client involvement in the design of a project team is illustrated in Figure 4.8. In this case the client and contractor may jointly design the project team, in which each staff position is carefully determined. Thus, Figure 4.8*a* indicates a numbered set of authorized staff positions and Figure 4.8*b* the agreed upon Project Manning Chart. The Project Manning Chart concept provides a fix on the project team growth and is essential for both estimating staff man-hours and labor costs, and the timely acquisition and release of project team personnel.

The actual implementation and control of a project team group requires the specification of formal lines of communication, authority, and responsibility for the project management effort involved. This organizational development must also be compatible with the management approach of the contractor. Thus, for example, in the engineering and design stages of the project a contractor may wish to pass the particular project through existing separate departments in a matrix formulation of the project team (see Figure 4.9*a*), or alternatively may prefer

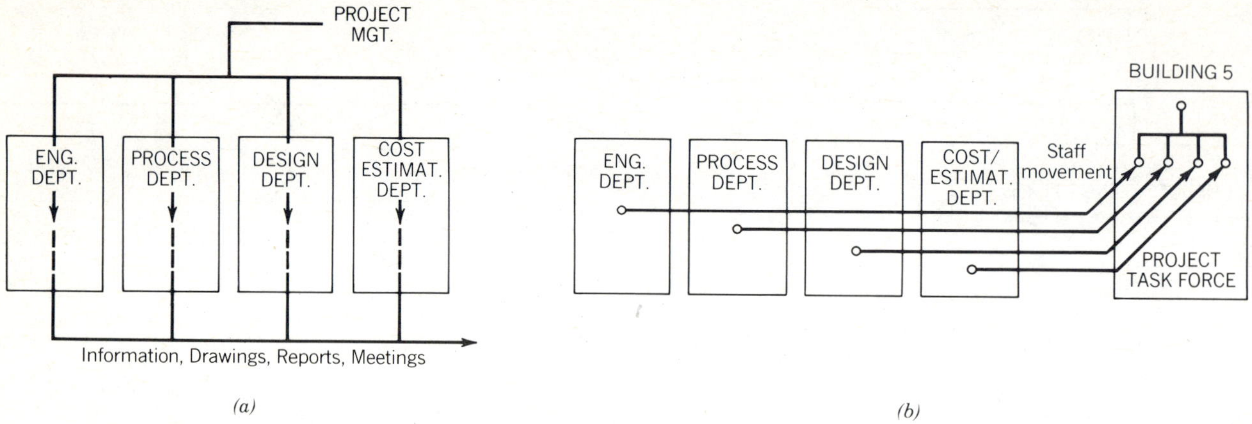

Figure 4.9 Management of project team in engineering and design phase. (*a*) Matrix approach. (*b*) Task Force approach.

to establish a more fully integrated project team effort by collecting a suitable staff from each department involved into a separately located group (see Figure 4.9*b*). The latter approach facilitates relocation in the field if the management approach of the contractor, project size, and client requirements so dictate. In this case the Initiate Project Action decision process merges into the Field Staffing decision process rather than involving a clear separation of head office and field concerns.

The overall objective of the Initiate Project Action decision process is to select, cull, and organize a project team that has the expertise for and capability of executing all phases of the construction project. This objective is accomplished through the implementation of standard company operating procedures combined with the knowledge of the relevant company data on prospective project team members. This decision process is used to produce project team organization charts, project focused operating procedures, staff man-hour estimates, staff manpower profiles, and the like. These data provide the basis for the timely transfer or allocation of company personnel to the project team and, when combined with specific project related inputs from other decision processes, such as project schedules and project control philosophies, become the basis for the often dynamic changes in the project team that occur over the life of the project. The integration and documentation of these data flows through the Initiate Project Action decision process constitutes the Management Information System (MIS) for this decision process. While this MIS is fully documented at the large firm level in the construction industry, at other levels it is an informal information process wherein normal project management roles are assumed for each project by a similar "cast of characters."

4.3 THE PROJECT TEAM COMPONENT APPROACH

At the small firm level, the construction market must be constantly surveyed for potential projects by monitoring bid advertisements, Dodge reports and the like, or by the simple *circuit riding* of the owner/partners. In the initial assessment of a project, consideration is given to its size, location, and complexity in relation to the firm's expertise and capabilities, especially as constrained by its available bonding capacity. Additional consideration is given to the volume of work in hand, the availability of key staff, and the responsible project contract manager. In this latter regard some contractors will refuse work from architect/engineer (A/E) firms who they believe have a reputation for providing inadequate drawings; delaying the authorization of progress payments; or who adopt, in their view, a too-direct interpretation of what is required with respect to quality and unspecified items. The seeking out of suitable projects is considered part of the normal business role of the owners, and is therefore not considered a manpower management problem for which they have to plan and manage the work of others.

The decision to bid can be considered a watershed decision for a contractor, because it commits the firm to a significant investment of scarce resources in a speculative proposition that may not be

TABLE 4.3 POLICIES OF THE SMALL FIRM

Policy	Description
P_S 1 NORMAL ROLES	Project team members assume normal project team roles.

successful. Nevertheless, most contractors must continue to commit and venture the time and skill of their management team in the preparation and submission of bids. In the competitive bidding environment common to the construction industry, the competitor must enter the race to win. The strictly limited and intimate management environment in which the personnel of the small firm operate ensures that the owner/partners are continuously cognizant of the status and activities of each member. Thus, once suitable projects are found, the central manpower management problem and focus is the determination of the project's labor content in terms of labor dollars.

The owner/project manager initiates project action by ALLOCATING OFFICE STAFF (A1) and RELEASING THESE STAFF TO THE PROJECT (W1).* No special techniques are involved in the performance of these management functions beyond the adoption of a tried and true standard project team structure such as that indicated in Figure 4.4, in which head office personnel assume *NORMAL PROJECT ROLES* (P_S1)† and involvement in the project. Thus, these functions are not executed formally and rarely constitute a significant manpower management decision problem. Each function is shown as part of the decision process for project initiation, because together they represent actions which must be taken prior to starting estimate preparation. If, for instance, there are only two estimators, a management decision would be made and one project estimator would be assigned to the proposed project.

*Whenever specific mention is made in the text to the performance of, or documentation for, a particular named manpower management function, the function name is capitalized. Thus, (referring to Figure 4.1), the manpower management function A1 is ALLOCATE OFFICE, ENGINEERING, AND DESIGN STAFF, and function W1 is RELEASE STAFF TO PROJECT. Notice that the function names are often slightly altered to fit the syntax of the textual sentence in which they are embedded.

†Policies that relate to how management functions are to be performed are shown in italicized capital letters. This distinguishes them from management functions. In addition, small firm policies are identified with a subscript $_S$ as shown above. Similarly, the subscripts $_M$, $_L$, and $_T$ are used to denote medium-sized and large firm policies and those associated with a firm using the total field autonomy management approach, respectively.

The formal decision process is illustrated in Figure 4.10, and the formal description of *NORMAL PROJECT ROLES* is provided in Table 4.3.

4.4 THE TRAVELING PROJECT MANAGER

The medium-sized firm is often solicited by potential clients to submit bids on selected projects. Depending on project size, the medium-sized contractor will often quickly evaluate the dollar magnitude and requirements of the project from the plans and specifications, in order to eliminate those projects which are undesirable because of their complexity, high risk nature, or which are too demanding on financial and cash flow resources. When selecting projects for which a bid will be prepared, executive management must be particularly cognizant of project merit versus profit potential. This consideration is critical, since each project must be successfully completed if the contractor is to maintain the dollar construction volume necessary to ensure the stability and continual growth of the firm. Consequently, projects which a small firm would eliminate simply because of their location or size, may be considered viable projects by medium-sized firms. This greater flexibility is mainly due to the range of project types available and the ability to move outside the local work region through satellite offices.

The project team is mobilized when the owners decide that a bid should be prepared for a project. An executive manager ALLOCATES OFFICE STAFF (A1) to the project according to the nature and complexity of the project, and a decision on the *PLANNED ROLES* (P_M1) (see Table 4.4) of staff and the desired project team structure is made. If the project is small and requires less management control, a project team structure similar to that of Figure 4.4 might be selected. On larger projects, where complexity might demand more management control, a standard project team organization like the one shown in Figure 4.11 might be implemented. Once the PROJECT TEAM IS RELEASED TO THE PROJECT (W1) by the executive manager and bid preparation commences, the project becomes the responsibility of the estimator and the project manager.

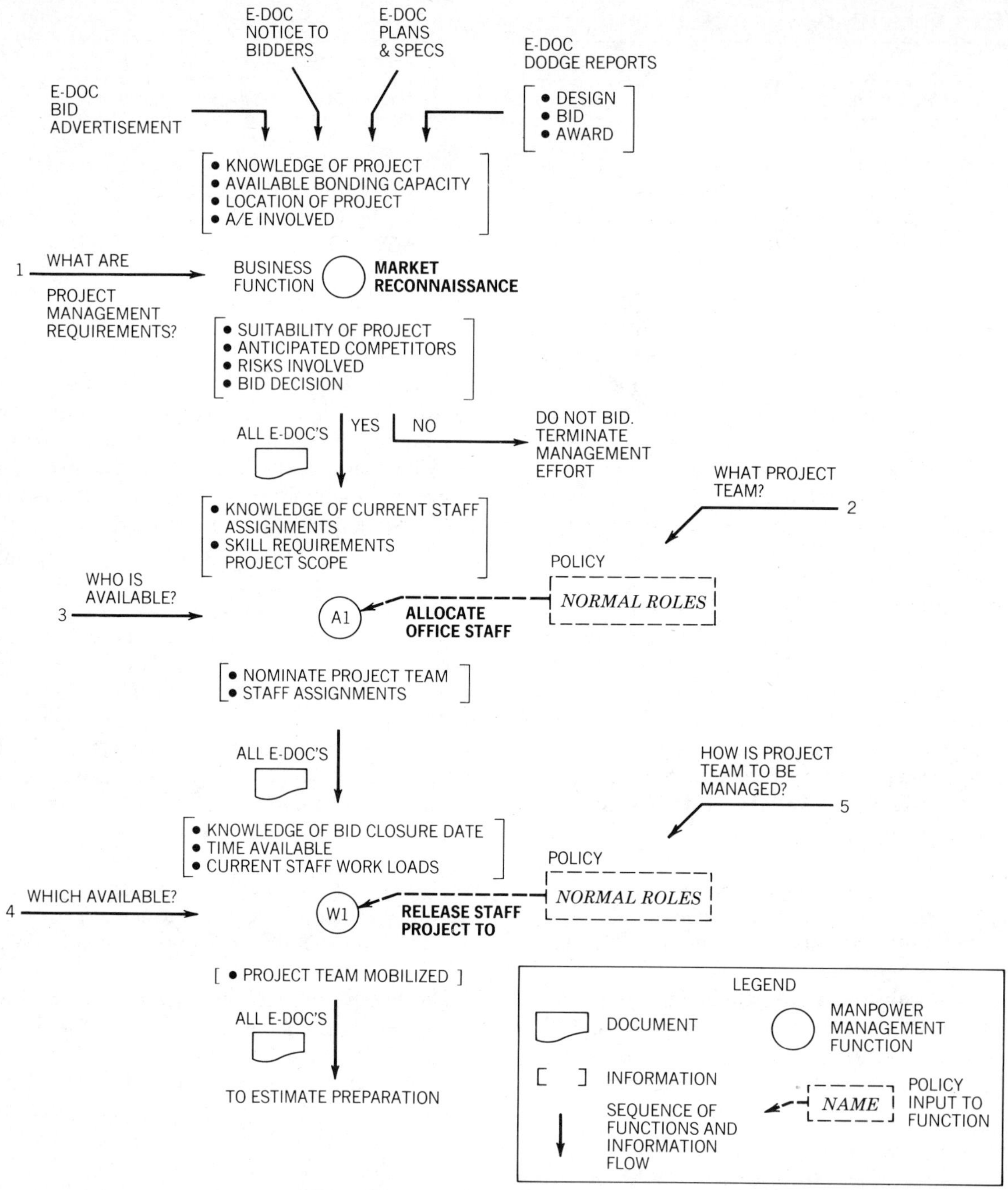

Figure 4.10 Initiate Project Action decision process (Project Team Component approach).

TABLE 4.4 POLICIES OF THE MEDIUM-SIZED FIRM

Policy	Description
P_M1 *PLANNED ROLES*	Project team structure and assignment of decision makers to a project depend on project complexity and size, and on the division of management functions desired to effectively manage the project.

Initial Project Action for the medium-sized firm differs from the approach of the small firm in that the key project field superintendent is selected and released to the project at this early time in order to gain the advantage of his participation in Project Scheduling and Field Staffing decisions. The decision process remains relatively simple except when large projects require a closer analysis of the project team's size and the functional roles of the staff involved. The decision process is summarized in Figure 4.12. Note its early interaction with the Field Staffing decision process and the emergence of a formal Project Organization Chart (P_M-DOC 1), (see Figure 4.11).

4.5 FIELD-BASED PROJECT MANAGEMENT

At the large firm level, the decision to evaluate a prospective project is often approached in a significantly different manner from that of the medium-sized firm. The large firm tends to solicit more business by selling their expertise to prospective clients. Consequently, these firms often create marketing and sales departments with full-scale programs for seeking out potential business. This approach requires the intensive and continuous surveillance of all available markets and clientele and, when appropriate, attempts to solicit requests from prospective clients to prepare and submit a

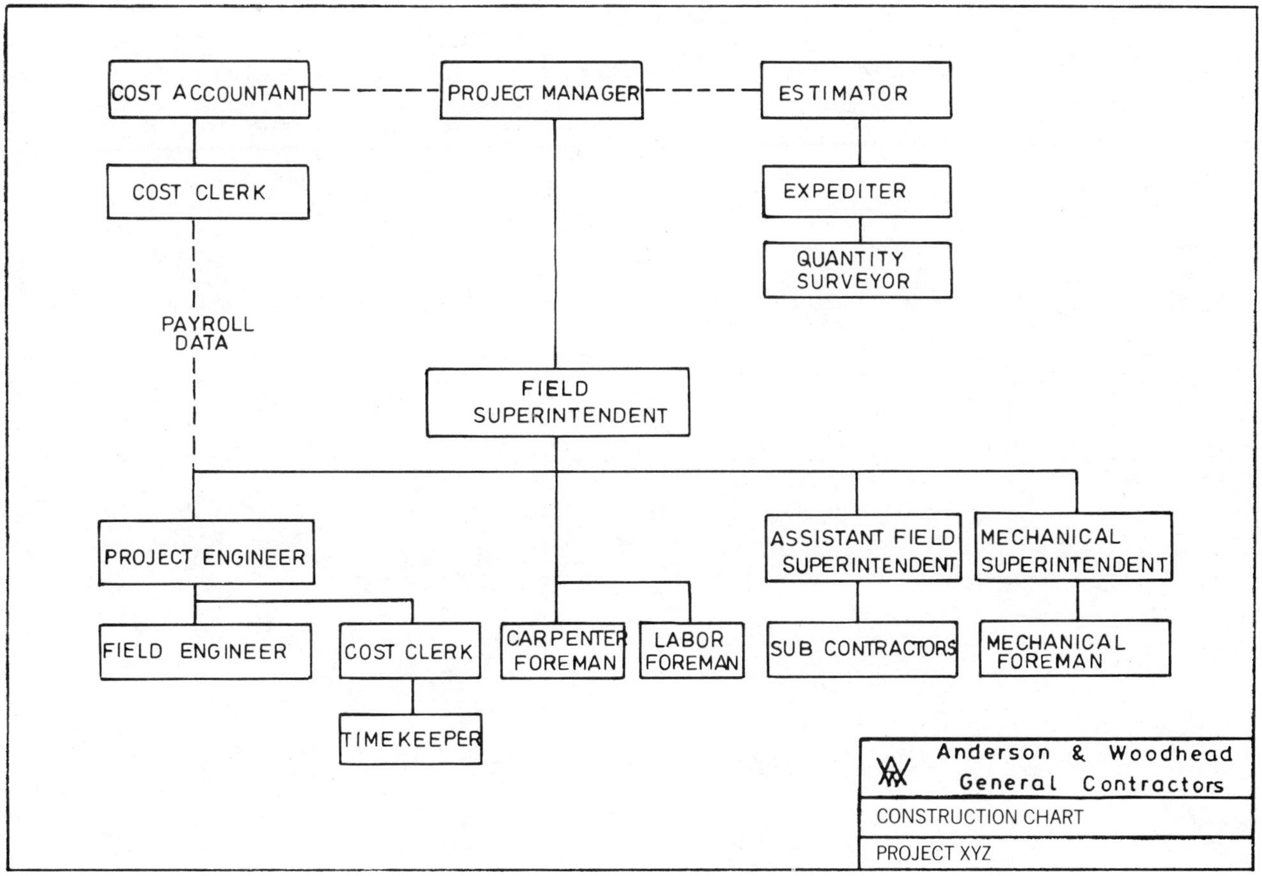

Figure 4.11 Project Organization (P_M-DOC 1).

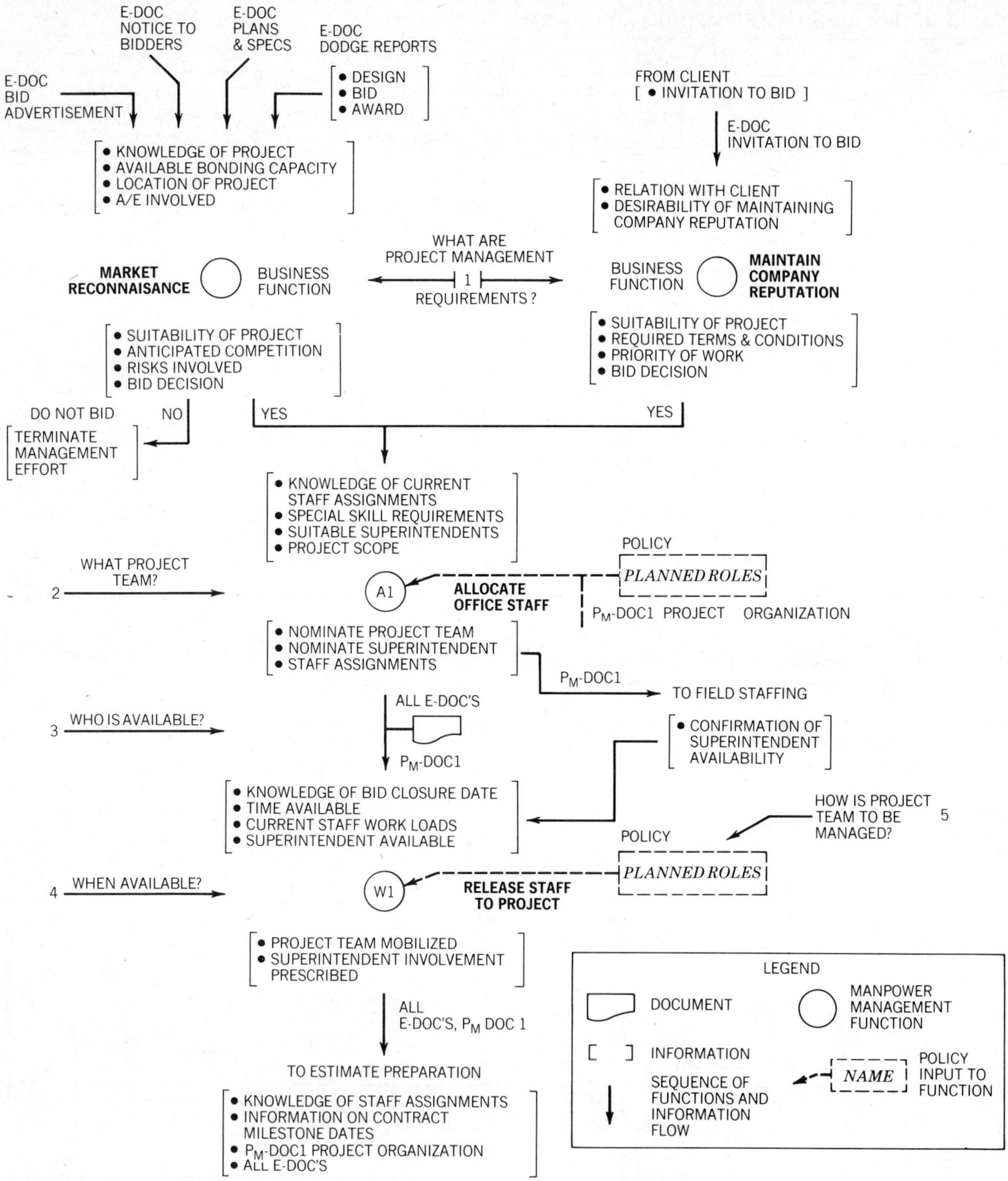

Figure 4.12 Initiate Project Action decision process (Traveling Project Manager approach).

bid, or proposal, for a project. In preparing a proposal, the effort may include, but not be limited to, the identification and potential commitment of key project management personnel; the presentation of typical project team organizational structures and control practices; and considerable project planning, estimate preparation, and project scheduling efforts. Consequently, in order to increase the potential for success, the contractor must make a considerable commitment to setting up, manning, and managing an information network to detect, screen, and evaluate potentially suitable projects for top management consideration. Figure 4.13 illustrates schematically this project solicitation decision approach.

When the contractor is preparing a fixed priced competitive bid based on a well defined scope, its Initial Project Action effort is similar to that of the medium-sized firm. The nature, complexity, and size of the project, however, will demand that the contractor take more care in the sizing and design of the project team structure than that required for the adoption of a standard team, as implied in Figure 4.11. In contrast, for design–construct projects to be compensated on a cost plus basis, the contractor, while relieved of the pressure of a competitive bid, must relate project team size to the demands for engineering and design information, which will change over the life of the project. Thus, various stages of project team structure and growth will occur, requiring the use of techniques such as historical factoring (see Figure 4.5) in initial staffing decisions.

The large firm, when operating in the Field-Based Project Management mode, is frequently involved in design–construct projects compensated on a cost reimbursable contract basis. Consequently, this project environment will serve as the basis for viewing Initiate Project Action.* While the project emphasis will be on the firm responsible for both the design and construction phases, the policies, functions, and techniques presented will apply for the most part to the contractors involved respectively only with either engineering or construction.

Full project execution is initiated when the firm is awarded a contract based on its proposal. This step is when project team mobilization can commence with OFFICE, ENGINEERING, AND DESIGN STAFF ALLOCATION (A1) and RELEASING THESE STAFF TO THE PROJECT (W1). Unlike small and medium-sized firms, large firms must treat Initiate Project Action as a critical manpower management problem area, because it poses the problems of the design and use of project teams comprised of large work forces. Figures 4.14*a*, *b*, and *c* illustrate the information flow structure and pertinent policy statements for this decision process, for a typical design–construct project. The initial mobilization effort phase is illustrated in Figure 4.14*a*, the staged development phase in Figure 4.14*b*, and the maintenance of the fully operational phase in Figure 4.14*c*. Figure 4.14*c*, for instance, displays the fully operating phase of project initiation, based on control budgets and schedules; the continuous analysis of project status; progress and labor performance; and the impact of changes in scope, design development trends, and overall project management directives that result from production engineering and design.†

At the inception of a project, the focus of the OFFICE, ENGINEERING, AND DESIGN STAFF ALLOCATION PROCESS relates to the design of the project team and the selection of key project personnel for both the home office and field, as shown in Figure 4.14*a*. These key project assignments are made by executive level management and include the personnel who will be most influential in the management process, such as the project manager, construction manager, project engineers, key lead engineers, and so on. The next level of project team assignments are those involving support functions like procurement, accounting, and cost and

*The design–construct project that is performed on a cost reimbursable basis will be the illustrative example for viewing all the decision processes as each relates to Field-Based Project Management.

†The detailed decision process for the management and control of the production of engineering specifications and construction drawings is outside the scope of this book. Decision processes for this area could be formulated provided function titles and descriptions are structured to focus on the detailed engineering and design process. As an example, Project Scheduling (S1, S2, S3, and A3) would cover engineering/design schedules from a general management perspective, as well as the detailed planning and control of the production of specific drawings. We believe that the management aspects of the design, development, and control of the engineering/design project effort, as it relates to the overall project management process, should be treated as part of the manpower management problem area. Thus, some decision processes will incorporate brief discussions of the engineering and design aspects of manpower management. The majority of the decision processes, however, will focus on the construction process rather than on the decision processes related to producing construction drawings.

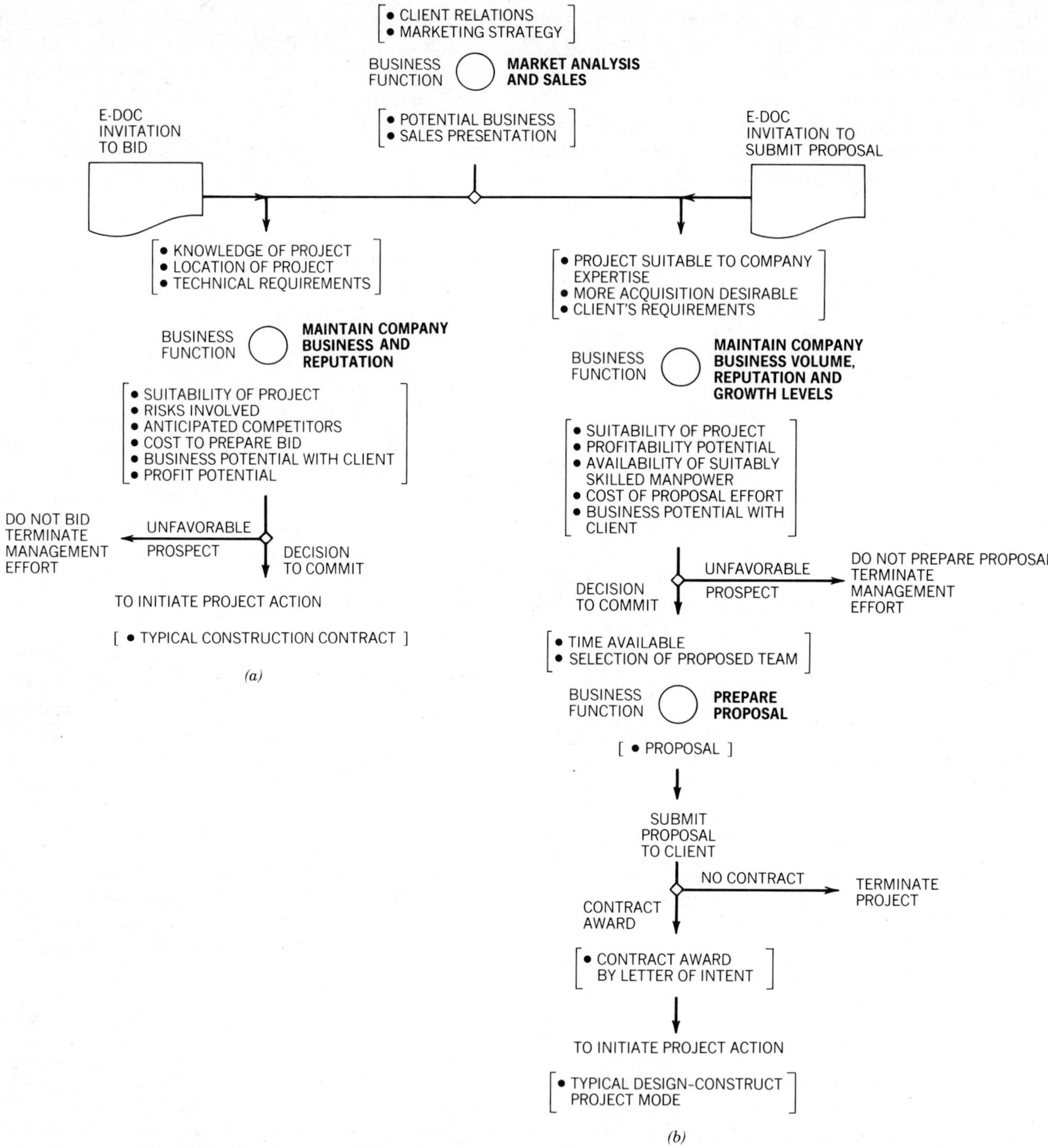

Figure 4.13 Project Solicitation decision process at the large firm level (business focus). (*a*) Fixed scope with detailed plans and specs for construction phase only on fixed price basis. (*b*) Broad scope concept for design–construct project on cost plus basis.

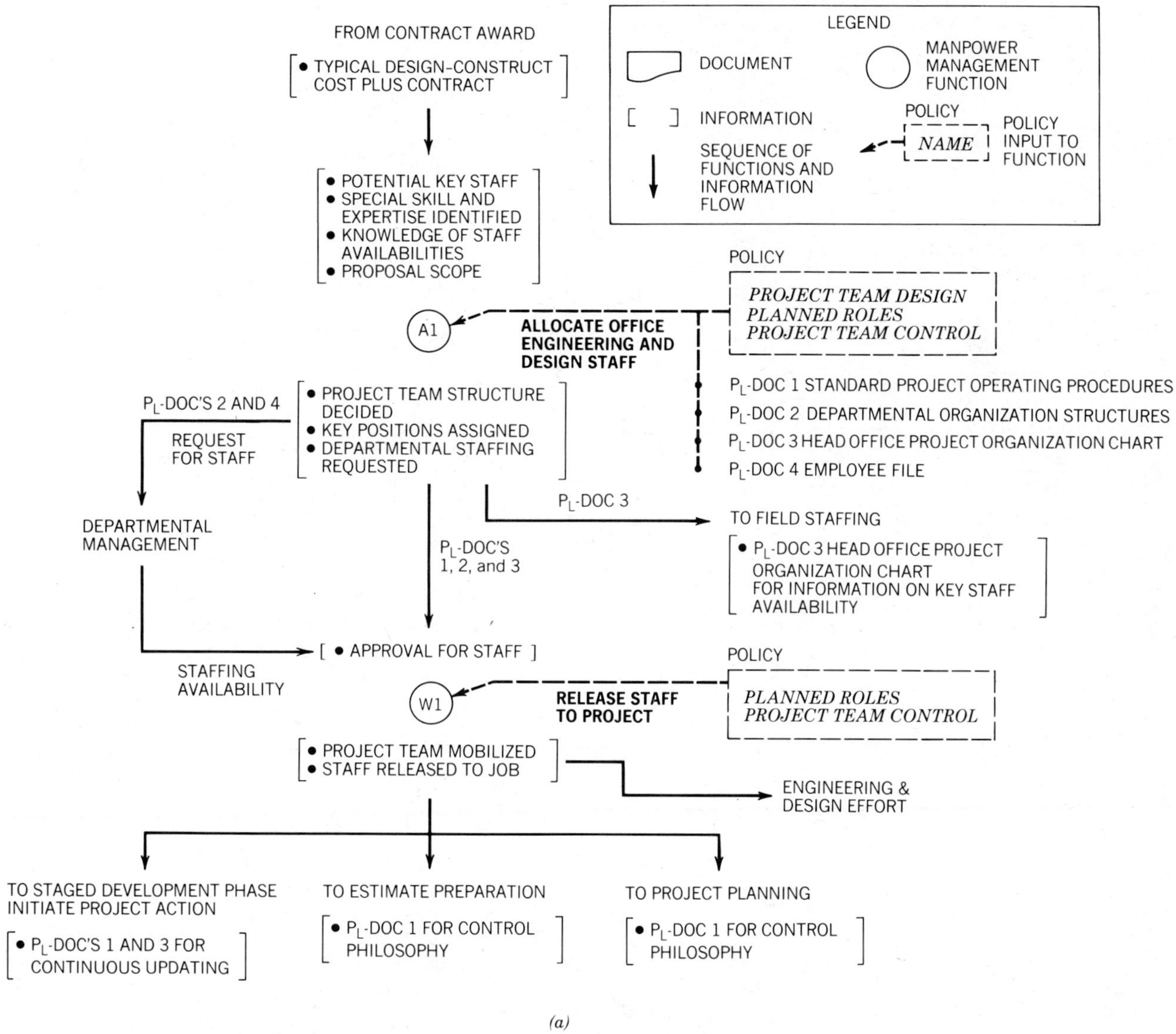

Figure 4.14 Initiate Project Action decision process (Field-Based Project Management). (*a*) Mobilization phase.

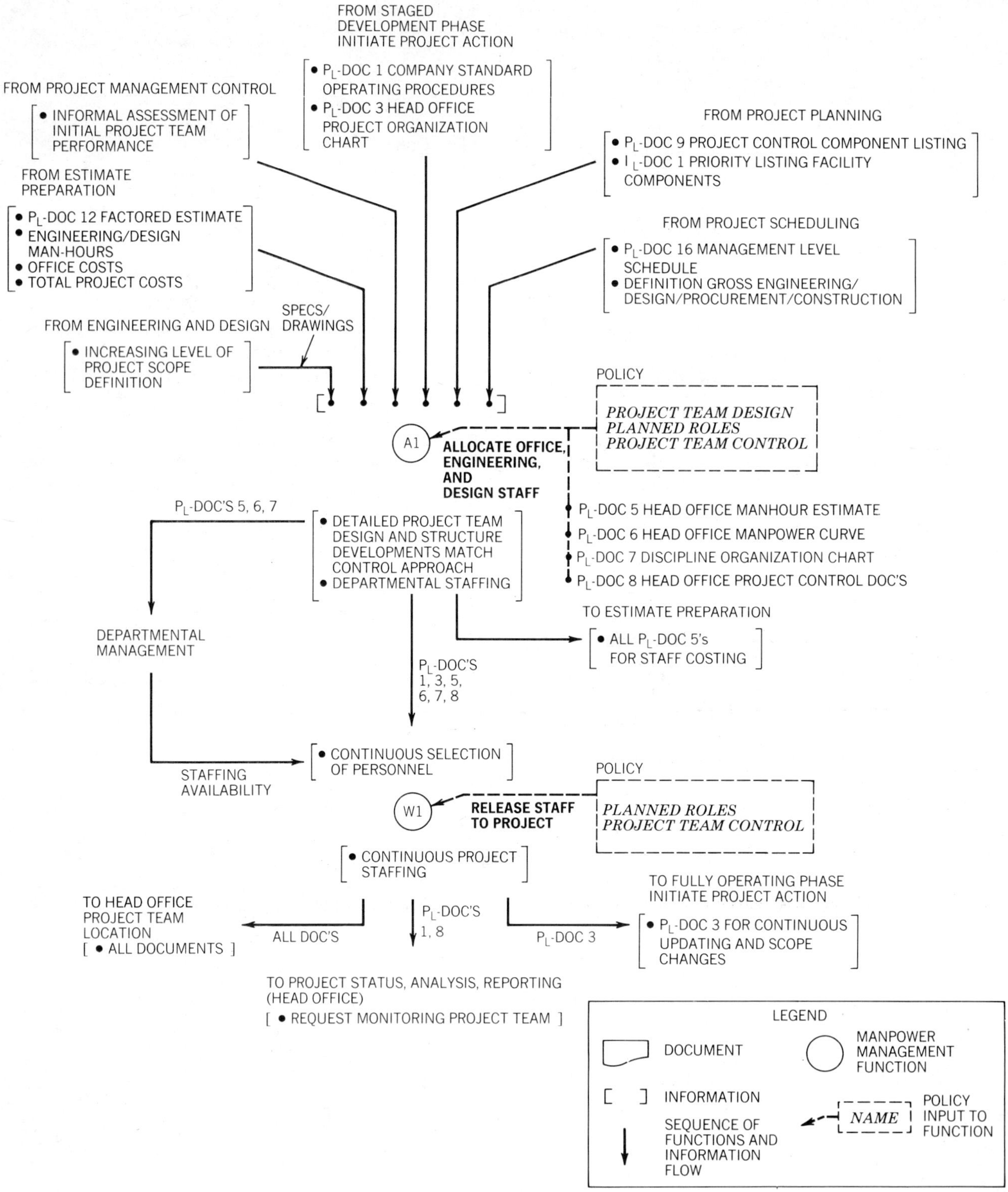

(b)

Figure 4.14 (*Continued*) (*b*) Staged development phase.

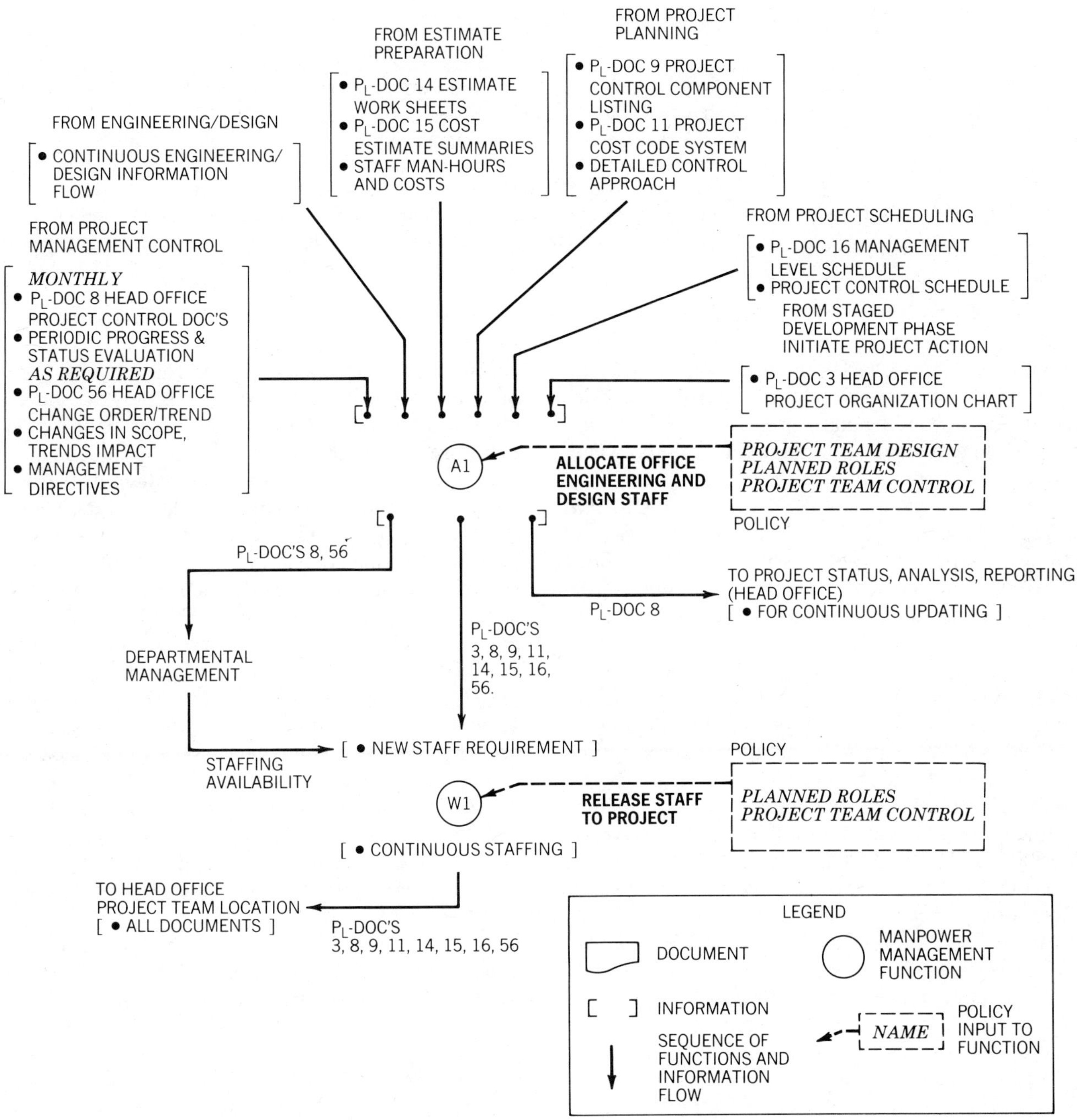

(c)

Figure 4.14 (*Continued*) (*c*) Fully operating phase.

TABLE 4.5 POLICIES OF THE LARGE FIRM

Policy		Description
P_L 1	*PROJECT TEAM DESIGN*	The project team structure is designed to maximize management control by giving the key managers full authority and responsibility for project execution.
P_L 2	*PLANNED ROLES*	The project team must have a complement of personnel so that all functional areas are assigned manager responsibility in order to achieve the desired level of management control as determined by standard company operating procedures.
P_L 3	*PROJECT TEAM CONTROL*	Project team assignments must be planned and monitored to ensure a smooth buildup as work intensifies and declines, and project management must be based on standard company project execution procedures.

planning engineering. The number of key management personnel assigned to the above categories could vary depending on the project's size and complexity, its technical nature, and the desired level of project management control.

The policies which govern the initial OFFICE STAFF ALLOCATION effort stress the project team concept (see Table 4.5). *PROJECT TEAM DESIGN* (P_L 1) provides the impetus for designing a project team with clear levels of key management responsibility and authority, and lines of reporting to these decision makers. The emphasis on *PLANNED ROLES* (P_L 2) ensures that the project team is initially staffed with personnel representing the requisite disciplines for initial project execution. Finally, because of the limited scope of the mobilization phase, project assignments must be planned to match only the initial requirements of the project for *PROJECT TEAM CONTROL* (P_L 3).

Project execution is guided by Standard Project Operating Procedures (P_L-DOC 1) as indicated by the *PLANNED ROLES* and *PROJECT TEAM CONTROL* policies. These procedures are written documents contained in manuals and are developed by each department within the company. As an example, the cost and planning department will have procedure manuals which describe separate but interrelated cost control techniques for both the head office and field locations. Company procedure manuals will influence all the decision processes in terms of the techniques and relevant documentation that emerge during project execution. The use of these manuals is first introduced during Initiate Project Action as stipulated by the above policy statements, and is hereafter considered as the basis for the documents generated in all the decision processes. Standard Project Operating Procedures (P_L-DOC 1), as schematically illustrated in Figure 4.15, will only be referred to as a source of specific project-oriented documents. Thus, for example, the Standard Company Cost Code is used as the source for the dissection of the Project Cost Code, and therefore will be referenced as a policy document.

A critical decision that must be made in the Initial Mobilization Phase (Figure 4.14*a*), is the determination of the organizational structure in terms of lines of reporting and staff relationship between key project decision makers. In many instances, the task force approach is selected. This means that project accountability is more project focused, and administrative links are maintained between the project team disciplines and their respective departments (see Figure 4.9 *b*). In the task force approach, project personnel are physically located together as much as possible. An alternative is the matrix approach, where functional accountability is maintained between the project team disciplines and their departments, while project accountability governs all relationships with the project manager (see Figure 4.9*a*). In this approach, the participants in the project may be located within their respective departments, and the project work is filtered through those departments. Whether the task force or matrix approach is selected, the various departments, which comprised the overall company operating structure, furnish the personnel for project-oriented work.* Departmental Organizational Structures (P_L-DOC 2) describes two company departmental organization structures typically seen in large design–construct firms (see Figure 4.16).

When the task force technique is selected, project personnel are provided to the project from each department, and normally the project occupies one floor or a specific building the firm may own. Thus,

*A combination of task force and matrix approaches is often used in today's competitive construction environment to lower project costs.

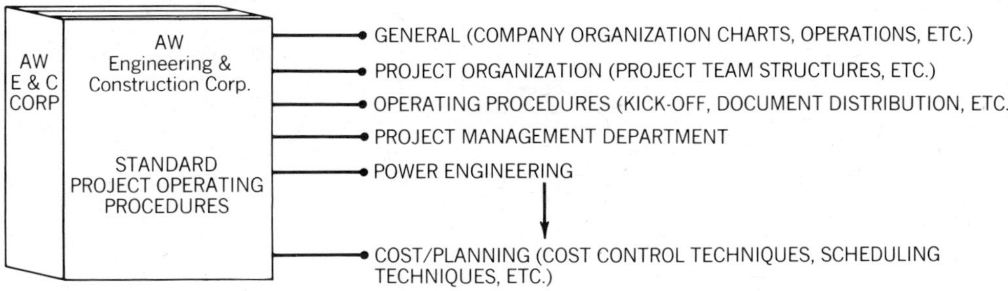

Figure 4.15 Standard Project Operating Procedures (P_L-DOC 1).

the initial structure of the Head Office Project Organization Chart (P_L-DOC 3), (Figure 4.17) reflects the breakdown of the departmental involvement in the project. The size of each functional area, such as engineering or planning, will be determined using historical factoring estimate techniques, based on project cost parameters, as presented in Figure 4.5. The results of this approach to estimating home office manpower can be formulated into total project office staff man-hours and labor dollars, and incorporated into factored estimates. This method also provides project management with an early indication of the magnitude of the head office effort needed, especially in relation to the project's direct field costs.

After establishing the basic project team structure and selecting the key project team personnel, the chosen decision makers can be RELEASED TO THE PROJECT according to *PLANNED ROLES*. Company personnel files, such as the Employee File (P_L-DOC 4) in Figure 4.18, illustrate a typical example of data available for matching company personnel with available projects.

Once the project moves into the Staged Development Phase of Initiate Project Action, Figure 4.14*b*, OFFICE STAFF ALLOCATION produces detailed organization charts based on scope estimates that reflect man-hour content and the planned manning to match Management Level Schedules and fit the desired project control approach. This scope estimate technique (see Figure 4.6) is the basis for gaining a better fix on the project team organization in terms of the documents (P_L-DOC 5, 6, and 7) shown in Figure 4.19. While the documents in Figure 4.19 portray only one functional area within engineering and design, if each functional area produces similar documents, a total engineering and design organization chart and manpower plan will result. When the other support groups (e.g., cost, planning, purchasing, etc.) are incorporated, a total office project manpower plan and staffing structure is defined. This phase of project initiation is reviewed continuously until final control budgets and manpower plans are approved by project management and by the client. Typical Head Office Project Control Documents (P_L-DOC 8), as related to the head office engineering and design effort, are illustrated in Figure 4.20.

When the client is involved in the approval of project organization charts, which is often the case for a cost reimbursable project, the project team may be modified in conjunction with client requests for a particular control emphasis in a given area. In addition, when the project moves into the Fully Operating phase of Initiate Project Action, Figure 4.14*c*, management may desire to modify the project team based on performance evaluations derived from the active monitoring of engineering/design progress and the status of a project, as well as management's assessment of the project team's effectiveness in managing the project. Techniques such as deleting and adding project positions (See Figure 4.7), may be utilized to expand the effort in one management phase while maintaining the original budget. Alternatively, the contractor may request additional project personnel through change orders based on a number of significant scope changes.

Both the Staged Development and Fully Operating phases of Initiate Project Action are dominately influenced by *PROJECT TEAM DESIGN* and *PLANNED ROLES* policies. Consequently, the project team structure and planned manning are continuously reviewed and revised as the project demands change. Furthermore, as capital cost estimates and project schedules are prepared based on well defined scope documents, a better definition of the office manpower effort can be made, which may result in modifications to the project team. In order

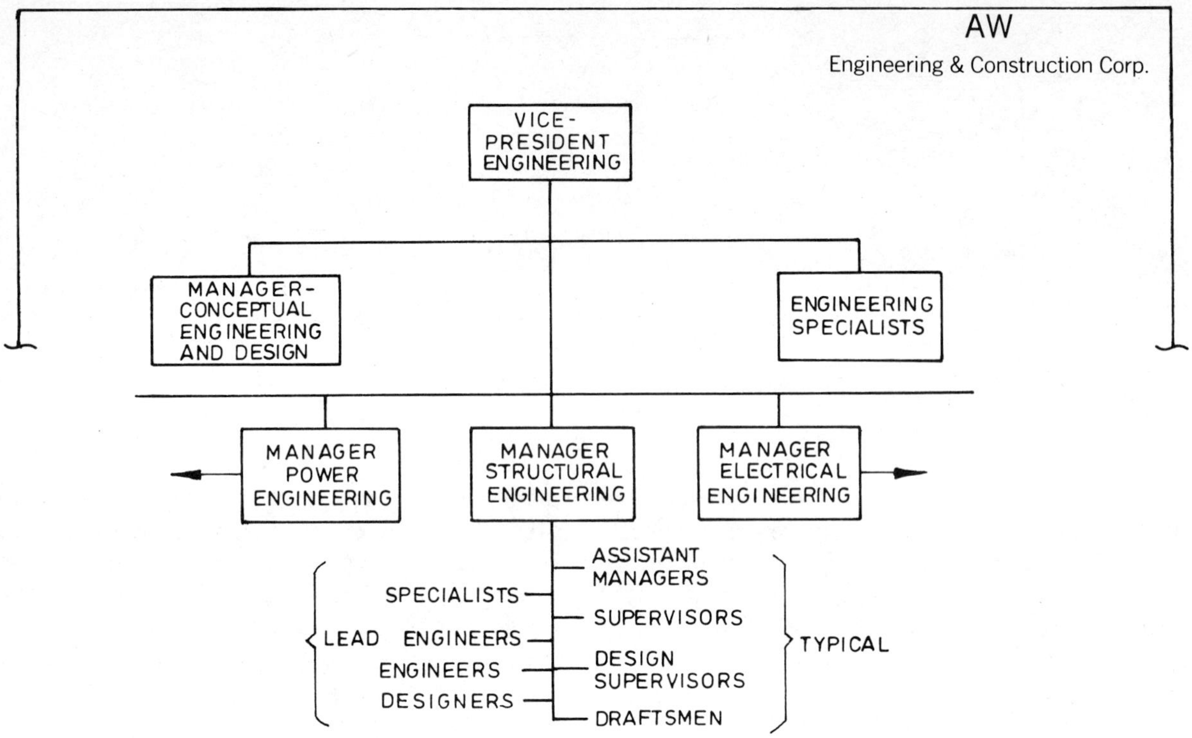

(a)

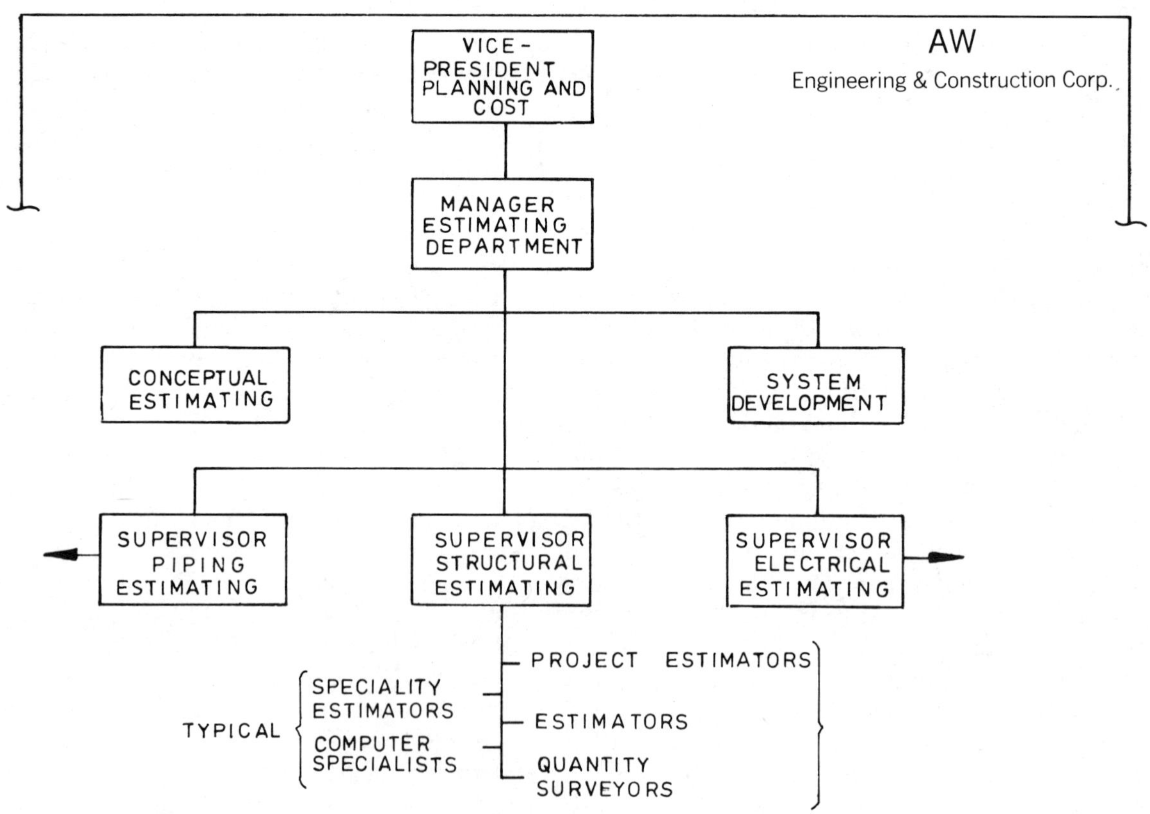

(b)

Figure 4.16 Departmental Organization Structures (P_L-DOC 2). (*a*) Engineering/design department. (*b*) Estimating department.

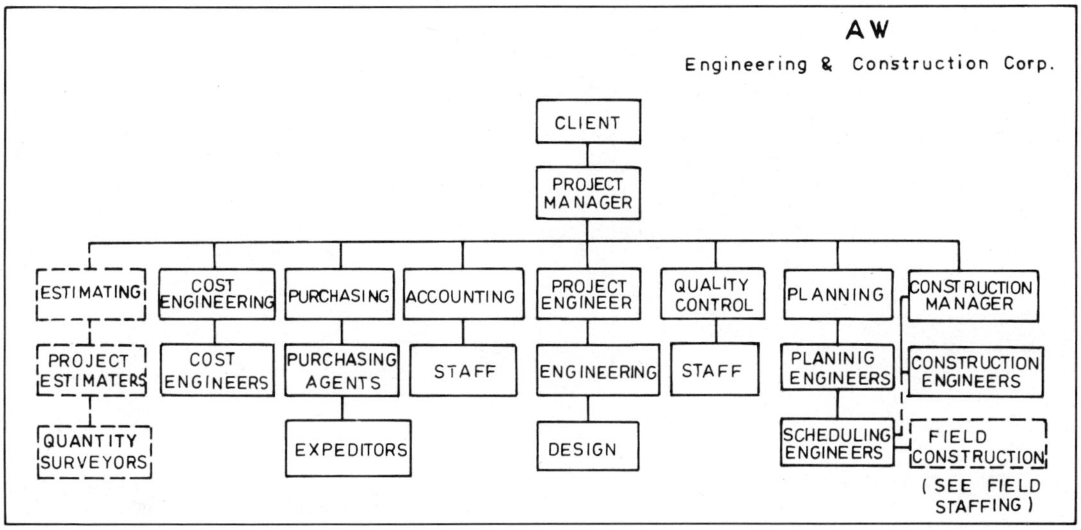

(a)

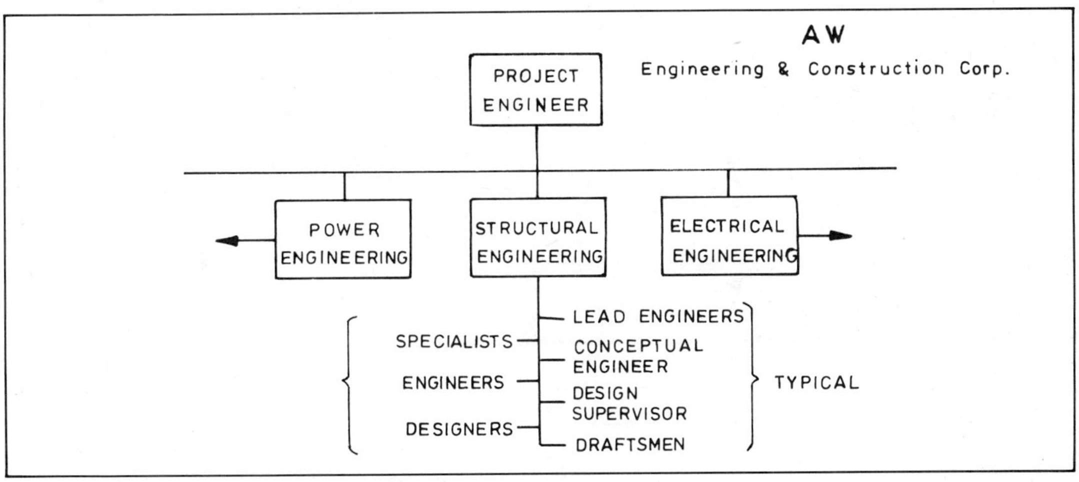

(b)

Figure 4.17 Head Office Project Organization Chart (P_L-DOC 3). (*a*) Overall head office project team. (*b*) Engineering and design project team.

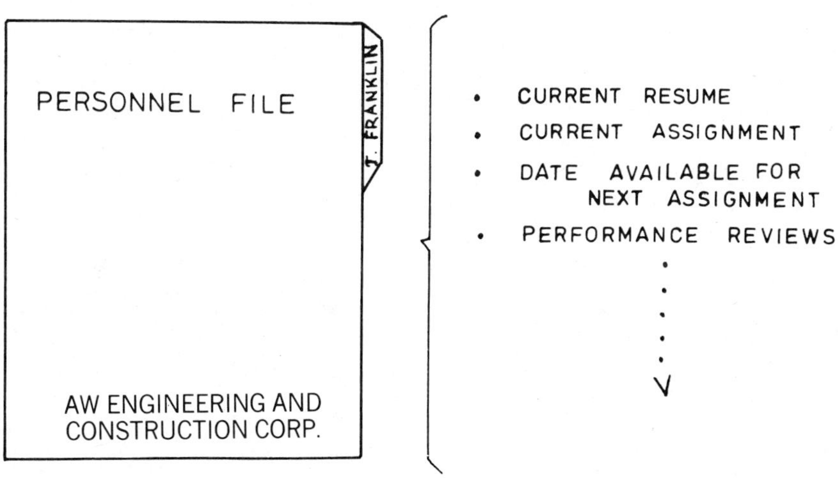

Figure 4.18 Employee File (P_L-DOC 4).

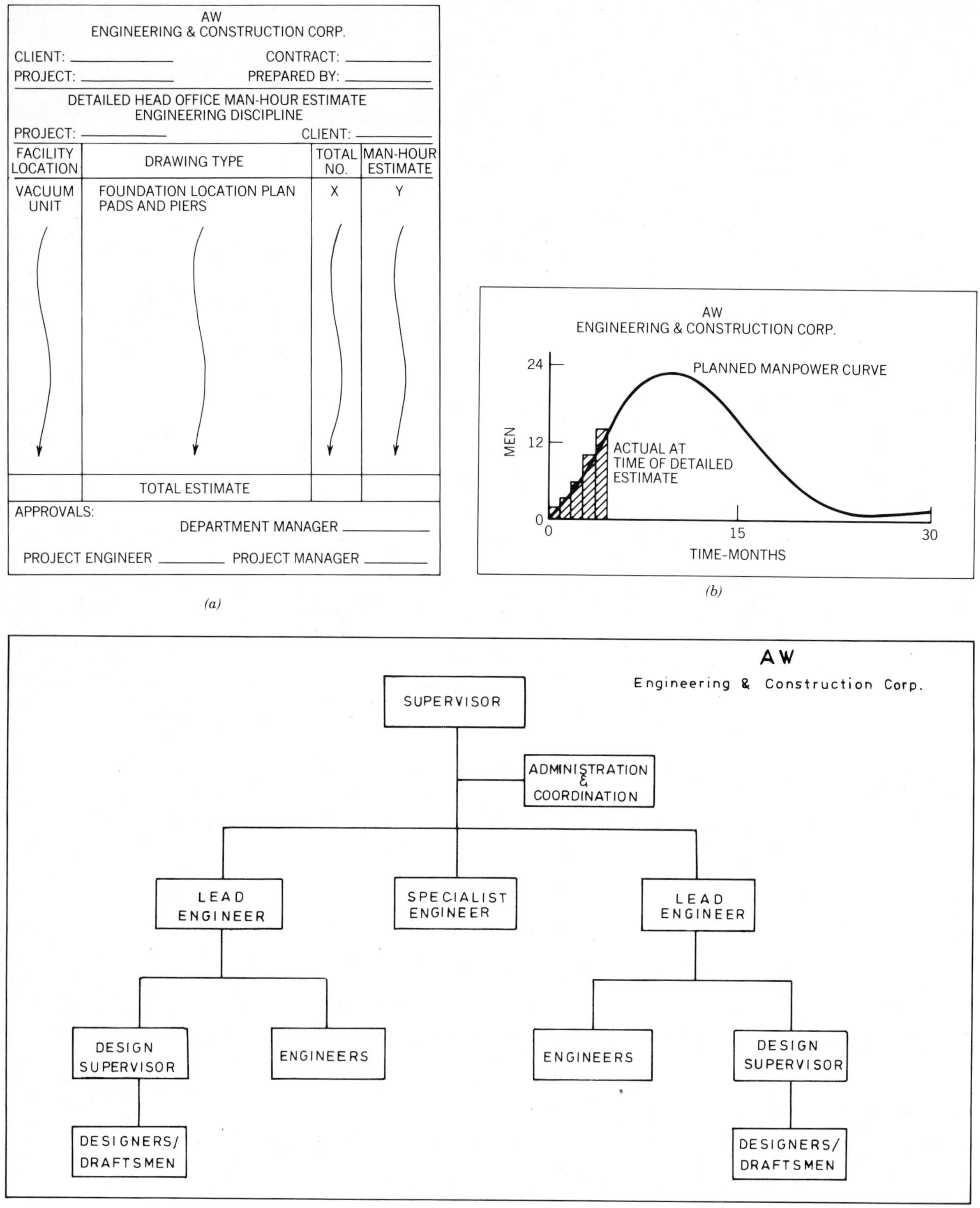

Figure 4.19 Scope engineering and design estimate documents. (*a*) Head Office Man-hour Estimate (P_L-DOC 5). (*b*) Head Office Manpower Curve (P_L-DOC 6) (*c*) Discipline Organization Chart (P_L-DOC 7).

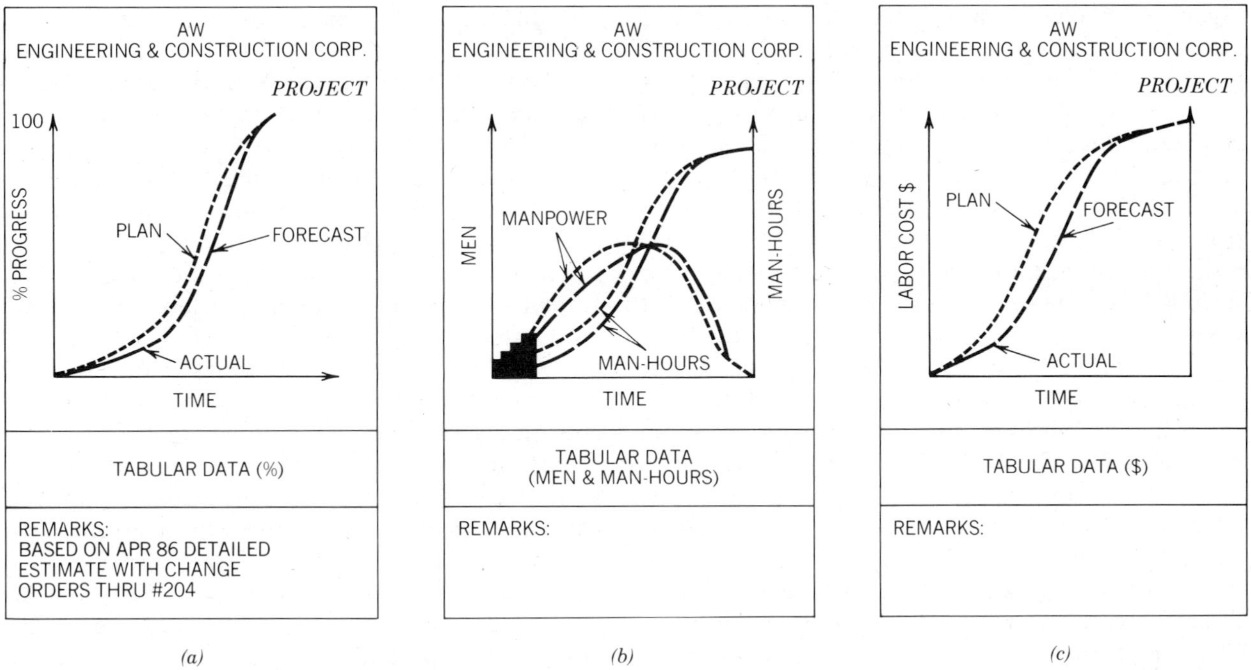

Figure 4.20 Head Office Project Control Documents (P_L-DOC 8). (*a*) Progress curves. (*b*) Manpower and man-hour curves. (*c*) Cost curves.

to control head office man-hour expenditures, project team manpower growth, head office progress, and head office labor costs, the Head Office Project Control Documents (P_L-DOC 8) in Figure 4.20 are frequently expanded upon. The statusing of these documents initiates an early interface with the Project Management Control decision process, wherein actual man-hour and labor cost expenditures are tabulated, engineering and design activities are statused, and related progress is evaluated. When the actual data is displayed against project plans on the Head Office Project Control Documents, deviations from the various project plans can be identified and analyzed, and corrective solutions decided on and enacted through the project team.

The effectiveness of the project team as a managing group can be assessed in part by evaluating its decision response to problems uncovered through the analysis of the data contained in project control documents like those shown in Figure 4.20. If these documents are not viable for identifying problems, then decisions may only be made when crisis situations arise. Alternatively, if these documents provide information which is not valid, wrong decisions may result. Thus, the project team's manipulation and use of project control documents is one indication of their management effectiveness. Other more subjective measures of project team effectiveness are the type, frequency, and timeliness of meetings and the actions which follow; project team morale; and the general flow of documents and whether these documents are utilized, encourage management action, and so on.

The most noticeable difference in the Initiate Project Action decision process when compared to that for the medium-sized firm, is the detailed planning of the project team structure, including the development of manpower loading plans. Furthermore, project team planning is focused on the head office phase. While the field construction organization is designed conceptually in terms of key personnel requirements, and their initial staffing often occurs at this stage, the detailed design of the organizational structure of the field project team and its planned staffing remains the problem of the Field Staffing decision process.

In summary, the Initiate Project Action decision process can be viewed as a documented Management Information System. The MIS structure is centered first around project mobilization efforts and then the staged development of the project team. In these phases, Standard Project Operating Procedures are guides to the definition of project team structure and size, and lead to formal project organization charts and the active growth of the project team through the acquisition of company personnel. The documents related to project team structures become more clearly defined as project control philosophies are established, schedules issued, and estimates of man-hours and costs evaluated. The MIS becomes fully operational when control base budgets and schedules are issued and project team positions are identified. The active staffing of the project team is a continuous effort which is conducted over the entire life of the project, through the use of employee files as manpower is requested. The subsequent de-staffing of the project will then occur as the project work load and volume declines and employees are transferred to other projects or to a department's central pool. At the large firm level, the acquisition of project personnel becomes a significant problem because of the often large number of personnel required for a project, especially for a design–construct project. Thus, the MISs become more formally documented in this project management environment.

4.6 TOTAL FIELD AUTONOMY

In the large firm environment, the Field-Based Project Management approach can be viewed as reflecting standard operating practices in project and construction management. While founded upon these standard management practices, the Total Field Autonomy approach enlarges its management focus to encompass the more specific management approaches required to achieve the level of management expertise expected and often demanded by the client. Hence, in this section, and in the sections entitled Total Field Autonomy in each succeeding chapter, only those *significant management problem areas* that are solved through either new or more integrated management approaches will be discussed, and only a selected number of documents, which relate to our selection of significant management problems, will be presented. Furthermore, the management solutions to these problem areas will also be described in relation to the policies that most dominately influence the particular management technique selected.

In summary, the description of the Total Field Autonomy approach will incorporate concepts that lead to possibly specialized management techniques, and also to the production of totally new documents or documents that may be slightly altered from that utilized by the large firm operating in the Field-Based Project Management mode.* For example, the Initiate Project Action decision process applied to the Total Field Autonomy approach will yield management approaches to project team design that reflect the pressures brought to bear on the contractor through client interaction in the decision process. Consequently, the structure and growth of the project team is often more carefully evaluated by the contractor because of the necessary authorization and approval of such matters required by the client. This frequently changes the documented format of the project team organization chart and the methods of executing a project. For instance, a contractor may produce project focused operating procedures designed for a specific client, in lieu of simply using standard project operating procedures, and organization charts tied to a specified number of client approved project positions. This approach impacts both the home office and field in such a way that a close interaction between the two is developed and maintained until the field assumes full management responsibility for the construction of the project.

The large firm operating in the Total Field Autonomy mode solicits business and prepares project proposals in a similar manner to that of the large firm operating in the Field-Based Project Management mode (see Figure 4.13). The proposal effort is often made more rigorous because of clients' desires to select the contractor with the best managing capability. Furthermore, the client is frequently well versed in current management techniques and will demand, and pay for, maximum project management effort from the contractor. Consequently, the client will actively participate in the proposal phase when selecting the contractor. In this situation, as depicted in Figure 4.21, client involvement increases after a project proposal is submitted. The contractor will often prepare presentations that focus on in-house technical expertise and on the project management systems proposed to manage the project. The contractor and client will negotiate the selection and commitment of key project man-

*The approach and format presented in Section 4.6 to describe the Total Field Autonomy approach will be utilized throughout the following eleven chapters (Chapters 5 through 15) for the section within each of these chapters that covers Total Field Autonomy.

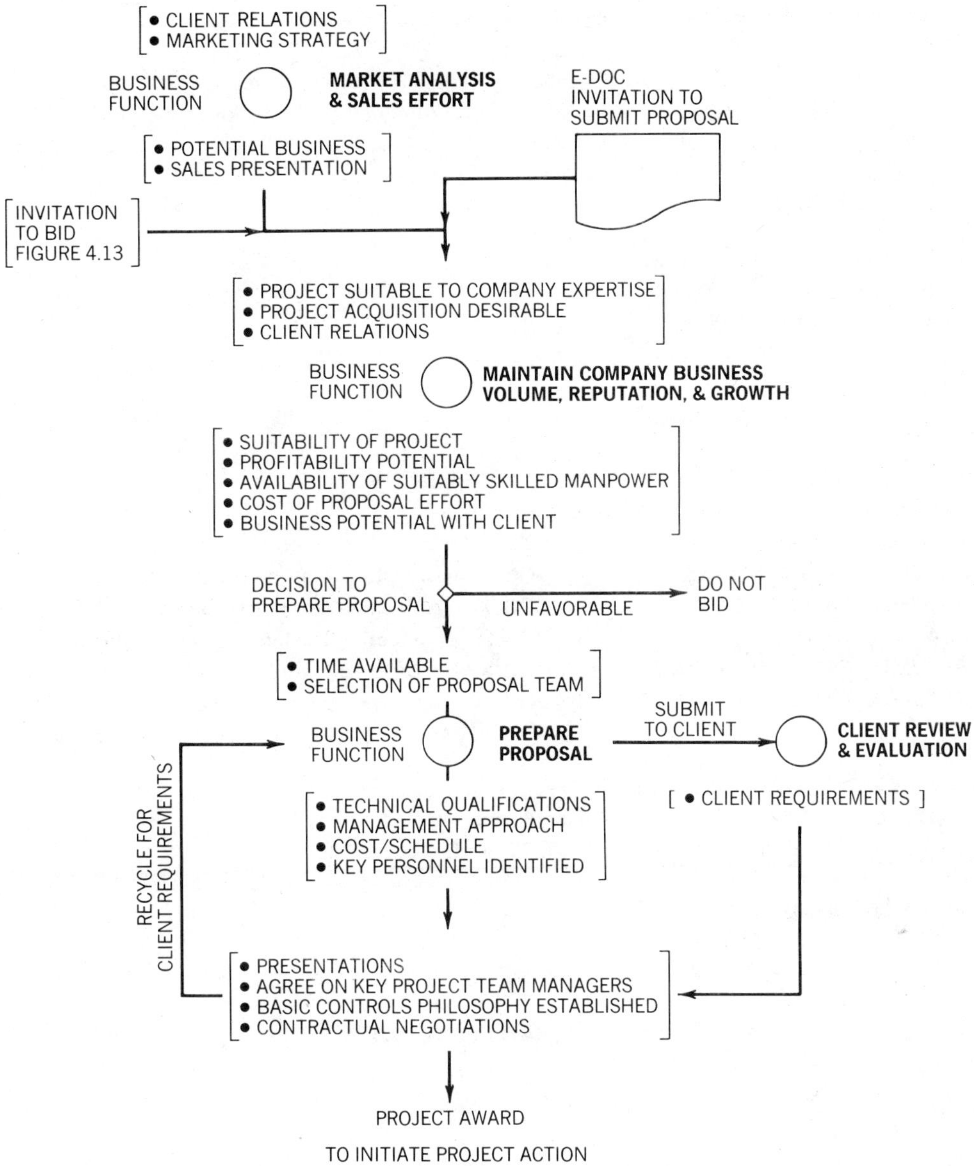

Figure 4.21 Project Solicitation decision process (Total Field Autonomy approach).

agers, establish and agree upon basic project control philosophies, and negotiate contractual relationships and responsibilities (e.g., construction management only, total design-construct responsibility, etc.).

The selection of the project manager (who is frequently a vice-president), and other key project managers is frequently made during the proposal stage and therefore the contract award immediately releases these personnel to the project to commence the Initiate Project Action decision process. This decision process, as described in Figure 4.22, is characterized by the three phases of effort including the initial mobilization (Figure 4.22*a*), the staged development phase of the project (Figure 4.22*b*), and the fully operating project phase (Figure

A1 ALLOCATE OFFICE, ENGINEERING, AND DESIGN STAFF

Policies	Documents
PROJECT TEAM FOCUS	P_T-DOC 1 Standard Project Operating Procedures
PROJECT TEAM CONTROL	P_T-DOC 2 Departmental Organization Structures
PROJECT TEAM DESIGN	P_T-DOC 3 Project Management Level Organization Chart
PLANNED ROLES	P_T-DOC 4 Employee File
PROJECT TEAM SPIRIT	

W1 RELEASE STAFF TO PROJECT

Policies

PROJECT TEAM CONTROL

PLANNED ROLES

(*a*)

A1 ALLOCATE OFFICE, ENGINEERING, AND DESIGN STAFF

Policies	Documents
PROJECT TEAM FOCUS	P_T-DOC 5 Project Execution Procedures
PROJECT TEAM CONTROL	P_T-DOC 6 Head Office Man-hour Estimate
PROJECT TEAM DESIGN	P_T-DOC 7 Head Office Manpower Curve
PLANNED ROLES	P_T-DOC 8 Engineering/Design Project Organization Chart
PROJECT TEAM MANAGEMENT	P_T-DOC 9 Head Office Manning Chart
PROJECT TEAM SPIRIT	P_T-DOC 10 Employee Availability Report
	P_T-DOC 11 Head Office Project Control Documents

W1 RELEASE STAFF TO PROJECT

Policies	Documents
PROJECT TEAM CONTROL	P_T-DOC 4 Employee File
PLANNED ROLES	P_T-DOC 10 Employee Availability Report
PROJECT TEAM MANAGEMENT	

(*b*)

A1 ALLOCATE OFFICE, ENGINEERING, AND DESIGN STAFF

Policies

PROJECT TEAM CONTROL

PLANNED ROLES

PROJECT TEAM MANAGEMENT

PROJECT TEAM SPIRIT

W1 RELEASE STAFF TO PROJECT

Policies

PROJECT TEAM CONTROL

PLANNED ROLES

PROJECT TEAM MANAGEMENT

(*c*)

Figure 4.22 Initiate Project Action decision process: Function, Policy, Document Plot (Total Field Autonomy approach). (*a*) Mobilization phase. (*b*) Staged development phase. (*c*) Fully operating phase.

TABLE 4.6 POLICIES OF THE AUTONOMOUS PROJECT TEAM

Policy		Description
P_T 1	*PROJECT TEAM FOCUS*	The project team is initially head office oriented until field operations are established, at which time there is a total transfer of management responsibility and control to the field under a top company executive acting as project manager.
P_T 2	*PROJECT TEAM CONTROL*	Key project team assignments, overall project team designs, and specifically designed project execution procedures as developed for project management are reviewed in consultation with and approved by the client.
P_T 3	*PROJECT TEAM DESIGN*	The project team structure is designed to maximize planning and management control by ensuring that the proper relationship exists between the level of manpower to be controlled in relation to the size of the project team.
P_T 4	*PLANNED ROLES*	The project team is a fully integrated group of managers combining their functional responsibilities through the project team structure to optimize planning techniques and the level of management control demanded by the client and company management, and agreed upon project execution procedures derived from standard project operating procedures.
P_T 5	*PROJECT TEAM MANAGEMENT*	Management evaluation of project progress, status, and performance on a continuous basis is essential to ensure that the project is being managed with maximum effectiveness by the project team.
P_T 37	*PROJECT TEAM SPIRIT*	Active participation of all project team members in the management process promotes goodwill and a positive management environment. It ensures the success of scientific management techniques and the total integration of the project team.

4.22*c*).* As illustrated in Figure 4.22 and Table 4.6, the key manpower management policy that guides the Initiation of Project Action is the *FOCUS ON PROJECT TEAM DESIGN* (P_T 1), (See Table 4.6 for description of this policy.)

This policy describes a project team headed by an executive project manager and stipulates that the total transfer of management responsibility and control to the field must occur when field construction activity is sufficiently established. Consequently, the selection and structuring of the project team must be formulated based on a control philosophy that will provide for compatibility between head office design and field construction efforts. The client is predominately involved in the design of the project team and in the selection of the key project team members as well as in pressuring the contractor to maintain a project team that is an effective managing body as seen through *PROJECT TEAM CONTROL* (P_T 2) and *PROJECT TEAM MANAGEMENT* (P_T 5). The project team structure and size may be redesigned as required until the most suitable project team, one which meets both company requirements and client approval, is achieved. Because of the dynamic nature of the project team structure at this stage, *PROJECT TEAM DESIGN* (P_T 3) becomes a critical policy of management. Planning the project team structure, size, and growth is correlated directly to the level of manpower to be controlled and the project control procedures to be implemented. This approach is ensured through *PLANNED ROLES* (P_T 4) and *PROJECT TEAM MANAGEMENT,* which promote project management through the use of a fully integrated project team. Thus, the staged design of the project team will often result in changes in key management positions, especially over the long-term life of the project.

The policies described in Table 4.6 and their impact on the decision process of Figure 4.22 in-

*It should be noticed that a tabular format depicting manpower management functions and their relevant policies and documents will be used in this section to describe the decision process, in lieu of the full decision network technique presented in Section 4.5. While a full decision process network could be developed for any Total Field Autonomy decision process, the authors believe the structure and flow of information would closely parallel that of the corresponding network for the Field Based Project Management approach. Thus, the authors believe that the identification of functions, their related policies, and the origin of each document (i.e., when a document is first used) in relation to the decision process, is sufficient to fully expose only those critical and unique management approaches associated with the Total Field Autonomy concept. The reader is encouraged to develop a network if desired.

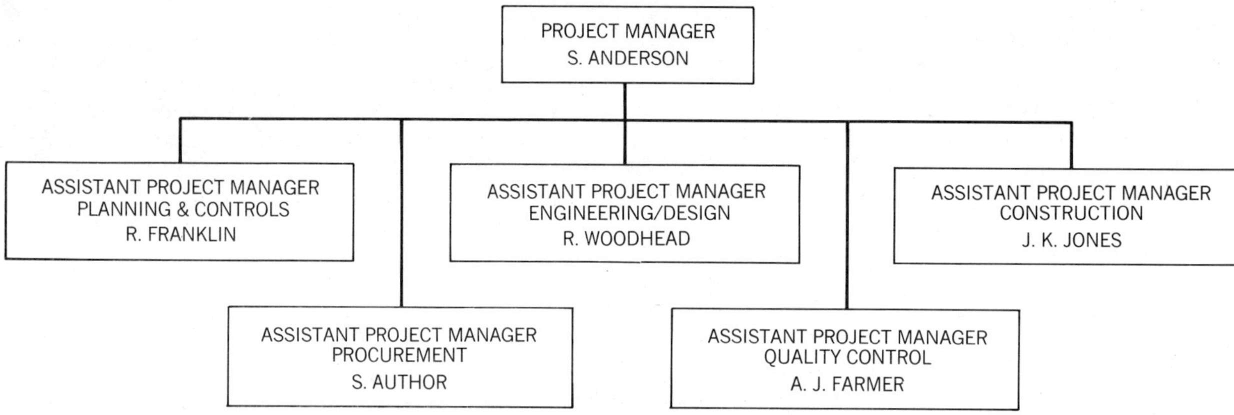

Figure 4.23 Project Management Level Organization Chart (P_T-DOC 3).

fluence the management functional performance techniques in a number of ways. For example, during OFFICE STAFF ENGINEERING, DESIGN AND ALLOCATION (A1), the departmental breakdown of functional involvement in a project is incorporated into the design of a top project management group that forms the basis for the project team structure. This Project Management Level Organization Chart (P_T-DOC 3) (depicted in Figure 4.23) limits the number of personnel reporting directly to the project manager. It alternately provides additional expertise and management strength and focus at the project management level, through the assistant project managers and through the establishment of their professional group required for project execution.* Consequently, the project manager need only be concerned with problems of a nature critical to the overall execution of the project as reported to him by the assistant project managers.

The basic project management level positions remain the same for both the head office and field locations. This allows for easy transfer of responsibility from the head office to the field. The particular managers occupying these positions may or may not transfer to the field, with the exception of the project manager, who normally relocates to the field. Once the project management level is designed and staffed, the full project team is designed and staffed.

An early function of the project team is to develop specific project-oriented execution procedures that meet the requirements and objectives of both project management and the client. These Project Execution Procedures (P_T-DOC 5) are based on the contractor's Standard Project Operating Procedures (P_T-DOC 1), but are modified for the requirements of the particular project. This evolutionary process of developing the Project Execution Procedures is illustrated schematically in Figure 4.24. These detailed project focused procedures are developed for most functional areas involved in the project during the Staged Development Phase of Initiate Project Action (see Figure 4.22*b*).

The detailed design and development of the full project team is directed by each assistant project manager for his or her respective functional area. The structure of the project team is based on a management control philosophy established by the project management staff. A common technique is to divide the project into discrete control areas based on geographical locations and/or production units. For each control area one of a subset of project teams is created, each having the necessary disciplines to perform all requisite functions pertinent to its area. Consequently, when the assistant project managers are designing the structure of their organiations, that structure must be consistent with the project control philosophy.†

The organizational design developed by any assistant project manager is subject to the approval of the project manager, department managers (head office only), and the client. This practice is relatively common when employing the Total Field Autonomy approach, because the design of the or-

*The terms project director and deputy project director are also used in practice on large projects to describe the project management level.

†A detailed discussion of the control area technique is included in Chapter 5, Project Planning.

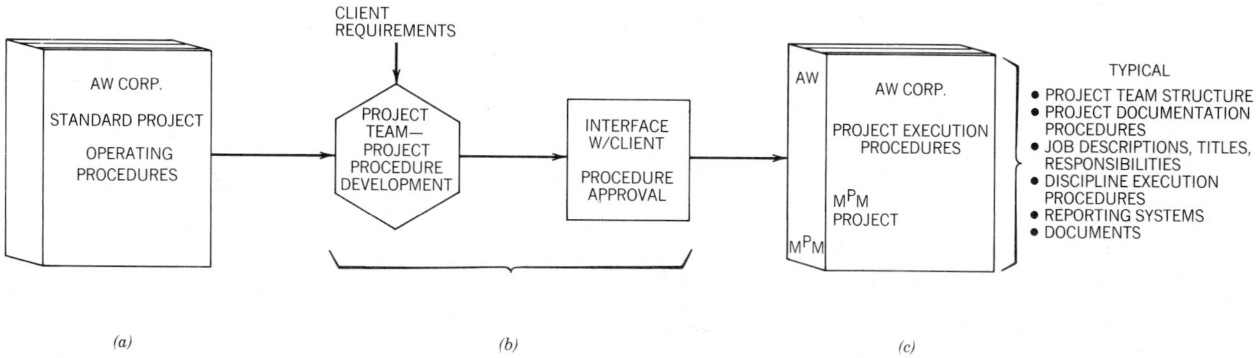

Figure 4.24 Project procedure development process. (*a*) Standard Project Operating Procedures (P_T-DOC 1). (*b*) Project procedural development with project team and client interaction. (*c*) Project Execution Procedures (P_T-DOC 5).

ganizational structure identifies the number of positions to be staffed. The total number of positions can be translated into project estimates of the man-hours required to engineer and design the project and manage the construction effort in the field. There are other factors that influence the size of the project team staff in any particular functional area, for example, the number of drawings required; the direct field man-hours to be controlled; or the number, type, and frequency of project management reports to be generated.

Scope estimating techniques (see Figure 4.6) are used in conjunction with increasing scope details to help define the project team required for the control of the production engineering and design effort. The final product of the Staged Development Phase (Figure 4.22*b*) is a project team that is fully authorized with client approved project team positions and manning charts. Figures 4.25 and 4.26 show a typical Project Team Organization (P_T-DOC 8), and Manpower Chart (P_T-DOC 9), respectively, for a section of the engineering and design group in the head office in terms of the project control area approach. Each project position is identified by a group number and an organization chart position number. These positions are often modified using the techniques illustrated in Figure 4.7, or through client-approved change orders resulting from additions to or decreases in the project scope.

RELEASING STAFF TO THE PROJECT (W1) requires the use of company Employee Files (P_T-DOC 4) (see Figure 4.18) and departmental reports (often computerized), and involves specifying the location and availability of company personnel to locate, acquire, and mobilize company personnel for a project (as demonstrated in Figure 4.27 Employee Availability Report (P_T-DOC 10)). In Figure 4.22, a number of policies are shown as influencing the manner in which personnel are released to the project. This is the result of continuous client interaction and approval requirements during all phases of the Initiate Project Action decision process. Finally, a policy of *PROJECT TEAM SPIRIT* (P_T 37) emerges at this management level as a means of actively involving key project team members in the project management effort, thereby creating a positive management environment and enabling the total integration of the project team.

Because of the heavy emphasis on maximizing the effectiveness of the project team via project progress and performance analysis as directed through policies like *PROJECT TEAM MANAGEMENT,* detailed head office progress and cost reporting systems will emerge initially in the Stage Development Phase of Initiate Project Action, and then will become fully operational during the production engineering and design project phases (see Figure 4.22*c*). Typical Head Office Project Control Documents (P_T-DOC 11), based on data from client-approved control budgets and schedules, are shown in Figure 4.28. When approved by the client, these control documents represent the control plans (i.e., progress, manpower, and dollars) for monitoring the project design effort. These documents are generated at a level of detail consistent with the project team structure and function groupings and control area concepts. Thus, each engineering/design discipline will have its own control base in total and/or for each control area as applicable to the particular discipline. Thus, control documents shown in Figure 4.28 will have a total project focus and an engineering discipline focus.

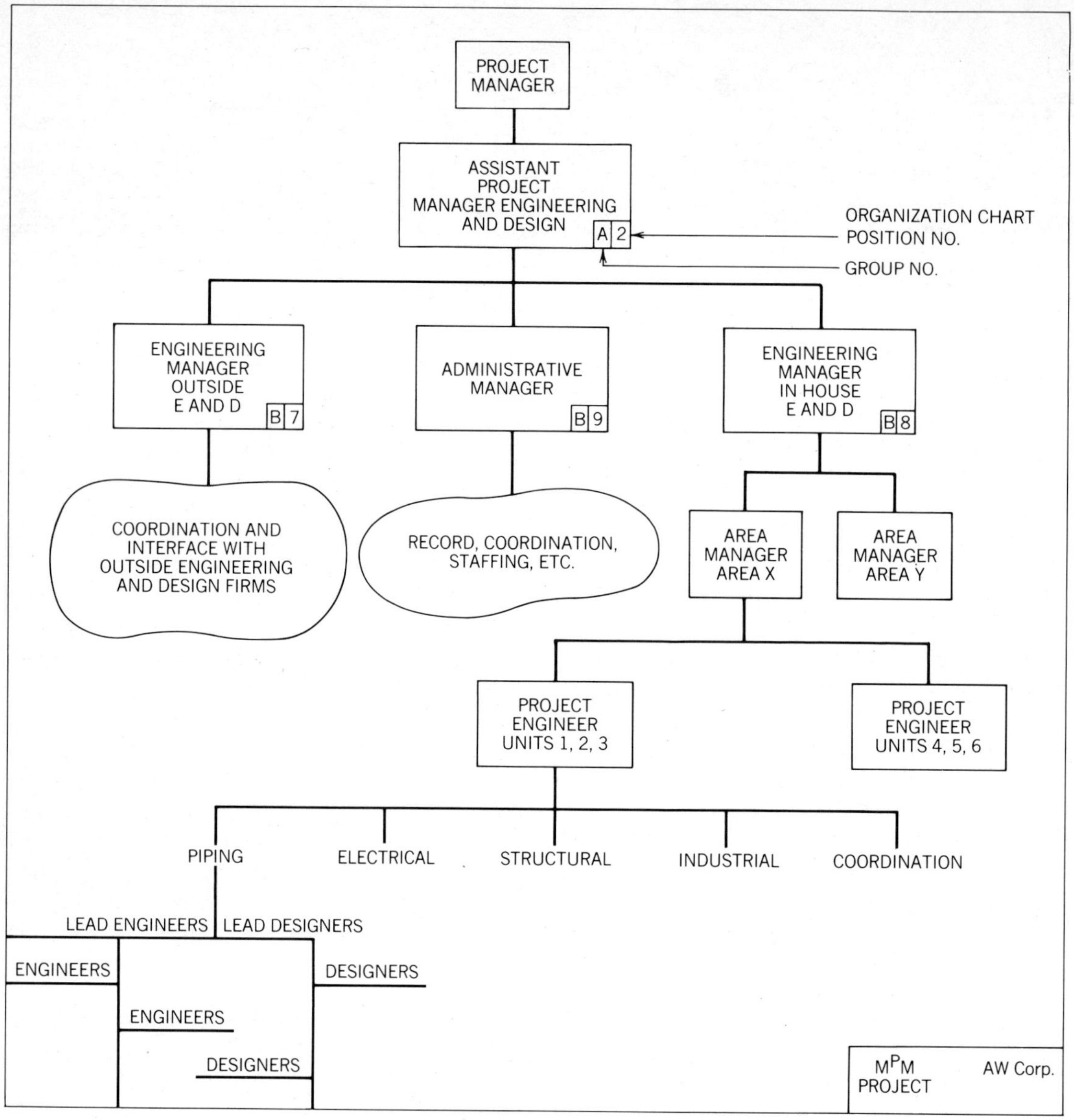

Figure 4.25 Engineering/Design Project Organization Chart (P_T-DOC 8).

M^P M PROJECT — AW CORP.

POSITION NO.	TITLE	NAME	CLASS CODE	DURATION M A M J J A S	TOTAL MAN-MOS.
GROUP A	PROJECT MANAGEMENT				
1	PROJECT MANAGER	S. ANDERSON		→	48
2	ASST. P.M. ENGR/DES.	R. WOODHEAD		→	40
3	ASST. P.M. CONSTRUCTION	J. K. JONES		→	46
4	ASST. P.M. PLAN/CONTROLS	R. L. FRANKLIN		→	48
5	ASST. P.M. PROCUREMENT	S. T. AUTHOR		→	48
6	ASST. P.M. QUAL. CONTROL	A. J. FARMER		→	36
	TOTAL				266
GROUP B	ENGINEERING/DESIGN				
7	MANAGER—ENGINEER	F. L. LYNCH		→	36
8	MANAGER—ADMIN.	G. L. LYDUM		→	32
9	MANAGER—ENGINEERING	R. M. GROVES		→	40

Figure 4.26 Head Office Manning Chart (P_T-DOC 9).

A W CORP.			
MANAGER: ____			
SUPERVISOR: ____			
EMPLOYEE	CLASS	CURRENT ASSIGNMENT	DATE AVAILABLE
S. D. ANDERSON	F213	REFINERY EXPANSION—WOODSON CORP.	MAR. 87
↓	↓	↓	↓

Figure 4.27 Employee Availability Report (P_T-DOC 10).

These control documents are frequently updated through the Project Status, Analysis, and Reporting and Project Management Control decision processes to reflect current progress and performance and forecasts of future progress and costs when deviations are apparent. Through this effort problems are identified and solutions determined through the Project Management Control decision process.*

The Initiate Project Action decision process can be summarized as an MIS, as illustrated in Figure 4.29. The initial selection of key project team managers establishes the Project Management Level. When these personnel are transferred or allocated to the project the full scale project initiation effort commences. As shown in Figure 4.29, the Initiate Project Action (DP 1) decision process is impacted by a number of documents that provide data input for the preparation of project team organization charts, manpower profiles, and man-hour estimates, in addition to specific project-oriented procedures. These documents establish the manpower requirements that can be used when staffing the

*The scope of this book is not large enough to cover the management of the engineering/design production effort. The Head Office Project Control Documents are introduced to illustrate a technique for monitoring the engineering and design effort.

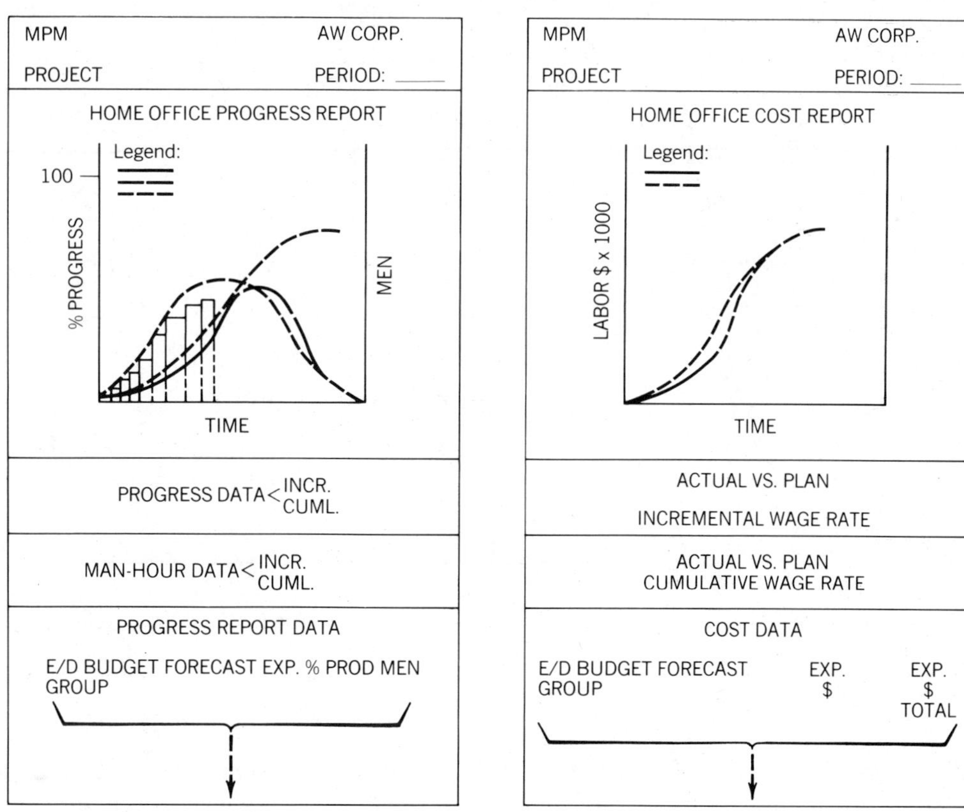

(a) (b)

Figure 4.28 Head Office Project Control Documents (P_T-DOC 11). (*a*) Progress report. (*b*) Labor cost report.

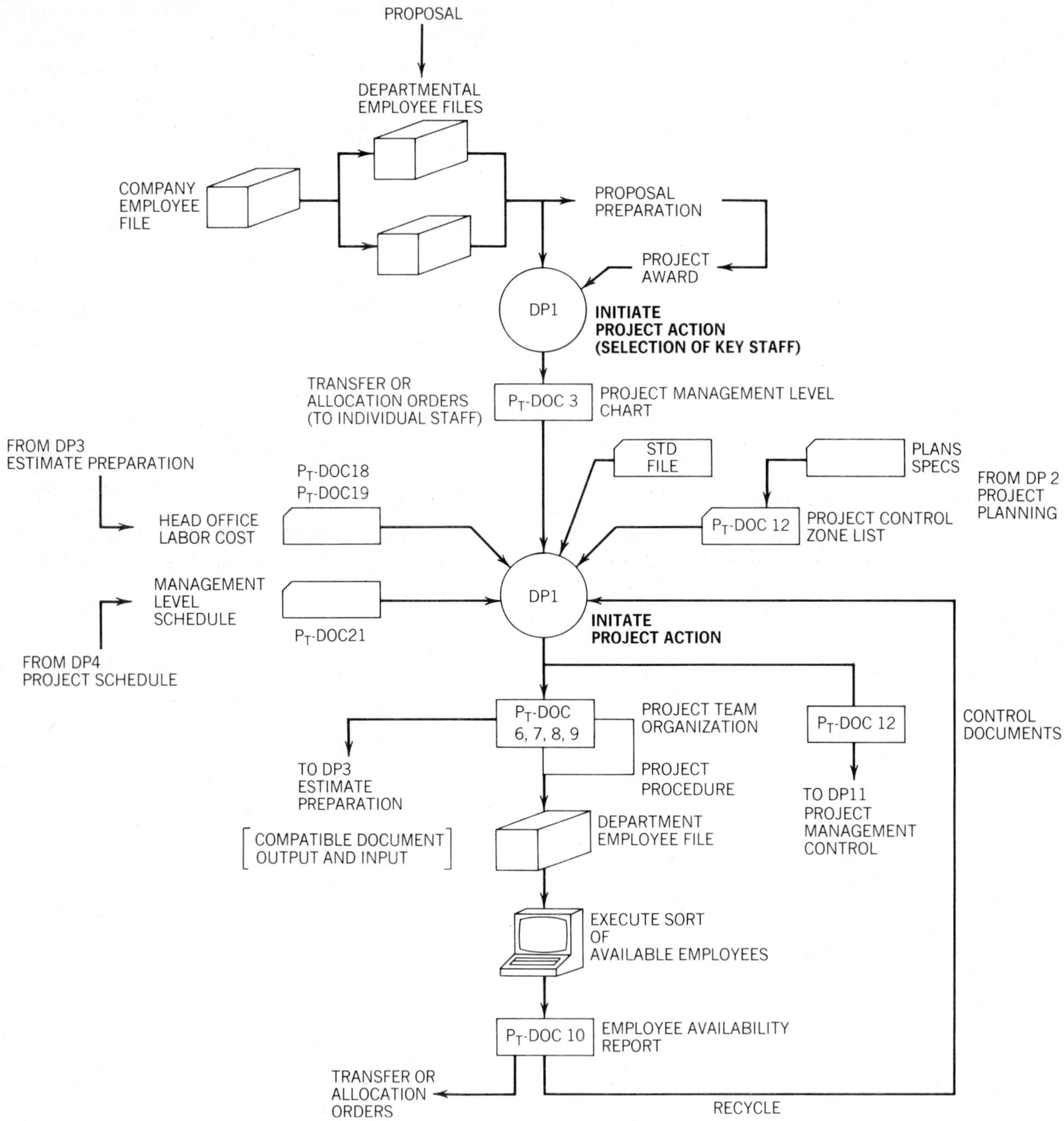

Figure 4.29 Initiate Project Action Management Information System (Total Field Autonomy approach).

project with suitable personnel who possess the requisite qualifications for each project team position. The department employee files can be searched by sorting on available employees as identified on the Employee Availability Report. The selection of suitable staff is made from this report, and transfer or allocation orders are given by management. The cycle is continuous over the life of the project and constitutes the MIS of Initiate Project Action. It is generally a fully documented system because of the large numbers of personnel assigned to projects in this management environment.

CHAPTER 5
Project Planning

Project Planning for the construction contractor commences with site reconnaissance and the determination of what has to be done. It involves prescribing how each construction activity is to be performed and enumerating the required resources. On major projects the client may request that the component facilities be built in certain sequences, in time periods established by contract milestones. In addition, on many design-construct projects and/or projects utilizing fast tracking techniques, the construction plan must reflect the order in which design and working drawings are to become available. Thus, Project Planning establishes the guidelines and sequence within which construction activity is to take place, and for which the contractor must mobilize and effectively use resources, including labor.

The reconnaissance and investigation of the labor environment is necessary for the determination of the availability of labor and skilled tradesmen, and for the assessment of their past and present performance on similar projects. This aspect of Project Planning is critical for the prediction of achievable manpower levels and productivities, unless the contractor has prior knowledge of the labor environment. The criticality of this management problem is such that the entire success of the project, in so far as cost, scheduling, and quality commitments are involved, can only be achieved by the correct evaluation of labor availability and the establishment of accurate performance targets. On small local projects contractors are well aware of labor availability and productivities and for similar projects can safely predict labor performance. On remote or foreign projects, however, where only scanty labor performance records are available, the project planning of manpower availability and labor performance is a risky undertaking.

The eventual attainment in the field of specific predetermined levels of productivity is dependent to some extent on the skill with which the contractor can manage the site area and work force with properly trained supervisors. On large, complex projects, however, work force management is a formidable problem and productivities are often very low, so that management must anticipate productivity levels during the planning process that are compatible with their field management capabilities. Project control procedures must be developed and later enforced to ensure that manpower planning objectives are achieved in the field. On most small local projects, the contractor can rely on company standard cost accounting and scheduling control methods as their procedure for manpower control. On large projects, however, where the size and complexity of the project site is significant and required manpower levels extend beyond past experience, specially designed manpower control procedures must often be developed and initiated in the field. With projects of this magnitude, consistent manpower reporting techniques are required to ensure that total project reports can be generated which are meaningful to top level field management. Consequently, developing manpower control procedures in the planning phase also becomes a critical problem.

The decision to initiate project action releases a project team to project planning. In many construction environments the science and practice of project planning is traditionally viewed in terms of planning construction methods and sequences prior to establishing the time frame in which construction activities will occur. The feasibility of the construction plan, however, depends upon the resolution of two critical and interrelated project manpower management problems, namely:

1. The determination of the location, size, and quality of actual and potentially available skilled labor pools, and a commitment to a planned work force size (by the adoption of suitable recruiting and hiring procedures), that is believed achievable.
2. The determination of acceptable levels of skilled labor productivities as the basis for planning, and the prescription of manpower control procedures that are believed capable of achieving these productivities in the work environment.

Thus, the manner in which these critical problems are addressed in the Project Planning decision process, and the quality of the decisions made then, directly impacts all time and cost aspects of a project and reflects the professionalism and experience of the contractor.

5.1 THE BASIC DECISION PROCESS

Project Planning encompasses a broad scope of functions that range from the complete investigation of all site factors that will influence the construction effort, through the broad sequence planning of major project components, the detailed planning of the sequence logic for construction op-

TABLE 5.1 SCOPE OF PROJECT PLANNING PROBLEM AREA

Major Decision Problem	Component Problem Statements
6. What is project composition:	What are the major components of the project? What is the nature of construction site? What flexibility exists in the sequencing of components? What are the desirable sequences?
7. What is project strategy:	What is the proposed project duration and what are established milestone dates? What priority of construction facilities meets contract and client requirements? What major project components must be subcontracted? What are project control procedures? What project breakdown is necessary to meet management control requirements for manpower, labor policies, work quality, and work progress? What techniques will be required to satisfy desired levels of project management control?
8. What are project manpower requirements:	What are project requirements for labor and skilled trades? What sequence of facility construction meets proposed labor and skilled trades profiles? Can effective use of labor be made in relation to proposed construction schedule sequence, material deliveries, and production of engineering drawings? Can short supply of labor be alleviated by adoption of more mechanized construction methods or by changes to construction schedule logic? How can subcontractor effort and labor requirements be integrated into the project effort?
9. What is manpower availability:	Where are labor sources? What is availability of labor and skilled trades? Can labor requirements be met by training and/or importation?
10. What are manpower policies:	Is project union or nonunion labor? What labor management policies have been established for the project? What labor organizations and agreements exist? What are prevailing craft wage rate structures in site areas? Are pay incentives required? When do labor agreements come up for renewal? Will a construction camp be required?
11. What is level of anticipated labor productivity:	How is productivity defined for control purposes? What impact will site conditions have on construction methods and labor productivities?
12. What will be Subcontracted:	What work is best contracted out of work force? What schedule and performance risks are associated with subcontracting work? What special management control procedures must be implemented to coordinate subcontracted work?

erations and, finally, to the development of project control procedures. In terms of project manpower management, Project Planning is specifically concerned with establishing feasible construction plans to meet key milestone dates that are compatible with anticipated manpower levels and performance targets. It is also involved with evaluating manpower management control procedures and formulating plans for their implementation. This effort entails the analysis of project requirements in terms of craft content, labor availability, labor management policies as well as the influence of subcontractor requirements and site conditions on total project manpower levels.

Project planning first establishes the desirable sequences of major project work areas of engineering (as applicable), and related procurement, subcontracts, and construction requirements. When the scope is well defined, these major work areas are further dissected into detailed project activities for which logical execution sequences can be developed. This formulation of the construction plan then becomes the basis for establishing the project schedule, and assists in defining the potential manpower involvement in the project. For instance, the contractor can determine the range of types of crews and crafts required and the approximate sequencing of these various crews and crafts over the life of the project. This will enable the contractor to formulate manpower requirements, to evaluate the impact of different levels of manpower availability in the site area, and to determine the potential subcontractor requirements for manpower. Finally, the contractor must design and be prepared to implement manpower control procedures which will ensure the attainment of various manpower goals established for field construction.

When the contractor delves into the Project Planning area, key personnel must be prepared to focus management attention on a variety of major problem areas, as presented in Table 5.1. The major problem areas related to project composition and strategy focus on generic problems which must be considered prior to entering into specific manpower-related major problem areas such as those for the determination of manpower requirements, labor availability, the development of effective policies, and the prediction of productivity levels. The degree of detail and intensity with which these major problem areas are addressed by the contractor depends on the type and nature of the project, its size and complexity, and its location.

The actions taken by management to solve each major problem area involved with Project Planning generally occur in the sequential order shown in Table 5.1. As indicated, these major problem areas can be further identified in terms of related management duties and tasks. When these management actions are placed in sequence as procedural steps, a general procedure for the Project Planning decision process emerges. This procedure is described in Table 5.2, and is considered typical of the range of Project Planning associated duties and tasks normally found in construction practice.

The management duties and tasks in Table 5.2

TABLE 5.2 GENERAL PROCEDURE FOR PROJECT PLANNING

Procedural Step	Management Duties and Tasks
1.	Evaluate construction, subcontracting, and procurement requirements of major project components; impact of site conditions on construction methods and proposed labor productivities; flexibility of sequencing major components in relation to best construction sequences.
2.	Determine project milestone dates and priority sequencing to meet client and/or contract requirements.
3.	Formulate project management control strategy in terms of level of detail, procedure for measurement of performance, work progress, and control techniques to be implemented by the project management team.
4.	Evaluate and establish labor and craft skill level requirements, local and/or regional sources of labor, craft wage rates, and so forth.
5.	Define site productivity goals and evaluate impact of site conditions on construction sequences and labor performance and costs.
6.	Establish construction sequence logic.
7.	Finalize areas of work which are best subcontracted, establish feasible durations, and integrate these into construction sequence.
8.	Determine whether union or nonunion labor will be used, whether training programs are required, and evaluate the impact of labor agreements and interfaces with unions.

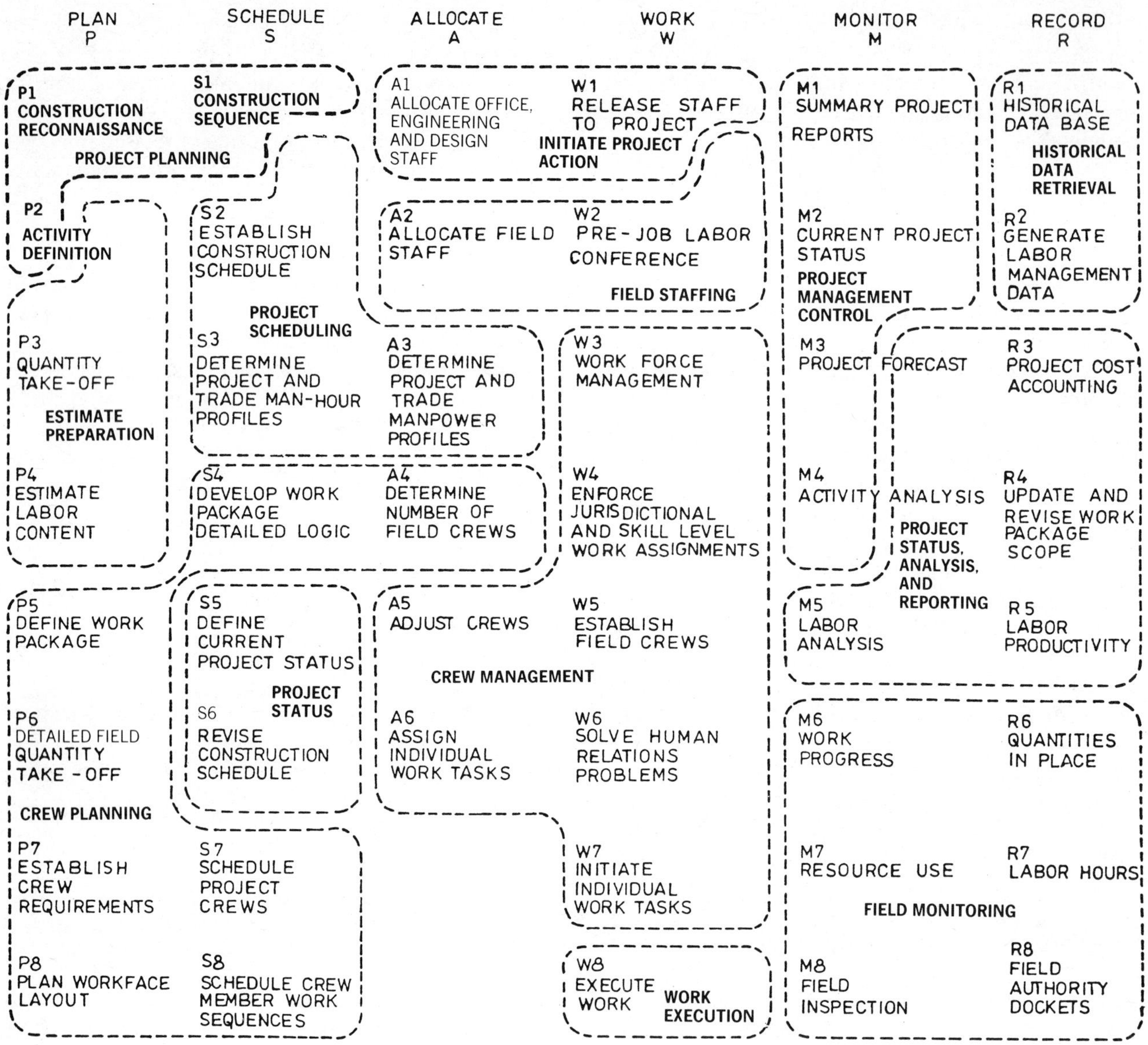

Figure 5.1 Functional plot of the Project Planning decision process.

are an expression of a decision process for Project Planning, and therefore can form the basis for identifying the manpower management functions which comprise the Project Planning decision process. The three functions involved are CONSTRUCTION RECONNAISSANCE (P1), ACTIVITY DEFINITION (P2), and CONSTRUCTION SEQUENCE (S1). The functional plot for the Project Planning decision process is shown in Figure 5.1 in relation to the other manpower management decision processes. In addition, a functional statement of this decision process, in terms of its component manpower management functions, is provided in Figure 5.2. As can be readily verified, the functional statements for each manpower management function generally correspond to one or more of the major problem areas or duties and tasks as delineated in Tables 5.1 and 5.2, respectively. The general relationship of Project Planning to other relevant decision processes is shown in Figure 5.3.

P1 CONSTRUCTION RECONNAISSANCE

- The assessment of project site conditions and local factors that may impact the project and which must be considered during project planning and management.
- Requires knowledge of project scope and specifications, construction methods, labor organizations and trade classifications, subcontractor availability,* and site inspection with checkout list for factors and facilities influencing site accessibility, mobilization, layout, and productivity.
- Considers availability of labor, location of union locals, union agreements and expiration dates, craft wage structure, local measures of productivity, restrictive work practices, local trade skill levels, attitudes to importation of labor, training mechanisms, subcontractor availability,* and impact of site conditions on construction methods and working conditions.

  *This applies to prime contractors who use subcontractors for specialized work.

P2 ACTIVITY DEFINITION

- Activity definition rationalizes project breakdown and control procedures as a means in some cases of developing the project scope (design-construct), and in all cases of identifying project scope and complexity and as a basis for project management control.
- Requires a detailed knowledge of the project scope, including preparatory work, proposed construction methods, and cost and scheduling control requirements, as well as an understanding of the level of breakdown detail required to match project team control strategy and expertise.
- Considers activities as recognizable portions of the finished product or work scope, correlation of activity description with estimate breakdown, schedule requirements, and cost control requirements and cost codes; implied crews and whether activity is worked via force account, direct hire labor, or by subcontract (if prime contractor).

S1 CONSTRUCTION SEQUENCE

- The identification of desirable sequences of major project work packages and the definition of the basic technological sequences of engineering, procurement, and field effort for each project activity.
- Requires knowledge of project scope, activity breakdown, and estimate details on work content and material, labor and equipment resources, subcontract items, and the impact on project and company resources of different sequences of major project efforts.
- Considers priority of facilities to be constructed in relation to contract requirements and decisions on availability of resources, drawings, and priorities of work effort, impact of construction methods on construction sequence logic; focuses on magnitude of effort required for integration of material deliveries with labor and equipment use to support proposed construction sequence and the impact of subcontractors (if prime contractor).

Figure 5.2 Functional statement of the Project Planning decision process.

5.2 FUNCTIONAL PERFORMANCE TECHNIQUES

The initial Project Planning effort generally focuses the contractor's attention on grasping the scope and nature of the project under evaluation. In many construction environments, this effort is relatively simple and often automatic because the magnitude of the project can be easily comprehended by the contractor as a result of its similarity to previous projects, or because of its limited physical size and functional complexity. On the other hand, when construction projects involve highly specialized technical processes, the project scope and nature is often virtually impossible to absorb and comprehend without a major Project Planning effort. This is especially evident when the project is located in a remote area. Project Planning is therefore concerned with the determination of what has to be done, of the sequence in which it will be done, the resources required to accomplish what has to be done, and the establishment of the technological construction processes required for the timely execution of planned construction activities.

In terms of Project Planning, the contractor must select and implement techniques specifically de-

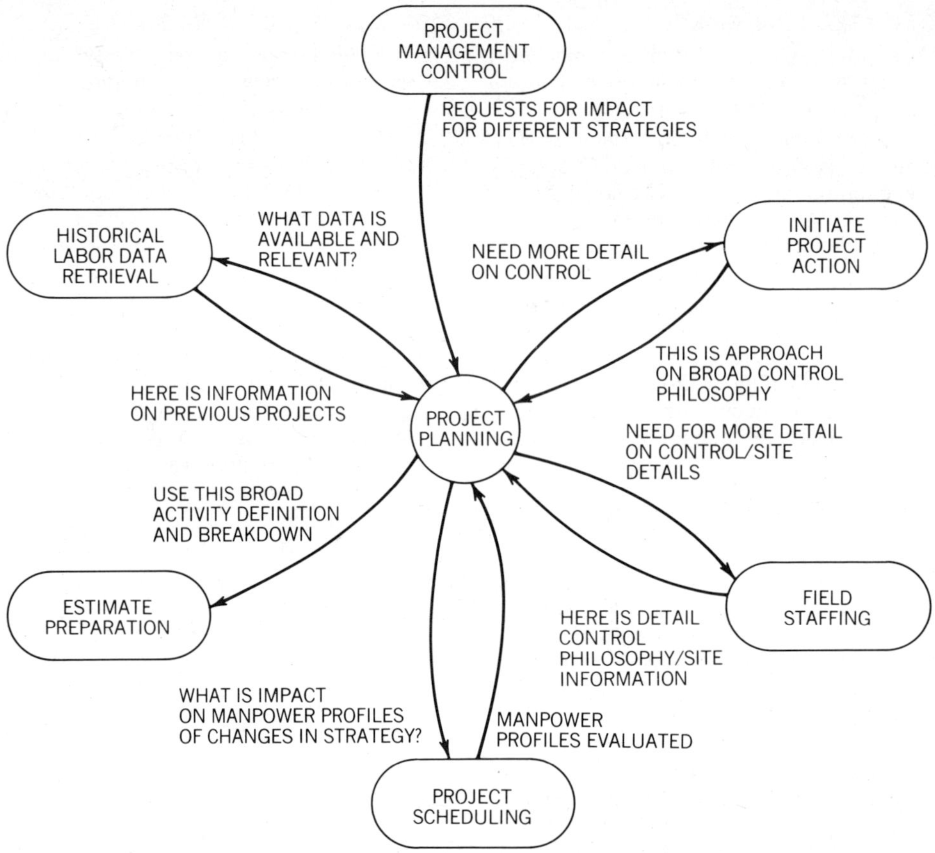

Figure 5.3 Schematic illustration of Project Planning decision process interfaces with other manpower management decision processes.

signed to solve the problems raised by the planning and sequencing of work, the analysis of anticipated labor conditions in the project site area, and the development of procedures designed to control manpower and to ensure that project manpower plans are achieved. When management focuses on these three areas, management concepts emerge that lead to the selection of performance techniques that are most relevant to the Project Planning decision area. Project Planning spans everything from the purely construction viewpoint of contractors involved only with the construction phase, to the overall project perspective of contractors responsible for complete management of an entire project including engineering/design and construction.

The contractor first considers how the project will be built in relation to broad overall sequences of engineering and/or construction. This effort requires the contractor to identify and sequence the physical order of appearance of individual facilities (or of component physical blocks of a single facility). This physical dissection of a project is easily accomplished for small to medium-sized projects, but becomes increasingly more difficult and critical for large projects where a vast number of physical components exist. A technique which can be used for this type of broad planning is the precedence network approach shown in Figure 5.4*a*.

The contractor must also integrate into the overall physical sequencing plans the client requirements that establish priorities of facility construction, and whatever facility completion milestones may already be fixed by the client. At times the contractor may have the operational flexibility to order and select the sequence for individual facility component construction. In this case it may be desirable to locate and plan the interface milestones that occur between sequences of work activity on each facility. This type of milestone planning is illustrated graphically in Figure 5.4*b*. Thus, these two types of broad planning approaches, while not directly manpower focused, begin to fix the se-

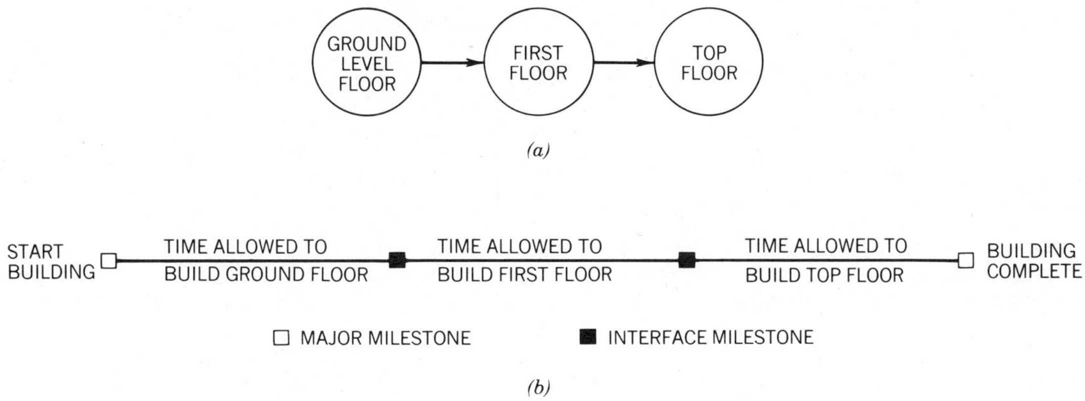

Figure 5.4 Broad planning concepts. (*a*) Physical precedence network (arrows represent sequences of construction). (*b*) Milestone planning.

quence in which manpower resources will be committed to the project.

In the small to medium-sized construction firm environments, the process of planning the broad dissection and sequencing of physical blocks of work is primarily focused on identifying potential bottleneck areas that, when completed, will release significant blocks of future work to construction crews. Figure 5.5*a* uses a physical precedence network to illustrate three typically critical blocks of work associated with the construction of a low-rise office building. In this case, the blocks designated as "Framing" and "Floors" are critical because each, when completed, opens up other critical blocks of work, as depicted in Figure 5.5*b*. Many of the additional blocks of work indicated in Figure 5.5*b*, such as the plumbing and HVAC systems, are typically subcontracted and thus become critical interface points for the contractor operating in this environment.

Physical precedence networks and milestone planning techniques can also be applied to the design-construct project environment. As an example, the precedence network shown in Figure 5.6*a* depicts physical work blocks which make up the overall project execution sequence from engineering to project start-up. At this level, the client could award major project execution blocks to specialist firms (for example, engineering to an engineering firm). These separately subcontracted work blocks might include design and procurement (i.e., blocks 1, 2, and 3), construction (i.e., block 4), or the entire process might be awarded to a single firm. By using milestone planning, both the gross relationships and key milestones between each major project execution process can be established, as illustrated in Figure 5.6*b*. In this situation, the milestone "start effective construction" would be controlled by the contractor, while "mechanical completion" would most likely be client-directed.

The overall broad dissection of work blocks begins to establish the basic control philosophy and the potential sequence for committing manpower

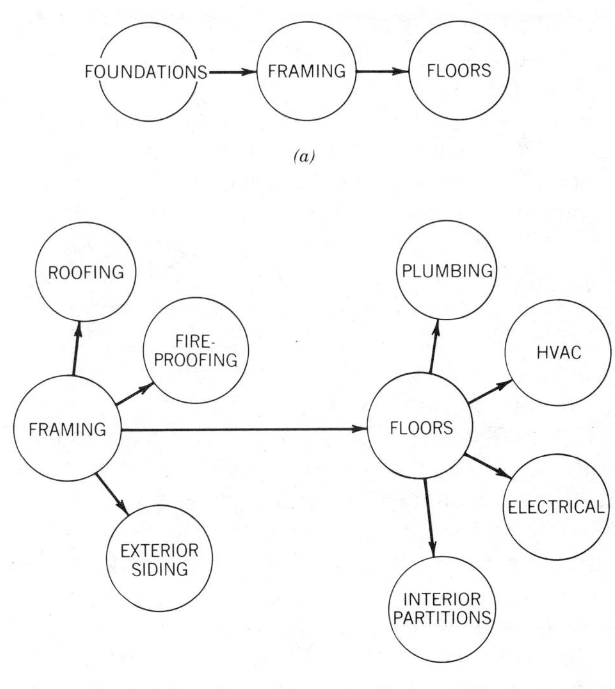

Figure 5.5 Physical precedence networks. (*a*) Physical sequences (arrows) for key blocks of work. (*b*) Completion of key blocks of work release other critical blocks of work.

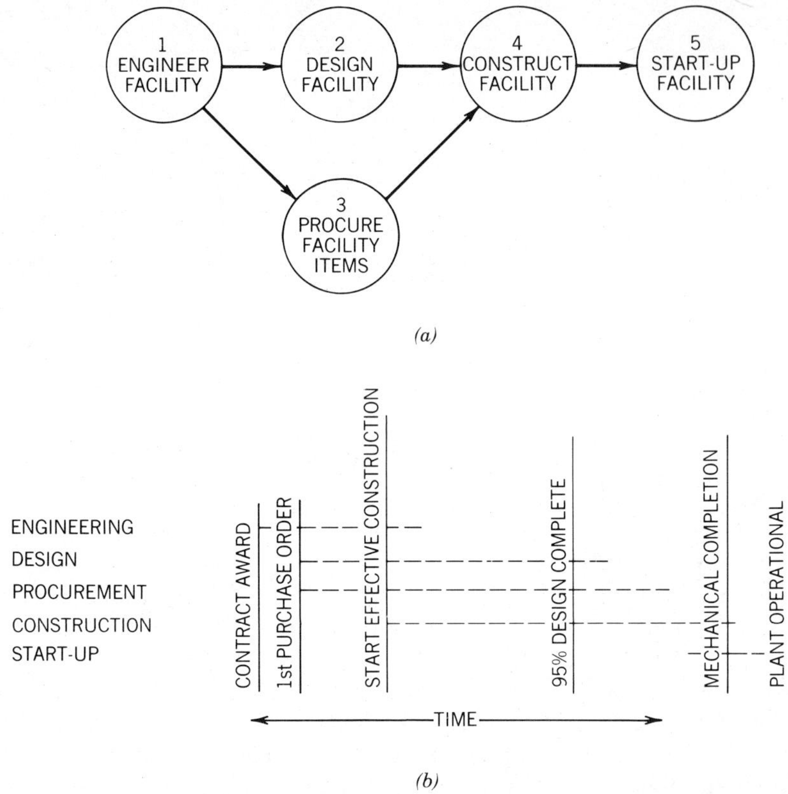

Figure 5.6 Broad planning concepts. (*a*) Project execution process by blocks of work. (*b*) Milestone planning.

resources to a project. For example, the construction process for a coal-fired power station can be divided into five physical blocks. The physical sequence of construction, as depicted in Figure 5.7, can clearly impact the utilization of manpower (and other) resources over time, depending on whether the physical blocks of work are performed concurrently (Figure 5.7*a*), sequentially (Figure 5.7*b*), or in a compromise sequence (Figure 5.7*c*). The sequence selected must integrate available resources, key milestones, and client-directed priorities. In addition, the five blocks of work in Figure 5.7 begin to define physical work to be controlled.

Two performance techniques which relate to the project control of major areas of work emerge at this level of planning. These two techniques are the Work Breakdown Structure (WBS) and the Control Area approaches. The WBS technique is particularly useful for a project that involves a number of different firms which are being managed by a single entity. As an example, level one of the WBS could be the total project program. Level two could focus on project assignments to firms for engineering/design and/or construction. Level three of the WBS is task oriented and could be the process engineering segment of a particular component facility within the overall facility. Subtasks, and finally, work packages would be defined as levels four through six and are determined as scope definition begins to develop. The Control Area approach is an alternate method of project control achieved by dissecting the project into manageable geographical or physical component areas. For example, the physical blocks of work shown in Figure 5.7 for the coal-fired power station could be designated Control Areas. Control Areas could easily be incorporated into the WBS if desired.

The broad construction planning described in Figure 5.7 represents initial steps in Project Planning, including the development of control procedures. By further dissecting a major block like the turbine-generator facility, physical pieces of this block can be identified and sequenced, thus establishing the next level of planning and control as delineated in Figure 5.8. Each major block can be identified and sequenced in a similar manner and, when integrated with project milestones, becomes a viable plan for top level management control. This level of detail, however, remains insufficient for detailed field planning, estimating, and scheduling.

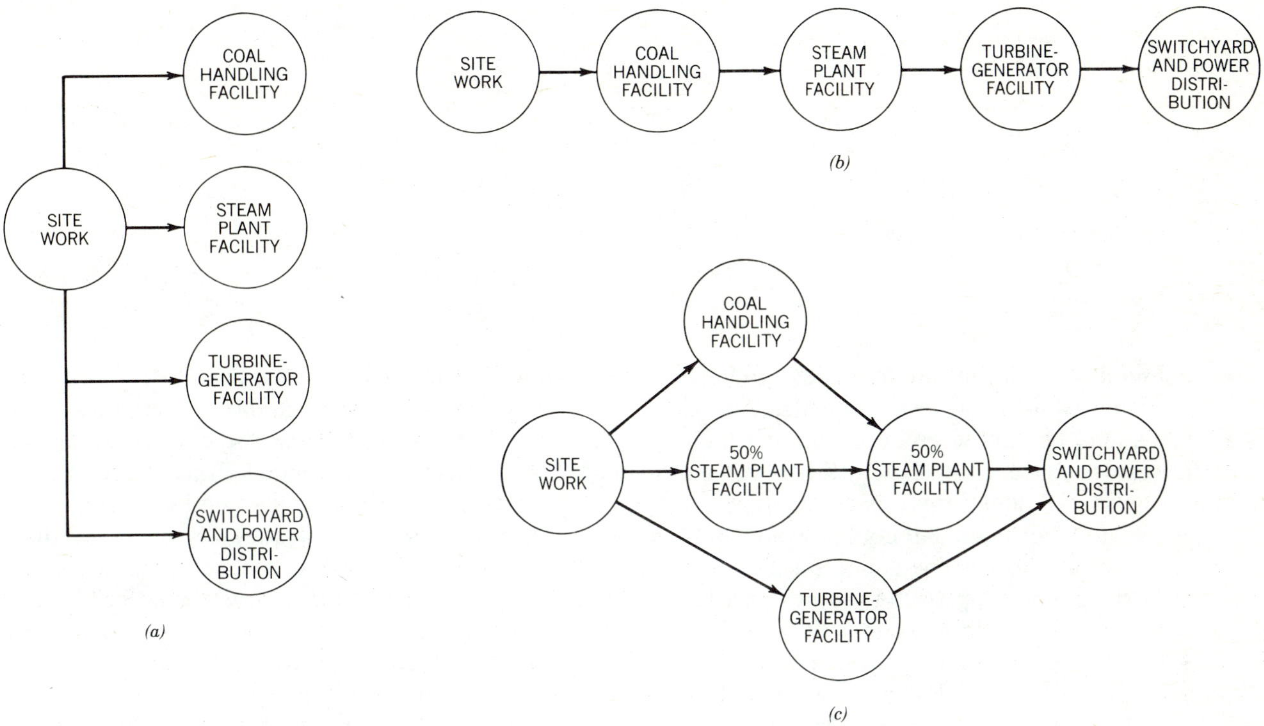

Figure 5.7 Overall physical sequence of construction. (*a*) Concurrent. (*b*) Sequential. (*c*) Compromise.

Cost Coding techniques are introduced at this level. This means that those physical pieces in Figure 5.8, for example, are coded into major cost categories such as 10xx for the "Foundation" block. It should be noted that the definition and sequencing of physical pieces corresponds to the normal construction order in which work is encountered. Cost codes normally reflect this same sequencing of construction work.

The contractor's concern now shifts to a level of planning which can be used by field managers and supervisors to control the project. This level of planning focuses on the management control of space, and serves as a means of establishing a compatible base for project control in relation to estimating, scheduling, and field planning and control. A finer application of the Work Breakdown Structure or Control Area (e.g., work packages) and Cost

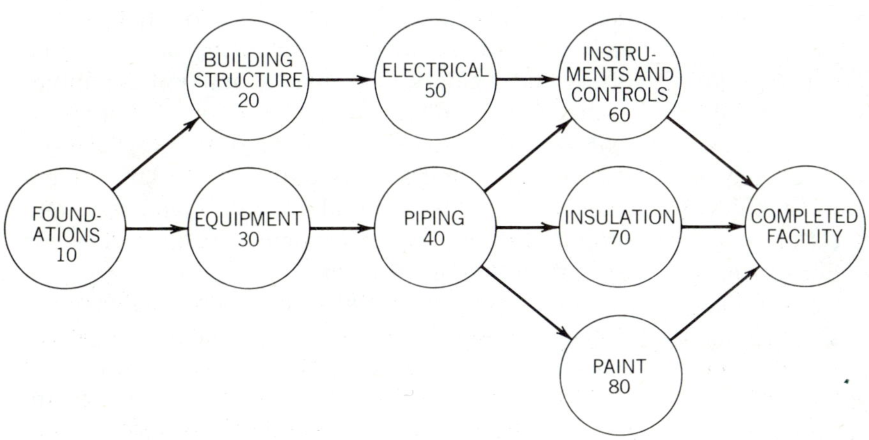

Figure 5.8 Physical components of turbine-generator facility.

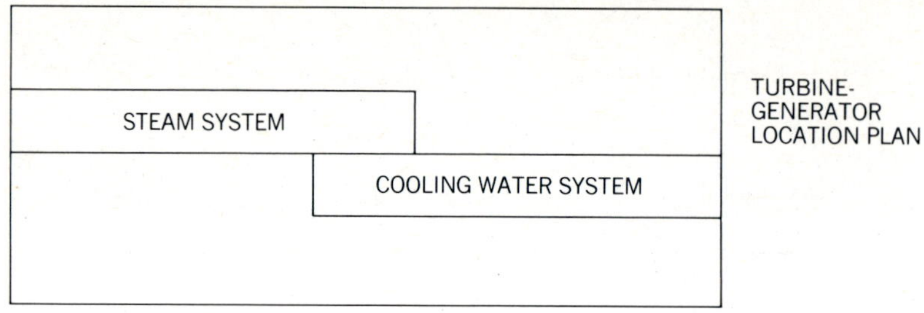

Figure 5.9 Management control of piping systems by work package: turbine-generator facility.

Coding techniques is appropriate to achieve the desired control of space. In a design-construct project, drawings may be numbered in a manner which ties directly into space control identifiers. As an example, Figure 5.9 illustrates space control by work packages for the steam and cooling water piping systems within the turbine-generator facility. While the work package approach provides an adequate level of detail for field management, a further breakdown of the work package into its component activities or tasks is essential in order to establish its anticipated manpower content. Figure 5.10*a* depicts a detailed activity dissection for the steam and condensate system work package within the turbine-generator facility. Cost coding techniques are applied to this detailed dissection of component activities and include designators for work packages. Figure 5.10*b* illustrates typical cost coding techniques for those items shown in Figure 5.10*a*. Finally, Figure 5.10*c* illustrates the concept of the dissection of common activities which forms a compatible basis for estimating, planning and scheduling, and field control. A greater level of detail is required to identify pipe size, for example, but if estimate data can be rolled up to the erect pipe activity level then sufficient information is available for scheduling and field planning.

A number of general concepts will emerge through the progressive analysis of Figures 5.7 through 5.10. First, the concept of rolling project data from the level of greatest detail (Figure 5.10) to the level of least detail (Figure 5.7) or vice versa ensures compatibility at all levels of project control. In addition, hierarchical levels of project control which are relevant to the various levels of project team decision making become clear. For example, top level management is interested in summary level data, while field supervision desires detailed activity component data. Finally, a common database for all levels of project control is desirable and can be achieved through the proper evaluation and dissection of the project using the WBS, Control Area, Cost Coding, and/or similar techniques.

As part of project planning the contractor must define the craft skills demanded by the nature of the project so as to anticipate project and craft requirements. This assessment is frequently aided by the broad dissection of physical blocks such as those illustrated in Figures 5.5 and 5.8. In the latter figure, for example, a piping component would indicate a need for pipe fitters, welders, plumbers, and the like, as is typical for the type of project under consideration. The assessment of craft requirements enables the contractor to evaluate problems related to labor availability in the site region, as well as establish wage rate structures for the required crafts, and estimate site productivities.

In construction environments where the work region is local and projects are small, the contractor generally knows the magnitude and mix of the craft requirements for the project and is familiar with local labor pools. Thus, the contractor can easily assess the impact of labor shortages or wage rate changes on the potential project if required. On large and/or remote construction projects, the contractor must not only identify but attempt to quantify the magnitude of craft requirements in a usable form such as that indicated in Figure 5.11. This Figure, for example, tells the contractor that historically 28% of the total project manpower are pipe fitters and welders, and that the maximum number of this type of craftsmen will be required at approximately two-thirds of the project duration. The contractor can combine this data with order of magnitude estimates of total project manpower and project duration to approximate the project needs for pipe fitters and welders.

Once the contractor has a basic knowledge of the project's nature, size, and anticipated manpower requirements, the firm must evaluate the site labor environment. The contractor must probe the site labor environment and pose questions which, when

PHYSICAL AREA DESIGNATION	WORK PACKAGE	MAJOR COST CATEGORY	DETAILED COST CODE	WORK PACKAGE DESCRIPTION	RESOURCE
TURBINE-GEN	STEAM	PIPING	ERECTION	ABOVE GROUND-CARBON STEEL	LABOR

(a)

30	42	4	5100	16" ABOVE GROUND-C.S. PIPE	1
30	42	4	5200	BUTT WELDING-C.S. PIPE	1

(b)

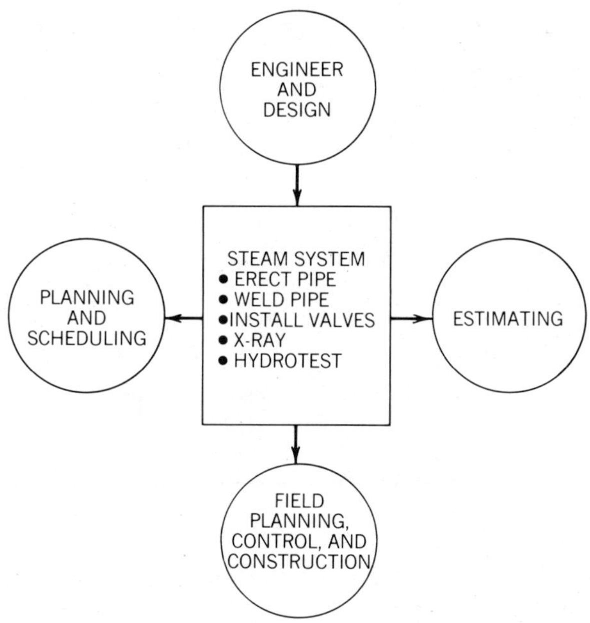

(c)

Figure 5.10 Detailed categorization of work. (*a*) Detailed dissection of work. (*b*) Cost coding applications to detailed dissection of work. (*c*) Detailed activity dissection which forms common basis for engineering/design, estimating, planning/scheduling, field planning, cost control, and construction.

answered, provide management with information in sufficient detail to make critical decisions. Thus, labor reconnaissance techniques emerge whereby the contractor can survey the project region to ascertain craft availability (number of available personnel), craft wage rate structures, work practices, and potential productivity levels. This survey could include a questionnaire, a search of company historical data from previous projects in the area, personal discussions with local unions and other contracting organizations in the area, other construction activity in the area, and so on. The results of site labor reconnaissance are frequently summarized in a report highlighting labor conditions the contractor is likely to meet. In any case, this type of analysis serves as input for decisions relating to camp requirements, whether to import and/or train labor, craft wage rates for estimating purposes, and the like.

Evaluating and quantifying the relationship between the contractor's labor estimating base and the anticipated site labor performance for the project raises several critical issues. The decision maker is first confronted with determining those

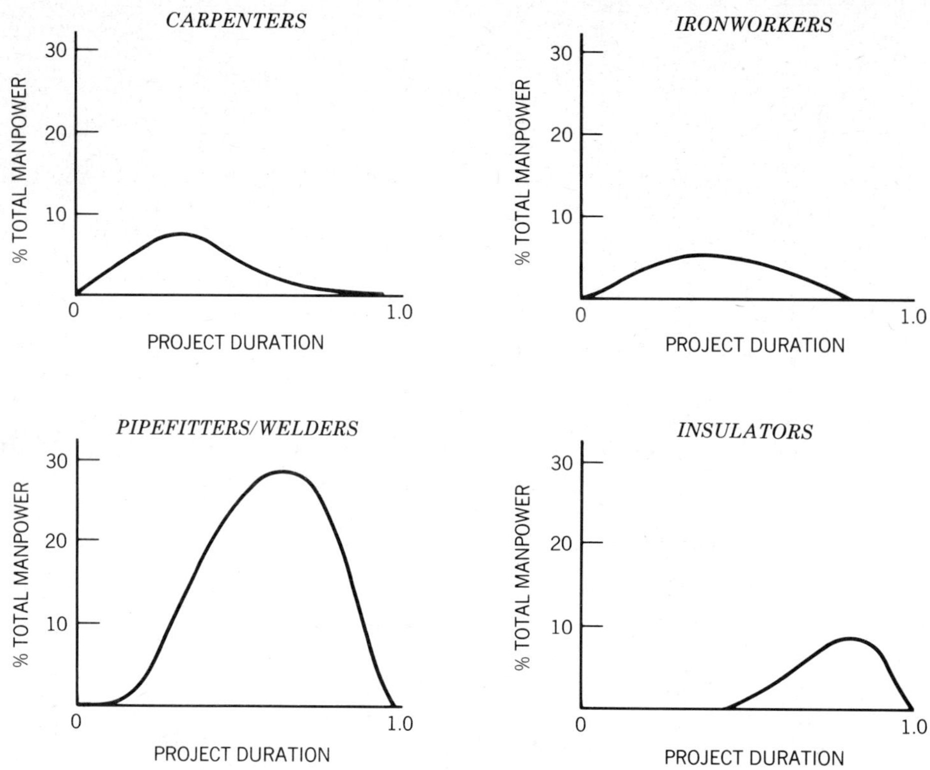

Figure 5.11 Typical standard craft profiles.

factors which are likely to make site labor performance for the current project different from the contractor's standard labor estimating base. For example, the contractor must consider the effect of a number of complex and often interrelated factors that influence productivity. These include:

1. Regional or location (often in foreign countries) variations.
2. Learning curve applications to the project.
3. Work schedule impacts (e.g., overtime).
4. Weather conditions.
5. The availability and acquisition of skilled craftsmen.
6. Work rules as applicable to union construction.
7. The contractor's ability to manage and motivate the work force.

While labor reconnaissance enables the contractor to qualitatively evaluate these factors, the contractor must also quantify this assessment, using experience, judgement, and common sense "gut" feelings. Thus, the contractor must determine the site value of standard labor units (e.g., 1 standard man-hour equals 1.15 site man-hours).

Once the contractor assesses the site labor efficiency in relation to the standard labor estimate base, it is necessary to determine whether or not site productivity (as related to the budget) can be improved through better management control techniques. In addition, the contractor must predict the anticipated rates at which productivity will occur over the life of the project. For example, productivity is normally lower during the early and late stages of the construction effort, with maximum productivity occurring somewhere between these periods. These concepts of productivity analysis are illustrated in Figure 5.12. Figure 5.12*a* depicts a brief tabular analysis of critical factors influencing labor performance, and the estimated percent adjustments as a result of those factors in relation to a standard labor estimate base for a project. This data is gathered from historical records, labor reconnaissance, analysis of weather conditions, typical craft production rates in the area versus standard rates, and so on. The composite adjustment of 10% is applied to the standard labor estimate base at the total project level, as shown in Figure 5.12*b*. This

LABOR AND SITE-RELATED IMPACTS	% ADD/DEDUCT TO STANDARD ESTIMATE BASE
• MAJORITY OF WORK IN WINTER	+15%
• LABOR SHORTAGES—KEY CRAFTS	+25%
• SKILL LEVEL LOWER	+ 5%
• ABOVE NORMAL REPETITIVE WORK	−10%
MANAGEMENT IMPROVEMENT APPROACHES	
• SELECT HIGH-QUALITY PROJECT TEAM	−10%
• USE METHOD IMPROVEMENT TECHNIQUES	−15%
TOTAL COMPOSITE ADJUSTMENT	+10%

(a)

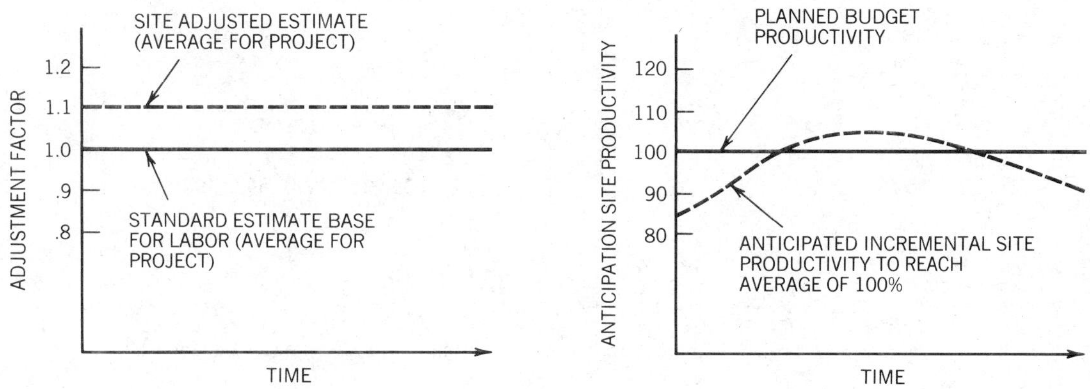

(b) *(c)*

Figure 5.12 Site productivity analysis. (*a*) Assessments of adjustments to standard labor estimate for anticipated site labor conditions. (*b*) Adjustment of standard base labor estimate. (*c*) Anticipated site productivity over time.

adjustment, however, could also be applied at different levels of detail to those work items associated with a craft that has lower production rates than the standard labor estimate base. Finally, Figure 5.12*c* predicts the rate of site productivity over the duration of the project that will reach the average overall 10% composite adjustment to the estimate base. This curve is particularly important when developing manpower plans for a project. The productivity analysis concepts described in Figure 5.12 commit the contractor to a target productivity profile over a given period of time. Failure to achieve the target profile carries both a cost and schedule impact. This failure, if it occurs, should be confirmed in the monitoring of actual productivities achieved when the relevant techniques are applied.

The goal of Project Planning in terms of a Management Information System can be described as the need to keep track of facility components and control area requirements. This need is fulfilled through the identification and definition of the facility components to be controlled when executing the project. When these facility components are integrated with cost code techniques as related to project control area concepts, the basis for controlling the project, and in particular the manpower component, is established. The documentation and information flow produced as a result of this Project Planning decision process provides the common data structure for the performance of Estimate Preparation and Project Scheduling, where control base budgets and schedules are established. In addition, this data structure provides the linkage between the control base plans and the actual results achieved as identified in Project Status, Analysis, and Reporting. In this way an integration of cost with schedule in the control of the project is ensured (see Chapter 16). Project Planning can be characterized as an MIS as shown in Figure 5.13. The fully documented MIS concept described in

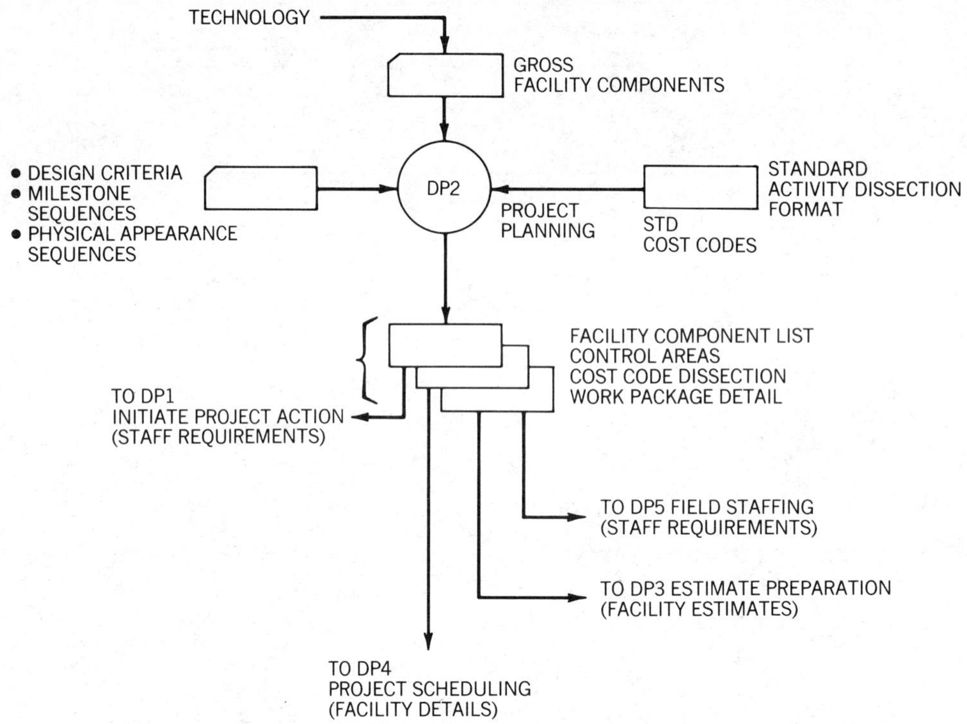

Figure 5.13 Project Planning Management Information System (MIS) concept.

Figure 5.13 is applicable to firms in the large firm environment. It is performed informally in the small and medium-sized firm environments, where traditional company cost codes are applied to projects, through Estimate Preparation. Otherwise, the physical size of the project and the related work volume can be readily grasped by most project team members.

5.3 THE PROJECT TEAM COMPONENT APPROACH

The planning process at the small firm level is straightforward and pragmatic, and is related directly to past experience as small firms rarely engage in construction activity outside of their area of expertise. The project manager and/or estimator will carefully examine the plans and specifications and make a site visit to locate boundaries, existing services, and other obstructions. The planner will immediately and intuitively grasp what is involved in the project. The project manager will review the entire construction process in his or her mind, taking note of items and features that are peculiar to the project. In addition, a determination will be made as to which areas of work will be handled in-house by the contractor's own limited number of crews, and which must be, or are better left to subcontractors.

At the small firm level the need for evaluating labor-related conditions within the site work region does not emerge, because an analysis of such factors as labor availability, union agreements, and local productivities is obviated by the limited (or localized) nature of the work region, small project labor forces, and utilization of a nucleus of highly skilled workers. Because the contractor works in a limited project environment, local productivity and available manpower levels, for instance, are well known to the project manager and estimators. Furthermore, the total project labor force, relatively small in actual numbers, rarely strains the local labor pool. Even when a shortage exists, the small firm continuously interacts with the same unions. Thus, the firm's knowledge of each union's policies and rules is quite extensive, and includes knowledge of all local restrictive practices. In this environment, therefore, construction reconnaissance focuses almost entirely on the evaluation of site conditions, and consequently is related directly to the Estimate Preparation decision process.

5.4 THE TRAVELING PROJECT MANAGER

The medium-sized firm operates in a highly competitive construction environment in which the contractor typically handles a portfolio of projects which vary considerably in total dollar magnitude and complexity. Individual projects can range from $1 million to $10 to $20 million, and require the resolution of a variety of engineering problems and the careful planning of critical construction operations. In addition, because fewer projects of this size exist, the contractor must be prepared to handle unfamiliar projects and operate in a strongly competitive bidding environment in which little time is given for the entire bidding process. Thus, the medium-sized firm begins to address engineering and planning problems.

The contracting firm must carefully evaluate alternative construction methods and quickly develop production, quality control, and cost comparisons of critical operations so that planning decisions can be made before estimating begins. However, specific engineering problems related to these decisions may be left for later, as considerable time pressures exist. Thus, for example, specification of required mobile crane sizes and location of lifting positions associated with a method of construction may be deferred once a decision has been made to use a crane rather than some other method.

While engineering and planning processes emerge at the medium-sized firm level, they are not formally addressed, but have a preliminary preestimating aspect with little labor or manpower focus. The contractor still operates in a local labor environment, and while occasionally prepared to move carefully outside a local region for an individual project, will endeavor to maintain the same management approach and knowledge of the local labor environment by establishing a branch office in a neighboring town or county. Hence, no manpower planning decision process is involved.

5.5 FIELD-BASED PROJECT MANAGEMENT

In the large firm environment the project's technical nature, size, and complexity significantly increase its manpower content. This creates a problem environment requiring a stronger management effort and a focus on Project Planning as a manpower decision area. This growth in the significance of Project Planning is apparent in both fixed priced contracting and cost reimbursable contracting, and is brought to bear most heavily on "grass roots" projects located in remote areas or foreign countries. Consequently, an individual project will require the resolution of many different and often complex engineering problems, and the careful preplanning of critical construction operations to match the often unique site conditions (e.g., tight lay-down areas, transportation of large and heavy permanent plant equipment, etc.). Field labor forces may extend into the hundreds or possibly thousands. Even when the contractor can to some extent grasp the project's nature and size, in addition to having a basic knowledge of the site location by virtue of past working experience and history in the area, the complexity of the large project will still produce Project Planning problems that the contractor must systematically solve prior to the final development of control budgets and schedules. Thus, Project Planning emerges as a critical manpower management focused decision process.

Project Planning commences when the project team is mobilized and released to the project. The Project Planning decision process is comprised of three manpower management functions, namely CONSTRUCTION SEQUENCING (S1), CONSTRUCTION RECONNAISSANCE (P1), and ACTIVITY DEFINITION (P2). The information flow structure, together with the relevant management functions and pertinent policy statements as presented in Figure 5.14, provides a basic description of the Project Planning manpower management process found in practice at large firms operating in the design-construct mode. Project Planning is similar to Initiate Project Action in that each phase of the decision process depicts an increasing level of effort and detail as the project scope is defined, and project control budgets and schedules are demanded by top management and the client. Furthermore, because Project Planning, as an independently recognizable manpower management decision process, initially appears in the Field-Based Project Management environment, it has a possibly limited field focus in CONSTRUCTION RECONNAISSANCE, and a predominately head office based orientation in CONSTRUCTION SEQUENCING and ACTIVITY DEFINITION.

Figure 5.14*a* depicts Project Planning during the staged development phase of engineering and design. In this phase, the basic project composition, major components and subcontracts, priorities of facility completion, and contractual responsibilities

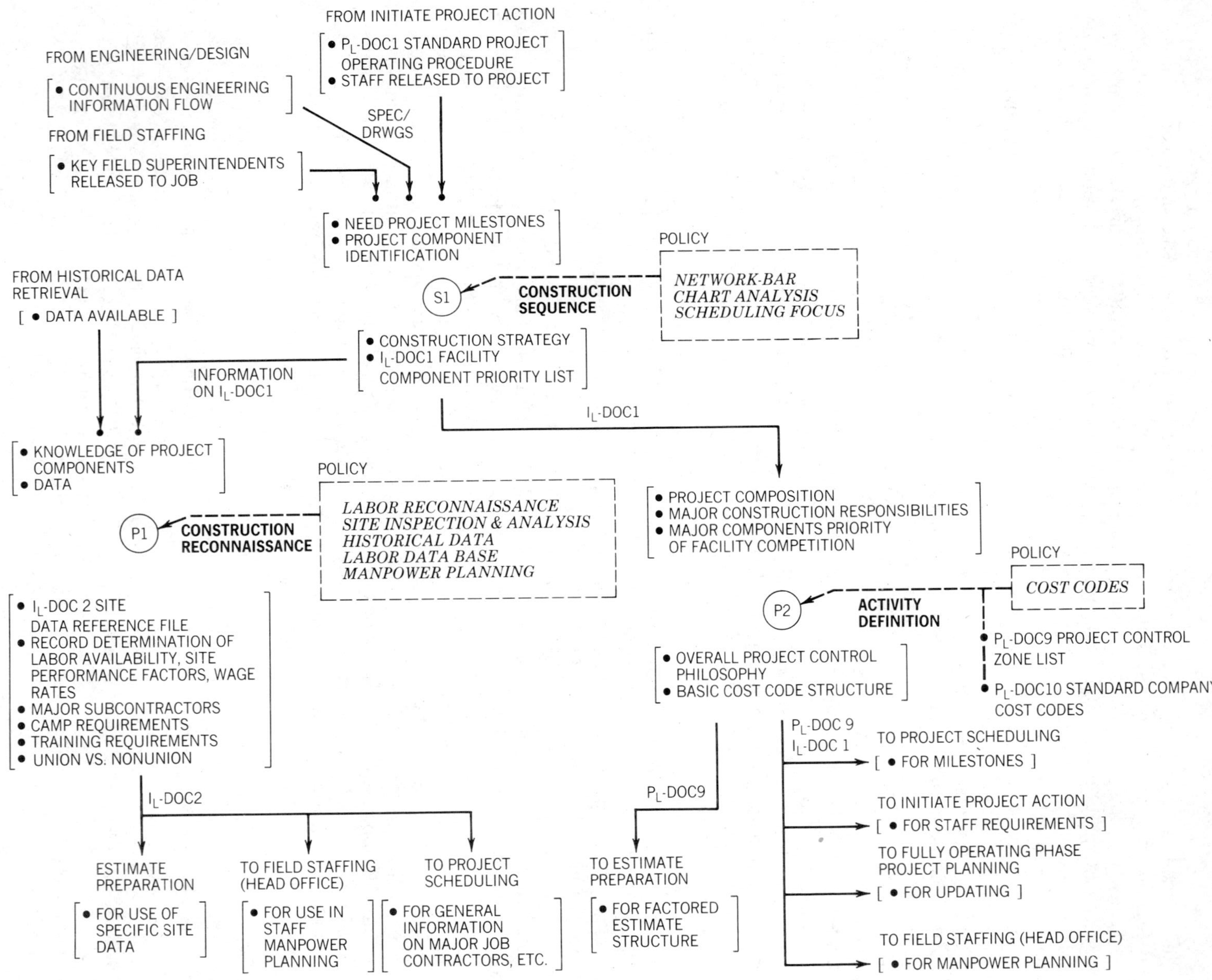

(a)

Figure 5.14 Project Planning decision process (Field Based Project Management). (*a*) Staged development phase.

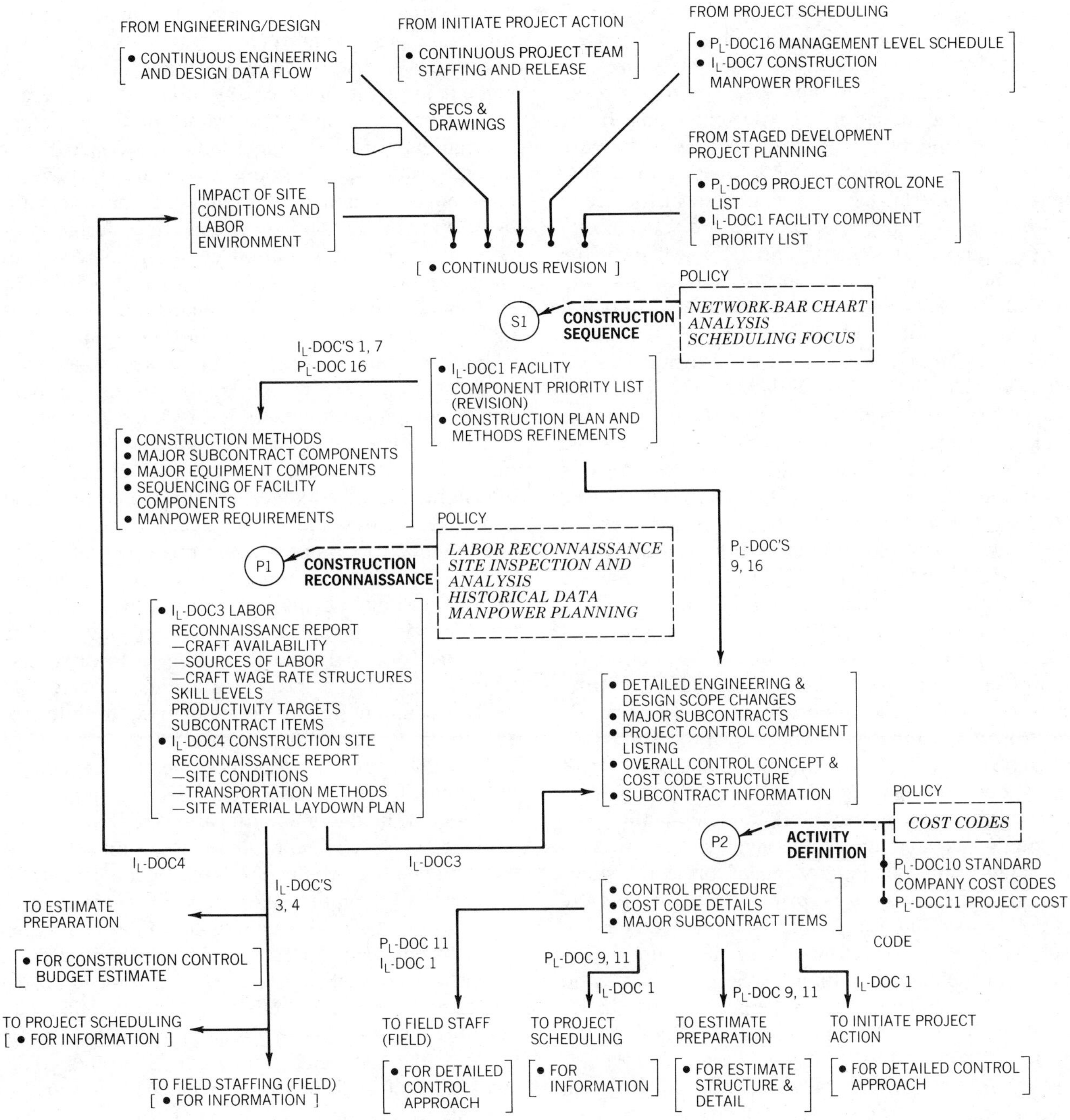

(b)

Figure 5.14 (*Continued*) (*b*) Fully operating phase.

are identified. This information provides the requisite input for establishing overall project control philosophies and major cost coding structures, as well as providing basic data essential for factored estimating and management level scheduling. In the fully operating phase of Project Planning, Figure 5.14*b*, when specifications and construction drawings are available, the data generated in the scope development phase becomes more specifically focused and documented. This data will form the basis for control budgets and schedules that will provide the field with the documents and information to plan and control the construction effort. Project Planning has a field orientation via CONSTRUCTION RECONNAISSANCE in the sense that site surveys, area labor reconnaissance, and so on are frequently performed through small field offices established shortly after the project's inception. Alternatively, CONSTRUCTION RECONNAISSANCE can also be performed from the head office, with frequent site visits to gather information. Whether or not a field office is established early in the project depends on the location of the project, its size and complexity, and the desires of the client. It may also be a function of permit requirements (i.e., obtaining appropriate construction, environmental, and other regulatory permits).

Project Planning commences with CONSTRUCTION SEQUENCING, wherein head office project team decision makers identify the desirable sequences of major project work, and define the basic technological sequences of engineering, procurement and construction for major project components. This function is primarily performed by construction and scheduling engineers under the supervision and direction of key field superintendents (particularly the general superintendent), who are selected and released to the project early in the effort. The key policies which dominate CONSTRUCTION SEQUENCING are *NETWORK AND BAR CHART ANALYSIS* (P_L 10) and *SCHEDULING FOCUS* (P_L 11).*

The initial planning focus of CONSTRUCTION SEQUENCING involves the identification of broad physical blocks of work. These blocks of work can then be related to key milestones either established by the client or the contractor. Both of these tasks can be graphically displayed, as shown in Figures 5.5 and 5.6, using Physical Precedence Network and Milestone Planning techniques.

*See Table 5.3, Policies of the Large Firm.

Selecting specific milestones depends on the nature of the project and typically includes such major milestones as start effective construction, start placement of the turbine generator mat, start erection of the structural steel frame, or facility commercial operation. On large industrial construction projects, start-up operations are included as client-directed major milestones, and are the dominant factor in the sequencing of engineering, design, procurement, and construction components. Start-up sequences will also impact the further dissection of each major physical process block (i.e., engineering, construction, etc.), and the determination of critical schedule contracts such as key major subcontracts (e.g., steam generation system), critical major equipment items (e.g., turbine generator), and so on. The small and medium-sized contractor rarely has to evaluate in detail project priorities, milestones, and the like (other than project start and completion dates), primarily because project size and complexity do not warrant this depth of analysis. The further dissection and sequencing of physical work blocks begins to reveal the nature of the project as is illustrated for a coal-fired power plant in Figure 5.7*c*. A Facility Component Priority List (I_L-DOC 1) is often produced at this stage to formally document facility components, milestones, and so on, as depicted in Figure 5.15.

When the project scope becomes better defined in terms of project specifications, drawings, and component facilities identification, CONSTRUCTION SEQUENCING begins to focus on construction engineering problems related to equipment usage, heavy lift requirements, lay-down space, and other technological sequences like the installation of the turbine and generator, or major chemical process reactor vessel. In addition, decisions are made regarding subcontract activities, and potential vendors for large key facility equipment (particularly those with long lead times) are identified. The evaluation of each of these problems leads to the final determination of the planned engineering and construction sequence logic.

Because of its size and technical complexity, the large project has a heavy requirement for manpower, and therefore CONSTRUCTION RECONNAISSANCE emerges as a predominant function in Project Planning. Projects often have field labor forces ranging from 200 to 2000 men. Obviously a project of this size can severely strain the local labor pool, creating potential craft availability problems which can impact both costs and schedules.

TABLE 5.3 POLICIES OF THE LARGE FIRM

Policy		Description
P_L 4	*LABOR RECON-NAISSANCE*	Comprehensive survey, analysis, and documentation of the project site labor environment is requisite to the accurate evaluation of the labor content and manpower requirements for the project.
P_L 5	*SITE INSPECTION AND ANALYSIS*	Comprehensive inspection, analysis, and documentation of project site conditions, transportation problems, etc., is requisite to accurately evaluating construction methods, plans and schedules, labor content, and to solve other unique problems.
P_L 6	*COST CODES*	A systematic classification of project work categories segregated by unique physical components and by cost codes is the basis for structuring the project scope into work tasks, categorizing estimated costs, reporting and controlling costs and manpower, and developing a historical database.
P_L 7	*UNIT MAN-HOUR*	Realistic evaluation of the labor content for a work task is essential for successful project planning and cost control. Therefore, labor costs must be based on work task man-hour content in relation to work volume, construction methods, local productivities, and labor availability.
P_L 8	*HISTORICAL DATA*	Historical labor data are basic to a realistic determination of the labor content for a work task as well as for major project components.
P_L 9	*LABOR DATABASE*	Historical labor data must be analyzed and documented in a meaningful way to be useful for estimating and planning manpower for future projects.
P_L 10	*NETWORK-BAR CHART ANALYSIS*	Broad network focused milestone planning is the best approach for identifying and sequencing major project components while a combination of the critical path method and bar chart model are best suited for project control level planning and scheduling.
P_L 11	*SCHEDULING FOCUS*	Scheduling definitions and constraints are: 1. A critical activity is one that, if delayed, will extend the completion date of the project. 2. Activity durations must be correlated to the estimated work content. 3. All constraints must be defined and incorporated into the sequence logic. 4. Manpower impact is an integral aspect of project scheduling. 5. Schedule must reflect optimum use of all resources. 6. Progress is based on the physical completion of work activities.
P_L 15	*MANPOWER PLANNING*	Long-range planning of manpower by craft and in total is a prerequisite to supporting the field effort, and must be performed in the planning and scheduling stages of the project.

Furthermore, when the site is located in an isolated area or in an area where skilled tradesmen are not available, the impact on labor costs and other ancillary costs can become significant, necessitating major decisions concerning labor productivity, training, housing, and so on. Because of these problems, CONSTRUCTION RECONNAISSANCE must provide information which is vital to evaluating the project's manpower content and the impact of manpower on the project as a whole.

CONSTRUCTION RECONNAISSANCE is performed by labor relations experts and construction engineers at different levels of detail depending either on the life stage of the project or its location. For example, if the contractor is constructing an expansion to an existing plant built by the same firm, certain types of analysis may not be required because of historical knowledge and available records from the previous project effort. Conversely, if the project is located in an extremely isolated area on a virgin site, and the contractor has limited knowledge of the area, detailed CONSTRUCTION RECONNAISSANCE is essential. The focus of this function is to gather data relating to both the site labor environment and the project site conditions, in so far as specific critical problems and labor conditions can be identified. This includes collecting information on the current availability of labor for each craft, prevailing labor practices (particularly unique restrictive practices in union construction environments), and prevailing wage rates. In addition, an analysis of existing or anticipated

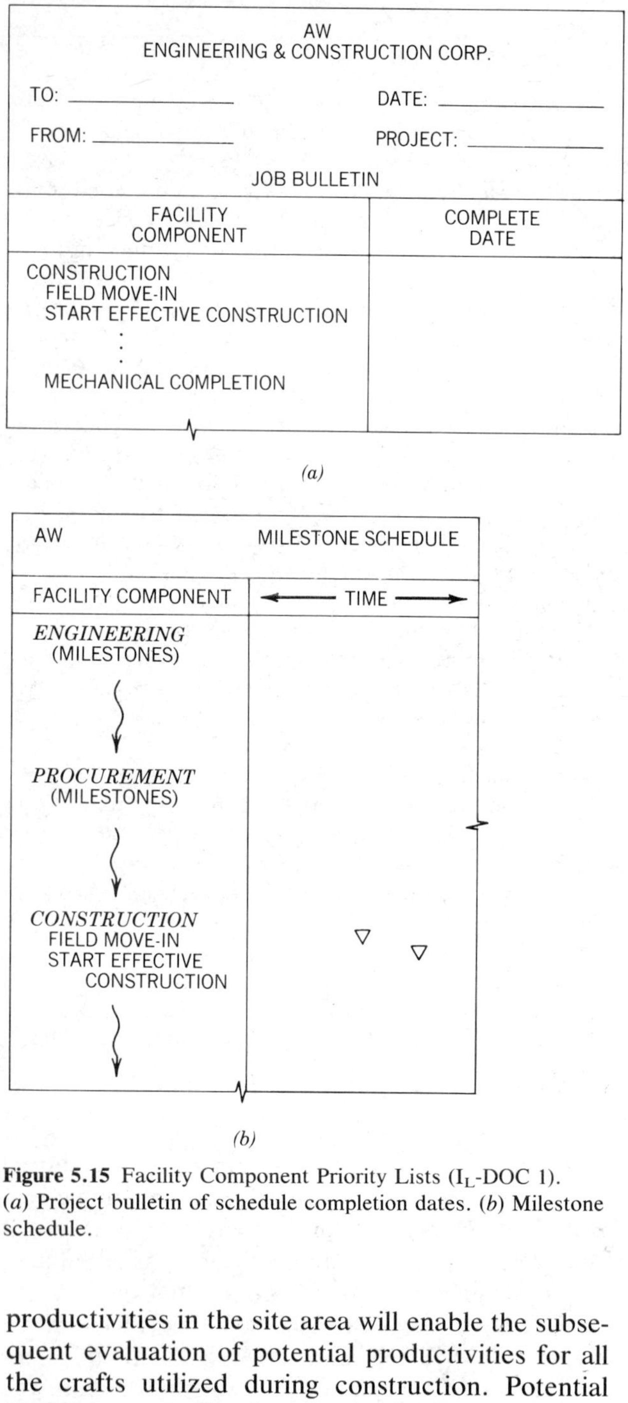

Figure 5.15 Facility Component Priority Lists (I_L-DOC 1). (*a*) Project bulletin of schedule completion dates. (*b*) Milestone schedule.

productivities in the site area will enable the subsequent evaluation of potential productivities for all the crafts utilized during construction. Potential problems regarding housing and training requirements are identified, and site visits are conducted to effectively evaluate possible problem areas and unique site conditions which may directly influence construction methods and techniques, procurement requirements, material handling, and the like.

During the design development phase (see Figure 5.14*a*), CONSTRUCTION RECONNAISSANCE is an in-house effort, wherein decision makers rely primarily on either *HISTORICAL DATA* (P_L 8) files or input from other sources. For example, the contractor might consult industrial relations specialists to gain labor-related data concerning the site. This type of input is useful for factored estmates and management level project scheduling, and can provide enough manpower data for initial decisions on camp requirements, training, manpower, and the like. A Site Data Reference File (I_L-DOC 2) might be a typical document compiled from the CONSTRUCTION RECONNAISSANCE effort (see Figure 5.16). During the fully operational phase, Figure 5.14*b*, data is collected via site visits or through temporary field offices* from which the respective decision makers contact local union representatives, other local contractors,† and vendors. The data they gather include the availability and quality of labor by craft, sources of labor, current productivities, restrictive work practices, and housing and training requirements. When analyzing this data, other sources of information such as *HISTORICAL DATA* from the *LABOR DATABASE* (P_L 9) on past projects located near the project under consideration or on projects currently under construction near the site location are frequently referenced. During the labor survey, other construction personnel will also visit the project site with the intent of documenting potential problem areas, unusual physical features, and site preparation problem areas, as well as determining the availability of local vendors.

According to the policies of *LABOR RECONNAISSANCE* (P_L 5), *SITE INSPECTION AND ANALYSIS* (P_L 6), and *MANPOWER PLANNING* (P_L 15) an extensive analysis of the information collected is documented in formal reports. The Labor Reconnaissance Report (I_L-DOC 3) in Figure 5.17 identifies such items as manpower totals by craft, current or estimated wage rates, detailed explanations of restrictive practices, and anticipated site productivities. (The anticipated site productivity is

*In the design–construct project environment, a field office may be established prior to the start of construction. In this case, site and labor reconnaissance are often handled almost exclusively from the field location. This provides the contractor with an additional advantage, because it allows a continuous flow of information to the head office as better site and labor information becomes available. Reports can be revised as necessary to reflect the latest field input.

†These contractors may be the potential subcontractors later used on the project.

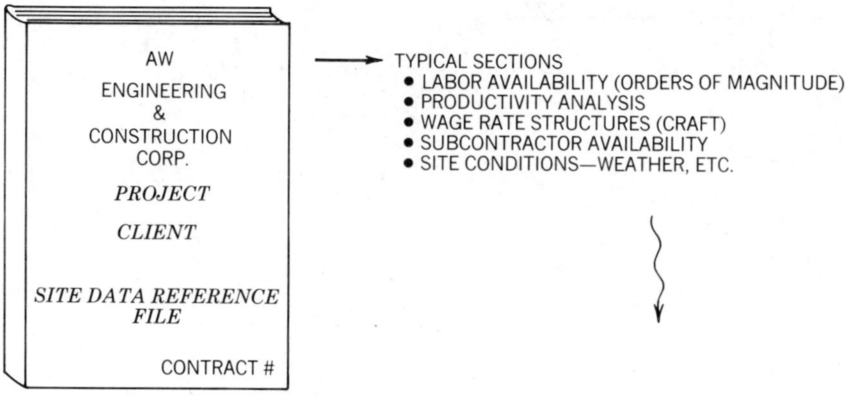

Figure 5.16 Site Data Reference File (I_L-DOC 2).

commonly referenced against a labor base productivity for a given region where productivities are essentially known from previous experience.) Techniques to evaluate and quantify both craft availability and labor productivity are described in Figures 5.11 and 5.12. The Construction Site Reconnaissance Report (I_L-DOC 4) in Figure 5.18 completely describes in detail all potential site problems, including access, storage, and other salient logistics factors. This descriptive documentation will often be accompanied by pictures, sketches, and/or other visual aids that clarify the written material. Further-

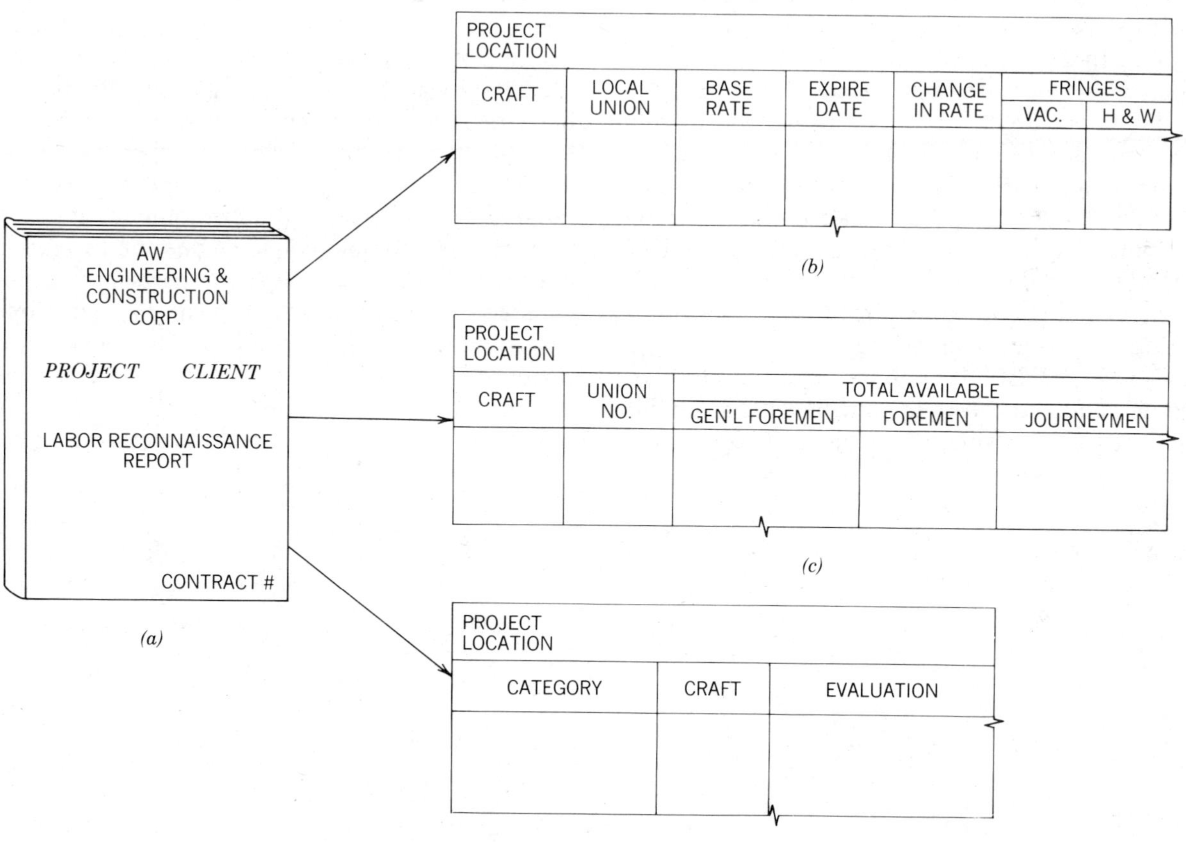

Figure 5.17 Labor Reconnaissance Report (I_L-DOC 3). (*a*) Report. (*b*) Wage rate data. (*c*) Craft availability data. (*d*) Productivity data.

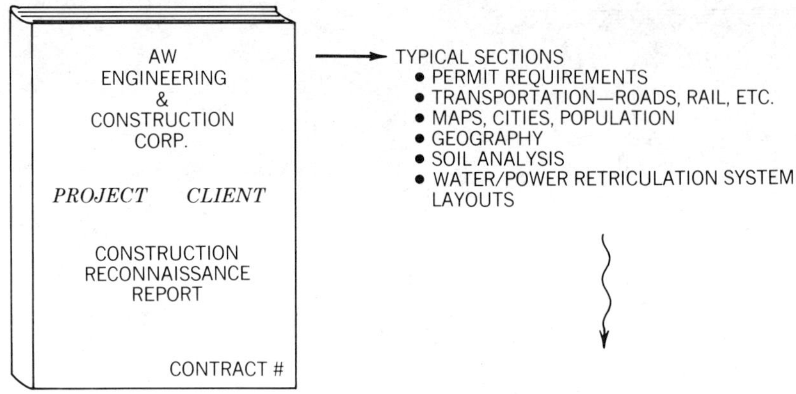

Figure 5.18 Construction Site Reconnaissance Report (I_L-DOC 4).

more, potential housing arrangements (where applicable) are described, including possible training requirements and potential camp facilities. In summary, these two reports address all aspects of the project that are influenced by site conditions and the labor environment.

In large firms, ACTIVITY DEFINITION has a completely different focus from that in small and medium-sized firms—especially in terms of Project Planning. In general, ACTIVITY DEFINITION in this environment attempts initially to define the overall project control philosophy and basic cost code structure (Figure 5.14*a*), and subsequently provide a detailed project control procedure and cost code structure based on that overriding control philosophy (Figure 5.14*b*). In addition, ACTIVITY DEFINITION determines both the major and minor subcontract activities for the purposes of estimating and project scheduling.

ACTIVITY DEFINITION is performed by cost engineers and estimators in concert with discipline engineers led by a project engineer. The cost engineer receives project scope information directly from engineering and design and combines this information with the project composition, major components, and so on, in order to form the basis for deriving an overall project control philosophy with the project manager's approval. The *COST CODE* policy (see Table 5.3) dictates that major project categories that are identified as physical components must first be given a cost code designation. For example, power plants have uniquely identifiable physical components, each of which can have its own scope definition. The turbine generator building, for instance, has excavation, concrete work, structural steel content, and the like. This building can be given a unique component designation that will completely segregate the turbine generator building from the other parts of the plant. Figure 5.19 illustrates a typical Project Control Zone List (P_L-DOC 9) document. Breaking the components of the plant down by a unique code designation provides a means for clearly defining the project scope of work, estimating the manpower

AW
ENGINEERING & CONSTRUCTION CORP.

CLIENT: ____________ LOCATION: ____________

PROJECT: ____________ CONTRACT #: ____________

CONTROL ZONE		DESCRIPTION
PRIME CODE	UNIT NO.	
10	15	SITE PREPARATION INCLUDES.....
20	25	BOILER BUILDING INCLUDES.....

Figure 5.19 Project Control Zone List (P_L-DOC 9).

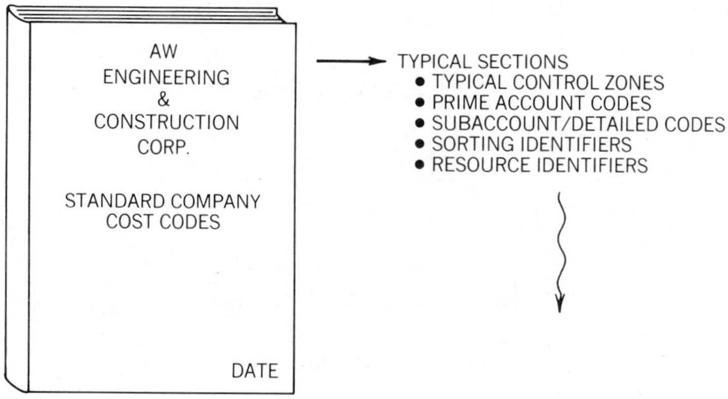

Figure 5.20 Standard Company Cost Codes (P_L-DOC 10).

content (and other costs), and monitoring and controlling construction labor costs and manpower in the field.

The cost code structure is further developed once the major physical components are identified and the project scope is increasingly more defined. The contractor normally has a standard system of cost codes which contains a classification of work categories typical of projects the firm traditionally designs and/or constructs. This document, referred to as the Standard Company Cost Codes (P_L-DOC 10), serves as the basis for developing the Project Cost Codes (P_L-DOC 11). Thus, the Project Cost Codes are a reflection of the Standard Company Cost Codes, only tailored to meet the requirements of the project in terms of control philosophy, scope definition, and specific client requirements. This approach is different from the small and medium-sized firms, because a unique project cost code is now generated as a document specifically defining the scope of work for a project. Small and medium-sized firms only reference a standard company cost code and do not create a document describing the cost code structure specific to a given project. This change in focus is the direct result of the increasing complexity and level of detail of the projects in the large firm environment. An illustration of a Standard Company Cost Code (P_L-DOC 10) is given in Figure 5.20. The cost code can be more than just a means of segregating the work into classifications like excavation or concrete. It can be an effective tool for sorting information into like categories of work from a variety of critical areas within the plant. Thus, when detailed cost codes are combined with physical component designators and other designators to identify types of work (e.g., elevated slabs), or activity operations (e.g., labor, material, subcontracts, etc.), the *COST CODE* system becomes a powerful base for many aspects of project management and control. An illustration of a Project Cost Code (P_L-DOC 11) structure, including physical components, sorting, and material/labor identifiers is shown in Figure 5.21*a,* with the detailed cost code format illustrated in Figure 5.21*b*.

The basic scope of the project is formally documented either in a written narrative based on the cost codes, or as reference material to the project estimate. In the former case, the narrative project scope, as part of the Project Cost Code, would typically appear as illustrated in Figure 5.21*c*. It might also include plant equipment lists, basic plot plans, or any other means of describing the project scope. The Project Cost Code is updated at various intervals over the life of the project, especially in the early phases when the project scope is being defined and project control philosophies are established for both design and construction.

The document structure of Project Planning is summarized in terms of an MIS in Figure 5.22. The initial Project Planning effort is based on Standard Company Cost Codes which are used in conjunction with gross facility components to define facility completion priorities, project control areas, and project cost codes. This cost code data forms the basis for a project cost code file that is continuously reviewed and updated as the project develops. The Project Cost Code provides the data structure and format for estimating, scheduling, and evaluating project team staff manpower and associated costs, and for the subsequent controlling of the project's cost and schedule.

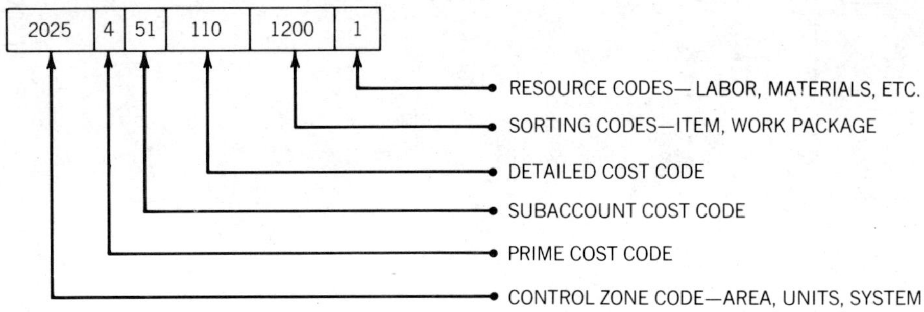

(a)

AW
ENGINEERING & CONSTRUCTION CORP.

PROJECT: ______________

CLIENT: ______________ PROJECT COST CODE: ______________

COST CODE						DESCRIPTION
ZONE	PRIME	SUB A/C	DETAIL	ITEM	RES.	
20 2025	10	14	000	1100	1	BOILER BUILDING SUBSTRUCTURE CONCRETE FOUNDATIONS REBAR LABOR

(b)

AW
ENGINEERING & CONSTRUCTION CORP.

PROJECT: ______________

CLIENT: ______________

COST CODE	SCOPE STATEMENT
2000 202500 202510000	BOILER BUILDING SUBSTRUCTURE CONCRETE—THIS ACCOUNT COVERS ALL CONCRETE FOUNDATIONS.....

(c)

Figure 5.21 Project Cost Codes (P_L-DOC 11). (*a*) Typical cost code structure. (*b*) Detailed cost codes. (*c*) Narrative description of cost codes.

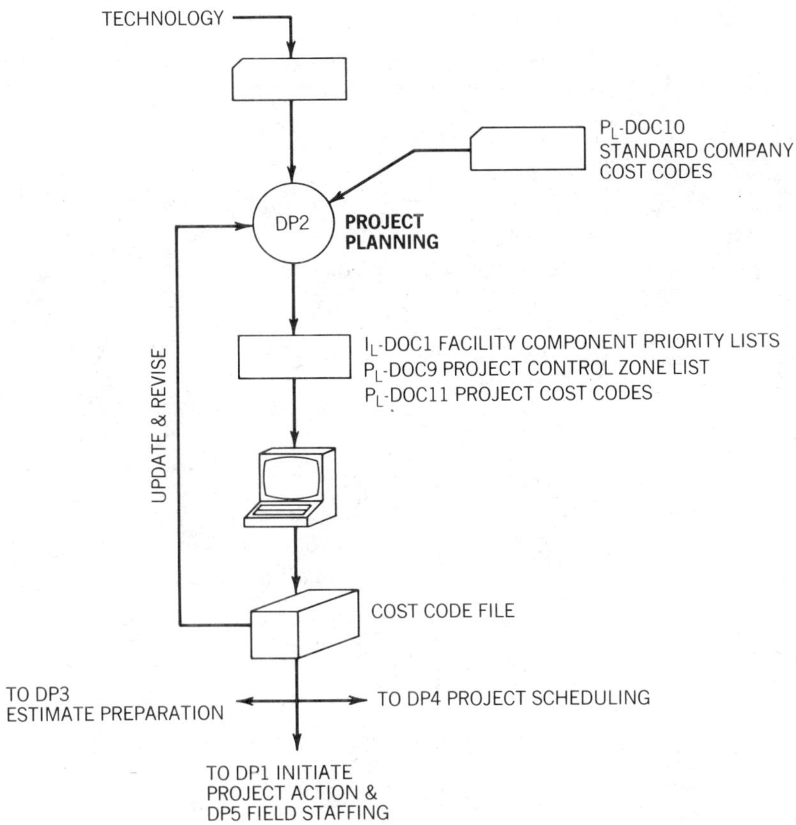

Figure 5.22 Project Planning Management Information System (Field-Based Project Management).

5.6 TOTAL FIELD AUTONOMY

Project Planning at the Total Field Autonomy level can be described in two phases, corresponding to the staged development phase (with increasing project scope definition) and to the fully operating phase (based on detailed scope definition). The functions, policies, and documents that characterize these two Project Planning phases are shown in Figure 5.23. Project Planning includes the manpower management functions of CONSTRUCTION SEQUENCING (S1), CONSTRUCTION RECONNAISSANCE (P1), and ACTIVITY DEFINITION (P2). The management effort and performance of these functions, as dictated by those policies shown in Figure 5.23, is increasingly penetrating and problem focused. While only one major new policy, *WORK BREAKDOWN* (P_T 8),* emerges at this management level, the other policy descriptions provide for an intensified management effort in guiding function performance. The change in focus is readily apparent when comparing the policy descriptions in Table 5.4 with those for the Field-Based Management approach in Table 5.3. The changing focus of manpower management policies at the Total Field Autonomy level is manifested in documents which, while similar to those of the large firm (see Section 5.5), are often more detailed and are oriented toward producing a totally integrated project management system.

The changing management focus of the Total Field Autonomy approach is reflected in three significant major problem areas. First is the integration of cost and schedule efforts at various management levels through the development of the integrated project management control structure. The second is the detailed analysis of site labor productivity in relation to a *standard base* man-hour. Finally, a large, complex project involving continuous interfacing with the client requires that the management approach be flexible enough to adapt to changes in control philosophies and planning techniques over the life of the project.

Project Planning commences with CONSTRUCTION SEQUENCING and focuses on the scope definition of the project in terms of the facility com-

*See Table 5.4, Policies of the Autonomous Project Team.

S1 CONSTRUCTION SEQUENCE	
Policies	Documents
WORK BREAKDOWN	I_T-DOC 1 Facility Component Priority List
NETWORK-BAR CHART ANALYSIS	
SCHEDULING FOCUS	
P1 CONSTRUCTION RECONNAISSANCE	
Policies	Documents
LABOR RECONNAISSANCE	I_T-DOC 2 Site Data Reference File
SITE INSPECTION AND ANALYSIS	
HISTORICAL DATA	
LABOR DATA BASE	
MANPOWER PLANNING	
P2 ACTIVITY DEFINITION	
Policies	Documents
WORK BREAKDOWN	P_T-DOC 12 Project Control Zone List
COST CODES	P_T-DOC 13 Standard Company Cost Codes
HISTORICAL DATA	P_T-DOC 14 Project Cost Codes (Preliminary)

(*a*)

S1 CONSTRUCTION SEQUENCE	
Policies	Documents
WORK BREAKDOWN	I_T-DOC 1 Facility Component Priority List (revisions)
COST CODES	
NETWORK-BAR CHART ANALYSIS	
SCHEDULING FOCUS	
P1 CONSTRUCTION RECONNAISSANCE	
Policies	Documents
LABOR RECONNAISSANCE	I_T-DOC 3 Labor Reconnaissance Report
SITE INSPECTION AND ANALYSIS	I_T-DOC 4 Construction Site Reconnaissance Report
HISTORICAL DATA	
LABOR DATA BASE	
MANPOWER PLANNING	
P2 ACTIVITY DEFINITION	
Policies	Documents
WORK BREAKDOWN	P_T-DOC 14 Project Cost Codes
COST CODES	P_T-DOC 15 Project Work Package Structure
HISTORICAL DATA	

(*b*)

Figure 5.23 Project Planning decision process (Total Field Autonomy approach). (*a*) Staged development phase. (*b*) Fully operating phase.

ponents to be constructed. The project control approach has its genesis through this function, because facility components are described and often designated by a unique unit numbering system. Each unit is then put in a sequence and assigned priority completion dates as defined by both the client and project management team. Thus, the *WORK BREAKDOWN* policy influences the manner in which CONSTRUCTION SEQUENCING is performed, and the subsequent documentation generated through this function takes on a slightly different focus from that of the Field-Based Project Management approach. For example, the Facility Component Priority List (I_T-DOC 1) in Figure 5.24

TABLE 5.4 POLICIES OF THE AUTONOMOUS PROJECT TEAM

Policy		Description
P_T 6	*LABOR RECONNAISSANCE*	Comprehensive survey, analysis, and documentation of the project site labor environment on a continuous basis is requisite to the accurate evaluation of labor content and planning of manpower requirements for the project at any time during the project's life cycle.
P_T 7	*SITE INSPECTION AND ANALYSIS*	A continuous comprehensive inspection, analysis, and documentation of project site conditions, transportation problems, etc., is requisite to the accurate evaluation of construction methods, schedules, and labor content and to solve other unique problems.
P_T 8	*WORK BREAKDOWN*	A work breakdown structure representative of the project scope and control philosophy is formulated to integrate costs and schedules with work packages for project management.
P_T 9	*COST CODES*	Cost codes are applied to the work breakdown structure and serve as a basis for categorizing estimated costs, reporting and controlling costs and manpower, developing a historical database, and integrating the schedule with the estimate.
P_T 11	*HISTORICAL DATA*	Historical labor data are basic to a realistic determination of the labor content for a work task or to identifying major project components and developing preliminary schedules.
P_T 12	*LABOR DATABASE*	Historical labor data must be normalized to a standard reference location and/or base and then documented to be useful for estimating and planning manpower.
P_T 13	*NETWORK-BAR CHART ANALYSIS*	Broad network-focused milestone planning is the best approach for identifying and sequencing major project components, while a combination of the critical path method and bar chart model are best suited for project planning and scheduling at the total project view level as well as the daily level.
P_T 14	*SCHEDULING FOCUS*	Scheduling definitions and constraints are: 1. A critical work package is one that, if delayed, will extend the completion date of the project. 2. Work package durations must be correlated to the estimated work content. 3. All restraints must be defined and incorporated into the sequence logic (i.e., material delivery, subcontract coordination, drawing receipt, etc.). 4. Manpower impact is an integral aspect of project scheduling. 5. Schedules must reflect optimum use of all resources relative to work package. 6. Progress is based on a realistic evaluation of the labor unit effort required to complete a component of work. 7. Schedules are geared to the daily work level and are prepared, analyzed, and reviewed weekly in relation to planned progress and productivity versus actual progress and productivities.
P_T 18	*MANPOWER PLANNING*	Long-range planning of manpower by craft and in total is a prerequisite to support the field effort and must be performed continuously throughout the project's life.

expands to incorporate a full listing and narrative description by unit number (Figure 5.24*a*) for each facility component, combined with a logic network diagram reflecting construction completion sequences (Figure 5.24*b*) and dates. The Facility Component Priority List provides the initial dissection of the facility that is used for assigning control zones and subsequent work packages for the project.

CONSTRUCTION RECONNAISSANCE addresses in full detail the descriptive scope statement of this function (see Figure 5.2). The key decision maker is the labor relations expert under the direction of the assistant project manager in charge of construction. Since CONSTRUCTION RECONNAISSANCE has both a heavy head office and field focus, the labor relations team will position personnel in both locations, in order to provide the advantage of continuous research and analysis of the local labor environment while having available a broad range of project team expertise to draw from for information.

M^PM PROJECT LOCATION ______			AW CORP.
UNIT NO.	AREA DESCRIPTIONS	PROCESS DESCRIPTION	SCHEDULED COMPLETION
	FACILITY A—GAS RECOVERY		
1	C_2 RECOVERY	THIS UNIT....	DATE

(a)

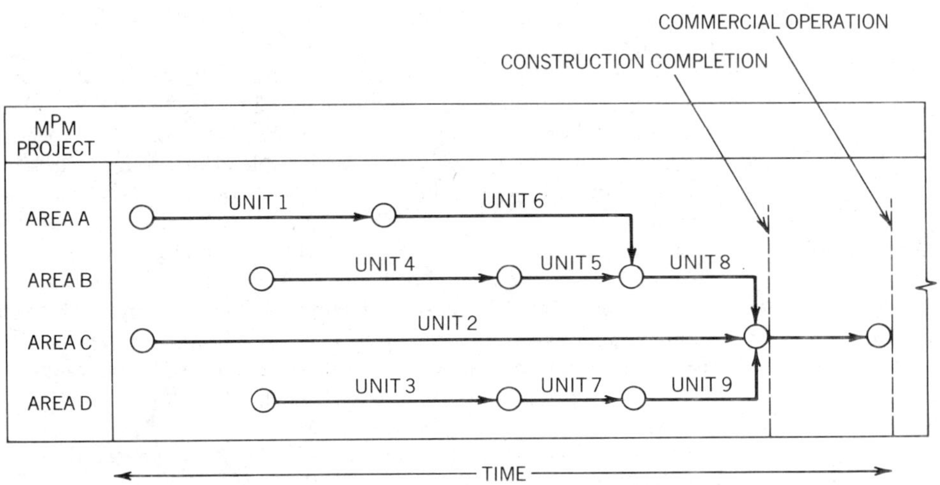

(b)

Figure 5.24 Facility Component Priority List (I_T-DOC 1). *(a)* Unit description and facility priority list. *(b)* Unit construction completion sequences.

The policies that affect the performance of CONSTRUCTION RECONNAISSANCE only change in emphasis as indicated in terms of their descriptive statement. Generally *LABOR RECONNAISSANCE* (P_T 6), *SITE INSPECTION ANALYSIS* (P_T 7), and *MANPOWER PLANNING* (P_T 18) focus on a *continuous* reconnaissance effort over the project's duration, as a means of consistently providing the most current manpower and site-related data. When using *HISTORICAL DATA* (P_T 11) from the contractor's *LABOR DATABASE* (P_T 12), the project team is constantly comparing project labor information that is referenced to a standard location or base. Thus, departures from this base can be fully documented and traced back to the standard base when changing project conditions and the labor environment create problems that require reevaluations of manpower content.

Consequently, it is critical to properly evaluate and document anticipated site efficiency factors in relation to the standard base man-hours. The Labor Reconnaissance Report (I_T-DOC 2) could contain an analysis of site productivity based on those techniques illustrated in Figures 5.11 and 5.12. Alternately, this document might include a comparative study that evaluates typical construction operations and the man-hour per unit effort commonly expended in the site area by local contractors, against the standard base man-hour per unit for the same operation. The relationship between these man-hour per unit rates will provide a basis for establishing the efficiency factor needed to convert one standard man-hour to equivalent site man-hours for a major work category, such as piping. An illustration of this type of comparative analysis is shown in Figure 5.25, and could result in an efficiency factor of 1.45 to 1.55 site man-hours to 1 standard man-hour for the piping operation. The client will often participate in these types of labor analyses by providing information on the site area (especially in

M^PM PROJECT	COMPARATIVE STUDY—SITE VS. STANDARD MAN-HOURS PER UNIT				AW CORP.
COST CATEGORY	WORK TASK OPERATION	AVERAGE MAN-HOUR PER UNIT	STANDARD BASE MAN-HOUR PER UNIT	EFFICIENCY FACTOR	REFERENCE
ABOVE GROUND PIPE	ERECTION—4″ C.S.	1.61 LF	1.22 LF	1.31 : 1.00	FIVE LOCAL CONTRACTORS
	6″ C.S.	2.53 LF	1.62 LF	1.56 : 1.00	
	8″ C.S.	3.18 LF	2.20 LF	1.45 : 1.00	
ABOVE GROUND PIPE	WELDING—4″ C.S.	3.46 EA	2.10 EA	1.65 : 1.00	
	6″ C.S.	4.14 EA	2.73 EA	1.52 : 1.00	
	8″ C.S.	5.70 EA	3.68 EA	1.55 : 1.00	

Figure 5.25 Comparative analysis of site man-hours versus standard man-hours based on site area labor reconnaissance (from I_T-DOC 3, Labor Reconnaissance Report).

foreign countries), and final approvals of productivity targets.

Project Planning focuses on ACTIVITY DEFINITION from a generic perspective, in terms of developing both the overall project control philosophy and the cost code level control approach for the project. This cost and scheduling function produces the common structure from which the entire project management information system is designed. Consequently, *WORK BREAKDOWN* structures and *COST CODES* (P_T 9) provide the project policy directives that ensure integration of cost and scheduling through work packages for all phases of project management and control.

A client approved Project Control Zone List (P_T-DOC 12), as shown in Figure 5.26, is the first stage in the development of common data structure for a project. The project plot plan is dissected into geographical zones as reflected in Figure 5.26*a*. Each zone is identified by an area designation and generally incorporates one or more project facility units, as shown in Figure 5.26*b*. The Project Control Zone List is combined with detailed cost codes selected from the Company Standard Cost Codes (P_T-DOC 13) to formulate an integrated Project Cost Code (P_T-DOC 14), as illustrated in Figure 5.27. Either the Work Breakdown Structure (WBS) or the Control Zone approach (or a combination of each), is

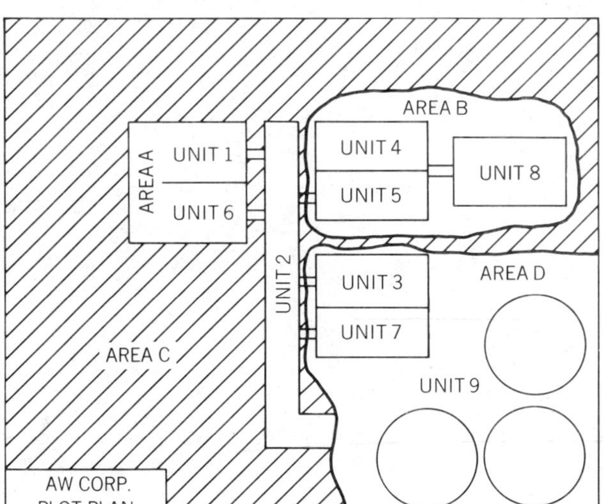

(*a*)

M^PM PROJECT		AW CORP.
CONTROL ZONE		DESCRIPTION
AREA	UNIT	
A	01	PROCESS UNIT.... ....

(*b*)

Figure 5.26 Project Control Zone List (P_T-DOC 12). (*a*) Project plot plan by control zone. (*b*) Control zone descriptions.

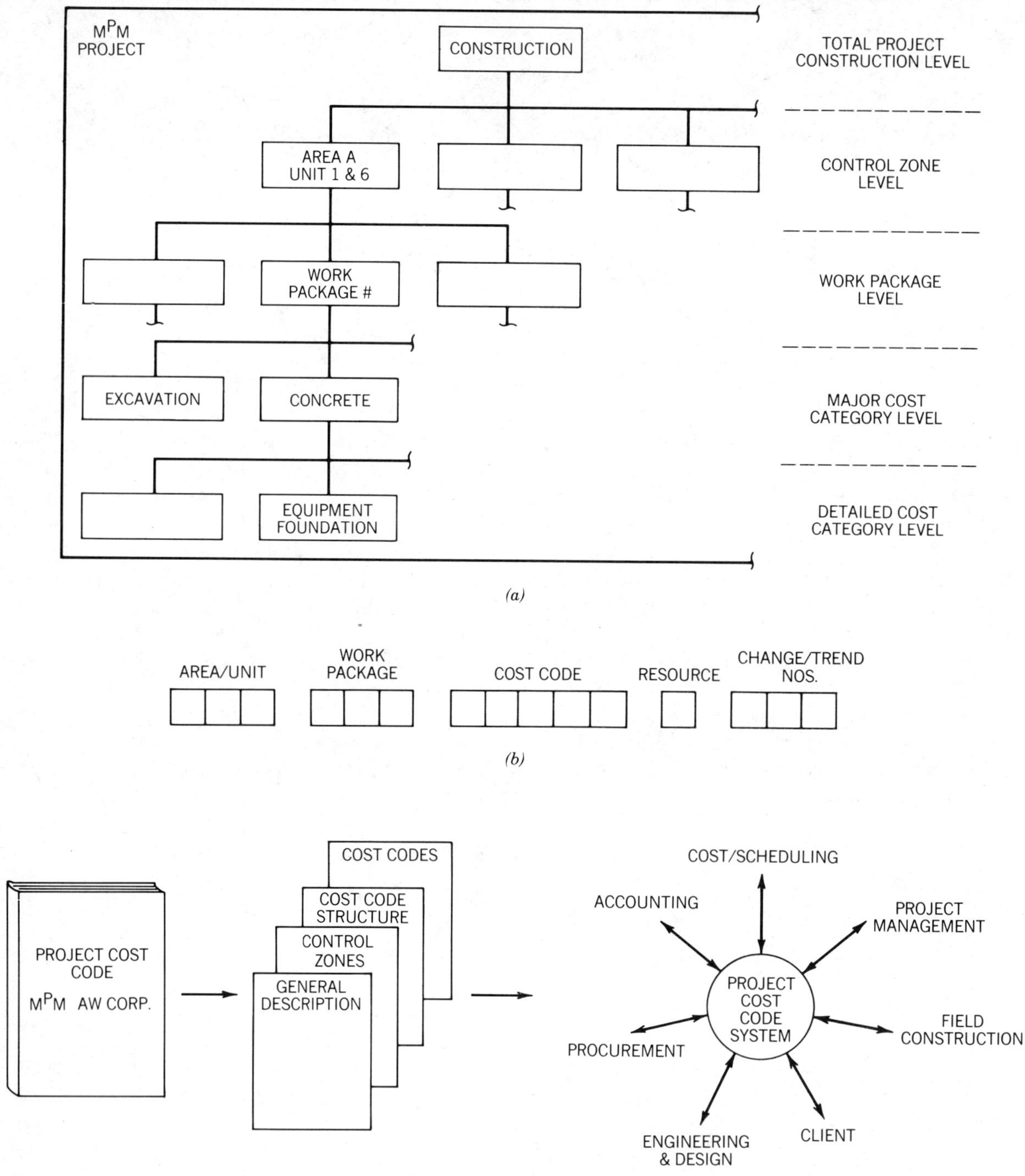

(c)

Figure 5.27 Project Cost Codes (P_T-DOC 14). (*a*) Work breakdown structure. (*b*) Cost code structure. (*c*) Cost code system interfaces.

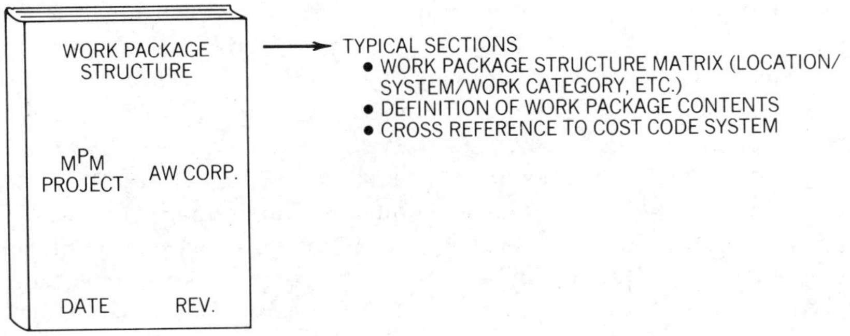

Figure 5.28 Project Work Package Structure (P_T-DOC 15).

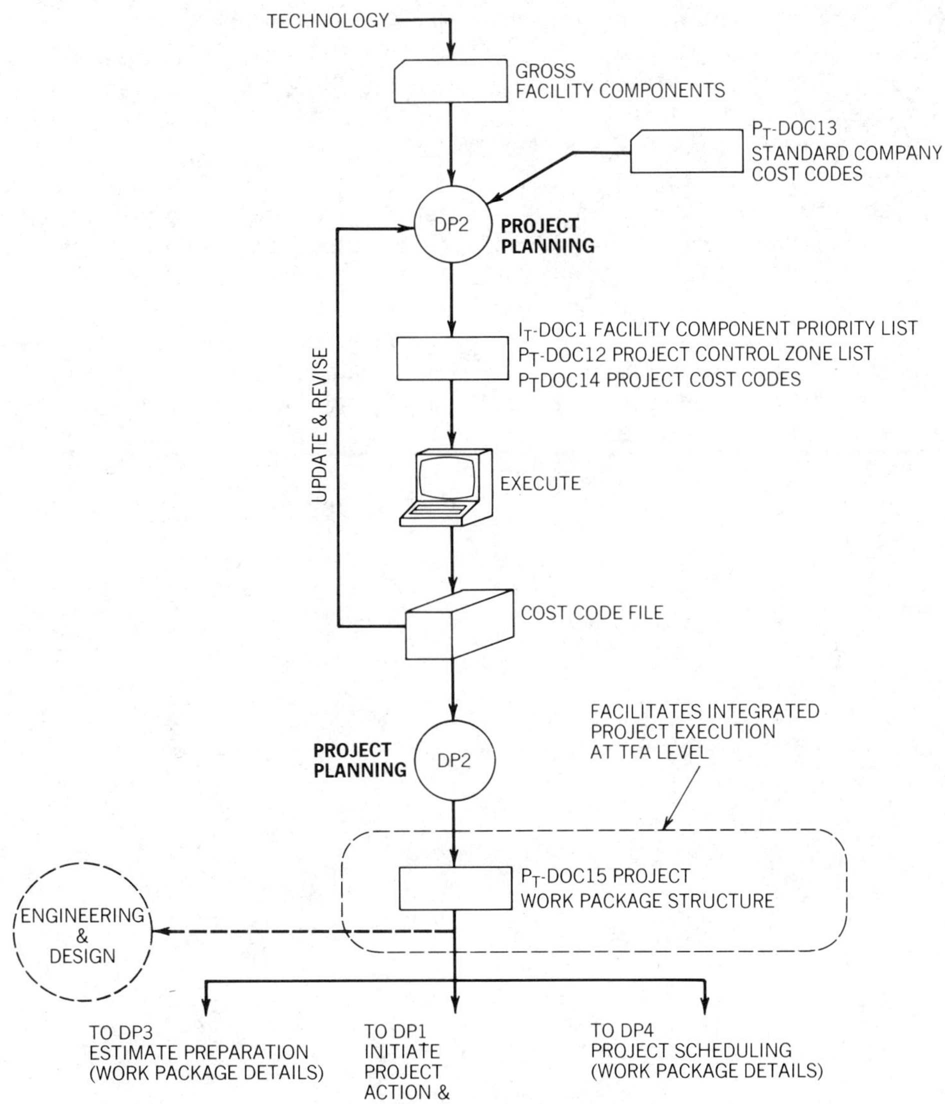

Figure 5.29 Project Planning Management Information System (Total Field Autonomy approach).

used in conjunction with Cost Coding techniques to formulate the project control structure. Thus, Figure 5.27*a* illustrates a typical WBS for the plant construction using a multi-level breakdown from the control zone level to the detailed cost category level. The cost code structure is further explained in Figure 5.27*b*, in which the components of the cost code are specified and given a project focus when unique areas, work package numbers, cost codes, and the like are assigned to each block. Because this system is computer based, the ability to sort and/or roll up data from any level of detail becomes an extremely important aspect of integrating costs and schedules with other project functions. Finally, Figure 5.27*c* presents a schematic illustration of the Project Cost Code System that becomes the focal point of data management for the project.

Once the project scope is defined, the Project Work Package Structure (P_T-DOC 15) in Figure 5.28 can be developed. The Work Package Structure defines the project work packages and relates them to the Project Cost Codes. These work packages not only become the basis for managing the project, but, if determined early in the project, can be tied directly into the engineering/design effort. In this way, construction drawings and other project data are all related through the work package concept, leading to an integrated project management information system.

The development of the Project Control Zone List, the Project Cost Codes, and the Work Package Structure are part of an evolutionary process, especially in the early phases of the project. The client's involvement in the project tends to maintain pressure on the contractor, hence the contractor is always focusing on means of upgrading and improving the management approach. This may result in the manipulation of control zones, cost codes, and the like. In addition, the dynamic nature of complex projects often forces the contractor into a creative management style whereby the firm may, for example, manipulate control zones to improve the management of the project.

The Management Information System describing the Project Planning decision process is only slightly different from that for the large firm using the Field-Based Project Management approach. As shown in Figure 5.29, the MIS reflects a more detailed dissection of the project scope into a WBS that defines work packages for both design and construction. This Work Package Structure, when incorporated into the Project Cost Codes, facilitates the integration of costs and schedules with management philosophies and the need for project team staff. Thus, the Project Planning MIS provides the database structure for project and construction management including estimating, scheduling, and project control.

CHAPTER 6
Estimate Preparation

Estimate preparation requires the careful study, compilation, and analysis of all the items, factors, and actions that influence and contribute to the capital cost of a construction project. It requires a detailed study of the contract documents to ensure a thorough knowledge of the requirements of the project, especially in relation to known and anticipated site conditions. Then, based on an intimate knowledge of the parameters associated with the firm's construction methods and support activities, the estimating effort requires the consideration of all items and activities leading to the completion of the project, and their systematic expression and analysis in a concise format for ready checking and evaluation. Consequently, Estimate Preparation has as its end objective the production of the approximate capital cost for a project based on a given project scope of work as defined by the client/owner or the architect/engineer (A/E) consultants.

Estimate Preparation is a critical management problem area and warrants significant management attention, because its product, the capital cost estimate, can directly impact the financial life of the construction company. For instance, in a fixed price contracting environment, the contractor is making an offer to furnish a given scope of work for the consideration of a single lump sum price, or else is paid for work done according to a schedule of fixed unit prices which reflect that scope of work. This offer and consideration legally exposes the contractor financially provided the client/owner accepts the lump sum bid or unit price schedule. Once the contractor enters into a contract, the firm is bound by a legal obligation to perform the work. Cost overruns, other than those generated by client/owner approved changes in scope, are absorbed by the contractor and will naturally impact the profit achieved on the project. Consequently, gross errors in estimate preparation can seriously impair a contractor's financial survival.

In other contracting environments the criticality of estimate preparation for the contractor assumes a slightly different focus, especially in a cost reimbursable contracting situation. In this environment, the client/owner will often base the economic analysis of a desired project on the contractor's capital cost estimate. If the economics clearly indicate a profitable venture, the client/owner will authorize the project and begin to commit large sums of money for its execution. If, as the project begins to develop, capital costs rise as a result of dubious estimating practices, the client/owner may choose to cancel the project on the grounds that the profitability potential of the project cannot be realized. Under these circumstances, apart from the potential reduction in the contractor's own profit picture, the contractor's reputation can be severely damaged and the ability to gain future work from that client destroyed.

Many construction estimates are prepared under extremely tight time schedules that allow little time for decisions or opportunity to seek advice or verify the results. Decision makers must often rely on prices quoted by subcontractors when compiling estimates for final presentation or submittal of the bid price. Furthermore, the scope of work may be insufficient in detail to produce a reliable estimate, yet the decision maker is often required to produce an estimate from which financial decisions can be made.

The Estimate Preparation process also establishes the basis for controlling project costs. Project management requires the establishment of an effective cost control system which is tailored to the budget estimate and the company and/or project cost code structure.

While estimate preparation becomes the basis for controlling project costs in most environments, it also provides information critical to manpower planning. This is especially significant where the size of the field manpower force may strain local labor pools and/or local housing markets. In these situations, estimate data can be combined with construction schedules to assess manpower peaks from which critical decisions can be made relative, for example, to the need for construction camps and the requirement to import and train labor. The contractor's professional ability and qualifications may be challenged, for instance, when a construction camp is sized for a peak of 750 men, and that peak is raised to 1000 when estimate omissions or errors are uncovered, site productivity analysis proves incorrect, or changes in the scope of work have not been properly documented.

In summary, Estimate Preparation is a decision problem area of critical management concern, because it helps determine the financial basis for the project. It establishes the framework for project cost control and assists in the determination of manpower requirements for the project. It can significantly impact the operation of a construction company in an environment that inherently produces pressures on the estimate preparation effort.

6.1 THE BASIC DECISION PROCESS

The Estimate Preparation process must necessarily focus on all factors influencing the cost of a project. However, the component of a construction project which predominantly influences the cost estimate of that project is the manpower or labor content. Because of its volatile nature, the labor component is frequently considered the most intangible and difficult cost to estimate. This salient characteristic makes the labor cost component possibly the most uncertain element within the cost estimate. However, the labor portion of the estimate must be sufficiently accurate to effectively define labor costs, man-hours, productivity, and the like, and also provides data for predicting manpower levels during the planning and scheduling of the project. This places a significant burden on the decision maker responsible for Estimate Preparation.

In terms of project manpower management, Estimate Preparation focuses on problems associated with the determination and evaluation of the anticipated manpower content for a given project scope. The manpower content of a project is expressed in a number of ways such as equivalent men, man-hours and/or labor dollars. In addition, the labor component must cover the labor costs associated with direct field construction, indirect field construction (e.g., field staff and support labor for such items as temporary construction facilities),* and head office support (e.g., project manager, estimators, engineers, designers, etc.). The magnitude of these labor-related costs will depend on the nature, size, and type of the construction project and the contractual responsibilities of the parties involved. Consequently, the labor component is an important estimate problem area in developing the total cost estimate for a project.

When the decision maker(s) address Estimate Preparation, the following major decision problem areas must be considered:

1. What are the estimate requirements?
2. What labor productivities can be expected?
3. What is the man-hour content?
4. What are the estimated labor costs?

The analysis of these problems is important because it requires the decision maker(s) to scan the broad range of factors and variables which can influence the quality and accuracy of the labor estimate, and thus can subsequently impact the bid price, financial analysis, and/or management control base. Table 6.1 presents a detailed compilation of the component decision problems associated with the labor component of Estimate Preparation.

The four key problem areas are important in Estimate Preparation as it relates to the manpower

*Labor wage related costs such as benefits (holiday, health, and welfare, etc.) and legislated burdens (FICA, workers' compensation, etc.) can either be included in the direct field labor costs or carried as indirect costs.

TABLE 6.1 SCOPE OF ESTIMATE PREPARATION PROBLEM AREA

Major Decision Problem	Component Problem Statements
13. What are estimate requirements:	What type of contract is involved? What type of estimate is required? What is required accuracy for estimate? What estimate techniques are required to meet accuracy? When is estimate required? What project breakdown is necessary to estimate labor and craft content? What is required material quantity content for each work item? Have all work items been considered?
14. What is labor productivity:	What crew sizes and mixes should be used? What labor productivities can be used?
15. What is man-hour content:	What labor estimating units are required for each item? What special labor units must be developed for unusual work areas? What engineering, design, and management effort is required?
16. What is estimated labor cost:	What labor and crew wage rates should be used? What are craft and staff benefits and legislated burdens? Where should these labor related costs be included in the estimate? What labor summaries and budget estimates are required for this project and construction environment? How does this estimate compare to current estimates? How does this estimate compare with historical ratios, guidelines, and the like?

content of a project. The decision maker must first determine the estimate requirements. This will establish the estimate techniques required to achieve the desired estimate accuracy within the time schedule for completing the estimate. The decision maker must understand the labor environment at the site location, and the impact it will have on the labor estimate. The labor estimating units, as established by the estimate techniques, must be carefully evaluated and applied to each component and detailed work item. In addition, the manpower content of the management support effort must be quantified. Finally, estimated labor costs per work item and in total must be calculated, compiled, analyzed, and checked against known factors prior to final estimate submittal. Successfully addressing these four major problem areas ensures that the labor estimate will provide management with the best quality estimate for the known scope and anticipated site conditions.

In the construction industry many different techniques and procedures exist for estimating the capital cost of a project, and, in particular, the labor component of that capital cost. The type of estimating approach is governed to a significant degree by the construction environment which surrounds and impacts this decision area. For instance, each component problem shown in Table 6.1 would be evaluated with different levels of intensity by a contractor estimating a small project in a local region where labor conditions are known, as opposed to a contractor who is estimating the labor content for a project located in a remote area or a foreign country.

In most construction environments, the basic structure of Estimate Preparation is similar and naturally corresponds to the order of the major problem areas and their component problem statements indicated in Table 6.1. Thus, commonly recognized management duties and tasks for estimating exist in construction practice. They are presented in Table 6.2 in roughly the order in which each is performed by the decision makers. There are estimate situations where various duties/tasks could be performed concurrently, and other situations where iterations between duties/tasks will be required, depending on the length of the estimate preparation cycle and feedback as estimate decisions are made.

The detailed duties and tasks described in Table 6.2 can be directly related to the three following manpower management functions:

1. ACTIVITY DEFINITION (P1)
2. QUANTITY TAKE-OFF (P2)
3. ESTIMATE LABOR CONTENT (P3)

These three functions associated with the Estimate Preparation decision process are shown in Figure 6.1 on the Manpower Management Function chart, in relation to the remaining manpower management functions and their associated decision processes. The composite functional statement for each Esti-

TABLE 6.2 GENERAL PROCEDURE FOR ESTIMATE PREPARATION

Procedural Step	Management Duties and Tasks
1.	Determine purpose and focus of estimate, whether for fixed price, control budget, or financial analysis.
2.	Evaluate the type of estimate and techniques to be used to achieve required estimate accuracy.
3.	Establish estimate milestones and completion dates, and determine the procedure for preparing estimate.
4.	Determine estimating scope, whether data consists of construction drawings and specifications, equipment lists, plant capacity requirements and/or the like.
5.	Dissect scope to the level of detail required by estimate techniques and cost and scheduling control system.
6.	Evaluate material quantity content for each item defined by estimate scope and check to ensure that all items are covered.
7.	Determine crew structures and estimate wage rates, establish burden and benefit patterns.
8.	Evaluate site productivity in relation to estimate base.
9.	Estimate direct field labor man-hours and/or labor dollars.
10.	Estimate field support labor man-hours and/or labor dollars, equipment costs, burden and benefit costs, home office man-hours, and labor costs.
11.	Summarize all estimate components costs including labor, material, subcontract, indirect support construction costs, office costs, fees, and contingency.
12.	Compare estimate to historical labor information in terms of typical ratios, factors, and so on.
13.	Reconcile new estimate with current cost forecasts.
14.	Review and approve estimate and release bid or estimate to client/owner.

mate Preparation manpower management function is described in Figure 6.2. Finally, the relationships between Estimate Preparation and other relevant decision processes are illustrated in Figure 6.3.

6.2 FUNCTIONAL PERFORMANCE TECHNIQUES

As a project definition component, Estimate Preparation often has a multi-level problem focus depending to a large extent on the project environment, the contractual relationship with the client/owner, and the intended use of a project's capital cost estimate. In the fixed price competitive bidding environment, the contractor's estimate would be a definitive estimate based on detailed project construction drawings and specifications, plus material and subcontract quotes. This is considered the most accurate cost estimate. Alternatively, an architect may develop a cost estimate which can be used to optimize design in relation to owner-desired cost projections, and also as a means of analyzing the quality of the competitive bids of several contractors. This type of estimate is normally a factored type of estimate, because it is based on facility parameters and is not necessarily supported by details from construction drawings. While the accuracy of such an estimate is less than that of the definitive estimate, this factored estimate is, nonetheless, an excellent management tool for the use intended.

On larger industrial projects and especially in the design-construct environment, the client/owner and contractor may together initiate several stages of cost estimates which correspond to the progressive scope development of the facility's technical processes and detailed design. This estimate cycle, when combined with an appropriate scope and/or cost trending procedure, provides both the client/owner and contractor with a means of controlling costs. This approach ensures that design development can be managed in a way that will maintain the client's overall dollar commitment and so keep the project economically viable. Several types of estimates emerge over the life of the project ranging from the conceptual order of magnitude estimate (based on cost-capacity factoring techniques), to the full definitive estimate (based on approved-for construction drawings and specifications *as well as* a percentage of the equipment and materials already purchased and/or committed). The estimate accuracy level and range is approximately ± 25 to 30%

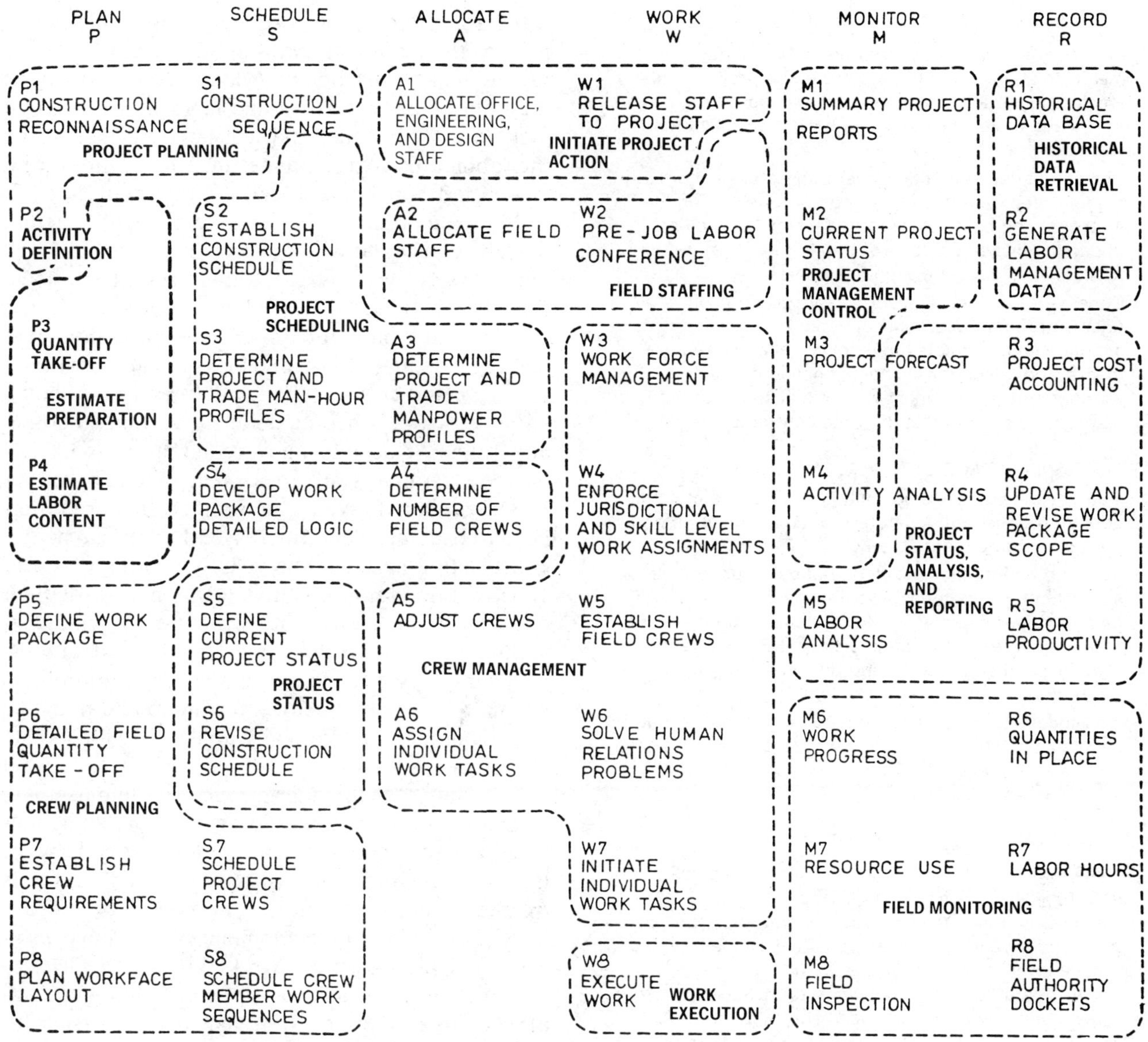

Figure 6.1 Functional plot of the Estimate Preparation decision process.

for the order of magnitude factored estimate to a ± 5 to 10% for the definitive estimate.

In terms of evaluating and defining a project's manpower content, the definitive estimate naturally produces the most accurate and detailed level of manpower information. However, in those project environments where the magnitude of the manpower content is significant (250 or more craft personnel), it is essential to know the manpower content as early as possible in the project's life in order to identify potential problems relevant to such areas as labor availability, camp requirements, training and importation of labor, and so on. Consequently, factored estimates, while not always readily identifying the labor component, are important estimates for the purposes of manpower management.

Estimate Preparation is governed by several general concepts that influence the various approaches utilized in cost estimating, including those approaches that are directly labor focused. The techniques and level of detail used for the estimate must be compatible with the desired level of project control, including the project's financial, scheduling, manpower, and field control aspects. For instance,

P2 ACTIVITY DEFINITION

- Activity definition rationalizes project breakdown as a means in some cases of developing the project scope (design-construct), and in all cases of identifying project scope and complexity and as a basis for project management control.
- Requires a detailed knowledge of the project scope, including preparatory work, proposed construction methods, and cost control requirements, as well as an understanding of the level of breakdown detail required to match project team expertise.
- Considers activities as recognizable portions of the finished product or work scope, correlation of activity description with estimate breakdown, schedule requirements, and cost control codes; implied crews and whether activity is worked via force account, direct hired labor, or by subcontract (if prime contractor).

P3 QUANTITY TAKE-OFF

- Quantity take-off systematically quantifies the material content (permanent, temporary, and consumable) of a project or component activity.
- Requires an intimate knowledge of construction practice, site conditions, material properties, and availability of manufactured items, a flair for systematic thinking and computation and the ability to work from architectural sketches and engineering and detailed drawings, and an awareness of implied labor content.
- Considers site conditions, construction methods, and the sequence in which the activity will be built; enumerates the material content and defines units of measure for each material required.

P4 ESTIMATE LABOR CONTENT

- Rationalizes the labor content for each project activity and the project as a whole.
- Requires an intimate knowledge of crew sizes and compositions for each proposed construction method and intended level of equipment use, as well as an extensive knowledge of labor productivities, craft wage rate structures, conditions, and historical labor database.
- Considers the impact of site conditions on labor productivity and the magnitude of activity work content in estimating activity duration and labor content. Extends direct labor content to include the labor burden as well as considering the labor content involved in the field management of labor and in engineering and other general office functions.

Figure 6.2 Functional statement of the Estimate Preparation decision process.

if the contractor intends to plan project manpower profiles the labor estimate should reflect man-hours rather than only labor unit costs. Alternatively, if the contractor collects actual labor unit costs that include craft benefits and legislated burdens, then the labor unit cost estimate must by definition include craft benefits and legislated burdens.

Estimate preparation, at whatever level of detail, must be consistent with the project scope (in that all items should be covered), should reflect site location aspects (based on knowledge of past productivities), should allow for time frame corrections (the correct indexing of historical records and cost rises for projected future scheduled work), and should allow for work schedule details (overtime and penalty rates). This relevancy requirement is critical because estimates are to a great extent based on historical data derived from projects that are different in terms of scope (increased facility capacity), construction periods (one year ago with a two year construction duration), and locations, (with their different unions, wage rate structures, work agreements, and skill levels). Thus, historical data must frequently be adjusted during the preparation of an estimate. This practice of referring to and updating historical data automatically applies to all estimate techniques.

A variety of techniques have been found useful in estimating practice that address the problems indicated above. Some estimators systematically work through drawings and color each item as it is treated. Others use a cost code system or prepare lists of activities (or equipment in individual projects) and progressively work through the list. In some cases, especially for scope estimating of complex industrial projects, it may be necessary to work off process flow diagrams as a first step in developing component and equipment item lists. When updating records, it is important to establish the equipment and productivity basis of previous construction methods (relative to improved equipment and mechanization methods), and to establish the basic parameters affecting costs (using index methods such as the ENR index). In some cases different parameters are involved and special care is required to either incorporate their effects on projections or to agree with the client upon a method to handle changes in cost due to inflationary price fluctuations.

As regards manpower management specifically, the labor estimate provides a measure of the manpower content and requirements for a project in

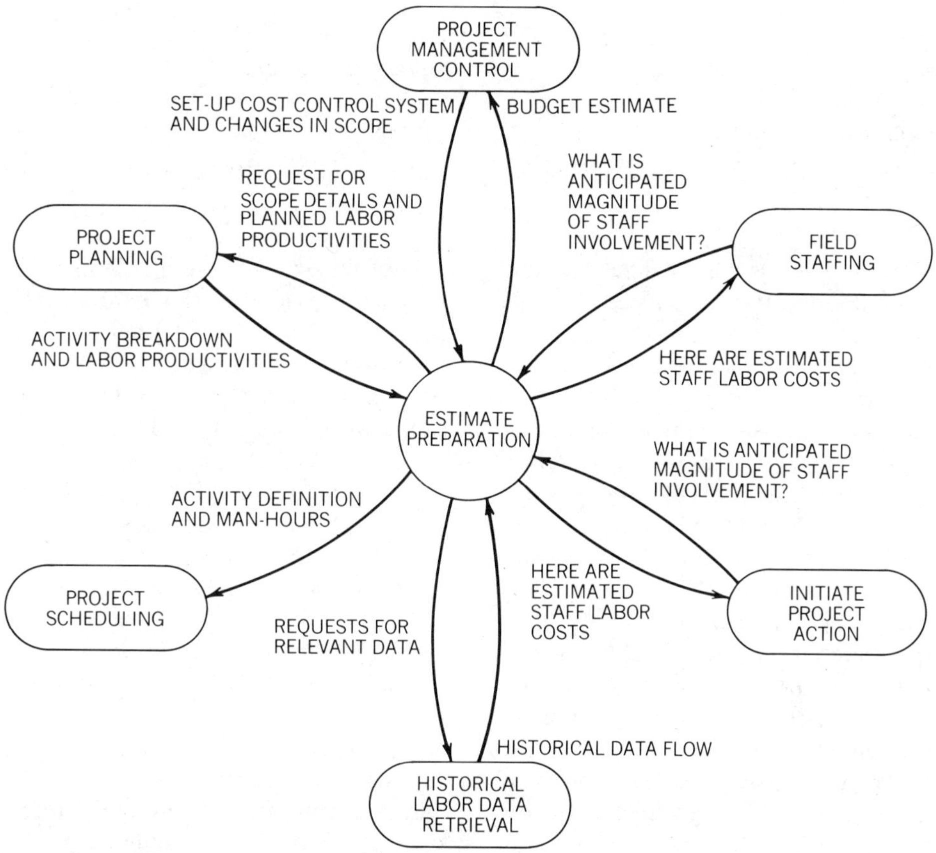

Figure 6.3 Schematic illustration of Estimate Preparation decision process interfaces with other manpower management decision processes.

terms of labor costs, man-hours, and the number of men to be involved in the project (particularly staff). Three major components determine labor costs for a construction project. The first is the money associated with hourly wages and related benefits, burdens, and overhead. The second factor, productive effort, is the quantity of work accomplished in a given period of time. The third factor, management, is the amount of support staff required to execute and control the construction project. Collectively, these factors focus on variables that influence the magnitude of a project's labor component, and which establish the control base estimate. Additionally, these variables should be monitored in both cost and schedule control efforts.

The maintenance of a standard labor estimate reference base which is easy to access and manipulate is essential for Estimate Preparation. This base is often manifested in the form of standard labor man-hour per unit charts or by a computer-aided labor data base, such as labor unit cost dissections by detailed work categories. These approaches provide consistent estimating techniques which are particularly important when a firm maintains a large estimating task force (e.g., 10 to 20 estimators).

In the construction industry, factored estimating models are most frequently associated with projects in the conceptual phase or during the engineering and design development phase. Under these conditions project scope detail is limited, and therefore the decision maker must employ techniques which can be applied to sketchy information. One estimating concept that emerges is the use of simple facility parameters, such as kilowatts for a power station or gross floor area in square feet for a building, to evaluate the order of magnitude of a project's cost. The estimate base is dollars per parameter which can be applied to the same parameter for the proposed facility, as shown in Figure 6.4. This technique produces order of magnitude costs and is naturally limited in its application, but can be useful for

FACILITY	HISTORICAL COST DATA—SIMILAR FACILITY	PARAMETER—PROPOSED FACILITY	ESTIMATE—PROPOSED FACILITY
Office Building:	$ per square foot	× Total square feet	= Capital cost ($)
Power Station:	$ per kilowatt	× Total kilowatts	= Capital cost ($)

Figure 6.4 Order of magnitude/conceptual estimate technique.

evaluating whether a project's cost will be one, ten, or one hundred million dollars. Another similar technique for estimating order of magnitude costs is simply to adjust the cost of a facility from one time frame to another time frame using cost indices. In this case, the base project must be nearly identical to the proposed project. If the manpower content is required, historical ratios identifying the percent labor dollars can be used to extract labor costs from the estimate and then converted to man-hours using an appropriate labor wage rate.

When the contractor has a better scope definition and a large number of facility components identified, estimates can be prepared based on the company's standard control philosophy. Typical examples are by major cost code category (e.g. structural steel), or by control areas (e.g., physical blocks of the facility). This basic estimate concept is illustrated in Figure 6.5. In Figure 6.5*a*, parameters typical of building construction projects are identified that are considered representative of projects of this size and scope. While only four parameters are provided in this example, fifteen to twenty parameters could be defined. When this technique is applied to a project like an office building, the requisite parameters are selected and the total unit of measure is established from preliminary sketches or drawings. The appropriate parameter and its measure is then applied to the details of the project, normally at the major cost category level, and a parameter cost is assigned as shown in Figure 6.5*b*. Naturally, historical data must be available for each major cost category in relation to the desired parameters. The labor component is incorporated into the parameter cost for each cost category. A second application of the factored estimate approach using the control philosophy concept is shown in Figure 6.6, using the control area for an example. This factoring technique is frequently referred to as cost capacity factoring, in which exponents are used to reflect the nonlinear increase/decrease in costs resulting from the economics associated with scale.

OFFICE BUILDING

TYPICAL PARAMETERS		MEASURE	QUANTITY
CODE	DESCRIPTION		
2	GROSS FLOOR AREA	SQUARE FOOT	ESTIMATED FROM SKETCHES
4	ROOF AREA	SQUARE FOOT	
12	INTERIOR PARTITIONS	LINEAR FOOT	
13	HVAC	TONS	

(a)

MAJOR COST CATEGORY	CODE	QUANTITY UNITS	UNIT COST	TOTAL COST
FOUNDATIONS	2	SF	$/SF	$
ROOFING & FLASHING	4	SF	$/SF	$
LATH & PLASTER	12	LF	$/LF	$
HVAC	13	TN	$/TN	$

(b)

Figure 6.5 Factored estimate based on major cost category control philosophy. (*a*) Typical parameters for office building. (*b*) Application of parameter data to major cost category.

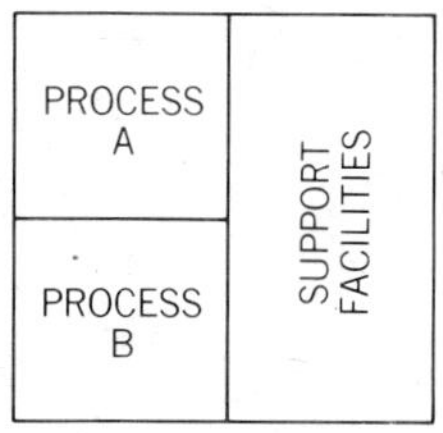

(a)

1. PROCESS UNIT A IS SIMILAR TO PROCESS UNIT X FROM A PREVIOUS PROJECT EXCEPT FOR A CHANGE IN CAPACITY
2. CAPACITY RELATIONSHIP
   PROCESS X 10,000 BARRELS PER DAY (BPD)
   PROCESS A 15,000 BARRELS PER DAY (BPD)
3. HISTORICAL DATA
   PROCESS UNIT X 250,000 LABOR MAN-HOURS (ACTUAL)
4. ESTIMATE OF MAN-HOURS FOR PROCESS UNIT A

$$\text{MAN-HOURS A} = \text{MAN HOURS X} \times \left(\frac{\text{CAPACITY A}}{\text{CAPACITY X}}\right)^{\text{EXPONENT}}$$

OR

$$319000 = 250000 \times \left(\frac{15000\ \text{BPD}}{10000\ \text{BPD}}\right)^{.6}$$

(b)

Figure 6.6 Factored estimate based on control area concept. (*a*) Proposed plot plan. (*b*) Estimate methodology.

This technique can be applied to total direct costs or specific components such as labor (see Figure 6.6). The exponent selected must be typical for the type of process unit being considered. It is generally arrived at through statistical analysis of past projects.

Another application of factored estimating relates facility machinery and equipment costs and capacities by means of a factor or ratio to other components of the direct field cost category such as labor man-hours.* In this case, engineering and design has proceeded to the extent that specifications for machinery and equipment are developed and can serve as a sound basis for equipment pricing. The prices are usually based on current day market conditions. Since the machinery and equipment component may represent a significant percent of the total direct field cost (e.g., 40 to 50%), this estimating approach may be more accurate than the cost capacity factored estimate, because a large element of cost is not based on history but on present market conditions for the equipment specified. Figure 6.7 illustrates this technique, which is commonly referred to as equipment, modular, or component ratio factoring.

Once the project scope is well defined, definitive or detailed estimates can be developed for fixed priced bids and/or as the project's control budget. Several common techniques are used in practice to develop definitive estimates.

The unit cost technique for labor estimating (e.g., dollars per cubic yard) is built on the following assumptions:

1. The implied labor content of the labor unit cost does not require labor dissection because the estimator has an extensive knowledge of the local labor environment and an inherent knowledge of project manpower content.
2. Labor unit costing often saves one to two steps in the labor estimating process. That is because local knowledge obviates the need for developing and converting man-hours to site location and applying labor wage rates.

*Factors or ratios can be comprehensive, covering all ancillary costs such as labor and material costs for concrete, steel, piping and so on, through a single factor or ratio.

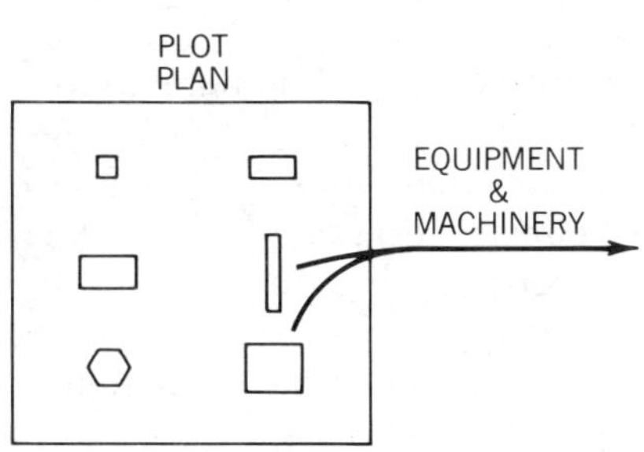

TYPE	PRICE	FACTOR	LABOR MAN-HOURS
VESSELS COMPRESSOR EXCHANGER PUMPS	(1)	(2)	PRICE × FACTOR EQUAL MAN-HOURS

(1) PRICE BASED ON CURRENT MARKET CONDITIONS AND EQUIPMENT DUTY SPECIFICATION

(2) FACTOR FROM HISTORICAL DATA FOR EQUIPMENT TYPES

Figure 6.7 Factored estimate based on cost of major plant equipment.

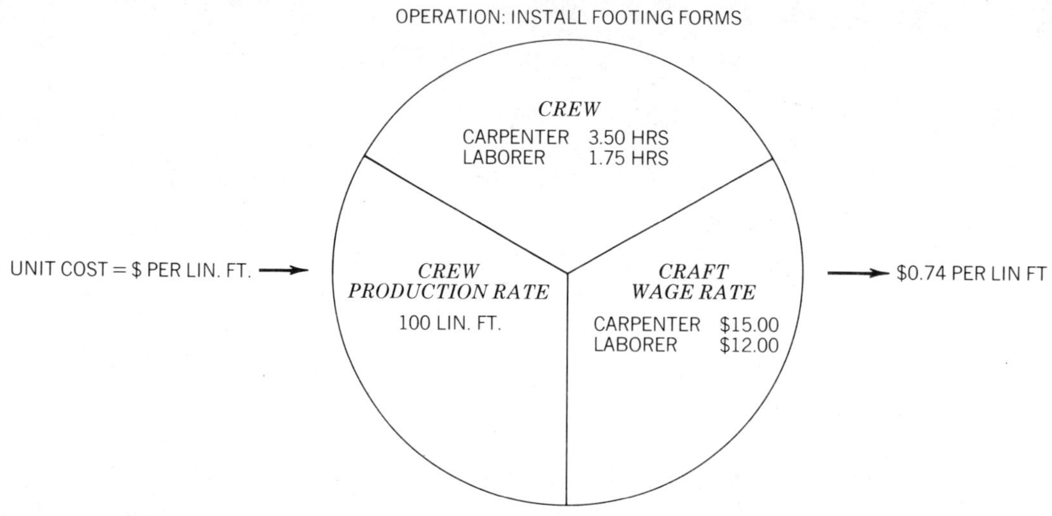

Figure 6.8 Implied labor content in unit cost technique.

This is particularly helpful to the contractor who is continuously operating in tight bidding situations. The contractor must maintain accurate and up-to-date historical labor unit prices to be successful using this technique. This concept is illustrated in Figure 6.8. Labor unit costs can be developed by evaluating those items inside the circle in Figure 6.8. If the contractor is estimating a schedule of unit prices for a project such as dams, tunnels, and so on, the unit price not only contains the labor component, as shown in Figure 6.8, but would also incorporate costs for material, construction equipment, staff, overhead, and profit.

The unit man-hour technique for labor estimating addresses two major requirements. First is the need to know the man-hour content because of project size and complexity, and often project location. Second, productivity, craft wage rate structures, and crew mixes are uncertain elements of labor costs and must be controlled. Figure 6.9 delineates the basic structure of the unit man-hour approach. A unit man-hour can be derived for any given activity based upon a detailed analysis of the basic crew size and composition, production quantities, and task duration components (inside the circle in Figure 6.9), or from field records normalized into standard labor man-hour per unit charts.

In some cases the nature of the estimating work is such that it lends itself to frequently repeated and structured dissections of physical components, equipment handling and installation, and a variety of construction processes. In these situations a computer-based approach may be useful. In such an approach, the estimator interacts with a terminal and need only enter critical parameters and answer pertinent questions (sometimes prompted by an interrogative program). The estimator has a journal printout suitable for checking and/or inclusion in a recapitulation or summary report. Many companies have extensive computer-based estimating systems which they use as a means of fully exploiting the knowledge and skill of their estimators and eliminating formal calculations and format considerations*.

The work package estimate technique is often used in situations where specialized documentation and analysis based on actual experience is required. This often occurs when a construction activity does not lend itself to historical labor unit costs or man-hour per unit data. This approach is illustrated in Figure 6.10, and can often serve as the basis for unit price estimates.

The estimating of direct field man-hours must be supported by estimates of field staff man-hours. This includes labor man-hours for the construction of temporary construction facilities and head office man-hours for engineering and design (especially in design–construct projects). At the factored estimate level, the concept of utilizing parameters that relate to the labor component areas referred to

*Many estimating systems are now available for use on microcomputers, and this is broadening computer usage in this area.

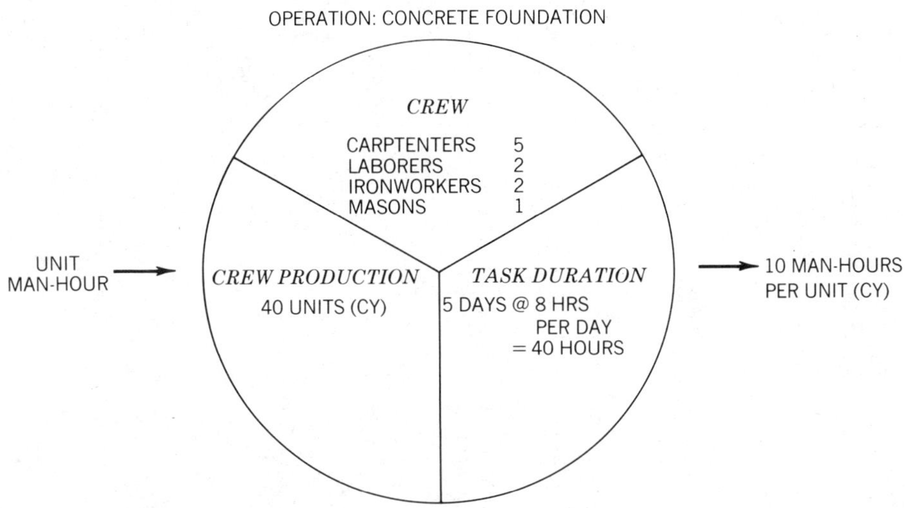

(*a*)

Add 10% or 1.10 × Base Man-Hours

(*b*)

Craft	Crew Size	Wage Rate	Total Dollars
Carpenter	5	15.00	75.00
Laborer	2	12.00	24.00
Ironworker	2	14.50	29.00
Mason	1	14.75	14.75
Total	10		142.75
Composite			14.28

(*c*)

1. Labor Man-Hours = 10 Man-Hours/CY × 100 CY × 1.10 (Site Adjustment) = 1,100 Man-Hours
2. Labor Cost ($) = 1,100 Man-Hours × $14.28 per hour = $15,700

(*d*)

Figure 6.9 Unit man-hour technique. (*a*) Implied labor content. (*b*) Site productivity adjustments. (*c*) Composite crew wage rate analysis. (*d*) Labor estimate.

above is effective in estimating these labor costs. For example, field staff man-hours normally relate closely to the direct field man-hours to be controlled. Figure 6.11*a* provides an illustration of this approach. It can be applied to other indirect field cost areas such as temporary construction facilities as well. Home office man-hours are often related to total direct field costs. Once the man-hours are determined appropriate wage rates must be selected so that the total labor costs for these indirect and office cost components can be estimated.

Definitive estimates must rely on estimate techniques which accurately portray the labor costs for project team staff requirements as they relate to the scope and control procedures. In this case, manning charts (as developed through Project Initiation) provide the best basis for estimating detailed labor costs for these categories. Figure 6.11*b*, for example, shows the manner in which manning charts are used to develop total field staff man-hours. By applying the appropriate wage rates in accordance with the positions on the chart, the labor costs can be estimated. Labor estimates for temporary con-

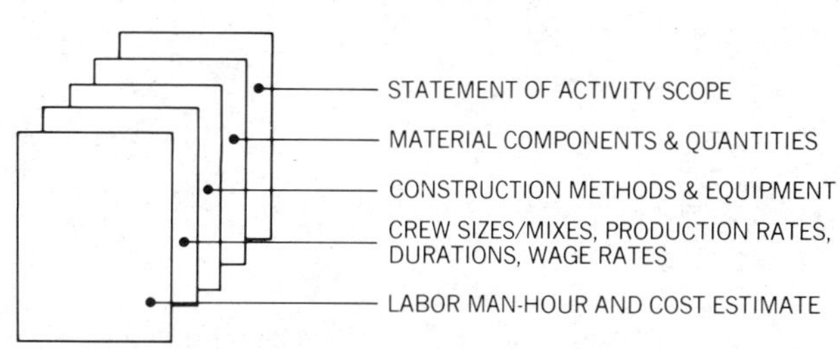

Figure 6.10 Specialized activity estimate using work package technique.

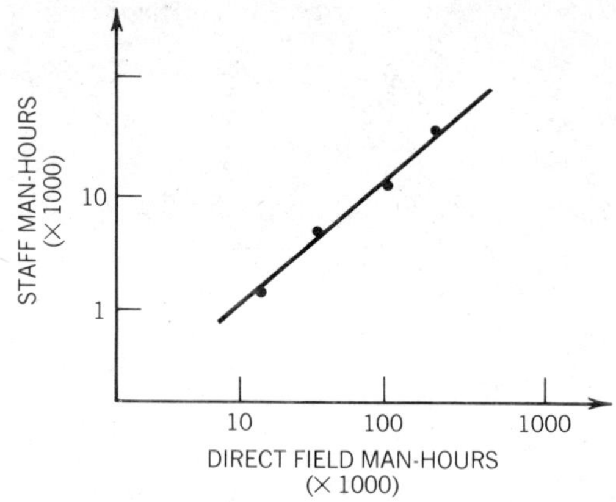

(a)

ORG. CHART POSITION	TITLE	DURATION-MONTHS	TOTAL MONTHS	HOURS/ MONTH	TOTAL HOURS
1	PROJECT MANAGER		12	170	2040
2	PROJECT ENGINEER		12	170	2040
3	ESTIMATOR		3	170	510
4	SUPERINTENDENT		9	170	1530

(b)

Figure 6.11 Field staff estimating techniques. (*a*) Factoring from history. (*b*) Estimate based on staff manning involvement.

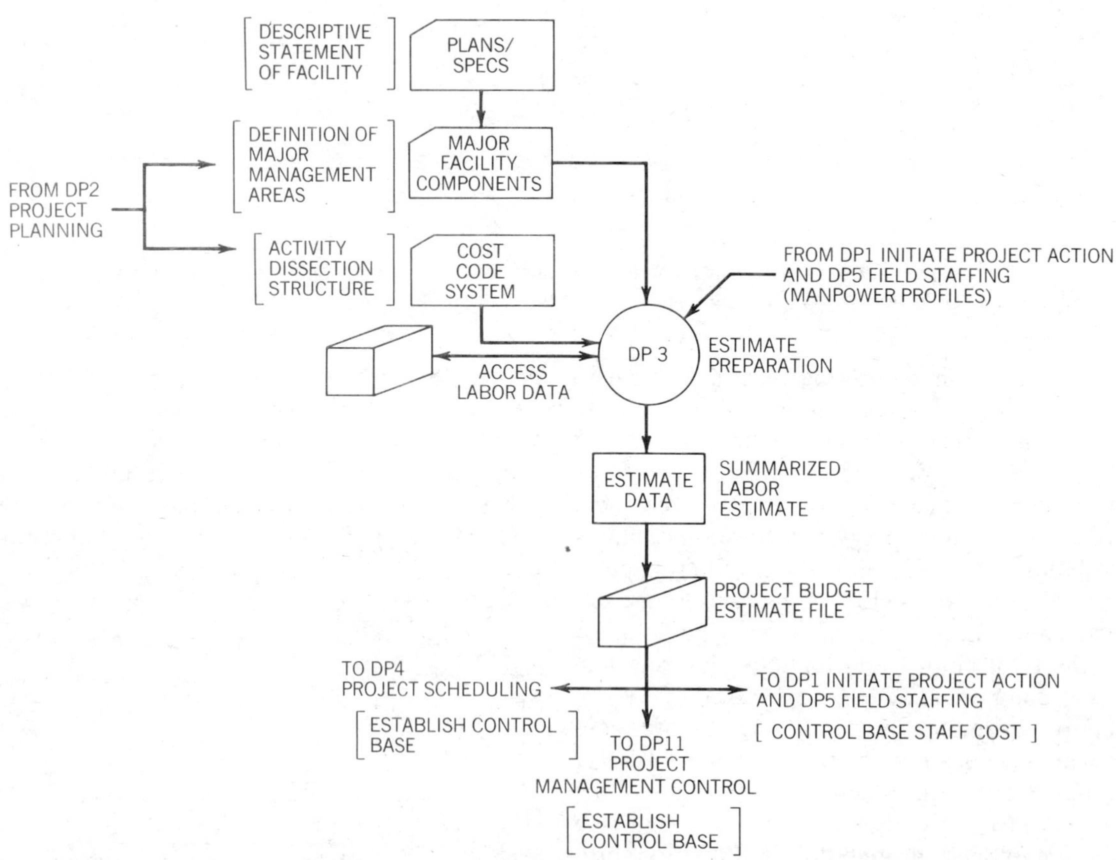

Figure 6.12 Estimate Preparation Management Information System (MIS) concept.

TABLE 6.3 POLICIES OF THE SMALL FIRM

Policy		Description
P_S 2	*VISUAL INSPECTION*	Visual site inspection is the method by which the impact of site conditions on project construction is evaluated.
P_S 3	*USE COST CODES*	Systematic classification of work categories segregated by unique cost codes is used to identify work tasks and collect costs.
P_S 4	*LABOR UNIT COSTS*	Labor unit costs are sufficient in project estimating provided that these costs reflect past experience from projects similar in scope and are adjusted, when necessary, to ensure compatibility with the current project scope.
P_S 5	*UNIQUE PROBLEM*	If unique manpower and crew aspects appear, labor unit costs are developed by evaluating crew size, mix, and productivity; otherwise traditional labor and crew situations are assumed.

struction facilities, for example, are frequently based on crew sizes, unit man-hour rates, and labor unit costs (if appropriate to the activity considered)*.

Estimate Preparation has as its primary goal the feeding of information into a project control budget estimate consistent with overall facility control approaches required to manage the project. This budget estimate data must be summarized in a manner that is useful to both top management (for approval and authorization to proceed), and other project team members (for implementation of detailed cost and scheduling control systems). Thus, the MIS that characterizes Estimate Preparation must produce critical documented information, which in turn becomes the requisite data input for the Project Scheduling and Project Management Control decision processes that establish the project cost and schedule control base. This MIS concept for Estimate Preparation is illustrated in Figure 6.12 and is applicable to all firm levels.

6.3 THE PROJECT TEAM COMPONENT APPROACH

At the small firm level, the manpower estimating focus is on the determination of the labor content of the project, and, for costing purposes, the duration of involvement of key field personnel. The project labor content is expressed in dollars and is usually evaluated by work categories. The accurate performance of Estimate Preparation is critical and ultimately determines whether or not the contractor is awarded the project.

Estimate Preparation requires the performance of the CONSTRUCTION RECONNAISSANCE (P1), ACTIVITY DEFINITION (P2), QUANTITY TAKE-OFF (P3), and LABOR CONTENT ESTIMATING (P4) functions.

The information required for CONSTRUCTION RECONNAISSANCE is contained in the plans and specifications (here identified as External Documents, E-DOC 1).† CONSTRUCTION RECONNAISSANCE is performed by a *VISUAL SITE INSPECTION* (P_S 2).‡ The estimator and project manager visit the project site and examine site conditions in light of the project scope to identify any special conditions which might influence the construction plan (such as relocation of services, access, and material storage areas, etc.) and the planned use of labor. CONSTRUCTION RECONNAISSANCE therefore provides specific site details for ACTIVITY DEFINITION.

Familiarity with site conditions aids the estimator in conceptualizing the structure to be built. He or she does this by dissecting the project scope into discrete work tasks or operations which are recognizable portions of the finished structure, and determining which items are to be subcontracted (if the small contractor is the prime contractor). While preparing an estimate, the estimator performs sufficient construction planning to ensure that the project can be completed within the scheduled time frame.

*The bibliography contains a list of references on construction estimating. These references provide greater detail on labor cost estimating and other estimating areas such as that related to material. The reader who is interested in the total estimating process is encouraged to obtain one or more of the references provided.

†Examples of plans and specifications are not shown but are only alluded to within the text.

‡See Table 6.3, Policies of the Small Firm.

Anderson & Woodhead Building Contractors	
	COMPANY COST CODES
WORK CODE	WORK CODE DESCRIPTION
A	Site Work and Excavation
B	Concrete
C	Masonry
D	Metals
E	Carpentry

(a)

Anderson & Woodhead Building Contractors	
B CODE Includes all labor costs for formwork, reinforcing steel and mixing and placing concrete work except...	
B.100	Columns and Wall Footings
.110	Grade Beams
.120	Slab on Grade
.130	Beams
.140	Columns
.150	Retaining Walls
.160	Unassigned
.170	Unassigned

(b)

Figure 6.13 Company Cost Codes (P_S-DOC 1). *(a)* Major work classification. *(b)* Detailed dissection of major work classification.

Cost codes help to establish a system of project work tasks organized in a logical construction sequence representing the general scope of work, with the activities segregated by unique code numbers. The need for *USING COST CODES* (P_S 3) requires the development of Company Cost Codes (P_S-DOC 1). Typical examples of this document are shown in Figure 6.13*a* and *b*. The ordered total listing of all project activities becomes the Project Scope List (I_S-DOC 1), a typical example of which is illustrated by Figure 6.14.

QUANTITY TAKE-OFFS and LABOR CONTENT ESTIMATING are the critical functions of Estimate Preparation. QUANTITY TAKE-OFFS determine the magnitude of all activities in terms of their material nature, units of measure, and content. Regardless of the level of detail, quantities are determined by: (1) direct take-off of items from the drawings (e.g., counting items); or (2) measuring dimensions for geometric shapes (e.g., rectangular column footings) representing a work task or its components, and calculating the quantities. By calculating all quantities according to the sequence of activities on the Project Scope List, the estimator can visualize the actual construction of the structure in a logical sequence from clearing and grubbing to final clean up.

Quantity development is basic to all levels of estimating in the construction industry, and varies only slightly from company to company. The manner in which quantities are recorded can vary significantly. Because the small firm generally estimates on a unit cost basis, estimated quantities are recorded to a level of detail consistent with the firm's historical database. The Quantity Sheet (P_S-DOC 2) is the document used for developing and recording quantities. A typical example of this document is shown in Figure 6.15.

During LABOR CONTENT ESTIMATION, *LABOR UNIT COSTS* (P_S 4) are developed for each work task and extended to labor dollars. The Estimate Recapitulation Sheet (P_S-DOC 3) shown in Figure 6.16 contains the recorded quantities, labor unit prices, and the total labor dollars for each work task. Historical cost data in the form of actual labor unit costs from previous work aid the estimator in developing new labor estimates for potential work. Historical costs can be extracted from completed projects or from projects currently under construction. Current labor unit costs best reflect prevailing

Anderson & Woodhead
Building Contractors

PROJECT: ____________ JOB NO.: ____________

ARCHITECT: ____________ DATE: ____________

LOCATION: ____________ SHEET NO.: 2 of 10

PROJECT SCOPE LIST

WORK ACTIVITIES	REMARKS
CONCRETE	
COLUMNS AND WALL FOOTINGS	DRAWING NO. 4510
SLAB ON GRADE	DITTO

Figure 6.14 Project Scope List (I_S-DOC 1).

Anderson & Woodhead
Building Contractors

PROJECT: ______ JOB NO.: ______
ARCHITECT: ______ DATE: ______
LOCATION: ______ SHEET NO.: ______
ESTIMATOR: ______
WORK CLASSIFICATION: CONCRETE CHECKED: ______

DESCRIPTION	NO.	DIMENSIONS L	W	D	CALC.	ESTIMATED QUANTITY	UNIT
DRAWING NO. 4510							
CONCRETE FOUNDATIONS	3000#	CONCRETE					
WALL FOOTINGS (B.100)		1206′-6″	2′-6″	1′-0″	3014		
		300′-6″	2′-0″	1′-0″	601		
TOTAL					3615	156	CY

Figure 6.15 Quantity Sheets (P_S-DOC 2).

wage rates and local productivities, however, and therefore are referenced most frequently. If a labor unit cost is not current, then it must be adjusted to reflect current labor conditions. The labor unit cost approach is quite meaningful because of the intimate knowledge the small contractor has of local labor conditions and productivities.

Actual manpower (crew size) for a work task is implicit in the labor unit price, because the newly evaluated labor unit price is based either on historical or present experience, both of which reflect the actual crew responsible for accomplishing similar work executed on recent or current projects. Consequently, the estimator performs little productivity analysis, but does recognize that various factors will influence labor performance and estimated labor unit costs. In order to reflect the impact of these factors, the estimated labor unit cost is often adjusted according to job conditions, the local labor market, the time of year, the bidding strategy, and the historical nature of the base data. The estimator often uses no systematic method or rationale for evaluating the adjustment factor. This evaluation is subjective and is based on past experience or "gut" feelings. Circumstances do arise, however, when the estimator can rationally evaluate these adjust-

Anderson & Woodhead
Building Contractors
ESTIMATE RECAPITULATION

PROJECT: ______ JOB NO.: ______
ARCHITECT: ______ DATE: ______
LOCATION: ______ SHEET NO.: ______
ESTIMATE BY: ______ CHECKED BY: ______

WORK CLASSIFICATION	QUANTITY	UNIT	UNIT COST	TOTAL ESTIMATED MATERIAL COST	UNIT COST	TOTAL ESTIMATED LABOR COST	TOTAL
CONCRETE							
COST CODE B.100							
3000#—WALL FTGS	156	CY	50.00	7800	20.00	3120	10920
COLUMN FTGS	122	CY	50.00	5199	22.50	2745	8845
FORMS—WALL FTGS.	3014	SF	.55	1658	2.60	7836	9494

Figure 6.16 Estimate Recapitulation Sheets (P_S-DOC 3).

Anderson & Woodhead
Building Contractors

ESTIMATE SUMMARY

PROJECT: ______ JOB NO.: ______

ARCHITECT: ______ DATE: ______

LOCATION: ______ SHEET NO.: ______

SUMMARY BY: ______ CHECKED BY: ______

WORK CLASSIFICATION	MATERIAL	LABOR	SUBCONTRACT	TOTAL
SITE WORK	1250	8700		9950
EXCAVATION AND BACKFILL	10100	12220		22320
CONCRETE	125352	69024		194376
MASONRY	46073	69045		115118
ELECTRICAL			225198	225198
PAVING, CURBING			56330	56330
CONSTRUCTION EQUIP.	18000	23400		41400
SUPERINTENDENT		15000		15000
JOB OVERHEAD—OTHER	24060	58270		82330
TOTAL COST	299428	128327	522810	950565
SALES TAX 4%				11977
PROFIT MARK-UP 8%				77003
BOND				6482
TOTAL BID				1046027

Figure 6.17 Estimate Summary (P_S-DOC 4).

ment factors. For example, a labor unit cost would be increased for anticipated escalating wage rates if, over the construction period, the wage scale for the crafts involved is likely to change. The percent increase is either calculated when increases in wage rates are known from union agreements, or estimated based on previous settlements or other sources such as published indices.

The estimator is only concerned with crew sizing during LABOR COST ESTIMATING when obvious manpower and crew aspects appear. Under normal circumstances, the estimated labor unit cost is based on construction methods using conventional tried and true construction techniques, such as a typical timber forming system for concrete column construction. Sometimes *UNIQUE PROBLEMS* (P_S 5) demand a detailed analysis of potential construction methods to support an estimated labor unit cost. These situations generally arise when extremely unfamiliar work is encountered.

The final step of LABOR CONTENT ESTIMATING is to summarize the estimated cost of all components. Labor dollars are transferred from the Estimate Recapitulation Sheets to the Estimate Summary Document (P_S-DOC 4) (see Figure 6.17). Also transferred to this document, as Figure 6.17 shows, are other costs such as (1) material dollars, (2) equipment costs, (3) subcontract costs, and (4) temporary construction costs. The company's overhead, bonding requirement, and profit are calculated, and the summary of all the items constitutes the total estimated cost of the project, as illustrated. The owners must develop their bidding strategy and formulate the final price submitted as their bid.

The Estimate Preparation decision process is illustrated in Figure 6.18. Estimate Preparation can be further depicted as a document handling system (see Figure 6.19). In this figure, input into Estimate Preparation (DP 3) includes plans/specification documents, standard company cost codes (which provide a means of categorizing estimated costs), and historical data from recent project files. The information is processed when performing the decision process (see Figure 6.18), the output being estimate data in document form. This estimate data becomes the project budget estimate file. Normally, the small contractor performs estimating on a manual basis and therefore the data would be retained in docu-

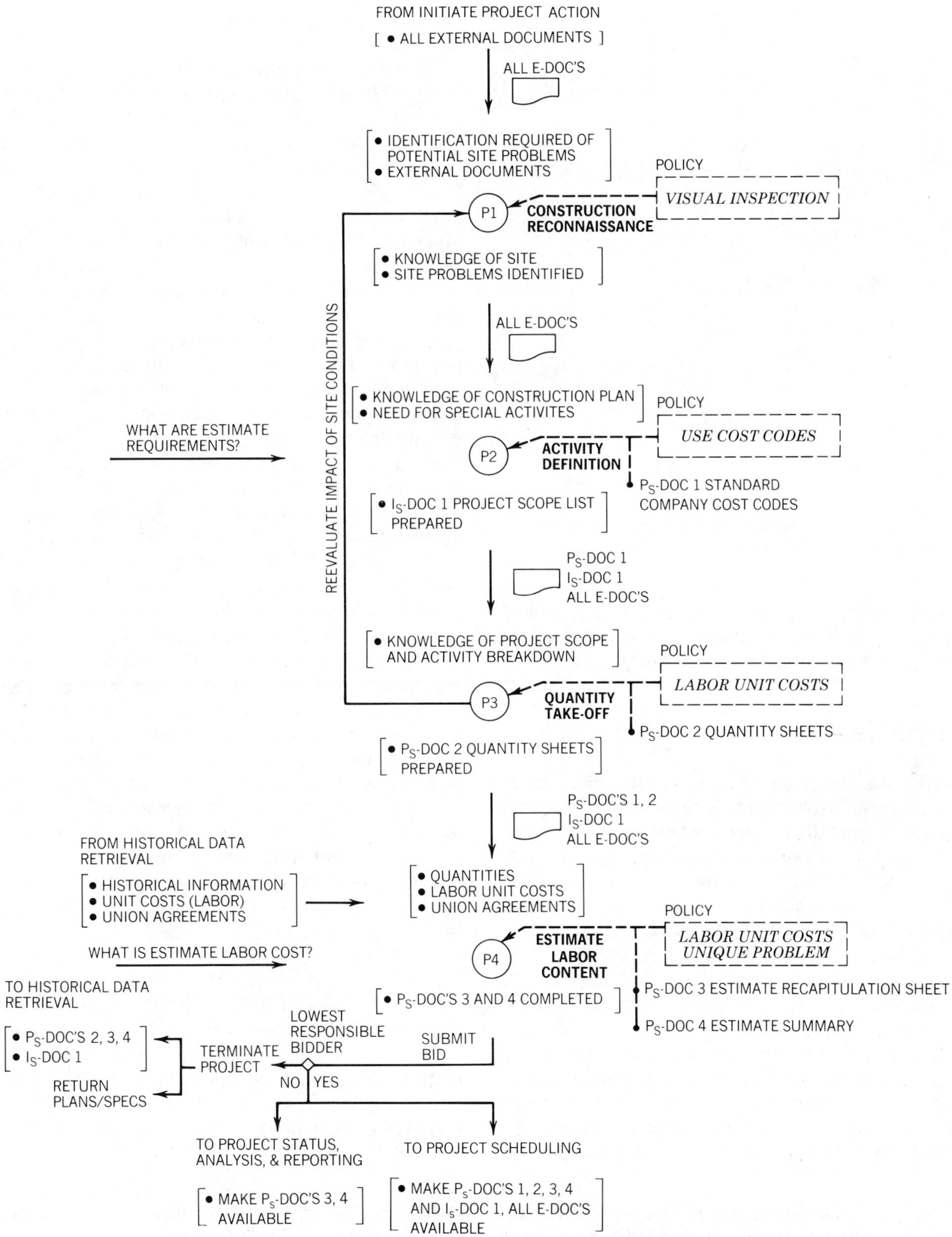

Figure 6.18 Estimate Preparation decision process (Project Team Component approach).

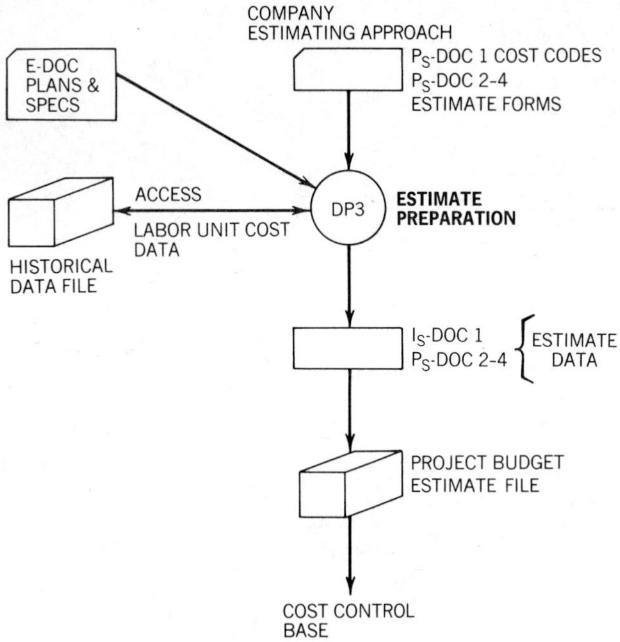

Figure 6.19 Estimate Preparation Management Information System (Project Team Component approach).

ment form and stored in a filing cabinet. The project budget estimate will then serve as the basis for cost control if the contractor is awarded the contract*.

6.4 THE TRAVELING PROJECT MANAGER

At the medium-sized firm level, the basic labor management problem of Estimate Preparation is similar to that found in small firms. The project labor content is expressed in dollars and is usually evaluated by work categories. The medium-sized contractor relies heavily on lump-sum competitive bidding and operates in a highly competitive market. Consequently, the estimator is still under continuous pressure to meet bid dates. While estimate preparation can be more detailed in this environment than for the small firm, the level of detail is still constrained by the pressure to meet bid schedules. As in small firms, however, the estimator does benefit from a limited work region, especially when branch offices are available. In many cases, this sensitivity to local productivities, labor availability, and union practices saves time in construction reconnaissance.

Estimate Preparation formally begins with CONSTRUCTION RECONNAISSANCE (P1), which involves both a *VISUAL SITE INSPECTION* (P_M 2)† and a *LABOR RECONNAISSANCE* (P_M 3) of the local labor environment. A Site Investigation Check List (P_M-DOC 2), as shown in Figure 6.20, is often prepared to document site conditions and other labor related data. Local union representatives are contacted to discuss and evaluate local labor availability and construction volume in the site area. In union construction, union officials are also queried regarding restrictive work practices applicable to their union.‡ The information derived from these discussions is evaluated in terms of its potential impact on the labor component of the project. While these conclusions are not formally documented, the estimator retains a knowledge of the local labor environment. In addition, the contractor will normally generate a file containing the local agreements of the unions in new work regions.§

Familiarity with site conditions aids in conceptualizing the facility to be built, and, when coupled with the plans and specifications, serves as basic input to ACTIVITY DEFINITION (P2). The focus of this function is to define the project scope and present it in a manner which is suitable for estimating the labor content and for the tracking of actual costs against estimates. The significant change in policy focus in this environment is the direct correlation between the project *COST CODES* (P_M 4) and the work categories within the estimate. Consequently, the estimator relies on company cost codes to assist in identifying work categories. In order to document the project scope, a standard check list is used for correlating the standard costs codes with the activities unique to the project. An illustration of Company Cost Codes (P_M-DOC 3) is shown in Figure 6.21. The check list of work categories is used in ACTIVITY DEFINITION, and becomes the Project Scope Check List (P_M-DOC 4) (see Figure 6.22), as the project activities are defined. The remarks column on this document is useful for denoting which activities will be subcontracted, or for identifying special requirements associated with an activity.

*Many of the estimating approaches described can be computerized at relatively low cost using microcomputers. Software programs are available which can be employed to accomplish this (see bibliography).

†See Table 6.4, Policies of the Medium-Sized Firm.

‡In non union construction practice, the estimator would need only to set a competitive wage rate structure.

§This would be in addition to files of those agreements covering the unions within the firm's normal work region.

TABLE 6.4 POLICIES OF THE MEDIUM-SIZED FIRM

Policy		Description
P_M 2	*VISUAL INSPECTION*	Visual site inspection is the method by which the impact of site conditions on construction is evaluated.
P_M 3	*LABOR RECONNAISSANCE*	Reconnaissance of the labor situation by interrogation of local labor officials is mandated for projects outside the normal work region.
P_M 4	*COST CODES*	Systematic classification of work categories segregated by unique cost codes is used to define work tasks, collect costs, and provide a means of comparing actual costs with estimated costs.
P_M 5	*LABOR UNIT COSTS*	A realistic value of labor dollars is sufficiently derived from labor unit costs for most tasks, provided that unit costs reflect current wages and potential production rates.
P_M 6	*CREW ANALYSIS*	Productivity and method analysis must supplement labor costs when evaluating cost differentials between a labor intensive operation and an equipment intensive operation.

QUANTITY TAKE-OFF (P3) commences once the project scope is fully defined and documented. The method the estimator or quantity surveyor employs for developing quantities consists either of counting directly the quantities on the drawings, or extracting dimensions from the drawings and calculating quantities. The document used to record quantities is the Quantity Survey and Estimate Work Sheet (P_M-DOC 5). A typical example of this document is shown in Figure 6.23, and is similar in most respects to the quantity sheets of the small firm (Figure 6.15). The key difference is the column allotted for inserting the cost code, which enables direct cost comparisons to be made with estimate data when actual costs are generated for each work task.

LABOR CONTENT ESTIMATING (P4) is generally based on the *LABOR UNIT COST* (P_M 5) technique. Several forms of this method are used, ranging from historical labor unit price to the development of a labor unit price based on anticipated crew sizes and mixes, craft production rates, construction methods, and prevailing wage rates. The method chosen will depend on the nature and criticality of each work task. For instance, historical labor unit prices are adequate if they reflect a similar activity from a previous or currently active project. These prices are adjusted as necessary to reflect existing conditions in a manner identical to that used in the small firm.

The estimator often investigates production rates and formulates labor unit prices by analyzing potential crew mixes and productivities. For example, the production rate is determined in labor hours per

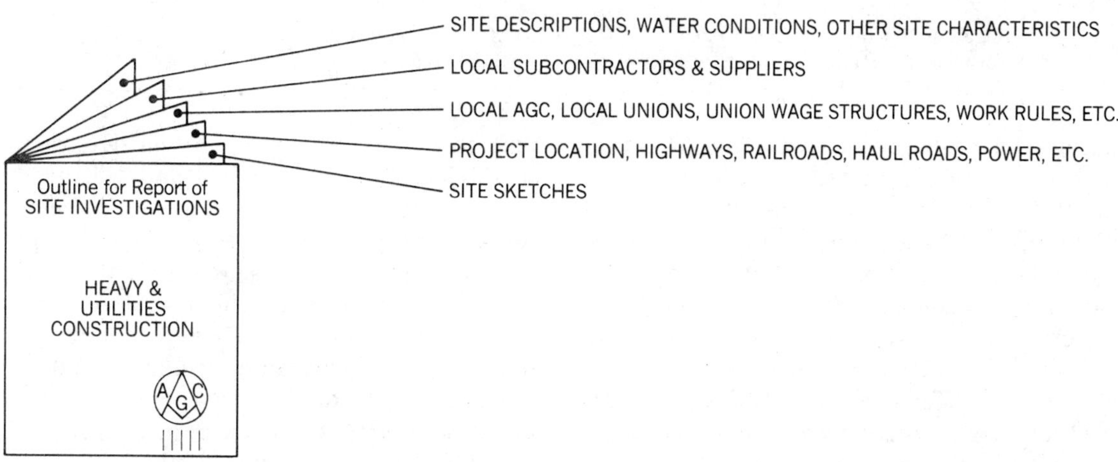

Figure 6.20 Site Investigation Check List (P_M-DOC 2); reprinted with permission of the Associated General Contractors of America, Inc.

WORK CODE	Anderson & Woodhead General Contractors GENERAL DESCRIPTION
1	GENERAL REQUIREMENTS
2	SITE WORK
3	CONCRETE
4	MASONRY
5	METALS
6	WOOD AND PLASTICS

(a)

02 9000 Landscaping	Lawns, trees, shrubs, and ground cover. Includes soil preparation and finish grading involving topsoil.

DIVISION 3—CONCRETE

	DESCRIPTION
03 0000 Concrete	NOTE 2.
03 0100 General Requirements—Concrete	NOTE 3.
03 1000 Concrete Formwork	All classes of concrete formwork. Does not include related ties or accessories.
03 2000 Concrete Reinforcement	Concrete reinforcing steel, accessories, welded wire fabric, stressing tendons and fibrous reinforcing.
03 3000 Cast-in-Place Concrete	Cast-in-place concrete material and placement.
03 4000 Precast Concrete	Concrete cast before placement (which may be on or offsite). Includes structural and architectural precast elements, erection accessories for connection, and related items.
03 5000 Cementitious Decks	Cast-in-place or prefabricated roof or floor decks including gypsum, cementitious, and wood fiber. Includes form boards, reinforcement, and support members when specified as a system.
03 5500 Concrete Finishes*	Finishes, such as flatwork, rubbing sandblasting. Includes curing.
03 6000 Grout	Metallic, epoxy, or other grouting.
03 6500 Concrete Accessories*	Accessory items used with the concrete, including angles, frames, anchor devices, rustification ties, anchors, expansion joints, waterstops, and similar items.
03 7000 Concrete Restoration and Cleaning	Cleaning, resurfacing, or rehabilitation of concrete.
03 8000 OPEN	NOTE 1.
03 9000 OPEN	NOTE 1.

AGC Document No. 1111 · Cost Classification Index, Broadscope Edition · February 1986
 Page 6

(b)

Figure 6.21 Company Cost Codes (P_M-DOC 3). (*a*) Major categories. (*b*) Detailed cost codes for major category; reprinted with permission of the Associated General Contractors of America, Inc.

unit measure for each craft in the crew. This rate reflects the construction method visualized to accomplish the work. The production rate (e.g., carpenter hours per 100 square feet of forms), is converted into labor dollars per unit measure (e.g., man-hours per 100 square feet × $16.00 per hour) for each craft, and summarized into a composite labor unit price for the work task.

Crew sizes are analyzed when an equipment-intensive activity is encountered. Possible crew sizes are evaluated against associated equipment needs to determine a cost per day for various construction methods. This *CREW ANALYSIS* (P_M 6) and cost trade-off approach will determine which method is most economical and will produce the best results. Developing labor unit prices as an alternate approach ensures that the unit price will reflect prevailing wages. It assists the estimator in selecting the method which best utilizes the resources of equipment, labor, and time. It also requires a more elaborate system of collecting and documenting field labor man-hour expenditures and quantity progress, because historical records of man-hours per unit of measure by craft are essential for developing labor unit rates.

The Estimate Recapitulation Sheets (P_M-DOC 6),

Anderson & Woodhead
General Contractors

PROJECT: ____________ JOB NO: ____________

LOCATION: ____________ DATE: ____________

PREPARED BY: ____________ SHEET NO.: ____________

COST CODE		WORK ITEM DESCRIPTION	√	X	REMARKS
3		CONCRETE			
0310.		CONCRETE FORMWORK			
0311.		FOOTINGS	√	X	
0312.		GRADE BEAMS	√	X	
0313.		SLAB ON GRADE	√	X	
0314.		BEAMS			
0315.		COLUMNS			

LEGEND:
√ REQUIRE ITEM OF WORK
X QUANTITY SURVEY COMPLETE FOR ITEM OF WORK

Figure 6.22 Project Scope Check List (P_M-DOC 4).

Anderson & Woodhead
General Contractors

PROJECT: ____________ JOB NO.: ____________

LOCATION: ____________ DATE: ____________

PREPARED BY: ____________ SHEET NO.: ____________

COST CODE	WORK ITEM DESCRIPTION	DIMENSIONS	UNIT MEASURE	QUANTITY
0310	FORMWORK			
0311	FOOTINGS			
	FOOTING	1′ x 3′ x 4 @ 105 EA	S.F.	1,260
	PEDESTAL	3′-6″ x 1′ x 4 @ 105 EA	S.F.	1,470
0311	TOTAL			2,730

(a)

LABOR UNIT COST	LABOR COST	MAT'L UNIT COST	MATERIAL COST	EQUIPMENT/ SUBCONTRACTS
1.22	3,330	.46	1,260	

(b)

Figure 6.23 Quantity Survey and Estimate Work Sheets (P_M-DOC 5). (*a*) Quantity survey sheet. (*b*) Estimate work sheet.

Anderson & Woodhead
General Contractors

PROJECT: ____________ JOB NO.: ____________

LOCATION: ____________ DATE: ____________

PREPARED BY: ____________ SHEET NO.: ____________

COST CODE	WORK ITEM DESCRIPTION	UNIT	TOTAL QUANTITY	LABOR		MATERIAL		SUBCONTRACT AND/OR EQUIPMENT	TOTAL COSTS
				UNIT COST	TOTAL $	UNIT COST	TOTAL $		
0310.	FORMWORK								
311.	FOOTINGS	SF	2,730	1.22	3,330	.46	1260		4,590
312.	GRADE BEAMS	SF	9,555	1.43	13,660	.70	6690		20,350
0300.	CONCRETE—TOTAL				276,045		512,655		788,700

Figure 6.24 Estimate Recapitulation Sheet (P_M-DOC 6).

summarize the detailed estimate information for each work task. As shown in Figure 6.24, the work tasks and their quantities are extracted from the quantity survey sheets, and, as the labor unit cost for each work task is evaluated, both quantities and unit prices are entered and identified by their unique cost codes.* The totals are transferred to an Estimate Summary (P_M-DOC 7) (illustrated in Figure 6.25), and insurance, overhead, and profit are calculated. The sum total of all costs, labor, material, subcontracts, equipment, overhead, and profit becomes the estimated cost to perform the work. The final price the contractor submits as the bid may vary according to the bidding strategy selected.

The complete formulation of the Estimate Preparation decision process at the medium-sized firm level is shown in Figure 6.26. When this decision process is described as a document handling system, its structure would be similar to that of the small firm, as shown in Figure 6.19. Most medium-sized contractors operate computerized estimating programs, however. A computerized information system for Estimation Preparation is illustrated in Figure 6.27. In this example, the input would include plans/specification documents, cost codes, and documented site details, as well as the forms used to input work categories and estimate quantity details to the computer. Quantity data, such as that shown in Figure 6.23*a*, would be developed and input to the computer labor estimate program file with a corresponding program command to execute the estimating program. The labor estimating program file accesses labor unit costs for each work category from a historical labor database file and applies the respective labor unit costs to generate Estimate Recapitulation Sheets and an Estimate Summary. The information contained in these documents is identical to that shown in Figures 6.24 and 6.25, respectively, except that it is formated by the computer. Computer-generated labor unit costs are not blindly accepted by the estimator but are carefully analyzed and reviewed, and adjusted if necessary to meet specific project conditions. This evaluation cycle will often result in several examinations of the labor estimate program file before the estimate data is incorporated into the contractor's bid.

The computerized estimating system will emerge in the medium-sized construction environment because business volume generally necessitates a computer system. In addition, if the contractor maintains computerized estimating, the firm is likely to have other computerized systems, such as one for cost control.†

*Calculations, sketches, and other support documentation would comprise the backup for the unit prices on these documents.

†Microcomputers provide a low cost means of implementing a computer-based system for estimating and cost control with current personal computer technologies.

Anderson & Woodhead
General Contractors

PROJECT: ____________________ JOB NO.: ____________________

LOCATION: ____________________ DATE: ____________________

CLIENT: ____________________ PREPARED BY: ____________________

COST CODE	DESCRIPTION	LABOR COSTS	MATERIAL COSTS	SUBCONTRACTS/ EQUIPMENT COSTS	TOTAL COSTS
0200.	SITE WORK	79,320	52,880		132,200
0300.	CONCRETE	276,045	512,655		788,700
0400.	MASONRY			1,410,000	1,410,000
0500.	METALS	327,725	989,175		1,318,900
0600.	WOOD AND PLASTICS	66,906	180,894		247,800
	TOTAL DIRECT COSTS	1,059,070	3,177,210	6,354,420	10,590,700
01	CRAFT BENEFITS, BURDENS		278,650		278,650
	EQUIPMENT COSTS		209,300		209,300
	TEMPORARY CONSTRUCTION	50,100	150,000		200,100
	TOTAL INDIRECT CONSTRUCTION COSTS	237,550	712,650		950,200
	PROJECT MANAGER (PART TIME)	25,000			25,000
	FIELD SUPERINTENDENT	37,500			37,500
	TOTAL PROJECT STAFF COSTS	327,835	140,465		468,300
	INSURANCE				131,100
	SALES TAX				206,520
	PROFIT/MARK-UP				897,880
	TOTAL PROJECT COSTS				13,244,700

Figure 6.25 Estimate Summary (P_M-DOC 7).

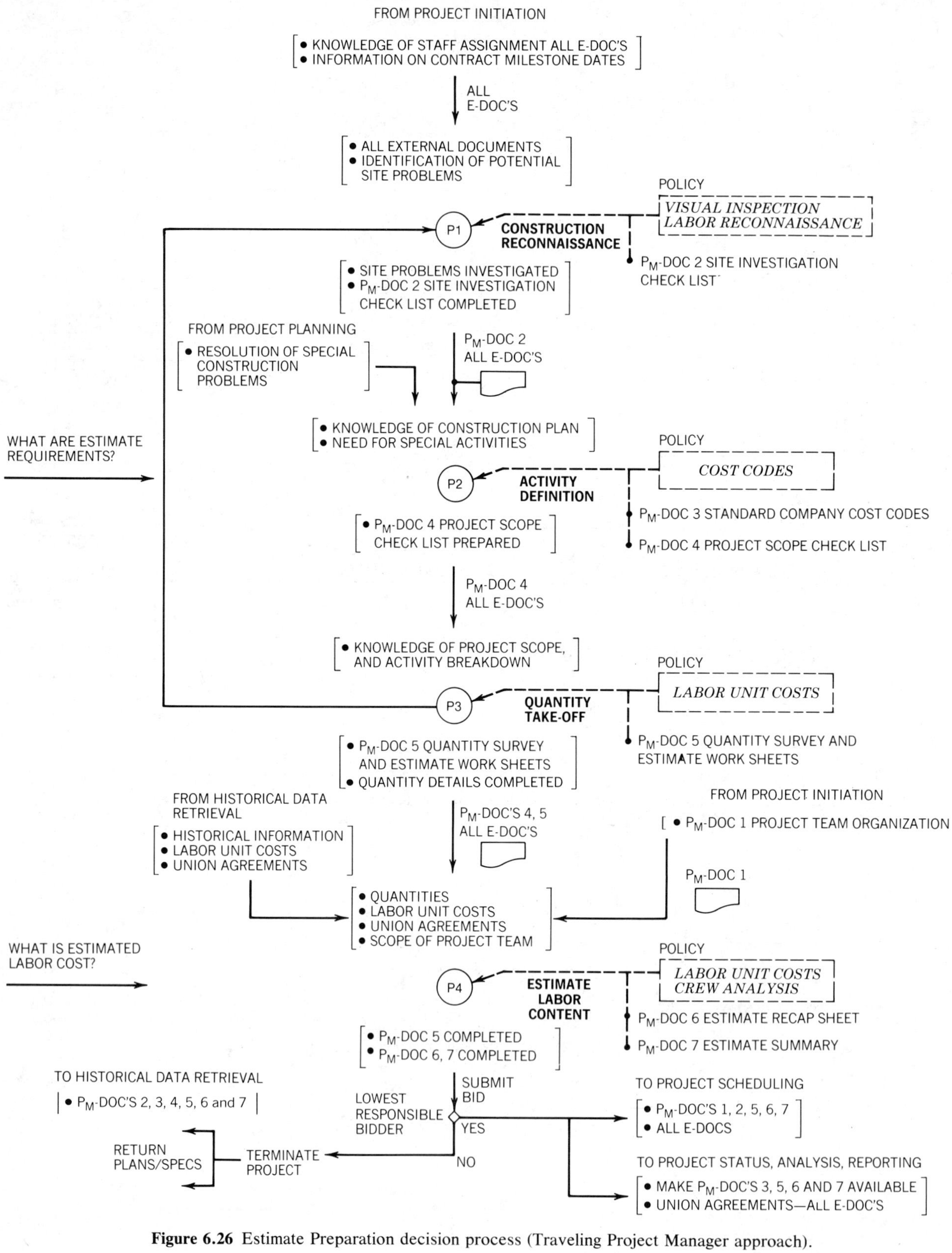

Figure 6.26 Estimate Preparation decision process (Traveling Project Manager approach).

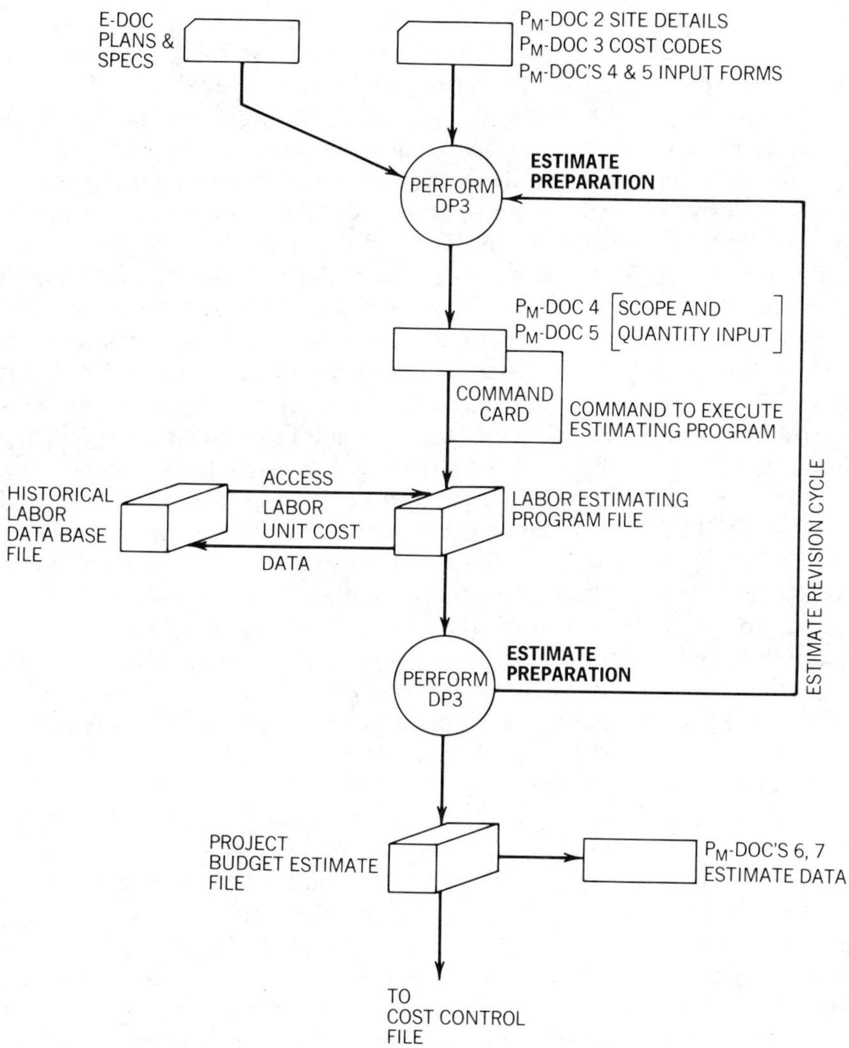

Figure 6.27 Estimate Preparation Management Information System (Traveling Project Manager approach).

6.5 FIELD-BASED PROJECT MANAGEMENT

Estimate Preparation at the large firm level for Field-Based Project Management follows several patterns, depending on the nature of the contract and the extent of the firm's responsibility for the project. If the contract is fixed price on a guaranteed maximum or lump-sum basis for construction only, and plans and specifications are essentially firm, a detailed control budget estimate is prepared and submitted as the firm's price for performing the work. Alternatively, under a design–construct contract several stages of Estimate Preparation may become necessary in order to fully define the project's labor content in terms of man-hour effort and unit man-hour rates as well as labor dollars. This represents a significant departure from the labor unit cost approach of the medium-sized firm, and reinforces the importance of quantifying the man-hour content of a project. The man-hour content is extremely important because man-hours are the primary ingredient for successful manpower planning and scheduling, for the determination of activity durations for all levels of project schedules, and for field crew planning. Furthermore, establishing the man-hour content and associated labor costs required for both the head office and field project team is essential, because when compared to those of the medium-sized firm, these manpower areas become significant percentages of the total capital cost of the typical large project, and are reviewed most carefully by perceptive clients.

Estimate Preparation is initiated either by a request from the Project Manager, or even top level management, by a predetermined estimate schedule, or by a directive from the client. At any time during the engineering and design phases of a project, an Estimate Preparation cycle can be generated in response to a number of factors. These include (1) a series of radical changes in scope, (2) apparent increasing costs which indicate significant cost overruns, (3) client cash flow problems, (4) predetermined stages in the development and definition of the project scope in terms of engineering and design detail, and so on. When contrasted with simpler procedures in the medium-sized firm, the large firm's frequently continuous Estimate Preparation cycle uses a broader spectrum of the various estimate techniques found in construction practice.

Estimate Preparation incorporates the three basic manpower management functions of ACTIVITY DEFINITION (P2), QUANTITY TAKE-OFF (P3), and LABOR CONTENT ESTIMATING (P4). The decision process structure of Estimate Preparation at the large firm level is similar to that of both small and medium-sized firms in so far as the three functions are performed sequentially. Conversely, at this level the focus of each function varies when compared with the medium-sized firm, depending on the type of estimate technique (e.g., factored versus detailed) utilized.

For example, ACTIVITY DEFINITION has minimal influence in Estimate Preparation because of an increasing emphasis on preparing a Project Cost Code System prior to Estimate Preparation, as the basis for scope definition and estimate structuring. This occurs as part of the Project Planning decision process at the large firm level. It is not a critical problem to the medium-sized firm, however, and therefore occurs simultaneously with the preparation of the detailed estimate for a fixed priced bid. Thus, the Estimate Preparation decision process can be broken down into two different phases, relating first to factored estimating during the engineering/design development phase as shown in Figure 6.28*a*, and then to construction control budget estimating as shown in Figure 6.28*b*, when the scope is defined by specifications and construction drawings.

The Factored Estimating Phase (Figure 6.28*a*), becomes significant in this environment, because it provides the means for establishing the approximate man-hour content of a project. When the man-hours are converted to craft manpower profiles, early indications of problem areas, such as shortages of certain skilled crafts, can be identified. The solution to these problems may lead to increasing costs for training programs or subsistence pay for travel to and from the project site.

In the design–construct environment, Estimate Preparation frequently passes through several stages, depending on the quality and quantity of the scope documents available, project management desires, and/or client demands. Normally, a first order of magnitude cost estimate is generated based on project parameters describing total facility (or facility component) capacities, such as kilowatts, barrels per day, and so on (see Figure 6.4). This Cost Capacity Factored Estimating technique produces rough approximations of facility cost and can be used to determine the gross labor content of a project both in terms of project team man-hours and field labor man-hours. Factoring techniques of this type are frequently based on the contractor's standard cost control methodologies, and rely heavily on historical cost data and historical manpower data in the form of curves, ratios, percentages, and so on. Figures 6.5, 6.6, and 6.11*a* illustrate typical examples of Capacity Factoring Techniques based on the concepts of company cost control methodologies.

As engineering and design proceeds, facility plot plans and general arrangement drawings emerge and specifications for facility equipment items are written. Similarly, as the procurement cycle for major equipment items (e.g., turbine generator) is progressing, some price quotes may be available. At this stage in the design, cost estimates based on Machinery and Equipment factored estimating techniques can be prepared (see Figure 6.7 for a typical example). Equipment factoring is historically based and therefore its accuracy depends to a large extent on the quality of the contractor's database.

Both cost capacity and equipment factored estimating techniques must often be supplemented with drawings and sketches of areas of a facility where factoring techniques will not produce reliable results. In this situation, preliminary quantity take-off for unit man-hour estimating is required. Consequently, the cost engineer/estimator must have a broad range of expertise and ability, not to mention good common sense and judgement, when tackling the problems raised by factored estimating approaches.

Project team man-hour content and labor dollars must be estimated in order to complete the manpower component of the total factored estimate for a project. These estimates are based either on historical cost curves (Figure 6.11*a*), historical factoring by departments (Figure 4.5), or approximate manning charts for each discipline furnished by the project team through the Initiate Project Action decision process. The man-hours are extended to labor dollars by wage rates determined through a wage rate analysis for each project team discipline. The cost engineer/project estimator is responsible for summarizing both the head office and field project team estimates.

The Factored Estimating Phase of Estimate Preparation is described in Figure 6.28*a*. All other cost components of the facility (e.g., permanent facility material and equipment, construction equipment, etc.) must be estimated and summarized into the total capital cost. A typical summary of a Factored Estimate (P_L-DOC 12) is shown by Component and Total Project in Figure 6.29.

The second phase of Estimate Preparation is the detailed, definitive or control budget estimate for construction control, as illustrated in Figure 6.28*b*. This estimate is based on a well-defined scope of work and is comparable in some respects to the estimates generated by small and medium-sized firms. For the design–construct project, however, engineering specifications and construction drawings are rarely entirely complete when the control budget estimate is required. Consequently, different types of quantity analysis emerge at the large firm level in addition to the contingency cost category necessary to cover known "soft" areas in the scope of work.

The management functions of the construction Control Budget Estimate Phase (Figure 6.28*b*), are performed primarily by an estimating group under the direction of the project cost engineer. The estimating group is comprised of estimators and quantity surveyors that are segregated into specific estimating specialties, such as structural, piping, and electrical. The actual size of the estimating team will vary with both project size and nature.

The primary focus of ACTIVITY DEFINITION by *COST CODES* (P_L 6)* is to ensure that the scope of work is consistent with project control requirements as determined by the Project Cost Code System (P_L-DOC 11). This is normally a cost engineering function, because the cost engineer coordinates the flow of scope documentation for the control budget estimate (e.g., equipment lists, specifications, drawings, etc.), and ensures that the estimate is produced according to the level of detail established by the Project Cost Code and in accordance with the Project Control Zone List (P_L-DOC 9). Finally, the flow of information must be consistent with the Estimate Preparation Schedule as prepared by the cost engineer and approved by the Project Manager.

QUANTITY TAKE-OFF commences as drawings and specifications become available. In many instances, the performance of this function is similar to that in small and medium-sized firms. That is, dimensions are measured from construction drawings and the quantities are calculated, or a physical count directly from the drawings is made. Because detailed quantities cannot be determined for every item, however, quantities are estimated based on specific parameters and factoring methods. For example, the tons of rebar for a type of concrete foundation is often estimated based on the density (in pounds) of rebar per cubic yard of concrete, or the frequency of pipe fittings and flanges is correlated to the linear feet of pipe for various sizes and ranges of pipe material classes. These frequency or density methods for estimating quantities are more common on large, complex projects where detailed quantity take-off is not practical for every item, and where design information is not 100 percent complete at the time of a control budget estimate. These methods are only effective if based on historical data statistically analyzed, documented, and incorporated into the company's standard estimating practices. This approach is a departure from techniques used by the medium-sized firm, and leads to more penetrating policies. The policies which influence QUANTITY TAKE-OFF are not only *COST CODES* and *UNIT MAN-HOUR ESTIMATING* (P_L 7), but also include *HISTORICAL DATA* (P_L 8) and *LABOR DATABASE* (P_L 9). The documents produced as a result of these policies vary in format but contain information similar to the Quantity Sheets previously discussed for the small and medium-sized firms. The main difference is that, at this level, quantities are tied to specific cost codes and component locations within the facility as dictated by the Project Cost Code structure. A typical example of a Quantity Work Sheet (P_L-DOC 13) is shown in Figure 6.30. Supporting sketches and

*See Table 6.5, Policies of the Large Firm.

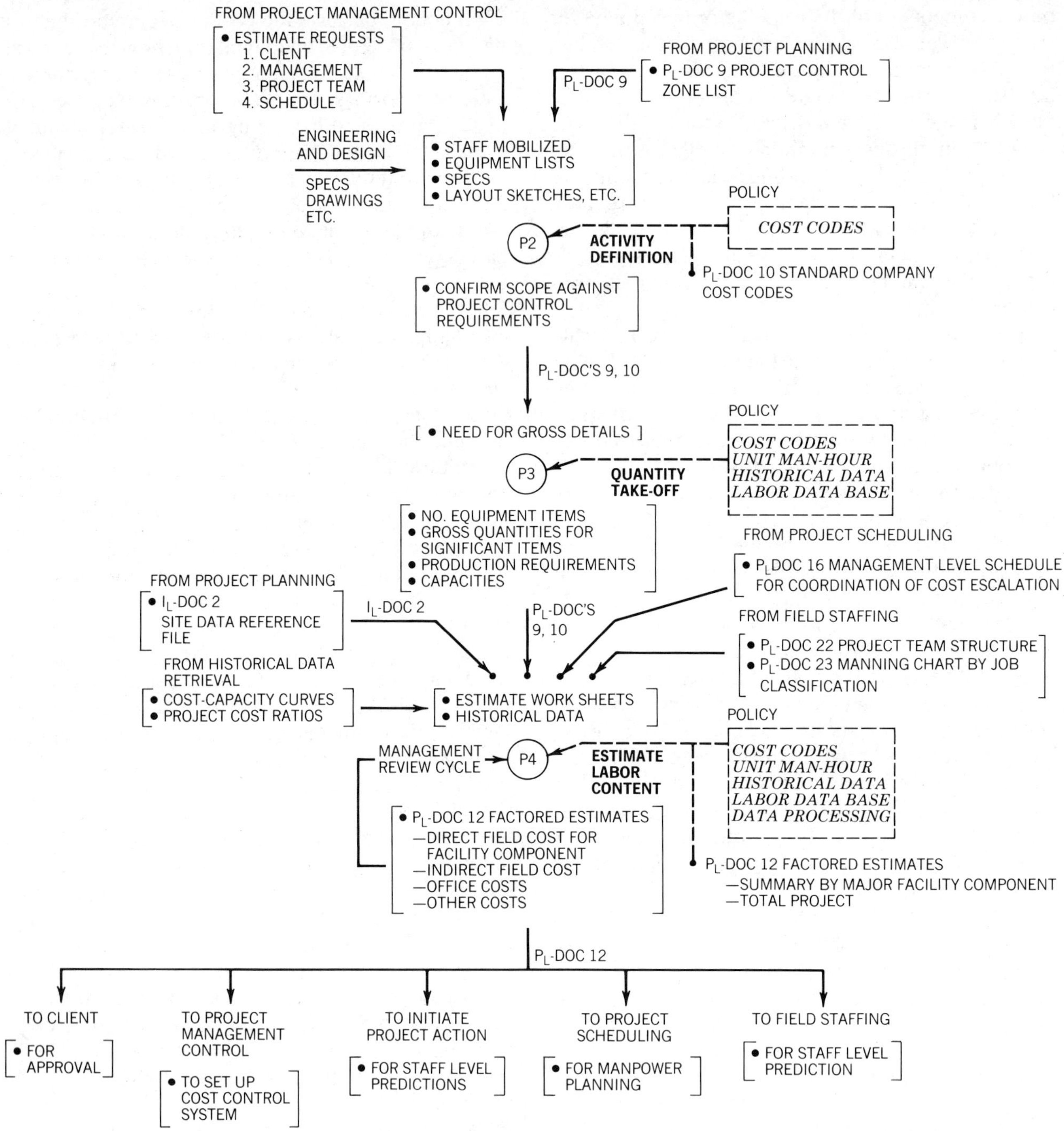

(a)

Figure 6.28 Estimate Preparation decision process (Field-Based Project Management approach). (*a*) Factored estimate phase.

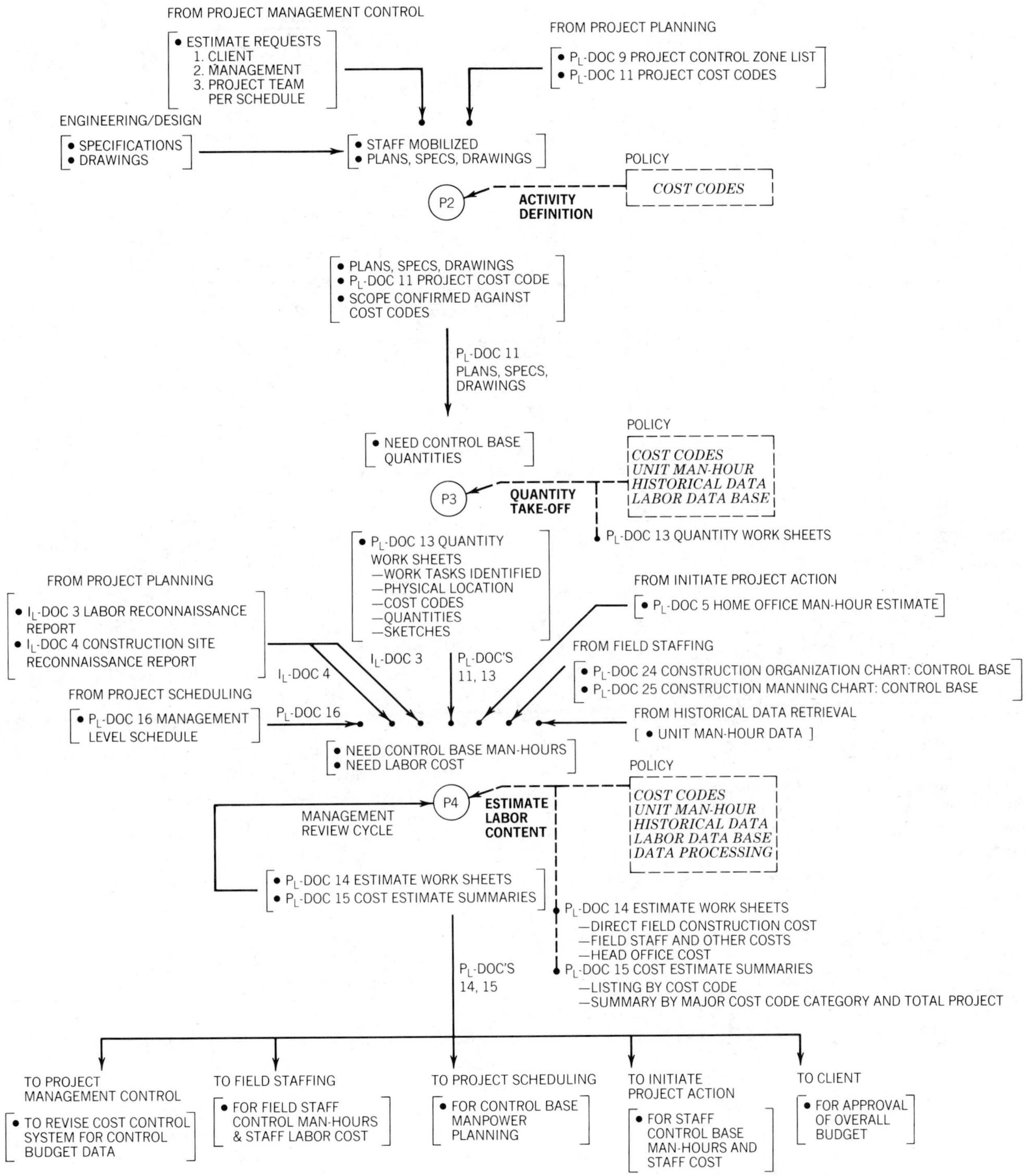

(b)

Figure 6.28 (*Continued*) (*b*) Construction control budget estimate phase.

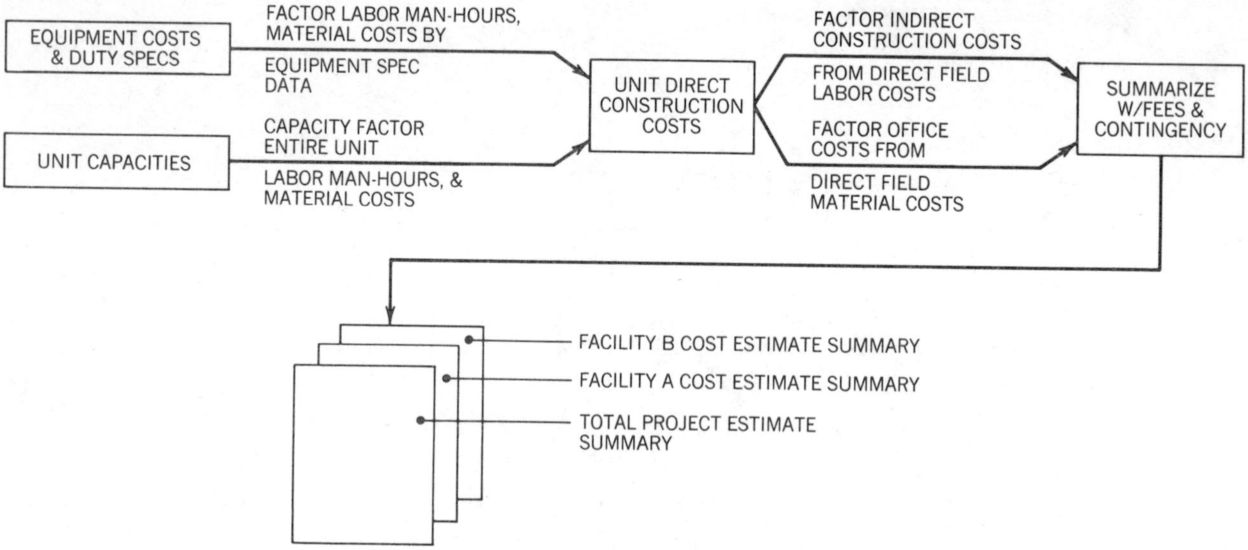

Figure 6.29 Factored Estimates (P_L-DOC 12).

other pertinent data used for the quantity take-off will accompany this document.*

As quantities are established for all the components of the project, the estimator begins LABOR CONTENT ESTIMATING. This function focuses on establishing the labor content of each item of work in terms of its labor man-hours and labor dollars. A different estimating approach is required to estimate man-hours, because the unit labor cost method normally does not reflect man-hours. Hence, the *UNIT MAN-HOUR* policy emerges as the common approach to estimating the man-hour content of work tasks. This estimating technique is described in Figure 6.9.

The unit man-hour model relies heavily on *HISTORICAL DATA*, either in the form of previous projects or a documented *LABOR DATABASE*. This data is generally referenced in terms of a common location and a known man-hour efficiency base such as a specific project or group of projects. The Project Cost Code system is utilized as the most reliable method of categorizing work. Thus, for any given work task which is described by a unique cost

*In design–construct projects where Computer Aided Design (CAD) Systems are used, quantity take-offs for estimating can be derived directly from the computer. Many CAD programs are rapidly approaching the capability to produce a total direct cost estimate, provided an estimate database is in the computer.

TABLE 6.5 POLICIES AT THE LARGE FIRM LEVEL

Policy	Description
P_L 6 *COST CODES*	A systematic classification of project work categories segregated by unique physical components and by cost codes is the basis for structuring the project scope into work tasks, categorizing estimated costs, reporting and controlling costs and manpower, and developing a historical database.
P_L 7 *UNIT MAN-HOUR*	Realistic evaluation of the labor content for a work task is essential for successful project planning and cost control. Therefore, labor costs must be based on work task man-hour content in relation to work volume, construction methods, local productivities, and labor availability.
P_L 8 *HISTORICAL DATA*	Historical labor data are basic to a realistic determination of the labor content for a work task as well as the labor content for major project components.
P_L 9 *LABOR DATABASE*	Historical labor data must be analyzed and documented in a meaningful way to be useful for estimating and planning manpower for future projects.
P_L 34 *DATA PROCESSING*	Maximum use of electronic data processing is often the most economical means of processing data into useable information.

AW
Engineering & Construction Corp.

CLIENT: ____________

PROJECT: ____________

DESCRIPTION OF WORK STATEMENT:

REFERENCES:
DRAWING NOS.
SPECIFICATION

COST CODE: _ _ _ _ _ _ _ _ _ _

QUANTITY ANALYSIS AND SKETCHES:

Figure 6.30 Quantity Work Sheet (P_L-DOC 13).

code, historical unit man-hour rates are available under that same cost code from previous projects (because Standard Company Cost Codes are the basis for all project-oriented cost code systems). Once the estimator establishes a reference unit man-hour for an activity/work task, he or she must adjust it to reflect any specific construction methods peculiar to the work task. This adjustment is normally based on experience, and on an analysis of the work task performed by the estimator.

Unit man-hour estimating often requires an analysis of the construction operation for a work task, as well as estimates of crew sizes and mixes based on specifically selected construction methods. Due to the size and complexity of the project, this approach is only performed for unique situations using the work package technique illustrated in Figure 6.10. For example, the estimator formulates a scope statement and enumerates the steps required to install a certain type of equipment. Next, the estimator selects the construction method, defines the crafts involved, and determines the construction equipment required. For each specific step the estimator establishes the crew size and mix and the hours needed to complete the operation. The sum of all the hours for each individual step provides the total estimated man-hours required to install the piece of equipment.

Subsequent to estimating the unit man-hours, these units must be adjusted to reflect anticipated site productivity in relation to the estimate reference basis. For example, 1.0 base man-hours may be equal to 1.2 man-hours at the project location. Consequently, the unit man-hour rate is adjusted to reflect the relative labor efficiency between the two locations. The efficiency factor is determined based on documented labor data from the Labor Reconnaissance Report (I_L-DOC 3) shown in Figure 5.18.

Converting work task man-hours into labor dollars requires an estimate of the wage rate associated with each work task. As described in Figure 6.9*c*, a composite wage rate is based on typical crew mixes associated with the type of work. Company standard estimating practices, previous experience, and knowledge of crew mixes are combined to formulate typical crews. The wage rate for each craft within the crew is derived from the Labor Reconnaissance Report and is weighted according to the percent of that craft to the total size of the crew. The sum of these weighted wage rates is the composite crew rate for the work tasks.

The unit man-hours, quantities, and wage rates, are recorded on Estimate Work Sheets (P_L-DOC 14), as demonstrated in Figure 6.31. The various extensions of man-hour rates and wage rates are calculated either manually or electronically, and the total man-hour and labor dollar estimate per task is determined. This process is repeated until the labor content of each work task is estimated. The sum of all the man-hours and labor costs for each component of the project equals the total direct field man-hour and labor dollar estimate for the facility. The other direct field costs, material and subcontract, are combined with the labor segment of the estimate

AW
Engineering & Construction Corp.

DIRECT FIELD CONSTRUCTION COSTS

CLIENT: ______________ DATE: ______________

PROJECT: ______________ CONTRACT: ______________

COST CODE	DESCRIPTION	QUANTITY	UNIT MEASURE	MAN-HOURS		WAGE RATE	UNIT COSTS			COSTS			
				UNIT	TOTAL		LABOR	S/C	MAT'L	LABOR	S/C	MAT'L	TOTAL

(a)

AW
Engineering & Construction Corp.

FIELD STAFF ESTIMATE

CLIENT: ______________ DATE: ______________

PROJECT: ______________ CONTRACT: ______________

COST CODE	STAFF CATEGORY	MAN-MONTHS	TOTAL MAN-MOS	TOTAL MAN-HOURS	WAGE RATE	LABOR COST

(b)

AW
Engineering & Construction Corp.

HOME OFFICE STAFF ESTIMATE

CLIENT: ______________ DATE: ______________

PROJECT: ______________ CONTRACT: ______________

COST CODE	ORG. NO.	STAFF CATEGORY	MAN-MONTHS	TOTAL MAN-MOS	TOTAL MAN-HOURS	WAGE RATE	TOTAL COST

(c)

Figure 6.31 Estimate Work Sheets (P_L-DOC 14). (*a*) Direct field construction costs. (*b*) Field staff costs. (*c*) Head office costs.

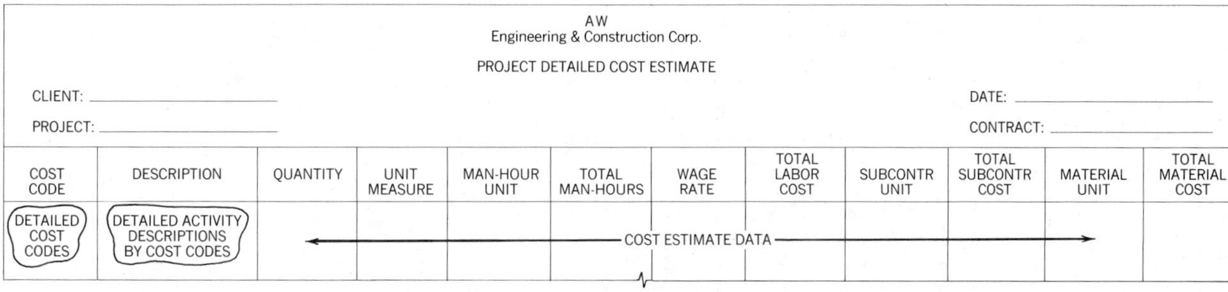

(a)

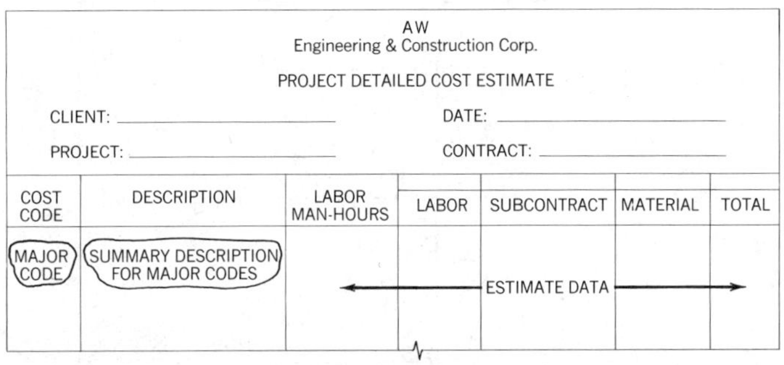

(b)

Figure 6.32 Cost Estimate Summaries (P_L-DOC 15). (*a*) Detail by cost code. (*b*) Summary by major cost code.

to provide the total direct field cost. These costs are shown on the Cost Estimate Summary (P_L-DOC 15) (see Figure 6.32).

Because of the nature, complexity, and lengthy duration of a large project, the man-hours required by the project team become a critical estimating factor. The detailed control budget manpower estimates for the field project management team are prepared during Field Staffing by the field office managers and key superintendents (see Chapter 8), and approved by the field project manager. These manpower estimates are input to Estimate Preparation, wherein the manpower totals are converted to man-hours and labor dollars in the head office by cost engineers/project estimators. The head office (engineering/design) man-hour estimate is prepared during Project Initiation by each individual discipline involved in the head office engineering and design effort. A labor wage rate analysis for each discipline is prepared by the cost engineer/project estimator, and is reviewed by the respective lead discipline engineers for their discipline. This wage rate is then applied to the total man-hours for the discipline and converted to labor dollars.

The summation of all project team man-hours and labor dollars is the responsibility of the cost engineer/project estimator. Once this step is completed, the appropriate benefit, burden, and overhead costs are estimated based on contractual percentages of labor dollars.

The project estimator completes LABOR CONTENT ESTIMATING by summarizing all project costs (materials, labor, subcontracts, etc.) and portraying them on a single document. This Estimate Summary can appear in many different forms, from hand-written to computerized documents. The supporting material can be extensive and encompass several volumes. Typical examples of estimates are shown in Figure 6.29 through 6.32. The estimate must reflect the appropriate time frame of the project and therefore will include escalated costs either incorporated into the details of the estimate or carried as a line entry in the estimate summary. Finally, at the large firm level, an evaluation of the probability of cost overruns or underruns is determined (often through risk analysis methods), and forms the basis for including a contingency cost category. While the definition of contingency varies drastically in the construction industry, it is normally considered as the probable percent of additional costs that arise due to unknown and/or unforeseen factors that have historically caused projects' costs to increase.

The final step in Estimate Preparation, whether it

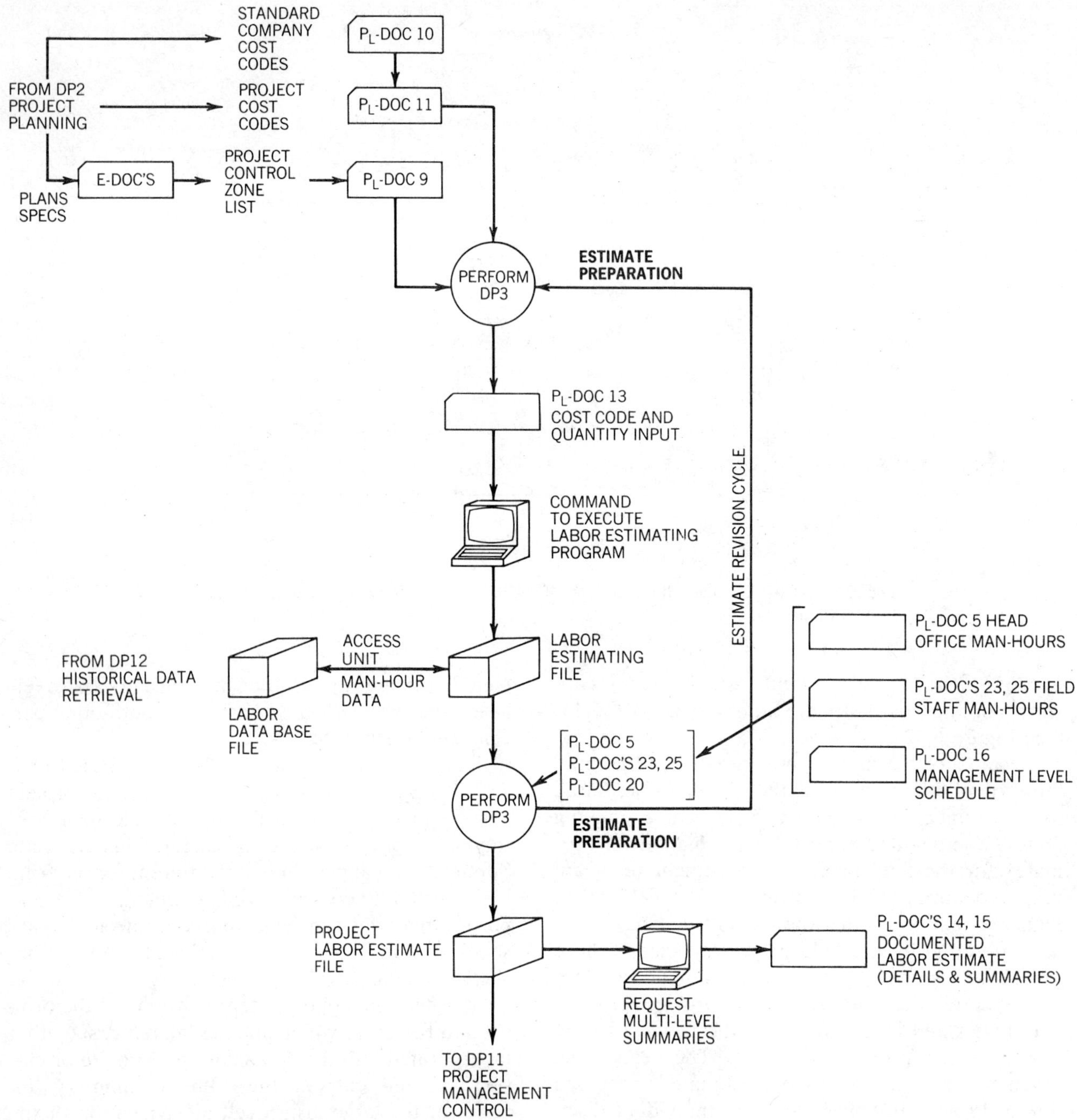

Figure 6.33 Estimate Preparation Management Information System (Field-Based Project Management).

TABLE 6.6 POLICIES OF THE AUTONOMOUS PROJECT TEAM

Policy		Description
P_T 8	*WORK BREAKDOWN*	A work breakdown structure representative of the project scope and control philosophy is formulated to integrate costs and schedules with work packages for project management.
P_T 9	*COST CODES*	Cost codes are applied to the work breakdown structure and serve as a basis for categorizing estimated costs, reporting and controlling costs and manpower, developing a historical database, and integrating the schedule with the estimate.
P_T 10	*UNIT MAN-HOUR*	A tight but realistic evaluation of the labor content for a work task is essential for successful project planning and management control. Therefore, estimates must be based on work task man-hour content in relation to work volume, construction methods, local productivities, labor availability, and site conditions.
P_T 11	*HISTORICAL DATA*	Historical labor data are basic to a realistic determination of the labor content for a work task or major project components and preliminary scheduling development.
P_T 12	*LABOR DATABASE*	Historical labor data must be normalized to a standard reference location and/or base and then documented in order to be useful for estimating and planning manpower.
P_T 21	*CREW MIX*	While impacted by union rules in some construction environments, the ratio of foremen to journeymen/apprentices within a crew is established to optimize both the foreman's level of supervisory control and the crew's performance.
P_T 36	*DATA PROCESSING*	A computer-based information system is the most feasible means of data processing for management control.

be a factored or control budget estimate, is internal company reviews from the project level, cost estimating department level, and so forth, all the way up to top executive management. These reviews cover many areas, such as verifying scope, checking typical project parameters and ratios, making comparisons against previous estimates, and so on. Once the estimate is approved by top management, it is released to the client and to other decision processes for project control purposes.

The Management Information System for the Estimate Preparation decision process (DP3) at the Field-Based Project Management level is summarized in Figure 6.33. The MIS approach is almost always computerized, due to the large volume of data and finer dissection of information dictated by project control philosophies. In addition, there must exist capabilities to summarize estimate data at various levels for project review and implementation as a control base information service. Thus, the MIS brings order to the handling of large volumes of estimate data and summary reporting of cost estimate results. It also produces estimate data in a format that is useful for other decision processes, such as developing manpower projections during Project Scheduling.

6.6 TOTAL FIELD AUTONOMY

Estimate Preparation for the Total Field Autonomy approach is similar to that for the Field-Based Project Management approach in terms of the basic decision process structure, information flows, and relevant manpower management functions (see Figure 6.28). However, subtle but significant differences emerge when comparing these two management approaches, in relation to policy formulation and the impact of policies on the nature and structure of the various documents generated. Three significant differences can be identified when contrasting the policies of the Large Firm (Table 6.5), with the policies of the Autonomous Project Team.* First, the focus on *WORK BREAKDOWN* (P_T 8) structures and *COST CODES* (P_T 9) ensures that the estimate is structured so that labor-related estimate

*See Table 6.6, Policies of the Autonomous Project Team.

P2 ACTIVITY DEFINITION	
Policies	
WORK BREAKDOWN	
COST CODES	
P3 QUANTITY TAKE-OFF	
Policies	
WORK BREAKDOWN	
COST CODES	
UNIT MAN-HOUR	
HISTORICAL DATA	
LABOR DATABASE	
P4 ESTIMATE LABOR CONTENT	
Policies	Documents
WORK BREAKDOWN	P_T-DOC 16 Factored Estimates
COST CODES	
UNIT MAN-HOUR	
HISTORICAL DATA	
LABOR DATABASE	
DATA PROCESSING	

(*a*)

P2 ACTIVITY DEFINITION	
Policies	
WORK BREAKDOWN	
COST CODES	
P3 QUANTITY TAKE-OFF	
Policies	Documents
WORK BREAKDOWN	P_T-DOC 17 Quantity Work Sheet
COST CODES	
UNIT MAN-HOUR	
HISTORICAL DATA	
LABOR DATABASE	
DATA PROCESSING	
P4 ESTIMATE LABOR CONTENT	
Policies	Documents
WORK BREAKDOWN	P_T-DOC 18 Estimate Work Sheets
COST CODES	P_T-DOC 19 Cost Estimate Summaries
UNIT MAN-HOUR	P_T-DOC 20 Management Estimate Summary
HISTORICAL DATA	
LABOR DATABASE	
CREW MIX	
DATA PROCESSING	

(*b*)

Figure 6.34 Estimate Preparation decision process—Total Field Autonomy approach. (*a*) Factored estimate phase. (*b*) Construction control budget estimate phase.

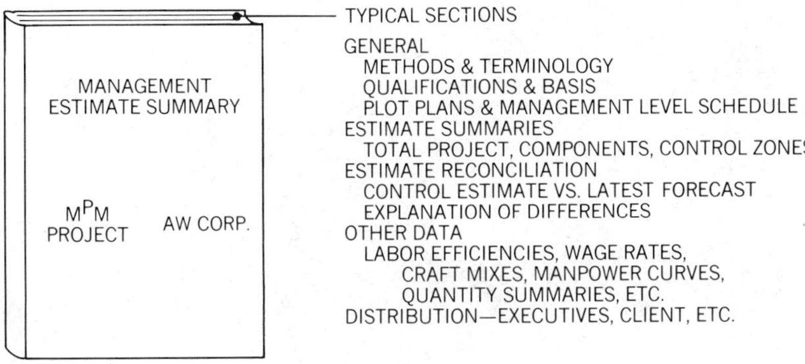

Figure 6.35 Management Estimate Summary (P_T-DOC 20).

data (e.g., quantities and man-hours) can be easily related to project schedules and field crew planning techniques. Thus, the estimate will have a structure consistent with the Project Control Zone List (P_T-DOC 12), Project Cost Codes (P_T-DOC 14), and the Work Package Structure (P_T-DOC 15), that includes and identifies cost components to the work package detail level. Second, the *LABOR DATABASE* (P_T 12) is referenced against a standard base location and time frame, and this base therefore remains fixed for all phases of labor man-hour estimating. Thus, a tight but realistic project site *UNIT MAN-HOUR* estimate (P_T 10) can be arrived at in relation to the standard base by applying efficiency factors that translate standard man-hours to the site location environment. This provides visibility for later adjustments, and control of productivity as actual labor conditions become known during the construction cycle. Finally, labor estimate assumptions are released for major decision making in relation to critical estimating base (e.g., efficiency, wage rates, craft mixes, etc.). These decisions, into which the client has input, facilitate later manipulation and changes that may be required due to actual site conditions in terms of labor efficiencies, wage rates, craft mixes, and so on.

The Factored Estimating and Construction Control Budget Estimating Phases of the Estimate Preparation decision process are shown in Figure 6.34 in terms of functions, policies, and documents. When comparing the Factored Estimating Phases of Total Field Autonomy (Figure 6.34*a*), with Field-Based Project Management (Figure 6.28*a*), the only differences that can be detected are related mostly to the structure of the estimate as defined by the Project Control Zone List, and to the development of Factored Estimates (P_T-DOC 16) utilizing the standard reference base man-hours. These two features of the Factored Estimating phase become extremely important, because they form the basis for tracking scope and cost trends as the design develops. Consequently, when current estimates are revised, the revised estimate can be reconciled against current estimates or forecasts through normalization of data, by control zone, into a standard estimate reference base. Provided these trends (and change orders) are well documented, the client will be fully apprised of current forecasted costs at the time of the control budget estimate. Consequently, notifying the client and getting his approval of deviations in costs is vital to the success of the Estimate Preparation process, in so far as a revised estimate will not dramatically change the current forecasted project cost.*

The Construction Control Budget Estimate Phase, as noted in Figure 6.34*b*, is also extremely similar to that shown in Figure 6.28*b* for the Field-Based Project Management decision process approach, with the exception of the addition of two new policies (*WORK BREAKDOWN* and *CREW MIX* (P_T 21)), and a generally more penetrating focus of the policy description, as previously discussed. The *WORK BREAKDOWN* policy coupled with *COST CODES* ensures that the labor estimate will match the level of detail of Project Cost Codes and the project Work Package Structure. Furthermore, basing the labor estimates on anticipated craft mixes aids in the future control of wage rates and site productivities.

The key documents produced during the control budget estimate are the Estimate Work Sheets (P_T-DOC 18) and the Cost Estimate Summaries (P_T-

*Change order/trending systems are discussed in detail in Chapter 13, Project Status, Analysis, and Reporting.

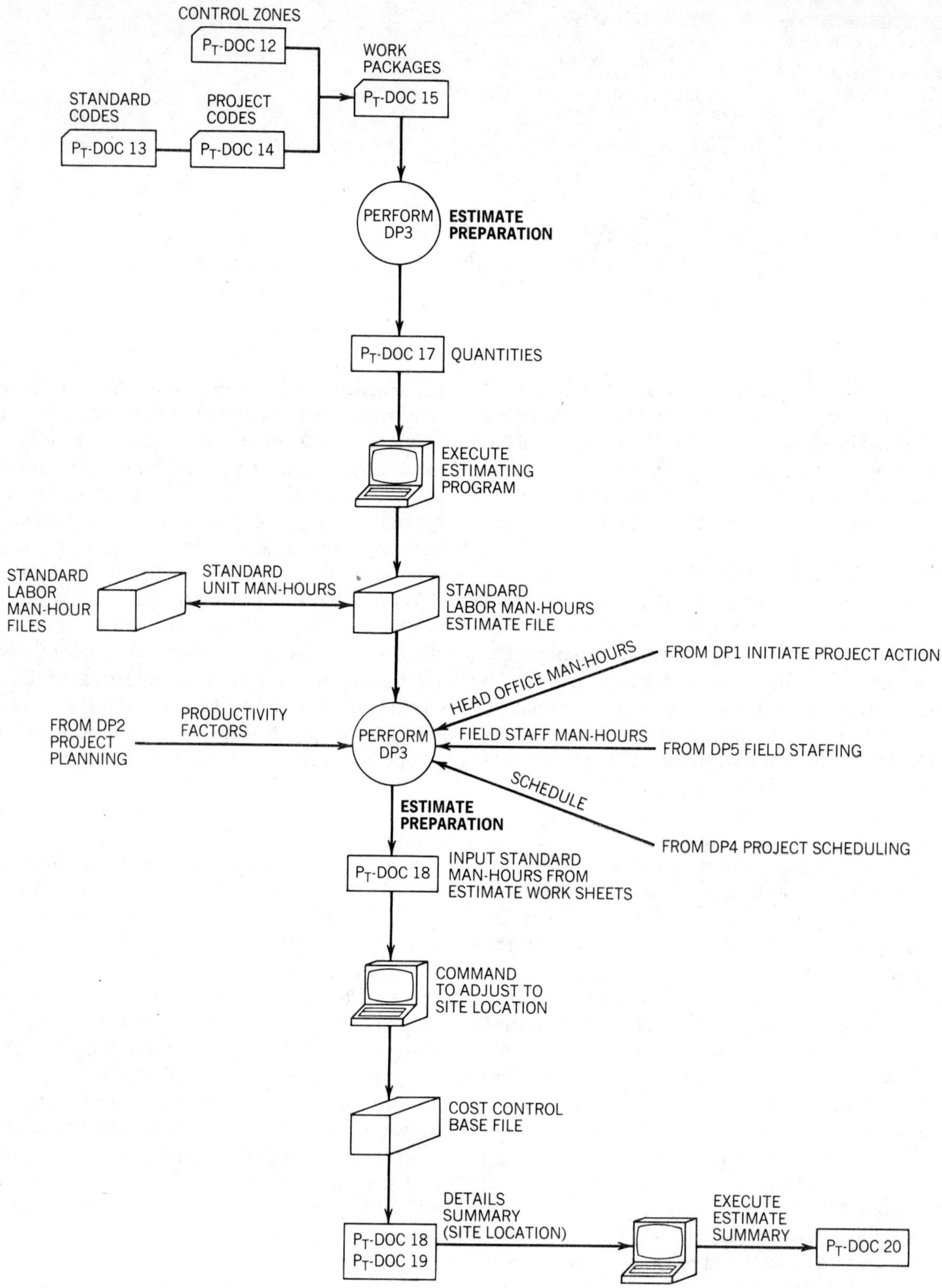

Figure 6.36 Estimate Preparation Management Information System (Total Field Autonomy approach).

DOC 19). Once these documents are generated, client approval of the estimate needs to be obtained. This approval should cover all components of the estimate, especially the head office and field project team budgets. Key estimate data are extracted from the control budget documents and collected together in a single document that essentially establishes a policy statement of the basic project approach adopted by management. This document contains the critical estimate bases, such as the estimate scope, which includes efficiency factors, craft wage rates and mixes, and the like. A schematic illustration of a typical Management Estimate Summary (P_T-DOC 20) is shown in Figure 6.35. The Management Estimate Summary provides a synopsis of the cost estimate and its basis, and leads automatically to the project control base system for construction and project management, from which estimate deviations can be monitored and controlled.

At the Total Field Autonomy level, the Management Information System is similar in focus and format to that for the Field-Based Project Management approach. For example, the labor estimate data is structured in accordance with Project Cost Code and Work Package Structures. However, as demonstrated in Figure 6.36, the MIS leads directly to the establishment of a project Labor Cost Control Base file, rather than beginning with a project labor estimate file which is subsequently loaded into the project cost control file. The MIS still maintains the capability to produce multi-level sorts of different types of estimate data that can be used to generate specialized reports and charts for top level management, such as, for example, the Management Estimate Summary shown in Figure 6.35.

CHAPTER 7
Project Scheduling

A contractor gets involved in project scheduling activities for two important reasons. First, it enables the contractor to coordinate many related activities over time with the procurement, availability, and planned use of resources. Second, it helps to evaluate the feasibility and soundness of a construction plan for meeting contract milestones and completion dates. Thus, scheduling efforts are directed to the coordinated and controlled use of resources. They are intended to minimize delays and reduce the incidence of idle resources, and to promote a smooth unfolding over time of the construction plan. In addition, scheduling provides the contractor with some assurance that the firm's obligation to provide the client/owner with a facility by a specified or agreed upon completion date can be met.

The overall scheduling requirements for a project are established when the various project milestones are identified, located in calendar time (or in terms of project weeks from some date), and tied into specific contract durations. Within this legal time framework, the contractor has to both formulate the construction plan and schedule the work effort so that all contract work will be completed within the allowed construction period. Project scheduling must locate in time the various component activities which comprise the plan for executing the project. This is accomplished by establishing durations for each project activity which are correlated to the availability of resources, work content, anticipated productivities, planned rates of progress, and project priorities. In addition, critical interface milestones between engineering, procurement, and major subcontractor activities must be integrated into the schedule to allow sufficient lead time for the contractor to meet these milestones. The summary schedule of its component activities must then reflect the effective utilization of resources while minimizing the demand on critical resources, and at the same time enable the contractor to meet key project milestones that are consistent with sequences of final testing and commissioning of project units, components, or systems.

The focus of project scheduling as a manpower management problem area is influenced considerably by the project environment. In certain project environments, the small size and relative simplicity of a project can obviate the need for significant scheduling documentation. The contractor can easily grasp the magnitude of the project and visualize the time commitment and manpower resources required to meet its completion dates. The contractor's knowledge of local labor conditions and of the ability of subcontractors in the area ensure that the schedule can be met.

As project complexity and size increases, the contractor must perform detailed scheduling to solve the problems raised by the need to comprehend what has to be done and how to manage it efficiently. The contractor cannot easily grasp the project scope, nor effectively procure special items and meld available resources with programmed activities without proper scheduling documentation. Finally, in the contracting environment where design and construction are being performed in a fast track mode, the contractor may be required to manipulate and schedule activities within a short-range planning horizon in which critical scheduling decisions must be made under management pressure to ensure that large work forces are productive and work activities are being completed in a timely manner.

Project scheduling is a critical management activity. It provides management with a knowledge of the magnitude and timing of the manpower and craft levels that must be committed to a project in order to meet contract and key project milestone dates.

7.1 THE BASIC DECISION PROCESS

Project Scheduling is the recognized decision area that commits the project plan and relevant resources to a calendar time table. It is a critical decision area, because it locks the contractor into a completion date that must be met if the firm is not to jeopardize its reputation and/or the client's return on investment.

In terms of project manpower management, Project Scheduling is concerned with the determination and effective use of project manpower. In this decision area, project manpower is frequently expressed in terms of the number of man-months, equivalent men, man-hour and manpower curves, craft curves and so on. Progress data that reflects the anticipated percentage completion rate will also emerge as the product of Project Scheduling.

In design–construct environments, the scheduled start of construction typically does not occur until forty to fifty percent of design is completed. While Project Scheduling is not directly concerned with manpower at this level, it is the necessary first step prior to evaluating the impact of the manpower resource on the schedule.

When project decision makers approach Project Scheduling, three major problem areas must be considered. First, the decision makers must produce the best schedule in relation to anticipated manpower availability, site productivity, potential bottlenecks within the contractor's own forces, and potential bottlenecks within subcontractors' work forces. Then, based on this best schedule, the decision makers must impact load the schedule with the man-hour content and manning strategies that will be implemented. Finally, these proposed manpower requirements must be converted into manpower profiles and leveled to reflect the availability of manpower and other resources, in order to establish the best use of total project manpower. These major problem areas and their respective component problem statements are given in Table 7.1. They describe in greater detail the scope of Project Scheduling.

When the contractor addresses Project Scheduling, these major problem areas and their component problem statements are generally evaluated in the order presented in Table 7.1. The breadth and detail of the contractor's approach to solving these problems will depend on the construction environment

TABLE 7.1 SCOPE OF PROJECT SCHEDULING PROBLEM AREA

Major Problem Area	Component Problem Statements
17. What are schedule requirements:	What is project execution plan? What levels of scheduling are required to meet this plan? What is desirable compatibility between levels of scheduling and breakdown of construction logic? What are best activity durations in relation to work volumes? Do activity durations reflect estimated labor content? Does overall schedule meet key milestones and client priorities?
18. What is best schedule:	What scheduling flexibility exists for this project? What specialist and craft bottlenecks can be eased by suitable scheduling? What is best project schedule for a given manpower availability, predicted labor productivities, and progress targets? What project requirements impact subcontractor scheduling?
19. What are manpower requirements:	What are implications of man-hour content of project work items? What scheduling and manpower profile strategies should be adopted for this project? What labor productivities are implicit in scheduling of project activities?
20. What is best use of manpower:	What are required craft and manpower levels for a given construction sequence and project milestones? Can these craft and manpower levels be met? How can scheduling craft and manpower conflicts be resolved? For any given planning period, what sequence and scheduling of project activities most effectively uses known labor resources?

TABLE 7.2 GENERAL PROCEDURE FOR PROJECT SCHEDULING

Procedural Step	Management Duties and Tasks
1.	Break down, depict, and analyse project construction logic to detailed activity level required for project scheduling control.
2.	Determine location in time of project activities as established by contract/project milestones.
3.	Evaluate the impact of site manpower availability, productivities, and progress targets in relation to meeting scheduled milestone dates.
4.	Assess impact of critical subcontractor interface points in relation to key milestones and completion dates.
5.	Determine detailed activity man-hour loadings of the construction schedule based on the estimated labor content and desired rates of progress needed to meet completion dates.
6.	Develop craft and total project manpower levels and associated progress curves.
7.	Level manpower resources for optimum use of labor in relation to given construction sequence, site conditions, labor availability, subcontractor interfaces, and project milestones.
8.	Develop short-range schedules which reflect known work scope, labor productivity, availability of manpower, and desired progress targets.

in which it operates, the required detail for the project schedules, and the magnitude of the manpower content of the contractor's typical project.

The management action required to solve the major problem areas described in Table 7.1 can be generally expressed as a sequential order of activities identified as management duties and tasks. The sum of these duties and tasks reflects the general procedure for Project Scheduling and is shown in Table 7.2. This procedure is considered typical of Project Scheduling as found in construction practice. The level of performance of each procedural step will naturally correspond to the constructor's mode of operation, as previously explained.

The performance of the detailed duties and tasks reflected in Table 7.2 is an expression of the decision process for Project Scheduling. Therefore, these duties and tasks form the basis for identifying the relevant manpower management functions which comprise the Project Scheduling decision process. The functions are:

1. CONSTRUCTION SEQUENCE (S1).
2. ESTABLISH CONSTRUCTION SCHEDULE (S2).
3. DETERMINE PROJECT AND TRADE MAN-HOUR PROFILES (S3).
4. DETERMINE PROJECT AND TRADE MANPOWER PROFILES (A3).

The functional plot for Project Scheduling in relation to the other major decision processes is shown in Figure 7.1. A functional statement for each management function involved in the decision process is given in Figure 7.2. The functional statements for each manpower management function generally correspond to one or more of the major problem areas and management duties and tasks, as delineated in Tables 7.1 and 7.2, respectively. Finally, the interface relationships between the Project Scheduling decision process and other relevant decision processes are illustrated in Figure 7.3.

7.2 FUNCTIONAL PERFORMANCE TECHNIQUES

Project Scheduling, for most contractors, is the initial effort required to establish the time frame in which project components will be completed. At this grass-roots level, the decision makers rely heavily upon their expertise, construction sense, and ability to visually comprehend the facility to be constructed and to grasp the resource commitment, especially manpower requirements. In these situations, the dominating objective is to complete the facility on time. A comparison of the project requirements with past or historical practice and experience confirms that this objective is achievable. In later decision-making situations, while timely project completion remains the dominant concern, the solution of ancillary problems such as the need to effectively use restricted manpower levels becomes critical to the determination of whether or not this primary objective can be successfully met. Consequently, decision makers must often assess, with some sort of detailed analysis, the impact of specific manpower levels on the attainment of

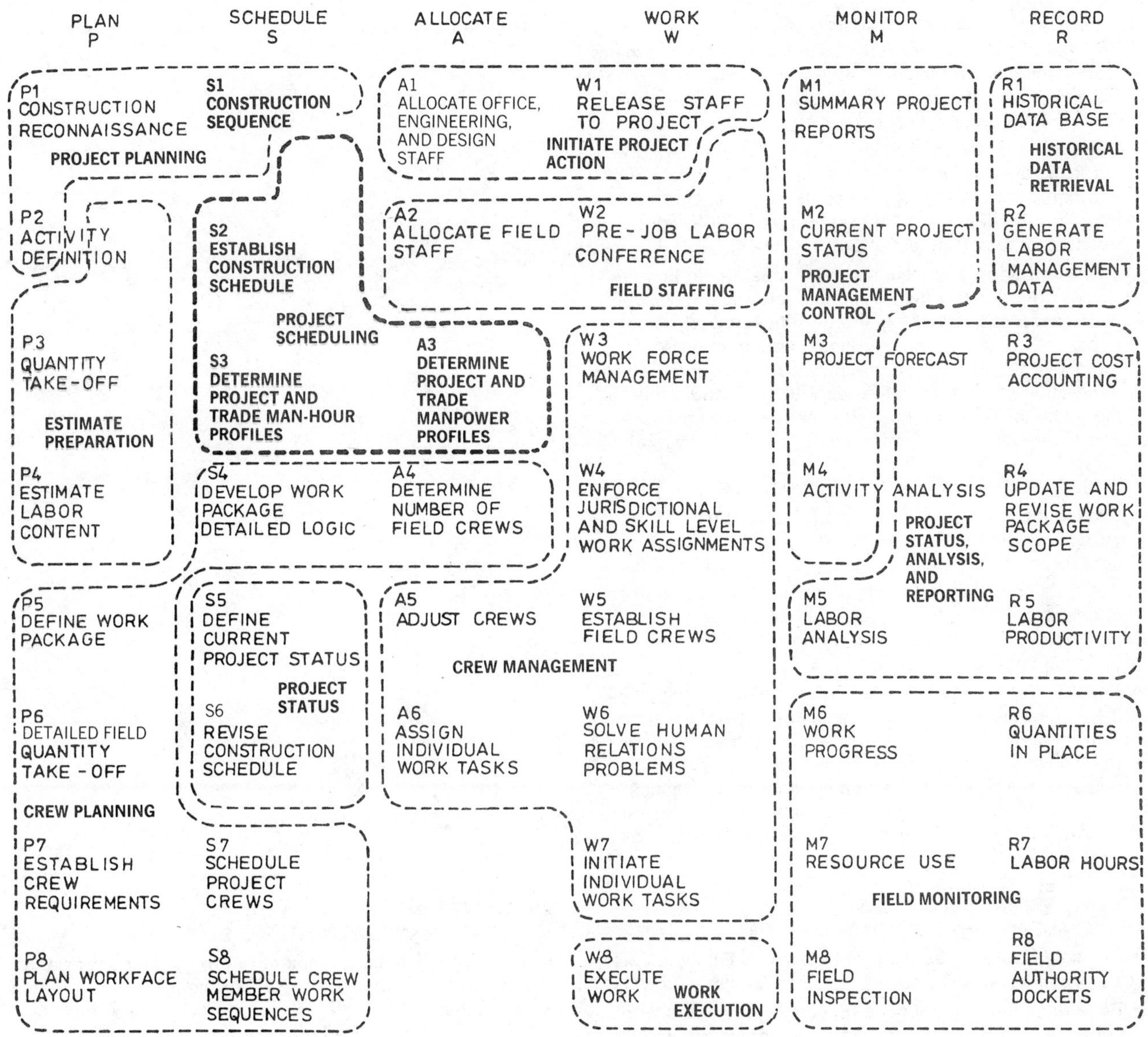

Figure 7.1 Functional plot of the Project Scheduling decision process.

scheduled completion dates. Because of this diversity in decision making during Project Scheduling, this book will introduce general concepts prior to presenting those that lead to specific functional performance techniques found useful in construction practice.

In the Project Planning decision mode, the contractor first defines the work to be accomplished and then the sequence in which it will be accomplished. When the contractor attempts to portray this plan in calendar time, the contractor checks to determine whether or not the volume of work can be accomplished in the available time, and thus verifies that the construction plan is achievable. Once a satisfactory schedule has been developed, the contractor must determine the implications the schedule has on available resources, including manpower, required to meet completion dates. This evaluation then leads to resource leveling, the objective being to optimize the effective use of resources. This may result in adjustments to the construction plan and its associated schedule.

Project Scheduling further provides the framework for managing and controlling the construction effort (including engineering, design, and procurement as applicable). This is manifested in control

S1 CONSTRUCTION SEQUENCE

- The identification of desirable sequences of major project work packages and the definition of the basic technological sequences of engineering, procurement, and field effort for each project activity.
- Requires knowledge of project scope, activity breakdown, and estimate details on work content and material, labor and equipment resources, subcontract items, and the impact on project and company resources of different sequences of major project efforts.
- Considers priority of facilities to be constructed in relation to contract requirements and decisions on availability of resources, drawings, and priorities of work effort, impact of construction methods on construction sequence logic; focuses on magnitude of effort required for integration of material deliveries with labor and equipment use to support proposed construction sequences and the impact of subcontractors (if prime contractor).

S2 ESTABLISH CONSTRUCTION SCHEDULE

- The location in time of planned project activity and its correlation with the availability of resources, activity work content, planned rate of progress, and project priorities.
- Requires intimate knowledge of planned construction sequences, activity work content, subcontract activities, realistic progress targets, and proposed construction methods and work durations.
- Considers the milestone dates for completion of key facilities; integrates construction logic and activity durations within the milestone time periods; correlates activity durations with work content, progress targets, size of proposed work force, and anticipated labor productivities based upon defined workweek and shift policies and subcontract activities; and establishes work sequences for float activities.

S3 DETERMINE PROJECT AND TRADE MAN-HOUR PROFILES

- The continuous assessment of best suited field work size and trade composition to meet planned level of field work.
- Requires knowledge of construction schedule and work content of project activities, and whether work is to be performed by union or nonunion labor, subcontractor (if prime contractor), and corresponding work classifications by trades and skill levels.
- Considers planned rate of work, desirability of maintaining high level of productive effort by matching planned work loads with work force size, and the need to minimize fluctuation in work force size; considers availability of labor in hiring halls and desirability of keeping highly skilled and motivated labor.

A3 DETERMINE PROJECT AND TRADE MANPOWER PROFILES

- The continuous establishment of project manpower and trade profiles to be mobilized over the life of the project. Establishes a fix on the project manpower and trade profiles to be mobilized over the life of the project.
- Requires a knowledge of project work requirements, labor man-hour estimates, and schedule of work areas requiring special skill levels and trade classifications.
- Considers man-hour estimates, availability of labor, training programs, necessity for establishing work camps, and correlation of project manpower size and fluctuations with the construction sequence, schedule, and anticipated work progress.

Figure 7.2 Functional statement of the Project Scheduling decision process.

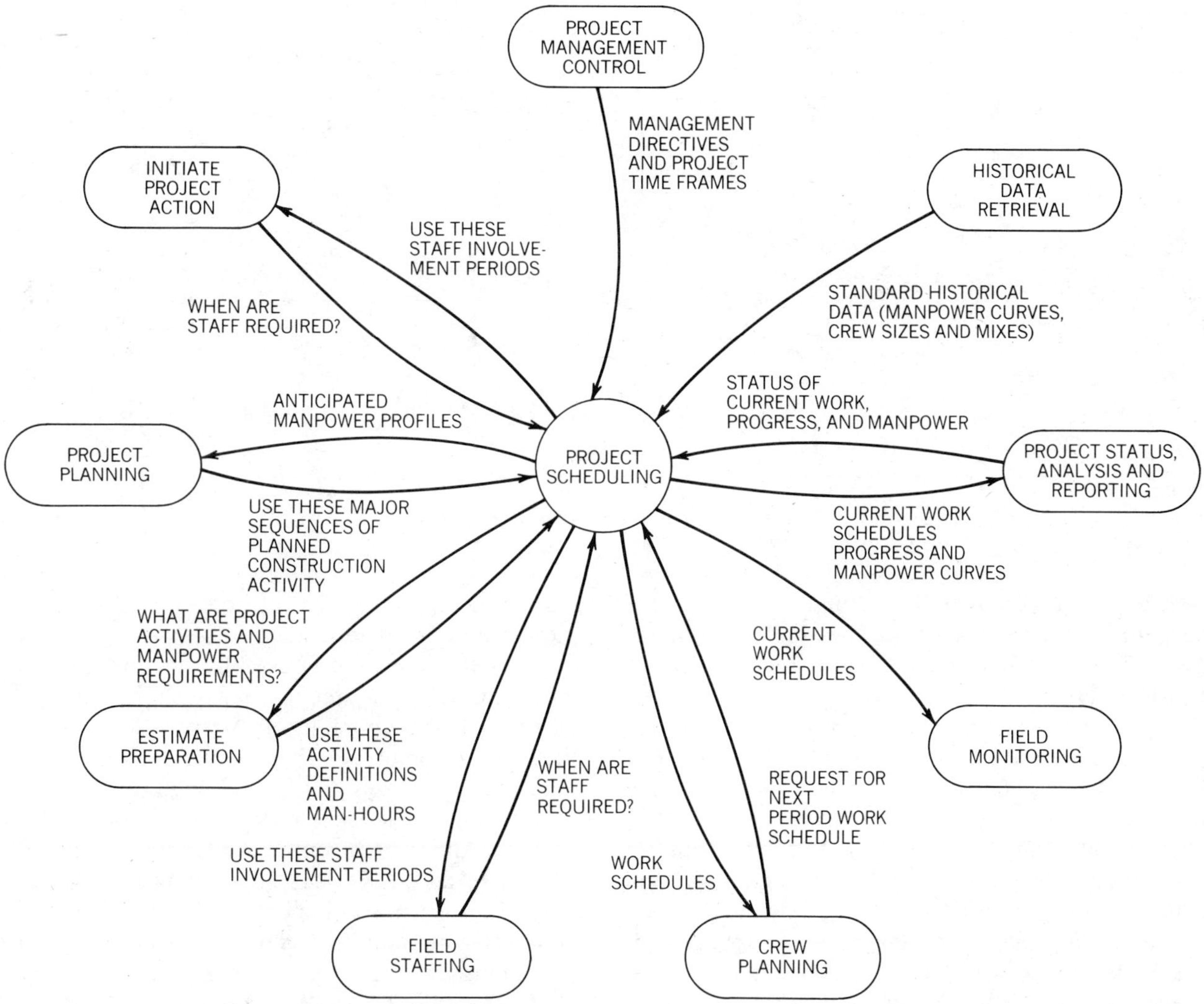

Figure 7.3 Schematic illustration of Project Scheduling decision process interfaces with other manpower management decision processes.

procedures, and in the ability of the contractor to coordinate design, procurement, and construction efforts, the acquisition of project equipment, and the effective use of subcontractors. Furthermore, in some construction environments the ability to schedule at various levels of detail provides management with multiple decision entry points, ranging from the greatest detail to the least detail, relevant to each scheduling problem being addressed. This approach is successful provided both the breakdown and/or roll-up activities are applied to similar project components (e.g., detailed steel erection activities, such as erect columns, bolt-up, weld, and so on, should roll up to a single activity; building steel).

The most basic concept in Project Scheduling is the identification in time of physical blocks of work by statements of start dates and/or completion dates, or of critical milestones such as delivery and subcontractor interface points. These statements could be verbal, or could appear as a simple written listing by major project activity. For example, a contractor may note that steel deliveries must be completed by March 26, 1986 if building framing is to be finished by April 23, 1986. An alternate technique is to list physical components and subactivities in a table identifying start dates, subcontractor requirements, and activities performed by the contractor's own crews. This technique is illustrated graphically in Figure 7.4. In this figure, the

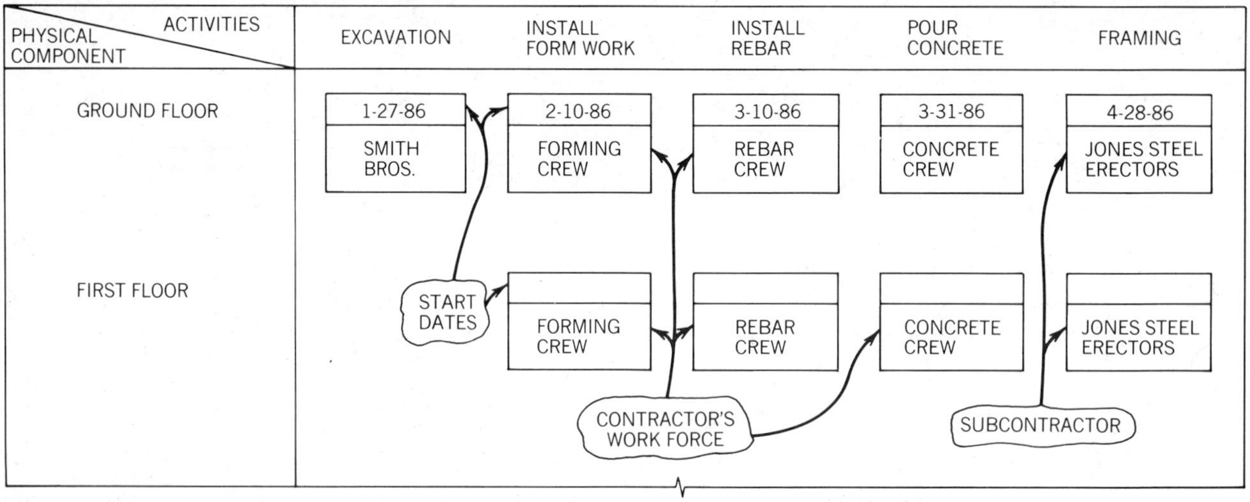

Figure 7.4 Tabular type schedule.

contractor must have the ground floor concrete poured by March 31, 1986 for the steel subcontractor "Jones Steel Erectors" to start steel framing on April 28, 1986.

In the construction industry, construction schedules are frequently based on the Time Scaled Bar Chart technique (Gantt Bar Charts). This technique reflects the concept of portraying physical sequences of major components, activities, and/or tasks over a time period (days, weeks, months), and is illustrated in Figure 7.5 It should be emphasized that the time scaled bar chart does not reflect logical ties between activities but only apparent relationships in terms of starting and completing each activity.

When network techniques form the basis for construction planning, project schedules depict the logical ties between construction activities in relation to time. This concept is illustrated in Figure 7.6. Time-scaled network scheduling can be applied to any level of detail, as dictated by project control procedures and the different management decision-making focuses the contractor may anticipate during a project. For example, time-scaled networks could be generated for specific construction control areas on a project, rather than for the project as a whole. This approach is acceptable provided the area schedules can be compressed into a project schedule with less detail for top management monitoring and control. Alternatively, a particularly critical project component could be scheduled by time-scaled network technique down to the task level of detail, such as, for example, a critical compressor station on a pipeline project, to ensure that its completion date is achieved.

After the schedule is established, the impact of the manpower resource is assessed. The concept of utilizing the components of the schedule, whether they be physical blocks, milestone activities, or detailed activities and tasks, applying a form of manpower content to these components, and summing up the manpower resources over time, emerges at this stage in the Project Scheduling decision process. In the most elementary form, crews (i.e., men) are estimated for each activity on a bar chart, and summarized for a total project manpower profile as shown in Figure 7.7. Other methods might be used on time-scaled networks. For example, the estimated labor man-hours from the labor cost estimate forms the basis for evaluating activity durations, as illustrated in Figure 7.8*a*. The man-hours are spread over the activity's duration and then summarized into a man-hour expenditure curve. The man-hour expenditure curves, both incremen-

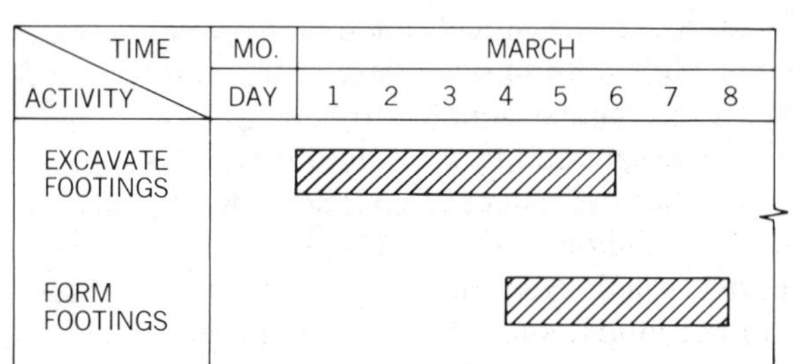

Figure 7.5 Bar chart type schedule.

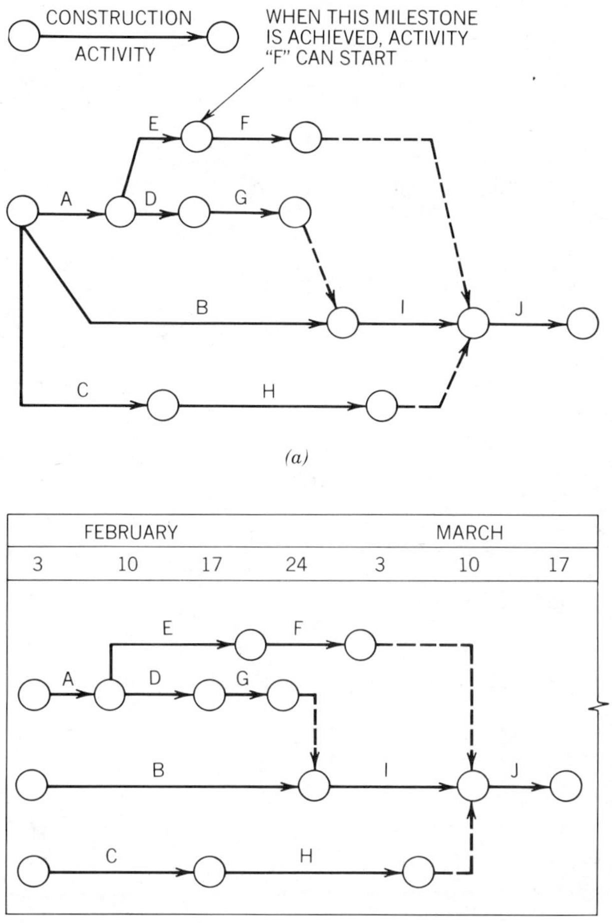

Figure 7.6 Time-scaled CPM schedule. (*a*) Logic network. (*b*) Time-scaled bar chart.

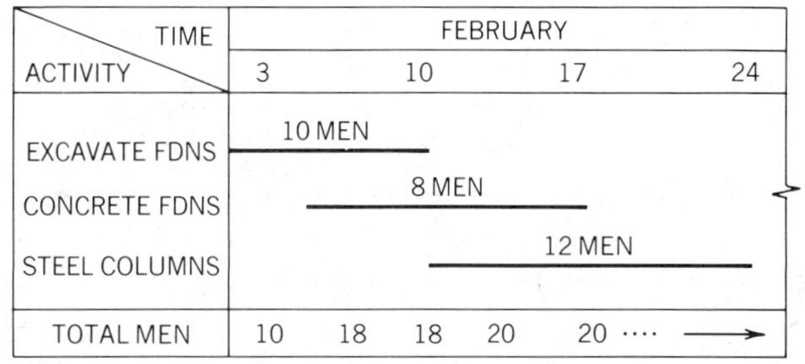

Figure 7.7 Development of manpower profile.

tal and cumulative, can be converted into a percent progress curve. If this operation is performed based on both the early start of each activity and the late start of each activity a percent progress envelope is derived, as depicted in Figure 7.8*b*. The planned progress curve should fall between the two curves as further illustrated in Figure 7.8*b*. Once the planned progress curve is determined, manpower profiles can be developed from the progress curve and the total estimated man-hours. Figure 7.9 reflects this method of arriving at manpower profiles. The manpower profile techniques presented in Figures 7.8 and 7.9 are particularly effective with networks having a significant number of activities (e.g., 500 to 1500), where activity man-hour spread averages produce a realistic progress curve through anticipated percent of man-hours expended. The approach presented in Figures 7.8 and 7.9 is intended to illustrate the concept of manpower profiles but is not the only method available to solve this problem. A limited application of these techniques could be useful on small networks.

The technique shown in Figure 7.9 is frequently employed during the early phases of a project, when only total man-hours from factored estimates are available and overall project level schedules have been developed. In this case, the planned progress curve would be a standard progress curve from the contractor's historical database files. Naturally, the project must be similar in scope to those in the contractor's history of projects for satisfactory manpower profiles to be produced. Manpower profiles developed at this phase in the project are important for evaluating the impact of site labor availability, potential camp and training requirements, field staff profiles, and so on.

On larger projects, man-hour and manpower applications frequently match control philosophies, such as control areas or major cost code categories, in which several control segments of the project must be summarized to reflect the total project manpower profiles. This concept is graphically demonstrated in Figure 7.10. The manpower profile for the chosen control segment could be uniquely determined using techniques similar to those presented in Figures 7.8 and 7.9.

Generally, initial manpower profiles rarely match the true needs of a project. Because of this, the contractor must "massage" schedules and related manpower profiles to meet anticipated site labor availability, to provide for a smooth growth and decline of manpower (thus avoiding excessive hiring, firing, and rehiring over short periods of time), and to achieve desired progress rates. Figure 7.11 is one example of a manpower profile where peaks and valleys are leveled to provide a smooth buildup to a desired level, and a smooth decline as the project reaches completion. In construction practice, man-

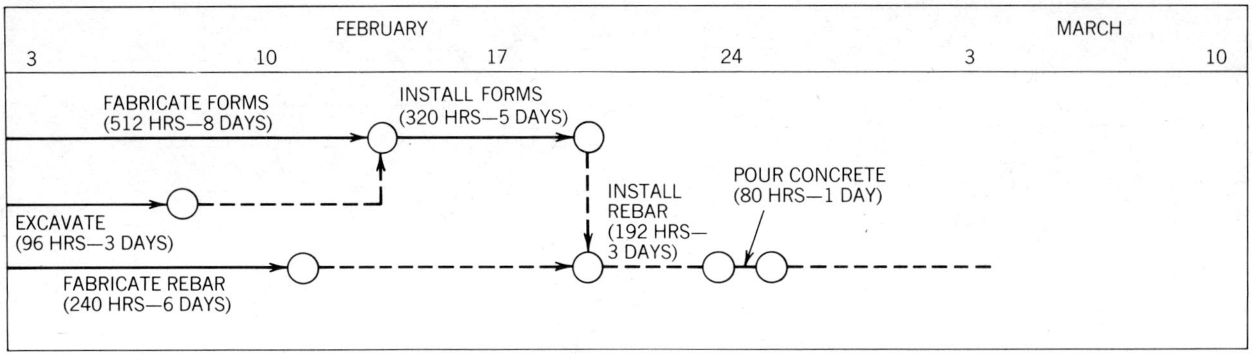

(a)

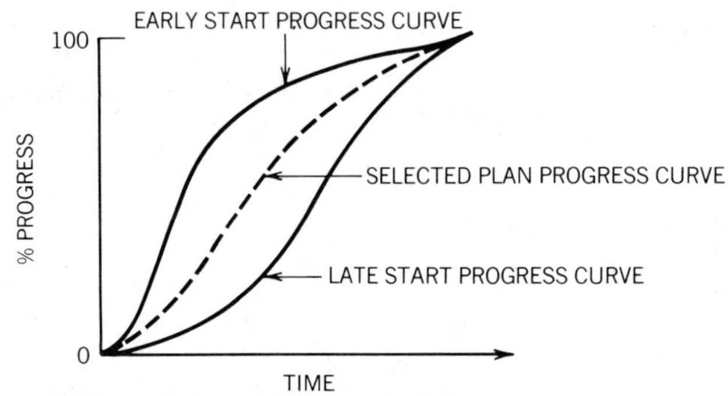

(b)

Figure 7.8 Progress curve development. (*a*) Weighted time-scaled network. (*b*) Progress curve based on man-hour weighted time-scaled network.

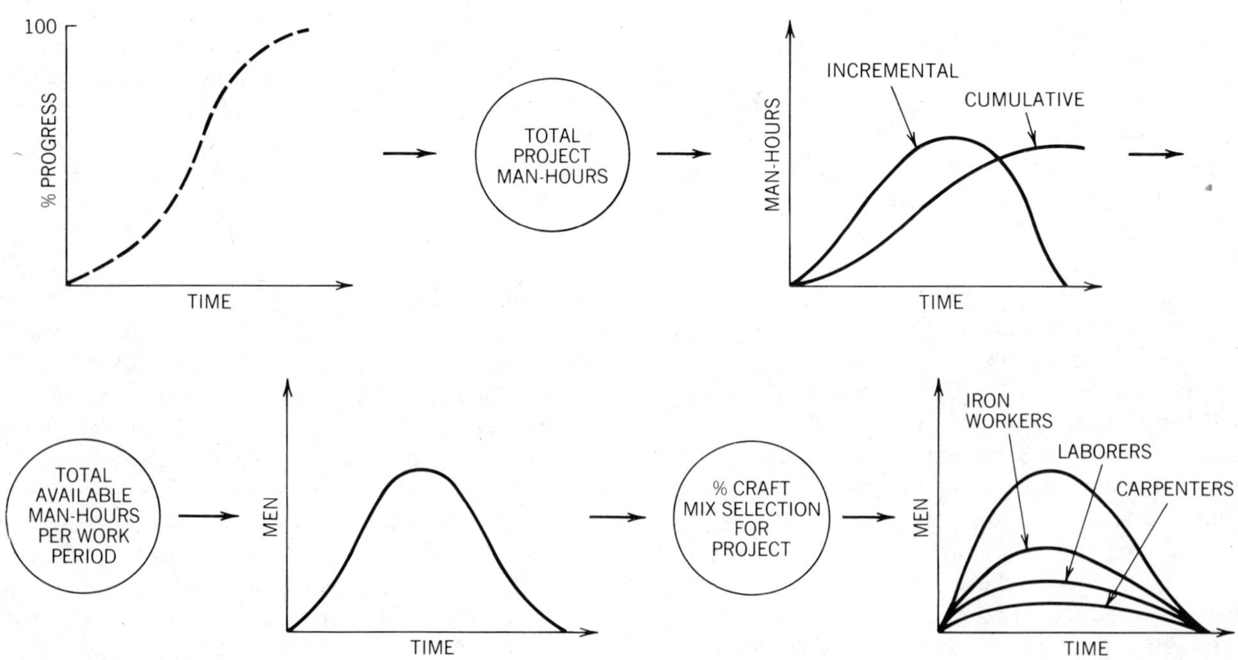

Figure 7.9 Man-hour and manpower profiles.

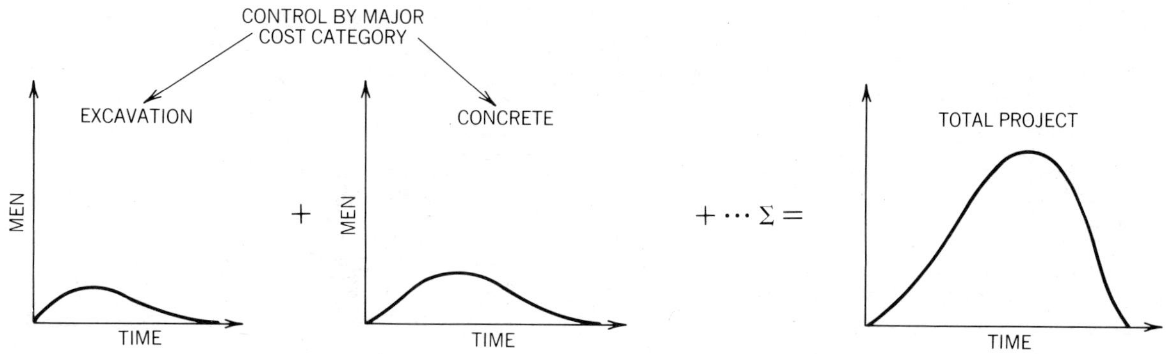

Figure 7.10 Total project manpower profile based on control philosophy.

power leveling can result in significant changes in both the plan and schedule, or require the utilization of float to adjust the location in time of activities. It can also be achieved by manipulating the manpower curve, provided the relevant progress curve falls within the scheduled early start/late start progress envelope (see Figure 7.8*b*).

On large and complex projects involving hundreds of field craft personnel, a critical problem is to ensure that project control level schedules are properly implemented and that they support Crew Planning. Since project control level schedules focus on the total project, it becomes necessary to prepare schedules for short-term planning horizons such as three to six months. The short-range schedule must contain enough work to at least meet or surpass the planned progress and planned manpower for the period considered. A short-range schedule technique emerges as a result of this concept, whereby each activity scheduled during the planning period is scrutinized in terms of its contribution to achievable percent progress and subsequent requirements (including other applicable resources such as construction equipment, materials, and the like). Figure 7.12*a* illustrates this concept in terms of a quarterly schedule where the first month contributes 4% progress and requires thirty men over the entire month. The first month's work schedule can be expanded into a listing of activities that will commence during that month. This monthly activity list, Figure 7.12*b*, is input to field supervisors for crew planning on a weekly or biweekly basis.

Another useful concept for short-range planning is to schedule more work (10–20% more) than is required to meet the progress plan for a period. This protects the field manager/supervisor from potential unanticipated activity work delays, whether these are due to material slippages or design problems, by arranging other work to fill the gaps. This approach also provides a forward look at activities available beyond the shortest planning time frame.

Project Scheduling focuses on three interacting objectives: the need to keep track of all facility components to be built, together with the detailed activities to produce these components and their logical sequences; the identification of time slots in which items will be built; and the enumeration of the resulting manpower levels required for their construction. Project Scheduling requires various data inputs in order to achieve these objectives. They include project control philosophies, labor estimate details (in addition to material and drawing delivery dates), and subcontractor interfaces, just to name a few. A Project Scheduling Management Information System must then integrate the

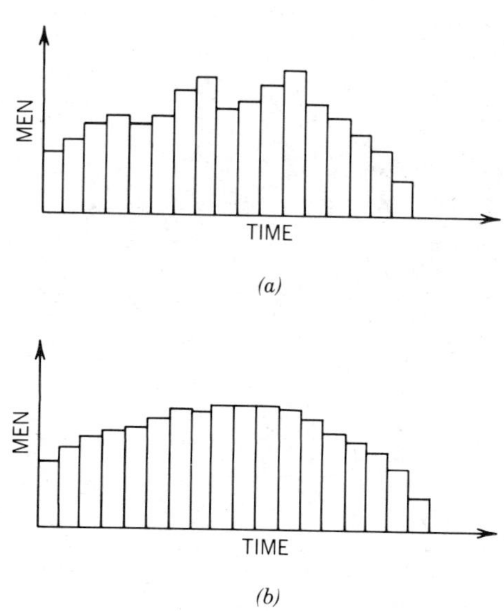

Figure 7.11 Manpower leveling. (*a*) Peaks and valleys. (*b*) Smooth buildup and decline.

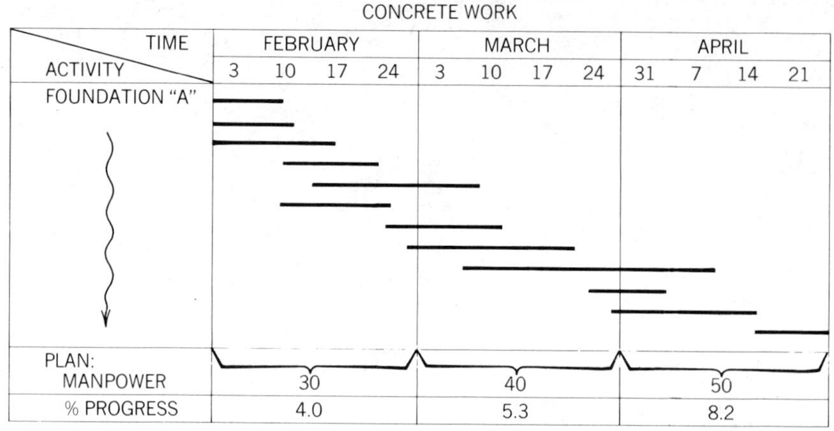

(a)

MONTH: FEBRUARY 1986

ACTIVITY	% COMPL	START		COMPLETE		MEN	PROGRESS UNITS
		PLAN	ACTUAL	PLAN	ACTUAL		
FOUNDATION A	0.0	FEB 3		FEB 7		12	50.0

(b)

Figure 7.12 Short-range planning tools. (*a*) Quarterly schedule. (*b*) Monthly activity list.

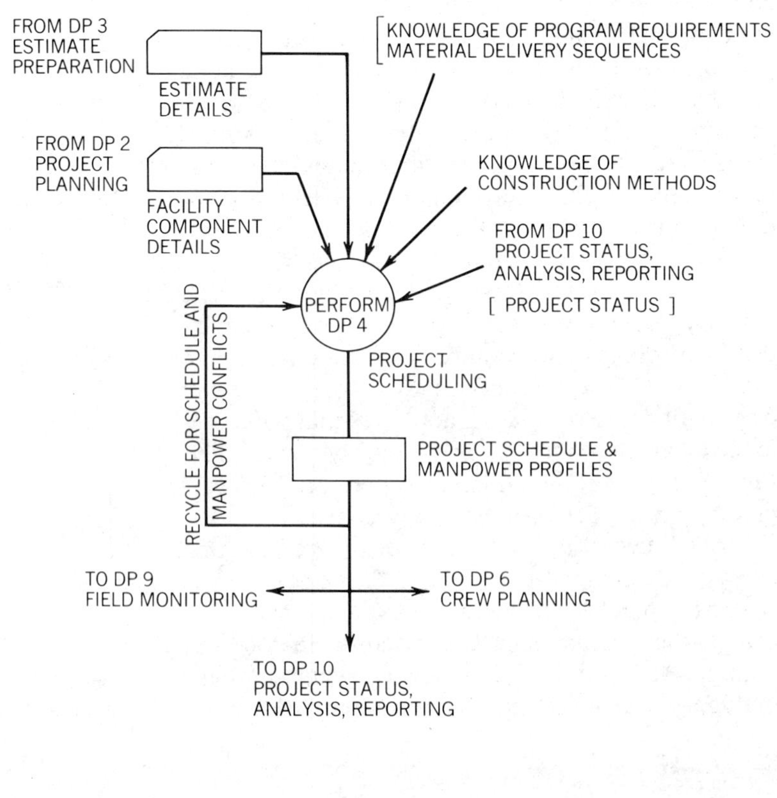

Figure 7.13 Project Scheduling Management Information System concept.

achievement of these objectives into realistic project control base schedules and manpower profiles against which the project's status and future direction will be measured.

The realization that schedule overruns on resource consuming activities, together with the enforced delays of inactive resources consuming activity float, imply additional costs is a basic concept for schedule control. This relationship is indicative of the nature of the interaction between planning the type of work and base productivity and estimating the work content and cost with the need to focus field monitoring attention and project status, analysis, and reporting in any coordinated schedule and cost control effort (see, for example, Chapter 16). The Project Scheduling MIS concept is illustrated in Figure 7.13. It is more frequently applied formally at the large firm levels, although small and medium-sized firms produce Project Schedule documents. In general, the MIS structure at the small and medium-sized firm levels remains relatively informal and is handled routinely by one or two key project team members (e.g., project manager and/or field superintendent), based on previous experience and "tried and true" construction techniques.

7.3 THE PROJECT TEAM COMPONENT APPROACH

At the small firm level, Project Scheduling effort is generally commenced on notification of a bid award. Its basic purpose is to portray (by work category areas) the various time involvements of effort necessary to complete the project within the contract period. Project Scheduling establishes both the feasibility of meeting the contract finish date and the consequent magnitude of required field effort. In this way, Project Scheduling establishes the work guidelines for field construction. Rarely does scheduling focus on, or raise, significant manpower management problems.*

The time frame available to the project manager for the scheduling effort is seldom sufficient to allow the performance of an in-depth analysis of the construction sequence. This is because, on the one hand, the uncertainty of the bidding process curtails any significant scheduling effort until the project is awarded, and on the other hand, because construction usually commences shortly after bid award. Thus, scheduling is often approached within a planning horizon just one step ahead of the project. The basic information available for scheduling is the project scope, estimate summary, estimate recapitulations, plans and specifications, and a knowledge of site details. These documents and data are sufficient to identify the basic activities and to manipulate them into a project construction sequence. The schedule visually depicts when, on a time scale, specific activities are to be performed, their construction durations, and their gross relationships with other activities. It does not represent the construction plan for accomplishing the work at the field level, since it only focuses on broad categories of activities associated with the physical components of the project, and excludes fine detail on field work sequences and support activities. Thus the schedule is only a management tool for coordinating project activity, monitoring progress, and identifying when subcontractors are required onsite. The construction schedule is usually developed in *BAR CHART FORM* (P_S 6)†, because project complexity rarely warrants more detailed construction planning techniques.

The scope of work defined in the estimate is restructured by the project manager into more encompassing types of activities. For example, the project estimate may segregate concrete foundations by wall footings, column footings, and foundation walls. In bar chart scheduling, these components might be combined into a single activity such as concrete foundations. However the construction activities are defined, a one-to-one correlation between activities on the schedule and on the estimate breakout is difficult to achieve. The construction sequence establishes the activity breakout and ordering. The activity list is not formally recorded because it is generally similar for most small projects. In any case, an intimate knowledge exists of the activities involved and their priorities. Thus, the project schedule is generally limited to less than thirty activities, and the project manager relies heavily on past experience and common sense to define and interpret the construction sequence. THE CONSTRUCTION SCHEDULE IS ESTAB-

*Thus the DETERMINATION OF PROJECT AND TRADE MAN-HOUR (S3) and MANPOWER PROFILES (A3) manpower management functions are not performed at the small firm level.

†The small firm policies relevant to the scheduling decision process are collected together in Table 7.3.

TABLE 7.3 POLICIES OF THE SMALL FIRM

Policy		Description
P_S 6	*USE BAR CHARTS*	The bar chart planning model is the approach best suited for project scheduling.
P_S 7	*SCHEDULING FOCUS*	Scheduling definitions and constraints are: (1) critical activity is any activity that opens up work areas for other crews and whose delay causes work volume bottlenecks. (2) Activity durations and sequence logic incorporate the impact of seasonal weather conditions. (3) Activity scheduling is based on the earliest possible start dates. (4) Manpower impact is neglected.

LISHED (S2) after determining the CONSTRUCTION SEQUENCE (S1) logic. The major factor influencing the construction schedule is the project completion date. This date is fixed by the owner and set forth in the contract between the owner or his agent and the contractor.

The work duration for each activity is estimated in conjunction with scheduling the activity. In this manner the activity is scheduled visually and within the overriding time constraints that are applied to the total project's construction life. The project manager is aware of subcontract considerations, material content and procurement constraints, and sequence logic, all of which influence activity start time and durations. For example, activity durations are formally developed in terms of material content versus scheduled project completion, and are independent of labor unit costs. The project manager knows from past experience the standard production rate for performing various activities, and can approximate activity durations from the estimated quantities available by dividing the quantities by the desired production rate. The only problem is the lack of a direct correlation between the activities on the schedule and the estimated quantities. This tends to lead to an ''eye-ball'' technique of plotting activities against a time scale, until a suitable schedule is obtained.

Several other factors are considered when determining the overall construction schedule. For instance, while the Notice-to-Proceed establishes the legal right to commence work, the actual start-up of work will depend upon the availability and expedited release of a specific superintendent/foreman. This, together with quick contact with the local hiring halls and/or availability of suitable tradesmen, constitutes the basic focus of the Field Staffing decision process, which will have an impact on the start date for the construction schedule.

As far as the actual scheduling process is concerned, a typical policy is to have the schedule reflect significant leeway at the end of the project, anticipating that some activities will consume more time than estimated. This *SCHEDULING FOCUS* (P_S 6) is based on commencing each activity at its earliest start time and reducing activity durations where feasible. Durations of critical activities are consistently reduced on the assumption that in the field, sufficient labor resources will be made available to ensure meeting the deadline. This is because the completion of critical activities directly influences the rate at which succeeding work becomes available. In this context, many activities are critical activities not in the CPM sense, but because their completion opens up work areas and work volume. Because the project life cycle of the small project is often short (less than a year), seasonal considerations are of paramount importance. Thus, projects are scheduled so that they are either completed prior to periods of inclement weather or so that a structure (if applicable) is completely enclosed should inclement weather occur. Finally, the project manager is rarely interested in correlating manpower requirements to the project schedule, mainly because the overall work force required for the project is rarely large enough to warrant formal planning of project manpower needs. Only when unusual or extremely unfamiliar work tasks are encountered is project manpower analyzed and tied into the project schedule during the scheduling phase.

THE ESTABLISHMENT OF THE CONSTRUCTION SCHEDULE generates the Bar Chart Construction Schedule (I_S-DOC 2), which is illustrated in Figure 7.14. It should be emphasized that the small contractor employing the bar chart model only schedules the work in gross detail and rarely describes the plan by which the activities on the schedule will be implemented. The bar chart does not show logical relationships between activities and is rarely used to delineate manpower needs over the life of the project. However, it is an effec-

Anderson & Woodhead
Building Contractors

PROJECT: ______ JOB NO.: ______

ARCHITECT: ______ STATUS DATE: ______

LOCATION: ______ DATE: ______

PREPARED BY: ______

CONSTRUCTION SCHEDULE

WORK CLASSIFICATIONS	P/A	STATUS		1986								PROGRESS	
		START DATES	COMPLETE DATES	FEB.	MAR.	APR.	MAY	JUN.	JUL.	AUG.	SEP.	% COMPLETE	50
SITE WORK	P/A	2-10/	2-26/										
EXCAVATION	P/A	2-17/	3-17/										
BACKFILL	P/A	3-12/	4-9/										
CONCRETE FOOTINGS	P/A	2-28/	3-19/										
SLABS	P/A	3-14/	4-2/										
COLUMNS	P/A	3-17/	4-22/										

LEGEND:
P = PLAN DATES
A = ACTUAL DATES
SCHEDULED ACTIVITY DURATION
ACTUAL STATUS
ACTIVITY DURATION EXTENSION

REMARKS:

Figure 7.14 Bar Chart Construction Schedule (I_S-DOC 2).

tive tool for scheduling and subsequently monitoring overall project progress for most firms engaged in construction in the small firm environment.

The Project Team Component (PTC) Project Scheduling decision process is illustrated in Figure 7.15.

7.4 THE TRAVELING PROJECT MANAGER

At the medium-sized firm level construction scheduling is normally only informally considered prior to contract award and during Estimate Preparation. The construction schedule is evaluated on an overview basis as labor unit costs are developed. This provides the firm with a means of evaluating the potential to meet the client's desired completion date. Depending on the nature, complexity, and size of the project, however, a more rigorous scheduling effort might occur before contract award. Thus, project size and complexity could influence the timing of the Project Scheduling decision process. In comparison with the small firm, the basic purpose of this decision process at the medium-sized firm level goes beyond simply portraying (by work category areas) the various time involvements of company effort needed to complete the job within the contract time period. Since *MANPOWER PLANNING* (P_M 13)* is considered critical to project control, the construction schedule is developed to best utilize the resources of time and manpower. Thus Project Scheduling not only establishes the feasibility of meeting the contract finish date, but also defines the approach for achieving project completion by determining the magnitude of the required field effort. In this way, Project Scheduling establishes the work force guidelines.

The project manager works closely with the key job superintendent when developing the project schedule. This interface is critical for the establishment of the TPM management team providing the job superintendent with adequate schedule information to properly plan his crew requirements. The job superintendent incorporates practical expertise and

*The medium-sized firm policies considered relevant to the Project Scheduling decision process are presented in Table 7.4.

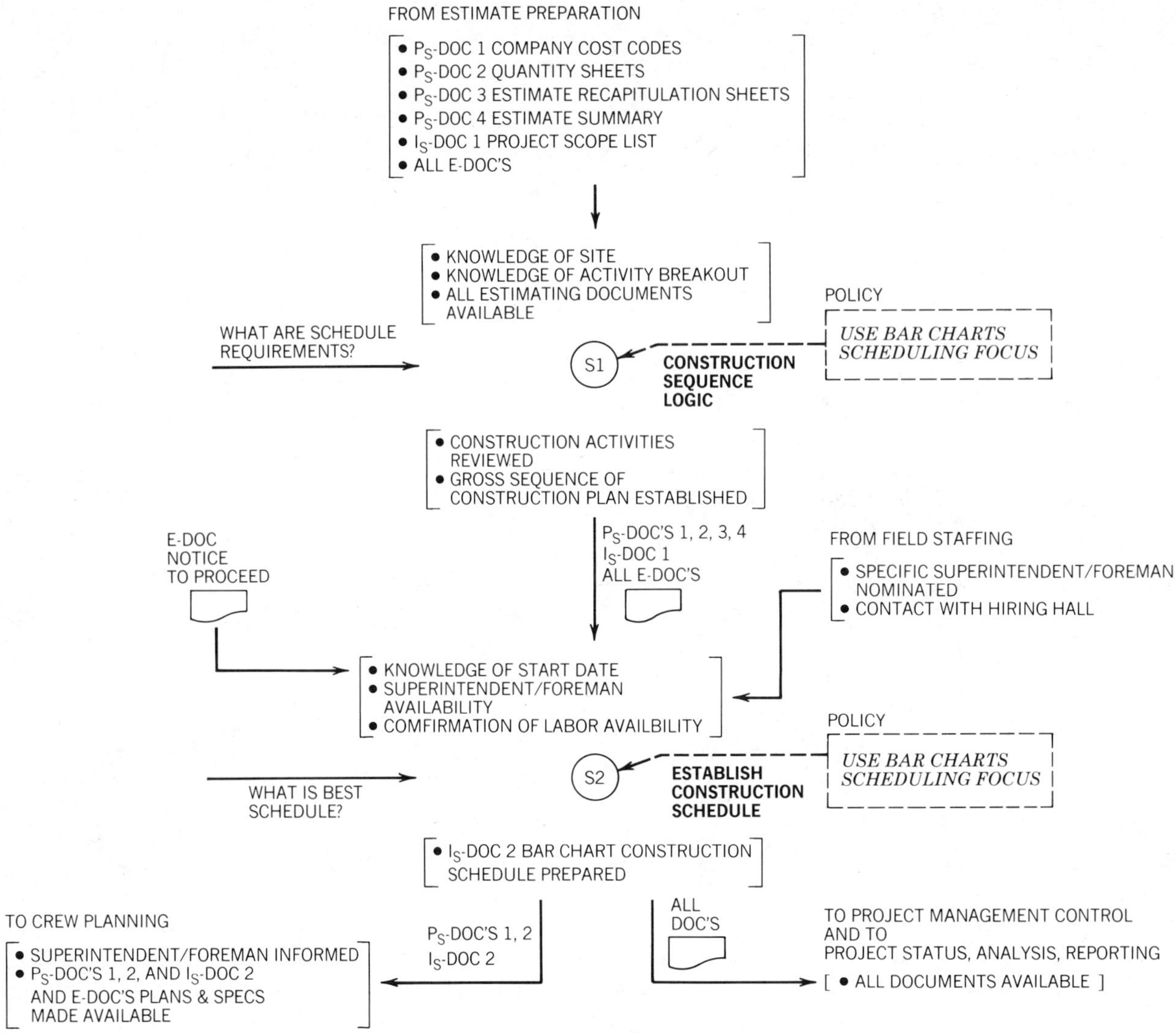

Figure 7.15 Project Scheduling decision process (Project Team Component approach).

knowledge of *UNION RELATIONS* (P_M 11), if applicable, into project scheduling, and the result is a viable construction schedule acceptable to both the field and head office. In conjunction with construction scheduling, the project manager and job superintendent make rough man-hour estimates in order to determine activity durations and short-range manpower levels. In this manner, short range work loads are identified and used to coordinate all facets of on-site activity during field construction, coordination of subcontractors, identification of key material deliveries, and project control, but are also used as a means of anticipating required field manpower levels and reporting progress. This implies that gross activity definition by major categories of work is sufficient for overall project management control purposes, but that a more detailed schedule and manpower analysis will assist the job superintendent in crew planning and the project manager in evaluating project status.

Project scheduling begins with the definition of the CONSTRUCTION SEQUENCE LOGIC (S1). The analysis of the construction activities and their sequence here is similar to the approach of the small firm. Accordingly, the *BAR CHART* (P_M 7) planning model is the common planning tool, but critical path methods are often introduced on a limited basis for suitable projects. If a CPM model is developed, it is viewed mainly as a planning tool to define the construction sequence and identify logical relationships

TABLE 7.4 POLICIES OF THE MEDIUM-SIZED FIRM

Policy		Description
P_M 7	*BAR CHARTS*	Bar chart model and limited application of the critical path method are best suited for project scheduling.
P_M 8	*SCHEDULING FOCUS*	Scheduling definitions and constraints are: (1) Critical activities are highly sensitive profit and loss activities and/or activities controlling the rate at which ensuing work can commence. (2) Activity durations and sequence logic incorporate the impact of seasonal weather conditions and restrictive work practices. (3) Manpower impact is considered on a conceptual basis. (4) Activity scheduling reflects the best sequencing of activities possible in order to effectively utilize all resources available.
P_M 11	*UNION RELATIONS*	In a union construction environment, maintaining good relations with each union, through the business agent, is helpful to successful project management.
P_M 13	*MANPOWER PLANNING*	Short-range planning of manpower at the head office pre-job level, coupled with field labor manipulation, is critical to project control.

between activities. It is rarely used for resource allocation or analyzing time/cost implications. The logic-based Network Plan (I_M-DOC 1) shown in Figure 7.16 could be either an *activity on node* or an *activity on arrow* application of the CPM approach and would probably be developed to focus on the interactive scheduling of survey, excavation, and forming crews or subcontractors once calendar dates are superimposed.

The logic-based Network Plan shown in Figure 7.16 is converted into a bar chart schedule by displaying the start and finish dates of each activity on a calendar time scale. By ESTABLISHING THE CONSTRUCTION SCHEDULE (S2) in this

IDENTIFIER	WORK DESCRIPTION
1	MOVE IN
2	ROUGH GRADE SITE
3	SURVEY-EXCAVATION
4	EXCAV-FTGS BAY A-C
5	EXCAV-FTGS BAY D-G
6	EXCAV-SEWER
7	FORM-FTGS A-C
8	FORM-FTGS D-G
9	PLACE SEWER

(a)

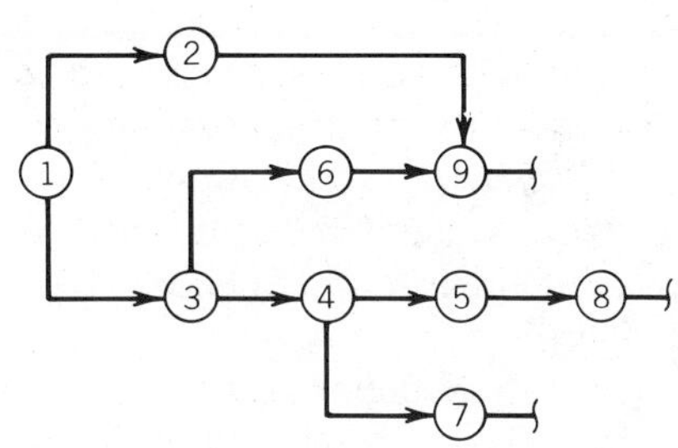

(b)

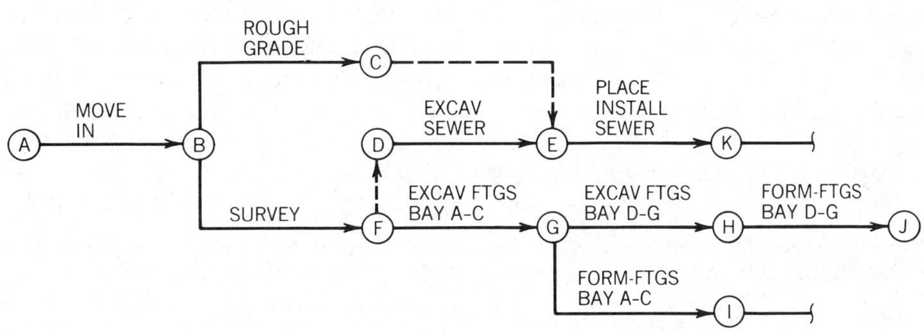

(c)

Figure 7.16 Network Plan (I_M-DOC 1). (*a*) Activity listing. (*b*) Precedence logic network. (*c*) Arrow logic network.

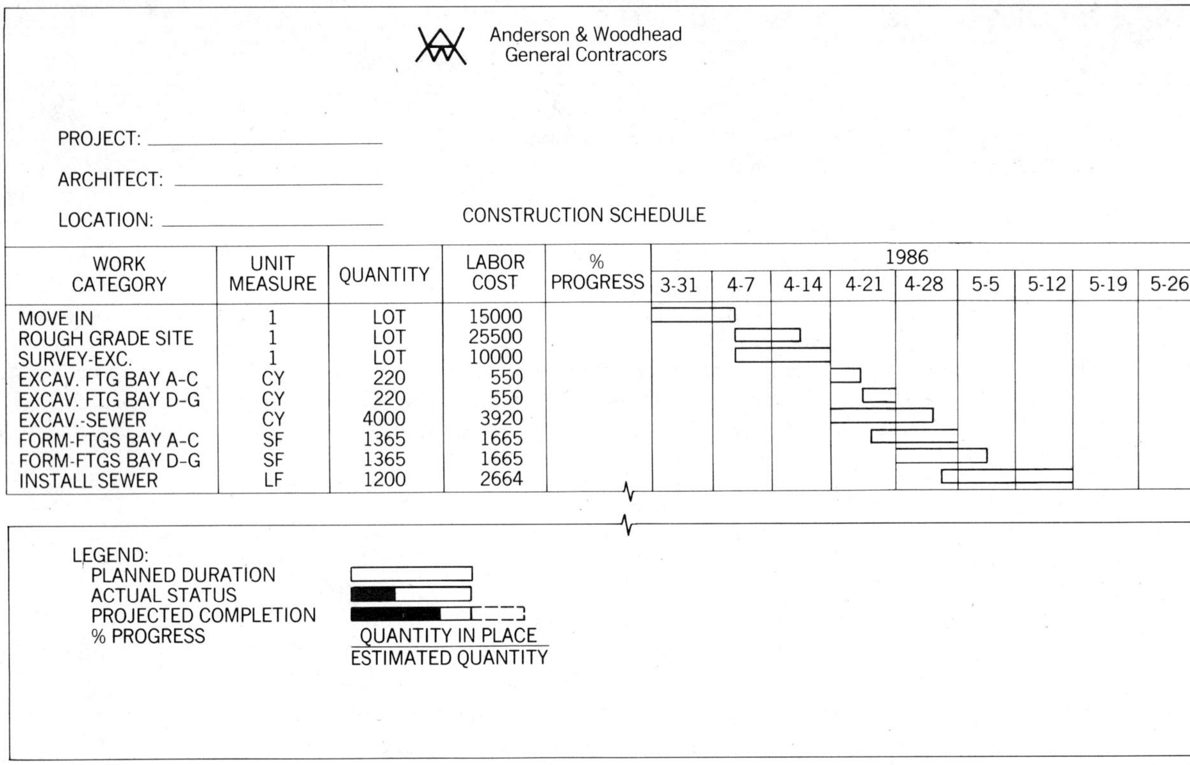

Anderson & Woodhead
General Contracors

PROJECT: ____________

ARCHITECT: ____________

LOCATION: ____________

CONSTRUCTION SCHEDULE

WORK CATEGORY	UNIT MEASURE	QUANTITY	LABOR COST	% PROGRESS	1986 3-31	4-7	4-14	4-21	4-28	5-5	5-12	5-19	5-26
MOVE IN	1	LOT	15000										
ROUGH GRADE SITE	1	LOT	25500										
SURVEY-EXC.	1	LOT	10000										
EXCAV. FTG BAY A-C	CY	220	550										
EXCAV. FTG BAY D-G	CY	220	550										
EXCAV.-SEWER	CY	4000	3920										
FORM-FTGS BAY A-C	SF	1365	1665										
FORM-FTGS BAY D-G	SF	1365	1665										
INSTALL SEWER	LF	1200	2664										

LEGEND:
PLANNED DURATION
ACTUAL STATUS
PROJECTED COMPLETION
% PROGRESS $\frac{\text{QUANTITY IN PLACE}}{\text{ESTIMATED QUANTITY}}$

Figure 7.17 Construction Schedule (I_M-DOC 2).

manner, the project manager and superintendent will establish a plan to meet the contract completion date. This type of *BAR CHART* scheduling is considered a satisfactory technique and might result in a document such as the Construction Schedule (I_M-DOC 2) shown in Figure 7.17.

The project manager and superintendent consider other factors when developing the overall construction sequence and project schedule. The *SCHEDULING FOCUS* (P_M 8) considers the criticality of each individual activity in terms of both the impact on total project cost and the scheduled completion dates; the impact of seasonal considerations;* the impact of restrictive work practices in union environments;† and the effective use of resources.

The main problem focus of Project Scheduling is activity sequencing and manpower allocation. While activity sequencing is conducted for each project, manpower allocations are more frequently considered on larger projects. This is a significant development in the Project Scheduling approach of the medium-sized firm. A rough analysis of manpower is evaluated through the determination of activity man-hours and PROJECT MAN-HOUR REQUIREMENTS (S3). This is accomplished by estimating the production rate and craft man-hour totals for each activity, in terms of *craft man-hours per quantity unit* and the total estimated quantity for the activity. Once a crew size and mix for the activity have been conceptualized, the man-hours are converted into a work duration. Standard production rates and typical crew sizes culled from past experience are the basis of the analysis. As the activity durations are evaluated, the activity can be integrated into the construction sequence. Finally, the project manager and job superintendent estimate the PROJECT MANPOWER REQUIREMENTS (A3) in order to properly staff their field management force, anticipate potential labor shortages in critical crafts, and optimize the construction schedule. This includes grossly summing up the crafts totals from the basic crews developed when analyzing activity durations. Formal documentation is not considered necessary unless the labor force

*Assuming the contractor is operating in an environment where weather conditions could significantly influence construction activities and/or crew productivity.

†This consideration is only an issue for the union contractor.

PROJECT: ______ DATE: ______

LOCATION: ______ JOB NO.: ______

BY: ______

ACTIVITY: FORMWORK FOOTINGS

QUANTITY: ESTIMATE 2730 SF

PRODUCTION: .07 MAN-HOURS/SF

CREW: 3 CARPENTERS PER DAY
1 LABORER PER DAY

ESTIMATED DURATION = $\frac{\text{2730 SF} \times \text{.07 MAN-HOUR/SF}}{\text{32 MAN-HOURS}}$

= 6 DAYS
OR 455 SF PER DAY

PROJECT MANAGER: JOHN SMITH

(a)

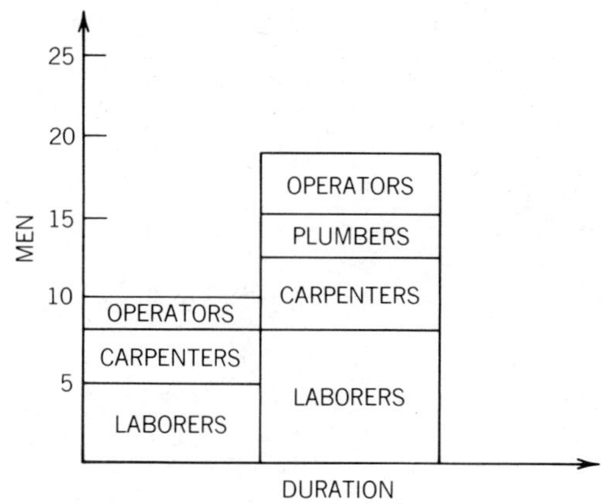

(b)

Figure 7.18 Manpower Summary (I_M-DOC 3). (*a*) Activity duration/crew size analysis. (*b*) Manpower loading.

size rises above normal past experience, in which case manning per activity may be depicted on either a bar chart schedule or in the form of a Manpower Summary (I_M-DOC 3) as shown in Figure 7.18.

Management focus is often concentrated on critical activities, because these activities can most noticeably influence total cost and project completion. *Cost critical* activities are those which are considered highly vulnerable to profit or loss and, if significant overruns occur, could cause total profits to suffer severely. *Schedule critical* activities are those which, if delayed, could deter project progress and have an impact on completion targets. In many instances, activities are both cost and schedule critical and consequently require special attention.

Cost critical activities will often be dissected into work tasks compatible with the estimate. For instance, the general category of site work may be comprised of work tasks such as cut, strip, and backfill, as shown in Figure 7.19. By evaluating labor dollars for each work task as a percent of the

	ACCOUNT	SITE WORK	LABOR COST	% TOTAL	SUBCONT. COST	% TOTAL
Cost Critical	0001	CUT	$15,865	39		
Cost Critical	0002	STRIP	3,750	9		
Cost Critical	0003	BACKFILL	5,941	14		
	0004	GRANULAR FILL	2,462	6	6,525	81
	0005	HAND EXCAVATION	2,229	5		
	0006	FOOTINGS	472	1		
	0007	FINISH GRADE	1,607	4		
	0008	CUT AT BLACK TOP	1,205	3		
Cost Critical	0009	COMPACT AT ROAD	2,812	7		
Cost Critical	0010	SPREAD TOP SOIL	2,700	7		
	0011	PRE CAST LOGS	620	2	1,550	19
	0012	FINISH GRADING	1,360	3		
	00	TOTAL	$41,023	100	7,075	100

Figure 7.19 Analysis of cost critical tasks for general categories of work.

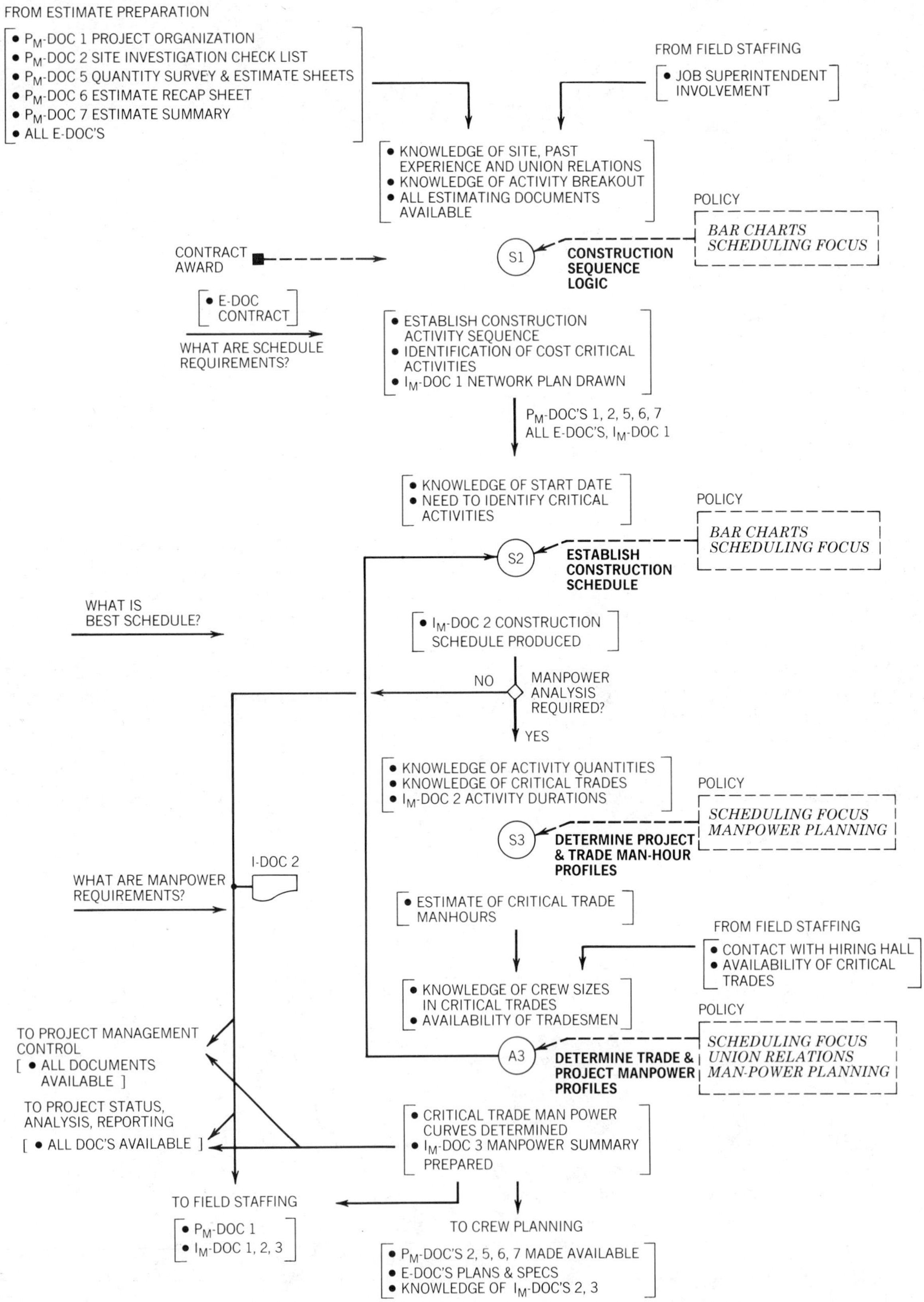

Figure 7.20 Project Scheduling decision process (Traveling Project Manager approach).

total labor dollars for the general category, the project manager can identify those work tasks which will have the most significant impact on the general category's profitability. This will determine the importance of those work tasks within the general activity category that should be carefully monitored from a cost and schedule perspective.

In summary, the Project Scheduling focus broadens and becomes more detailed at the medium-sized firm level. Solving the activity sequencing and manpower problems leads to an optimization of the construction schedule in terms of resource (time and manpower) allocation. The decision process for Project Scheduling at this level is illustrated in Figure 7.20.

The informal nature of Project Scheduling naturally leads to an informal structure that is often characterized by a single-pass, gross network logic scheduling approach. This approach provides the contractor with a simple means of checking activity relationships and their logical sequences, and identifies main tie-in points for interfaces with project subcontractors. It also provides an indication of the schedule's critical activities that must be carefully monitored and controlled by the contractor. While scheduling is mainly performed on a manual basis, many contractors are using computer-based project scheduling applications*. However, due to time constraints and pressure of work, the computerized project schedule is often not properly statused or updated over the life of the project; rather, this effort is performed manually.

7.5 FIELD-BASED PROJECT MANAGEMENT

At the large firm level, Project Scheduling is approached in a number of ways depending on the nature of the contract and the extent of the large firm's responsibility for the project. If the contract is fixed priced, and the plans and specifications are detailed, broad but focused scheduling is performed concurrent with Estimate Preparation. This enables the contractor to devise a realistic plan and schedule for the proposed work and analyze manpower requirements to ascertain the magnitude of the field effort the contractor must manage. Once the project is secured, the contractor proceeds to the detailed planning, scheduling, and field manpower analysis needed to support the full construction management requirements of the project. When the plans and specifications are not well defined, such as in a design–construct situation, the staged development of Project Scheduling often occurs in conjunction with Estimate Preparation. Each scheduling stage provides a better definition of the plan and schedule for construction, and the level of manpower detail increases with the better definition and the contractor's knowledge of the site labor environment and labor estimate details. This fully documented multilevel focus of Project Scheduling for the Field-Based Project Management approach is a major departure from the Traveling Project Manager approach of medium-sized firms.

In terms of manpower management, Project Scheduling for large projects becomes a critical decision process. This is because the magnitude of the man-hour effort involved in the large project, when translated into manpower profiles, can significantly impact construction plans and schedules if the contractor cannot achieve the anticipated levels of skilled craftsmen required to meet estimated site productivity and progress targets. At the small and medium-sized firm levels, these problems are rarely a consideration, because the size of the field labor force is not a dominant factor.

The manpower management functions of the Project Scheduling decision process at the large firm level include SEQUENCING THE CONSTRUCTION LOGIC (S1), ESTABLISHING THE CONSTRUCTION SCHEDULE (S2), DETERMINING PROJECT AND TRADE MAN-HOUR PROFILES (S3), and DETERMINING PROJECT AND TRADE MANPOWER PROFILES (A3).† In contrast to the medium-sized firm, these functions are performed not only in the head office but in the field as well. This requires the first movement of functions that were previously head office based to a field location in support of the Field-Based Project Management approach. Thus, the management problem focus of Project Scheduling is two-fold. First, Project Scheduling has a top management orientation and develops the broad calendar schedule that meets client (and contractual) time constraints, as well as the client's re-

* Many scheduling programs are available for microcomputer application, which can provide a cost-effective means of implementing computerized scheduling (see bibliography).

† This section will focus primarily on Project Scheduling as related to the construction phase of a design–construct project. The interdependencies between design, procurement, and construction will be discussed as necessary to describe their impact on construction planning and scheduling.

quested priority sequencing of facility components. When combined with estimated man-hour data, gross manpower profiles are established for the total project and by craft, and provide the requisite input for site manpower availability analysis. This management level scheduling phase is illustrated in Figure 7.21. Second, head office Project Scheduling has a project level orientation and expresses the detailed construction plan and schedule for accomplishing work activity in logical sequences over time. At this level, the work force is quantified in terms of craft man-hours and manpower based on construction control budget labor estimate details from which percent progress plans are derived. Project level scheduling attempts to outline the best utilization of manpower and equipment to achieve satisfactory field construction progress, in relation to anticipated engineering and design progress and the predicted delivery of materials and permanent plant equipment. This project level scheduling phase is illustrated in Figure 7.25. Project Scheduling in the field is concerned with short-range planning and scheduling of detailed work activities, specifically to support construction crew planning and match planned progress targets and manpower levels with actual field requirements. This phase of Project Scheduling is described in Figure 7.28.

The primary project team members involved in Project Scheduling and responsible for its performance are construction engineers and scheduling engineering specialists. Key construction managers and/or field superintendents also provide their expertise in a review and approval mode, while the project manager normally impacts Project Scheduling from a total project basis through policy directives such as project completion dates, key milestone dates, priority sequencing of facility components and desired progress rates, and field manpower levels. Client involvement in Project Scheduling increases, especially for design–construct projects produced on a cost reimbursable basis. The client frequently reviews and approves schedules and progress and field manpower plans.

Project size and complexity make it difficult to operate with only one-level of schedule, as is done in the small and medium-sized firm environments. Thus, it is not uncommon for the contractor to generate several schedules, with the detail of each successive schedule being more specifically focused and detailed than the last as activities are further subdivided into their component parts. Accordingly, *NETWORK-BAR CHART ANALYSIS* (P_L 10)* dictates that a combination of the critical path method and bar chart models forms the basis for planning and scheduling at all levels. For example, the bar chart is a basic management level schedule technique, while CPM-based scheduling in conjunction with bar charts is best suited for project and field construction control scheduling.

The initial Project Scheduling effort focuses on management level scheduling as described in Figure 7.21. In this phase, ESTABLISHING THE CONSTRUCTION SCHEDULE begins to fix the project scope (e.g., project composition, major components, and priority of facility completion, etc., as defined during Project Planning) in calendar time in relation to contractual arrangements and other key milestone dates determined by the project manager as *SCHEDULING FOCUSED* (P_L 11) constraints. The scheduling and construction engineer relies heavily on past history for standard construction durations (or engineering durations as applicable) and standard procurement cycles, and relies on current market conditions for delivery times of major equipment items and facility materials. The durations determined for major categories of work are integrated with the key milestones and are adjusted as needed to meet the project completion date. The schedule is reviewed with the project manager and other key project team personnel until an approved Management Level Schedule (P_L-DOC 16) is generated, as illustrated in Figure 7.22.†

The Management Level Schedule serves as the basis for the DETERMINATION OF PROJECT AND TRADE MAN-HOUR AND MANPOWER PROFILES. To do this, the scheduling engineer analyzes manpower levels to determine if the Management Level Schedule is realistic with respect to initial readings of available manpower in the site region. This objective is stipulated in *MANPOWER PLANNING* (P_L 15) and is also a critical *SCHEDULING FOCUS*. The basic information required to evaluate manpower profiles at this level of scheduling is the estimated project field man-hours from the Factored Estimates (P_L-DOC 12), and the basic time frames established by the Management Level Schedule. In accordance with the *LABOR DATABASE* (P_L 9) policy the scheduling engineer relies on historical data such as Standard Construction Progress Curves (P_L-DOC 17) and Standard Con-

*See Table 7.5, Policies of the Large Firm.

†These schedules are often referred to in practice as master schedules, directive schedules, and so on.

TABLE 7.5 POLICIES OF THE LARGE FIRM

Policy	Description
P_L 9 *LABOR DATABASE*	Historical labor data must be analyzed and documented in a meaningful way to be useful for estimating and planning manpower for future projects.
P_L 10 *NETWORK-BAR CHARTS ANALYSIS*	Broad network focused milestone planning is the best approach for identifying and sequencing major project components, while a combination of critical path method and bar chart model are best suited for project planning and scheduling.
P_L 11 *SCHEDULING FOCUS*	Scheduling definitions and constraints are: 1. A critical activity is one that, if delayed, will extend the completion date of the project. 2. Activity durations must be correlated to the estimated work content. 3. All constraints must be defined and incorporated into the sequence logic. 4. Manpower impact is an integral aspect of project scheduling and scheduling control. 5. Schedule must reflect optimum use of all resources. 6. Progress is based on the physical completion of work activities.
P_L 15 *MANPOWER PLANNING*	Long-range planning of manpower by craft and in total is a prerequisite to support the field effort, and must be performed in the planning and scheduling stages of the project.
P_L 16 *CREW PLANNING*	Field planning, scheduling, and allocation of manpower requires a precise definition of work tasks and a combination of past experience, knowledge of work content, man-hour estimates, and interaction between field project team members.
P_L 34 *DATA PROCESSING*	Maximum use of electronic data processing is often the most economical means of processing data into useable information.

struction Craft Mixes, (P_L-DOC 18) as a basis for manpower analysis.

The Standard Construction Progress Curve in Figure 7.23*a* shows a standard construction duration, expressed in equal intervals of time (horizontal axis), for which a percent progress curve is plotted representing typical rates of progress for a construction project commonly built by the company. The Standard Construction Progress Curve is applied to the specific project duration but expanded or contracted as specified by the project duration of interest. The resultant curve is the Construction Progress Curve (I_L-DOC 5) illustrated in Figure 7.23*b*. Because there exists a direct relationship between labor man-hour expenditures and progress, the percent complete at any increment (or period) of time, multiplied by the total estimated man-hours, provides the man-hours required to achieve that percent progress for the increment. In this approach (see Figure 7.9), man-hours are portrayed as a function of time from which labor man-hour curves are plotted. These Man-hour Curves (I_L-DOC 6) are models that predict the rate of man-hour expenditures needed to achieve a desired progress rate. The total project man-hour curves (Figure 7.23*c*), are further divided into their component craft man-hours by applying percentages from the Standard Construction Craft Mixes (P_L-DOC 18) (Figure 7.23*d*). Craft curves are similar to those shown in Figure 7.23*c*. The man-hour profiles shown on I_L-DOC 6 (Figure 7.23*c*), are translated into equivalent manpower profiles by dividing the predicted man-hour expenditure per period by the total available man-hours to be worked in the period. Project and Craft Manpower Curves (I_L-DOC 7) are illustrated in Figure 7.24.

While Figures 7.23 and 7.24 represent only total project curves, curves could be developed for major categories of work such as excavation or concrete, and then summarized into total project curves. This technique is illustrated in Figure 7.10. Finally, Manpower Curves are leveled to provide a smooth buildup and decline in manpower. This is done by simply massaging the curve to remove "peaks and valleys" in the manpower profiles. This massaging technique is acceptable at the management scheduling level, provided the profiles still reflect the construction durations and typical manpower peaks in relation to project durations. Finally, computer *DATA PROCESSING* (P_L 34) is a necessity when developing manpower profiles, especially when significant manpower leveling is required.

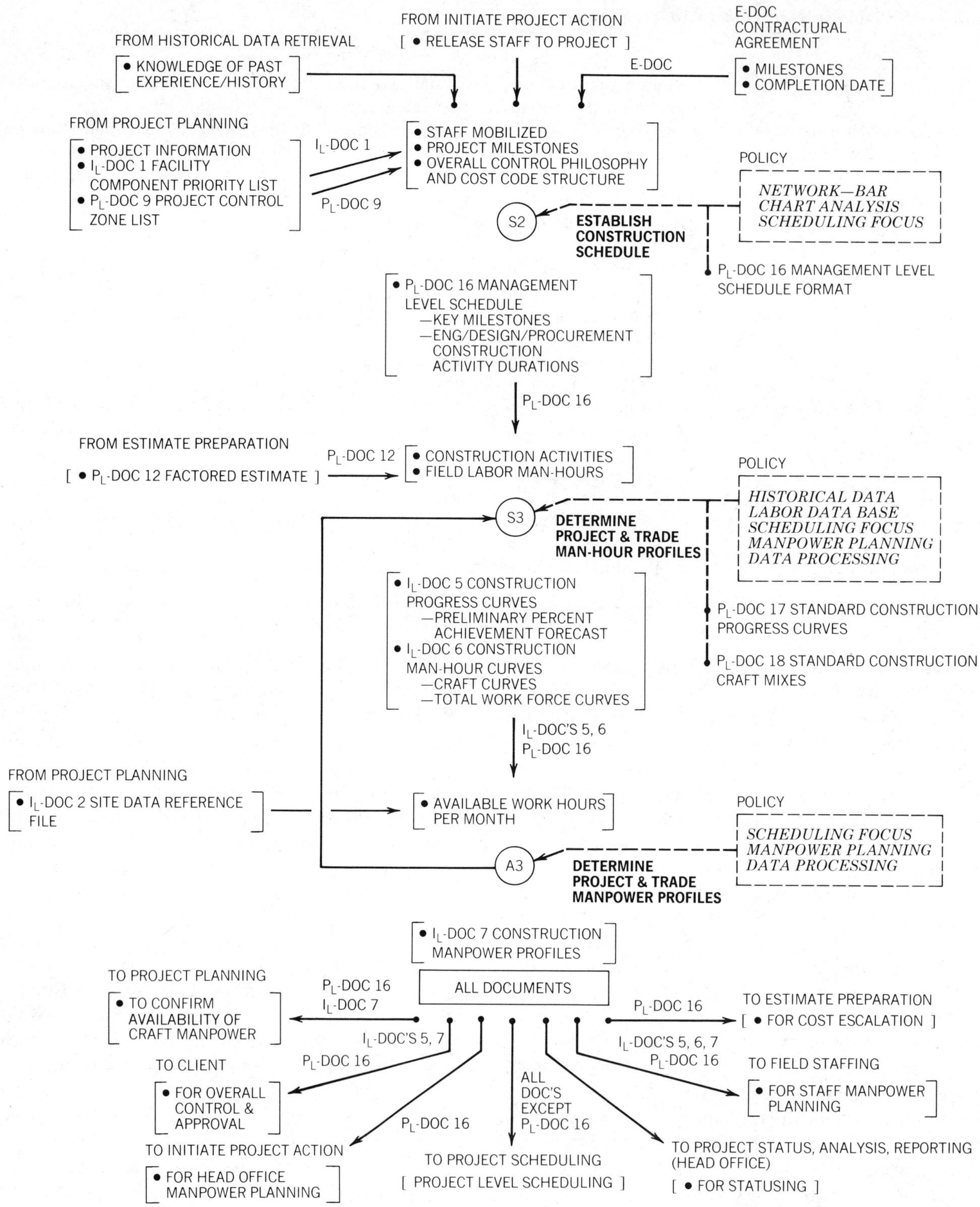

Figure 7.21 Project Scheduling decision process: management level scheduling (Field-Based Project Management approach).

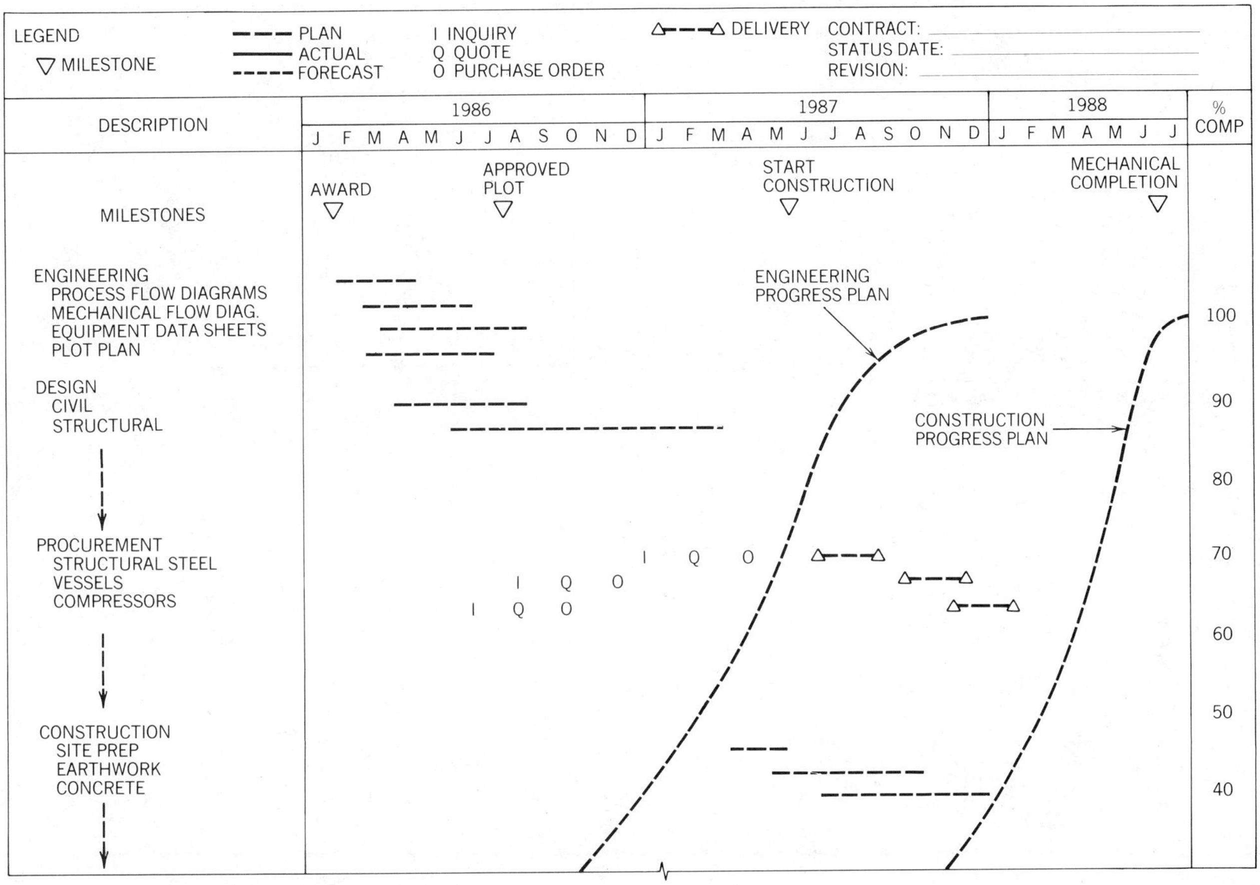

Figure 7.22 Management Level Schedule (P_L-DOC 16).

The establishment of manpower profiles by craft at the management level stage in Project Scheduling is critical, because it provides the basis for analyzing the availability of various crafts. This analysis can lead to decisions relative to camps, training, and so on.

When the project approaches the start of construction, Project Level Scheduling commences, as illustrated in Figure 7.25. This phase incorporates those functions of the Management Level Scheduling phase and begins with CONSTRUCTION SEQUENCING defining the detailed construction activities and the points at which they interface with material deliveries and drawing issue activities. Then, logical relationships between all of these activities are established. These logical relationships are displayed through CPM network techniques and in construction plans for accomplishing the work. When specific activity durations are established and start and completion dates are selected, the CPM logic network becomes the plan for completing the project. The CPM schedule must be consistent with the Management Level Schedule in that it reflects the same total project duration, major milestones, and priority sequence for completing the facilities.

The scheduling engineer is primarily responsible for Project Level Scheduling. He or she works closely with construction engineers and relies on input from key superintendents and construction managers. Depending on the size and complexity of the project, there may be a team of scheduling engineers whose responsibilities are dispersed according to the types of components that comprise the project. For example, one scheduling engineer might have responsibility for all mechanical construction.

The scheduling engineer gathers information for CONSTRUCTION SEQUENCING from a variety of sources, as depicted in Figure 7.25. Information and experience from past projects is used to establish the initial listing of activities and their interdependencies. Working in concert with construc-

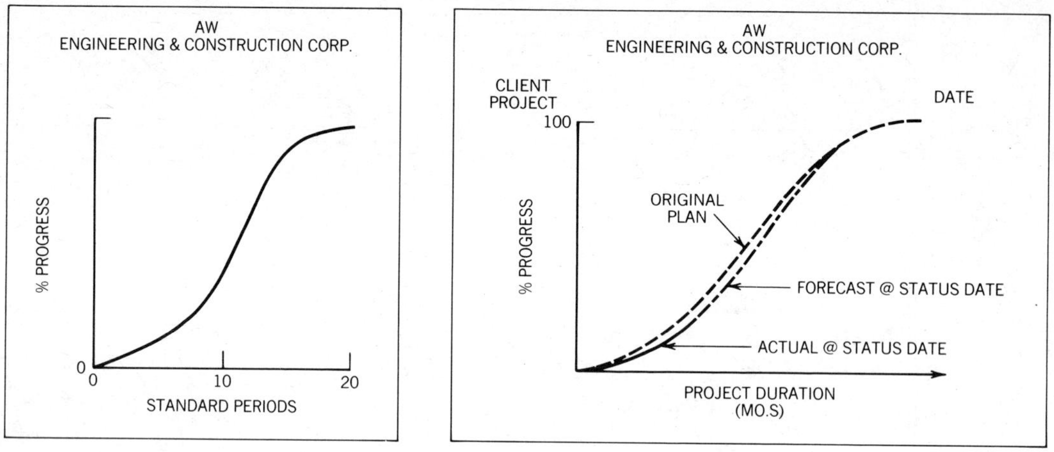

(a) *(b)*

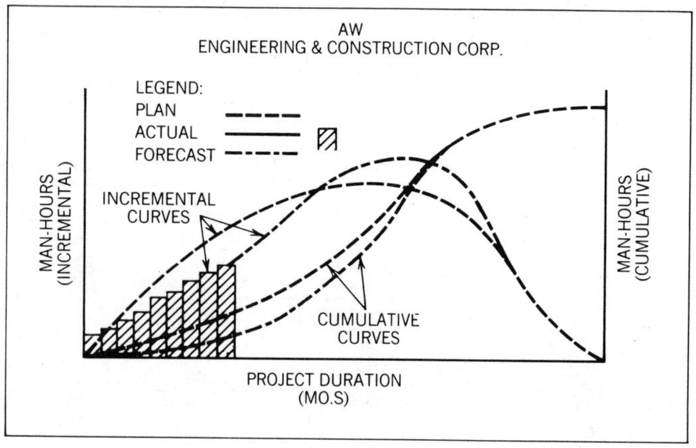

(c)

CONCRETE

CRAFT	% MIX
LABORERS	15
CARPENTERS	55
IRONWORKERS	20
CEMENT MASONS	5
OPERATORS	3
TEAMSTERS	2
TOTAL	100

(d)

Figure 7.23 Progress and manpower data. (*a*) Standard Construction Progress Curve (P_L-DOC 17). (*b*) Construction Progress Curves (I_L-DOC 5). (*c*) Project Man-Hour Curves (I_L-DOC 6). (*d*) Standard Construction Craft Mixes (P_L-DOC 18).

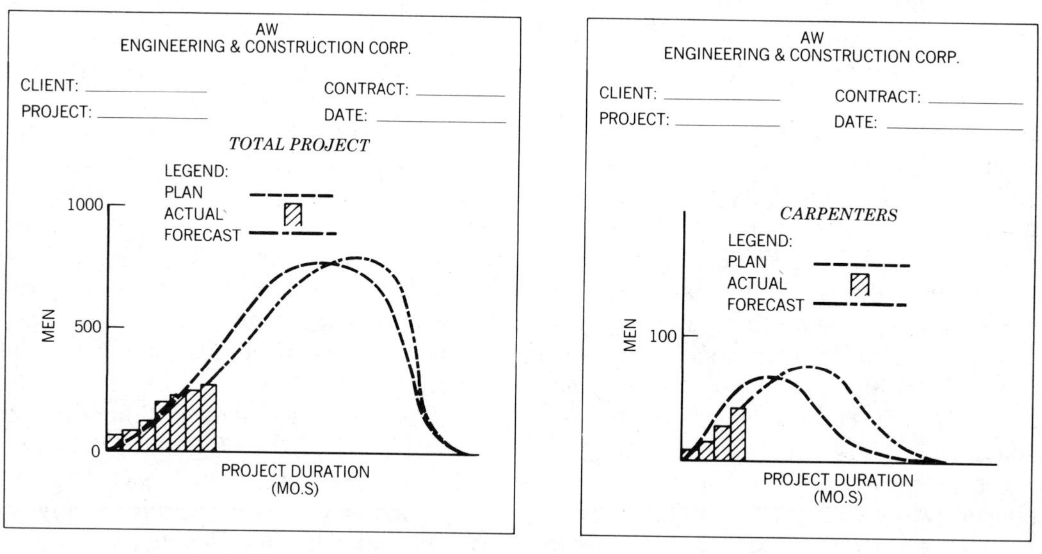

(a) *(b)*

Figure 7.24 Construction Manpower Profiles (I_L-DOC 7). (*a*) Project. (*b*) Craft.

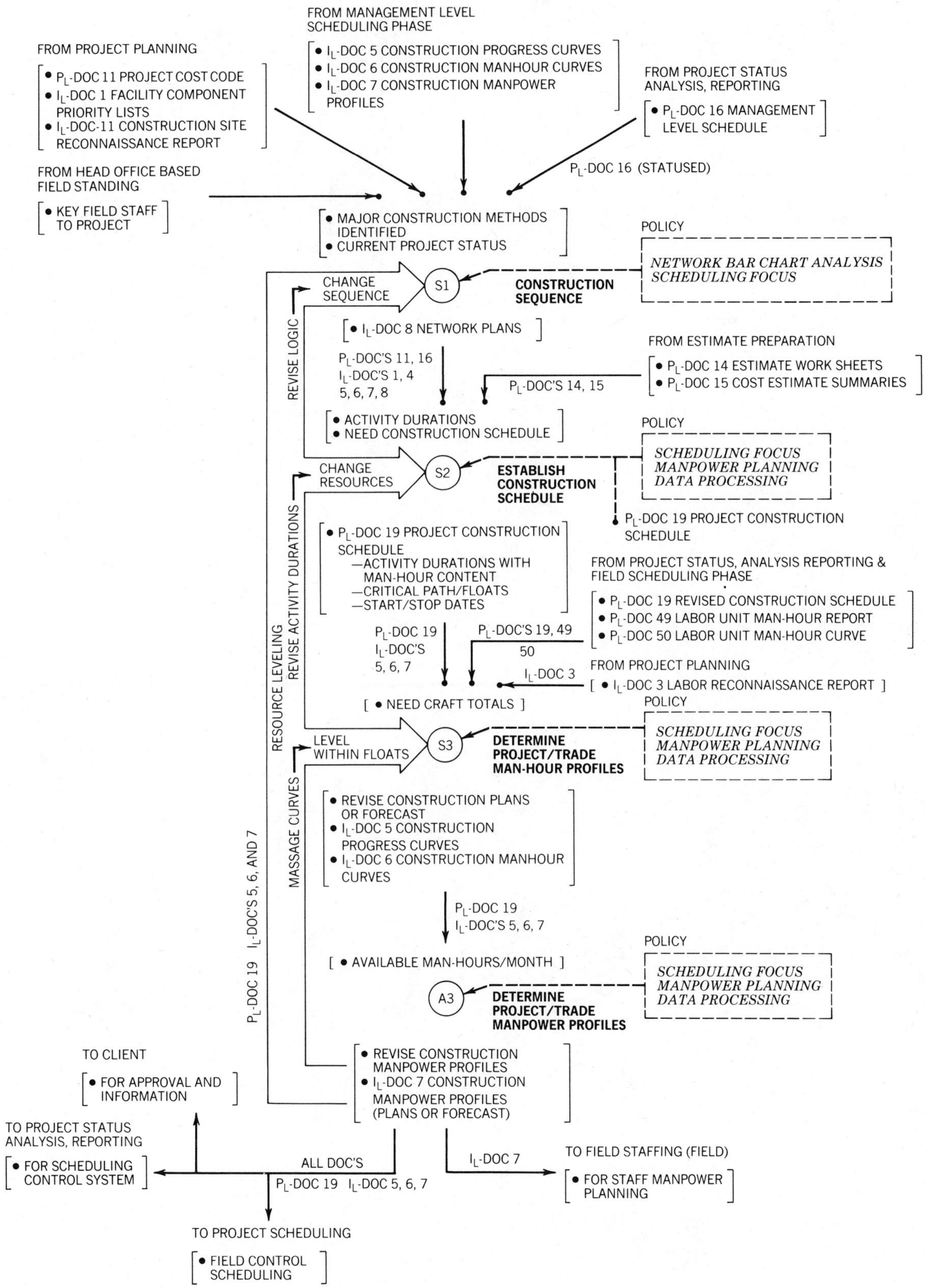

Figure 7.25 Project Scheduling decision process: project level scheduling (Field-Based Project Management approach).

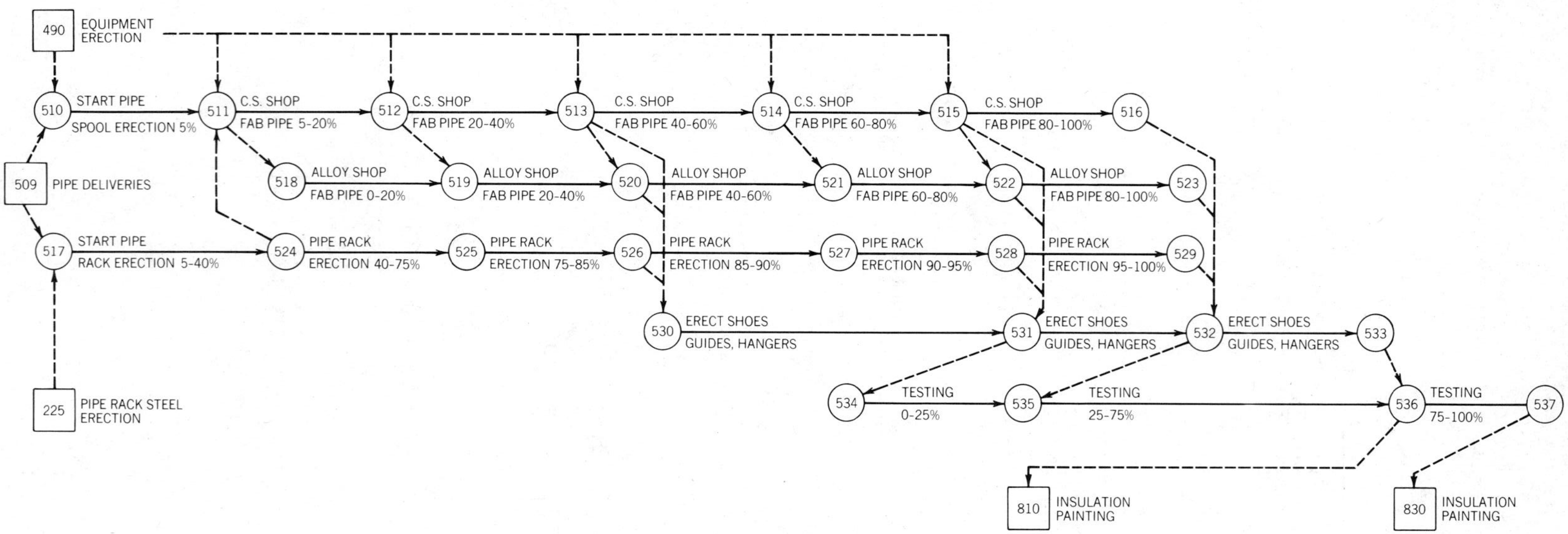

Figure 7.26 Network Plans (I_L-DOC 8).

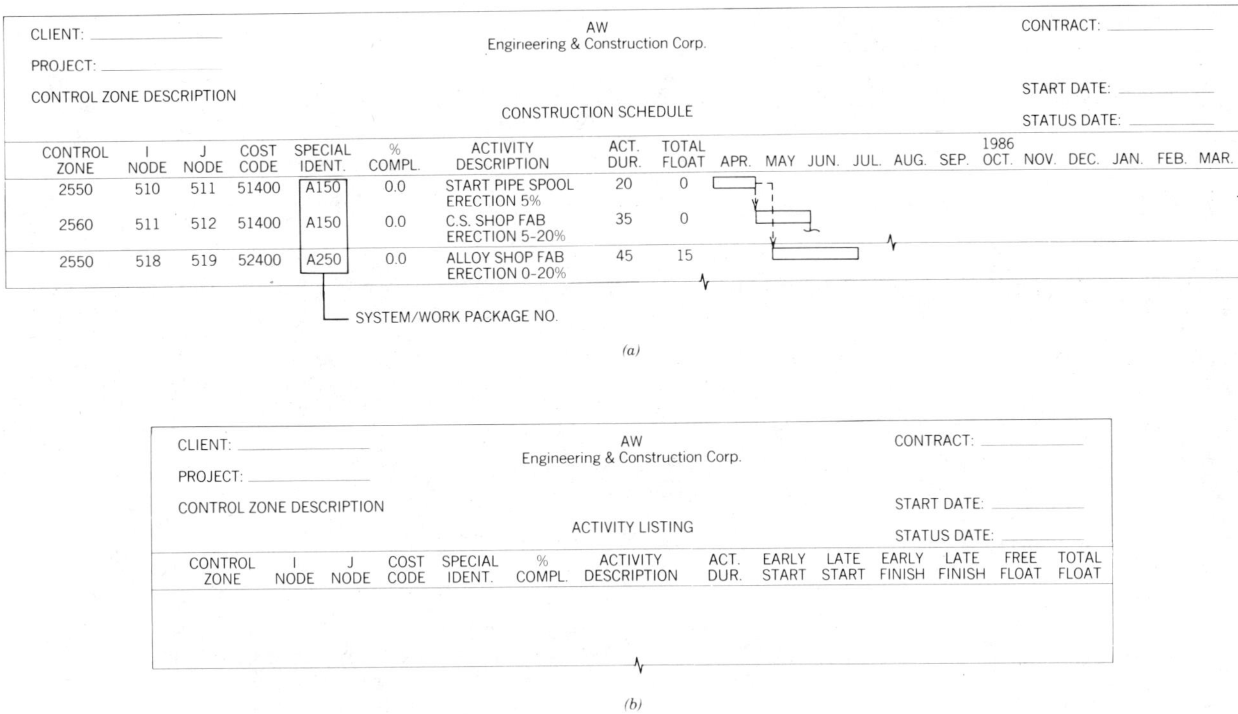

Figure 7.27 Project Construction Schedule (P_L-DOC 19). (*a*) Logic-based time-scaled schedule. (*b*) Activity listing.

tion engineers, a CPM logic-based Network Plan (I_L-DOC 8) (either a precedence or an arrow diagram) is generated (see Figure 7.26).

Prior to having detailed estimating information such as man-hours and quantities for each activity, the scheduling engineer utilizes scaled activity durations from similar projects, while being restrained by key project milestone dates and general completion dates from the Management Level Schedule. When control budget estimate labor man-hours and activity quantities (from P_L-DOCs 14 and 15) become available, activity durations are adjusted to reflect the impact of this revised information. Production rates and crew sizes for each activity are analyzed in light of estimated labor man-hour and activity quantity content, and from this a realistic activity duration is established.

When the CPM model is used, the actual *DATA PROCESSING* of the network logic, activity start, and completion dates, as well as the calculation of the critical path and the like, are executed electronically by a computer. The computer will often draw the CPM network either as a bar chart without displaying logical ties, or with logic ties only referenced against a time scale, or will use both techniques in combination. A segment of a Project Construction Schedule (P_L-DOC 19) is shown in Figure 7.27*a*, with its corresponding activity listing illustrated in Figure 7.27*b*.

The man-hour loaded CPM schedule becomes the basis for revising the progress and manpower curves using those techniques presented in Figures 7.8 through 7.11. As reflected in Figure 7.25, the process is normally repeated several times before the final construction plans are reviewed and approved by the project manager and key field managers. These progress and manpower plans will also incorporate the detailed input from the Labor Reconnaissance Report (I_L-DOC 3) and the Construction Site Reconnaissance Report (I_L-DOC 4).

The documents which emerge from head office project scheduling set the control plan and schedule for the project. The Project Construction Progress Curves and Construction Manpower Curves provide a control base from which actual progress can be tracked and compared against planned progress. These documents become the basic guides for field project scheduling, crew planning, and management control.

Field Construction Control Level scheduling, Figure 7.28, is the transitional link between project definition and field construction. The problem focus

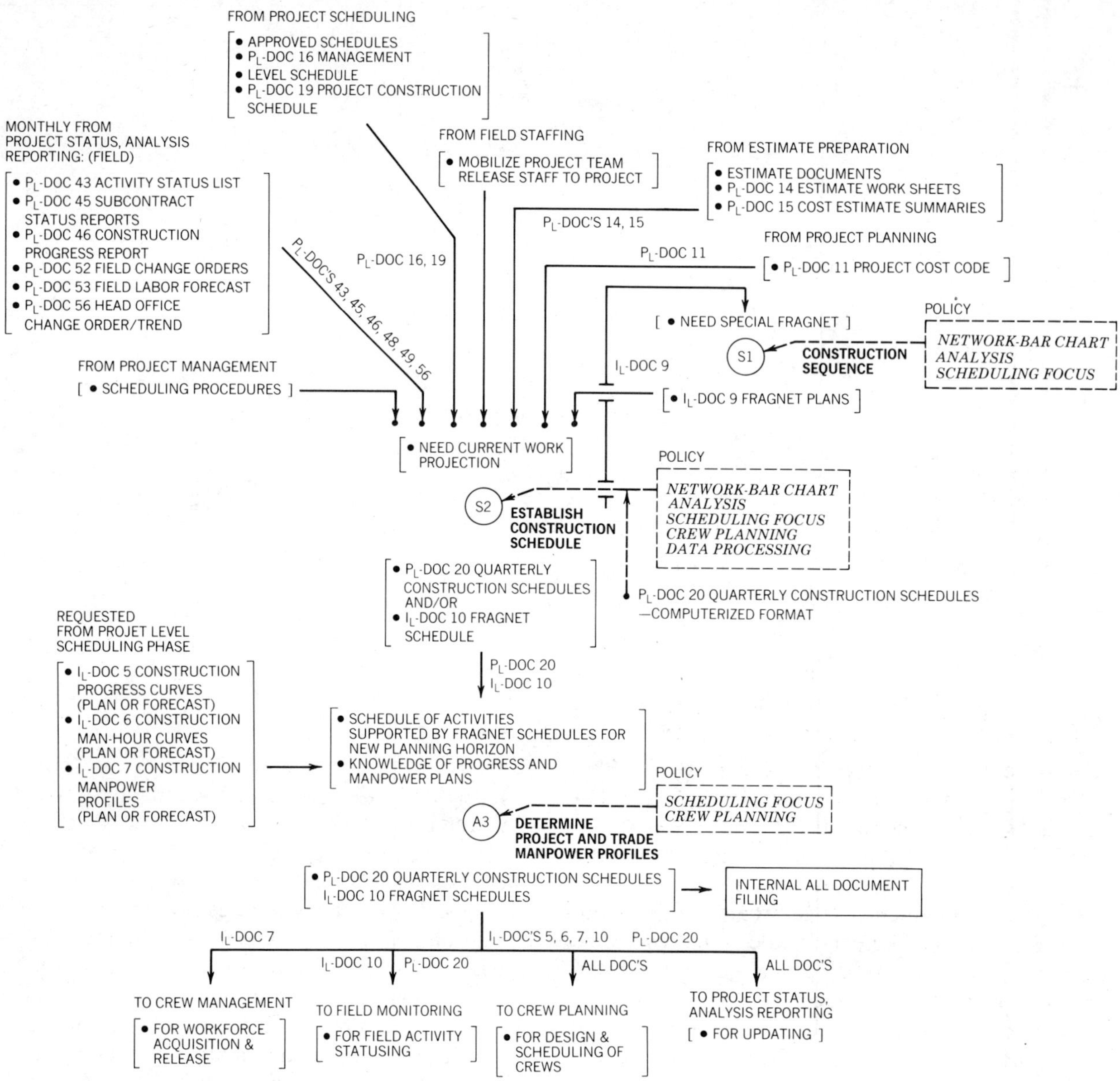

Figure 7.28 Project Scheduling decision process: field construction control level scheduling (Field-Based Project Management approach).

of field scheduling is short-range. It extends to a level of detail sufficient to support crew planning, as performed by the assistant and/or craft superintendents and general foremen. This decision phase, as shown in Figure 7.28, is comprised of the same functions as the Project Level Scheduling phase, with the exception that MAN-HOUR PROFILES are not analyzed.

The function progression of field scheduling parallels Project Level Scheduling, but the scope is limited to a shorter planning horizon such as the one indicated in the Quarterly Construction Schedule (P_L-Doc 20) shown in Figure 7.29. The level of detail is normally far greater because activities on the Project Construction Schedule are further divided, where requested, into their finite work task components in order to support crew planning. Finally, a Fragnet CPM can be drawn to depict the sequence logic and plan for constructing work task components as required. A typical Fragnet Logic Network

CLIENT: ______ AW Engineering & Construction Corp. CONTRACT: ______

PROJECT: ______

CONTROL ZONE DESCRIPTION START DATE: ______

QUARTERLY CONSTRUCTION SCHEDULE STATUS DATE: ______

CONTROL ZONE	I NODE	J NODE	COST CODE	SPECIAL IDENT.	% COMPL.	ACTIVITY DESCRIPTION	ACT DUR	TOTAL FLOAT	APRIL 7 14 21 28 / MAY 5 12 19 26 / JUNE 2 9 16 23 30
2550	510	511	51400	A150	0.0	START PIPE SPOOL ERECTION 5%	20	0	XXXXX XXXXX XXXXX XXXXX
2550	511	512	51400	A150	0.0	C.S. SHOP FAB ERECTION 5-20%	35	0	XXXXX XXXXX XXXXX XXXXX XXXXX XXXXX XXXXX
2550	518	519	52400	A250	0.0	ALLOY SHOP FAB ERECTION 0-20%	45	15	XXXXX XXXXX XXXXX XXXXX XXXXX XXXXX

SYSTEM/WORK PACKAGE NO.

Figure 7.29 Quarterly Construction Schedule (P_L-DOC 20).

(I_L-DOC 9) is illustrated in Figure 7.30*a*. A similar approach is applied to each new activity in the planning period.

The Fragnet Logic Network is referenced against the planning period to ESTABLISH THE CONSTRUCTION SCHEDULE for the Short-Range Bar Chart Schedule (I_L-DOC 10), in Figure 7.30*b*. The cumulative construction durations, start dates, and finish dates for all work task component identified must be consistent with those of the general activity from the Project Construction Schedule.

TRADE MAN-HOUR AND MANPOWER PROFILES for Fragnet Schedules are determined by estimating the number of crafts required for a detailed work task as defined by normal crew sizes or the numbers of crews available, and then establishing rough crew ratios for a detailed work task based on its estimated man-hour content. This pro-

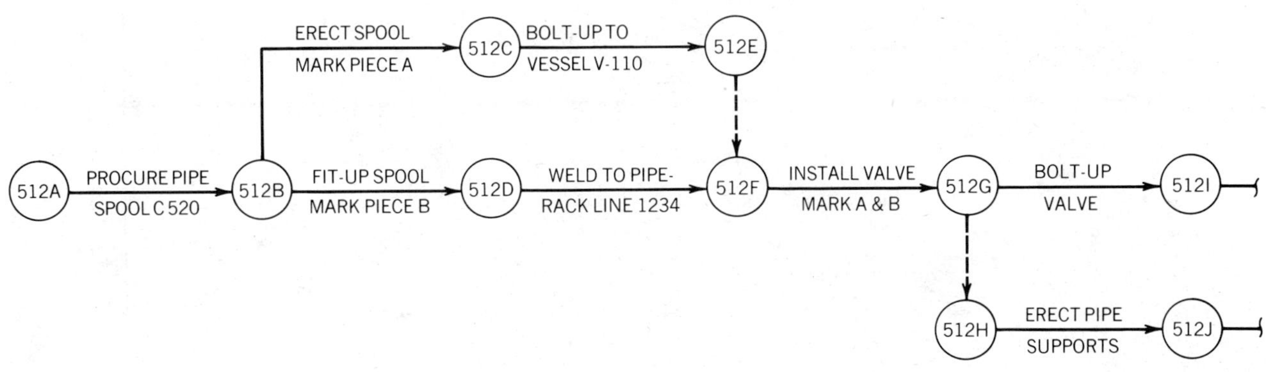

(*a*)

CLIENT ______

PROJECT ______

I NODE	J NODE	ACTIVITY DESCRIPTION	ACT DUR	CREW REQ'MTS	EQUIP REQ'MTS	DURATION M T W T F S S M T W T F
512A	512B	PROCURE PIPE SPOOL C 520	1			
512B	512C	ERECT SPOOL MARK PIECE A	2	PIPEFITTING CREW 15	CRANE	
512B	512D	FIT-UP SPOOL MARK PIECE 8	4	PIPEFITTING CREW 16/ WELDING CREW 10	CRANE WELDING MACHINE	

(*b*)

Figure 7.30 Fragnets for short-range scheduling. (*a*) Fragnet Plan (I_L-DOC 9). (*b*) Fragnet Schedules (I_L-DOC 10).

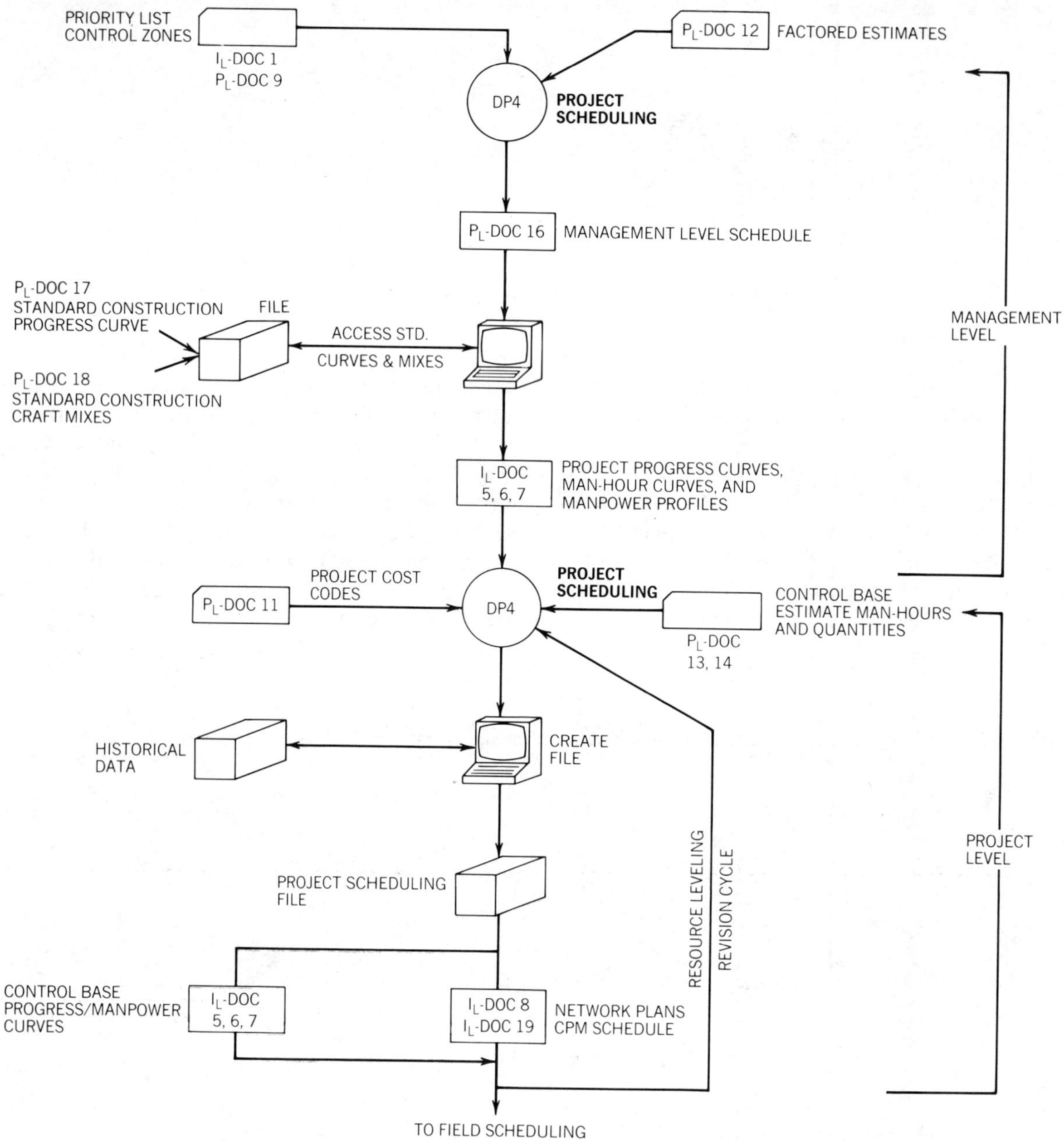

Figure 7.31 Project Scheduling Management Information System (Field-Based Project Management approach).

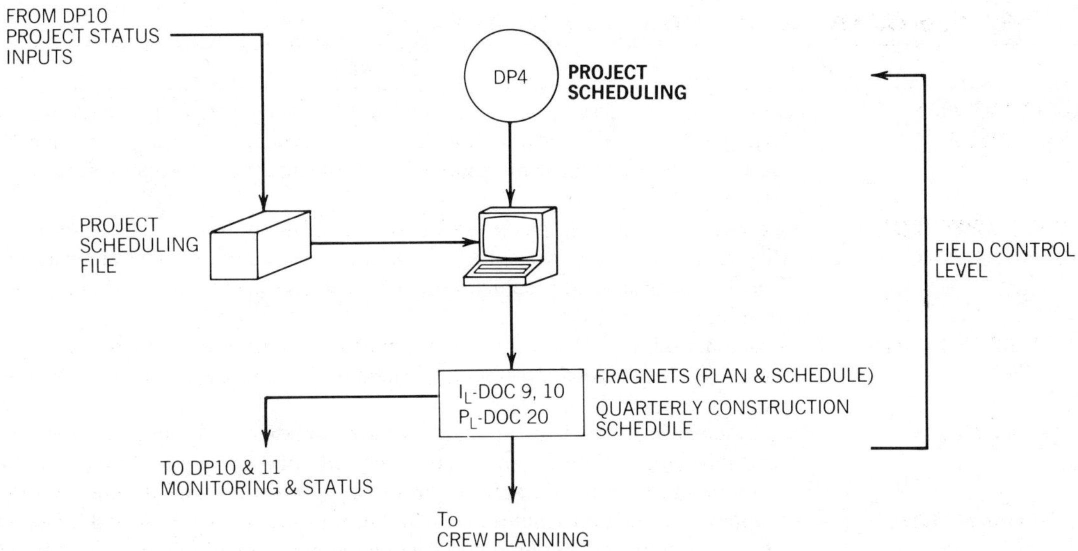

Figure 7.31 (*Continued*)

vides the field construction team with a prediction of the short-range manpower requirements. It must be compared to planned manpower profiles to ensure that planned progress is achievable over the short-range planning period.

At the large firm level, the Project Scheduling decision process is supported by a complex Management Information System, as illustrated in Figure 7.31. This complexity results from the need to handle a larger volume of project data and to accomplish a finer dissection of this data. It also results from the uniquely focused nature of the scheduling requirements for management, project, and control level components. When coupled with project control philosophies, the MIS produces many schedules and fragnets based on iterative decisions on time slots, manpower levels in relation to material deliveries, drawing issues, and other key interface areas. Consequently, most contractors using Field-Based Project Management design their Project Scheduling MIS based on a computer application. The computer approach provides the flexibility to access reports at many different levels based on management requirements and specific project needs. Thus, as Figure 7.31 shows, the Project Scheduling MIS has a multi-level entry focus.

7.6 TOTAL FIELD AUTONOMY

Total Field Autonomy Project Scheduling is the decision area where the project group begins to operate fully as a team. This decision area maximizes the integrated team approach that is established by project team design and by the control philosophy decisions made during the Initiate Project Action and Project Planning decision processes. The integrated project team concept leads to an integrated Project Scheduling decision process. The decision process structure, general information flows, and relevant management functions are comparable to the Field-Based Project Management approach (see Figures 7.21, 7.25, 7.28). Total Field Autonomy differs in that an increasing number of policies, and the more penetrating focus of these policies, impact Project Scheduling by promoting a fully integrated system that interrelates cost and schedules. For example, the cost and scheduling engineering functions are closely integrated through the project *WORK BREAKDOWN* (P_T 8)* structure and *COST CODING* (P_T 9) system, thus establishing a common database of major work packages. In this way, items in the estimate will correlate directly to activities in the schedule through work packages, and when given a *SCHEDULING FOCUS* (P_T 14), *MANPOWER PLANNING* (P_T 18) can be more readily tied to estimated labor man-hours and quantities. Furthermore, when changes in scope or trends impact project man-hours and work scope, these changes can be readily incorporated into the schedule through the work breakdown structure. This requires continuous updating and reevaluating

*See Table 7.6, Policies of the Autonomous Project Team.

TABLE 7.6 POLICIES OF THE AUTONOMOUS PROJECT TEAM

Policy		Description
P_T 6	*LABOR RECONNAISSANCE*	Comprehensive survey, analysis, and documentation of the project site labor environment is requisite on a continuous basis to accurately evaluate labor content and plan manpower requirements for the project at any time during the project's life cycle.
P_T 7	*SITE INSPECTION AND ANALYSIS*	A continuous comprehensive inspection, analysis, and documentation of project site conditions, transportation problems, etc., is requisite to accurately evaluate construction methods, schedules, and labor content, and to solve other unique problems.
P_T 8	*WORK BREAKDOWN*	A work breakdown structure representative of the project scope and control philosophy is formulated to integrate costs and schedules with work packages for project management.
P_T 9	*COST CODES*	Cost codes are applied to the work breakdown structure and serve as a basis for categorizing estimated costs, reporting and controlling costs and manpower, developing a historical database, and integrating the schedule with the estimate.
P_T 10	*UNIT MAN-HOUR*	A tight but realistic evaluation of the labor content for a work task is essential for successful project planning and management control. Therefore, estimates must be based on work task man-hour content in relation to work volume, construction methods, local productivities, labor availability, and site conditions.
P_T 11	*HISTORICAL DATA*	Historical labor data are basic to a realistic determination of the labor content for a work task or major project components and preliminary scheduling development.
P_T 12	*LABOR DATABASE*	Historical labor data must be normalized to a standard reference location and/or base and then documented to be useful for estimating and planning manpower.
P_T 13	*NETWORK-BAR CHARTS ANALYSIS*	Broad network focused milestone planning is the best approach for identifying and sequencing major project components, while a combination of the critical path method and bar chart model are best suited for project planning and scheduling at the total project view level as well as at the daily level.
P_T 14	*SCHEDULING FOCUS*	Scheduling definitions and constraints are: 1. A critical work package is one that, if delayed, will extend the completion date of the project. 2. Work package durations must be correlated to the estimated work content. 3. All restraints must be defined and incorporated into the sequence logic (i.e., material delivery, subcontract coordination, drawing receipt, etc.). 4. Manpower impact is an integral aspect of project scheduling and scheduling control. 5. Schedules must reflect optimum use of all the resources relative to the work package. 6. Progress is based on a realistic evaluation of the labor unit effort required to complete a component of work. 7. Schedules are portrayed to the daily work level and are prepared, analyzed, and reviewed weekly in relation to planned progress and productivity versus actual progress and productivities.
P_T 18	*MANPOWER PLANNING*	Long-range planning of manpower by craft and in total is a prerequisite to supporting the field effort, and must be performed continuously throughout the project's life.
P_T 34	*SCIENTIFIC MANAGEMENT*	Techniques of scientific management, such as work sampling, time lapse, activity analysis, multiple activity charts, and prediction models, can contribute to improving crew performance.
P_T 36	*DATA PROCESSING*	A computer-based information system is the most feasible means of data processing for management control.

S2 ESTABLISH CONSTRUCTION SCHEDULE

Policies	Documents
WORK BREAKDOWN	P_T-DOC 21 Management Level Schedules
NETWORK-BAR CHART ANALYSIS	
SCHEDULING FOCUS	

S3 DETERMINE PROJECT AND TRADE MAN-HOUR PROFILES

Policies	Documents
WORK BREAKDOWN	P_T-DOC 22 Standard Construction Progress Curves
COST CODES	P_T-DOC 23 Standard Construction Craft Mixes
HISTORICAL DATA	I_T-DOC 5 Construction Progress Curves
LABOR DATA BASE	I_T-DOC 6 Construction Man-hour Curves
SCHEDULING FOCUS	
MANPOWER PLANNING	
DATA PROCESSING	

A3 DETERMINE PROJECT AND TRADE MANPOWER PROFILES

Policies	Documents
WORK BREAKDOWN	I_T-DOC 7 Construction Manpower Curves
COST CODES	
SCHEDULING FOCUS	
MANPOWER PLANNING	
DATA PROCESSING	

Figure 7.32 Project Scheduling decision process: management level scheduling (Total Field Autonomy approach).

of manpower by facility components through the manipulation of work package activity sequences. Total integration emerges at all levels of scheduling through *NETWORK BAR CHART ANALYSIS* (P_T 13), wherein weekly work schedules are tied to the planned work depicted on short-range planning schedules, which, in turn, are directly tied into comprehensive overall project schedules. Thus, the management and control of detailed field activities and manpower will be consistent with the overall project plan as it is viewed by top management and the client. The schedule then becomes a live and dynamic tool that can be used effectively at different management levels as dictated by management need. Finally, a detailed schedule package for the next planning period emerges and identifies the basis (e.g., manpower, progress, other requirements, etc.) for accomplishing the major work packages considered in the next period. This package places the client and top field management on notice as to current plans and requirements, and when performed continuously provides a track record for scheduling control of the project. These packages also provide the broad input for detailed Crew Planning (DP6).

The management functions, relevant policies, and documents for the three levels of Project Scheduling in the Total Field Autonomy approach are introduced in Figures 7.32, 7.35, and 7.39. These three levels are called Management Level, Project Level, and Field Construction Level Scheduling respectively. The Management Level Scheduling phase, Figure 7.32, establishes the basic time frames in which major project components will be executed, and displays these components in relation to major project milestones. While this scheduling focus is similar to that of the Field-Based Project Management approach, the introduction of the *WORK BREAKDOWN* policy leads to Management Level Schedules (P_T-DOC 21) that correlate to control zones as determined by the Project Control Zone List (P_T-DOC 12). Based on the nature of the project and its composition, Management Level Schedules are developed by ESTABLISHING THE CONSTRUCTION SEQUENCE from scope definitions available for each control zone. This task is performed by the cost/scheduling engineer, working closely with other project team personnel. The sequence in which control zone schedules are developed follows the natural priority sequence of facility completion, as defined by the Facility Component Priority List

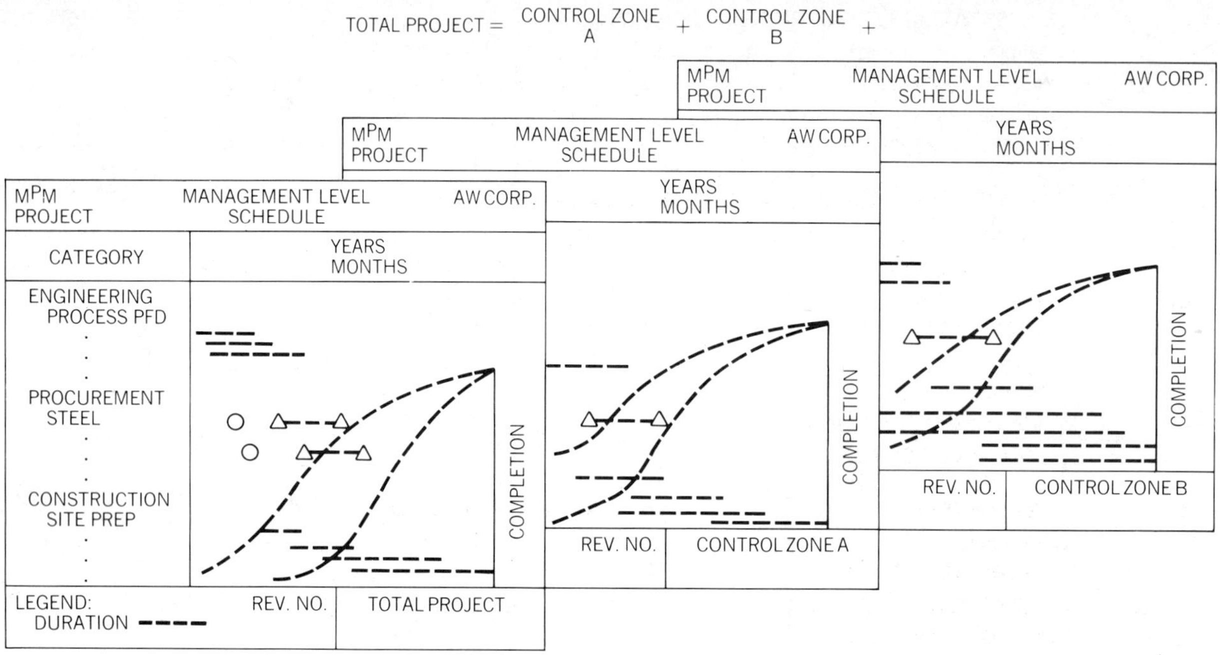

Figure 7.33 Management Level Schedules (P_T-DOC 21).

(I_T-DOC 1), and thus an orderly approach to manning and executing engineering, design, and construction is established. This scheduling effort is cyclical as the scope is defined, and requires the continuous interface with the client until each control zone schedule is approved. These control zone schedules must roll into the overall project Management Level Schedule, as illustrated in Figure 7.33, and must reflect not only engineering, procurement, and construction, but also contractual obligations such as completion of the project by a specified date.

As project Management Level Schedules are generated, manpower analysis is initiated first at the total project level, and eventually by control zone as warranted and necessary to support construction.* The methodology and techniques for manpower planning are the same as those introduced in Section 7.5. For example, Standard Construction Progress Curves (P_T-DOC 22) and Standard Construction Craft Mixes (P_T-DOC 23) form the basis for deriving manpower curves that are adjusted to match Management Level Schedules and factored estimate man-hour data. In addition, the *COST CODING* policy frequently leads to the preparation of progress curves and manpower profiles by major cost category (e.g., steel, piping, etc.) as well as by control zones. Provided control zones and major craft category manpower profiles and progress curves can be summarized in a number of ways, a considerable range of manpower data can be provided to the field for CONSTRUCTION RECONNAISSANCE and craft availability analysis. Typical examples of various progress and manpower curves are illustrated in Figure 7.34 in relation to control zone roll-ups. In the Total Field Autonomy environment, the client interacts closely with the contractor and reviews Management Level Schedules and overall project manpower and progress plans as these plans are issued.

Project Level Scheduling, Figure 7.35, commences once engineering and design progress makes sufficient scope detail available. The focus on project *WORK BREAKDOWN* analyses through control zones and *COST CODING* techniques has a pervasive influence in this phase of scheduling. Project work composition begins to be considered in terms of major work packages as defined by the Project Cost Code (P_T-DOC 14) and the project Work Package Structure (P_T-DOC 15). This staged development of construction logic networks for each control area is frequently based on Standard

*This detailed manpower analysis effort is initially focused on the engineering/design teams and is also performed on a control zone basis (see Chapter 4, Initiate Project Action).

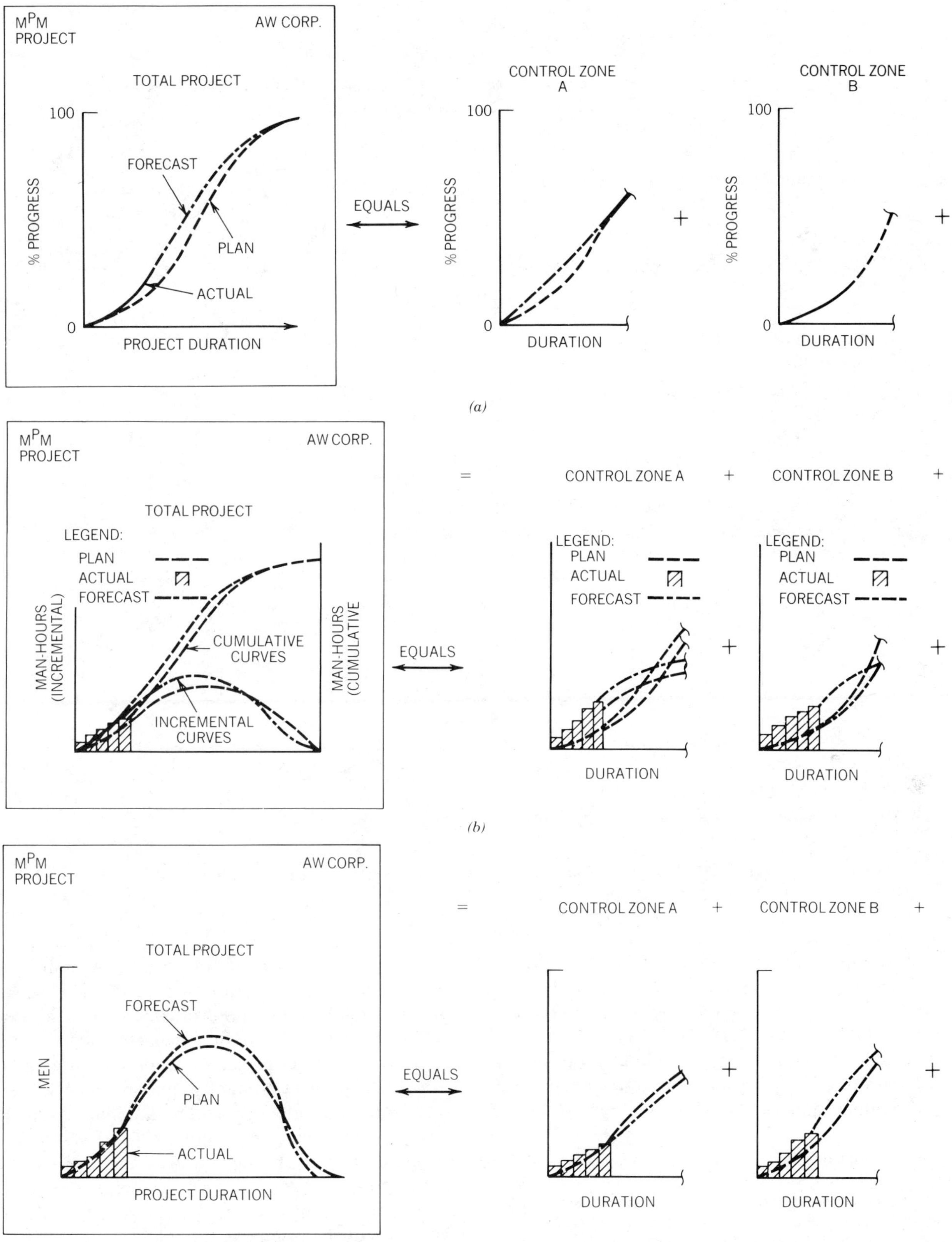

Figure 7.34 Construction progress and manpower curves. (*a*) Construction Progress Curve (I_T-DOC 5). (*b*) Construction Man-hour Curves (I_T-DOC 6). (*c*) Construction Manpower Curves (I_T-DOC 7).

S1 CONSTRUCTION SEQUENCE

Policies	Documents
WORK BREAKDOWN	P_T-DOC 24 Standard Construction Network Modules
COST CODES	I_T-DOC 8 Network Plans
HISTORICAL DATA	
LABOR DATA BASE	
NETWORK-BAR CHART ANALYSIS	
SCHEDULING FOCUS	
DATA PROCESSING	

S2 ESTABLISH CONSTRUCTION SCHEDULE

Policies	Documents
WORK BREAKDOWN	P_T-DOC 25 Project Construction Schedules
COST CODES	
UNIT MAN-HOUR	
NETWORK-BAR CHART ANALYSIS	
SCHEDULING FOCUS	
MANPOWER PLANNING	
DATA PROCESSING	

S3 DETERMINE PROJECT AND TRADE MAN-HOUR PROFILES

Policies	Documents
WORK BREAKDOWN	I_T-DOC 5 Construction Progress Curves
COST CODES	(Control Base Plan or Current Forecast)
SCHEDULING FOCUS	I_T-DOC 6 Construction Man-hour Curves
MANPOWER PLANNING	(Control Base Plan or Current Forecast)
DATA PROCESSING	

A3 DETERMINE PROJECT AND TRADE MANPOWER PROFILES

Policies	Documents
WORK BREAKDOWN	I_T-DOC 7 Construction Manpower Curves
COST CODES	(Control Base Plan or Current Forecast)
SCHEDULING FOCUS	
MANPOWER PLANNING	
DATA PROCESSING	

Figure 7.35 Project Scheduling decision process: project level scheduling (Total Field Autonomy approach).

Construction Network Modules (P_T-DOC 24)*, as depicted in Figure 7.36. These standard modules reflect *HISTORICAL DATA* (P_T 11) from the contractor's *LABOR DATABASE* (P_T 12), and are used by the cost/scheduling engineer on a first pass basis for identifying logical sequences of construction. Each module must be massaged to reflect the correct scope and project work breakdown structure. In addition, the best plan for constructing the facility (or facilities) for a control zone must be devised. Thus, the cost/scheduling engineer must have a solid grasp of the project's scope and all available scope and policy documents, and must work closely with other specialist project team members, including key construction personnel. When a standard module does not properly reflect the scope of a facility, the cost/scheduling engineer must develop the construction logic network based on a detailed analysis of the composition of the facility.

Once a Network Plan (I_T-DOC 8) is defined for a control zone and activity durations, and estimated

*The Standard Construction Modules are a logic-based network which describes a typical plan for construction work packages/activities for a project discipline or component, such as piping or turbine generator equipment. Standard modules represent historical practice and depict typical construction sequences for projects commonly constructed by the firm. Chapter 15 provides further information on standard network modules.

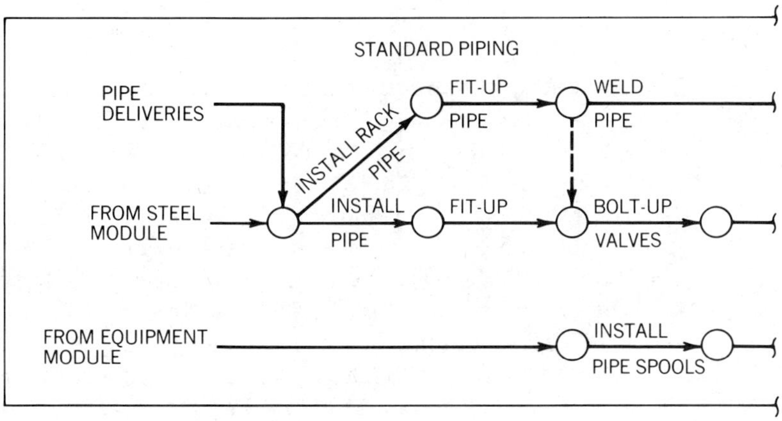

Figure 7.36 Standard Construction Network Module (P_T-DOC 24).

start dates and other relevant data are established, computer based *DATA PROCESSING* (P_T 36) is utilized to provide Project Construction Schedules (P_T-DOC 25).* However, work package activity durations must be analyzed in relation to *UNIT MAN-HOUR* (P_T 10) data for the work packages, as defined by Estimate Work Sheets (P_T-DOC 18) and Estimate Summaries (P_T-DOC 19). These durations are adjusted as necessary to reflect estimated man-hour content, work volume, construction methods, site productivities, and so on. The Project Construction Schedules are then manipulated by reevaluating construction logic sequences and revising work package activity durations, until these schedules are generally considered compatible with key milestones, drawing issue dates, and material delivery forecasts, as determined by the control zone Management Level Schedules and the best utilization of anticipated resources.

The man-hour loaded Project Construction Schedules serve as the basis for manpower analysis that will establish the client approved Construction Progress Curves (I_T-DOC 5) and Construction Manpower Curves (I_T-DOC 7). The techniques presented in Figures 7.8 through 7.11 are used to generate project and control zone Construction Progress Curves and Construction Manpower Curves. These plans must reflect the anticipated site labor environment and represent realistic targets in relation to the estimated site productivity (see Figure 5.12*c*), while meeting key facility milestone dates and fulfilling the priority sequence of the control zones. These progress and manpower curves would be identical in format to those illustrated in Figure 7.34, except that they are revised and based on the detailed construction schedule and control base estimate labor man-hours.

Project Level Scheduling has both a head office and field orientation. While this phase may commence in the head office it frequently culminates in the field once the field project team is fully operating. This shift to the field is accomplished by the relocation of the cost/scheduling engineering team (consisting of the same personnel as much as is possible), to the field. This helps ensure, through project team interaction, that the construction schedule

*This document would be similar to the document illustrated in Figure 7.27, with the exception that sorting techniques would assist in capturing all activities in a work package.

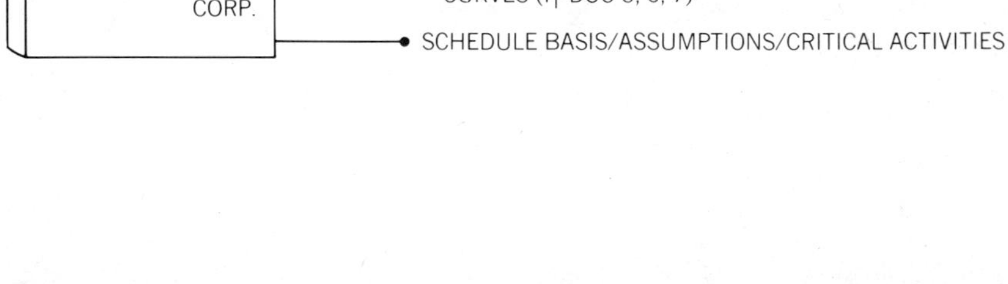

Figure 7.37 Project Construction Schedule (P_T-DOC 25) for control zone.

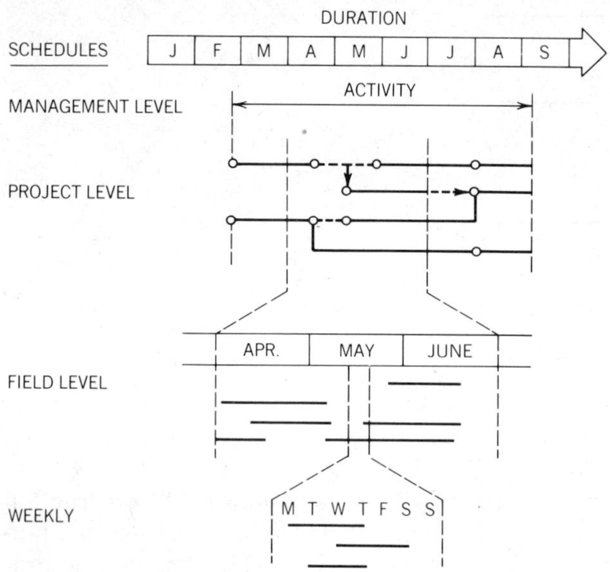

Figure 7.38 Roll-up/roll-down technique for scheduling compatibility and integration.

accurately reflects site conditions and major construction processes and methodologies. Consequently, construction personnel participate in the development, analysis, and review of each control zone schedule and its associated manpower and progress plans, thereby becoming fully indoctrinated to schedule and manpower requirements as approved by the client. The final Project Construction Schedule for a control zone becomes a fully documented schedule package, as shown in Figure 7.37. The contents of these schedule packages serve as the control base schedule from which short-range plans are derived and actual progress, status, and manpower will be measured and compared.

A general scheduling philosophy begins to emerge clearly during the Project Level Scheduling phase, as indicated by the interrelationship between the detailed scheduling data portrayed on Project Construction Schedules with the broad scheduling data on the Management Level Schedules. This

S1 CONSTRUCTION SEQUENCE

Policies	Documents
WORK BREAKDOWN	I_T-DOC 9 Fragnet Plans
NETWORK-BAR CHART ANALYSIS	
SCHEDULING FOCUS	
DATA PROCESSING	

S2 ESTABLISH CONSTRUCTION SCHEDULE

Policies	Documents
LABOR RECONNAISSANCE	P_T-DOC 26 Construction Schedule Package
SITE INSPECTION AND ANALYSIS	P_T-DOC 27 Monthly Activity Plan
COST CODES	I_T-DOC 10 Fragnet Schedule
NETWORK-BAR CHART ANALYSIS	
SCHEDULING FOCUS	
DATA PROCESSING	

S3 DETERMINE PROJECT AND TRADE MAN-HOUR PROFILES

Policies	Documents
WORK BREAKDOWN	
COST CODES	
SCHEDULING FOCUS	
MANPOWER PLANNING	
DATA PROCESSING	

A3 DETERMINE PROJECT AND TRADE MANPOWER PROFILES

Policy	Documents
LABOR RECONNAISSANCE	
WORK BREAKDOWN	
COST CODES	
SCHEDULING FOCUS	
MANPOWER PLANNING	
DATA PROCESSING	

Figure 7.39 Project Scheduling decision process: field construction control level scheduling (Total Field Autonomy approach).

TYPICAL SECTIONS

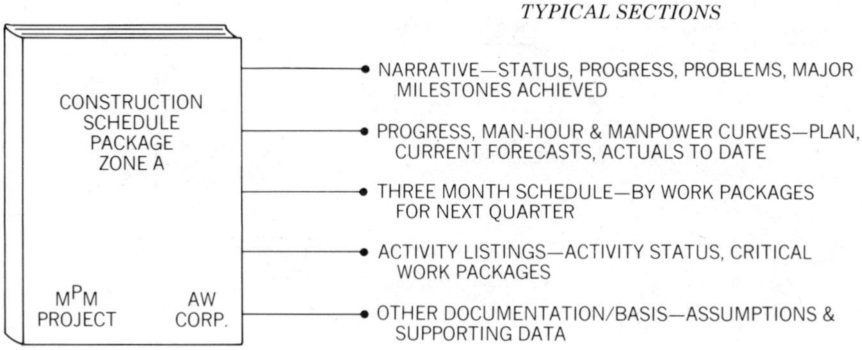

(a)

M^PM PROJECT — CONTROL ZONE A — AW CORP.

DESCRIPTION	COST CODES	% COMPL	MONTHS MAY	JUN.	JUL.
WORK PACKAGE S					
.					
.					
.					

	MAY	JUN.	JUL.
PLANNED PROGRESS UNITS PER MO.	800	1000	1000
PLANNED % PROGRESS PER MO.	8	10	10
PLANNED MANPOWER PER MO.	50	70	70
PLANNED MAN-HOURS PER MO.	8650	12110	12110

(b)

Figure 7.40 Construction Schedule Package (P_T-DOC 26). (*a*) Summary Schedule Package—Zone A. (*b*) Three-Month Schedule.

roll-up or roll-down technique, as specified by the *NETWORK-BAR CHART ANALYSIS* and *SCHEDULING FOCUS* policies, and, when combined with project work breakdown structure, provides management with the flexibility to locate problem areas at any level of scheduling detail. Furthermore, scheduling compatibility is maintained through this technique, because it enables Project Construction Schedules to meet key project milestones and control zone priority sequences, while requiring adjustments to Management Level Schedules to ensure that the detailed work package activities roll-up to the Management Level Schedule categories. Finally, the roll-up technique is carried into the Field Construction Control Level Scheduling phase. An illustration of the roll-up concept is presented in Figure 7.38.

The Field Construction Control Level Scheduling phase, represented in Figure 7.39, is a critical problem area because it melds the project control structure with labor man-hour estimates and quantities, construction CPM schedules, current site information, current drawing issues, and material delivery status, thus creating a short-range plan (e.g.,

M^PM PROJECT — AW CORP.
LOCATION: ______ CONTROL ZONE A PERIOD: ______

WORK DESCRIPTION	COST CODE	% COMPL.	PLAN START	ACTUAL START	PLAN COMPL.	ACTUAL COMPL.	PROGRESS UNITS
WORK PACKAGE S							
ACTIVITY 1							
ACTIVITY 2			←	DATES		→	
.							
.							

Figure 7.41 Monthly Activity Plan (P_T-DOC 27).

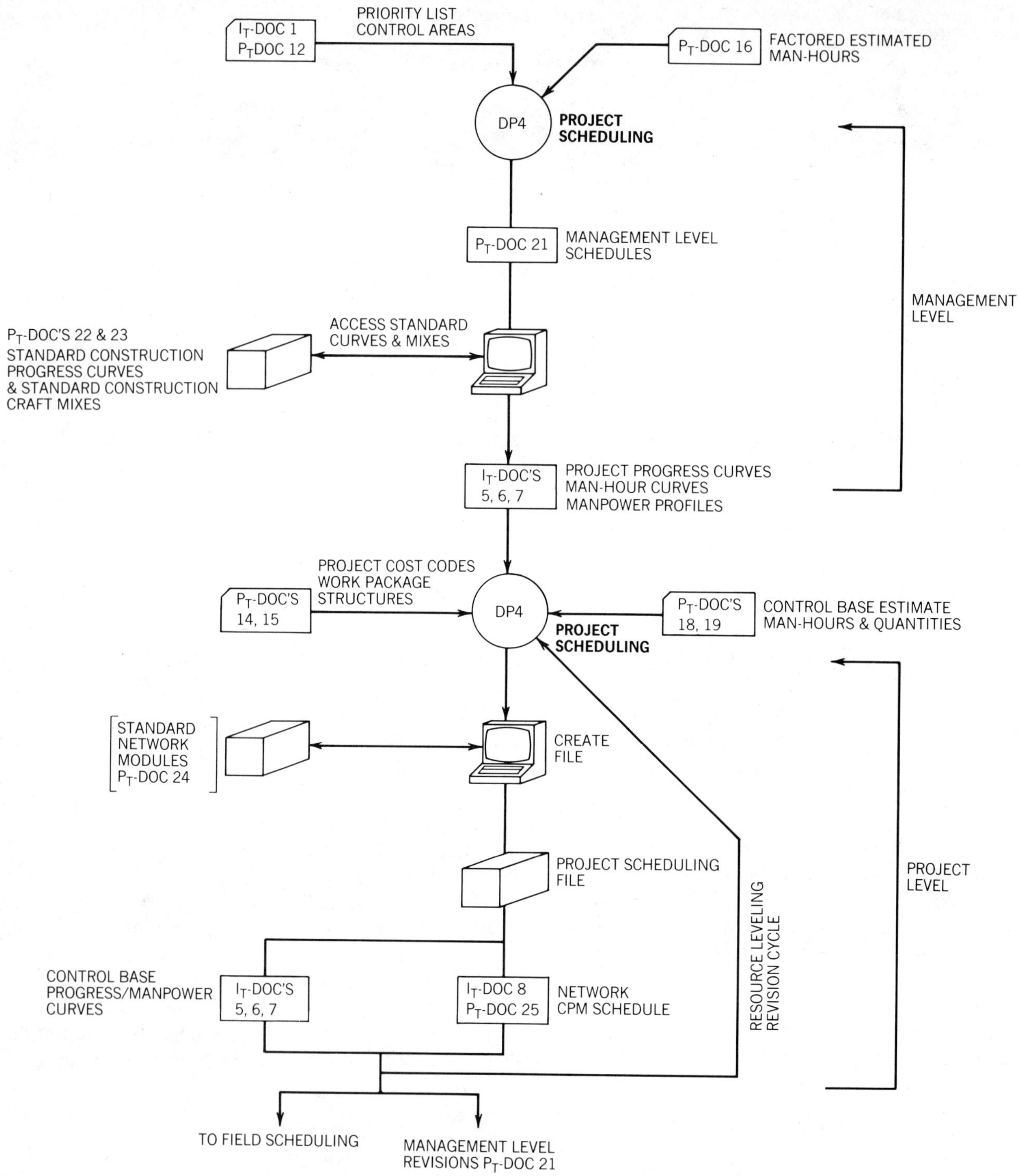

Figure 7.42 Project Scheduling Management Information System (Total Field Autonomy approach).

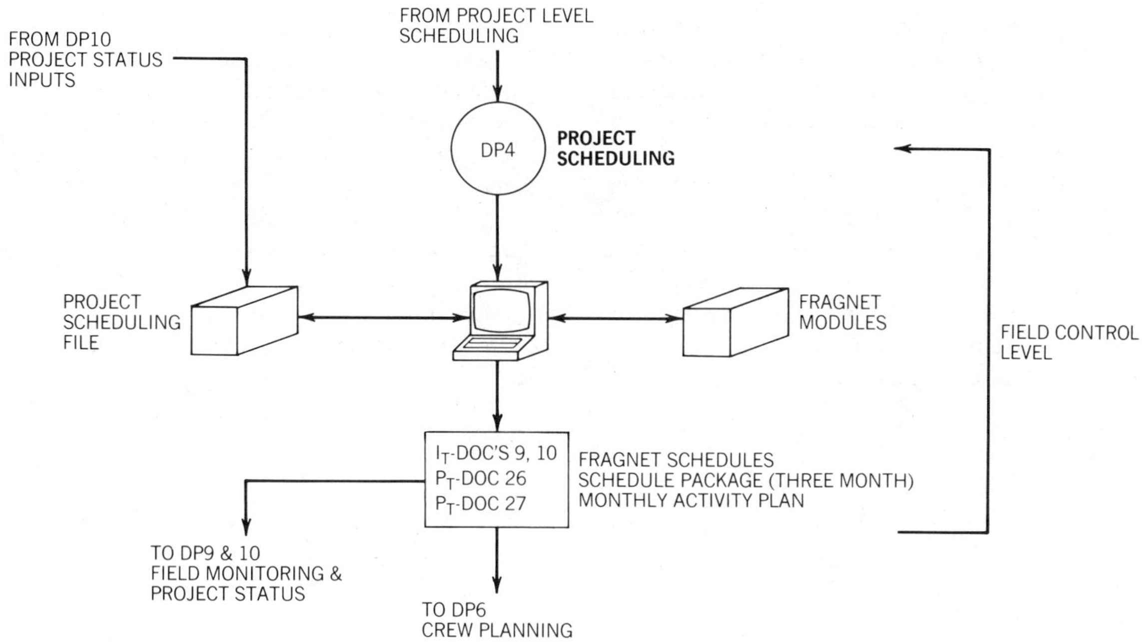

Figure 7.42 (*Continued*)

three months) that becomes the basis for both monthly and weekly crew planning. The *SCHEDULING FOCUS* of this decision phase attempts to ensure that project and control area construction progress plans and construction manpower plans are achieved within acceptable productivity targets by scheduling sufficient work in the current planning horizon to maintain planned and/or forecasted progress and manpower. The success of short-range planning lies in the ability of the project team to identify the drawings, equipment, materials, craft manpower, and other related items needed to support the work packages scheduled in the current planning horizon, usually a period of three months. This is accomplished by preparing and issuing a comprehensive Construction Schedule Package (P_T-DOC 26). As is schematically illustrated in Figure 7.40*a*, this document contains a considerable level of detail, which specifically describes the requirements for the current planning horizon in addition to including overall progress and manpower targets. A key document included in the Construction Schedule Package is the three-month schedule (see Figure 7.40*b*). The three-month schedule focuses on work packages in terms of progress and manpower requirements and start/completion dates. The work packages scheduled in the three-month planning period will be further dissected for detailed crew planning and scheduling. Short-range planning is normally performed by control areas and naturally results in continuous *LABOR RECONNAISSANCE* (P_T 6), *SITE INSPECTION AND ANALYSIS* (P_T 7), and *MANPOWER PLANNING* by *WORK BREAKDOWN* and *COST CODING* approaches. The short-range planning technique described in Figure 7.12, coupled with specially selected fragnet planning analyses, produces the Construction Schedule Package. As part of the package, a Monthly Activity Plan (P_T-DOC 27) (seen in Figure 7.41) is provided to the respective control area discipline superintendents (e.g., civil, structural, piping, etc.), for crew planning.

Short-range planning is a cycle that continues throughout the life of the project. As new work areas (e.g., piping, then electrical) commence, the Construction Schedule Package expands to cover a broader spectrum of construction activity. Furthermore, data from the Project Status, Analysis, and Reporting decision process, along with major project directives/strategies from Project Management Control, must be incorporated into future issues of the Construction Schedule Package. This package will highlight existing and potential problem areas. The impact of these problems on schedules and the appropriate remedial action (possibly through the Project Management Level) that is necessary to resolve problem areas is also stated in the schedule package. The Construction Scheduling Package also includes revised progress and manpower forecasts (when different from plans), and modification to construction logic and work package durations, as dictated by current project status, revised schedules, and changes in scope.

The Project Scheduling Management Information System at the Total Field Autonomy level expands to incorporate the increasing use of standard scheduling modules for fragnet and project level planning and scheduling. This scheduling effort is based on project control philosophies that allow for integration of the schedule with project labor man-hour estimates through common work packages. Furthermore, using computer roll-up techniques, the greatest level of detail can be summarized to the least level of detail, thus ensuring that all levels of scheduling data are compatible for either top level project management at one extreme, or field supervisors at the other extreme. This MIS application is equally useful for project schedules, progress curves, and manpower data. In all other aspects, the MIS summarized in Figure 7.42 is similar to that of the Field-Based Project Management approach presented in Figure 7.31.

CHAPTER 8
Field Staffing

The decision to prepare for and manage field construction activity requires the contractor to consider the field management requirements for a project team, and then to design and subsequently staff that field project team. This effort usually requires the contractor initially to select key personnel as the nucleus of the field construction and management group, and subsequently to expand and develop the project team. This team will then proceed under the management group's direction to meet project, site, and client requirements. The establishment, control, and management of the field project team is a continuing and dynamic management problem for the contractor, who must therefore react to the changing needs of technical areas of the project and meet the operating problems for each construction phase. Thus, addressing and solving field staffing problems absorbs a large amount of top management's consideration and effort.

In many cases, the reputation and public image of the contractor is built through contact with field personnel alone, and on small projects through dealings with one company representative. Often top management must delegate full authority to one or two members of the project team, and, as a consequence of this delegation, become responsible for the problems raised by their decisions, attitudes, and positions on sensitive issues. Thus, the top management's handling of subcontractors or efforts to remove troublesome but influential workers may have repercussions on the professional image of the contractor. In addition, the success of the project and the magnitude of the profit earned depends, to a large extent, upon the skill and dedication of a few people.

On large projects the selection of a suitably qualified project team is considerably more difficult. This is especially true when a highly specialized team capable of working with a sophisticated client must be quickly established by a contractor who has access to a large pool of technical and professional people. In this latter case, the decision maker cannot simply rely on a personal knowledge of a limited number of staff as can small firms. Instead, staff selection requires the meaningful use of personnel files and a personal evaluation and selection process that is sensitive to the needs of the client, to the perceptions of staff, and matches staff with the needs of a project in a professional effort that is expected to enhance the reputation of the firm. Thus, the selection of key personnel and the field staffing of the project team is a critical management problem for the contractor.

8.1 THE BASIC DECISION PROCESS

The Field Staffing decision process is similar in many respects to the Initiate Project Action decision process in that it is also concerned with the manpower aspects of the contractor's involvement in the construction project. Initially it is concerned with the selection of a suitable "cast of characters" chosen mostly from the so-called permanent staff. These staff, however, rarely hold top management positions in the company in the small and medium-

TABLE 8.1 SCOPE OF FIELD STAFFING PROBLEM AREA

Major Decision Problems	Component Problem Statements
21. What field team:	What magnitude of field staff effort is required for the project? What is best project team structure? What is best size of field project team and what support staff are required?
22. Who is available:	What is location and availability of suitable staff? Where can suitable support staff be found?
23. When available:	When can suitable key staff be released to the project? When can suitable support staff be released/assigned to the project?
24. How is field team to be managed:	How should field project team organizational structure be developed and staffed? Do field staff have function overloads?
25. What are work force requirements:	What jurisdictional areas are involved? What are established work rules? What are hiring/firing procedures? What are subcontractor management procedures? What are pre-job conference arrangements for project team staff and subcontractors?

sized firm environments because they have a dominant field function to perform. It is only in large firms that the field staff come from departmental areas whose managers are in the contractor's top line of managers. Once the field management and supervisory team has been chosen, however, its main concern is with the selection, acquisition, and maintenance of the field work force itself via hiring and firing procedures. It is not surprising then to find that the scope of the Field Staffing problem area, as shown in Table 8.1, is very similar to that of Initiate Project Action (see in Table 4.1). The only difference in terms of major decision problems, as can be seen in a comparison of these tables, is that the Field Staffing problem area does not address Project Management Requirement issues, but instead substitutes Work Force Requirement issues (labeled 25 in Table 8.1).

Field Staffing is a decision area that commits staff to work on the construction site. The field project team is the very visible project management and engineering staff acting in the contractor's behalf. Staffing this team is a very critical area for the contractor, because the field team establishes the contractor's ability to perform, displays the professionalism and effectiveness of the firm's planning and management efforts, and especially demonstrates its experience in melding together a project work force that is capable of achieving planned progress with quality work within time and cost constraints. Thus, the selection of suitable staff for key positions such as the general and craft superintendents and general foremen directly impacts the quality of field work and labor relations, while the

TABLE 8.2 GENERAL PROCEDURE FOR FIELD STAFFING

Procedural Step	Management Duties and Tasks
1.	Evaluate desirable field project team structure.
2.	Establish field project management guidelines, policies, and procedures.
3.	Identify key project team positions and assess availability of personnel for these positions.
4.	Select key staff and release them to the project.
5.	Design project team structure and identify desirable growth pattern of project team.
6.	Assess availability of other potential key and support staff, select personnel, and release them to project.
7.	Develop and clarify working agreements with relevant labor organizations.
8.	Brief field staff on relevant company policies.
9.	Establish lines of communication with relevant labor organizations.
10.	Monitor management effectiveness of project team, levels of performance, and quality of effort.
11.	Adjust project field team as project conditions, scope, and client/owner requirements change.

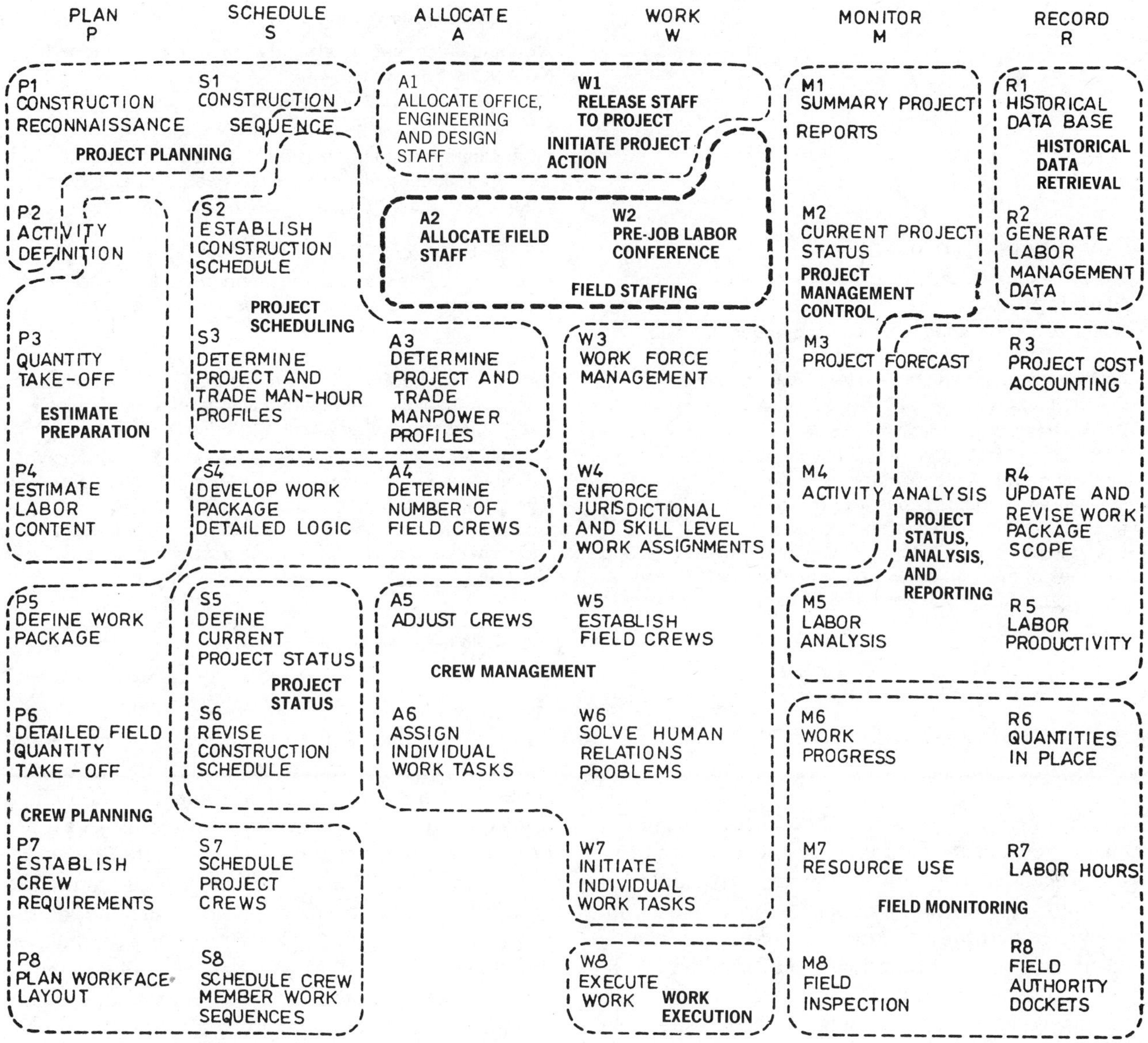

Figure 8.1 Functional plot of the Field Staffing decision process.

selection of the project and methods engineers will impact the cost-effectiveness of field construction methods. Finally, the selection of the cost and scheduling engineers will impact the quality and accuracy of the project's cost, status, and progress reporting system.

The general procedure for performing the Field Staffing decision process is given in Table 8.2. The procedural steps describing the various management duties and tasks are similar to those for the Initiate Project Action decision process (see Table 4.2), and lead to a similar functional plot of the involved manpower management functions, as shown in Figure 8.1. A full functional statement of the decision process is given in Figure 8.2. The relationship between the Field Staffing decision process and other decision processes with which it interfaces is presented schematically in Figure 8.3.

WI RELEASE STAFF TO PROJECT

- Directs office, engineering, design, and field staff to begin project work.
- Requires authorization from top management and release from previous assignments.
- Considers availability of project decision makers, relevant contract documents, office space, and desirable rate of buildup of head office and field project teams.

A2 ALLOCATE FIELD STAFF

- The selection, allocation, and commitment of staff best suited to the project.
- Requires a knowledge of project scope and complexity and the qualifications and skill level required to perform field construction and the desired level of field project management.
- Considers past experience, suitability, and availability of potential key staff, and their ability to develop a well-motivated project team, and work force in relation to the special requirements peculiar to this project; defines and finalizes field management functions, nature of project team structure, desirable management control levels, and methods of managing the project team's growth, redeployment, and effectiveness as a managing body.

W2 PRE-JOB LABOR CONFERENCE

- The development and clarification of working agreements with relevant labor organizations and the establishment of lines of communication.
- Requires thorough knowledge of company manpower management policies and procedures, labor organizations and ability and experience in labor negotiations, and subcontract coordination (if prime contractor).
- Considers project labor requirements and importance of establishing agreements, strives for clarification and solution of jurisdictional dispute areas, and during negotiations knows desirability of reaching compromises, when to draw the line, and knowledge of subcontract administration.

Figure 8.2 Functional statement of the Field Staffing decision process.

8.2 FUNCTIONAL PERFORMANCE TECHNIQUES

Field Staffing, as indicated in Section 8.1, is similar in many respects to the Initiate Project Action decision process in terms of the manpower aspects of the contractor's involvements in the construction project. It naturally follows that similar concepts and their relevant functional performance techniques influence the Field Staffing decision process area as well. Different applications of the techniques introduced for Project Initiation (see Section 4.2) emerge, however, when they are applied to the resolution of Field Staffing problems. These variations most frequently occur in the large firm design–construct project environment. For example, while order of magnitude project costs are often used with historical factoring techniques to evaluate the total manpower* for a particular organizational area such as construction, other techniques which focus on the task force approach to the design and size of the project team itself provide a more accurate fix on the field staff manpower content of a project.† With respect to the latter technique, the development of an organizational chart of the project team would not necessarily be required. A chart delineating its general structure and identifying the classifications considered to be involved in the project (e.g., project manager, superintendents, field engineers, etc.,) and the estimated number of man-months required of each classification over the life of the project, would suffice. This field staff planning, estimating, and scheduling technique is illustrated in Figure 8.4. Both the man-hour curve and rough manning chart approaches to Field Staffing are acceptable in the early phases of a design–construct project, because the immediate concern is with the proper design and growth of the engineering, design, and procurement project teams. The key field manager, however, will be selected early in the project, along with several other key field staff members such as area managers and/or general superintendents.

At the other end of the spectrum, when construction control budget estimates are required and construction is commencing, project team design and growth models based on the acceptance and modification of standard project team structures may be used for the Field Staffing analysis. Similarly, the development and implementation of authorized project team positions and agreed upon

*See Section 4.2 and Figure 4.5.
†See Figure 4.9*b*.

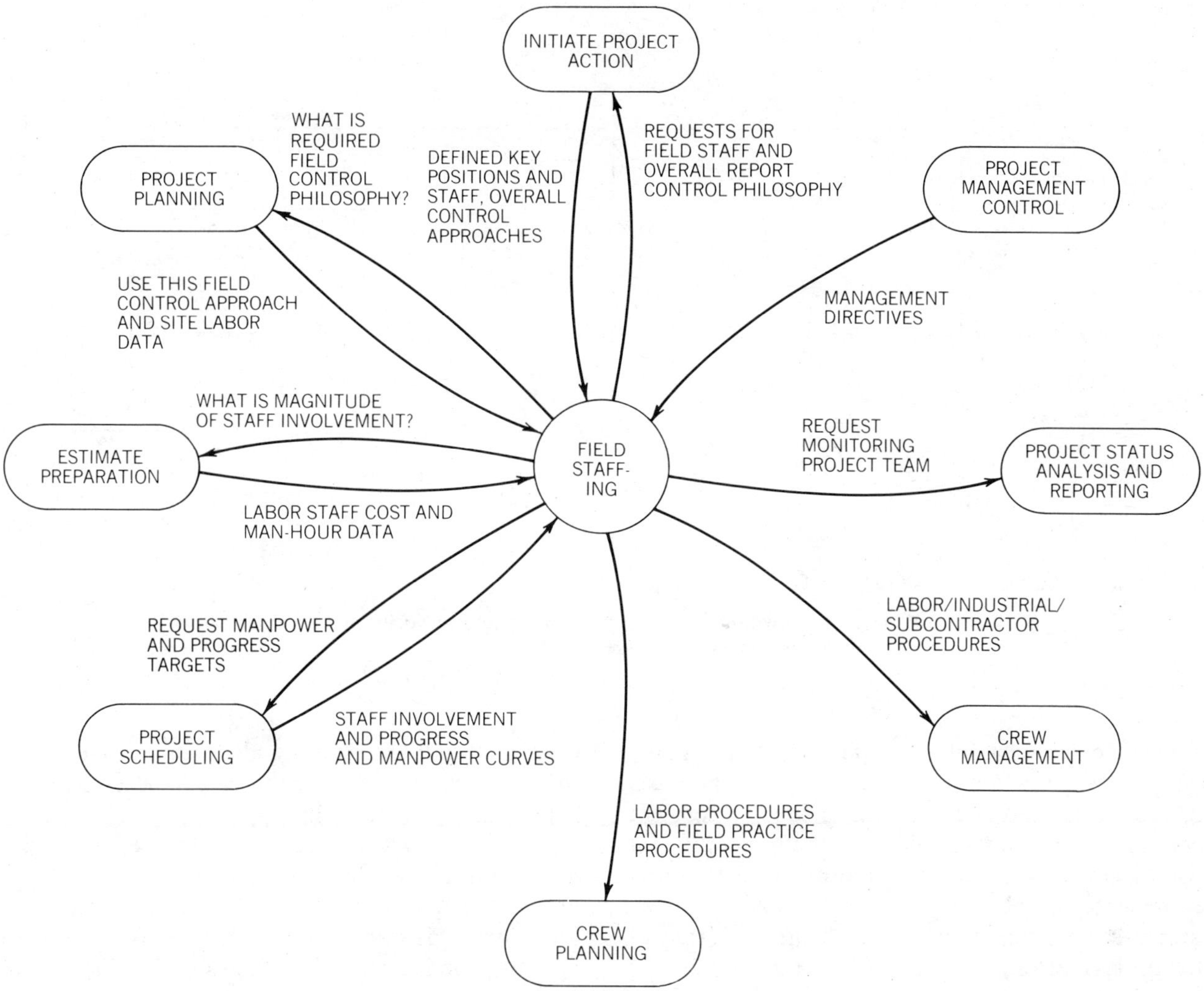

Figure 8.3 Schematic illustration of Field Staffing decision process interfaces with other manpower management decision processes.

CLASSIFICATIONS	DURATION-MOS.												TOTAL MAN-MOS.	AVE HRS/ MO.	TOTAL MAN-HOURS
	1	2	3	4	5	6	7	8	9	10	11	····→			
PROJECT MANAGER	1	1	1	1	1	1	1	1	1	1	1	····→	24	173	4150
PROJECT ENGINEER	1	1	2	2	2	2	2	2	2	2	2	····→	·	·	·
FIELD ENGINEERS			1	1	1	2	2	4	4	4	4	····→	·	·	·
SUPERINTENDENTS		1	1	1	3	3	3	5	5	5	5	····→	↓	↓	↓
TOTAL FIELD STAFF															

Figure 8.4 Field staff analysis by classification and required man-months.

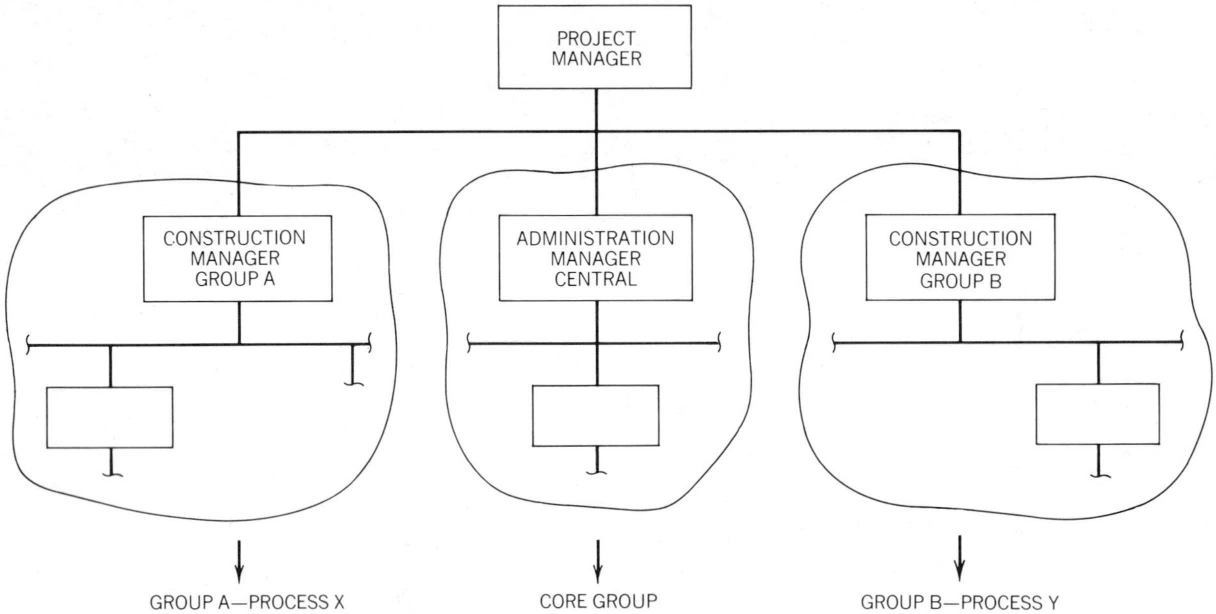

Figure 8.5 Project team with core management group (project focus) and geographical groups for specific project facility processes.

manning chart schedules (as described in Figures 4.7 and 4.8) are often relevant. However, the basis for the design and development of the field project team is far more intangible than the concepts and techniques used in the Initiation of Project Action as presented in Section 4.2. For example, the factored estimate approach for head office design of the project effort (i.e., based on the number of drawings, etc.)* rarely provides a solid base for field project team design. It may be more useful to relate field project team design and growth to the project control philosophy in relation to the project's nature and the field labor man-hours to be controlled. This approach is illustrated in the construction situation where the project is separated into distinct geographical groupings, often simply resulting from physical distance separating the components of a facility. In order to ensure common control procedures and practices, a core group is established and functions as the coordinator of overall project management and control. Each geographical group could have its own semiautonomous project team, as depicted in Figure 8.5.

Apart from these differences in the focus and relevance of otherwise common concepts in the Initiate Project Action and Field Staffing decision processes, the Field Staffing decision problem area raises many new issues that must be considered. These issues are equally applicable to all levels of construction and are mostly concerned with staff location and availability models, labor/industrial relations techniques, and subcontractor interface techniques (as applicable to the construction manager or prime contractor). These issues and considerations will be the focus of the remaining discussions of this section.

The construction industry is unique in that it employs labor on a temporary and fragmented manner for the sole purpose of constructing one-of-a-kind facilities at locations and time periods determined by the needs of isolated clients. A contractor attempts to overcome this labor utilization problem by building up and maintaining a nucleus of highly skilled craftsmen which can (through the skill and efforts of the contractor) be continuously employed from project to project. Similarly, the contractor's front-line supervision, especially general foremen and foremen, are vital but are not permanent management resources. Thus, it behooves the contractor to keep them on the payroll if at all possible, so as to be able to maintain high standards of quality and performance. The contractor is forced to depend on the capability of front-line supervisors to meet and satisfy the skill levels and management

*See, for example, Figure 4.6.

requirements demanded. The capabilities of these supervisors are often unknown. The contractor must achieve productivity levels through the efforts of front-line field managers/supervisors and a field team comprised mostly of employees who are transients and who frequently work for several contractors in different work regions and locations in any one year. Nevertheless, this composite field management team often represents the visible public image of the contractor, on which a reputation is built. A simple and practical solution to these industry-wide problems is to cultivate, through continuous employment, a nucleus of foremen loyal to the contractor, while at the same time maintaining these employees in a nonpermanent employment status. In a union construction environment, nonpermanency is ensured, from the contractor's point of view, in part because of the foreman's continued affiliation with his or her local union. The nonunion contractor has greater flexibility in this respect and can more readily ensure permanent positions for such key people if desired. This is because the contractor is not hindered by trade and/or work rules, and can reclassify jobs frequently. This problem area is particularly critical to small and medium-sized contractors, as these contractors rarely have the overhead capital to continue to employ foremen and skilled craftsmen during slow construction periods. Small contractors, and often medium-sized contractors as well, are often forced to utilize the same employment strategies with superintendents unless the contractor is large enough to maintain permanent superintendents on a full-time basis. In many respects, the retention of loyal foremen is easier for the small to medium-sized contractor to achieve, because of the localized nature of the work region.

The large union contractor attempts to retain loyal foremen especially in work regions where the contractor is established as a permanent construction firm, and therefore has high expectations of gaining additional projects. When a large contractor enters a particular region on an infrequent or one time basis, however, difficulties in employing top quality foremen are often encountered. On projects where the construction cycle spans several years, the large contractor has the opportunity to cull for retention the foremen who demonstrate high skill levels and good management potential. The large nonunion contractor has a considerable advantage, because there is no need to abide by potentially restrictive work practices. Foremen can be reduced temporarily to the laboring level, or can be used outside of their craft, and thus be maintained in some form of permanent employment status. Large union and nonunion contractors will both maintain permanent superintendents on the firm's payroll, because these personnel are essential on projects with large manpower forces (which therefore require many foremen), especially in work regions where the availability of suitably qualified foremen may be inadequate. In this construction environment, the superintendents (or assistant superintendents) participate heavily in the Crew Planning, Crew Management and Field Monitoring decision making areas.

In the construction industry, as with other industries, the employment of labor naturally results in labor relations problems and, in the case of union labor, in industrial relations problems as well. A most essential aspect of the Field Staffing decision process is the establishment of policies, and therefore techniques, for solving labor and industrial relations problems. Some contractors, especially those in the medium-sized range, enjoy the benefits of craft loyalty to a greater degree than others. This is due to their concern for field managers and their management approach to the resolution of work force problems and issues. However, this rather personal management approach is not particularly viable on extremely large projects. The nonunion contractor has an additional advantage in that labor can be moved from one work location to another, and from one type of work to another, without serious repercussions.

When dealing, with unions or site committees, the contractor usually establishes a hard-line position on site benefits and conditions from which there will be no compromise or movement. The contractor will frequently have a list of "give away" items, however, that can be used to bargain with when dealing with union agents or site delegates while cooperating on peripheral dispute matters. The union contractor must also have an extensive working knowledge of local union agreements and work rules, including craft wage rate structures and margins, and general foremen/foremen to journeymen ratios.

The contractor's basic concern with site subcontractor interactions is to establish proper and effective lines of communication in order to ensure the integration of the overall project work plans and schedules, and to facilitate the coordination of the work force guidelines and practices of each subcontractor on the construction site. The recognized technique is to set up pre-job conferences with the selected subcontractor(s) prior to their required involvement in the project. At these conferences, agreed procedures and policies are adopted. This approach is particularly significant to the prime/general contractor who is often responsible for many subcontractors.

In summary, the Field Staffing decision process is directed towards the achievement of two major independent but interrelated objectives. One objective is the selection, culling, and organizing of a field construction management project team with the expertise and capabilities needed to execute all phases of the field construction effort. The second objective of Field Staffing is to establish procedural approaches to labor relations, craft hiring policies, and subcontractor interfaces. Together, these two objectives are accomplished through the implementation of standard operating procedures combined with the knowledge of company data on prospective project team members. The results are project team organization charts, project focused procedures, staff man-hour and cost estimates and manpower profiles, and the like. The timely transfer or allocation of company personnel to the project team depends on data generated through Field Staffing decisions in relation to data input from other decision processes such as project schedules, and on particular decisions concerning project control philosophies. The integration and documentation of these data flows through the Field Staffing decision process constitutes the Management Information System for this decision process. While this MIS is fully documented in construction practice at the large firm level, at other levels the MIS is an informal information process wherein normal construction management roles and construction practices are assumed for each project by a similar "cast of characters."

8.3 THE PROJECT TEAM COMPONENT APPROACH

Before the project is scheduled and issued to the field for actual construction start-up, the project manager selects the field superintendent. At the small firm level this selection is not difficult and rarely requires more than a mental assessment of the personnel on the company's payroll, and possibly a discussion with a partner if the firm is jointly owned. FIELD STAFF ALLOCATION (A2) is governed by the policies of *COMPANY ALLEGIANCE** (P_S 8), in which the company relies on loyal superintendents and foremen and on a simple field team in which each member carries out *NORMAL ROLES* (P_S 1). On extremely small projects of simple construction complexity, a foreman would likely be assigned the key field management position. In this capacity, the foreman would assume all major field responsibilities. If a foreman is the key field manager, the project manager would probably keep a close contact with the field and assume minimal job superintendent type functions. Larger and more complex projects involving a larger labor force would normally require that a job superintendent be assigned key field management responsibility. Thus, while Field Staffing is not a significant or formal decision process, the contractor must coordinate the field start schedule with the availability of a superintendent/foreman, and must check the local hiring halls for the ready availability of craft.

Because in certain environments the superintendent and/or foreman are rarely permanent employees of the firm, many owners are timid about the details of labor cost and work progress information they afford to their field managers. The ramifications are obvious as this attitude can create distrust between field and head office management, producing an antagonistic relationship between these two management levels. A relationship of this nature can severely affect the job satisfaction level of the field manager, and ultimately may negatively influence his managerial capabilities. The smaller firm strives to alleviate this problem by hiring superintendents and maintaining them on the payroll, thus developing pseudo-permanent positions, and with them, company loyalty and mutual trust. The problem escalates at the foreman level because the first journeyman employed on the job becomes a foreman via the union agreement. These managers can thus be foremen on one project and journeymen on the next. Under such circumstances, their loyalty will generally remain with the union and not with the contractor. If enough work is available, however, the small contractor will attempt to retain key foremen by extending their employment with the firm, thereby promoting better communication and better relationships between the head office and the field.

*The small firm policies relevant to the Field Staffing decision process are collected in Table 8.3.

TABLE 8.3 POLICIES OF THE SMALL FIRM

Policy		Description
P_S 1	*NORMAL ROLES*	Project team members assume normal project team roles.
P_S 8	*COMPANY ALLEGIANCE*	Field management is staffed from men loyal to the company without hiring on a permanent basis.
P_S 10	*UNION RELATIONS*	In a union environment, maintaining good relations with each union (through the business agent) is helpful to successful project management.
P_S 11	*UNION PRACTICE*	Field management must subscribe to local union practices when operating in a union environment.

The small contractor in a union environment rarely deals directly with unions in any PRE-JOB CONFERENCE (W2), but gains its strength and bargaining power through affiliation with a local chapter of contractors such as The Associated General Contractors, who collectively bargain with individual unions to establish labor control. Accordingly, during Field Staffing the small firm follows the pattern and guidelines established by the labor contract, and strives to maintain good relations with local union officials. In nonunion situations the contractor would tend to develop close ties with a small group of skilled craftsmen by continuity of employment, and in a union environment, might notify the various business agents of an intention to use their manpower. Formal meetings, however, are obviated because the contractor is well acquainted with the local union officials. The contractor strives to maintain good relations with each union's business agent, as the policies on *UNION RELATIONS* (P_S 10) and the monitoring of local *UNION PRACTICE* (P_S 11) suggest. Good relations with each union's agent may circumvent problems during the field construction process.

Field staffing does not raise any significant problems in terms of manpower management for the small firm, mainly because the work force is small. Consequently, no formal decision process network is presented at this level of operation.

8.4 THE TRAVELING PROJECT MANAGER

Field Staffing first emerges as a structured decision problem area at the medium-sized firm level, because of the nature and emphasis of the Traveling Project Manager management approach. This management approach requires that the field team be manned with staff capable of achieving a high level of management control for the given project. In keeping with this philosophy, the medium-sized firm establishes a standard organizational chart for the project team in which each member assumes *PLANNED ROLES* (P_M 1).* As previously mentioned, the standard project team expands to include, where appropriate, field personnel such as project and/or field engineers, assistant and specialty superintendents, and timekeepers†. The project/field engineer is involved in many aspects of the construction process. His or her functional role is related to field analysis in support of crew planning, work progress, and performance analyses. In addition, the project/field engineer often assists the project manager and superintendent in the field scheduling of work activities. This scheduling responsibility ensures that proper coordination occurs with the project subcontractors. The timekeeper's role in manpower management is primarily the monitoring and recording of labor hour expenditures.

Medium-sized firm partners support and actively participate in local chapters of contractor organizations, and in negotiations with unions on labor contract terms. Since they have a bigger share of labor usage they stand to gain (or lose) the most. Hence, it is in their interest to be in the forefront of establishing relationships between increased wages and improved conditions with the elimination of troublesome work rules. In addition, because of their need for trouble-free production, they maintain close contact with all relevant business agents.

During the Initiation of Project Action, a decision is made to support the Traveling Project Manager by providing him, as early as possible, with a job superintendent. Thus a key superintendent is RELEASED TO THE PROJECT (W1) and assists the

*The medium-sized firm policies relevant to the Field Staffing decision process are illustrated in Table 8.4.

†See, for example, Figure 4.11.

TABLE 8.4 POLICIES OF THE MEDIUM-SIZED FIRM

Policy		Description
P_M 1	*PLANNED ROLES*	Project team structure and assignment of decision makers to a project depend on project complexity and size, and on the division of management functions desired to effectively manage the project.
P_M 10	*COMPANY ALLEGIANCE*	Key craft foremen are staffed from foremen loyal to the firm. Foremen other than loyal company foremen are selected on the basis of experience, managerial capability, and desire.
P_M 29	*PROJECT TEAM SPIRIT*	Active participation of all field managers in the management process promotes goodwill and a positive management environment.

project manager in project scheduling, in determining the project team size, and in selecting and ALLOCATING FIELD STAFF TO THE PROJECT (A2) to meet the established construction schedule. Field staff positions are assigned to foremen who are *LOYAL TO THE COMPANY* (P_M 10), and a significant attempt is made to get these field staff on site and motivated by fostering *PROJECT TEAM SPIRIT* (P_M 29).

On small projects, a key superintendent can maintain a close watch over foremen, often obviating the need for placing one of the firm's own key foremen on the project. On projects requiring tighter management, key foremen are more likely to be used because of the increased administrative responsibilities involved. The remaining field personnel assigned to the project are chosen according to the number required to effectively manage the labor force and meet the construction requirements. Since the field management force is not large, project staffing can be carried out informally. The majority of field personnel are permanent employees, and the key foremen are normally the company's loyal foremen. As field staffing is completed, the project can be released to the field for construction start-up, and the field construction phase commences.

The Traveling Project Manager Field Staffing decision process is illustrated in Figure 8.6.

8.5 FIELD-BASED PROJECT MANAGEMENT

Field Staffing for the small firm is virtually a nonissue and only pertains to the selection of the key field supervisor, which occurs *after* bid award at the inception of construction start-up during Project Scheduling. Furthermore, a heavy reliance on the contractor's intimate knowledge of the local labor environment virtually eliminates most decision problems concerning labor and subcontractor relations. At the medium-sized firm level, however, Field Staffing begins to emerge as a problem area in so far as the Traveling Project Manager approach requires the *early* involvement of the key field superintendent in Project Scheduling *prior* to bid submission. While some projects in this environment are large enough to warrant decisions on field organizational structures, in most cases the medium-sized firm relies on the company standard organizational approach, and therefore determines only the types of positions required and the availability of personnel for those positions. These approaches are no longer viable in the large firm environment where the contractor must control hundreds or even thousands of craftsmen over a significant period of time. This situation is compounded when the project is located in an isolated area and the contractor has only a limited knowledge of the local labor environment. Thus, the Field Staffing decision process emerges as a significant management problem. As portrayed in Figure 8.7, Field Staffing for Field-Based Project Management reflects an increasing level of complexity by its double layered decision process structure. That is, it comprises both a head office located phase, Figure 8.7*a*, and a field focused phase Figure 8.7*b*. Also reflective of the increased complexity in this environment is policy formulation and the manner in which the policies impact the performance of the various manpower management functions of Field Staffing.

Field Staffing is initially concerned with the design and staffing of the field project construction organization. Because of project size and complexity, and the magnitude of the field labor force, the planning and manning of the field management team is an integral facet of project team development. The size and structure of the project construction

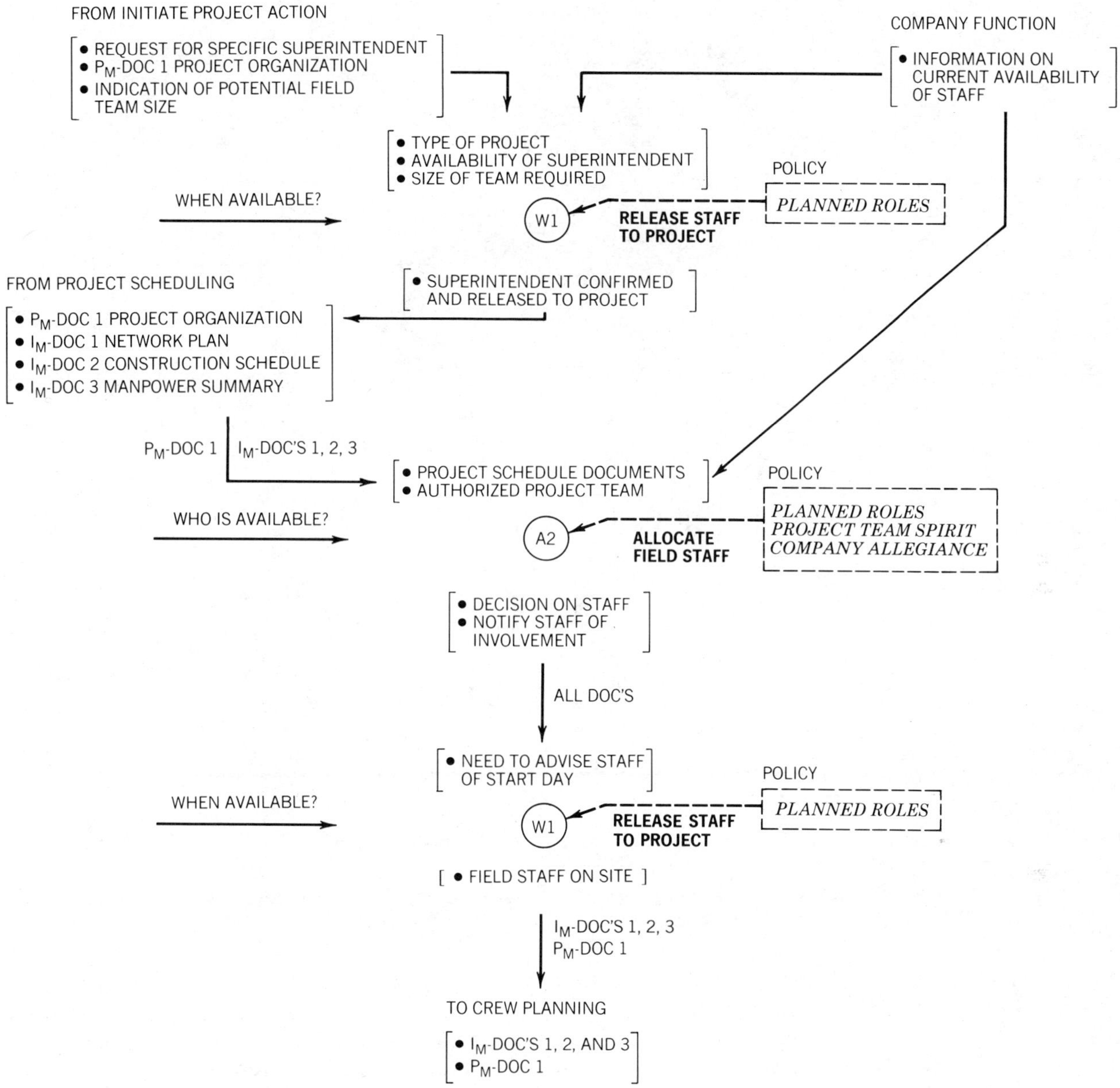

Figure 8.6 Field Staffing decision process (Traveling Project Manager approach.)

organization is influenced primarily by three factors, namely the management control philosophy as it relates to desired levels of management control, the size of the field labor force to be managed, and the technical complexity of the project. These factors are carefully considered when defining the functions that field management must perform and when identifying the authority and corresponding responsibility that each field manager is to have with respect to those functions.

The manpower management problem of Field Staffing includes first the identification and selection of the key construction management nucleus of the field team, and then the full design of the project construction organization (see Figure 8.7*a*). The key construction managers (e.g., resident field project managers), and superintendents (e.g., general superintendents), are selected as part of the Initiate Project Action decision process when the basic overall project control philosophy is established.

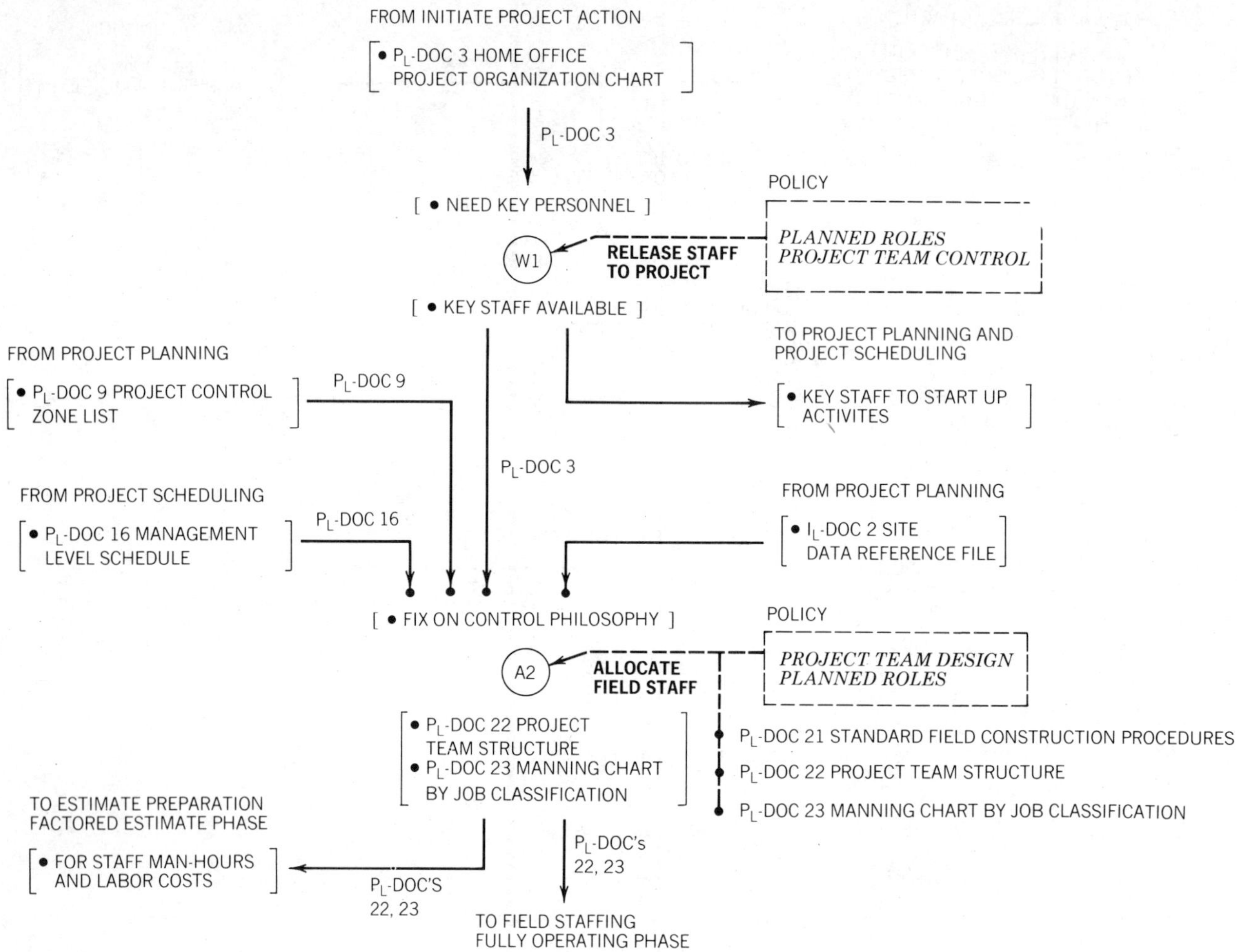

Figure 8.7 Field staffing decision process (Field-Based Project Management approach). (*a*) Head office located field staffing phase.

These key construction managers are RELEASED TO THE PROJECT (W1) to take up *PLANNED ROLES* (P_L 2)* and to ensure that effective *PROJECT TEAM CONTROL* (P_L 3) is achieved. This means that the primary field construction personnel will be available to participate in critical office project planning and scheduling decisions, as indicated in Figure 8.7*a*. When the Project Control Zone List (P_L-DOC 9) is defined, and the Management Level Schedule (P_L-DOC 16) and preliminary field manpower curves become available through head office Project Scheduling, the nominated resident field manager, in conjunction with the head office construction manager, begins FIELD STAFF ALLOCATION (A2). At this stage in Field Staffing decisions are normally guided by Standard Field Construction Procedures (P_L-DOC 21), and an initial Project Team Structure (P_L-DOC 22) is selected and modified to match the basic control philosophy and the specific technical requirements of the project.†

A commitment to *PROJECT TEAM DESIGN* (P_L 1) and *PLANNED ROLE* policies dominately influences the approach to the nature and make-up of the field project team structure. This team design process includes the identification of basic project team roles and of their functional responsibilities and resulting lines of authority. The number of personnel assigned to each job classification depends on the desired level of management control, the

*See Table 8.5, Policies of the Large Firm.

†In the design-construct project environment early release and involvement of key field staff also provides an opportunity for constructability input during the design of the facility.

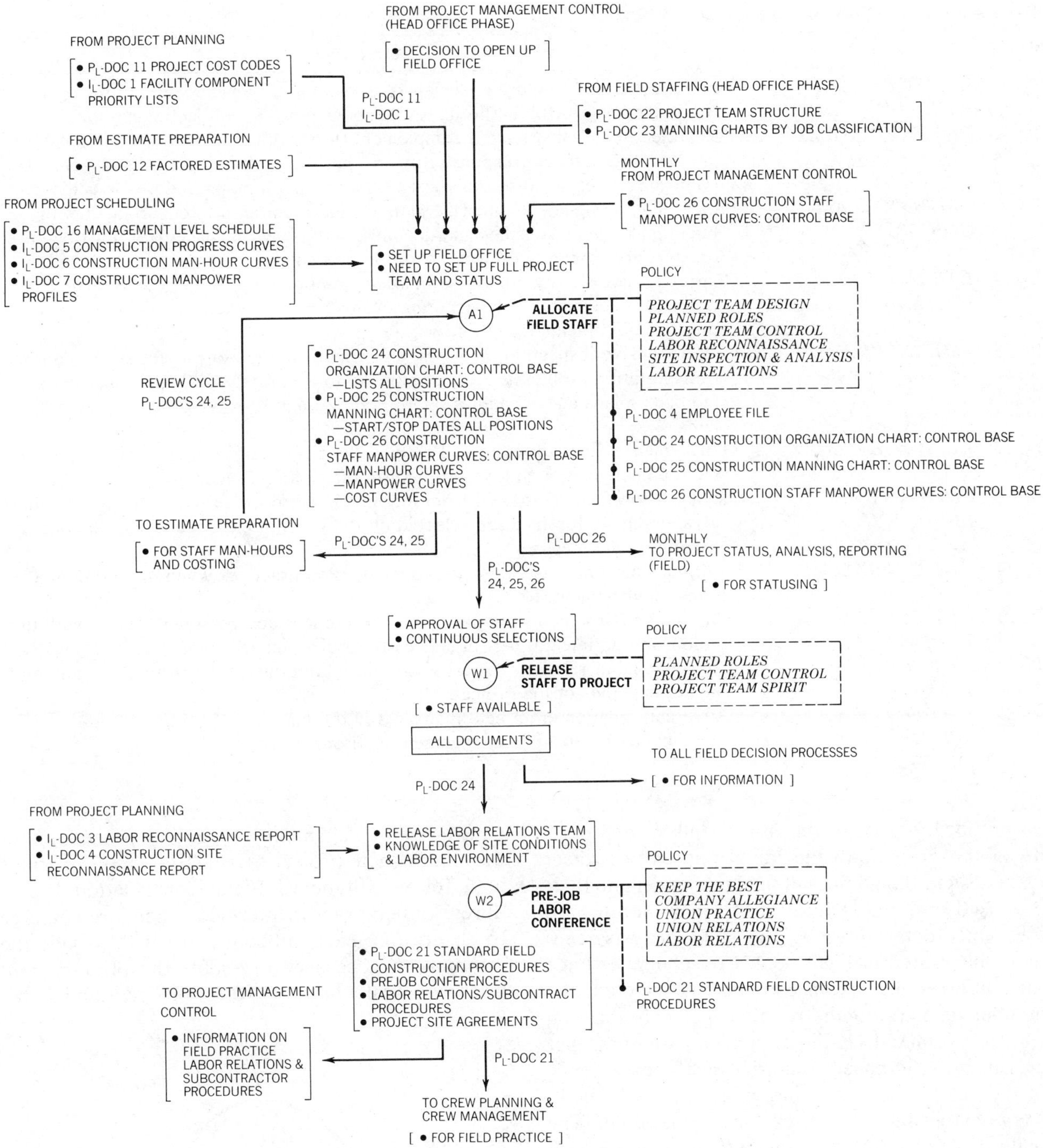

Figure 8.7 (*Continued*) (*b*) Fully operating field phase.

TABLE 8.5 POLICIES OF THE LARGE FIRM

Policy		Description
P_L 1	*PROJECT TEAM DESIGN*	The project team structure is designed to maximize management control by giving the key managers full authority and responsibility for project execution.
P_L 2	*PLANNED ROLES*	The project team must have a complement of personnel so that all functional areas are assigned manager responsibility, so that the desired level of management control as determined by standard company operating procedures is achieved.
P_L 3	*PROJECT TEAM CONTROL*	Project team assignments must be planned and monitored to ensure a smooth buildup as work intensifies and declines, and project management must be based on standard company project execution procedures.
P_L 4	*LABOR RECON-NAISSANCE*	Comprehensive survey, analysis, and documentation of the project site labor environment is requisite to the accurate evaluation of the labor content and manpower requirements for the project.
P_L 5	*SITE INSPECTION AND ANALYSIS*	Comprehensive inspection, analysis, and documentation of project site conditions, transportation problems, etc., is requisite to accurately evaluate construction methods, plans and schedules, labor content, and to solve other unique problems.
P_L 12	*KEEP THE BEST*	Management strives to retain and employ the best qualified men available, and to build a nucleus of men loyal to the firm through this security.
P_L 13	*COMPANY AL-LEGIANCE*	Key craft foremen are staffed from foremen loyal to the firm. Foremen other than loyal company foremen are selected on the basis of experience, managerial capability, and desire.
P_L 14	*UNION PRACTICE*	Field management must subscribe to local union practices when operating in a union environment.
P_L 21	*UNION RELA-TIONS*	In a union environment, pre-job union conferences and personal contact with the unions are an integral part of successful project management.
P_L 23	*LABOR RELA-TIONS*	A full-time labor relations expert is often required due to the complexity and manpower demands of the projects.
P_L 35	*PROJECT TEAM SPIRIT*	Active participation of all field managers in the management process promotes goodwill and a positive management environment.

magnitude of the field labor force, and the complexity of the project. In the initial phases of project team design, the number of field personnel and their proposed assignment durations (in man-months) are frequently determined by the staff analysis technique illustrated in Figure 8.4. Consequently, once the Initial Project Team Structure is selected, the number of man-months required per classification can be estimated. Estimation of man-months per classification is based primarily on the experience of the field managers as individuals, as well as on "brain storming" between these personnel. The buildup of the field management team generally precedes but can overlap with the buildup of the field labor force. It is performed by man loading field positions over the project duration. A typical example that illustrates the growth of the field management team in relation to that of the field labor force is shown in Figure 8.8.

Some critical issues during the initial field staffing activities are industrial relations and subcontractor management. A general procedure for dealing with a project's industrial/labor relations area follows Standard Field Construction Procedures. Subcontractor interface procedures emerge to ensure that each subcontractor will provide the information necessary to enable the primary contractor to evaluate total project progress and status

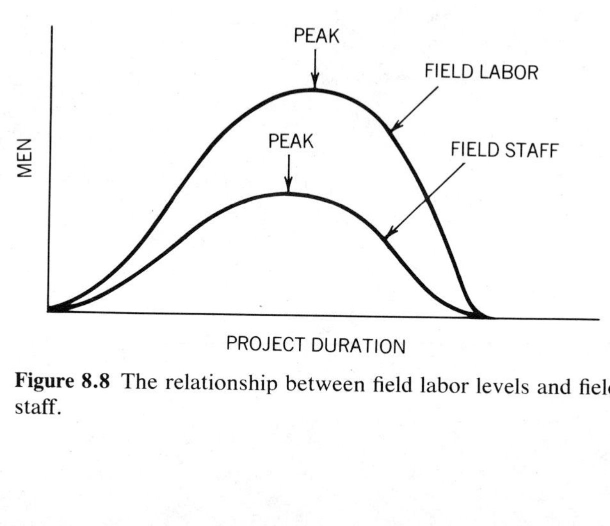

Figure 8.8 The relationship between field labor levels and field staff.

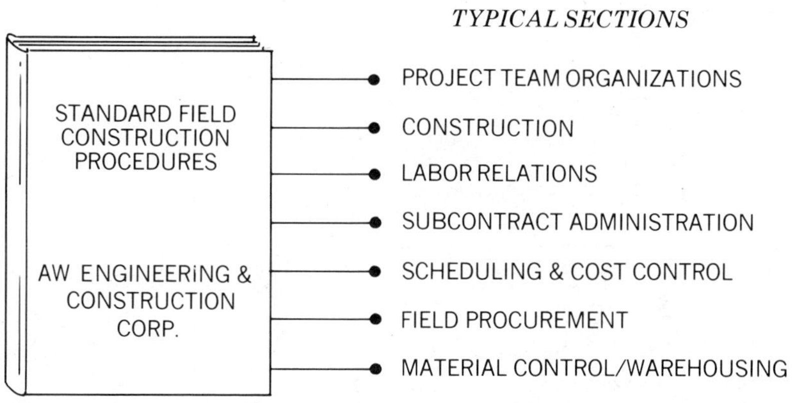

Figure 8.9 Standard Field Construction Procedures (P_L-DOC 21).

in relation to cost and schedules. Normally, Standard Field Construction Procedures are available for formulating specific procedures relevant to subcontract administration. A typical document which illustrates the contents of these procedures is presented in Figure 8.9.

The design of the project team for early project budget control purposes is the dominant focus of FIELD STAFF ALLOCATION. As depicted in Figure 8.7*a*, the resident field manager and key management level personnel begin to massage the Initial Project Team Structure developed in the head office into a usable organization for managing the construction effort. At the large firm level, various relationships exist which are critical factors impacting the design of the field project team structure and the number of personnel required. For example, the level of field supervision (front-line superintendents) versus the size of the field labor force to be managed is a critical problem. Because the size and number of field crews becomes rather large, analyzing supervisory ratios is imperative to the design and manning of the project team. Thus, the relationships between superintendents in terms of the number of general foremen, foremen, and typical crew sizes are established. These relationships, in the form of ratios, are normally determined by experience and are influenced by the size of the project's labor force, anticipated site productivity levels, and labor availability. At this stage, the analysis is rough because the field managers are working with rough manpower levels. The next level of supervision (i.e., craft superintendents), is a function of the types and magnitudes of crafts involved. Finally, a more discrete breakdown of functional responsibilities depends on the management approach in terms of the project and site control measures to be implemented. A project may be subdivided into functional control areas with a superintendent responsible for each specific function, for instance, for all the mechanical work. The mechanical superintendent would have authority over those craft superintendents within the mechanical area.

A final consideration in the design and manning of the field project team is the determination of the kind of support to be provided to the superintendents in such areas as cost, scheduling, and field engineering. The number of support personnel is a function of the management philosophy and varies widely depending on project size, complexity, the project's management information system, among other factors. The field project team normally contains field engineers, cost and scheduling engineers, timekeepers, and so on, all of whom support the superintendents by furnishing cost and scheduling data, progress information, quantity data, manhours, and field drawings. The ratio of construction superintendents to support personnel will depend on the contractor's control system and view of cost and scheduling as an important area.

Typical examples of the initial Project Team Structure (P_L-DOC 22) and Manning Chart By Job Classification (P_L-DOC 23) that result from FIELD STAFF ALLOCATION are shown in Figure 8.10.

The client/contractor decision to establish a field office and relocate field personnel permanently to the field location initiates the Fully Operating Field Staffing Phase, as shown in Figure 8.7*b*.* While field staff are involved with the problems and func-

*The size, location, and timing of the establishment of a field office may be constrained by permit requirements. Therefore, the permit process often controls this client/contractor decision.

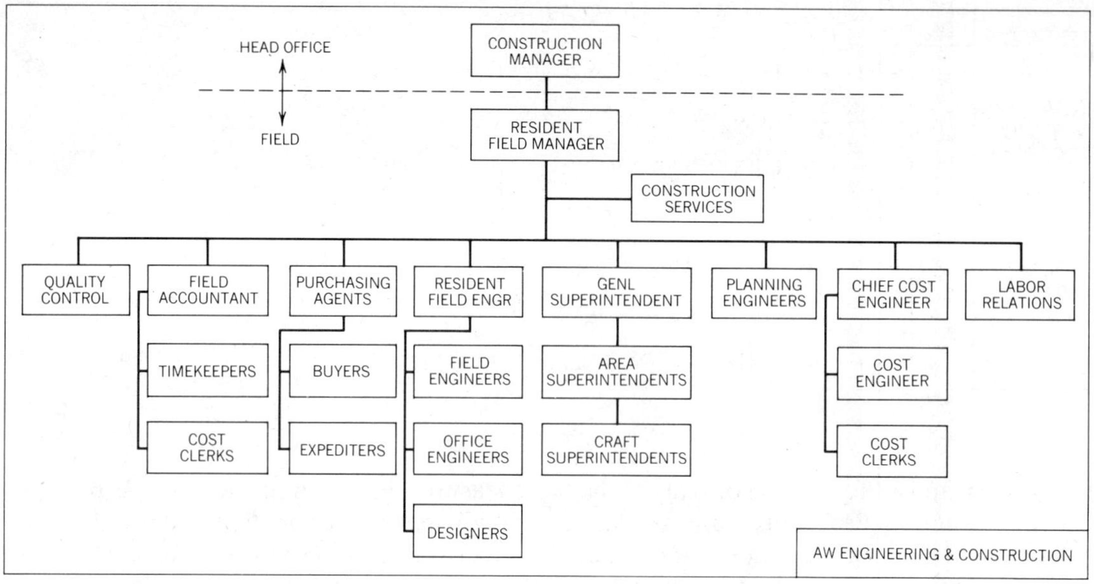

(a)

AW
Engineering & Construction Corp.

CLIENT ______________ CONTRACT ______________

PROJECT ______________ DATE ______________

STAFF MANPOWER ESTIMATE

COST CODE	STAFF CLASSIFICATION	DURATION MAN-MONTHS									TOTAL MAN-MOS.
		J	F	M	A	M	J	J	A	····→	
93110	RESIDENT FIELD MANAGER		1	1	1	1	1	1	1	····→	36
93120	GENERAL SUPERINTENDENT			1	1	1	1	1	1	····→	34
93140	RESIDENT FIELD ENGINEER			1	1	1	1	1	1	····→	36
93120	AREA SUPERINTENDENTS					1	1	2	2	····→	103

(b)

Figure 8.10 Field project team organizational development. (*a*) Project Team Structure (P_L-DOC 22). (*b*) Manning Chart by Job Classification (P_L-DOC 23).

tions related to establishing a field office (e.g., securing temporary facilities, building temporary buildings, etc.), the main manpower management focus of this phase of Field Staffing is two-fold. It involves: (1) the design, implementation, and control of a fully operating project team in order to manage construction; and (2) the definition and implementation of labor/industrial relations programs and subcontractor interface procedures. Thus, Field Staffing not only incorporates FIELD STAFF ALLOCATION and the timely RELEASE OF STAFF TO THE PROJECT functions, but also the PRE-JOB CONFERENCE (W2) management function.

The product of FIELD STAFF ALLOCATION is the Construction Organizational Chart: Control-Based (P_L-DOC 24) and the Construction Manning Chart: Control Base (P_L-DOC 25), examples of which are shown in Figures 8.11*a* and *b* respectively. The Construction Manning Chart serves as the basis for establishing curves used to monitor and control staff man-hours and labor costs as shown in Figure 8.11*c*, Construction Staff Manpower Curves: Control-Base (P_L-DOC 26). The size and structure of the field project team is continuously reevaluated as better information on project characteristics becomes known, and site labor conditions change.

As the construction manager designs the project team and the other key positions are identified, personnel are selected and their availability determined through documents such as Employee Files (P_L-DOC 4). Thus, the orderly RELEASING OF STAFF TO THE PROJECT (W2) can proceed via *PLANNED ROLES* and *PROJECT TEAM CONTROL* in preparation for setting up a field office.

Once the construction effort gains momentum, the PRE-JOB LABOR CONFERENCE function of Field Staffing, under the direction of a Labor Relations team, begins to emerge. This function is easily performed by small and medium-sized contractors, because of the local nature of the region in which they operate. Its performance is occasionally demanded in the large firm environment, however, because of the size of the contractor's own field labor force and the problem of coordinating this labor force with subcontractor labor forces. Pre-job conferences with local union representatives are probably more frequent in a union construction environment, because a common understanding of restrictive work practices, jurisdictional areas, and other union rules by all project participants is vital to circumventing problems, and provides a forum for implementing the fundamental management philosophies of the project. Problems can be discussed, evaluated, and resolved prior to commencing the construction effort. Thus, *UNION PRACTICE* (P_L 14) and *UNION RELATIONS* (P_L 21) become critical policies for the large firm.

When viewed as a Management Information System, Field Staffing first focuses on the initial development of the field project team through a head office based Field Staffing phase. In this phase, Standard Field Construction Procedures are guides to the definition of the project team structure and its size. This leads to the selection and acquisition of key field project team personnel through the company information system. As the project moves toward construction, Field Staffing becomes a fully operational field located effort. In this phase, considerable data flow from other decision processes, such as control base schedules and estimates, enters the MIS of Field Staffing. As the project team structure, size, and manpower plans become defined, the continuous growth and management of the project team becomes the key decision focus of Field Staffing. In addition, decisions relating to project labor relations and subcontractor interface procedures are made and implemented for Crew Planning and Crew Management. The acquisition of field staff becomes a critical problem in the large firm environment because of the often remote location of projects, and the need for more specialized field staff personnel. Thus, the MIS becomes a more formally documented system in large firms.

8.6 TOTAL FIELD AUTONOMY

The large firm operating in the Total Field Autonomy management mode approaches Field Staffing from the same perspective as it approaches the Initiate Project Action decision process. Specifically, its management approach is to transfer total management responsibility, decision making, and control to the field. Thus, the design and structure of the field project team must incorporate all the functional areas that emerge in the head office. This interest requires the duplication of the Project Management Level group (see (P_L-DOC 3)) in the Field Staffing decision process. This is a significant departure from the Field-Based Project Management approach, wherein the resident field manager, in conjunction with the head office construction manager, controls the number and size of the functional groups involved in the project subject to the project manager's approval. Thus, Field Staffing can be viewed as a repeat or extension of the Initiate Project Action decision process (DP1), but now transplanted to the field location for the construction management phase of the project.

Field Staffing is further influenced by the control philosophy adopted by the contractor for the project. This follows because, to effectively transfer total management authority and responsibility to the field, a common control base must exist in so far as engineering/design data (e.g., specification, construction drawings, and material identification) integrates with the construction management approach. As a result, construction control zones must match, to a significant degree, engineering and design control zones. For example, construction drawings will specify control zone and work package locations. This approach also increases the effective use of project schedules and labor estimates that are tied into common project control philosophies.

The Total Field Autonomy project environment places increasing demands on Field Staffing in terms of the need to acquire, train, and motivate large work forces. This results in a strong focus on the industrial relations procedures and practices selected for a project. In addition, the project may involve a number of specialist contractors. The management of these contractors must follow over-

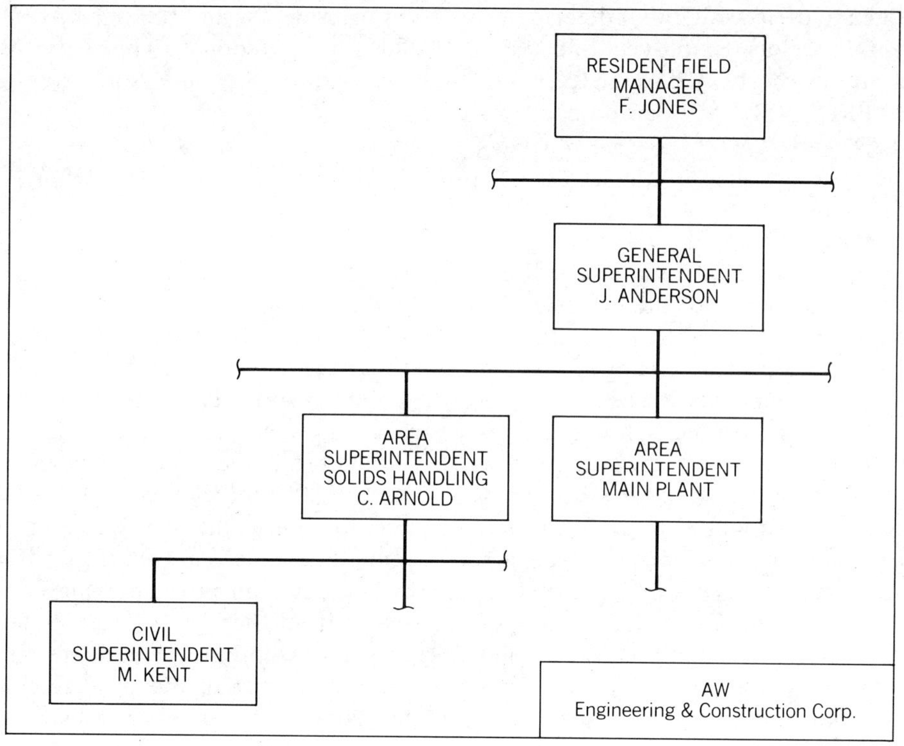

(a)

AW
Engineering & Construction Corp.

CLIENT ____________________

PROJECT ____________________

STAFF MANPOWER ESTIMATE

COST CODE	STAFF CLASSIFICATION	NAME POSITION	DURATION MAN-MONTHS									TOTAL MAN-MOS.
			J	F	M	A	M	J	J	A	····→	
93110	RESIDENT FIELD MANAGER	F. JONES PROJECT MANAGER		1	1	1	1	1	1	1	····→	36
93120	GEN'L SUPERINTENDENT	J. ANDERSON			1	1	1	1	1	1	····→	34
93121	AREA SUPERINTENDENT	C. ARNOLD SOLIDS HANDLING					1	1	1	1	····→	24
	CIVIL SUPERINTENDENT	M. KENT SOLIDS HANDLING					1	1	1	1	····→	24

(b)

Figure 8.11 Project team organization documents. (*a*) Construction Organization Chart: Control Base (P_L-DOC 24). (*b*) Construction Manning Chart: Control Base (P_L-DOC 25).

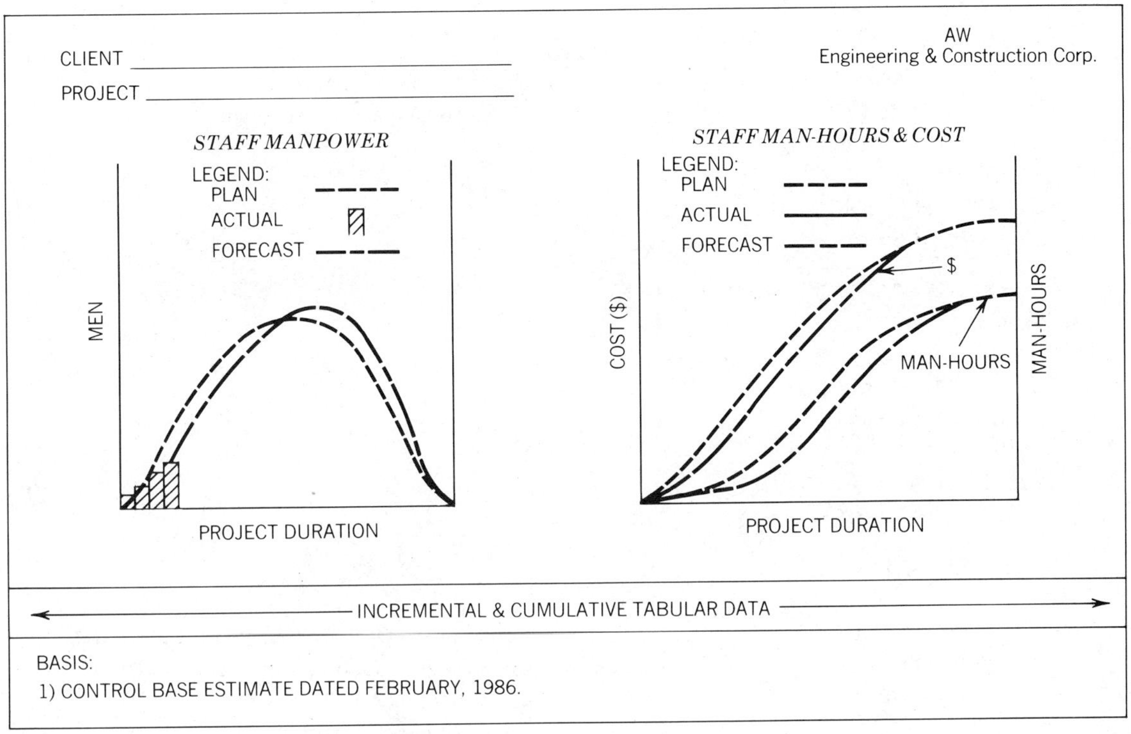

(c)

Figure 8.11 (*Continued*) (*c*) Construction Staff Manpower Curves: Control Base (P_L-DOC 26).

all project control approaches so as to minimize competition for personnel and crafts, to organize access to critical site areas, and to measure the contractor's ability to achieve desired project performance targets. These industrial relations and subcontractor interface problems lead to specially prepared project field construction procedures, often developed in conjunction with the client.

The manpower management functions, policies, and documents that characterize the Field Staffing decision process for the Total Field Autonomy approach are shown in Figure 8.12 for both the Head Office Phase and the Fully Operating Field Phase. During the head office phase of Field Staffing (see Figure 8.12*a*), the Project Team Structure (P_T-DOC 28) is developed. It is based primarily on *PROJECT TEAM FOCUS* (P_T 1),* *PROJECT TEAM CONTROL* (P_T 2), and *PLANNED ROLES* (P_T 4) policies and serves as a means of ALLOCATING FIELD STAFF (A2) personnel to key positions and establishing a field staff estimate basis for Factored Estimating. Since the client interfaces frequently with project personnel during the performance of this function, approvals of key staff are obtained quickly and STAFF ARE RELEASED TO THE PROJECT (W1). A typical Project Team Structure (P_T-DOC 28) is depicted in Figure 8.13*a*. Simultaneously, project team personnel, under the direction of the management level group, determine the size of each field group by the classifications and man-months required to support construction for each control zone. A Manning Chart by Job Classification (P_T-DOC 29) like that in Figure 8.13*b* will be produced and input to the Estimate Preparation decision process for conversion into staff man-hours and staff labor dollars. The client interfaces heavily with the contractor during the Field Staffing decision process. Once the client approves the Factored Estimate (P_T-DOC 15), the field Project Team Structure is approved and the contractor enters into the Fully Operating Field Phase of Field Staffing (see Figure 8.12*b*).

When Field Staffing moves to a field location, as indicated in Figure 8.12*b*, the design of a fully operating field management team begins. At this stage, *PROJECT TEAM DESIGN* (P_T 3) becomes a criti-

*See Table 8.6, Policies of the Autonomous Project Team.

A2 ALLOCATE FIELD STAFF

Policies	Documents	
PROJECT TEAM FOCUS	P_T-DOC 28	Project Team Structure
PROJECT TEAM CONTROL	P_T-DOC 29	Manning Chart by Job Classification
PROJECT TEAM DESIGN		
PLANNED ROLES		
PROJECT TEAM MANAGEMENT		
PROJECT TEAM SPIRIT		

W1 RELEASE STAFF TO PROJECT

Policies

PROJECT TEAM FOCUS
PROJECT TEAM CONTROL
PROJECT TEAM DESIGN
PLANNED ROLES
PROJECT TEAM MANAGEMENT
PROJECT TEAM SPIRIT

(*a*)

A2 ALLOCATE FIELD STAFF

Policies	Documents	
PROJECT TEAM FOCUS	P_T-DOC 30	Construction Organization Chart: Control Base
PROJECT TEAM CONTROL	P_T-DOC 31	Construction Manning Chart: Control Base
PROJECT TEAM DESIGN	P_T-DOC 32	Construction Staff Control Curves
PLANNED ROLES	P_T-DOC 33	Standard Field Construction Procedures
PROJECT TEAM MANAGEMENT		
LABOR RECONNAISSANCE		
SITE INSPECTION AND ANALYSIS		
PROJECT TEAM SPIRIT		

W1 RELEASE STAFF TO PROJECT

Policies

PROJECT TEAM FOCUS
PROJECT TEAM CONTROL
PROJECT TEAM DESIGN
PLANNED ROLES
PROJECT TEAM MANAGEMENT
PROJECT TEAM SPIRIT

W2 PRE-JOB LABOR CONFERENCES

Policies	Documents	
LABOR RECONNAISSANCE	P_T-DOC 34	Project Field Construction Procedures
SITE INSPECTION AND ANALYSIS		
KEEP THE BEST		
COMPANY ALLEGIANCE		
UNION PRACTICE		
LABOR RELATIONS		
PERSONAL CONTACT		
SCIENTIFIC MANAGEMENT		
PROJECT TEAM SPIRIT		

(*b*)

Figure 8.12 Field Staffing decision process (Total Field Autonomy approach). (*a*) Head office located field staffing phase. (*b*) Fully operating field phase.

TABLE 8.6 POLICIES OF THE AUTONOMOUS PROJECT TEAM

Policy		Description
P_T 1	*PROJECT TEAM FOCUS*	The project team is initially head office oriented until field operations are established, at which time there is a total transfer of management responsibility and control to the field under a top company executive acting as project manager.
P_T 2	*PROJECT TEAM CONTROL*	Key project team assignments, overall project team designs, and specifically designed project execution procedures as developed for project management are reviewed in consultation with, and approved by, the client.
P_T 3	*PROJECT TEAM DESIGN*	The project team structure is designed to maximize planning and management control by ensuring that the proper relationship exists between the level of manpower to be controlled in relation to the size of the project team.
P_T 4	*PLANNED ROLES*	The project team is a fully integrated group of managers combining their functional responsibilities through the project team structure to optimize planning techniques and the level of management control demanded by the client, company management, and agreed upon project execution procedures derived from standard company operating procedures.
P_T 5	*PROJECT TEAM MANAGEMENT*	Management evaluation of project progress, status, and performance on a continuous basis is essential to ensure that the project is being managed with maximum effectiveness by the project team.
P_T 6	*LABOR RECONNAISSANCE*	Comprehensive survey, analysis, and documentation of the project site labor environment is requisite on a continuous basis to accurately evaluate labor content and plan manpower requirements for the project at any time during the project's life cycle.
P_T 7	*SITE INSPECTION AND ANALYSIS*	A continuous comprehensive inspection, analysis, and documentation of project site conditions, transportation problems, etc., is requisite to an accurate evaluation of construction methods, schedules, and labor content, and to solve other unique problems.
P_T 15	*KEEP THE BEST*	Management strives to retain and employ the best qualified men available and to build a nucleus of men loyal to the firm through this security.
P_T 16	*COMPANY ALLEGIANCE*	Key craft foremen are staffed from foremen loyal to the firm. Foremen other than loyal company foremen are selected on the basis of experience, managerial capability, and desire.
P_T 17	*UNION PRACTICE*	Field management must subscribe to local union practices when operating in a union environment.
P_T 23	*LABOR RELATIONS*	When union labor is utilized, pre-job union conferences and personal contact with the unions are an integral part of successful project management. Regardless of the construction environment, a full-time labor relations expert is required because of the complexity and manpower demands of the projects.
P_T 37	*PROJECT TEAM SPIRIT*	Active participation of all project team members in the management process promotes goodwill and a positive management environment. It ensures the success of scientific management techniques and the total integration of the project team.

cal policy influencing the decision-making approach to FIELD STAFF ALLOCATION. Careful analysis and consideration is given to the detailed study of the size and magnitude of field manpower to be managed and controlled. Project control zones and work package structures become an important factor in determining the management span of control of project team personnel, as reflected in the project team design and structure. This determination includes the analysis of construction supervisory ratios and other pertinent factors like those considered by the large firm in Section 8.5. Because of a policy focus on effective project control (see Table 8.6), and active client involvement in the project management process, support groups are more carefully designed. Their structure parallels the project control zone approach of the construction group which these groups support. Finally, to achieve consistency in project reporting from all groups, and to provide the project management

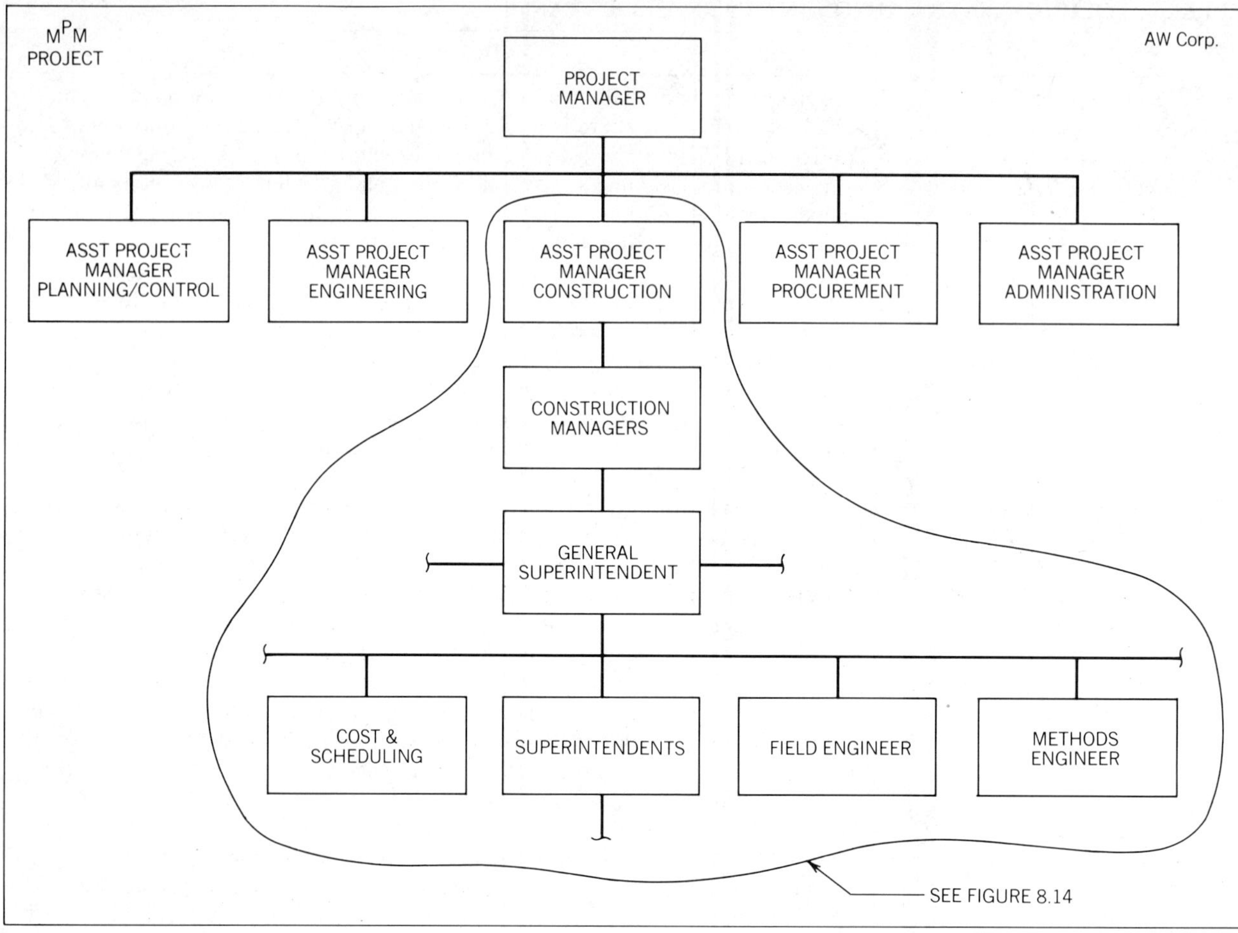

(a)

M^PM
PROJECT

AW CORP.

AREA ______ DESCRIPTION ______

POSITION	CLASS	DURATION-MOS.					TOTAL MAN-MOS.	TOTAL MAN-HRS.
		J	F	M	A	M		
CONSTRUCTION MGR.	M225	1	1	1	1	1	36	7920
GENERAL SUPT.	S215	1	1	2	2	3	90	19800
SUPERINTENDENTS	J312	1	1	3	3	5	200	44000

(b)

Figure 8.13 Project team organizational development. (*a*) Project Team Structure (P_T-DOC 28). (*b*) Manning Chart by Job Classification (P_T-DOC 29).

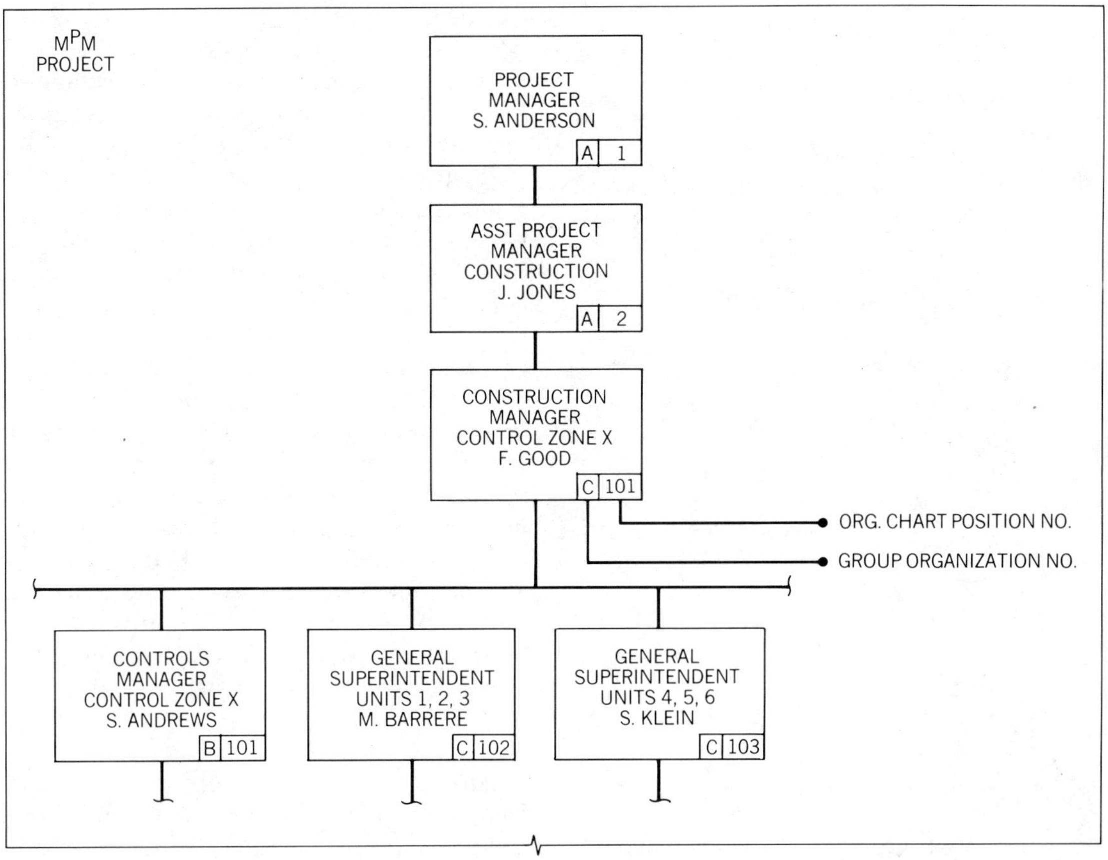

(a)

M^P M PROJECT

AREA ____________ DESCRIPTION ____________

AW CORP.

GROUP ORG. NO.	NAME-POSITION	CLASS	DURATION-MOS. J	F	M	A	M	J	TOTAL MAN-MOS.
C101	F. GOOD-CONSTRUCTION MANAGER	M225	X	X	X	X	X	X	36
C102	M. BARRERE-GENERAL SUPERINTENDENT	S215	X	X	X	X	X	X	36
C103	S. KLEIN-GENERAL SUPERINTENDENT	S215			X	X	X	X	30
	TOTAL CONTROL ZONE X								1520

(b)

Figure 8.14 Control base project team organization. (*a*) Construction Organization Chart: Control Base (P_T-DOC 30). (*b*) Construction Manning Chart: Control Base (P_T-DOC 31).

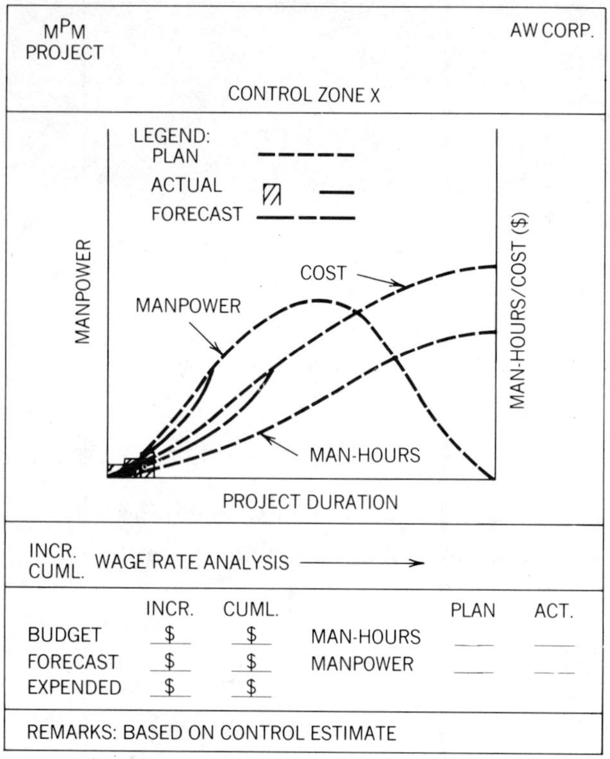

Figure 8.15 Construction Staff Control Curves (P_T-DOC 32).

level with sufficient progress and status information to make major project-oriented decisions, a core (or central) group is often established (especially on large projects) using similar techniques to those used in the one iilustrated in Figure 8.5. The products of this evolution of the project team design and structure are a client-approved Construction Organization Chart: Control Base (P_T-DOC 30), and a Construction Manning Chart: Control Base (P_T-DOC 31) as shown in Figure 8.14. The technique of identifying organizational chart positions by numbering them and relating the numbers to the Manning Chart permits the client to approve personnel for a specific position, provides the contractor with a method of managing the growth and release of personnel over the project's life, and provides a basis for processing change orders and trends to the control base estimate. The man-months derived from the approved organizational structure are incorporated into the control base budget estimate.

The Construction Manning Chart is converted into Construction Staff Control Curves (P_T-DOC 32), as demonstrated in Figure 8.15. These curves become an additional tool for monitoring staff man-hour expenditures and manpower levels against plans. When coupled with budget cost data they also provide a basis for evaluating actual staff wage rates and labor dollar expenditures against plans or current forecasts. This actual expenditure information is gathered and documented through Project Status, Analysis, and Reporting, wherein deviations are evaluated. These deviations become input for project management decisions concerning changes in the Construction Organization Chart, such as adding or deleting positions or personnel, or modifying the chart's structure to improve management control in a given area of the project.

Because project conditions such as project scope, site conditions, the labor environment, and so on are constantly changing, the project team design process is a continuous undertaking. Thus, Field Staffing is a dynamic decision area throughout

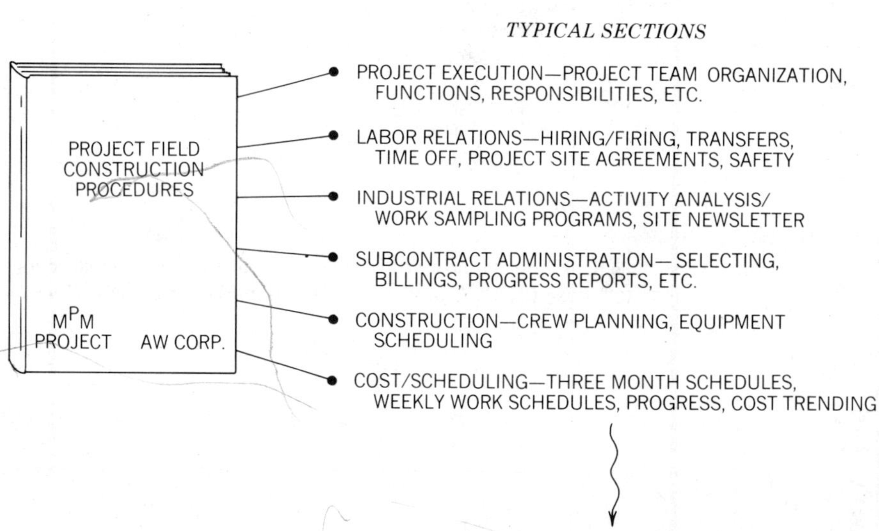

Figure 8.16 Project Field Construction Procedures (P_T-DOC 34).

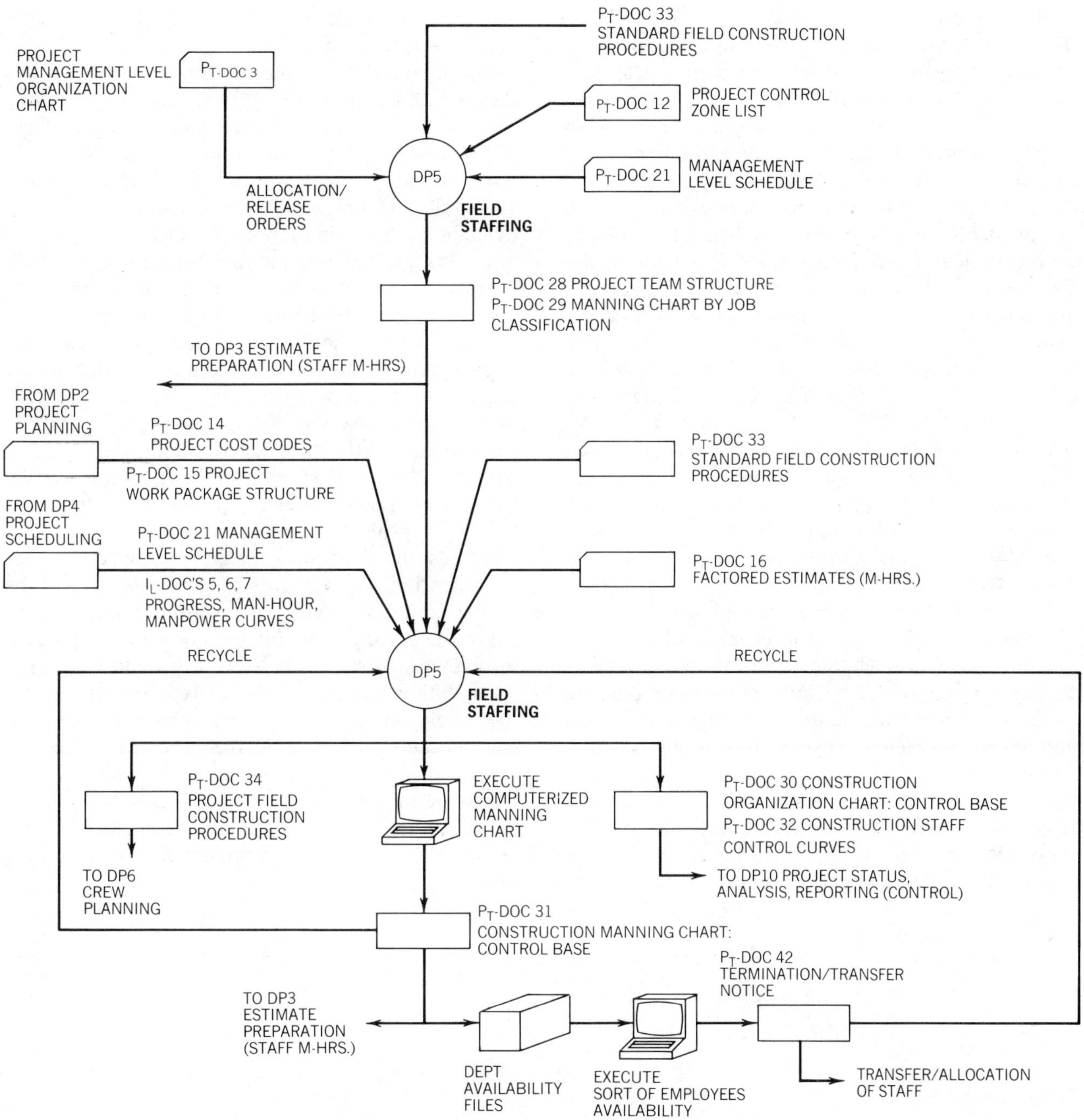

Figure 8.17 Management Information System Field Staffing decision process (Total Field Autonomy approach).

the life of the project. *PROJECT TEAM MANAGEMENT* (P_T 5) alerts project and construction managers to potential problem areas. A continuous flow of *LABOR RECONNAISSANCE* (P_T 6) and *SITE INSPECTION* (P_T 7) information often provides the impetus for changing project team designs and staffing requirements. For example, a critical shortage of welders may necessitate increased training programs and thus more training staff. In a dynamic project team environment, the approved Construction Organization and Construction Manning Charts may reflect a number of major revisions over the life cycle of a project. Thus, *PROJECT TEAM SPIRIT* (P_T 37) becomes a critical policy in the constantly changing project management environment to promote good project morale.

A critical facet of the Fully Operating Phase of Field Staffing is the development and implementation of field construction procedures. While Standard Field Construction Procedures (P_T-DOC 33) are available, they are only used as guidelines to derive Project Field Construction Procedures (P_T-DOC 34) designed specifically for the project and approved by the client. These procedures formally establish and document project-wide productivity improvement programs, training policies, hiring and firing practice, and the like, as directed by such policies as *PERSONAL CONTACT* (P_T 24) and *SCIENTIFIC MANAGEMENT* (P_T-DOC 34). These project-wide programs emerge as a result of the need to control and manage large work forces by those control zones unique to the project. These procedures provide a common method and basis from which a construction manager responsible for several control zones can operate.

Project Field Construction Procedures also provide guidelines for subcontractor interfaces. This is particularly important because procedures that are properly implemented help to minimize congestion and access problems between the various contractors on the site. They also establish a common basis by which all site subcontractors will report progress, status, and productivity. This uniformity of reporting will aid the managing contractor when rolling up total project progress and status, and will provide a mechanism for evaluating each subcontractor's performance. This can be successfully accomplished if the client fully supports and approves the subcontractor's procedures. A schematic illustration of typical Project Field Construction Procedures (P_T-DOC 34) is shown in Figure 8.16.

In summary, Field Staffing can be characterized as a fully documented Management Information System, as shown in Figure 8.17. Once the selection of key project team personnel is made, initial project team organization and manning charts are developed. This leads to the full-scale Field Staffing decision process, wherein a number of documents impact field staffing and are critical to the proper evaluation of control base project team organization charts, manpower profiles, and project-focused construction procedures. The manpower chart is computerized so that staffing requirements can be frequently updated over the project's life to reflect the latest project scope. The manning chart also identifies the type of project position, the basic classification requirements of the position, and the timing commitment of the position. These criteria aid in the selection of suitable company employees through a search of Employee Files and Employee Availability Reports. This approach provides for the timely release of company personnel to a project and constitutes the Field Staffing MIS. The cycle is repeated throughout the project's life as illustrated in Figure 8.17, and is critical to providing effective field construction management for large technical projects in often remote and isolated locations.

CHAPTER 9

Crew Planning

Crew Planning is concerned with the matching of proposed work activity with the scheduled availability of suitably qualified crews. It therefore involves the planning and scheduling of crew work sequences and the planning and mobilization of all preparatory work, such as field procurement of materials and movement of construction equipment to the work face, and ensuring the availability of drawings, working tools, and the procurement of all ancillary working materials relevant to the construction method involved and essential to the start of field activity.

Crew planning is a vital concern for the construction contractor, because it is the critical phase which converts a technical problem into a plan for its construction and resolution. It portrays the manner in which the job is going to be built (that is it establishes constructability), exposes the competency of field staff, and establishes the feasibility of meeting planned progress and scheduled milestones. In addition, the number of crews and their proposed scheduling to work areas around the site must be compatible with the control procedures adopted to ensure that field labor productivities are met. Crew Planning is of course the key decision area that establishes whether field productivities, progress targets, and schedules are attained or not. Therefore, it is the first field management area to be investigated when work performance is poor and work is not proceeding as planned.

On small projects requiring a limited number of general purpose crews, Crew Planning is simply a matter of keeping sufficient ready work ahead of each crew. As project size, complexity, and work volume increase, however, the emergence of a large number of specialists, labor, and material handling crews makes the Crew Planning process very complicated and all the more essential to the achievement of planned progress. On very large projects it is a major problem to schedule, mobilize, and coordinate crew activity and it is often impossible to relocate crews once they have been committed to a certain location on a specific day. Similarly, as the number of subcontractors involved in a project increases, the coordination of their work activity and their need for access to work areas significantly complicates the scheduling of the contractor's own crews.

Thus, the smooth coordination of field operations reflects the proficiency with which the Crew Planning process is performed and reflects positively on the competency of the field project team. Conversely, not only does poor Crew Planning impact project costs due to the nonattainment of planned productivities, but the resulting confusion and enforced delays to crew activity is demotivating to crew members and can lower project morale. Thus, Crew Planning is a critical problem area in Project Manpower Management.

9.1 THE BASIC DECISION PROCESS

Crew Planning is the recognized management area for the superintendent and the foremen, and involves a decision process that is basically intuitive

TABLE 9.1 SCOPE OF THE CREW PLANNING PROBLEM AREA

Major Decision Problem	Component Problem Statements
26. What is to be done:	What is the type and scope of work? What are requirements of chosen construction method? What drawings, types of materials, construction equipment/tools, and other support items are required for work package and proposed construction method? What is specific material content (quantities) of work package activities? What are impacts of site conditions and work environments? What quality control criteria are involved? What skill levels are required? What is best size for work package definition of labor content? Should work packages be defined for crew requirements and to suit actual crews? Does a fully integrated work package control exist? What is acceptable level of detail for work task breakdown of work package? What are crew requirements for the work package?
27. Who can do it:	What is best crew size and craft skill mix? Who is best suited for this work package assignment? Is the crew size compatible with the foreman's ability to manage and with the work face layout, site conditions, and accessibility? Can the crew size be tailored to suit this work package?
28. How will it be done:	Can work face layout assist in raising crew productivity? Can construction method be detailed and sequenced so that field personnel can achieve desired productivity and meet production milestones? Can crew member work patterns be developed to advantage on this work package? How can field personnel be briefed on need for this level of detail? Are field personnel adequately briefed on scope of work, job requirements, and construction method details for this work package?
29. When will it be done:	How many crews can be managed on work site? How can crews be located and accurate predictions made on their availability for new work packages? What is the availability of suitably experienced foremen and crews? When are crews scheduled to commence work packages? What are best crew member work sequences?

and conditioned by past experience and observation. Most field managers adopt traditional crew sizes and modify these slightly on a trial-and-error basis to suit local conditions and the manner in which they plan to manage the crew and the work area. This pragmatic and eminently practical approach to this decision area is a natural consequence of the fact that field supervisors at the foreman level have rarely had any formal management training and generally dislike paperwork of any kind. Field studies have established, however, that traditionally sized crews are not always the most productive. This is especially true for industrial projects involving large work forces, congested areas, and highly skilled labor and specialized work. This situation is aggravated on large projects where a shortage of experienced foremen causes a significant drop in field productivities. Management concern with the magnitude of these problems has resulted in the Crew Planning decision problem (in this project environment) becoming the province of specialist engineers and methods planners. In these situations the Crew Planning process becomes more

TABLE 9.2 GENERAL PROCEDURE FOR CREW PLANNING

Procedural Step	Management Duties and Tasks
1.	State scope of work, select construction method, and identify quality control requirements.
2.	Investigate work area, local site conditions, and accessibility.
3.	Check on work quantities and technical aspects of construction method.
4.	Establish best work approach and required level of trade skills for work activity.
5.	Nominate crew(s) and predict work progress.
6.	Establish manner in which crew(s) will be put to work at work face and predict start day for activity.
7.	Schedule crew(s) and crew member work sequences.

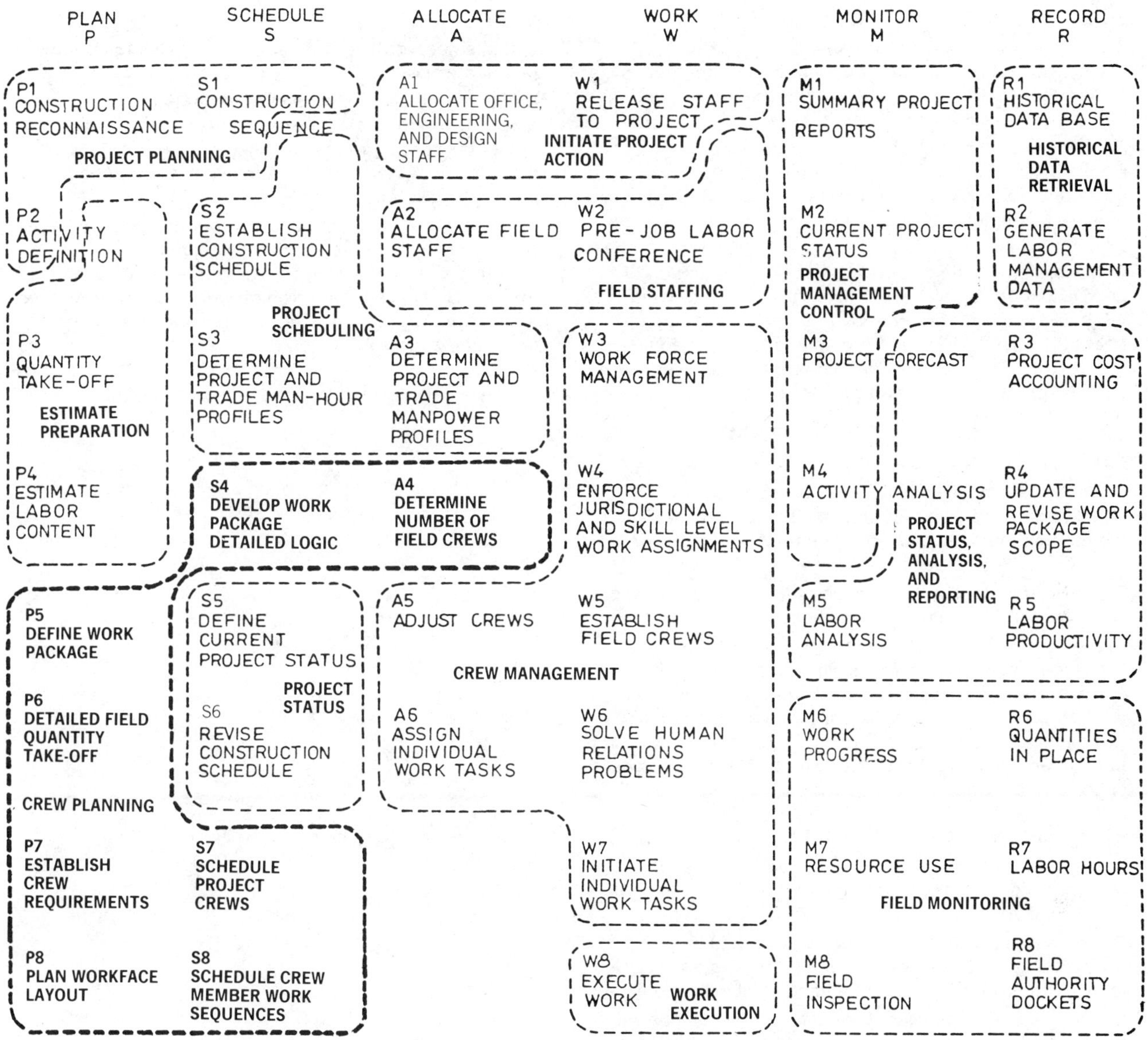

Figure 9.1 Functional plot of the Crew Planning decision process.

objective and is the subject of constant input from field monitoring and analysis. It is also more fully documented and planned. It is in this typically cost reimbursable design–construct contract environment that the work package approach emerges. This approach involves the preparation of fully documented plans for crew activity, and the almost total departure from the use of standard traditional crew sizes.

In contrast to the more project-focused problem areas and decision processes (such as Project Scheduling), Crew Planning and the other field-oriented management decision processes become more directly manpower and labor oriented. This specific focus is illustrated in Table 9.1, which lists the component problem statements and major decision problem breakout of the scope statement of Crew Planning. Similarly, the breakdown of the general procedure for the Crew Planning decision process (shown in Table 9.2) reflects a more specific and obvious list of management duties and tasks, with a corresponding more direct equivalence to the related basic manpower management functions involved in the Crew Planning decision process (see Figures 9.1 and 9.2). The factors considered by the foreman/ superintendent decision makers, as given

S4 DEVELOP WORK PACKAGE DETAILED LOGIC

- Establishes full scope and work details of project work segment based upon field requirements and conditions as basis for the briefing and scheduling of field construction managers/supervisors and crews.
- Requires knowledge of project and contract requirements, basis of estimate, specific details of drawings and field conditions, construction methods, cost effectiveness of work techniques, methods improvement, and value engineering.
- Considers effective use of resources and elimination of needless effort, breakdown of work package into work tasks and work assignments by trade; detailed logic (or fragnet) for work package from construction schedule; field estimate of quantities and labor hours involved, and material procurement, equipment location, and material flow routes to workface.

P5 DEFINE WORK PACKAGE

- In some cases develops the project scope (design-construct), and in all cases documents the scope of work, job requirements, and construction method details for specific work to be done.
- Requires an extensive knowledge of the cost impact and effectiveness of construction methods and trade practices and equipment and material handling processes.
- Considers activities as recognizable portions of the finished product or work scope, current field conditions, the skill and level of understanding of field construction managers and supervisors, and develops best construction method and sequence of activity that minimizes cost, effort, and duration of work activity.

P6 DETAILED FIELD QUANTITY TAKE-OFF

- Prepares work and material quantity details for specific work packages as basis for briefing of those field personnel to be involved in the work.
- Requires constant awareness of field conditions in relation to the accuracy and detail of engineering and working drawings, and awareness of implied labor content.
- Considers initial estimates, contract documents, current field status, performs detailed quantity take-off correlated with field measurements and work package definition, and prepares detailed list of work tasks.

P7 ESTABLISH CREW REQUIREMENTS

- Matches proposed crew size and trade skill mix with job requirements.
- Requires knowledge of trades, scope of work, field conditions, schedules, estimates, and acceptable ranges of crew sizes and mixes in relation to anticipated and required production.
- Considers magnitude of work content, production rate, scheduled durations, available labor and skill levels, and impact of site conditions on workface accessibility, material supply routes, and extent to which equipment can be made available.

A4 DETERMINE NUMBER OF FIELD CREWS

- Establish best-suited number of field crews to cope with planned work.
- Requires intimate knowledge of project site, work areas, types of work and planned rate of work, project requirements for special trades or skilled crews, agreements made on crew size, and relationship with working and nonworking foremen.
- Considers availability of suitable foremen, desirable span of control levels for key project management staff in instructing and supervising foremen on work details, need to maintain special trade or highly skilled crews, and variation in number of crews to suit workface layout, conditions, and accessibility.

S7 SCHEDULE PROJECT CREWS

- Schedule work for project crews.
- Requires knowledge of crew sizes and composition, current assignments and status, inventory of available work packages for execution, knowledge of project priorities.
- Considers current work status of each crew, productivities, outstanding work, anticipated work completion time, required skill levels and quality of finish requirements, and matching of crews with different types of work.

Figure 9.2 Functional statement of the Crew Planning decision process.

P8 PLAN WORKFACE LAYOUT

- Locates men, equipment, tool boxes, material dumps, and supply routes to workface to best advantage.
- Requires intimate knowledge of workface area, working conditions, equipment availability, and job requirements.
- Considers workface maneuvers over life of work activity and the location and interaction of the pacing and support trades. Minimizes work effort by elimination of needless walking and secondary movement of materials.

S8 SCHEDULE CREW MEMBER WORK SEQUENCES

- Establishes desirable work sequences for crew members for a specific work package.
- Requires intimate knowledge of work package scope, construction methods to be used, and required quality control specifications and skill levels.
- Considers the composition and size of the crew, skill levels, work capacity and motivation of individual crew members, and relates to the work content and workface layout of the operations.

Figure 9.2 (*Continued*)

in Figure 9.2, are almost self-explanatory and are equivalent to those in the scope statement of Crew Planning (see Table 9.1). Finally, the relationships of the Crew Planning decision process to those decision processes with which it interacts are illustrated in Figure 9.3.

9.2 FUNCTIONAL PERFORMANCE TECHNIQUES

Crew Planning at the foreman/superintendent grassroots level is generally an ad hoc process in which these field supervisors synthesize their visual and

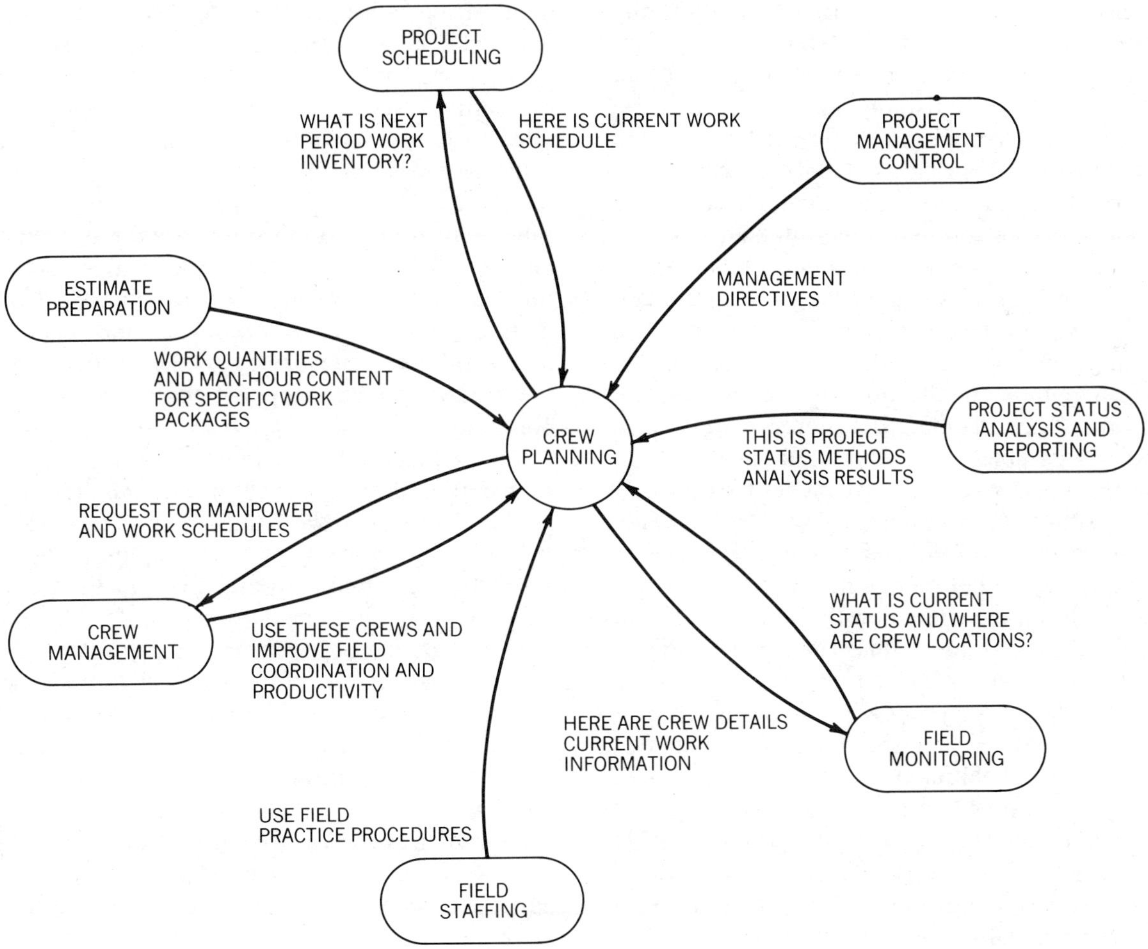

Figure 9.3 Schematic illustration of Crew Planning decision process interfaces with other manpower management decision processes.

(a)	(b)	(c)
Masonry: 5 Masons 2 Laborers ½ Carpenter	Where three (3) or more Carpenters are employed on a project, one (1) shall be designated as foreman and receive foreman's pay. Maximum crew for each Carpenter Foreman shall not be more than ten (10) journeymen and apprentices combined. Only a craft foreman who normally works with tools of the trade during the straight time period will work with tools during overtime periods.	Use standard piping crew for all pipework less than 6 feet above work floor. Add 1 fitter to standard piping crew for pipework above 6 feet.
Piping: 6 Pipefitters 3 Welders ½ Equipment Operator	Where four (4) or more Carpenter Foremen are employed on a project, there shall be a designated General Foreman, and he shall receive General Foreman's pay. General Foremen shall supervise and give orders to Carpenter Foremen and shall not supervise over six (6) foremen on a project. When any workman in the trade is given foreman's responsibilities, he shall receive foreman's pay. Provided qualified men are available, all foremen shall have worked at least one year within the last three years within the jurisdiction of one or more Local Carpenter Union or Unions within this work region.	

Figure 9.4 Historical crew sizes approaches. (*a*) Standard crews. (*b*) Agreement clauses. (*c*) Impact of work conditions.

mental assessment of many factors into an internal vindication of their decisions. Rarely do such decision makers have any clear or objective decision process. Instead, they usually take recourse in specific statements of opinion supported by general norms of historical practice to justify their decisions to others. In most cases, the common sense and experience of the foreman/superintendent is such that effective crew planning results. In this management environment it is difficult to explain the concepts involved in the decision process in other than an intuitive manner. The following paragraphs introduce a number of the dominant concepts relevant to the Crew Planning decision process.

An obvious objective in Crew Planning is to establish the crew size that best meets both desired work rates and the economies of work volume, and which is capable of being effectively managed by the foreman or superintendent. In many cases in practice this is accomplished by simple trial-and-error modifications to standard crew sizes. However, the crew size is limited by the nature of the work face, the congestion of space, the complexity of the work involved, and the skill and management approach of the field manager involved. In this respect, the continued consideration of safety aspects, the concern for the minimization of crew delays of all kinds, the effort required to achieve desired levels of control and labor productivity, as well as the normal requirement to direct and monitor crew activity, places a considerable management strain on foremen. In some cases the management strain will limit the effective span of control and smaller crews than normal may be essential.

The foreman will try to plan a crew according to the nature of the work involved, so that there is an effective matching of trade and skill levels. Thus, in order to make the crew as productive and cost effective as possible, the foreman will tend to separate the unskilled supportive work for assignment to laborers from the skilled work that is to be performed by the generally much higher paid tradesmen. It then becomes a much simpler problem to match the crew, according to the productivities of the tradesmen and supporting laborers, to work content.

In most cases the foreman will rely on past experience and adopt historical or *standard* crew sizes, as these represent tried-and-true practical solutions to crew selection. Typical examples of historical standard crew make-up are given in Figure 9.4*a*.* When working in areas which have formal contractual agreements between unions or organized labor and contractors, the foreman will rely on special contract clauses which impact the planning of the crew size. These clauses often focus specifically on jurisdictional solutions to the number of support groups, and even on such matters as whether or not the foreman is to be a working member of the crew itself (see Figure 9.4*b* for typical examples). Finally, on some projects the contractor will have es-

*See, for example, *Building Construction Cost Data,* Robert Snow Means Co., Inc., Kingston, Mass. Published annually.

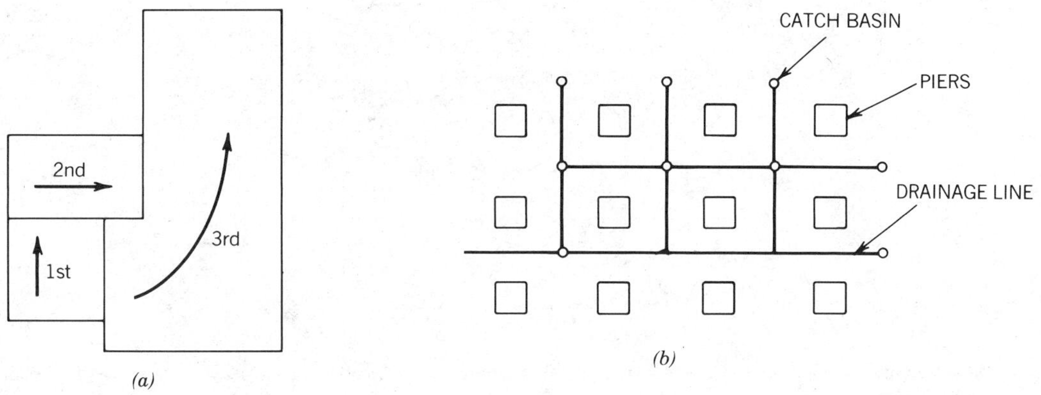

Figure 9.5 Work layouts. (*a*) Roof decking erection sequence (single crew). (*b*) Plan showing pier and drainage line locations (foundations and piping crews).

tablished, by previous field studies, *optimal* crew sizes for different work conditions. Thus, for example, Figure 9.4*c* indicates different piping crew sizes and shows that the need for supporting labor is a function of the working height of the pipework itself. If this concept is carried further and the work volume warrants it, special *support activity* crews will be planned. It is always good practice to keep foremen aware of the importance of maintaining quality standards and productivity levels, and the importance of thorough preplanning and mobilization efforts. Foremen must be able to rely on field management for the supply of good quality drawings and information, and on an adequate provision of support resources. Thus, field procurement, material handling, and special consideration of the location, clearance, and management of supply routes to the work face will be considered in Crew Planning. It may even become necessary to sketch the proposed order of development of work areas in order to ensure the smooth delivery of materials and minimize double handling of material, and site hazards resulting from material lying loosely around the work face. Figure 9.5*a* shows a typical work layout sketch of a roof decking erection sequence. In this case work proceeds from the lowest roof elevation to the highest with its implied changes in material flows.

In terms of the immediate preplanning for crew activity, the field supervisor(s) must check out availability of drawings, the location of material and necessary equipment, and must ensure that these have been authorized and made available for the work activity. Finally, the foreman/superintendent must carefully plan the actual field mobilization prior to work start-up. On large and congested sites it will be necessary to see to it that free access to the work site itself is possible. This may require coordination with other superintendents and crew foremen. One practical approach on congested sites is to have the work area plans developed and color coded for individual crew access priorities on a weekly or even daily basis. An example of this is shown in Figure 9.5*b*, in which a concrete crew will form and pour a series of concrete piers while a piping crew installs drainage lines and precast concrete catch basins in close proximity. If the schedule is such that these two crews must be involved in the same area at almost the same time, then it may be necessary to establish the order in which the piers and the drainage lines are installed simply to avoid congestion and crew interference.

In more difficult or critical cases it may be necessary to plan a crew size and detail crew member work sequences for an activity for which either no previous experience exists, or where it is essential to ensure that a feasible crew size can meet required production rates or imposed equipment availability time constraints. In this *work task analysis* approach to crew sizing, it is necessary to determine all the work tasks that are involved in the activity, and to plan the crew size and activity sequence in such a way that the design of the crew can be checked *before* field activity commences. The Crew Balance Chart is one technique which enables this to be done. It is illustrated in Figure 9.6. Simple time-scaled bars are used to denote the planned activity of each crew member and the resources critical to these activities. Thus, crew member 1 has a work sequence of work tasks 1, 2, and 4, but work task 4 must be performed in conjunction with crew member 2 and a crane. In this way, once simple

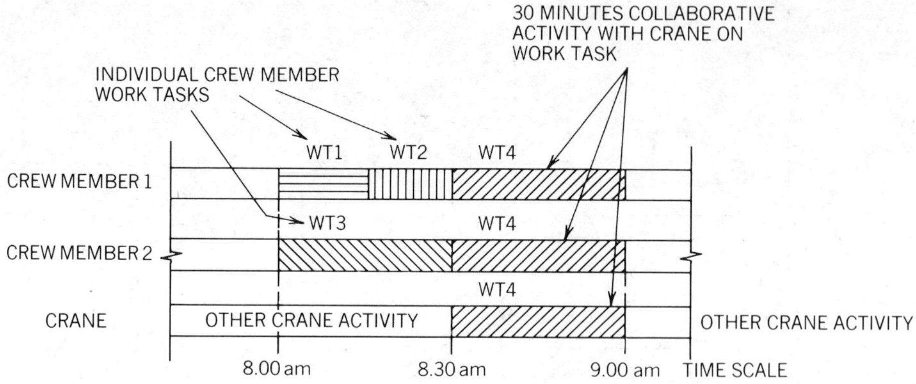

Figure 9.6 Crew balance chart.

work task duration estimates are available, a plan of crew activity can be prepared and decisions relating to work sequences, crew size and mix, and crane use schedules can be verified before field activity commences.

A practical approach that integrates formally most of the previous concepts, and which is commonly found in large industrial type construction projects, is called the *work order* approach. The work order approach is a field planning technique which is used to break out work for crew activity from what is otherwise only previously scoped out cost and estimating data for work packages of major items of construction. In this respect the work order must define the work involved in terms of drawings and material requirements, and must indicate quantities, construction methods, optimal crew size, and construction sequence in sufficient detail that it can be farmed out by the relevant area superintendent to specific general foremen and their crews. A schematic of the scope and functional breakdown of a fully documented work order is illustrated in Figure 9.7. The work order approach is very useful when the volume of work is such that crew planning responsibilities are removed from the foreman, who is then free to devote all his talents to crew management. In this crew planning environment, work order definition must be performed by experienced construction engineers, superintendents, and methods engineers, and care must be taken to ensure that the documented material is presented in a manner that makes it easy for the foreman to understand the scope of work involved, the resources that are to be made available, and the manner in which field activity can proceed.

A useful technique and an aid to both superintendents and foremen in Crew Planning is the development of a detailed work schedule, either weekly or bi-weekly, as shown in Figure 9.8. This schedule is a combination of a simple bar chart schedule for the current week's activities for a specific crew or crews, and a reminder to the assistant superintendent/foreman of next week's planned activities. Thus, the foreman is gently led into an objective planning mode while continually being reminded of his or her management requirement to prepare for the next week. If, in addition, each bar is tagged with manpower, equipment, and material requirements and, if desirable, drawing numbers and so on, the schedule becomes a very powerful manpower planning and preplanning document, especially when distributed to field engineers, warehousing personnel, material control engineers, and other personnel responsible for supporting the construc-

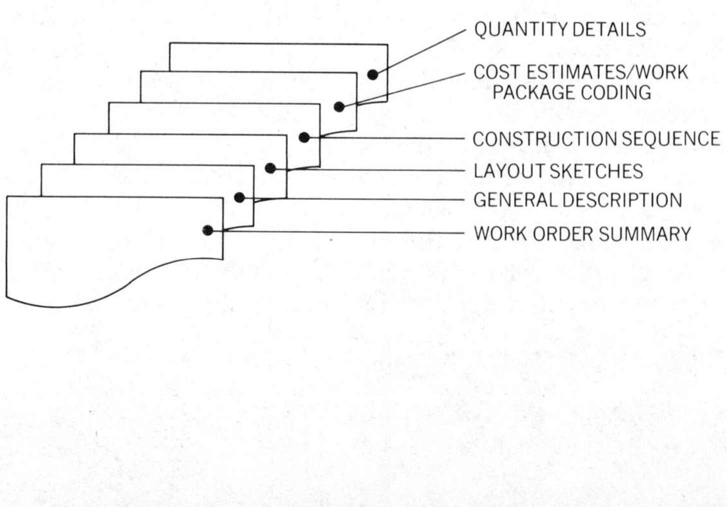

Figure 9.7 Work order approach.

COST CODE	ACTIVITY	MATERIAL QUANTITY	CREW SIZE	EQUIPMENT	M	T	W
110001	EQUIPMENT FOUNDATION						
	FORMING	9680SF	20				
	REBAR	44000#	5				
	POURING	176 CY	10	VIBRATORS 2 EACH			
	DRAWING #10-000-24						

Figure 9.8 Detailed work schedule.

tion effort. In many cases the work schedule is a very successful technique for involving the foremen in specific quantifiable management functions.

An extension of the crew balance chart can be applied to the situation where many crews are involved and compete for shared major resources. A technique that is derived from this extension is called the Multiple Activity Chart (see Figure 9.9). It is very useful for the scheduling of crews by floor on a high-rise building, or in a management control area approach for congested industrial or power-oriented construction projects with hundreds of crews. Its format can be readily changed to a time-scaled bar form similar to that of Figure 9.7 or Figure 9.8. As indicated in Figure 9.9, each crew (and if necessary, each shared major equipment item), is considered a major resource and is given an individual entry. This distinguishes it from the crew balance chart of Figure 9.6, in which each crew member is uniquely identified. The time scale of the Multiple Activity Chart may reflect project weeks as shown in Figure 9.9, or two-week periods only as shown in Figure 9.8, or be designed especially for the cycle time required to produce a repetitive constructional unit such as the floors of a high-rise building.

A number of formal external documents are involved in the Crew Planning decision process. Technical drawings prescribe what is to be built and where. Therefore, the superintendent/foremen must carefully study the requirements for the next crew activity, noting measurement, material needs, connection details, and so on. Similarly, quality requirement specification clauses must be related to field conditions. Finally, the identification of special materials and the check on performance specifications for specialized tools or equipment must be made. Thus, the superintendent/foremen must be familiar with all the technical information covering the facility component to be built by each crew and the associated resources such as material, equipment, and tools.

During the preplanning of the crew's next task, the superintendent/foremen must evaluate the physical aspects of the job, such as its location, access, working environment, as related to the location of

		PROJECT WEEKS		
CREW		WEEK 58	WEEK 59	WEEK 60
SCAFFOLDING CREW 2		AREA 4	AREA 10	AREA 5
·				
PIPE HANGING CREW 8		AREA 6	AREA 4	AREA 10
·				
·				
WELDING CREW 3		AREA 2	AREA 6	AREA 4
·				
·				
LABORING CREW 5		AREA 7	AREA 2	AREA 6
·				

Figure 9.9 Multiple activity chart: control of work areas and scheduling of crews.

PIPE HANGING CREW 8 INVOLVED IN PIPING WORK ORDER ACTIVITY IN AREA 10 DURING PROJECT WEEK 60.

TECHNOLOGICAL SEQUENCE OF CREWS IN AREA 6.

tool boxes, material dumps, and so on for the selected construction method. They must also relate individual skills of crew members to the various basic and cooperative tasks involved in the performance of the construction method, and see to it that the proper steps are taken to minimize needless effort so that the crew can perform in an efficient manner. This process must take place during all stages of the crew's planned and scheduled activity. This considerable effort (usually mental) of definition and analysis constitutes the so-called informal Crew Planning Management Information System (MIS).

When the volume of work and the technical complexity of a project involves a large number of specialized crews, more formally documented information is required for crew planning. This information frequently includes guidelines to work simplification and practical suggestions for the elimination of needless work. In addition, there is the need to know current status, crew availability, and projected manpower requirements, and to have access to short-range (bi-weekly) schedules highlighting anticipated crew activity. In such an environment the Crew Planning decision process relies more extensively on documented information inputs from other decision processes, such as detailed project scheduling, man-hour budgets, and current status and progress. As a result, Crew Planning produces an increasing level of documented data for input to other decision processes, such as Crew Management and Field Monitoring. In this way, Crew Planning becomes a more formally documented MIS.

9.3 THE PROJECT TEAM COMPONENT APPROACH

Crew Planning is a critical decision process for the small firm. It establishes the method of performing the work, the magnitude of the field labor effort, and the plan by which the work will be accomplished. In this way, crew planning focuses on two key management problems: (1) crew sizing and work face layout; and (2) crew scheduling.

In order to address the crew sizing problem, the job superintendent must establish the work content involved by performing a DETAILED FIELD QUANTITY TAKE-OFF (P6) if the desired level of quantity detail is not available in the estimate documents. The selection of the most desirable construction method essentially enables the superintendent and foreman to select a suitable crew and establish the basic CREW REQUIREMENTS (P7) which, in conjunction with the specifics of the location of material dumps and access routes to the work face, allows them to PLAN THE WORK FACE LAYOUT (P8) and determine achievable production rates and activity durations. The Crew Planning decision process is completed once the CREW SCHEDULE (S7) or activity start-up date is established. The job superintendent can act autonomously because of policies conferring *FIELD RESPONSIBILITY* (P_S 12)* on him. In this environment he can plan, schedule, allocate, and monitor project activity according to preferred methods based on *PAST EXPERIENCE* (P_S 13). In this crew planning activity the job superintendent will act within the *UNION PRACTICE* (P_S 11) and *CREW MIX* (P_S 15) policies adopted by executive management through its relationships with local unions and commitment to jurisdictional findings.†

The superintendent has several methods available for *CREW SIZE SELECTION* (P_S 14). As indicated by the policy statement, the approach most commonly adopted by field managers is to base crew size on *historical crew sizes and mixes*. This approach is best suited to the small firm project environment because the work is conventional and tried-and-true construction methods prevail. Thus *standard crews*, which are consistently used with these methods, are applicable. The standard crew size and mix will vary slightly depending upon the estimated material content and the desired production level. The *trial-and-error crew buildup approach* is applied less frequently than the historical crew size and mix method. In some cases, and especially in situations where the contractor is concerned about the level of field control and the rate of labor dollar expenditures, a smaller than normal crew will initially be used. In this crew sizing approach the contractor has more time to monitor crew performance, to determine its best size and mix, and to add crew members later if required. In terms of seasonal impact on crew size, if the project must be completed, or, in the case of a building, the

*The small firm policies relevant to the Crew Planning decision process are presented in Table 9.3.

†Restrictive work practices are all but eliminated for the nonunion contractor. Thus, the policy *UNION RELATIONS* (P_S 11) would not be required for the small nonunion contractor.

TABLE 9.3 POLICIES OF THE SMALL FIRM

Policy		Description
P_S 7	*SCHEDULING FOCUS*	Scheduling definitions and constraints are: (1) critical activity is any activity that opens up work areas for other crews and whose delay causes work volume bottlenecks. (2) Activity durations and sequence logic incorporate the impact of seasonal weather conditions. (3) Activity scheduling is based on the earliest possible start dates. (4) Manpower impact is neglected.
P_S 11	*UNION PRACTICE*	Field management must subscribe to local union practices when operating in a union environment.
P_S 12	*FIELD RESPONSIBILITY*	Actual planning, manipulation, and general management of labor is a total field responsibility.
P_S 13	*PAST EXPERIENCE*	Field planning, scheduling, and management of manpower is based on a free-wheeling ad hoc approach, with heavy reliance on past experience.
P_S 14	*CREW SELECTION*	Crew selection is based on these models: (1) historical crew sizes and mixes, (2) trial-and-error crew buildup, (3) maximum crew size.
P_S 15	*CREW MIX*	While impacted by union rules in some construction environments, the ratio of foremen to journeymen/apprentices within a crew is established to best optimize both the foreman's level of supervisory control and the crew's performance. Foremen other than loyal company foremen are selected on the basis of experience, managerial capability, and desire, in that order.

basic structure enclosed prior to periods of inclement weather, it may become necessary to increase the size of the crew in order to ensure the attainment of these objectives. Conversely, if it is deemed necessary to continue the work during the winter months, and the structure cannot be adequately enclosed prior to this period, the net effect on crew size might be an overall reduction of manpower.* The primary purpose behind the latter approach is to minimize the need for show-up or attendance pay when inclement weather prevents effective work.

The job superintendent/foreman must also be aware of those activities which are deemed critical in terms of the project schedule. Field managers will have the overall schedule as established by the head office, and will be responsible for both maintaining and updating the schedule. Hence, because critical activities influence the rate at which succeeding work becomes available (see policy P_S 7, *SCHEDULING FOCUS*), the field superintendent will generally expand the crew size for those activities in an attempt to expedite their completion. The increase in crew size may be unprofitable, and is therefore acceptable only if the increase will ultimately make a profit for the project as a whole (by expediting the scheduled completion of the project in relation to overhead and liquidated damage cost overruns).

*This naturally applies to geographical areas where the weather, especially in the winter, is a particularly significant problem.

Various Crew Planning policies come about as byproducts of union rules that directly influence the crew selection process. For example, the *CREW MIX* policy suggests that the foreman/journeyman ratio be optimized within the constraints imposed by the union's working rules. In nonunion construction, the contractor has the distinct advantage of being able to establish the desired crew mix ratios of foremen to journeymen and helpers. In addition, this contractor is not concerned with jurisdictional issues during crew sizing.

Crew sizing and scheduling are extremely critical problems at the small firm level, and their resolution provides the keys to successful field construction planning. It is within Crew Planning that the actual details of construction execution are determined. Poor crew planning can jeopardize the successful completion of the project, and can create crew management problems even before crew mobilization. The burden of Crew Planning is carried entirely by the field superintendent and the foremen as the project manager rarely enters this decision process. The Crew Planning decision process network is shown in Figure 9.10, with the crew sizing and layout and crew scheduling problems emphasized on the network.

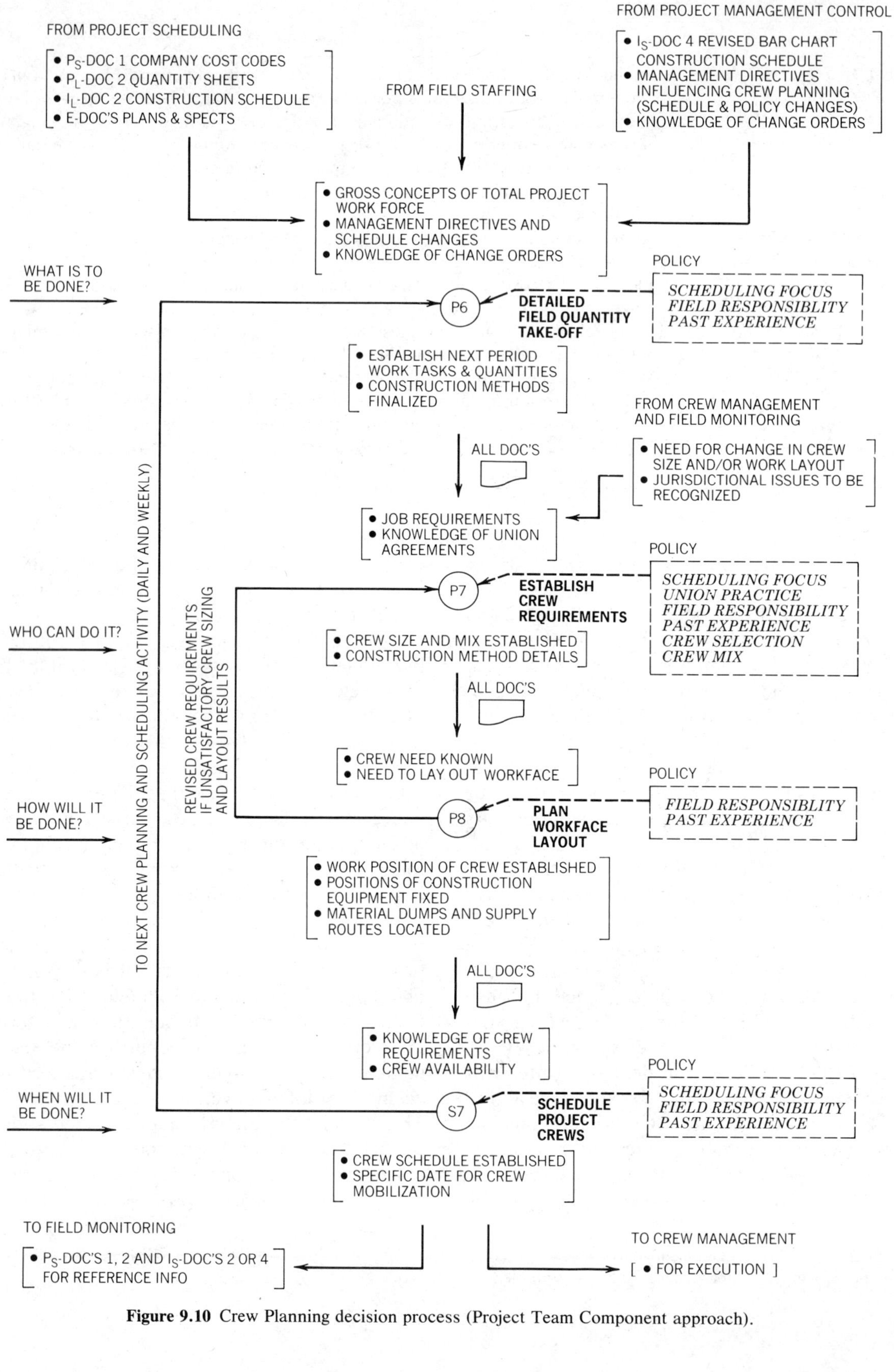

Figure 9.10 Crew Planning decision process (Project Team Component approach).

TABLE 9.4 POLICIES OF THE MEDIUM-SIZED FIRM

Policy		Description
P_M 12	*UNION PRACTICE*	Field management must subscribe to local union practices when operating in a union environment.
P_M 13	*MANPOWER PLANNING*	Short-range planning of manpower at the head office pre-job level, coupled with field labor manipulation, is critical to project control.
P_M 14	*PAST EXPERIENCE*	Field planning and allocating of manpower is based on a freewheeling approach, past experience, and close interaction between the project manager and field management.
P_M 15	*CREW SELECTION*	Crew selection is based on three models: (1) historical crew size and mix, (2) trial-and-error buildup, and (3) maximum crew size.
P_M 16	*CREW MIX*	While impacted by union rules in some construction environments, the ratio of foremen to journeymen/apprentices within a crew is established to best optimize both the foreman's level of supervisory control and the crew's performance.
P_M 18	*PERSONAL CONTACT*	Manpower management begins with head office scheduling and allocation of manpower with field input, but the primary emphasis remains on-site good relations via personal contact and recognition.

9.4 THE TRAVELING PROJECT MANAGER

While the Crew Planning decision process of the medium-sized firm is similar in many respects to that of the small firm, there are subtle differences in the methods of solving these problems. This is due mainly to the character of the information inputs into the decision process and to the size of the field work force. For instance, as distinct from the small firm, the continuous (and often daily) cost and schedule data that the Traveling Project Manager brings with him to the field from the continuous consideration of the Project Scheduling and Project Status decision processes often provides detailed inputs relative to basic manpower requirements, the changing needs of a detailed construction schedule, and the attainment of labor budget objectives. This continuous knowledge of project cost/schedule status enhances the Crew Planning process, because the impact of schedule changes in terms of time and manpower can be readily incorporated into the daily consideration of crew sizing and scheduling problems. Similarly, the sizing and scheduling of crews is affected by the level of field effort as ascertained from daily monitoring of crew performances. A congested work area, for example, may ultimately signal a need for reexamination of the planned crew size and mix, construction method, and other details affecting the crew's structure.

Because some medium-sized firm projects are larger and more complex, the planning horizon for the planning and scheduling of manpower to various work tasks, when referenced against stipulated project completion dates, can become a crucial consideration. If the project is particularly complex, the crew planning horizon must contain enough lead time for the coordination and planning of crew sizes and crew activity around the work site so that all crews interface efficiently, especially with subcontractors' crews, without creating congestion. Thus, pragmatic *MANPOWER PLANNING* (P_M 13)* begins to appear naturally in this field environment. Harmonious relations between the project manager, job superintendent, and foreman through constant *PERSONAL CONTACT* (P_M 18) are seen to be the best approach to achieving maximum crew planning efficiency.

A DETAILED QUANTITY TAKE-OFF (P6) is the first function initiated in Crew Planning. The project schedule provides the time segment and planning horizon for each activity, and the plans and specifications coupled with the labor estimate enable the field staff to break an activity into definable work tasks. Critical activities are given close attention in crew sizing. The work tasks and their quantities from construction drawings are determined for the next planning period. Then, based on *PAST EXPERIENCE* (P_M 14), the schedule, and the estimate of the work volume, CREW REQUIREMENTS ARE ESTABLISHED (P_7). The *historical crew size and mix* approach is used by the craft foremen and superintendents to plan crews for less critical work tasks, or work tasks which inher-

*The medium-sized firm policies relevant to the Crew Planning decision process are collected in Table 9.4.

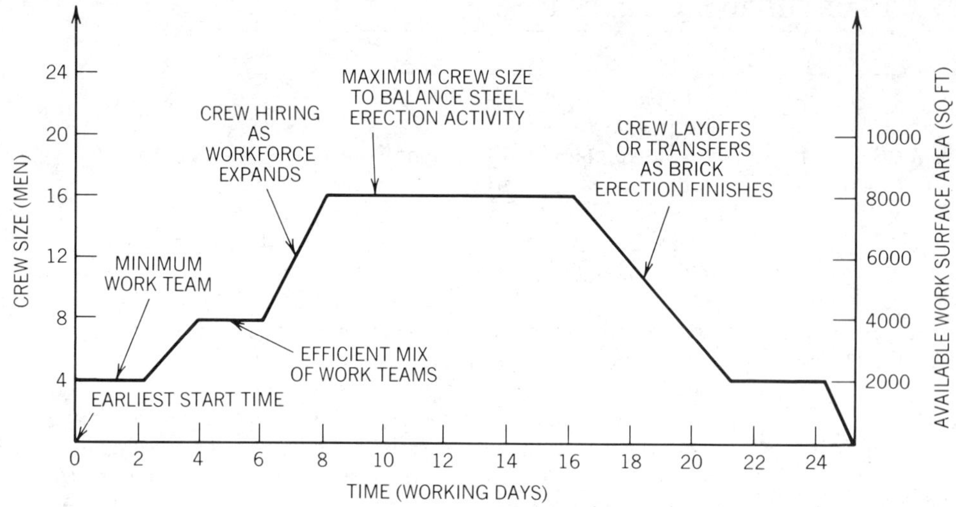

Figure 9.11 Exterior brick activity: crew growth-decay curve as a function of available work surface.

ently require small crews. The foreman continuously informs the job superintendent of *MANPOWER NEEDS*. On larger projects, the craft foremen would work more closely with the superintendent when sizing crews using the historical crew size and mix approach.

The *trial and error crew buildup approach* is most applicable to work tasks having a significant material placement volume, a standard construction method, and the necessity to match work progress with parallel and interlocking activities. Thus, this approach is most useful for those work tasks where work volume fluctuates with project progress (e.g., exterior brick masonry operation where the work face expands in proportion to the completed superstructure). The crew is initially sized to commence work on the activity at the earliest start time, is built up as the work face expands, and then reduced as the work task nears completion. Crew buildup is characterized by a crew growth-decay concept, as illustrated in Figure 9.11. For example, a masonry activity such as laying the exterior face brick will commence at the earliest possible time, that is, when a significant enough percentage of the superstructure is completed to allow this masonry activity to commence. The initial crew size is small because the available work surface is limited. As the available surface area for masonry work expands, the crew is enlarged accordingly. The rate of buildup is dependent upon the rate at which the work surface expands, and upon both the desired and actual production rates of the crew. These variables are considered, along with the activity's work duration and construction method, to determine crew size at any point in time. In addition, the superintendent/foreman must continuously observe activity progression and crew production in order to modify crew size and anticipate crew buildup over time. Thus, the crew size increases as the work surface expands and, as the work is completed (denoted by a decreasing work surface), the crew size slowly decreases.

The *maximum crew size approach* calls for commencing a work task with a full crew and adjusting crew size only when the desired production rate is not achieved. This approach can be applied to a variety of work tasks, and requires knowledge of the production rate, construction method, and activity duration. It is from these that the crew size is determined. After the crew size and mix is evaluated, the work face layout established, and the crew is scheduled to start, the work task is manned with the full crew as determined by the variables specified.

The WORK FACE LAYOUT (P8) for each major construction activity must be carefully planned regardless of the crew sizing approach. The location of material dumps and equipment layout is an essential feature of the crew planning problem. Normally, formal documentation is not required, and previous experience, common sense, and judgment are the basis for solving these problems. Careful attention should be given to an activity if crew buildup is designed to fluctuate with the available work area. In this case, the location of the material dumps and material flow routes must be reevaluated continuously during field monitoring to ensure that a smooth crew growth curve is achieved.

PROJECT CREW SCHEDULING (S7) provides the approximate date of crew mobilization defined by the latest project schedule, and by manpower, material, and equipment availability. The job superintendent and foremen assess these status and resource criteria to determine the most advantageous date for crew mobilization on new project activities. Project activities that are critical, as well as resource-heavy activities, are normally scheduled for the earliest possible start dates because their early completion will open up work areas and release resources for other activities. Schedule changes can also impact the timing of crew mobilization either by slipping crew start-up in order to utilize available manpower more advantageously, or by requiring an earlier than anticipated start-up, because of the ready availability of major equipment.

Crew sizing and scheduling problems are considered critical management problems of the medium-sized firm. These problems are keys to the effective planning of field construction, and are solved by continuous interaction between the project manager, superintendents, foremen, and other field support personnel. The decision network for the Crew Planning decision process is illustrated in Figure 9.12.

9.5 FIELD-BASED PROJECT MANAGEMENT

The size and technical complexity of the large project and the general manner in which its planning and scheduling are developed, together with the often large field work force involved, make the Crew Planning decision process in this environment significantly different from that of the small and medium-sized firms. This difference is due not only to the considerably larger crew planning problem involved on large projects, but also to the increased technical management requirements associated with such projects.

At the large firm level the magnitude of the work content and technical complexity of projects usually require the segregation of common work activities, during the project scope and planning phases, by the use of project cost codes. The need to have specific groupings of work activities leads to the practice of defining work packages of related work tasks. These work packages are often only identified by scope statements in the early planning phases of the project, and their detailed analysis and construction method determination is consequently usually left to the crew planning stage. In addition, the logistical problems associated with the provision of field drawings, material, equipment, and specialist tools on large technical projects is such that the Crew Planning process for specific work packages must pass through a number of resource check milestones before these work packages can be released to crews. Finally, because of the size of the field work force and the large number of crews, many of whom are specialist and high-skill focused, it becomes necessary to ensure that crews are scheduled to work systematically in their areas of expertise. Thus, the freewheeling short-range planning cycle approach to crew planning commonly associated with the small and medium-sized firms becomes increasingly difficult to perform at the large firm level. Consequently, the Crew Planning decision process that emerges begins to focus on the detailed scheduling of work task sequences to correlate with the desired number of crews, and the effective crew member sequences in individual crews. Thus, the decision process no longer has only a crew sizing and crew scheduling emphasis, but purposely addresses the matching of individual crew design to the required work activity.

In this more complex decision environment, the number of manpower management functions involved with Crew Planning, and the level of documentation required for their performance, increases. Because the work package approach to work definition becomes part of the crew planning process, several new manpower management functions emerge which are not formally performed at the medium-sized firm level. These additional functions focus on the DEFINITION OF THE WORK PACKAGES (P5) involved and the DEVELOPMENT OF THE DETAILED LOGIC OF EACH WORK PACKAGE (S4), together with the DETERMINATION OF THE NUMBER OF FIELD CREWS (A4) required in any period. The SCHEDULING OF CREW MEMBER SEQUENCES (S8) also becomes essential for many critical work tasks. In addition, the nature of the work involved in the formulation of detailed work packages and crew size decisions is such that the Crew Planning decision process can be readily split into a Crew Analysis and Design phase (see Figure

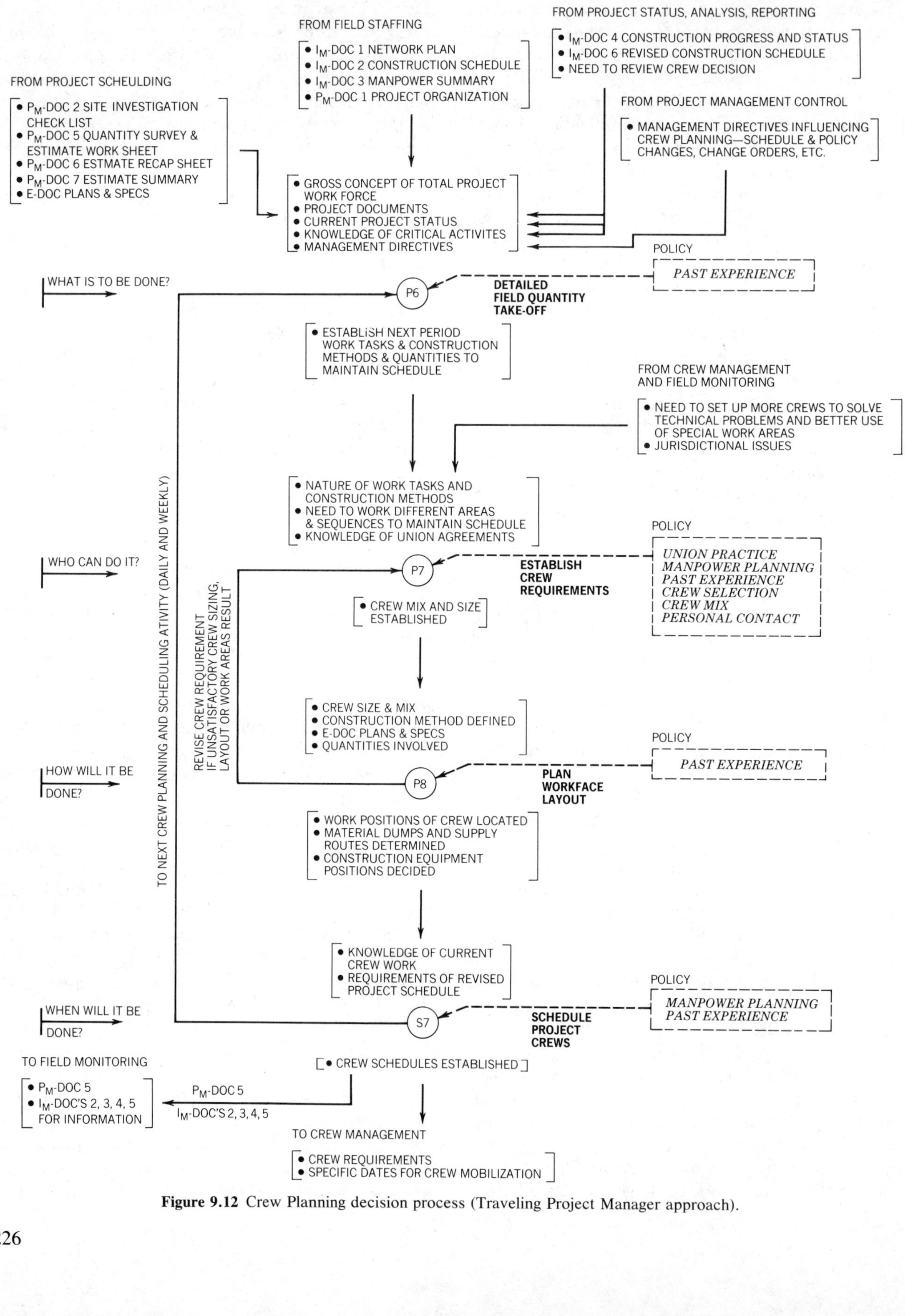

Figure 9.12 Crew Planning decision process (Traveling Project Manager approach).

9.13) involving only planning functions, and a Crew Scheduling phase (see Figure 9.17) which focuses on crew related allocation and scheduling functions. This split focus means that more field project team members become involved in the decision process. Thus, crew analysis and design for a work package is the responsibility of a combination of field managers/supervisors. The general superintendent, field engineers, and assistant or craft superintendents are primary participants in Crew Analysis and Design, and are supported by field scheduling engineers. The general foremen and foremen are also influential in crew design, however, their focus is different from that in the medium-sized firm, where they have a more dominant role to play. Crew Scheduling is most frequently performed by superintendents, general foremen, and scheduling engineers, as required. The primary documents which stimulate Crew Planning are the short-range Quarterly Construction Schedules (P_L-DOC 20), the relevant Fragnet Schedules (I_L-DOC 10), and in addition, the various documents indicating current activity status and project reporting, such as Weekly Work Status (P_L-DOC 40) and Percent Complete Reports (P_L-DOC 42) emanating from Project Status, Analysis, and Reporting. These documents, together with technical drawings, provide the level of detail needed to define work packages. This information, combined with the relevant field managers' knowledge of field activity status, is used in foremen meetings to inform these managers of the potential work available in the next planning horizon. These scheduling documents, however, are still too broad for daily and weekly crew planning, and therefore the Quarterly Construction Schedules become a guide for assigning activities to work packages as each activity is scheduled to commence.

Crew Planning begins with the Crew Analysis and Design phase as shown in Figure 9.13. The craft or assistant superintendents, together with field engineers, use *COST CODE* (P_L 6)* techniques to *DEFINE THE WORK PACKAGE* content in terms of those activities scheduled for the next period and those that are currently in progress. Once a work package is defined by a set of activities, these activities are dissected into work tasks representative of the work to be performed. Work requirements for these work tasks are then defined and identified. The superintendents and field engineers, working together as a *PROJECT TEAM UNIT* (P_L 35), determine the drawings, construction methods, equipment, and materials required for a work task by analyzing each operation of that work task in detail and establishing the best construction approach. These personnel rely on general construction drawings and specifications, combined with past experience and knowledge of labor estimate details, to establish the *CREW PLANNING* (P_L 16) and *CREW SIZING* VARIABLES (P_L 18) associated with each work task. Once the components of an activity are defined, the field engineer must first determine if the requisite detailed field drawings are available for issue to crew foremen, as illustrated by this decision loop in Figure 9.13. On large design–construct projects with overlapping design and construction efforts, drawing availability often controls the crew design effort. This is not a major problem for the medium-sized firm because the contractor is usually operating with approved drawings and specifications, most of which are always available.

A DETAILED QUANTITY TAKE-OFF (P6) is performed by the superintendent and/or field engineer for the work tasks within the particular work packages. This may be in the form of a final detailed field take-off or may be an analysis of a bill of materials. The quantity take-off defines the material content of work package activities and leads to a material availability and location check. While this decision loop is performed infrequently at the medium-sized firm level, it becomes a critical problem on large projects where special items and significant bulk materials, such as structural steel, re-steel, piping, electrical, and so on, are required and often stored in warehouses or laydown areas located at considerable distances from the proposed work face. It is best to have all materials available for an activity prior to proceeding to the full-scale crew design and activity layout. However, pressures of schedule often require that work be scheduled to proceed without all available materials on the premise that the missing materials will arrive on site and at the work face before scheduled work starts.

ESTABLISHING CREW REQUIREMENTS (P7) remains the critical function in Crew Analysis and Design. The ingredients which predominantly influence this function are *CREW SIZING VARIABLES* such as the estimated activity labor man-hours, quantities, and scheduled activity durations. The labor man-hours are derived from Estimate Work Sheets (P_L-DOC 14), and serve as a guide for crew design, while activity durations are available

*See Table 9.5, Policies of the Large Firm.

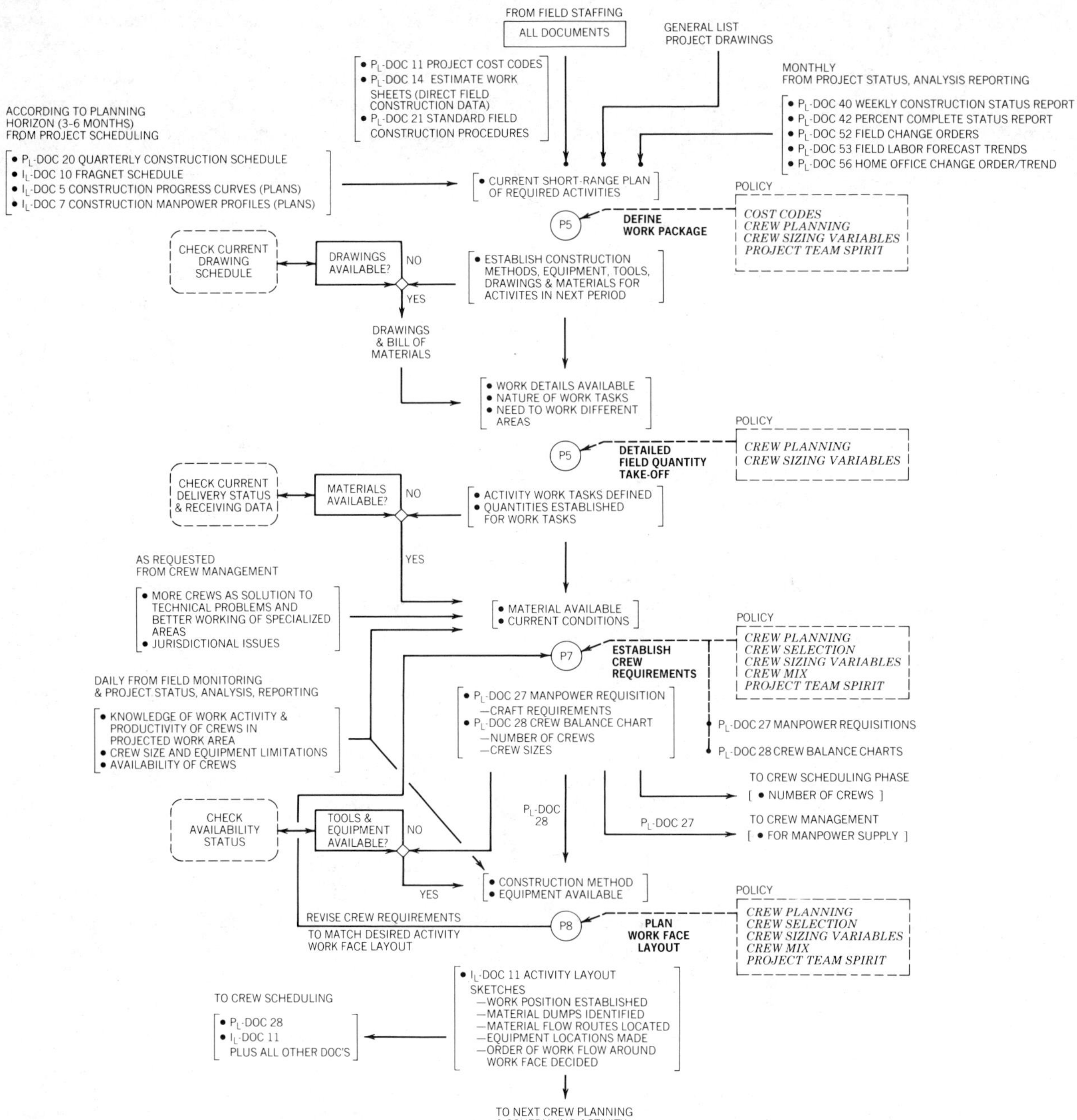

Figure 9.13 Crew Planning decision process: crew design phase (Field-Based Project Management approach).

from the Quarterly Construction Schedule. In addition, Construction Progress Curves (I_L-DOC 5) and Construction Manpower Curves (I_L-DOC 7) provide further guidelines in terms of progress targets (percent complete) and anticipated manpower levels for the next planning horizon. The need to meet these guidelines often leads to requests for additional labor and in the preparation of Manpower Requisitions (P_L-DOC 27), as shown in Figure 9.14.

A variety of crew sizing techniques are in common use. The specific method chosen by the field personnel depends on the situation and the general size of the field labor force. For instance, on proj-

TABLE 9.5 POLICIES OF THE LARGE FIRM

Policy		Description
P_L 6	*COST CODES*	A systematic classification of project work categories segregated by unique physical components and by cost codes is the basis for structuring the project scope into work tasks, categorizing estimated costs, reporting and controlling costs and manpower, and developing a historical database.
P_L 10	*NETWORK-BAR CHART ANALYSIS*	Broad network focused milestone planning is the best approach for identifying and sequencing major project components, while a combination of critical path method and bar chart model are best suited for project planning and scheduling.
P_L 11	*SCHEDULING FOCUS*	Scheduling definitions and constraints are: 1. A critical activity is one that, if delayed, will extend the completion date of the project. 2. Activity durations must be correlated to the estimated work content. 3. All constraints must be defined and incorporated into the sequence logic. 4. Manpower impact is an integral aspect of project scheduling and scheduling control. 5. Schedule must reflect optimum use of all resources. 6. Progress is based on the physical completion of work activities.
P_L 15	*MANPOWER PLANNING*	Long-range planning of manpower by craft and in total is a prerequisite to supporting the field effort and must be performed in the planning and scheduling stages of the project.
P_L 16	*CREW PLANNING*	Field planning, scheduling, and allocation of manpower requires a precise definition of work tasks and a combination of past experience, knowledge of work content and man-hour estimates, and interaction between field project team members.
P_L 17	*CREW SELECTION*	Crew selection is based on these models: 1. Maximum crew size 2. Number of crews and crew buildup 3. Work task analysis 4. Historical crew
P_L 18	*CREW SIZING VARIABLES*	Labor dollars, man-hours, and quantities are primary ingredients in crew selection and must be weighed against scheduled activity durations.
P_L 19	*CREW MIX*	While impacted by union rules in some construction environments, the ratio of foremen to journeymen/apprentices within a crew is established to best optimize both the foremen's level of supervisory control and the crew's performance.

ects with large labor forces, the project-wide focus on the number of crews that should be established at any one time is a major decision and corresponds in concept (total number of crews rather than just one crew) to the crew buildup technique often used by the medium-size firm for several individual critical crews. When the labor force is smaller it becomes more practical to determine the number of crews and their individual sizes using a maximum crew size approach. The work task analysis approach becomes a common tool, however, because the complexity of specific work often confronts the field managers/supervisors with unique operations which do not compare to previous experience. In these cases, the crew will be sized based on an in-depth method analysis.

When the *historical* and *maximum crew size* techniques are used by the large firm, the specific approach would be much the same as that described for the medium-sized firm. However, these two approaches are rarely suitable for large projects. Rather, the *number of crews buildup* approach is used. This approach is based on determining a standard crew for a work task type, and then evaluating the total number of that type of crew required for the duration of the work task. If the duration carries over several quarterly planning periods, then only the number of crews required for the current quarterly period would be evaluated. Because this approach is often applied to the crew manning of activities with longer durations, the actual timing and buildup of the crews is important. The likelihood

AW
Engineering & Construction Corp.

MANPOWER REQUISITION

REQUESTED BY: ______ TITLE: ______

CONTROL ZONE: ______ DATE: ______

CRAFT CODE	CRAFT DESCRIPTION	GEN'L FOREMEN	FOREMEN	JOURNEYMEN	DATE REQUIRED
06	CARPENTERS		1	10	MAY 15
08	CEMENT MASONS		2	25	MAY 30
16	LABORERS	1	3	20	JUN 10
	.				
	.				
	.				
	.				

APPROVALS:

PROJECT MANAGER ______ GEN'L SUPERINTENDENT ______

Figure 9.14 Manpower Requisitions (P_L-DOC 27).

that other crews will be added depends on the availability of work space, material content, and availability of support construction equipment. This is also impacted by the desired rate of progress as given by the Construction Progress Plan or the Current Progress Forecast. Thus, the number and buildup of project activity crews over the life of a project matches the planned or forecasted progress targets. In this crew buildup approach, the basic standard crew size is often formulated from historical crew sizes and mixes, but with more careful analysis given to quantities, labor man-hour and unit rates as established by the labor estimate, proposed construction methods, and material logistics and layout problems.

The complexity and highly technical nature of large projects often confronts the superintendents and foremen with unique work operations to which the *work task analysis* crew selection model is applied. This approach attempts to model the crew operation for a work task on a method analysis of the various operations involved. This might require layout drawings of equipment locations, material dumps, and flow routes to the work face. These aspects are combined with a step-by-step method of construction for which crew sizes and mixes are determined using quantities, labor unit man-hour rates, quantity output per day, and expected construction durations. The *work task analysis* approach frequently produces Crew Balance Charts (P_L-DOC 28) which serve as aids to crew design and crew member instruction, as illustrated in Figure 9.15. In this approach, the size of a crew is determined by relating small component work team groups to specific sequences of work tasks in the typical component work cycles in the work to be done. Once this is done, effort is directed to establishing the crew size which maximizes the productive *basic job* percentages of crew time on the activity, and minimizes as far as possible the unproductive idle times associated with lack of coordination between small work teams. This planning effort eliminates needless work.

The basic ESTABLISHMENT OF CREW REQUIREMENTS must be carried out in relation to the work face conditions associated with the activity and its work tasks, and those critical problems of material layout, material movement, and equipment location that are common to large projects. Simple visual observations and discussions of material and equipment logistics problems with respect to crew locations are not always possible. Each work task must be considered individually, depending on its nature, location on the project, and criticality. The main reason for an increased awareness of PLANNING THE WORK FACE LAYOUT (P8) is the technical complexity of the project and the characteristics of the site. The construction site is often large, and in refinery or power plant projects may cover one-half of a square mile densely packed with facilities. In this environment, large quantities of bulk materials (valves, piping, cable, etc.) must be identified, and their planned transportation to the appropriate location integrated into the crew de-

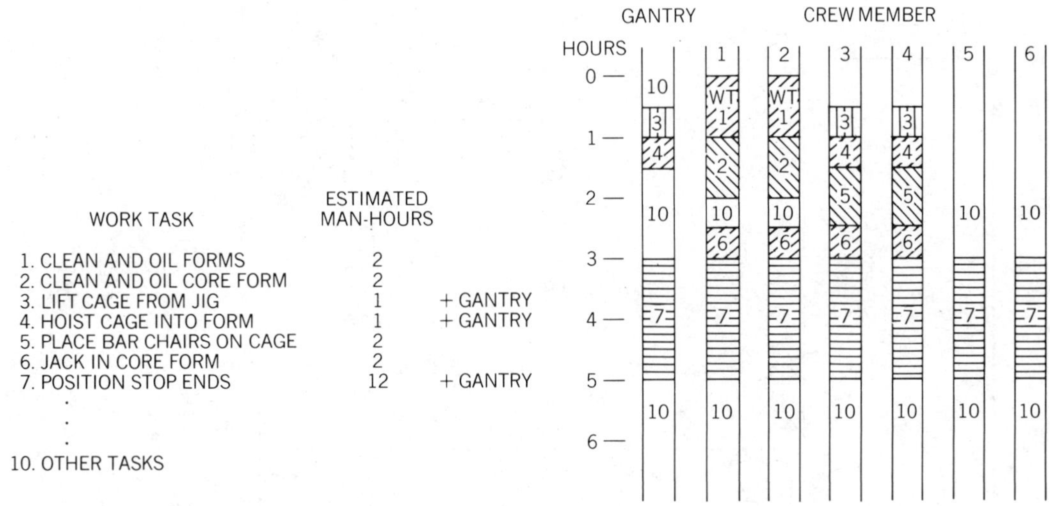

Figure 9.15 Crew Balance Charts (P_L-DOC 28): Bridge Unit Casting Crew inserting steel cage into casting mold.

sign. Alternatively, in multi-story high-rise office building projects work is often spread over many floors so that considerable ingenuity is required to plan subcontractor and crew activities which maximize the use of limited access to and through the building, especially if the facility is located in a major city.

The information which impacts PLANNING THE WORK FACE LAYOUT consists of the work package activity content, crew size and mix, number of crews, anticipated productivity, planned manning of the work task, and anticipated field progress targets. The rate of expected field progress has the most significant influence on this function, however, because it controls the level of manpower which can be added to the work area without causing congestion or saturation. The superintendent and foreman together plan the location of the crews in relation to the material dumps, material flow routes, and equipment locations. Depending on the location of the material dumps (i.e., warehouse, stockpiles, etc.), a crew (or a work team component of a crew) may be assigned to function solely as a material handling crew to supply materials to the work face. Formal policy documents are not required, but material and equipment location sketches and material flow routes may be roughed out as needed to ensure that the crews have the support they need in order to achieve maximum productivity and progress. Thus, Activity Layout Sketches (I_L-DOC 11) might be produced, as illustrated in Figure 9.16.

Crew analysis and design for each work package leads to crew scheduling in terms of work package items. This scheduling problem is handled informally by the superintendents, field engineers, and/or scheduling engineers. It involves the development of the work package sequence logic in terms of the sequencing of detailed activity work tasks. The total number of crews required for the work package is estimated, and actual crew mobilization dates are established. This decision phase is shown in Figure 9.17.

Crew scheduling begins with DEVELOPING THE WORK PACKAGE DETAILED LOGIC. This management function takes the detail of the work package activities and develops a logical sequence plan of construction for those activities. This activity sequencing may be based on Fragnet Schedules for a specialized work operation, or a specific fragnet may be drawn from the details derived through crew work task analysis. In most cases, however, when the superintendent performs this function he relies on past experience or suggestions from scheduling engineers, and rarely documents the construction sequence for detailed work package activities. While these decisions are made quickly and without documentation, they are dominantly influenced by *SCHEDULING FOCUS* (P_L 11) and *CREW PLANNING* policies. These policies stress knowledge of details, clear definitions of work, and consideration of schedule impact. When scheduling especially critical work packages, the project field engineers, scheduling en-

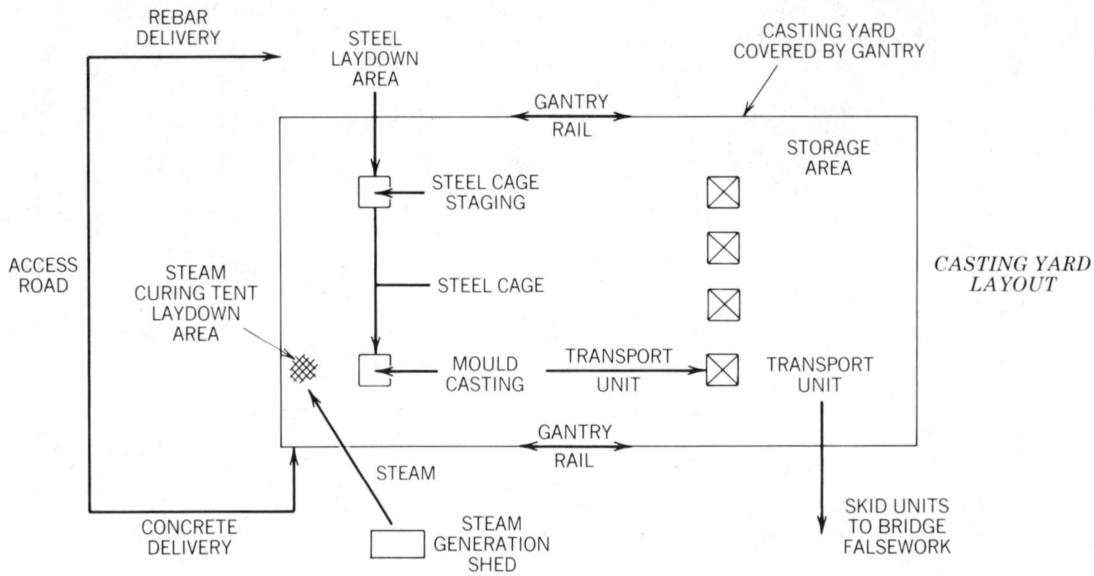

Figure 9.16 Activity Layout Sketch (I_L-DOC 11).

gineers, and other project team members may enter the decision process to assist the superintendent with final detailed scheduling of work task logic sequences. In any case, the short-range scheduling of work activity to specific crews is established at this point. This schedule principally focuses on the next work period, and incorporates the work package activities that are currently in progress with available new work package activities as defined by DETAILED WORK PACKAGE LOGIC plans. A work period usually consists of one or two weeks. Consequently, knowledge of current activity status and its percent complete, coupled with the volume of work left to complete for a particular activity, is vital when scheduling the next week's work effort. An example of a Bi-weekly Work Schedule (P_L-DOC 29), is illustrated in Figure 9.18. At this stage a final check of material availability and construction equipment/tool availability is performed, and locations are verified. A Construction Equipment Schedule (P_L-DOC 30) is utilized to portray the timing of equipment needed in the next period. Figure 9.19 depicts a typical Construction Equipment Schedule.

The superintendent can DETERMINE THE NUMBER OF CREWS that will be required for the next planning period based on a detailed assessment of the construction logic. The specific crew requirements are established for each activity during the crew analysis and design phase, and, when summarized by work package, become a projection of the estimated total number of crews for the work package.

The coordination and scheduling of multi-crew situations can be accomplished through Multiple Activity Charts (P_L-DOC 31), of which Figure 9.20 is a typical example. In this crew planning and scheduling tool, the sequence of work activities for each crew over a work period cycle is plotted to a time scale on strip bar charts. Each bar represents an individual crew and/or major resource that is to be shared amongst crews. Thus, it is a simple matter to ensure that resource use conflicts or difficulties of access to limited critical areas are eliminated by examining the planned use of crews and resources at any given time.

The detailed construction plans for available work packages, and their associated estimated crew manpower requirements, are formally documented and presented to specific foremen in weekly foreman meetings during the SCHEDULING OF PROJECT CREWS (S7) function. The Bi-weekly Work Schedule forms the basis for SCHEDULING CREW MEMBER WORK SEQUENCES prior to mobilizing and then releasing the crew to the various work package activities. Crew member work sequences are reviewed daily as new activities commence and in-progress work activities are completed. This review is carried out by both the general foreman and foremen under the superintendent directly responsible for the general foreman.

At the small and medium-sized firm levels, the

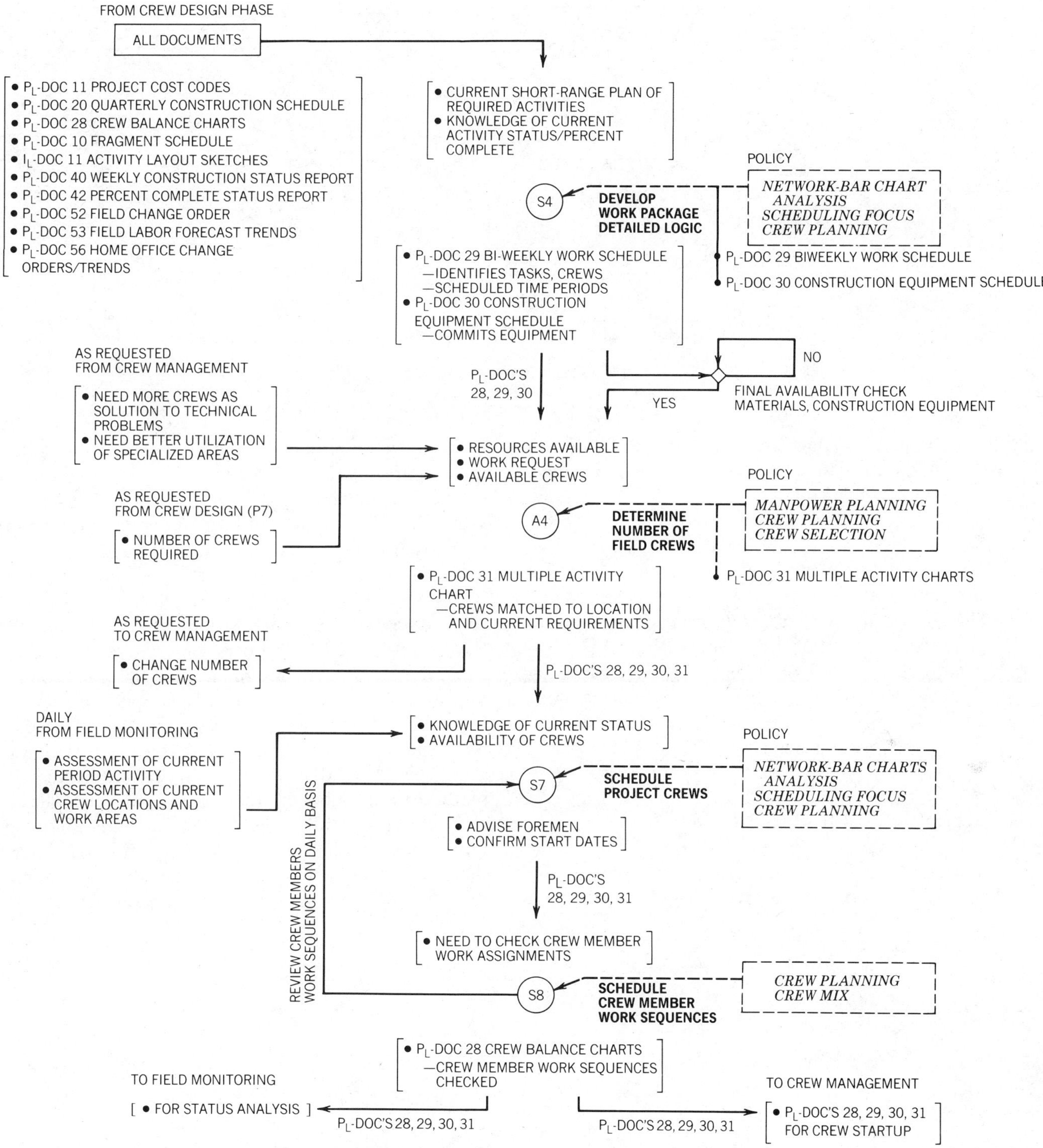

Figure 9.17 Crew Planning decision process: crew scheduling phase (Field-Based Project Management approach).

AW
Engineering & Construction Corp.

PROJECT ______________ CONTRACT ______________

LOCATION ______________ WEEK ENDING ____________ DATE ______________

PREPARED BY ______________

COST CODE	ACTIVITY DESCRIPTION	MATERIAL QUANTITY	CREW SIZE	JULY					
				2	3	4	5	6	7
	CONCRETE: FOUNDATION C-110 DRAWING 100020								
2015.2110	FORMWORK (FAB & INSTALL)	1000 SF							
2015.2120	REBAR	250 lbs							
2015.2130	CONCRETE POUR & FINISH	100 CY							
	⋮								

Figure 9.18 Bi-weekly Work Schedule (P_L-DOC 29).

Crew Planning Management Information System is performed informally at best. At the large firm level informal practices still occur, but there emerges an increasing level of documentation to support crew planning decisions, including detailed short-range schedules, manpower requirements, and current activity status and progress reports. This more formalized documented information provides the field managers/supervisors with an increasing level of information from which crew planning and scheduling can occur. Thus, supporting documentation does emerge, as shown on the Crew Planning decision process networks in Figures 9.13 and 9.17.

9.6 TOTAL FIELD AUTONOMY

In the final form of project team development, the Total Field Autonomy approach to Crew Planning maximizes the project team concept by enlisting the active participation of many project team personnel in the Crew Planning decision process. This integrated project team focus is often directed by a construction methods engineer specializing in the analysis of construction operations and planning techniques. Thus, Crew Planning at this level tends to produce more documentation with greater detail, that is integrated with costs and schedules. However, the increase in documentation and the growth in project team size necessary to support this documentation will only occur *provided* the client demands and pays for the effort, and/or the contractor convinces him that, through more sophisticated methods improvement techniques, the contractor can reduce overall project costs and maintain or potentially accelerate project completion dates. In the above philosophy, the changing management effort toward Crew Planning will become apparent

AW
Engineering & Construction Corp.

PROJECT ______________ CONTRACT ______________

LOCATION ______________ DATE ______________

WEEK ENDING ____________ PREPARED BY: ______________

EQUIP. NO.	EQUIPMENT DESCRIPTION	SERIAL NO.	DATE IN SERVICE	DATE OUT SERVICE	PROJECT LOCATION	RESPONS. PARTY	AVAILABILITY	
							REQ'D	RELEASE
102	12 TON CHERRY PICKER	25 x 2400z	2-24-86		BOILER BUILDING	J. SMITH	3-24-86	5-25-86

Figure 9.19 Construction Equipment Schedule (P_L-DOC 30).

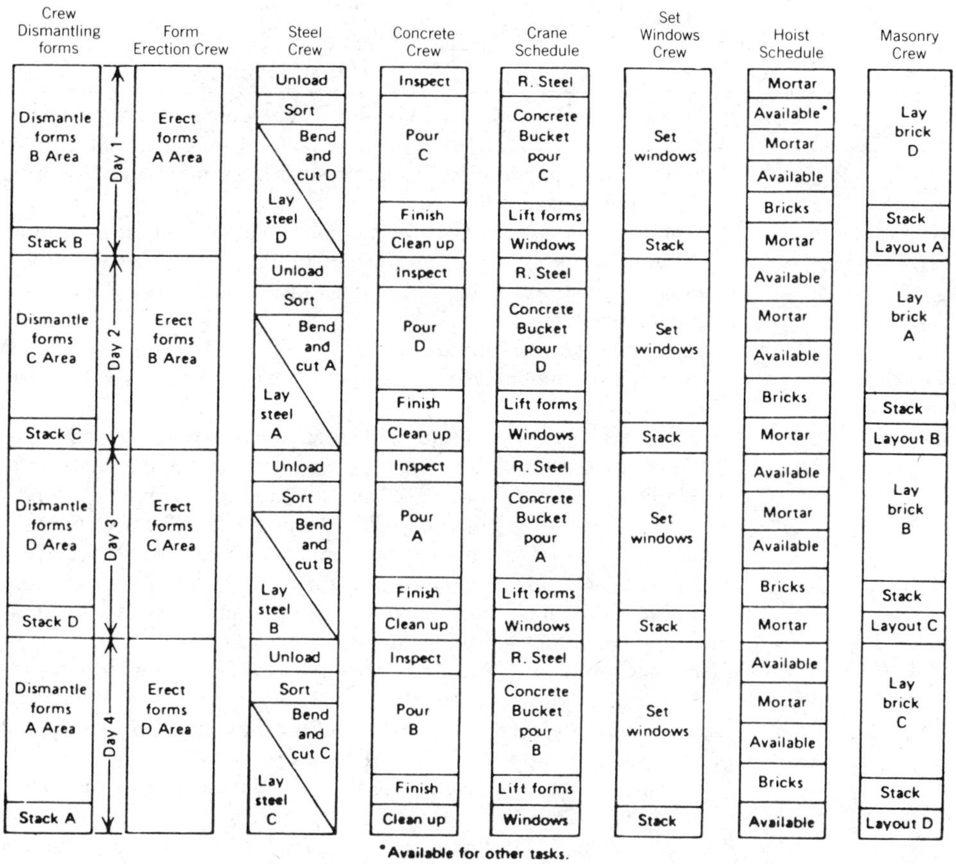

Figure 9.20 Multiple Activity Charts (P_L-DOC 31).

when contrasted with that of the Field-Based Project Management approach. The decision network structure of Crew Planning, while the same as that presented in Figures 9.13 and 9.17 for the Field-Based Project Management approach, will clearly focus on the *WORK BREAKDOWN* (P_T 8)* and *COST CODE* (P_T 9) systems which are broken down by work package, and for which detailed *WORK ORDERS* (P_T 19) can be prepared based on field estimates of *UNIT MAN-HOURS* (P_T 10), *CREW PERFORMANCE* (P_T 30), and *PRODUCTIVITY* (P_T 31) targets. These field work order estimates are performed at the control zone level, and incorporate construction methods analysis and determination. Crews are sized as each of these components are integrated with material, drawing, and equipment requirements to fulfill the work order. This comprehensive work order approach places increasing management pressure on the field project team involved with their preparation.

* See Table 9.6, Policies of the Autonomous Project Team.

Crew Planning is initially based on a three-month scheduling horizon. In this environment, work orders must be planned to fit resource availability and priority sequences in order to avoid simply working crews on those work package activities that have all the material. Thus, the work order approach tends to correlate procurement and other resource needs to a specific order of work packages. It also impacts crew sizes and the number of crews by establishing crews that correspond directly to work content and to the most economical construction methodologies. This approach is supported by more detailed weekly scheduling of crews, and frequent updating of progress and work order status. The weekly planning effort has a multi-level priority focus wherein sufficient work orders are planned per week to achieve desired rates of progress. In addition, twenty to thirty percent more work per week is planned to cover problems that interrupt high priority work orders (e.g., unexpected material shortage or drawing holds).

The responsibility for Crew Planning is shifted to

TABLE 9.6 POLICIES OF THE AUTONOMOUS PROJECT TEAM

Policy		Description
P_T 6	*LABOR RECONNAISSANCE*	Comprehensive survey, analysis, and documentation of the project site labor environment on a continuous basis is requisite to the accurate evaluation of labor content and the planning of manpower requirements for the project at any time during the project's life cycle.
P_T 7	*SITE INSPECTION AND ANALYSIS*	A continuous comprehensive inspection, analysis, and documentation of project site conditions, transportation problems, etc., is requisite to the accurate evaluation of construction methods, schedules, labor content, and to solving other unique problems.
P_T 8	*WORK BREAKDOWN*	A work breakdown structure representative of the project scope and control philosophy is formulated to integrate costs and schedules with work packages for project management.
P_T 9	*COST CODES*	Cost codes are applied to the work breakdown structure, and serve as a basis for categorizing estimated costs, reporting and controlling costs and manpower, developing a historical database, and integrating the schedule with the estimate.
P_T 10	*UNIT MAN-HOUR*	A tight but realistic evaluation of the labor content for a work task is essential for successful project planning and management control. Therefore, estimates must be based on work task man-hour content in relation to work volume, construction methods, local productivities, labor availability, and site conditions.
P_T 13	*NETWORK-BAR CHART ANALYSIS*	Broad network focused milestone planning is the best approach for identifying and sequencing major project components, while a combination of the critical path method and bar chart model are best suited for project planning and scheduling at the total project view level as well as the daily level.
P_T 14	*SCHEDULING FOCUS*	Scheduling definitions and constraints are: 1. A critical work package is one that, if delayed, will extend the completion date of the project. 2. Work package durations must be correlated to the estimated work content. 3. All restraints must be defined and incorporated into the sequence logic (i.e., material delivery, subcontract coordination, drawing receipt, etc.). 4. Manpower impact is an integral aspect of project scheduling and scheduling control. 5. Schedules must reflect optimum use of all resources relative to the work package. 6. Progress is based on a realistic evaluation of the labor unit effort required to complete a component of work. 7. Schedules are geared to the daily work level and are prepared, analyzed, and reviewed weekly in relation to planned progress and productivity versus actual progress and productivities.
P_T 18	*MANPOWER PLANNING*	Long-range planning of manpower by craft and in total is a prerequisite to supporting the field effort, and must be performed continuously throughout the project's life.
P_T 19	*WORK ORDER*	The work order is a key tool for field construction planning and crew sizing because it focuses on detailed estimates based on field measurements, describes construction methods, defines material and equipment mobilization, evaluates the most economical approach, and considers schedule impact, all in relation to work site conditions and work package.
P_T 20	*CREW SELECTION*	Crew selection is based primarily on the work order model, but it does incorporate all other models, such as the following: 1. Maximum crew size 2. Number of crews and crew buildup 3. Work task analysis 4. Historical crew
P_T 21	*CREW MIX*	While impacted by union rules in some construction environments, the ratio of foremen to journeymen/apprentices within a crew is established to optimize both the foreman's level of supervisory control and the crew's performance.

TABLE 9.6 (*Continued*)

Policy		Description
P_T 30	*CREW PERFORMANCE*	Crew performance is a function of labor productivity, job conditions, and management effectiveness, and must focus on a tight but realistic performance indicator.
P_T 31	*PRODUCTIVITY*	Productivity is an essential concept and must be rationally defined for office and field utilization. An awareness of the components that affect productivity and their relation to man-hours, activity durations, and progress is essential, as well as that all these integrate and are manageable.
P_T 36	*DATA PROCESSING*	A computer-based information system is the most feasible means of data processing for management control.

the methods engineers and assistant superintendents and away from the foreman. Thus, the foreman carries out the work order and can fully concentrate on its implementation, and on the management of those crews involved with each work order that he or she must manage. The foreman still has input into the preparation of work orders, but the final decision making, in regard to crew size and mix, is left to the field engineers and superintendents.

The manpower management functions, policies, and documents which describe Crew Planning are summarized in Figures 9.21 and 9.23 for the Crew Design and Crew Scheduling Phases respectively. At the Total Field Autonomy level, the Crew Planning decision process revolves around the three key milestone checkpoints of (1) drawing availability, (2) material availability, in terms of location of existing materials and potential location of materials near desired work areas, and (3) construction equipment/tool availability and potential location near desired work areas. These milestones are critical decision points, and can significantly impact the effectiveness of the Crew Planning effort if they are not fully analyzed and reviewed by the project team when evaluating work orders and scheduling them for execution. Thus, careful consideration must be given to these milestones by the project team, especially as they relate to time pressures brought to bear because of fast-tracking project execution philosophies and heavy client demand to meet component facility start-up sequences and completion dates.

Crew Design for work packages, Figure 9.21, is tied into the project *WORK BREAKDOWN* and *COST CODE* system, through both Construction Schedule Packages (P_T-DOC 26) and Monthly Activity Plans (P_T-DOC 27), including current progress and activity status of these documents. Thus, DEFINING THE WORK PACKAGE (P5) content begins with examining predetermined work packages supported by complete craft manpower and progress targets from either the control base estimates and/or current forecasts. These work packages must be further dissected into activities for which *WORK ORDERS* (P_T 19) can be assigned and evaluated. The work order technique, illustrated in Figure 9.7, naturally increases the level of documentation and project team involvement in Crew Planning. This technique provides a methodology for fully scoping an activity, that is, documenting drawing, material, and construction equipment requirements and their availability for an activity as required for the formulation of the work order package. While the work order technique increases field project team costs, it can reduce direct field labor costs through better productivity, to the extent of possibly overcoming the additional cost associated with increasing the management effort. Furthermore, because the work order documents an activity's content, material shortages, for example, can be identified in a timely manner for immediate resolution. Thus, the work order not only optimizes construction methods for cost effectiveness, but also provides a vehicle for coordinating material, drawing, equipment and manpower requirements for work package activities. When the work order system is applied to the project as a whole, efficient Crew Planning results, producing a means of meeting construction completion dates. The work order system need not be incorporated into the Total Field Autonomy approach, however, unless the client demands it and/or will pay the costs of the system. If the work order is not an integral part of the Project Field Construction Procedures (P_T-DOC 34), then techniques similar to those presented for

P5 DEFINE WORK PACKAGE

Policies	Documents	
LABOR RECONNAISSANCE	P_T-DOC 35	Work Order
SITE INSPECTION AND ANALYSIS		
WORK BREAKDOWN		
COST CODES		
UNIT MAN-HOUR		
SCHEDULING FOCUS		
WORK ORDER		

P6 DETAILED FIELD QUANTITY TAKE-OFF

Policies	Documents
UNIT MAN-HOUR	
WORK ORDER	

P7 ESTABLISH CREW REQUIREMENTS

Policies	Documents	
UNIT MAN-HOUR	P_T-DOC 36	Manpower Requisitions
WORK ORDER	P_T-DOC 37	Crew Balance Charts
CREW SELECTION		
CREW PERFORMANCE		
PRODUCTIVITY		

P8 PLAN WORKFACE LAYOUT

Policies	Documents	
WORK ORDER	I_T-DOC 11	Activity Layout Sketches
CREW SELECTION		
CREW MIX		
CREW PERFORMANCE		

Figure 9.21 Crew Planning decision process: crew design and analysis phase (Total Field Autonomy approach).

the Field-Based Project Management approach will be utilized, with the exception that the control zone/work package approach will still be maintained.*

A typical set of Work Order documents (P_T-DOC 35) is schematically illustrated in Figure 9.22. These cover each management function performed in the Crew Design phase (see Figure 9.21). Although the work order does not recognize ad hoc crew planning, it does recognize the need for good judgement, common sense, and past experience. This is reflected in *UNIT MAN-HOUR* and *CREW SELECTION* (P_T 20) policies, wherein a variety of field estimating and crew selection models are potentially incorporated into the work order planning process. It should be noted that the work order documents cover a broad range of tasks and therefore require the composite efforts of many project team members.

A focus on *CREW PERFORMANCE* and *PRODUCTIVITY* policy formulation when ESTABLISHING CREW REQUIREMENTS (P7) and PLANNING THE WORK FACE LAYOUT (P8) promotes crew designs supported by tight but realistic performance targets. These targets can be achieved through effective field management, provided proper planning methods are utilized. These policies are critical regardless of whether or not the work order system is maintained, because they reflect a positive management attitude and concern for the control of labor productivity beginning with the *planning* stages of a construction activity.

The DEVELOPMENT OF THE WORK PACKAGE DETAIL LOGIC (S4) initiates the Crew Scheduling of Work Packages, as illustrated in Figure 9.23. This phase of Crew Planning is also focused on the work package concept through the project *WORK BREAKDOWN* and *COST CODE*

*If work order planning is not used on a project-wide basis, it may be effectively used for highly critical activities in selected applications.

CONTROL ZONE ____________
WORK PACKAGE ____________

WORK ORDER 210
SCOPE:
1.
2.
REQUIREMENTS:
1.MATERIALS
2. TOOLS AND EQUIPMENT
3. WORK AREA
4. DRAWINGS
CONSTRUCTION METHODS/SEQUENCES
1.
2.
3.

(a)

CONTROL ZONE ____________
WORK PACKAGE ____________

WORK ORDER 210
1. COST CODE W/WORK PACKAGE ID.
2. QUANTITY TAKE-OFF ANALYSIS
3. LABOR MAN-HOURS
a. STANDARD MAN-HOURS
b. CREW SIZE & MIX
c. SITE MAN-HOUR
d. PERFORMANCE TARGET $\frac{\text{SITE M-HRS}}{\text{STD. M-HRS}}$
e. WAGE RATE ANALYSIS
4. MATERIALS
1.
2.
5. SUBCONTRACTS

(b)

MPM PROJECT — WORK ORDER 210 SUMMARY — AW Corp.

DESCRIPTION

CONSTRUCTION METHOD/SEQUENCE

DESCRIPTION	COST CODE	STANDARD M-HRS	ESTIMATE M-HRS	PERFORMANCE TARGET	WAGE RATE	LABOR $	S/C $	MAT'L $	TOTAL $
WORK PACKAGE X 1. 2. 3.									
TOTALS									

SCHEDULE ANALYSIS:
SCHEDULED START: ____________
SCHEDULED COMPL: ____________

KEY MILESTONES:
1. RECEIPT OF

(c)

Figure 9.22 Work Order (P_T-DOC 35). (*a*) Definition. (*b*) Estimate. (*c*) Summary.

system. Consequently, each management function of the Crew Scheduling process revolves around the work packages considered in the next planning horizon. When the *WORK ORDER* system is employed as part of the work package, the various work orders are identifiable components of the crew plan and schedule for the next period.

The *SCHEDULING FOCUS* (P_T 14) of the Crew Scheduling phase contains the following major stipulations: (1) progress is based on a realistic evaluation of the labor unit effort required to complete components of work; and (2) work activity is prepared, analyzed, and reviewed weekly in relation to planned progress and productivity versus actual progress and productivity. These two policy directives are implemented through a detailed Weekly Work Schedule (P_T-DOC 38). As illustrated in Figure 9.24, the Weekly Work Schedule considers all major resources involved with the work orders planned for the next week, and identifies crew requirements and progress units to be achieved in relation to targets for the periods. The document contains sufficient data to evaluate progress and productivity on a weekly basis in accordance with *CREW PERFORMANCE* and *PRODUCTIVITY* policies.

S4 DEVELOP WORK PACKAGE DETAILED LOGIC

Policies	Documents	
WORK BREAKDOWN	P_T-DOC 38	Weekly Work Schedule
COST CODES	P_T-DOC 39	Construction Equipment Schedule
UNIT MAN-HOUR		
NETWORK-BAR CHART ANALYSIS		
SCHEDULING FOCUS		
WORK ORDER		
DATA PROCESSING		

A4 DETERMINE NUMBER OF FIELD CREWS

Policies	Documents	
MANPOWER PLANNING	P_T-DOC 40	Multiple Activity Charts
WORK ORDER		
CREW SELECTION		

S7 SCHEDULE PROJECT CREWS

Policies	Documents
SCHEDULING FOCUS	
WORK ORDER	
CREW MIX	
CREW PERFORMANCE	
PRODUCTIVITY	

S8 SCHEDULE CREW MEMBER WORK SEQUENCES

Policies	Documents
WORK ORDER	
CREW MIX	
CREW PERFORMANCE	
PRODUCTIVITY	

Figure 9.23 Crew Planning decision process; crew scheduling phase (Total Field Autonomy approach).

Apart from the detailed Weekly Work Schedule, which is based on work packages defined in terms of work orders, the Crew Scheduling phase remains essentially the same as in the Field-Based Project Management approach, in terms of the methods and techniques used in its performance. Thus, as shown in Figure 9.23, Construction Equipment Schedules (P_T-DOC 39) and Multiple Activity Charts (P_T-DOC 40) are all documents incorporated into the Crew Scheduling of work packages. The project team members involved in this phase are principally the superintendents and cost/scheduling engineers, with the support of other project team personnel, as the ***PROJECT TEAM SPIRIT*** (P_T-DOC 37) policy requests.

At the Total Field Autonomy level, Crew Plan-

MPM PROJECT — WEEKLY WORK SCHEDULE — AW Corp.

CONTROL ZONE: ______ UNIT: ______ WEEK ENDING: ______

ACTIVITY DESCRIPTION	COST CODE	PROGRESS UNITS	% PROG.	PERFORM. TARGET	CREWS	MEN	RESOURCES	PRIORITY	CRITICAL NEEDS
WORK PACKAGE WITH WORK ORDER DETAIL	IDENTIFIER	PROGRESS UNITS AVAILABLE	% OF TOTAL PROGRESS UNITS	EST M-HRS / STD M-HRS	PIPEFITTERS WELDERS OPERATORS	10 5 2	CRANE WELDING RODS WELDING MACHINE	1 OR 2 OR 3	1. DRAWING NO.

Figure 9.24 Weekly Work Schedule (P_T-DOC 38).

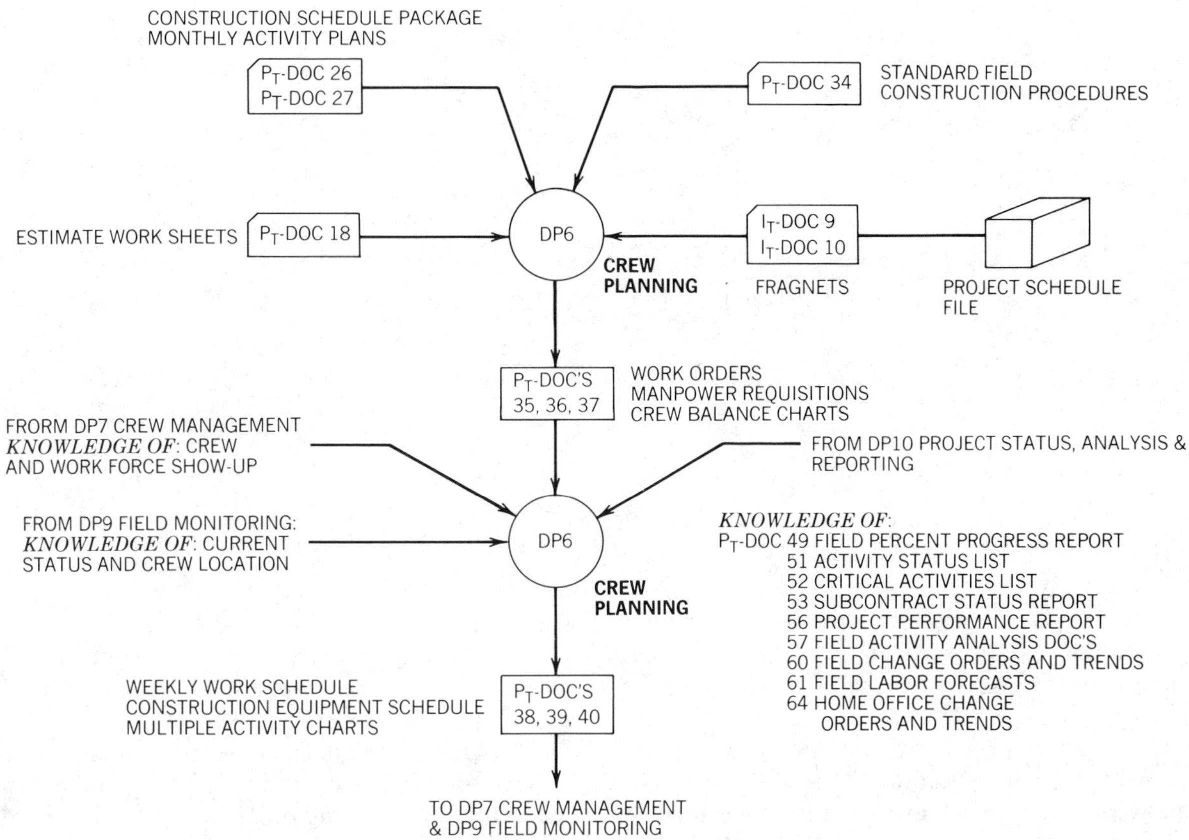

Figure 9.25 Crew Planning Management Information System (Total Field Autonomy approach).

ning can be characterized as a formal Management Information System, as illustrated in Figure 9.25. This figure depicts the increasing level of support documentation and other information required for Crew Planning in the Total Field Autonomy environment. It further demonstrates that many more documents are required to facilitate both the planning of crew activity (Work Order), and scheduling of crews (Weekly Work Schedules). The type of MIS described in Figure 9.25 is often performed manually, but with the advent of microcomputers most of this process could be computerized. In addition, most of the information input to Crew Planning is derived from a computer-based management information system utilized in the other Total Field Autonomy decision processes.

CHAPTER 10
Crew Management

Crew Management is generally considered to be the professional role of the foreman/superintendent and the grass-roots level of the field management of labor and tradesmen. Crew Management thus involves the assignment of work tasks to individual crew members and the communication of instructions on how these work tasks are to be performed so that full advantage is taken of each individual's skills and abilities, and an integrated crew activity results. It also involves the continuous supervision and direction both of individuals and of the crew as a whole, to maintain productive work at acceptable quality levels. In addition, the foreman/superintendent must be a good leader, skillful in the motivation of workers. He or she must be alert to the early detection and resolution of personal problems as they relate to the quality of work performed, work attitudes in general, and in particular the relationships between crew members.

On small projects in a local environment, the contractor's key field personnel have more control of the quality of a crew because they know who is in the local union, and, through past contact, have already evaluated most of the journeymen's skills, work attitudes, and performance. Thus, the contractor moves very quickly to remove incompatible workers. On large projects, however, it becomes necessary to pay more attention to the skill evaluation of the potential work force. In this case the hiring process must effectively screen out undesirable individuals. The hiring position changes significantly on giant projects, however, because of the need for large work forces to meet production schedules, and the often limited pool of available labor, which frequently forces the contractor into accepting gate show-up labor and even workers possessing inferior skills. Thus, a significant problem on giant projects is the acquisition and training of suitably skilled labor in an environment where the contractor is being pressed on schedule matters and on the cost inefficiencies associated with low labor productivities. On these projects, the interviewing, skill evaluation, hiring, and assignment of labor to a specific foreman and crew is the basic management action involved in the establishment and maintenance of a crew. Finally, at this level of operation the Crew Management area will also require continuous consideration of the number of crews required at any one time on the site, in relation to the work programs for each project control area. This issue is related to the supervision and productivity problems raised by the general management of large field work forces.

The Crew Management area is very important to the contractor because it is the management interface of the project team with labor. Thus, it is the responsibility of the superintendent and foremen, by suitable leadership and management styles, to motivate individual crew members and the crew as a whole to make their best effort to meet work quality and performance requirements, and to encourage them to identify themselves with project goals. Field management has the obligation to support foremen/superintendents in their work by proper instructions and encouragement in the implementation of field procedures that ensure productive work, in the elimination of useless work and needless tool box trips, and in the provision of adequate resources so that productive activity can result. Crew Management is therefore the initial area to investigate if field progress and labor performance is not satisfactory.

10.1 THE BASIC DECISION PROCESS

The labor-oriented aspect of the Crew Management decision process is common to all projects and is the basic responsibility of the foreman. Its performance depends upon the skill and experience of the foreman and his or her management style, since, as mentioned before, it focuses on the actual instruction and continuous direction of the crew members in productive work, and the detection and resolution of individual conflicts between crew members. Thus, the dynamic manipulation and regrouping of the crew into effective work teams from time to time according to the work needs requires an ability to look ahead, a thorough technical knowledge of the construction operation, and a keen perceptive assessment of crew members. Hence, Crew Management, for the foreman, is concerned with crew direction and the resolution of personal conflicts and problems within the crew.

The manpower-oriented focus of the Crew Management decision process, however, is concerned with the hiring and firing of labor and the setting up of project crews. Thus, the determination of the required number of project crews is a critical management decision on large projects. It directly impacts the volume of work that can be handled at any one time, and the number of work faces at which crews can be located for work around the site. In addition, the number of project crews impacts (through the quality of foremen involved) attainable levels of productivity and hence, ultimately affects project costs. Thus, for the superintendent or field project manager, Crew Management has both a crew mobilization focus and a stress on the manipulation of crew size and craft compositions to meet the changing needs of the project.

The natural division between the manpower and labor management aspects of Crew Management, however, only begins to emerge clearly at the large firm level when large field work forces are required. This is because, on projects requiring less than one

TABLE 10.1 SCOPE OF THE CREW MANAGEMENT PROBLEM AREA

Major Decision Problem	Component Problem Statements
30. How many crews:	How many specialist crafts are involved in the project? How many different work areas are needed at any one time? How many qualified foremen are available? What is the desirable level of flexibility and size of work groups?
31. What are hiring and firing practices:	Who should interview? What information is required concerning new applicants? Should proficiency tests be carried out? What checks should be made on past employment and experience? Is applicant work motivated? What working conditions should be prescribed? How should the worker be cautioned that performance is unsatisfactory? Who should fire? How and when should a worker be fired?
32. What crew manipulation:	What are desirable crew sizes? How many are under or over sized? Can variations in work face conditions be best managed by changes in the number of crews, or in the sizing of individual crews? How can a new worker be best incorporated into the work force? Can field personnel problems be resolved by transferring members to other crews?
33. What crew direction:	Who are skilled workers? Who can best perform these particular work tasks? What work teams should be set up? What distribution of work tasks should be made to individual workers? How much instruction is needed? Does the foremen show them how? Who needs help?
34. What personal management:	What personal guidance is necessary? Does the foreman show them how to do it? How can the foreman help or cope with this particular problem? Should the foreman correct them? What can the foreman do to make the work more productive and satisfying? How can individual crew members be motivated? Should the foreman select some for praise and if so which ones? Has the foreman motivated the entire crew?

TABLE 10.2 GENERAL PROCEDURE FOR CREW MANAGEMENT

Procedural Step	Management Duties and Tasks
1.	Review work load volumes, in terms of manpower requirements broken down when necessary by crafts and specialist groupings, over the planning horizon.
2.	Determine location of proposed work on the project and acknowledge the sequence of work packages and scheduled milestone deadlines.
3.	Determine the number of available foremen and relate them to the work volume, control areas, and desired work sequences.
4.	Establish the number of crews to operate in this work period, the requirement to split or combine existing crews, and/or the need to initiate new crews.
5.	Select foremen for each crew.
6.	Hire workers and allocate them to suitable crews when necessary, or transfer in workers from other crews.
7.	Establish and maintain crew size and mix of trades and skills.
8.	Brief crew members on proposed work activity, good work rules, required quality standards, expected performance, and the need for productive cooperative efforts.
9.	Avoid jurisdictional disputes by making correct assignments of work tasks to crew members.
10.	Direct crew members to work tasks in suitable work team groups.
11.	Instruct crew members as required on technical details and methods of ensuring achievement of required quality standards.
12.	Strive to motivate the crew to meet work deadlines.
13.	Detect problem areas and resolve disputes.
14.	Transfer or fire those crew members who are unresponsive to direction, help, or correction.
15.	Reassign crew members to other work tasks or crews on completion of work.

hundred workers, it is still possible to maintain close personal contact with each crew, and to develop recognition of and interest in the individual. The scope of the Crew Management problem area, as given by Table 10.1, clearly indicates this threshold change in terms of the major decision problems of determining how many crews are needed, outlining hiring and firing practices, and managing crew manipulation. These decision problems are primarily manpower management focused, whereas the crew direction and personal management major decision problems have a labor management focus.

The general procedure for the Crew Management decision process is shown in Table 10.2. This table attempts to portray formally that process which for many foremen is an heuristic and subjective consideration. The functional plot for Crew Management, in terms of basic manpower management functions, is shown in Figure 10.1 and can be readily related to the tasks and duties of Table 10.2.

The full functional statement of the Crew Management decision process is shown in Figure 10.2, and while describing the decision process in still more detail, establishes a specification for the qualities of the decision maker and lists a variety of factors that he or she should consider when making decisions which have a close relationship to the problem statements listed in Table 10.1. Finally, Figure 10.3 presents a schematic illustration of Crew Management interfaces with other relevant manpower management decision processes.

10.2 FUNCTIONAL PERFORMANCE TECHNIQUES

Crew Management has both a manpower management and a labor management focus. The building up of a field work force and its division into crews, is a manpower management area concerned with the balancing of the work force and crews against the scheduled work volume needs of the project. Thus, it addresses the general hiring and firing of labor; the selection of foremen, and, if necessary, of superintendents; and the implementation of policies aimed at raising field productivities and minimizing conflicts and jurisdictional disputes. The leadership and management of a specific crew, however, has a labor management focus. The crew must be directed by a person who is competent and who has the ability to direct and lead men in the performance

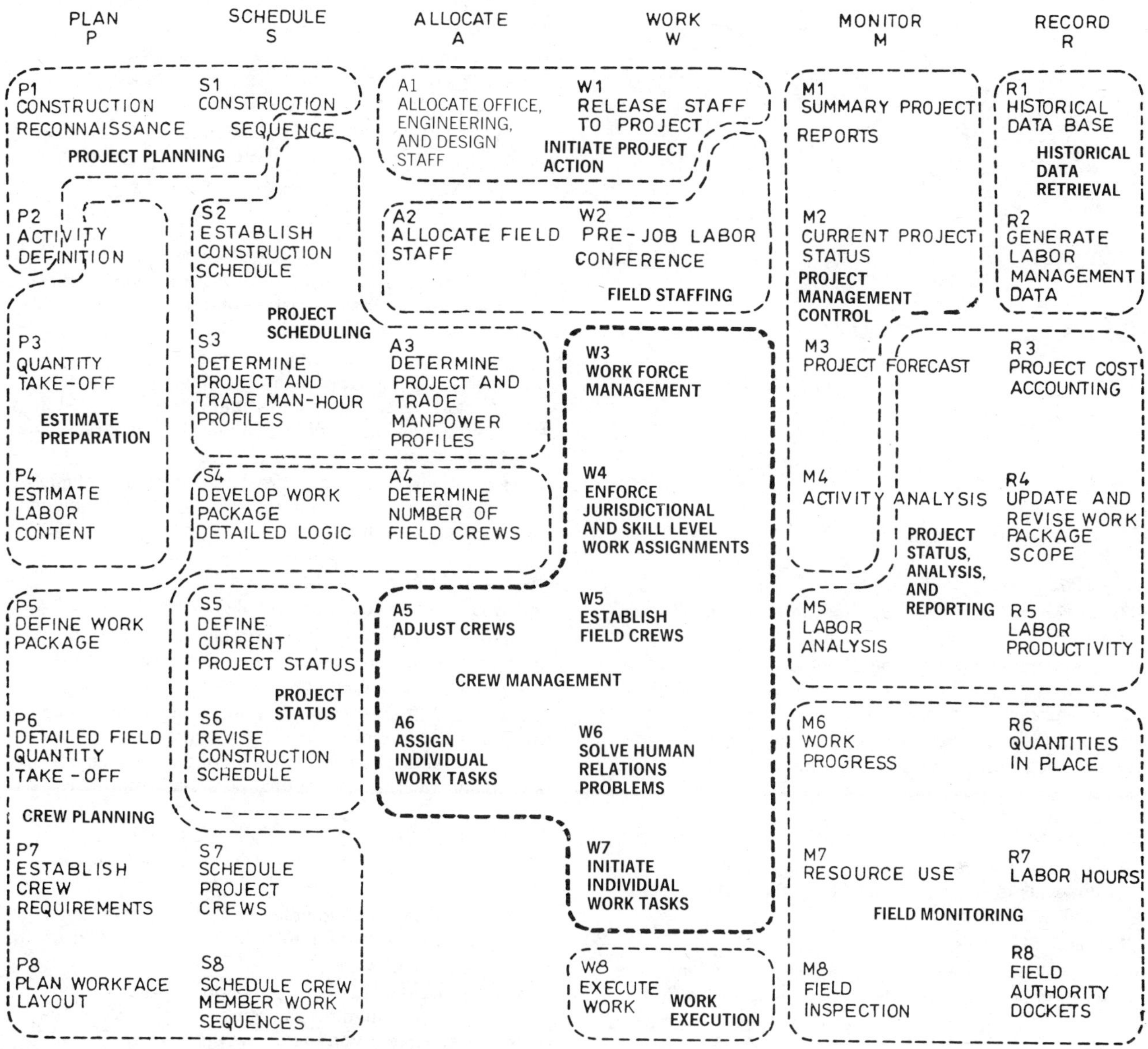

Figure 10.1 Functional plot of the Crew Management decision process.

of efficient, productive work, and, if required, the expertise to perform skillfully any work task that is demanded of a crew member.

A major concern of many contractors when starting field operations is the quality of their field supervisors, especially at the foreman and superintendent levels. Consequently, once experienced and capable foremen have been found, the contractor will endeavor to keep them on the payroll and move them from job to job. The position of job superintendent is a much more critical field position to fill than that of foreman. In general, if the contractor needs a superintendent and does not have a foreman already earmarked for such a job, the firm will hire in a superintendent. The hiring of foremen is still a critical issue, however, and the contractor will always strive to place a few loyal experienced men on a job first. In this way, the contractor can hire without the necessity of making the first craft union hired worker on the job the craft foreman. This strategy enables the contractor to gain time to carefully assess the potential of a new worker for the foreman position. In addition, the contractor is interested in the management style of the potential superintendent/foreman and will often need people with a management style to match the field environ-

A5 ADJUST CREWS

- Maintenance of well-balanced, motivated and productive crews.
- Requires intimate knowledge of working conditions and work loads for each construction operation and field situation, and desirable crew mixes and sizes.
- Considers desirability of adding and deleting new members to match work loads or solve human conflict problems.

A6 ASSIGN INDIVIDUAL WORK TASKS

- The matching of individual crew member skill and motivation levels to requirements of the construction operation and work load.
- Requires intimate knowledge of men.
- Considers best approach to instructing, controlling, and motivating crew members in the workface environment.

W3 WORK FORCE MANAGEMENT

- The active hiring, redirection, and firing of all types of project labor to ensure that field work force size matches available and projected work load.
- Requires knowledge of hiring and firing procedures, and projected work loads by trade and skill areas, and ability to assess characters and personalities of workers.
- Considers availability of labor, size of labor pools, desirability of using in-house work force; checks prospective worker credentials and attitudes toward work.

W4 ENFORCE JURISDICTIONAL AND SKILL LEVEL WORK ASSIGNMENTS

- The recognition of quality control requirements or skill levels and enforcement of established jurisdictional areas.
- Requires intimate knowledge of jurisdictional areas established by union agreements or implied by nonunion job site skill classifications, the ability to handle men in tough situations, and proper briefing of key field management personnel.
- Considers worker attitudes, areas of potential conflict, and attempts clarification and solution of field problems with minimum conflict.

W5 ESTABLISH FIELD CREWS

- Involves foreman selection and allocation of labor.
- Requires knowledge of crew's proposed work assignments, crew size and composition, and desired manner of crew buildup to full strength.
- Considers suitability and willingness of individual field personnel for foreman's job, and attempts to establish crews that strengthen work ties among crew members and minimize areas of potential personality conflicts.

W6 SOLVE HUMAN RELATIONS PROBLEMS

- Strives for happy, warm, well-motivated work environment among crew members.
- Requires perceptive understanding of work relations and ability to anticipate or tactfully handle situations involving real or imagined conflicts between men.
- Considers ways of defusing situations before they become problems, attempts to define and discuss the nature of a specific human relations problem and solution methods in nonthreatening ways that maintain human dignity, and in the last resort considers reassignment of individuals to other areas or terminating the individual's employment.

W7 INITIATE INDIVIDUAL WORK TASKS

- Briefing and direction of crew members on their work roles and contributions to the crew effort.
- Requires intimate knowledge of the construction operations and assignments compatible with skill levels and seniority of individual crew members, and best combinations of crew members to form work groups.
- Considers physical nature of required work in relation to crew member physique and attitude to work, attempts strengthening of work bonds between individual crew members, and implicit recognition of leadership roles of individual men.

Figure 10.2 Functional statement of the Crew Management decision process.

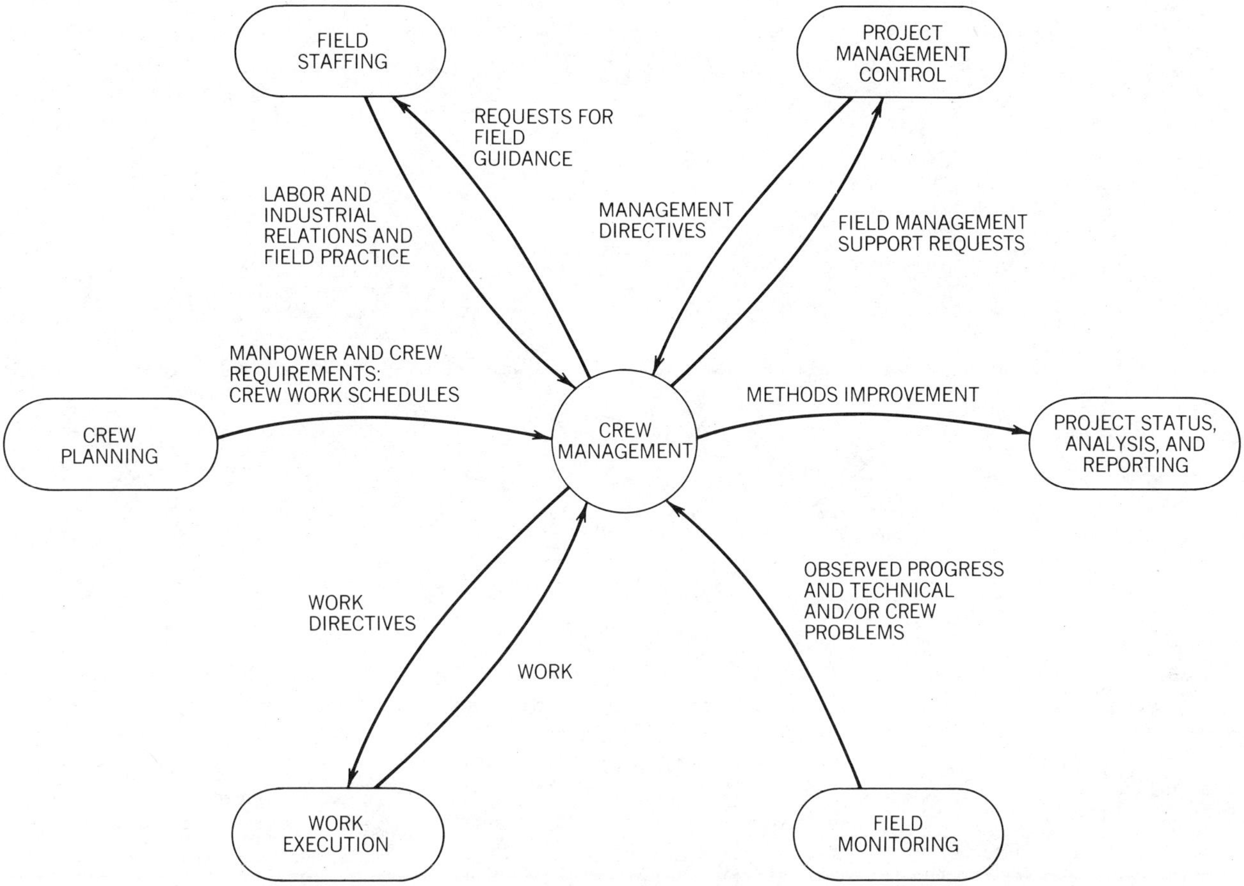

Figure 10.3 Schematic illustration of Crew Management decision process interfaces with other manpower management decision processes.

ment and the status of the project on a whole. Thus, a forceful foreman, one who is fair and who can "dish it out" as well as "take it," may be needed in some situations, while the talents of a persuasive foreman may be more successful in others, for example, where a single word at the right time may be all that is required to solve a problem.

Foremen must be technically competent in their craft area and have leadership qualities. Since only properly instructed workers can competently perform what is required, the foreman must be able to clearly explain ideas and be observant to the needs of the crew and the job. The foreman must be skilled at establishing work groups within the crew structure that take advantage of the special skills of individuals and compensate for the lack of skills in others.

A good foreman is always in control of the crew and knows when it is best to direct, to persuade, or to lead. When necessary, the foreman will find the opportunity to have a personal word of advice with an individual without the other crew members being aware. Generally, however, the foreman will use the field shack for conversations affecting the crew as a whole, such as how to tackle the next assignment, or practical aspects of a new safety concept. The foreman must strive for the avoidance, elimination or, if this is impossible, the minimization of conflict within the crew. The foreman must also, within the limits of the job at hand, establish achievable goals for each crew member, since success is always a good motivator. The foreman must be capable of giving praise when it has been earned and of reprimanding individuals when necessary, but in such a way that it can be seen as helpful. Over and above everything else, however, the foreman must be capable of maintaining an effective crew, of implementing the chosen work plan in an efficient work manner, and in achieving planned work progress and productivities.

Interest
Self Motivating
Technical Competence
Experience
Attention to Detail
Recognizable Skill
Forward Thinking
Analytical Sense
Financial Sensitivity
Sense of Values
Moral Judgements
Honesty
Tolerance
Tact
Caring
Ability to Cope with Pressure
Ability to React Quickly
Sound Judgements
Adaptability
Ability to Convey Ideas
Ability to Manage Subordinates
Ability to Motivate

(a)

Theory *X*: man has an inherent dislike of work and will avoid it if he can. Therefore, he has to be bullied and driven toward achievement, but he prefers to be directed anyway, and wants security rather than responsibility.

Theory *Y* makes completely different assumptions about people, and facilitates improvements in productivity. External control and threats are not the only means of getting men to work toward objectives, as men who are committed to objectives will be self-directed; commitment is related to rewards and achievements (including esteem and self-actualization); effort in work is as natural as play or rest; people can learn to accept responsibility; many people are underutilized intellectually, and could contribute more.

Management must somehow tap the hidden potential of its work force; by creating an environment in which self-motivation and control may develop and flourish.

(b)

A variety of theories regarding managerial behavior can be identified. These theories, or sets of assumptions, are based on the connections between three organization universals: *concern for production*, or the amount of emphasis supervision places on achieving production; *concern for people*, the productive unit of organization; *hierarchy*, or the boss aspect. Whenever a person acts as a manager, he/she is in some way making assumptions about how to solve problems of achieving the organization purposes of production through people.

The Managerial Grid®

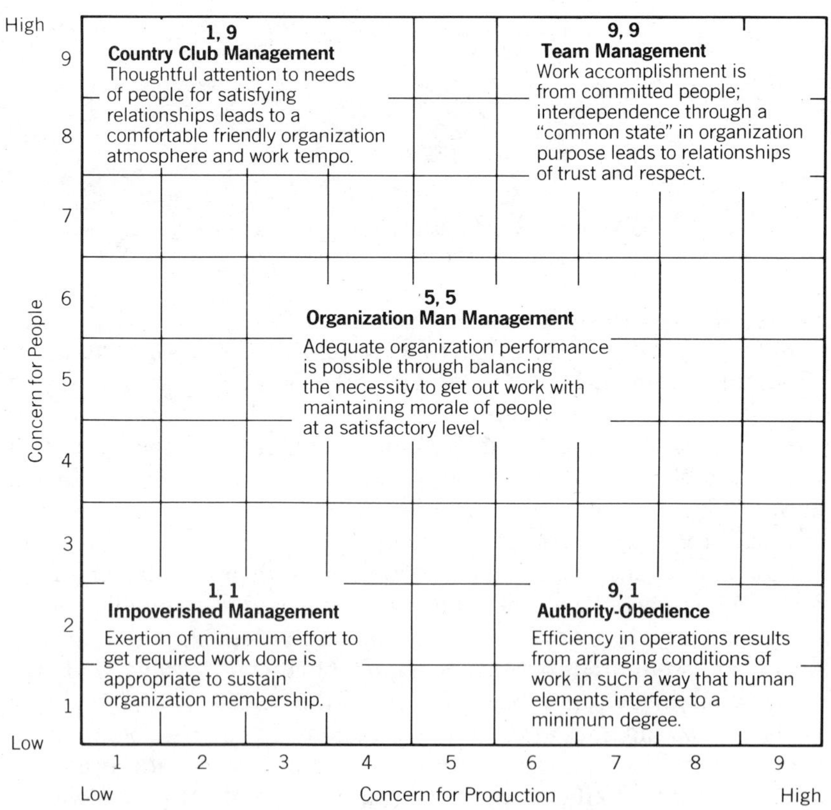

(c)

Figure 10.4 Leadership styles. (*a*) Leadership qualities. (*b*) Theory X–Theory Y of management styles. (*c*) Grid® concept of management style. The Managerial Grid figure from *The Managerial Grid III*, by Robert R. Blake and Jane Srygley Mouton. Houston: Gulf Publishing Company, Copyright © 1985, page 12. Reproduced by permission.

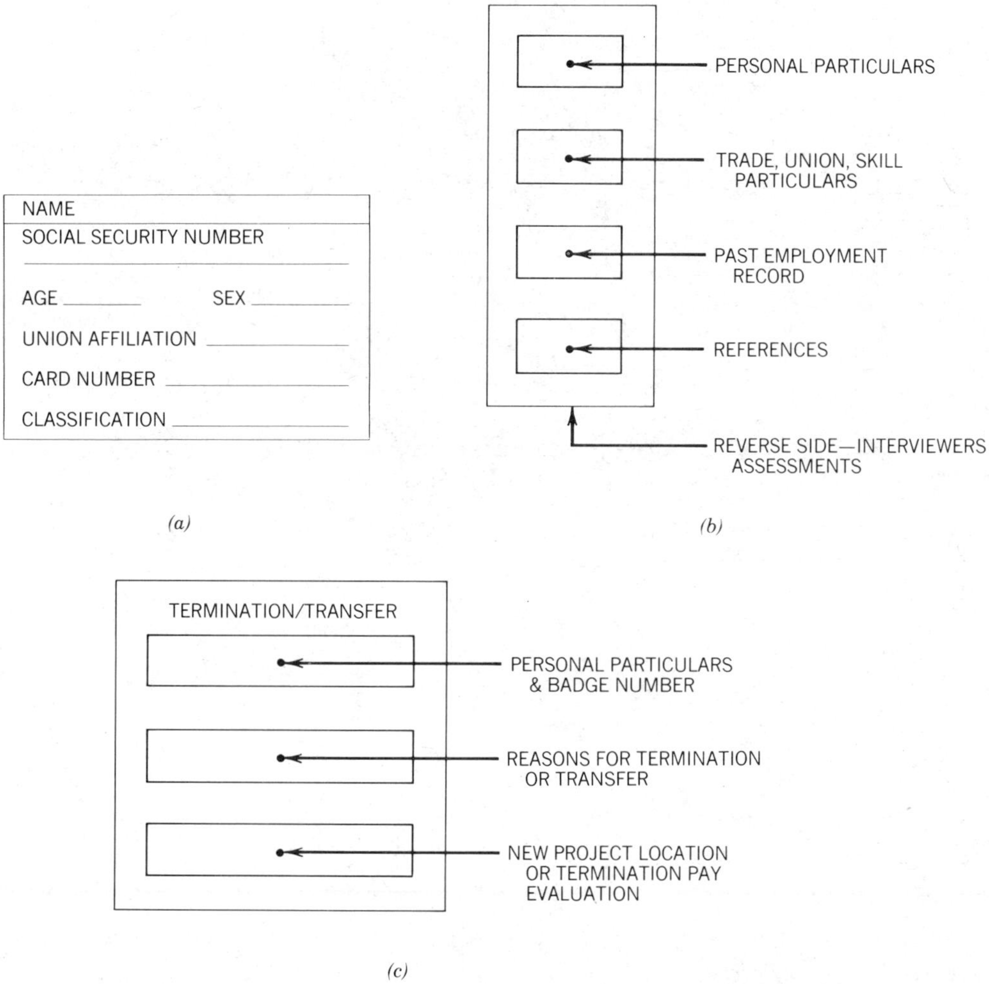

Figure 10.5 Hiring, firing, and transfer documentation. (*a*) Informal standard data. (*b*) Formal hiring interview document. (*c*) Formal termination or transfer document.

Key field managers such as superintendents and foremen must work at improving whatever natural talents and abilities they possess. Thus, by suitable exposure to work simplification techniques, productivity improvement, and management behavioral seminars, they can enhance their technical competency, leadership qualities, and management styles. Some of the indicators of leadership qualities are listed in Figure 10.4*a*. Several theories have been developed to describe and explain management styles. The fairly simple Theory X Theory Y view of management attitude is illustrated in Figure 10.4*b*, and the more comprehensive and flexible concept of the Management Grid® Theory is illustrated in Figure 10.4*c*.

The overall size of the field work force is built up and managed by means of the hiring and firing of labor and skilled tradesmen. In its most elementary form, this is done simply by telephoning the relevant local hiring halls or taking in people who apply at the work site gate. These formalities may require only the most simple personal details to be recorded (See Figure 10.5*a*). Generally in the local environment, the contractor knows from past contact the name, skill level, and work attitude of each member of the local unions, and may only need to see a union card (see Figure 10.6*a*). Consequently, the contractor/superintendent may reject the top-of-the-list candidate, or agree to pay show-up money only, if the particular craftsman is known to be shoddy in work quality or is a firebrand troublemaker. In cases where the skill level is unknown or the worker is suspected of falsely claiming a trade, the foreman/superintendent can devise a sim-

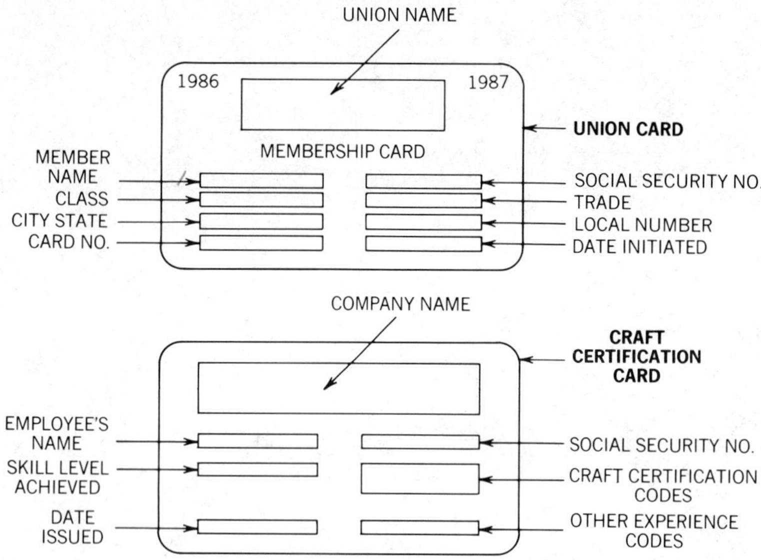

(a)

(i) Reflective evaluation of workers statement of past experience, work periods with each contractor, and frequency of job changes and on this basis question the applicant on details of work activity and work face procedures.

(ii) Close inspection of worker during trial work period on tool and equipment handling techniques. The manner in which the worker sets out a job, touches up to acceptable quality finish standards, or "walks the iron" if an equipment operator.

(iii) The assignment of the worker to a specific task requiring competency for its completion.

(b)

(i) When premiums are payable on skill levels the employer may request practical proof of skill to be demonstrated at the work face on jobs that the applicant will be employed.

(ii) The applicant may be requested to repeat union/trade tests relevant to the certificates presented as proof of skill.

(iii) Dozer Operator Skills Test as a sample of a Formal Skill Test.

A rigorous, quantitative, test of a bulldozer operator's ability to cut and move material has been produced by J. O'Brien of Department of Engineering Construction & Management, University of New South Wales and Loren Wilson of Central Arizona State College and validated in the U.S.A. and Australia.

The test involves requiring an operator to cut a flat bottomed slot in flat terrain. A profile is specified that consists of a down ramp beginning and ending at specified pegs, a flat bottom and an upramp again between beginning and ending pegs. The test takes about 5 minutes and is scored on the operator's ability to:

(a) Maintain plane surfaces i.e., roughness
(b) Natural twist
(c) To start and stop ramps at the predetermined spots.

The test has been shown to discriminate well between experienced drivers and will rapidly detect gross levels of ability.

(c)

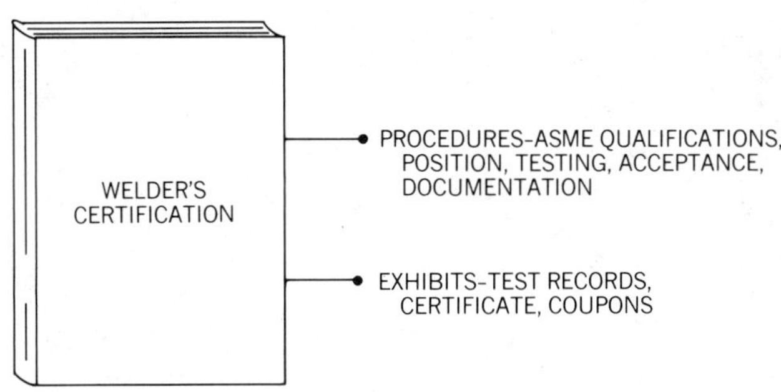

(d)

Figure 10.6 Skill evaluation. (*a*) Union and craft certification cards. (*b*) Informal test. (*c*) Formal test. (*d*) Quality assurance test.

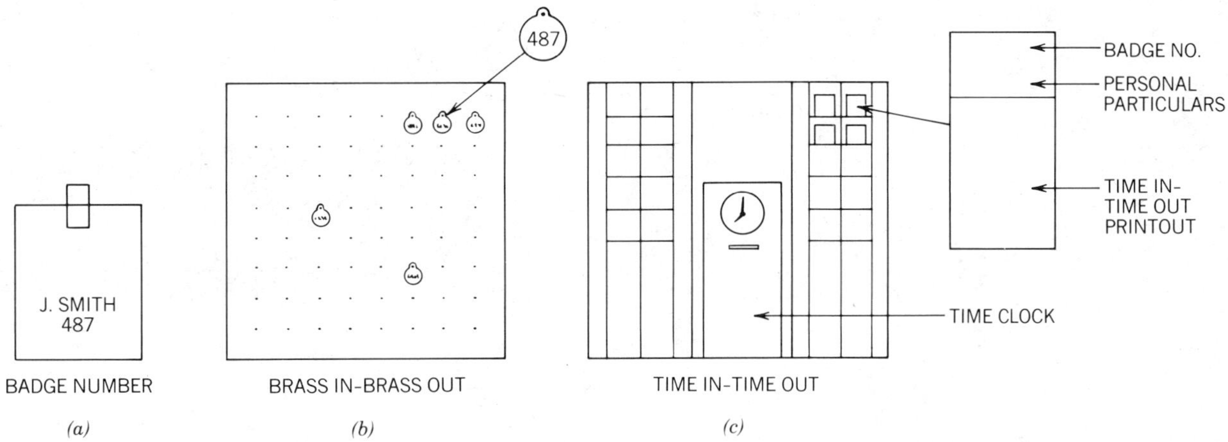

Figure 10.7 Work force time keeping control. (*a*) Badge number. (*b*) Brass in–brass out. (*c*) Time in–time out.

ple skill test or carry out a check (see Figure 10.6*b*), and on the worker's performance pass judgement as to whether he or she possesses an acceptable skill level. In other cases, if the applicant is a carpenter, for instance, a simple toolbox check for tool condition and care will give an indication of how he or she may perform.

On larger projects it may be company policy never to hire from the gate but to have a fact-finding interview complete with an interrogation on past experience (see Figure 10.5*b*), with or without a formal skill test evaluation (see Figure 10.6*c*),* and on this basis initiate a follow-up phase or make an offer by letter. In some construction environments, a craft certification card can be used to identify skill levels of potential employees who have previously worked for the company (see Figure 10.6*a*). On very large projects there may be a "show-up-at-the-gate-take-anyone" policy backed up by an inaugural job indoctrination briefing and follow-up training programs for those with inadequate skills. Some contractors make a formal contact by letter invitation to all those previously employed by the company. In many highly skilled areas, quality control and assurance requirements may require the constant testing of skilled tradesmen with consequent verification (see Figure 10.6*d*).

The firing procedure may consist of a simple statement, may involve severance pay, or even may require that a formal document be processed indicating the reasons for termination (see Figure 10.5*c*). In all cases, however, the contractor must strictly follow the procedures that are spelled out in labor agreements, and conform, where appropriate, to the requirements of civil rights legislation. In many cases, by following a warning procedure indicating causes for dissatisfaction with worker performance or attitude to work, the contractor may gain union support for a dismissal action.

The actual management control of the work force, insofar as show-up and time on site aspects are concerned, is a vital matter for both payroll and job morale purposes. Methods can be based on personal recognition by the foreman or superintendent on a daily basis, or on a badge number assigned to each work force member (see Figure 10.7*a*), or on simple gate entry controls, such as the "brass in brass out" system indicated in Figure 10.7*b*. It can also be handled by an automatic time clock system, as indicated by Figure 10.7*c*.

The field location of a crew is determined by the work sequence of activities, the availability of materials and equipment, and safety considerations. On small projects there is little need to adopt any formal method of keeping track of crews. On large projects involving many crews and control areas, however, crew location becomes a major problem. One simple approach is to issue colored hats so that different trades or crews can be readily identified. On congested sites it may become necessary to set up a site location board in the field office and to post crew markers on this site location board. A typical example of this system of portraying crew location on a project site is shown in Figure 10.8.

*"Operating tests for track type dozer operators" J. O'Brien and L. Wilson, Department of Engineering Construction and Management, University of New South Wales, New South Wales, Australia, August 1983.

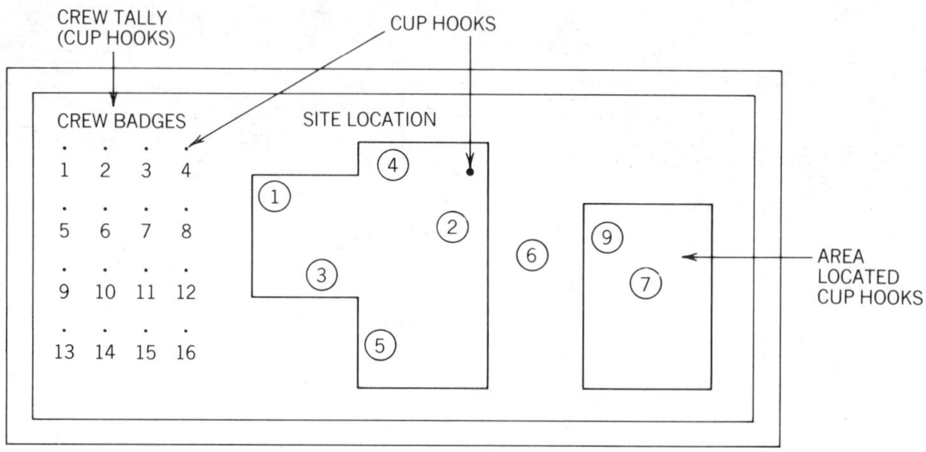

Figure 10.8 Wall display crew location chart.

Field management personnel should seek every opportunity to encourage foremen in their efforts to achieve productive work. Suggestions should be made on the elimination of needless time-consuming activities, such as frequent trips to tool boxes and other supportive activity that can be minimized through better field instruction to crew members. It is often helpful to provide foremen/superintendents with work simplification and elimination guidelines, to give them the opportunity to attend management workshops that focus on quality circle or productivity improvement issues, and/or to expose them to results or case study conclusions from other jobs. If top management intends to initiate a productivity, work study, or motivational improvement program, the foremen/superintendents should be thoroughly briefed on the manner, techniques, and conclusions to be drawn from such efforts, so that they can brief their crews and individual crew members on the issues involved from a knowledgeable position and in a convincing manner.

The active assignment of work tasks to individuals and their subsequent direction in these efforts is intimately tied to the superintendent/foreman assessment of the individual skills and work attitudes. In this respect, the field supervisors constantly call to mind their observations of the past performance of these crew members (Field Monitoring). Thus, the superintendent/foreman must continuously upgrade their personal assessments and judgements of the capabilities of individual crew members, and of their own ability to effectively build an efficient crew with high morale in every work situation that it addresses. This effort constitutes the life blood of the work of the superintendent/foreman, and is eventually tied into the Field Monitoring effort. It represents the informal Management Information System of the Crew Management decision process.

10.3 THE PROJECT TEAM COMPONENT APPROACH

The Crew Management decision process, which at the small firm level is essentially labor management oriented, and the *FIELD RESPONSIBILITY* (P_S 12)* of the superintendent and foremen, operates in four different modes that correspond to the different life stages of the crew. These modes are crew start-up, crew direction, the detection and resolution of crew problems, and the subsequent transferring of crew members to other work situations or disbandment.

The action of supplying crew manpower for an activity initiates WORK FORCE MANAGEMENT (W3). Once a CREW HAS BEEN PROPERLY ESTABLISHED (W5) (meeting the necessary JURISDICTIONAL AND SKILL REQUIREMENTS (W4)), and the foreman ASSIGNS INDIVIDUAL WORK TASKS (A6) to crew members and DIRECTS THEM TO START WORK (W7), the crew has entered the work execution phase. On completion of its work tasks the crew reenters Crew Management at CREW ADJUSTMENT (A5), for firing or transfer to another activity or project. If crew problems arise, they are generally detected by the

*The small firm policies relevant to the Crew Management decision process are collected in Table 10.3.

TABLE 10.3 POLICIES OF THE SMALL FIRM

Policy		Description
P_S 8	*COMPANY ALLEGIANCE*	Field management is staffed from men loyal to the company without hiring on a permanent basis.
P_S 9	*KEEP BEST*	Field management must retain or employ the best qualified men available and strive to build a nucleus of men loyal to the firm (i.e., create a permanent work force).
P_S 10	*UNION RELATIONS*	In a union environment, maintaining good relations with each union (through the business agent), is helpful to successful project management.
P_S 11	*UNION PRACTICE*	Field management must subscribe to local union practices when operating in a union environment.
P_S 12	*FIELD RESPONSIBILITY*	Actual crew planning and management and manipulation of labor is a total field responsibility.
P_S 14	*CREW SELECTION*	Crew selection is based on these models: (1) historical crew sizes and mixes, (2) trial-and-error crew buildup, and (3) maximum crew size.
P_S 15	*CREW MIX*	While impacted by union rules in some construction environments, the ratio of foremen to journeymen/apprentices within a crew is established to optimize both the foreman's level of supervisory control and the crew's performance. Foremen other than loyal company foremen are selected on the basis of experience, managerial capability, and desire, in that order.
P_S 16	*BEST MIX*	The best mix of men is guided by skill level and personal compatability, and is reflected in crew assignments.
P_S 17	*PERSONAL CONTACT*	Field labor management is not formally considered beyond maintaining good on-site relations via personal contact and recognition, coupled with hiring and firing practices to weed out undesirable labor.

foreman or the superintendent during field inspection, when PERSONAL PROBLEMS ARE RESOLVED (W6).

When crews are initially manned, the job superintendent must tap available sources of manpower. To do this, he or she will try to obtain crew members from the company's nucleus of loyal employees, which could mean transferring employees from other company projects, reassigning some from within the existing project labor force, or hiring workers who normally work for the contractor, provided these craftsmen are available. These *PERSONAL CONTACTS* (P_S 17) are the preferred means of supplying manpower for project crews, because they are the best method of creating a permanent work force through *COMPANY ALLEGIANCE* (P_S 8). The other alternative is to hire from the local labor pool, in which case the job superintendent is not always assured of obtaining the best qualified individuals. Regardless of the source of manpower, WORK FORCE MANAGEMENT strives to *KEEP THE BEST* (P_S 9) craftsmen available. This means that the superintendent attempts to employ the most highly skilled workers available, and to provide work for them continuously.

While highly skilled craftsmen are constantly sought by contractors at all levels in the construction industry, the small contractor must always obtain the best men available. Crews in this environment are normally small, and any "dead wood" within a crew would adversely affect crew productivity and activity progress. As a result, skill level is the primary quality to be considered when evaluating employee credentials. It must be recognized that the employer often has minimal control over craftsmen furnished by the union, which is one shortfall the union contractor must overcome. The various union halls are contacted and the desired number of workers are hired. Each will be given an employee number for payroll purposes, and will be assigned to a foreman.

Prior to establishing field crews in a union environment, the job superintendent/foreman must ensure that jurisdictional areas are defined and work assignments are made according to those jurisdictional areas. When ENFORCING JURISDICTIONAL WORK ASSIGNMENTS, it is essential that each field manager keep aware of the respective unions' work rules, because in a union environment the contractor must subscribe to local *UNION PRACTICES* (P_S 11). While the foremen are eminently versed in their unions' work rules, the job superintendent must have a good knowledge of the working rules of many unions. This cannot be emphasized enough, because an incorrect *CREW MIX* (P_S 15) or WORK ASSIGNMENT could cause a needless loss of work days due to walkouts over a jurisdictional dispute. Maintaining *GOOD RELATIONS* (P_S 10) with each union's business agent is a policy which is helpful in preventing such delays, because problems may be resolved rapidly in consultation with these agents. Generally, however, the small contractor relies on a local chapter of a contractor organization such as the AGC to establish union contracts and work rule agreements.

The ESTABLISHING OF FIELD CREWS implies the selection of crew foremen and of the workers assigned to them. When a firm does not employ regular foremen, the job superintendent must select a foreman from the journeymen that are hired. In a union environment, if a firm employs several regular foremen, then the crews must reflect the requirements stipulated in the union agreement. For instance, a firm's regular carpentry foreman will supervise all carpentry work on the job. Each carpentry crew, however, will reflect the proper ratio of foremen to journeymen. Accordingly, the crew member selected as foreman will be the man with previous experience in that position. If no one has the experience, the person with the most leadership potential and basic supervisory knowledge will often be selected as foreman. In either case, the foreman selected in this manner achieves this position by virtue of the union agreement, and often assumes little direct management responsibility.

The job superintendent is aware of the skill level of the loyal workers and will have little problem achieving a good mix of craftsmen. The job superintendent must use his own judgment when assigning new employees to crew foremen. The superintendent may personally interview new hires to determine their qualifications and personalities.

On-site labor relations, as part of Crew Management, is primarily the responsibility of field supervision. Head office involvement occurs only when either major negotiations with a union are necessary as a result of strikes and contract negotiations, or special problems arise which require owner decision making. The job superintendent or foreman on smaller projects most probably is the key decision maker who will execute the majority of labor relations related functions. They must interface frequently with crew members, and in union environments with union business agents and other foremen, to develop successful on-site labor relations. The foremen/superintendent must help to SOLVE HUMAN RELATIONS PROBLEMS and ENFORCE JURISDICTIONAL WORK ASSIGNMENTS. This approach tends to promote good relations at the work face level, thus preventing a job from going sour.

On-site labor relations is a key management function which can have a significant impact in determining the success of a project. Personal contact and continuous communication with each employee afford field supervision with the best environment in which to manage labor. If, for instance, the job superintendent knows the names and personalities of all crew members under his direct supervision on a site, he can then more readily communicate work objectives and anticipate, discuss, and solve personal or job-oriented problems. Furthermore, the superintendent can develop crews which reflect the best mix of workers in terms of interpersonal compatibility and skill level. This approach can greatly improve productivity.

The foreman is responsible for ASSIGNING INDIVIDUAL WORK TASKS to crew members. Work assignments must be compatible with the skill level of each crew member and each assignment must be consistent with union jurisdiction rules. In addition, the foreman must recognize potential personal problems and resolve them, if possible, before making crew assignments. This requires daily planning of the work and directing crew members to START INDIVIDUAL WORK TASKS.

When FIELD INSPECTION (M8) (through field monitoring) reveals a potential problem area, the field manager must react and attempt to resolve the problem. The severity of the problem will determine which field staff are involved in solving it. Most daily problems are successfully administered by the foreman and/or the job superintendent. Only if the problem has serious project implications will head office management be alerted. Thus, field staff retain full authority to implement corrective action. Depending on the nature of the problem and its cause, the decision maker will analyze the crew selection variables involved. For example, if low productivity, resulting from a saturated work space (limited site accessibility) is a problem, the foreman/superintendent could reduce overall crew size or change the physical location of the crew members to relieve this congestion. If these alternatives are not feasible, the construction method might be

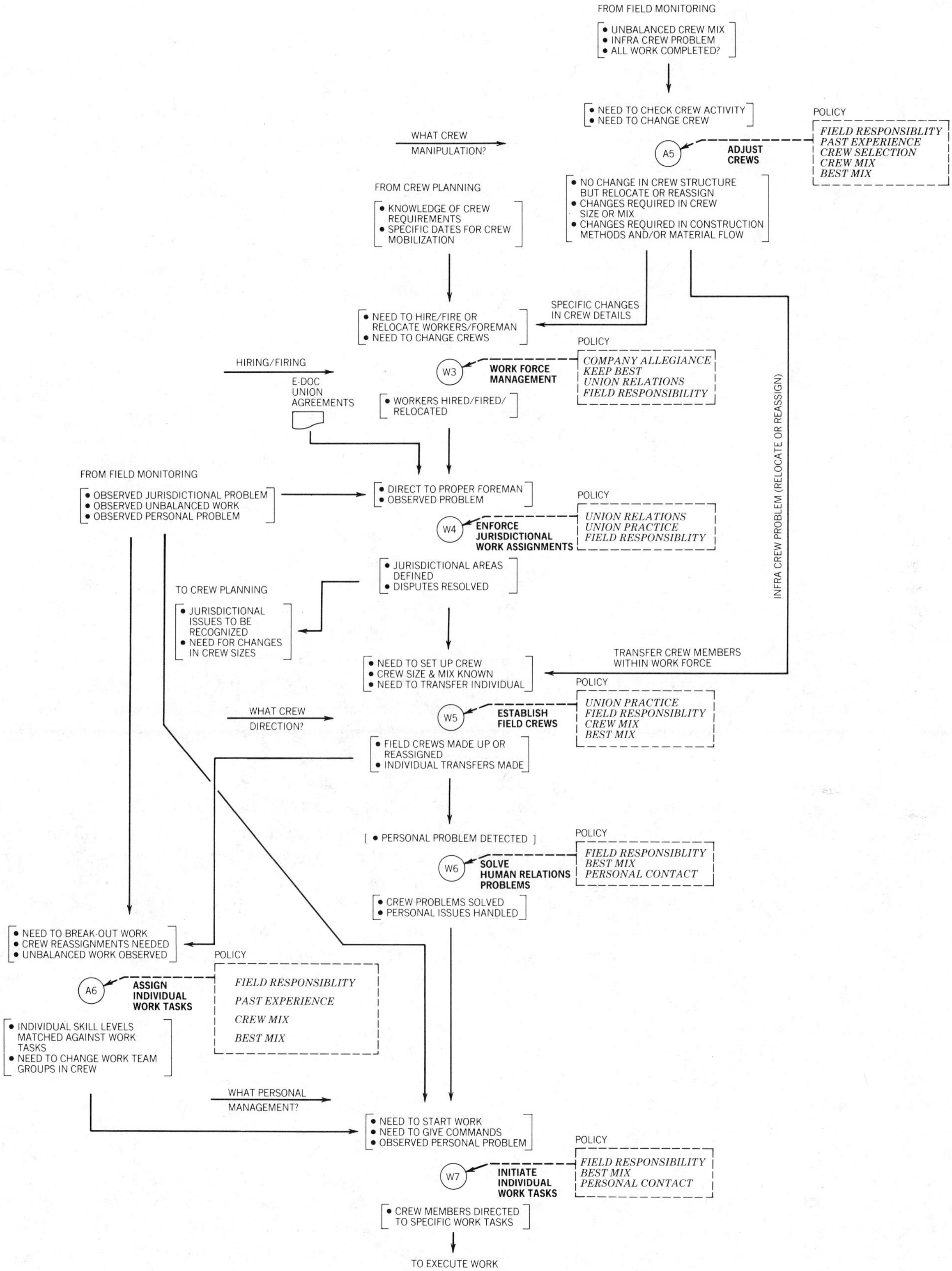

Figure 10.9 Crew Management decision process (Project Team Component approach).

altered or the material flow routes changed. The best solution among the available alternatives would be selected and implemented expeditiously in order to circumvent a decrease in productivity. The timely ADJUSTMENT OF CREWS is a critical management function and the total *FIELD RESPONSIBILITY* of the superintendent and foreman.

Crew management is another critical labor management problem of the small firm. It involves the day-to-day interaction between field managers and workers, while attempting either to start-up crews, modify crews, or solve crew problems. On small projects, for small contractors, successful crew management can mean the difference between profit and loss. An illustration of the Crew Management decision process is shown in Figure 10.9.

10.4 THE TRAVELING PROJECT MANAGER

Crew Management, as mentioned in Section 10.3, addresses four different issues, namely, crew start-up, crew direction, solving crew problems, and crew manipulation. The most noticeable difference in Crew Management between the Traveling Project Manager concept and that of the small firm does not appear in the methods of performing each function, but rather in the management roles of the personnel responsible for these functions. In this approach, the project manager assumes partial responsibility for WORK FORCE MANAGEMENT (W3) and the SOLVING OF HUMAN RELATIONS PROBLEMS (W6). His functional responsibility is a result of *PERSONAL CONTACT* (P_M 18)* policies which promote a heavy project management role in labor relations. The primary vehicle for implementing this policy of personal contact is to recognize, and strive for good relations with, the superintendents/foremen and the project labor force. Because the job superintendent is also a permanent employee, the project manager attempts to promote this policy in the field. Constant site visits by the project manager also help to promote good relations, provided time is taken to "chat" with key foremen and others on the job on a first name basis.

Personal contact, coupled with endeavors to *KEEP THE BEST* (P_M 9) foremen and craftsmen, enhances *COMPANY ALLEGIANCE* (P_M 10) and provides the impetus for effective employee motivation. In addition, the project manager, where possible, should strive to incorporate the key superintendent and foremen into the decision-making process by seeking their expertise at both the office level during the manpower allocation stages, and in the field as problems are encountered. The results of such an approach are obvious: The labor force acquires a high spirited attitude toward work, and the existence of a cooperative atmosphere between office and field management ensures an extremely effective on-site labor management program.

The larger labor forces encountered in the medium-sized firm environment have a salient impact on the composition of the work force and the responsibilities of the foreman. A subtle change emerges with respect to the percent of loyal company employees to the total project labor force. While this percent is relatively high for the small firm utilizing a nucleus of key people, it decreases markedly for the medium-sized firm, especially on larger projects, simply because it is difficult to maintain that many key workers on a continuous basis. Thus, the project manager and superintendent have to rely more frequently on the local labor pool for their supply of manpower. Hence, the Crew Management decision process begins to take on a *MANPOWER PLANNING* (P_T 13) focus, and relies more heavily on the *PAST EXPERIENCE* (P_T 14) of both the Traveling Project Manager and field job superintendent. New hires will often be screened carefully by the project manager or superintendent, and are frequently observed by these staff members during an initial trial period so that the quality of their work can be judged. Finally, the key craft foremen are assigned more responsibility in Crew Management because the superintendent simply cannot devote as much time to this area.

Crew Management is a critical labor management problem because it involves the daily interaction of the project manager, field manager, and work force, while attempting either to start-up crews, modify crews, or solve crew problems. These issues more frequently influence the project manager, particularly when costs and schedule implications are involved. The Crew Management decision process is highlighted in Figure 10.10, and except for an enlarged set of more specific policies, is identical to the small firm's approach (see Figure 10.9).

*Medium-sized firm policies relevant to the Crew Management decision process are listed in Table 10.4.

TABLE 10.4 POLICIES OF THE MEDIUM-SIZED FIRM

Policy		Description
P_M 9	*KEEP THE BEST*	Management strives to retain or employ the best qualified men available and build a nucleus of men loyal to the firm.
P_M 10	*COMPANY ALLEGIANCE*	Key craft foremen are staffed from foremen loyal to the firm. Foremen other than loyal company foremen are selected on the basis of experience, managerial capability, and desire.
P_M 11	*UNION RELATIONS*	In a union environment, maintaining good relations with each union through the business agent is helpful to successful project management.
P_M 12	*UNION PRACTICE*	Field management must subscribe to local union practices when operating in a union environment.
P_M 13	*MANPOWER PLANNING*	Short-range planning of manpower at the head office pre-job level, coupled with field labor manipulation, is critical to project control.
P_M 14	*PAST EXPERIENCE*	Field planning and allocating of manpower is based on a freewheeling approach, past experience, and close interaction between the project manager and field management.
P_M 16	*CREW MIX*	While impacted by union rules in some construction environments, the ratio of foremen to journeymen/apprentices within a crew is established to optimize both the foreman's level of supervisory control and the crew's performance.
P_M 17	*BEST MIX*	The best mix of men is guided by skill level and personal compatability and is reflected in crew assignments.
P_M 18	*PERSONAL CONTACT*	Manpower management begins with head office scheduling and allocation of manpower with field input, but the primary emphasis remains onsite good relations via personal contact and recognition.
P_M 26	*MANAGEMENT-BY-EXCEPTION*	Manpower management by exception as problems arise is an acceptable decision approach. However, management decisions are made when the impact will result in a significant cost/scheduling savings.
P_M 29	*PROJECT TEAM SPIRIT*	Active participation of all field managers in the management process promotes goodwill and a positive management environment.

10.5 FIELD-BASED PROJECT MANAGEMENT

The Crew Management decision process, especially in its labor management focus, is not significantly influenced by the changing construction environments currently found in construction practice. The decision process structure, information flows, and policy statements that decide Crew Management issues at the large firm level are similar or often identical to those of the medium-sized firm (see Figure 10.10). The Crew Management decision process for the Field-Based Project Management approach is shown in two separate decision process networks, so that details can be represented conveniently. The overall Project Crew Management focus is illustrated in Figure 10.11*a*, and the more personal focus of Individual Crew Management is illustrated in Figure 10.11*b*.

Many of the policies adopted by the large firm for Crew Management, Figure 10.11, are the same as those of the medium-sized firm. Typical examples are *KEEPING THE BEST* (P_L 12), *COMPANY ALLEGIANCE* (P_L 13), *UNION PRACTICE* (P_L 14), *CREW MIX* (P_L 19), *BEST MIX* (P_L 20), *PERSONAL CONTACT* (P_L 22), and *PROJECT TEAM SPIRIT* (P_L 35).* These policies are applied to Crew Management in the same way as described for the medium-sized firm. However, one new policy, *LABOR RELATIONS* (P_L 23) is a response to the demands placed on Crew Management as a result of increasing field manpower levels and long-range manpower planning, as stipulated by the changing focus of the *MANPOWER PLANNING* (P_L 15) policy. These policies create a need for a labor relations team of experts solely responsible for the project site labor relations and industrial relations programs. The functions which are most noticeably impacted by this labor relations focus are WORK FORCE MANAGEMENT (W3) and ENFORCING JURISDICTIONAL WORK ASSIGNMENTS (W4).

*See Table 10.5, Policies of the Large Firm

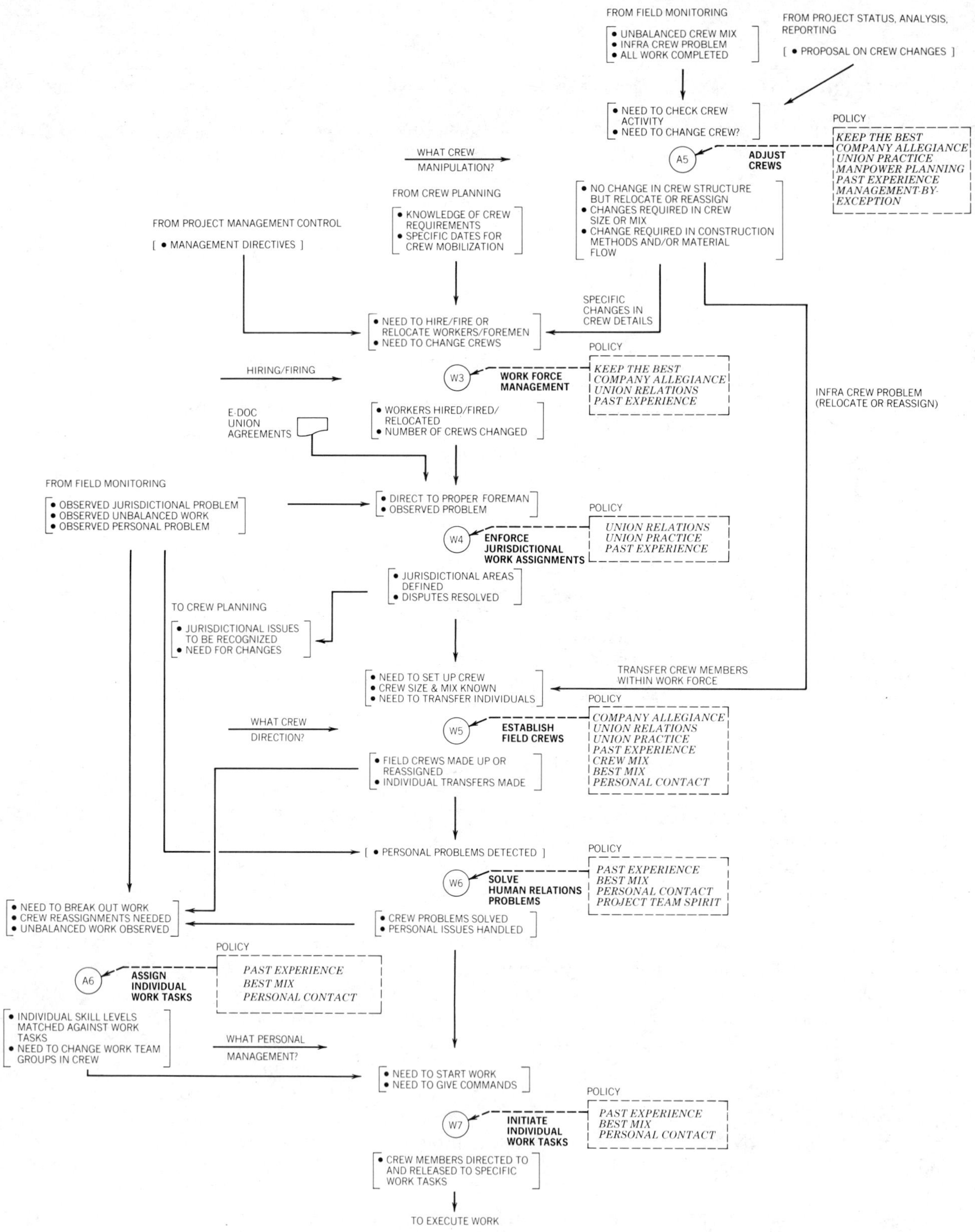

Figure 10.10 Crew Management decision process (Traveling Project Manager approach).

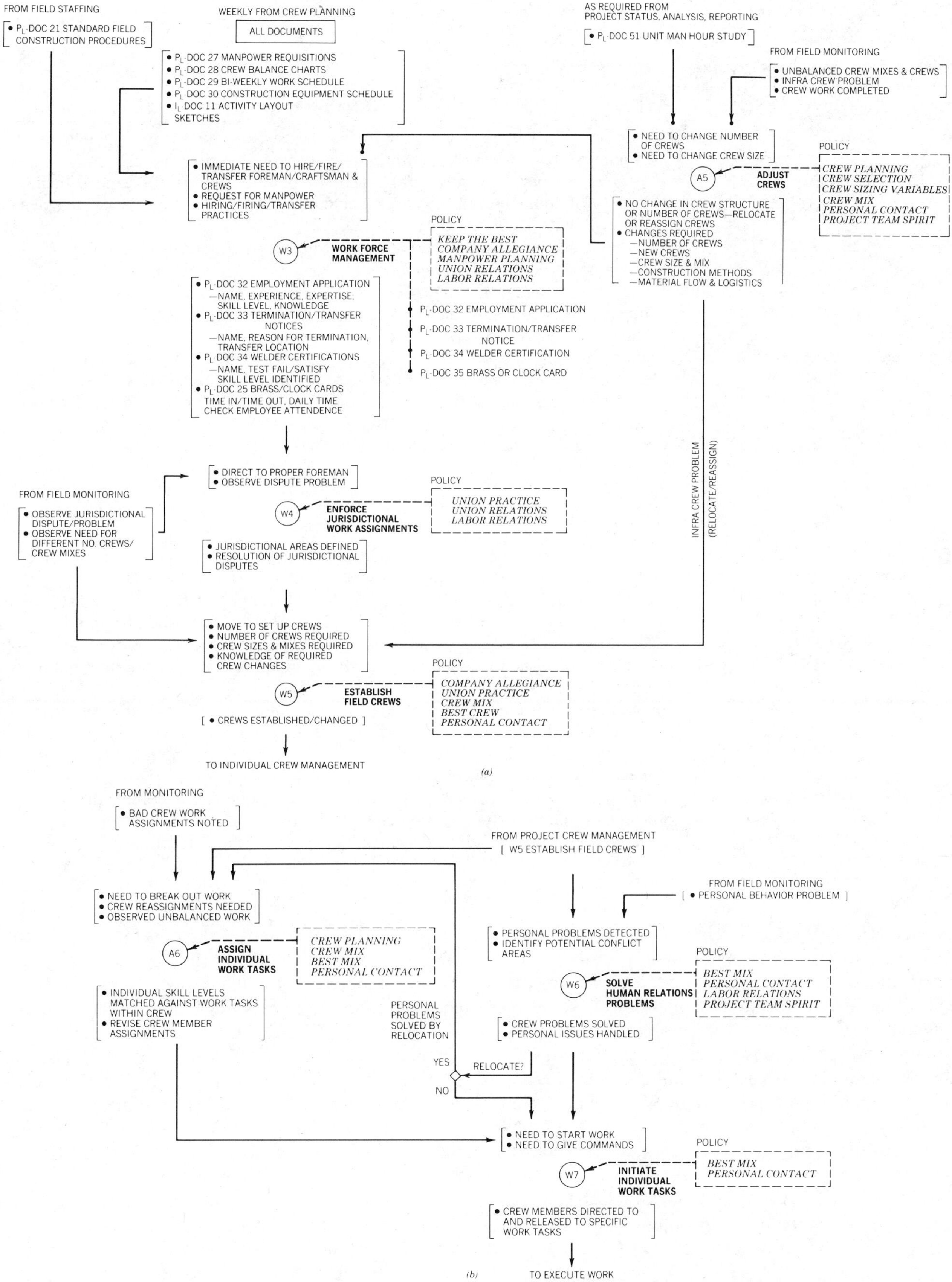

Figure 10.11 Crew Management decision process (Field-Based Project Management approach). (*a*) Project crew management. (*b*) Individual crew management.

TABLE 10.5 POLICIES OF THE LARGE FIRM

Policy		Description
P_L 12	*KEEP THE BEST*	Management strives to retain and employ the best qualified men available and build a nucleus of men loyal to the firm through this security.
P_L 13	*COMPANY ALLEGIANCE*	Key craft foremen are staffed from foremen loyal to the firm. Foremen other than loyal company foremen are selected on the basis of experience, managerial capability, and desire.
P_L 14	*UNION PRACTICE*	Field management must subscribe to local union practices when operating in a union construction environment.
P_L 15	*MANPOWER PLANNING*	Long-range planning of manpower by craft and in total is a prerequisite to supporting the field effort, and must be performed in the planning and scheduling stages of the project.
P_L 16	*CREW PLANNING*	Field planning, allocation, and scheduling of manpower requires a precise definition of work tasks and a combination of past experience, knowledge of work content, man-hour estimates, and interaction between field project team members.
P_L 17	*CREW SELECTION*	Crew selection is based on these models: 1. Maximum crew size 2. Number of crews 3. Work task analysis 4. Historical crew
P_L 18	*CREW SIZING VARIABLES*	Labor dollars, man-hours, and quantities are primary ingredients in crew selection, and must be weighed against scheduled activity durations.
P_L 19	*CREW MIX*	While impacted by union rules in some construction environments, the ratio of foremen to journeymen/apprentices within a crew is established to optimize both the foremen's level of supervisory control and the crew's performance.
P_L 20	*BEST MIX*	The best mix of men is guided by skill level and personal compatibility and is reflected in crew assignments.
P_L 21	*UNION RELATIONS*	In a union construction environment, pre-job union conferences and personal contact with the unions is an integral part of successful project management.
P_L 22	*PERSONAL CONTACT*	Crew management must be coupled with personal contact and good relations between office field management and construction workers.
P_L 23	*LABOR RELATIONS*	A full-time labor relations expert is often required due to the complexity and manpower demands of the project.
P_L 35	*PROJECT TEAM SPIRIT*	Active participation of all field managers in the management process promotes goodwill and a positive management environment.

In the former, the labor relations expert's responsibilities concern the hiring of manpower from local labor pools, releasing manpower from the project, recruiting and acquisition of manpower as determined by Project Manpower Curves (I_L-DOC 7) and Manpower Requisitions (P_L-DOC 27), and possibly developing job programs for training employees and improving employee motivation. When operating in a union environment, the labor relations team will work closely with union representatives and will be heavily involved in ENFORCING JURISDICTIONAL WORK ASSIGNMENTS, which requires knowledge of many union agreements,* negotiating work disputes on behalf of the contractor, and settling jurisdictional issues as they arise.

The manpower-oriented aspects of Crew Management, and the manpower-focused functions like WORK FORCE MANAGEMENT, ENFORCING JURISDICTIONAL WORK ASSIGNMENTS, and ESTABLISHING FIELD CREWS, involve different functional performance techniques. For example, the recruiting and acquisition of labor may require advertising in local newspapers within the region, or a constant communication with local union business agents to sustain a continuous flow of manpower until the project manpower peak is realized. In addition, formal hiring, transfer, and termination documents will be used in an effort to determine the quality and work history of crafts-

*This need is obviated if the union is a party to national agreements often signed with large construction firms.

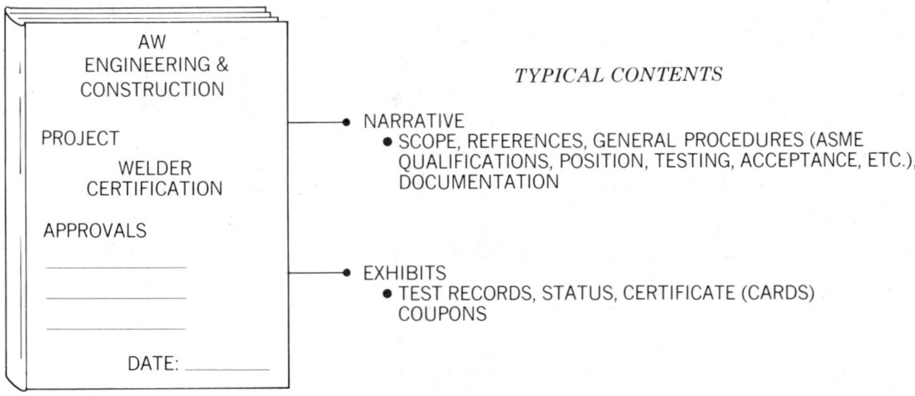

Figure 10.12 Welder Certification (P_L-DOC 34).

men, and to document specific reasons for termination in the event a craftsman's skills are not sufficient to meet the requirements of the job. The data typically included on the Hiring Interview Form (P_L-DOC 32) and Termination and Transfer Notice (P_L-DOC 33) are illustrated in Figures 10.5*b* and 10.5*c*, respectively. The acquisition of manpower for large projects is dependent upon accurate and current Manpower Curves, supported by a continuous flow of Manpower Requisitions to cover the short-range planning periods.

Because of the increased technical complexity normally associated with large industrial or heavy construction projects, formal testing techniques resulting in certification documents begin to emerge. Examples are shown in Figures 10.6*c* and 10.6*d*. These documents are often necessary to ensure that the project meets the requisite quality assurance standards either imposed by the client, and/or by engineering and design codes as prepared by professional organizations and accepted as standard practice. A Welder's Certificate (P_L-DOC 34) is an example of the type of documentation arising out of formal project site testing procedures. A typical Welder's Certificate is shown in Figure 10.12.

Large firms commonly have the problem of monitoring and controlling craft movement on a daily basis as craftsmen arrive and depart from the job site. This problem is intensified as the labor force grows and the area of the job site is enlarged or is located in several distinct physical facility areas. Typical solutions employed by the large firm are the Brass system or a Clock Card (P_L-DOC 35) approach. While either of these techniques can emerge on larger projects in the medium-sized firm environment, they are more frequently associated with large projects employing significant labor forces. The Brass and Clock Card systems are illustrated in Figures 10.7*b* and 10.7*c* of Section 10.2. More sophisticated electronic time and attendance systems that serve to control craft tool requisitions and camp facility usage, not to mention daily attendance recording of craft personnel, are also emerging in construction practice.

Apart from those functions in which the labor relations team participates in Crew Management, this decision area remains the domain of the superintendent and foremen. Both of these field personnel must, however, carefully devise management and leadership styles that can effectively operate in a labor environment that is becoming more impersonal at the work task level. This situation is due to the need for a larger number of crews comprised of a significantly increased number of transient tradesmen. Furthermore, the superintendent/foremen must devise effective methods for personally keeping track of their crews as project complexity and geographical size often makes it impossible for the superintendent/foremen to simply know where every crew is located and where each crew member is working on the job site.

10.6 TOTAL FIELD AUTONOMY

Crew Management for the large firm operating in a Total Field Autonomy mode changes only in terms of policy formulation and in the increased involvement of a variety of construction personnel in the Crew Management decision process. The increasing involvement of the project team group is most noticeable in relation to the manpower-oriented functions, and those decisions involved in modifying crews through CREW ADJUSTMENT (A5) de-

W3 WORK FORCE MANAGEMENT

Policies	Documents	
KEEP THE BEST	P_T-DOC 41	Employment Applications
COMPANY ALLEGIANCE	P_T-DOC 42	Termination Transfer Notice
UNION PRACTICE	P_T-DOC 43	Welder Certification
MANPOWER PLANNING	P_T-DOC 44	Brass or Clock Cards
LABOR RELATIONS		

W4 ENFORCE JURISDICTIONAL AND SKILL LEVEL WORK ASSIGNMENTS

Policies	Documents
UNION PRACTICE	
CREW MIX	
LABOR RELATIONS	

W5 ESTABLISH FIELD CREWS

Policies	Documents
COMPANY ALLEGIANCE	
UNION PRACTICE	
WORK ORDER	
CREW MIX	
BEST MIX	
PERSONAL CONTACT	

A5 ADJUST CREWS

Policies	Documents	
UNION PRACTICE	I_T-DOC 12	Crew Location Chart
WORK ORDER		
CREW SELECTION		
CREW MIX		
BEST MIX		
PERSONAL CONTACT		
CREW MANAGEMENT		
PRODUCTIVITY		
PROJECT TEAM SPIRIT		

(*a*)

A6 ASSIGN INDIVIDUAL WORK TASKS

Policies

CREW MIX

BEST MIX

PERSONAL CONTACT

W6 SOLVE HUMAN RELATIONS PROBLEMS

Policies

CREW MIX

BEST MIX

PERSONAL CONTACT

WALK JOB

PROJECT TEAM SPIRIT

W7 INITIATE INDIVIDUAL WORK TASKS

Policies

CREW MIX

BEST MIX

PERSONAL CONTACT

(*b*)

Figure 10.13 Crew Management decision process (Total Field Autonomy approach). (*a*) Project crew management. (*b*) Individual crew management.

TABLE 10.6 POLICIES OF THE AUTONOMOUS PROJECT TEAM

Policy		Description
P_T 15	*KEEP THE BEST*	Management strives to retain and employ the best qualified men available and build a nucleus of men loyal to the firm through this security.
P_T 16	*COMPANY ALLEGIANCE*	Key craft foremen are staffed from foremen loyal to the firm. Foremen other than loyal company foremen are selected on the basis of experience, managerial capability, and desire.
P_T 17	*UNION PRACTICE*	Field management must subscribe to local union practices when operating in a union construction environment.
P_T 18	*MANPOWER PLANNING*	Long-range planning of manpower by craft and in total is a prerequisite to supporting the field effort and must be performed continuously throughout the project's life.
P_T 19	*WORK ORDER*	The work order is a key tool for field construction planning and crew sizing because it focuses on detailed estimates based on field measurements, describes construction methods, defines material and equipment mobilization, evaluates the most economical approach, and considers schedule impact, all in relation to work site conditions and work packages.
P_T 20	*CREW SELECTION*	Crew selection is primarily based on the work order model, but it does incorporate all other models, such as the following: 1. Maximum crew size 2. Number of crews and crew buildup 3. Work task analysis 4. Historical crew
P_T 21	*CREW MIX*	While impacted by union rules in some construction environments, the ratio of foremen to journeymen/apprentices within a crew is established to optimize both the foreman's level of supervisory control and the crew's performance.
P_T 22	*BEST MIX*	The best mix of men is guided by skill level and personal compatibility, and is reflected in crew assignments.
P_T 23	*LABOR RELATIONS*	When union labor is utilized, pre-job union conferences and personal contact with the unions are an integral part of successful project management. Regardless of the construction environment, a full-time labor relations expert is required because of the complexity and manpower demands of the project.
P_T 24	*PERSONAL CONTACT*	Crew management must be coupled with personal contact and good relations between field management personnel and construction workers if scientific management techniques are to be applied successfully.
P_T 25	*WALK JOB*	In the field, daily activity progress and crew performance is monitored by visual inspection, and crew management decisions are made on the basis of a subjective evaluation of observed progress and productivity versus planned progress and productivity. This decision process often incorporates a number of decision makers.
P_T 29	*CREW MANAGEMENT*	Field crew management is based on past experience coupled with scientific management techniques.
P_T 31	*PRODUCTIVITY*	Productivity is an essential concept and must be rationally defined for office and field utilization. An awareness of the components that affect productivity and their relation to man-hours, activity durations, and progress, and that all these integrate and are manageable, is essential.
P_T 37	*PROJECT TEAM SPIRIT*	Active participation of all project team members in the management process promotes goodwill and a positive management environment. It ensures the success of scientific management techniques and the total integration of the project team.

cision making. For example, methods engineers are often called in to solve problems that arise when changes in construction techniques or relocation of material supply points might mean better support of the pacing crew's production rates. The Crew Management decision process, together with its relevant management functions, policies, and documents is shown in Figure 10.13 in terms of Project Crew Management (Figure 10.13*a*), and Individual Crew Management (Figure 10.13*b*). Except for policy deviations, this decision process is identical to that of the Field-Based Project Management approach as presented in Figure 10.11.

The most significant change in Crew Management at the Total Field Autonomy level concerns the policy focus on scientific field management techniques such as work sampling for improving field performance and productivity. This change can be difficult to manage because many past applications of scientific management have been viewed as "spying" on project foremen and individual craft members to detect "screw-ups and poor performers." Thus, scientific management techniques will only be successful if field supervisors and craftsmen are involved in the program to the extent of understanding and gaining a full awareness of its objectives. From their perspective, its purpose is to assist them in the productive application of their skills.

An examination of many of the relevant policies will reveal the close interaction and cooperation at all levels that ensures successful implementation of scientific management approaches. For example, *PERSONAL CONTACT* (P_T 24),* *CREW MANAGEMENT* (P_T 24) and *PROJECT TEAM SPIRIT* (P_T 37) are critical policies of Crew Management with respect to scientific management techniques and their influence on the performance of many functions (see Figure 10.13). Consequently, at the Total Field Autonomy level field construction managers become more sensitive to the management of craft and time relationships through scientific management techniques.

Because of the large number of complex multiple work face environments, and the increasing field management pressures brought to bear by the Total Field Autonomy approach, field supervisors often devise techniques to control the location and work task involvement of their crews and crew members. As a result of these pressures, Crew Location Charts (I_T-DOC 12) similar to the one presented in Figure 10.8 begin to emerge. These charts may take a number of different forms, from marking facility plot plans with colored pins representing crews, to using an individual pin to represent a specific crew member. This type of technique is particularly effective on projects with large crews operating either in large physical areas or congested areas, such as a process unit with a high density of process equipment and piping located at many different elevations.

*See Table 10.6, Policies of the Autonomous Project Team.

CHAPTER 11
Work Execution

Work is the necessary means by which construction progress is achieved. Work requires the physical application of human effort and skill, albeit often augmented mechanically by the effective operation of plant and equipment. All construction effort requires human work.

Work itself to some is a burden to be endured, to be exchanged for the means of acquiring whatever the individual wants, be it food, shelter, pleasure, or rest. To others work is a challenge and an opportunity to develop and apply skills. Some view it as a physical workout rather like a gym training exercise, and they gain pleasure in hard physical effort and in the development of good physical condition. Others also see it as an opportunity to work with a team in an outdoor environment with team members who have similar skills. Many see work, especially construction work, as the means of producing facilities that are essential to society, national security, and so on, and thereby identify themselves with the project and its role in society. Some see it as the necessary role of man to subdue and order the earth. In this view, work has a moral, philosophical, or even a religious significance. Whatever the individual worker's attitude may be, it helps provide stimulus for physical activity, as well as the basic work site ingredient for the initiation, development, and maintenance of project tone and work relations on the site.

Thus, different attitudes to work usually exist in any collection of workers. The basic ingredients in the behavioral make-up of each worker are founded on the individual's cultural, moral, and religious background. Nevertheless, the individual's attitude to work is strongly influenced by the work environment, peer pressure from other crew members, and perceived management attitudes.

While these conditions vary from place to place, significant factors that mold the work site environment include the size of the construction firm involved, its management approach, and the size of the field work force itself. For instance, as work force size increases it becomes more difficult to maintain personal contact with and recognition of individuals. Thus, communication barriers may develop in a multi-tiered contractor organization unless perceptive and effective measures are taken to counteract them.

By recognizing these factors in the development of mutual respect, and in the implentation of better communication channels, the contractor can significantly improve the workers' attitude to work. This will, in turn, positively impact the field management of the construction project. Thus, Work Execution is a critical decision area that the contractor must appreciate if there is a desire to establish an effective project team and improve project tone.

11.1 THE BASIC DECISION PROCESS

A number of factors significantly impact work effort and its effectiveness. They can be classified into three main groups, as follows:

1. Those affecting the physical environment in which work is to be performed.

2. Those which establish the management environment under whose constraints work is to be performed.
3. Those which affect the mental hygiene and aspiration that basically form the individual's attitude and drive to work.

The actual physical environment in which work is to be performed often determines the nature of the construction method to be employed, and affects the discomfort and dangerous positions in which the construction worker is expected to execute work. In many cases, work postures demanded by a confined area of operation, coupled with the inconveniences associated with noise, dust, water, cold, and heat, severely limit productivity. These factors are aggravated by the natural caution associated with self-preservation in such dangerous situations, and the physical impediments to work efficiency associated with protective clothing, goggles, helmets, and large waterproof boots. Thus, the physical environment imposes upper limit constraints on what can be humanly achieved, unless their impact can be eliminated by the use of suitable construction methods, or can be minimized by perceptive planning and remedial and preventative work.

The construction worker is very much aware of the effectiveness of the contractor's planning, procurement, and field management efforts since these continuously impact the local work environment. If the craftsman or laborer senses that management is indifferent to work conditions, appears to be incompetent in the direction of work, and is subject to frequent changes of mind leading to the continual replacement of work already in place for whatever reason (whether improper drawing control, inefficient foremen, and so on), then the natural human result is for the workers to lose their original work drive. If these poor management conditions are not quickly rectified, field morale and the work productivity will plummet, with the consequential disastrous impact on construction quality, schedule, and costs.

The final area influencing work effort resides in the individual worker's mind. While it is very difficult, if not impossible, to change the mental attitudes of individuals, it is often possible to revive and maintain a worker's attitude to work by simple actions that in fact reflect a change in attitude by management to the work force in general, and individuals in particular. Thus, when management seeks suggestions on work simplification or method improvement, and gives praise for work well done, it is obvious to workers that management is responsive to field work conditions. These management actions provide the threshold factors that can change a worker's attitude to the manner in which

TABLE 11.1 SCOPE OF WORK EXECUTION PROBLEM AREA

Major Decision Problem	Component Problem Statements
35. What is work environment:	Is the work face clear? What is work face access? What hazards are involved? Must I/should I wear protective boots, clothing, goggles, helmet, respiration device, etc.? What safety rules must be enforced? What physical postures are required by congested work area? What physical efforts are required? What difficulties are involved in performing this work activity?
36. What is management environment:	Where is required material and equipment? Are relevant drawings available? Are these drawings accurate and self explanatory or must field decisions be made? Have I been adequately briefed? Does field management really know what has to be done? Are they going to change their minds? Who really makes the decisions? Is the foreman capable of effectively working under these conditions? What is job steward's view on this activity?
37. What do I have to do:	What has to be done now? How am I going to do it? Who will help me? Where are the tools, equipment, and material that I will use? What is to be done next? Will we keep the same work teams? Can I move onto the next work task as soon as I am ready, or must I wait for the foreman's direction?
38. What is the work motivation:	Am I working automatically? To do just enough to avoid disciplinary action? To exercise skills? To show that I can do a first-class job? Because I am interested in working? To gain approval from my workmates, foreman, and others? Because I am a member of a good crew? Because I am building this job and identify my effort with the finished product?

TABLE 11.2 GENERAL PROCEDURE FOR WORK EXECUTION

Procedural Step	Management Duties and Tasks
1.	What is the next job and location?
2.	What are site access and work face conditions like?
3.	What hazards are involved and what safety procedures must be followed?
4.	How is crew to be worked and managed?
5.	What is my involvement?
6.	Do I know what to do next and how to do it?
7.	What do I need to carry out the work?
8.	Have I laid out my tools and material competently around the work face?
9.	Am I working skillfully, safely, and with consideration of others?
10.	Am I thinking about how I am working?

work is addressed. The outward signs of a good worker are clear. They include careful preparation for work, proper care of tools, promptness under work direction, pride in skill and quality of work achieved, and a cheerful, willing, and cooperative team attitude. It is this work environment that ensures success.

The scope statement of the Work Execution problem area is given in Table 11.1, and the General Procedure for the Work Execution decision process in Table 11.2. The functional plot for Work Execution is shown in schematic form in Figure 11.1, the full detail in Figure 11.2, and related to other decision processes in Figure 11.3.

11.2 FUNCTIONAL PERFORMANCE TECHNIQUES

Work is the cutting edge of construction. It is the threshold for productive activity, backed up by a complex technical activity-oriented organizational system in which a whole series of interacting and compounding influences affect and limit the attainable efficiency with which work can proceed. This situation is not unlike the proverbial chain, which is as strong as its weakest link. It is often the case that work effort is unduly constrained by inadequate preplanning, and tardy and incompatible supportive efforts, or that it is misused by inefficient field management direction. Thus, every examination of work functional performance techniques must consider the environment in which the individual worker is expected to perform.

The work performance of an individual depends first upon the intrinsic difficulty of the task to be undertaken by the crew, the role given the individual within the crew, the natural talents he or she possesses, the degree to which the individual subscribes to his or her allotted role, and the extent to which the individual applies him or herself to the chosen course of action in preplanning the work task. The first two factors are dependent upon the skill of the contractor in designing and planning the operation, and on the foreman's ability to match crew members to the necessary roles implied by the operation. The third factor is a function of the contractor's skill in recruiting suitably trained personnel and in allocating them to suitable crews. Thus, only the last two factors relate to the individual's acceptance of the role chosen for him or her, and the extent to which he or she is instructed in its performance.

Construction work is inherently dangerous. Consequently, considerable attention must be given to the minimization of hazards, and to the adoption of safe working practices. The contractor has an obligation to all involved to avoid accidents. Thus, the subject of safety is one of noncontroversial mutual interest between management and the work force. Nevertheless, personal discomfort is often the excuse given when safety procedures are not adopted. Recent OSHA legislation, however, has implemented a formal program of mandatory safety practices to which all firms are bound. Figure 11.4 illustrates typical areas in which workers can be advised and reassured concerning management's intention to comply with safety procedures.

Generally speaking, most of the problems encountered on sites relate to the task environment and the frustrations generated by absence of material, incorrect or inadequately detailed drawings, and large amounts of rework resulting from changes, equipment breakdowns, and so forth. There is little evidence to suggest that workers are not prepared to apply themselves and do a fair day's work for a fair day's pay. In fact, the opposite seems to be true. If this view is correct, the problems of low productivity are external to the worker and relate to organizational contexts, reward systems, and the relative disposition of rewards, communication flows, and the like. These factors, being

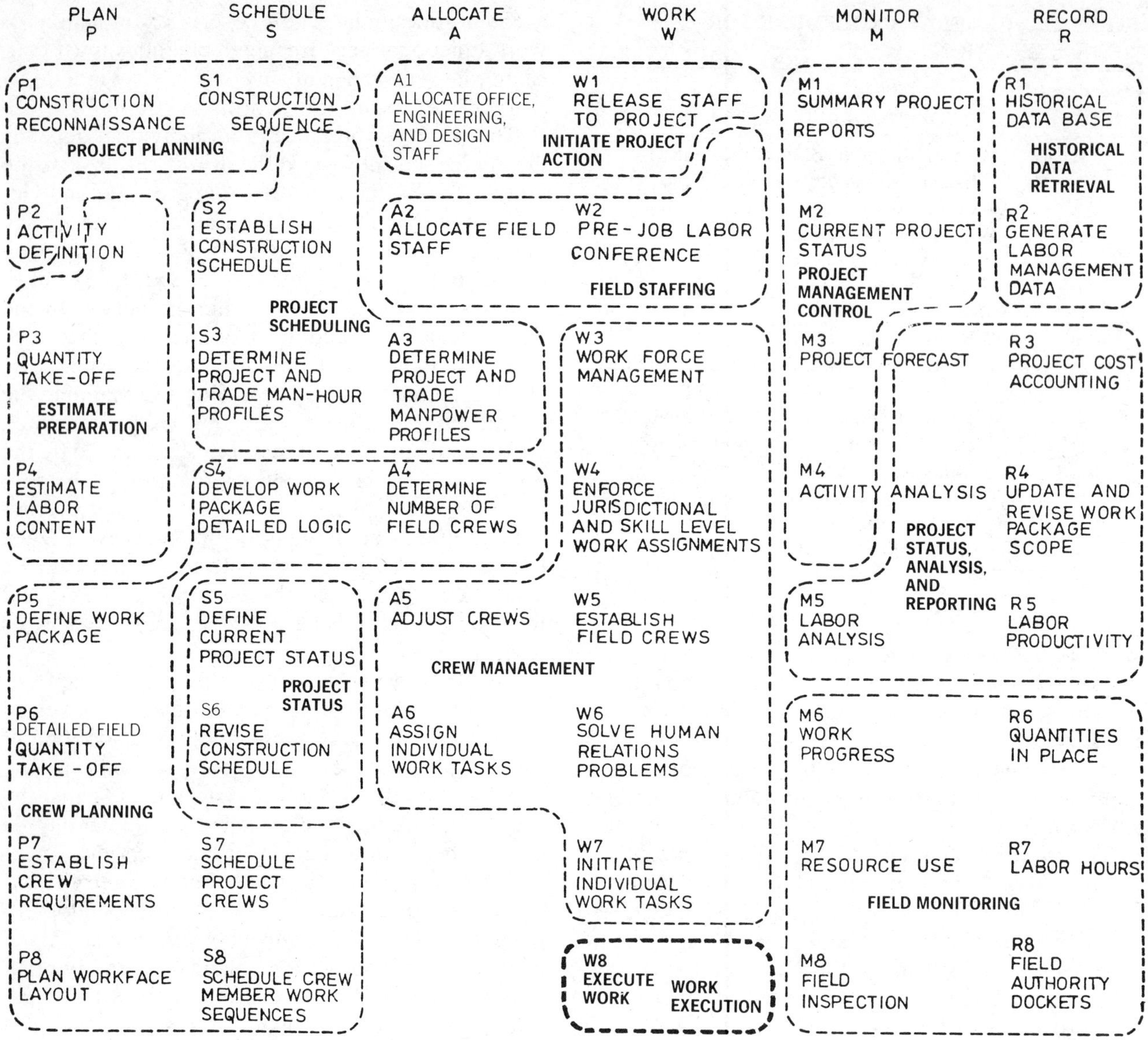

Figure 11.1 Functional plot of the Work Execution decision process.

man made, are much easier to change than the attitudes, values, or motivations of an individual worker. Thus, if construction workers appear to get more satisfaction from working in a well-run and effective organization than they do for one that is poorly run, their satisfaction in work performance may well be found to be a consequence of, rather than a prerequisite to, good performance. Thus, management interested in work performance improvement may find it more profitable to view the construction planning, mobilization, and work activity phases of construction in terms of how to improve productivity, rather than focusing on the behavioral aspects of motivating people as a solution to low work performance. Nevertheless, much can be done to motivate people on construction sites. This especially favorable situation exists in the construction environment because each worker at each level can see the results of personal effort, and can thereby readily identify with the constructive effort.

The study of human behavior is the focus of behavioral scientists. Many theories have been brought forward and elaborate experiments conducted to explain human behavior. Since man is a complex animal, theories of behavioral science rarely have a rigorous formulation, are generally

W8 EXECUTE WORK

- The achievement of high quality productive work in an efficiently directed, safety conscious, cooperative manner.
- Knowledge of construction method performance, the proper use of tools and equipment, the careful handling of materials, a consideration for the work of others and a care for finished work, the awareness of dangerous situations and of effective safety procedures, a willingness to act under direction, to work with others in a cooperative manner, requires motivated and dedicated labor.
- Consider personal involvement in work sequences, nature of the physical environment in which work is proceeding, constant awareness of hazards and locations of moving equipment, need to implement safety procedures, demands of the crew and work team groups, advantages of hard hats and protective gear, positive aspect of giving good example, awareness of good practice and of fatigue levels in self and others, advantages of cooperative companies in avoidance of personal conflict situations in crew.

Figure 11.2 Functional statement of the Work Execution decision process.

subject to contrary interpretations, and at best give qualitative concepts. Maslow's theory that man is a wanting animal and his need to satisfy these wants is hierarchical in nature is an attempt to explain some facets of human behavior. Figure 11.5 illustrates the basic premises of the theory, and also contains a hierarchical statement of man's needs as Maslow sees them. Herzberg has proposed a mental-hygiene theory (see Figure 11.6) based on factors that lead to job satisfaction and dissatisfaction. This theory proposes that improved worker motivation hinges on the elimination of factors that lead to worker job dissatisfaction, while at the same time factors that lead to job satisfaction are strengthened.

Skilled workers gain satisfaction from applying their skills to a challenging problem. This is especially true if they are required to work in front of observers or with others in a close-knit work group. The group situation offers a sense of belonging and of being able to work well with others. Group members earn one another's praise, and so gain the security and fulfillment that comes from knowing that others can rely on one. These are positive motivators, and the generally healthy environment of

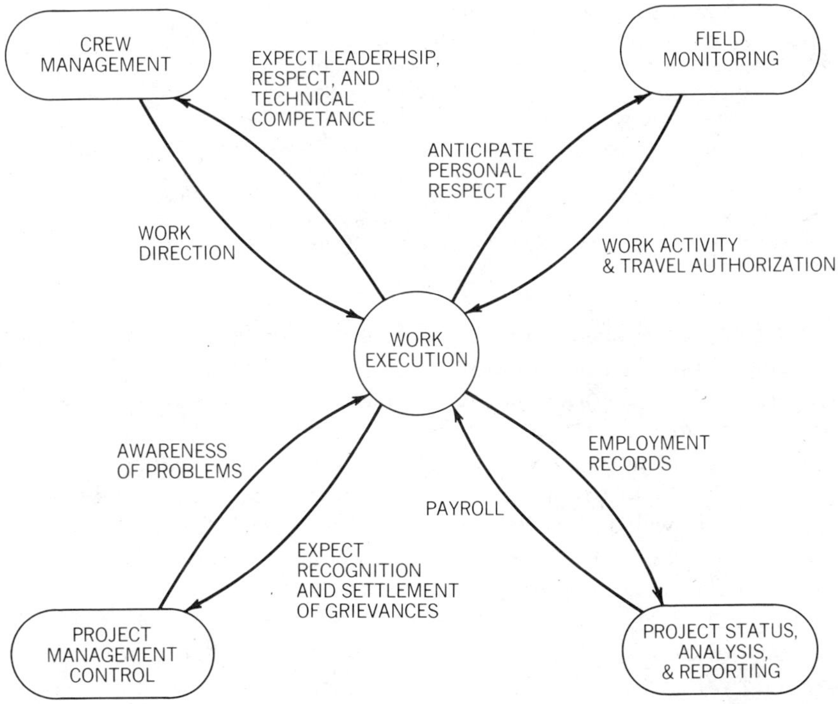

Figure 11.3 Schematic illustration of Work Execution decision process interfaces with other manpower management decision processes.

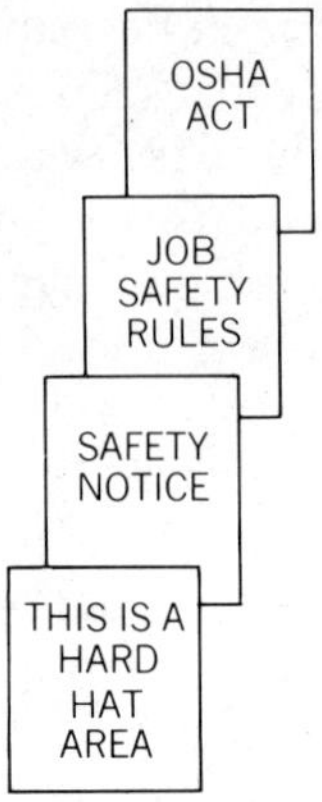

(*a*)

Re: OCCUPATIONAL SAFETY & HEALTH ACT 1970 (Construction) (OSHA)

Employers, Owner, Contractors, Subcontractors, Superintendents or Foremen in charge shall not direct or permit an employee to work under conditions which are not in compliance with the above code.

Every employee shall observe all provisions of the above codes which directly concern or affect his conduct. He shall use the safety devices provided for his personal protection and he shall not tamper with or render ineffective any safety device or safeguard.

Overhead Hazards—All employees shall be provided with *HARD HATS* and shall use HARD HATS.

Falling Hazards—Every hole or opening in floors, roofs, platforms, etc., into or through which a person may fall shall be guarded by a barrier sufficient to PREVENT FALLS.

Projecting Nails—Projecting nails in boards, planks and timbers shall be *removed,* hammered in or *bent over* in a safe way.

Protective Eye Equipment—Eye protection shall be provided by employers and *shall be used* for cutting, chipping, drilling, cleaning, buffing, grinding, polishing, shaping or surfacing masonry, concrete, brick, metal or similar substances. Also for the use and handling of corrosive substances.

Protective Apparel—Waterproof Boots where required shall have safety insoles unless they are the overshoe type. *Waterproof Clothing* shall be supplied to the employee required to work in the rain.

(*b*)

Cranes & Derricks—All cranes and derricks shall be equipped with a properly operating boom angle *indicator* located within the normal view of the operator.

Every derrick and crane shall be operated by a designated person.

A copy of the *signals in use* shall be posted in a conspicuous place on or near each derrick or crane. Cranes and derricks shall have a *fire extinguisher* attached.

Trucks—Trucks shall not be *backed* or dumped in places where men are working nor backed into a hazardous location unless guided by a person so stationed on the side where he can see the truck driver and the space in back of the vehicle.

The above items do not encompass all the construction safety regulations as they pertain to OSHA, but are intended as a guide to the ever present hazards and primary causes of accidents in our industry.

(*c*)

Figure 11.4 Typical safety notices. (*a*) General safety notices. (*b*) Specific personal and work area safety standards. (*c*) Specific equipment safety standards.

outdoor construction activity provides an ideal environment in which each worker has the opportunity to achieve these goals. Basic theories that relate to group behavior, the emergence of group leaders, and the qualities of leadership are presented in Figure 11.7 (see also Figure 10.4).

Indicators that a worker is well motivated and responsive to the work environment include positive physical behavior patterns such as prompt arrival on site, prompt response to direction, efficient layout, use, and care of tools, constant production of good quality work, and willingness to give assistance to others. Obvious negative behavior patterns include sullenness, slowness to respond to all manner of requests, generally lax behavior in the care of materials and approach to work, and unwillingness to initiate new action or offer suggestions.

Considerable management attention is being focused on the need for management to get involved with the work force, to open up channels of communication, and to find ways of encouraging the workers to identify themselves more with the project. In many cases, workers will respond positively to such a simple but powerful stimulus as personal recognition and attempts to seek information or suggestions on work improvement by a supervisor.

Man is a "wanting" animal. And wants in turn become "needs" that man tries to satisfy. These needs are not only physical, to be satisfied by material things, but also psychological, that is, they act through the mind.

Highest Need	Need to self-actualize, to find self-fulfillment, to maximize one's potential
	Need for beauty, order, and symmetry
	Need for information, to know, to understand, to explore
	Need for self-esteem, respect, importance, independence, to be competent, to gain approval and recognition
	Need for belonging and love, to affiliate, to be accepted
	Need to be safe, secure, out of danger
Lowest Need	Physiological Needs—hunger, thirst, pain, etc.

Maslow's theory proposes that man first satisfies his/her most basic needs; once these are satisfied they are replaced by new needs and he/she is motivated to attain these. Needs once satisfied are no longer motivators, but there is always a new need to replace the old.

Figure 11.5 Maslow's hierarchy of needs.

A typical technique directed to this end, illustrated in Figure 11.8 is the project news sheet. A recent development aimed at improving communication channels on a work site and on reviving worker motivation and attitude to work on projects that have gone sour, is the so-called Productivity Improvement Program. A schematic outline of such an effort is given in Figure 11.9. These programs have been successful and very effective in highlighting basic management control defects on large sites, and in preparing and implementing methods for their resolution.

The intensely intimate and personal aspects of work are obvious. The individual crew member must bring his or her mental knowledge of what has to be done to fruition. The manner in which actual physical work results will depend on the physical ability of the individual, and his or her skill and motivational drive. While the individual's skilled application of effort to the work activity is in progress, he or she must be constantly aware of the location of other crew members, and of the continual movement of the (possibly) heavy equipment involved in the work task being undertaken, and be sensitive to the consequential hazardous aspects of the work environment. Thus, human action requires an intense informal and intimate integration of the many cognitive and sensory processes involved in construction activity. This mental effort constitutes the informal Management Information System at the Work Execution level.

11.3 THE PROJECT TEAM COMPONENT APPROACH

The small firm usually operates in a local region with projects that require small field work forces. Even when subcontractors are required, their field work effort rarely extends beyond that of a single crew. Thus, the work environment is conducive to the development of close personal relationships between the members of the field work force and field management.

The small firm is often a family business in which a small nucleus of tradesmen develop strong bonds with the company to the extent that the firm can often have nonunion labor in a strong union neighborhood. In most cases, the contractor and the job superintendent/foreman know virtually all members of the local unions by name, past performance, and reputation. Thus, the contractor has considerable control over the hiring of workers. Local contracts often include clauses allowing a contractor to call for specific tradesmen by name provided they have a previous employment history with the firm. In

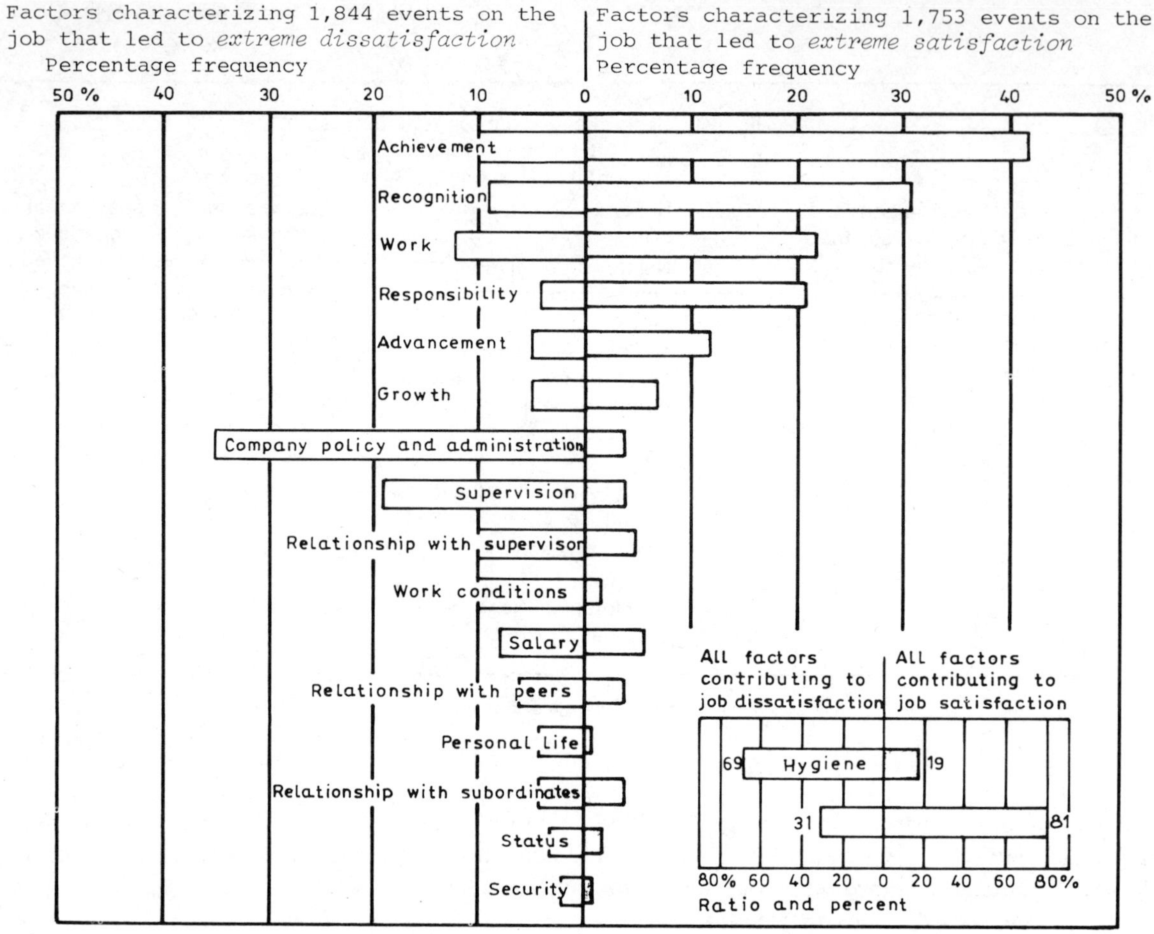

Figure 11.6 According to Frederick Herzberg's Motivation–Hygiene Theory, job-related factors that produce worker satisfaction are different than, and separate from, factors that produce worker dissatisfaction. Thus, worker needs are divided into two categories: motivators that produce job satisfaction through job content variables and hygiene factors that relate to the job environment and produce dissatisfaction. Reprinted by permission of the Harvard Business Review. An exhibit from "One More Time: How to Motivate Employees?" by Frederick Herzberg (January/February 1968). Copyright © 1968 by the President and Fellows of Harvard College; all rights reserved.

GROUP NORMS	Are standards for acceptable behavior within a group. Group norms are either developed or adopted by group members to a reasonable degree and provide criteria for guiding an individual's behavior in relation to what is expected. Group norms influence individuals to conform. Whilst attitudes apply to individuals these attitudes and behavior patterns must closely equate to group norms if the individual wishes to conform to the group and be accepted or feel belonging.

Figure 11.7 Typical behavioral science theories for group behavior.

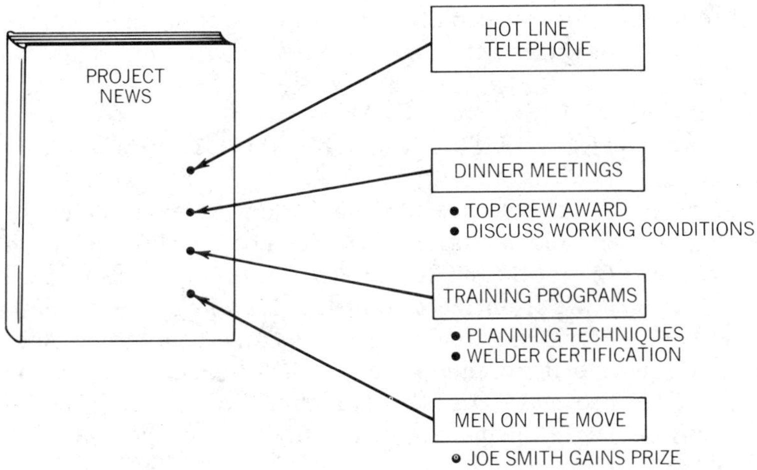

Figure 11.8 Typical items in project news sheet.

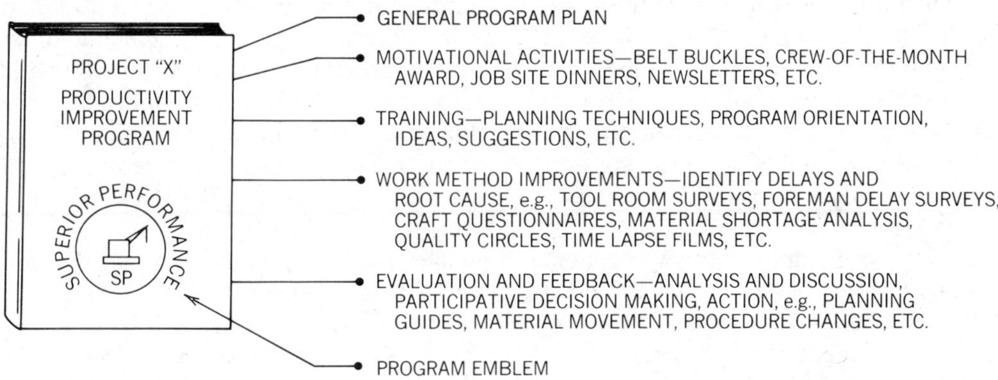

(a)

DELAY QUESTIONNAIRE	
DELAY CAUSE	*LOST TIME*
WORK OUT OF PHASE	0.5 HOURS
DRAWING ERROR	1.0 HOURS
CHANGE BY SUPERVISOR	0.0 HOURS
. . .	

(b)

PLANNING SHEET						
ACTIVITY	DRAWING		MATERIALS		CREW	
	ISS.	REC.	ORD.	REC.	SIZE	MIX

(c)

Figure 11.9 Productivity improvement program. (*a*) Program details. (*b*) Typical delay survey. (*c*) New crew planning document.

addition, with the exception of the occasional transient tradesman, all workers live locally, intend to remain in the neighborhood or county, and generally share the same community mores and attitudes to work. Individual tradesmen are therefore not likely to jeopardize future job opportunities, will be prepared to work consistently to quality standards, and will refrain from wildcat actions unless a strong union-backed grievance is involved. However, in this intimate environment it may be difficult for an outsider to gain access to jobs. Individuals who are loners must therefore rely on higher skills and performance to acquire continuous work.

The small firm work environment offers many inducements for the skilled journeyman. Since small work forces are involved, there is a variety of work over a wide spectrum of experience in which the journeyman may have full control and work on his or her own initiative. There are frequent opportunities to interact with other trades, especially with other subcontractors, who will similarly be present in small numbers working under their own initiative with little supervision. Thus, there is a high sense of achievement. A natural consequence of this work environment is that the journeyman is known personally by name to the entire work force and to the contractor himself. Hence, skilled journeymen gain recognition and praise and can usually win steady work with the same firm.

Most small projects are run by a superintendent or a foreman with little interference from the head office, except for the occasional management-by-exception visit by the project manager. Hence, the field supervisory staff will be able to mold the field work environment to suit their management style in an intimate personal labor management mode. This intimate work environment ensures that many opportunities for the development of easy communications will exist. Provided the superintendent/foremen are sensitive to the individual needs of work force members, many of the factors that lead to worker dissatisfaction will not occur.

The small contractor therefore has many things going for the firm in the work environment. This is often reflected in higher levels of productivity and increased confidence in meeting work progress milestones. The small firm attempts to maintain continuity of work and to minimize layoffs and rehiring situations if at all possible. Finally, small contractors cannot afford to be complacent in their attitude toward local workers since the firm's future professional existence is intimately tied to the skills, availability, and attitudes of a small number of tradesmen.

11.4 THE TRAVELING PROJECT MANAGER

A number of factors make the work environment of the medium-sized firm significantly different from that of the small firm. The most important of these flows from the role of the Traveling Project Manager, who continuously visits the field project site, and there enters into the field management of the project. Thus, the Traveling Project Manager injects top management authority and policies into the work environment and frequently will insist on the implementation of management decisions that can be readily seen as being due to his or her action. Immediate consequences of these situations are that the superintendent/foreman are no longer sole authorities on the work site. Attempts at establishing friendly relations with the foreman may be in conflict with head office's concept of discipline and of the correct project environment. In addition, unless the project manager has an interest in establishing personal contact with individual foremen and workers, his decisions may be seen to be impersonal and threatening. The project manager's role, however, is to ensure that head office policies and decisions are implemented. Provided this is accomplished tactfully and in a manner that maintains the authority and respect of field staff and produces results that are clearly seen to be effective and positive, the professional tone of the project team is strengthened.

Since the medium-sized firm can handle much larger and more varied projects than the small firm, the required size and skill level of the work force significantly increases. Thus, the work becomes more interesting and exacting, and calls for more experienced foremen than those normally found on small firm projects. A consequence of this situation is that individual workers may be attracted to such projects because of the greater learning opportunities they present, and also because they may desire work requiring higher levels of skill and quality control. Hence, it is much easier for a very skilled tradesman to pick and choose the type of work he or she desires. Another advantage for workers in this environment is that these projects may still be managed in a personal and intimate fashion, since it may still be possible for every person in the work force to be known individually. While the composition of the work force may no longer reflect a local environment, the contractor will still be able to seed the field work force with favored foremen and workers who have strong company allegiance.

The work environment of the medium-sized firm offers similar inducements for the skilled journeyman as does that of the small firm. Additionally, the size and complexity of the projects met at this level presents the journeyman with a greater variety of skilled work in his or her craft area, with occasional work in other areas. The craftsman has more security since the company, having more projects, can often readily schedule work activities from job to job. This increased continuity of work, coupled with the knowledge that the individual is personally wanted on a variety of jobs within the county or region, gives him or her an enhanced sense of achievement. Finally, since the company is much larger in size than the small firm, more opportunities exist for advancement to the foreman and superintendent levels.

The ability of the field management team to control work in the field situation will be influenced by the availability, experience, skill, and motivation of the field workers, and the extent to which they can be integrated into the project team. The project work environment can be significantly improved if mutual respect between head office management (in the person of the Traveling Project Manager in cooperation with the superintendent and foremen) and labor is established, and if management effort is directed to the elimination of needless conflicts and to the improvement of the work environment. In most cases, labor responds to perceptive and effective field management effort that is based on respect. Thus, the role of the Traveling Project Manager and the skill with which he or she implements head office policies and decisions are major determinants of the quality of the work environment that will exist on the project site.

11.5 FIELD-BASED PROJECT MANAGEMENT

The typical project in the large firm's portfolio requires a large field work force. Work forces of between 100 and up to 3000 are not uncommon. The immediate consequence of this work environment is that personal and labor management focuses cannot extend far beyond the crew level. In addition, since a considerable field management effort must be mounted, the project team will require many foremen and more tiers of management. Thus, effective barriers against communication concerning the work force environment emerge between those at the work face and the top field managers. Unless great care is taken, management decisions will often be misinterpreted and workers will view tighter field authority as unresponsive to their needs, and will see decisions affecting their work sequences and productivity as being impersonal and coldly automatic.

On large projects, the work force will often be comprised of workers from many different cultural backgrounds, and will be affected by the frequent replacement of transitory tradesmen. Thus, there is a tendency for the crew to lose cohesion and common interest unless it is fortunate enough to have a competent and perceptive foreman. The combination of the above factors means that it is difficult for the foreman to establish a crew team spirit, and must therefore be continuously involved in efforts directed to strengthening team cohesion.

The work environment for the skilled journeyman at the large firm level is significantly different from those at the small and medium-sized firm levels. While many projects offer job security for periods of up to three years, this is not a dominant concern if dissatisfaction arises because of poor field management. On fast-track industrial construction projects it is often difficult to keep the work up to each individual crew. Thus, the journeyman must spend more time getting materials and tools, and may even have to look busy when in fact their is nothing to do. If the journeyman is highly skilled and motivated, he or she will leave and gain ready employment on another job with the more satisfying work environment of a smaller project, probably in the medium-sized firm category. Thus, a natural result of large projects may be that the skill level of crews diminishes. This is because the workers remaining on the job may be those who do not mind slack time, and maybe would have difficulty in getting work elsewhere. While job security exists for considerable periods of time on large projects, there is little chance of being transferred to other projects because these may be located in geographically remote areas. Often the skilled journeyman is well educated, more financially secure, and established in a region from which he or she will not readily relocate. Thus, the journeyman does not always identify with the large contractor nor look to it for future employment or security.

The large work forces engaged in projects normal

to the large firm provide a highly visible and vulnerable target for militant unions and unionists in some countries. In these cases, unless issues are quickly resolved, a hotbed situation can easily develop that can so disrupt the working environment as to make effective work almost impossible. Some contractors have reacted to this environment by extensively subcontracting out work through brokering practices so that their own work forces are much smaller than normal. The transitory nature of the subcontracting crews diminishes the target for militant unionists. In addition, if the individual subcontracting crews have intimate ties with their company due to common ethnic backgrounds or because they are shareholders, then it may in fact be the case that their field productivity, company allegiance, and resulting pride in their work provides such a working environment that the workers themselves will eject the poor quality craftsmen or individuals who do not meet the standards of work established by the crew.

A more significant factor impacting the work environment of large projects is the apparent inability of management to solve crew support problems. This is especially true for fast-track projects where design detail changes often slow down construction effort. In many cases management is unaware of the extent of their loss of management control of the work force, and of the impact of repeated dismantling and reerection of field work on worker motivation. The inability of competent tradesmen to achieve production because of errors and problems that are beyond their control leads to lower productivity and morale at the same time that management is becoming aware of the impact of low productivities on costs and schedules.

On large projects, management faces the significant problem of efficiently supporting the work force effort of many crews, because productivities and worker motivation quickly deteriorate when support systems are not effective. This problem area requires a great deal of management attention. Also, in order to counteract to some extent the impersonal working environment that often results, it is necessary for management to get involved and to devise ways in which the individual worker can get involved with the project, and can feel that there exists fairly responsive channels of communication concerning important issues. If management wishes to be effective in this regard the way is open for the elimination of needless conflict in confrontation situations, which if anticipated early enough, require only a little understanding and initiative for their resolution.

11.6 TOTAL FIELD AUTONOMY

Management of the work environment on very large projects is possibly the most difficult and critical of all problems faced by field management. The huge impact of poor work face management on cost and schedules directly affects the client and lowers the client's confidence in the contractor's ability to perform. Consequently, contractors working in this project management approach strive to be constantly aware of indicators that reflect inefficient field support systems, lowered productivities, and reduced worker motivation. Managers are aware that by the nature of complex projects in the fast-track environment, managers themselves are often to blame for inefficient field operations. Thus, in this field management environment management is usually committed to constant surveillance of the problem area, and may even canvas the work force through questionnaires as to what remedial actions they deem appropriate and possibly effective. It is important, therefore, that workers be aware that management is endeavoring to rectify matters, and is striving to inject a concern for the working environment of individuals into its management approach.

The work environment for the skilled journeyman on very large projects is such that it is easy to lose personal identity, motivation, and productivity. The sheer magnitude of the project, the size of the work force, and the multi-level field hierarchy are intrinsically demotivating to the individual craftsman who takes pride in his or her work. In addition, the nature of the work itself is often very repetitive, so rework is often needed through no fault of the journeyman. Also, the individual's contribution can hardly be located or identified to others on the project as something personally built. This demotivating work environment can to a great extent be remedied by the adoption of geographic or zone control areas. In this way, it is possible to imitate the intimate work environment of the small or medium-sized firm by assigning the journeyman to one control area only. If this personal atmosphere can be maintained and a variety of work assigned in lieu of the same old repetitive activity, then renewed motivation and improved productivities result and the skilled journeyman can regain a sense of achievement and pride in good workmanship.

As mentioned previously, many public-relations focused and motivational improvement techniques are relevant to the work environment problems commonly met on large scale projects. Project news sheets, project outings (picnics), family visits to the project, regular meetings of top management with crews, hot lines for worker suggestions and complaints, are some things that management can provide in order to initiate action to improve the work environment and raise worker motivation. Whether these methods are successful or not hinges to a large extent on the ability of the contractor to exhibit competence in the preplanning of field operations, and on the effectiveness with which these operations are established and supported at the work face. In the last resort, the individual worker responds to the opportunity of using acquired skills and in proving that he or she can perform.

CHAPTER 12
Field Monitoring

The Field Monitoring of construction activity encompasses the day to day observation and evaluation of crew performance and activity progress, and the field management responses to daily problems that are subsequently detected. In addition, Field Monitoring is the decision area where critical project information relevant to field resource utilization is first documented for later analysis during project status and reporting. This decision area is considered the first level of overall project field control, and is an area in which the contractor must rely heavily on the skill and technical abilities of field managers and supervisors, including general and working foremen. A critical management problem that Field Monitoring addresses is project management's need for continuous and timely technical feedback on the effectiveness of the work effort.

On small projects, where the small number of field crews can be easily located and observed, Field Monitoring is based heavily on the personal contact that field managers and supervisors have with craft labor. These employees are normally on a first name basis. On these projects field management is normally immersed in the project, and personally monitors field activity at all times. As project size increases, field managers can no longer maintain personal contact with each crew, and must therefore work hard at being informed of project activity status and labor performance. Once the project grows to a level where it is impossible for one or two managers to effectively comprehend all field work activity, the project is often split into manageable areas such as control zones. The key field managers must then rely heavily on other field personnel to inform them of project activity. They will be able to inspect and review critical problem areas on a management-by-exception basis only. It is in this field environment that more perceptive Field Monitoring efforts are required, since cost and schedule impacts of poor planning, worker dissatisfaction, and productivity drops can be significant. Thus, management must rely on special techniques and suitably trained staff to maintain an effective contact with field conditions and field performance.

The documentation of field activity is relatively straightforward on projects with small labor forces, few geographical areas, and limited complexity. The need for documentation on these projects is generally limited, provided that the supportive knowledge the field managers and supervisors obtain through visual observation of the work in progress and labor performance is sufficient to enable them to successfully control the project. On larger and more complex projects such personal and interpretive understanding is difficult. Thus, the controlled use and movement of equipment and labor around the site, and the formalized documentation of its progress, becomes essential. Consequently, there is an increasing reliance on documented data, and integrating and summary systems emerge to make the control of field labor and work progress more effective.

Field Monitoring is critical for three major reasons. First, it is the decision phase which alerts those responsible to take corrective action to resolve technical, resource use, productivity, attitudinal, or personal problems that may impact the work effort. That is, it has an important troubleshooting focus. Second, by monitoring field performance top field management can assess the technical competence and effectiveness of the field managers and supervisors directly responsible for the field work effort. Finally, it provides top management with a means of assessing the competence of the field management staff and their ability to cope with unforeseen field situations.

12.1 THE BASIC DECISION PROCESS

Field Monitoring is a decision area where the skill, technical ability, past experience, and "gut" reactions of field managers and supervisors characteristically play a critical role in problem solving as work effort takes place. These decision makers are continuously reviewing work quality and progress, crew performance, crew morale, construction methods, and so on, on a daily or even hourly basis, via visual observation of field performance and other indicators of resource use. Their assessment of the work effort and ability to detect problems and initiate corrective action is the essence of Field Monitoring for most contractors.

The Field Monitoring process is also the source for documentation of current resource utilization. This aspect of Field Monitoring does not often receive the attention it warrants. Superintendents, and especially foremen, generally have little interest in the paperwork associated with Field Monitoring. In some construction environments, this attitude is probably acceptable because the size of the project warrants only those types of documents necessary for payroll purposes and progress payments. Top management, through periodic site visits, is generally well aware of the status of a project. On larger projects, where a large complement of manpower is used, the documentation of field data becomes essential to any understanding of the effectiveness of project activity. This documentation is the first line of defense for the early detection of changes in general performance indicators that can alert management to emerging problems. It is the basic input to top management decision processes. This is because it is no longer sufficient to walk the job to fully comprehend the project's position. Consequently, field data, once processed, provides top management with the only means of ascertaining the position of the project against relevant planned progress and status indicators. Thus, on large projects, establishing an effective Field Monitoring process built around an integrated

TABLE 12.1 SCOPE OF FIELD MONITORING PROBLEM AREA

Major Decision Problem	Component Problem Statements
39. What is being done:	Who is doing what? Is it being done correctly? Is sufficient skill evident? Are materials, construction equipment, and tools sufficient to keep the crews busy? Are workers idle?
40. What has to be done:	Just how much has been done? Has it been done correctly and at an acceptable quality? Who has been doing this work? Is it sufficient? How widespread is fall-off of site activity and performance? How can the desired output and quality of work be achieved? How will I assess the work that has been done? Can I remember the details, or will I write it down?
41. What field data:	Where are the raw data? How do I get them? How accurate are these data? What will I do with them? What level of detail is required to support project statusing, analysis, and reporting?
42. What documentation is required:	What records must I keep? Who worked on each task? How many hours were spent on each task? What cost code is required? Are work quantities in place reports adequate for labor productivity analysis and work progress evaluation? How will labor productivities and work progress be reported in the field? Are worker activity or attitudinal surveys required?
43. What workforce is available:	What is today's workforce? How many are absent? How many are late arrivals or early departures? How many have called in sick? Who has been terminated?
44. What work site control is required:	How many man-hours are to be charged to work packages? Who authorizes movement from work locations? What manpower effort was required to relocate?

TABLE 12.2 GENERAL PROCEDURE FOR FIELD MONITORING

Procedural Step	Management Duties and Tasks
1.	Visual inspection of individual crew member activity, work progress, and quality of finished product.
2.	Visual inspection of crew performance and resource use.
3.	Troubleshooting detection of problems.
4.	Preparation of time cards, plant hire reports, quantities in place reports, and percent progress reports.
5.	Control of field work force movement, equipment movement and materials requisitions.
6.	Consideration of effectiveness of location of material dumps, tool boxes, and construction equipment, and factors affecting smooth work progression.
7.	Observation of crew activity and attitude to work, and the effectiveness of the foreman's control of the crew.

documentation system is a critical problem for management.

Field Monitoring is a highly manpower- and labor-oriented decision area in contrast to other decision areas. The more specific manpower and labor focus of Field Monitoring is illustrated by the component problem statements and major problem area breakout depicted in Table 12.1. Similarly, the breakdown of the general procedure for Field Monitoring in Table 12.2 reflects a more specific and obvious listing of management duties and tasks that bear on manpower and crew output. The summary of management duties and tasks shown in Table 12.2 compares almost directly to the six manpower management functions which comprise the Field Monitoring decision process, as shown in Figure 12.1. The functional scope statement of each manpower management function in this decision process, and the factors generally considered by frontline field decision makers as given in Figure 12.2, are almost self-explanatory. Taken together, these are equivalent to the scope of Field Monitoring as described in Table 12.1. Finally, the interface relationships between Field Monitoring and other relevant decision processes is illustrated schematically in Figure 12.3.

12.2 FUNCTIONAL PERFORMANCE TECHNIQUES

In the small to medium-sized construction firm environment, Field Monitoring dominantly focuses on field observations and visual assessments of crew activity and work progress as the primary management areas of importance to total project control. This aspect of Field Monitoring is classically the domain of superintendents and foreman, and the project manager in cases where he or she interacts closely with the field through frequent visits from the head office. As project size and complexity increase, the importance to top management of the data gathering function of Field Monitoring is paramount. Such data auditing becomes the grassroots level sensing apparatus and vehicle for overall project management and control. Because of this difference in perspective, Field Monitoring for large projects is not necessarily viewed as a direct hands-on management area from the overall project standpoint, but rather it encompasses the recording of work activity status and progress, and often many different aspects of resource utilization. In this capacity it forms the basis for project management decision making. Thus, the Field Monitoring decision process can be dissected into two different but interrelated phases, and the importance of each phase to overall project management can be directly related to project size and complexity, as illustrated in Figure 12.4. The concepts related to functional performance techniques of each of these phases of Field Monitoring will be presented separately in the remainder of this section.

Field Monitoring to most foremen and some superintendents, particularly on small projects, is related to the belief that direct field observation is the most reliable means of determining activity work progress. The acceptance and implementation of this belief provides foremen with almost complete autonomy when monitoring the field work effort. The approach, referred to here as the Direct Management Role of the Foreman, is schematically illustrated for a typical brick masonry activity in Figure 12.5. The foreman relies on visual observations to check actual progress which, when combined with recollection of past experience, determines whether or not satisfactory work progress has been achieved.

The field observation phase begins to uncover and penetrate problem areas when observations of

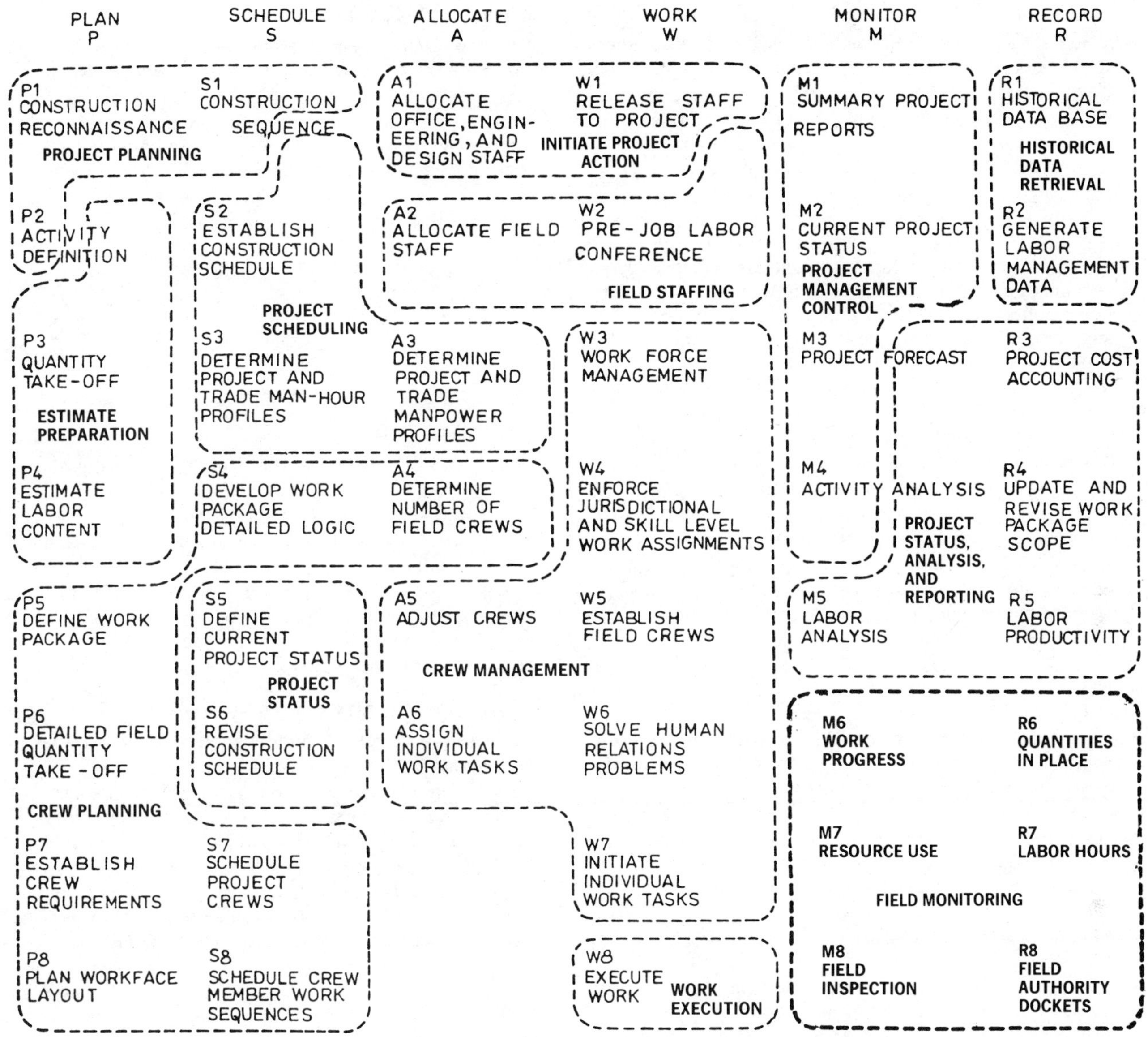

Figure 12.1 Functional plot of the Field Monitoring decision process.

work progress are combined with techniques used to evaluate work effectiveness of crews. Walking the job for troubleshooting purposes and inspecting recognized work activity problems are techniques used by superintendents and foremen. These techniques are both practical and effective for solving problems and evaluating crew and labor performance. As an example of this approach, the schematic depiction of a brick wall activity is provided in Figure 12.6. As indicated, crew performance is satisfactory early in the day (Figure 12.6*a*), but by 11:00 a.m. crew performance problems are observed by the superintendent (Figure 12.6*b*) as he or she walks the job. The foreman and superintendent interact when inspecting the work activity problem, and thereby evaluate the problem and select the appropriate solution (Figure 12.6*c*). The solution is implemented and by late afternoon crew performance is once again satisfactory (Figure 12.6*d*), as observed by both while walking the job. While one such problem-solving situation may not influence the success of the overall project, an accumulation of such walk-the-job troubleshooting incidents could have a significant impact.

M6 WORK PROGRESS

- Evaluate the percent progress and activity status for each work package and the project as a whole.
- Requires an intimate knowledge of project cost codes and project scope documents, including drawings and specifications, schedules, work package structure, construction materials, and work in progress.
- Considers either the actual quantities placed by measuring the physical output via direct quantity take-off of the material placed, logging quantities, counting quantities, or visually assessing quantities placed or by developing a system of labor units of work that represent the complexity of an activity and weights the activity according to its difficulty; activity status in relation to scheduled start dates and completion dates and percent progress.

R6 QUANTITIES IN PLACE

- Documents the physical quantities placed and/or percent progress achieved by work task and/or work package.
- Requires an intimate knowledge of project cost codes; project scope documents, including drawings and specifications; work package structure; construction materials; estimates; and pertinent recording documents.
- Considers the actual quantities placed through physical measurement and/or percent progress evaluation.

M7 RESOURCE USE

- The assessment of the utilization of field labor.
- Requires a knowledge of the work in progress, project cost codes, employees assigned to each work task, and the work performed by each employee.
- Considers the types of work performed, the actual hours expended to perform the work, the number of employees assigned to the work, craft type, and classification.

R7 LABOR HOURS

- Documents the manpower and man-hours assigned to work packages.
- Requires a knowledge of the work tasks, employees assigned to each work task, the project cost codes, and the pertinent recording documents.
- Considers the actual man-hours expended against a work task by each employee, the number of employees assigned to the work task, and the craft type and classification (e.g., foreman, journeyman).

M8 FIELD INSPECTION

- Visual assessment of crew performance, construction methods, quality of work, daily activity, progress, material movement and supply, and project morale.
- Requires an intimate knowledge of working conditions and work loads for each construction operation and field situation, desirable crew mixes and sizes, anticipated production rates and progress, union work rules and restrictive work practices, method improvement techniques and the ability to detect human conflict problems.
- Considers actual crew performance and progress in relation to planned crew performance and progress, optimizing crew design, problems in material movement, needless waste of effort, congestion at the workface, crew adjustments, and changes in construction methods.

R8 FIELD AUTHORITY DOCKETS

- Controls the movement and location of labor, material, and equipment.
- Requires an intimate knowledge of employees, their identification numbers, material warehouses and laydown areas, equipment locations, and pertinent documents.
- Considers the location and movement of each employee under a field manager, the type and location of material and equipment required for each work task, and the personal and company business of each employee.

Figure 12.2 Functional statement of the Field Monitoring decision process.

Field observations and assessments begin to encompass a broader management focus and interest when both project managers and general superintendents are involved. While the direct field observation of field activity becomes less frequent for these field personnel, the focus of their observations is specifically directed toward field management control in operation. The most common techniques in this area are simply for the field manager to observe and mentally assess how the assistant superintendents/foremen are performing as managers of the daily/weekly work activity, in terms of their timely identification and rectification of crew performance problems. Alternatively, project managers and general superintendents can analyze the information produced by field supervisors as a means of assessing their management effectiveness and control of manpower. Figure 12.7 presents several areas where the skill of assistant superintendents/foremen can be evaluated in terms of the in-

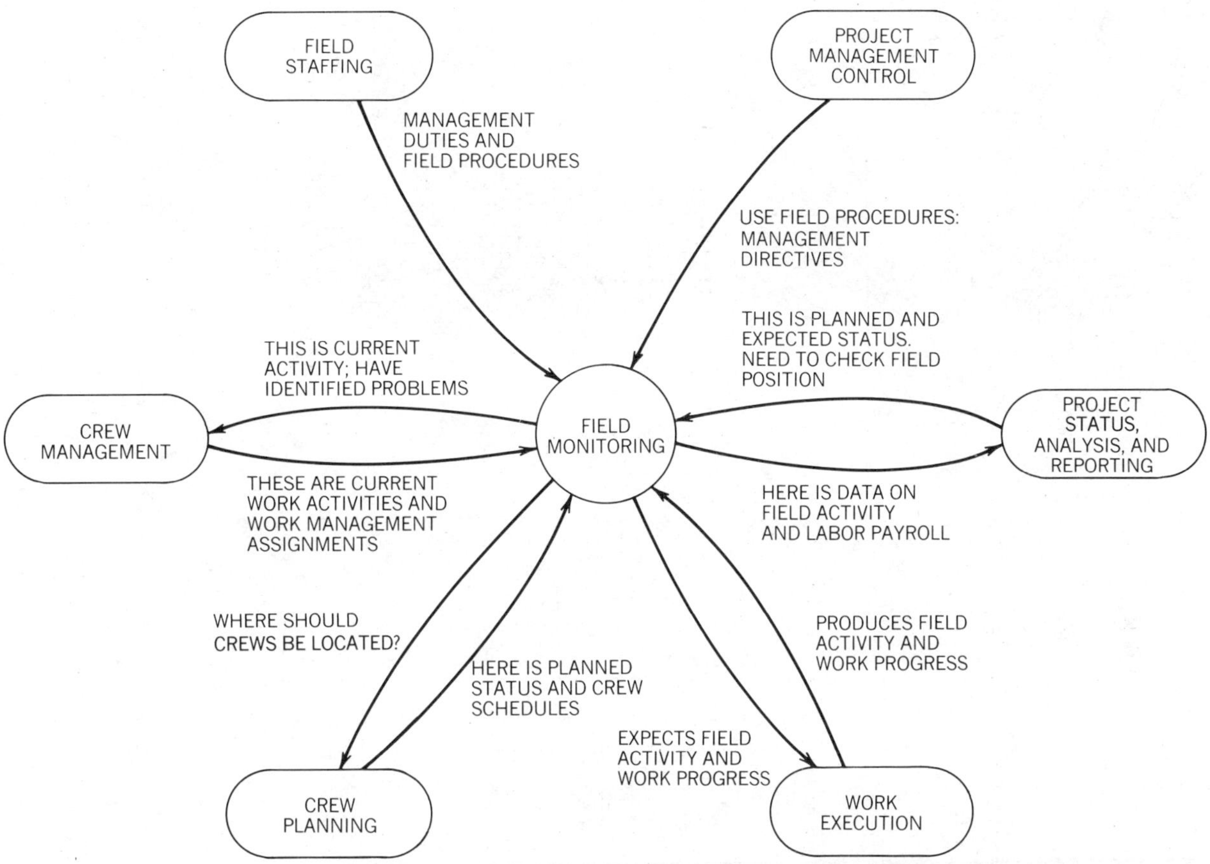

Figure 12.3 Schematic illustration of Field Monitoring decision process interfaces with other manpower management decision processes.

formation they produce in their daily and weekly Field Monitoring of work activity. Figure 12.7*a*, for example, depicts the statusing of activities in progress during the week. The bars are colored according to the estimated percent completion. This graphic display quickly indicates to the field manager which activities are ahead of or behind schedule, and shows the remaining work to be completed for any given activity. The weekly status report is often supported by a supervisor's delay log, as shown in Figure 12.7*b*, that identifies man-hours lost due to delays outside the direct control of the assistant superintendent and foreman. Finally, on projects with large work forces, travel authorization, as demonstrated in Figure 12.7*c*, is requisite to keep track of craft labor movement. Other techniques used to control the location of craft and measure total craft manpower on large projects include colored hard hats representing individual crafts or specific work areas, different colored employee badges, and so on.

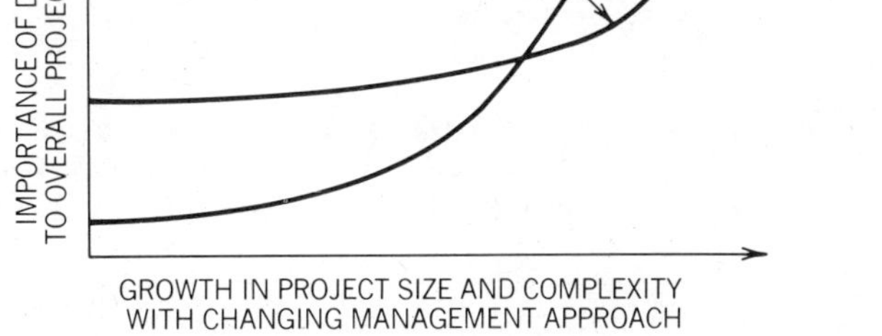

Figure 12.4 Field Monitoring decision process: importance of component decision phases in relation to the growth of project size and complexity.

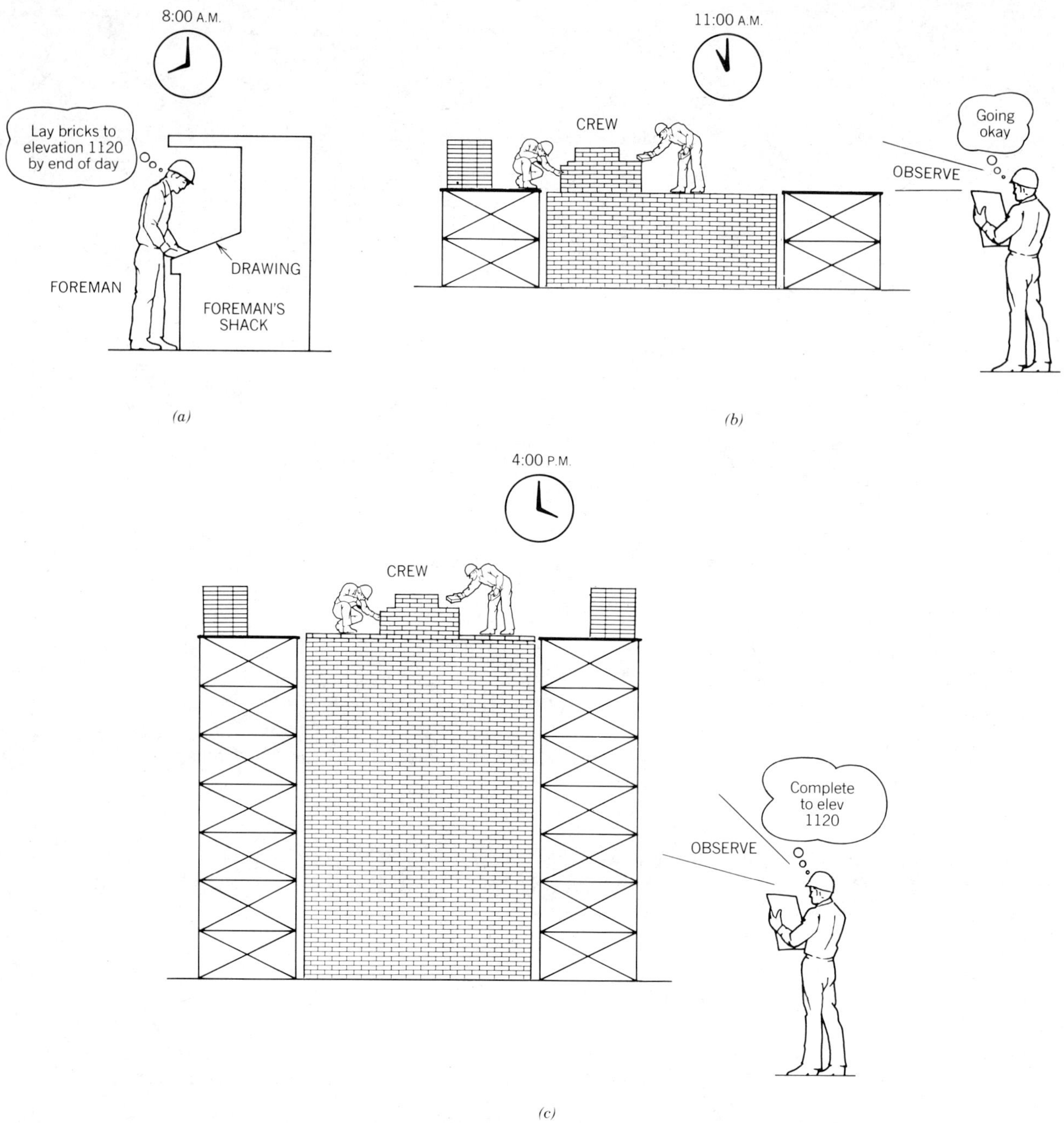

Figure 12.5 Direct management role of foreman in Field Monitoring. (*a*) Foreman determines progress requirements for crew on brick wall. (*b*) Foreman observes that work is proceeding properly. (*c*) Foreman observes that work was completed as planned.

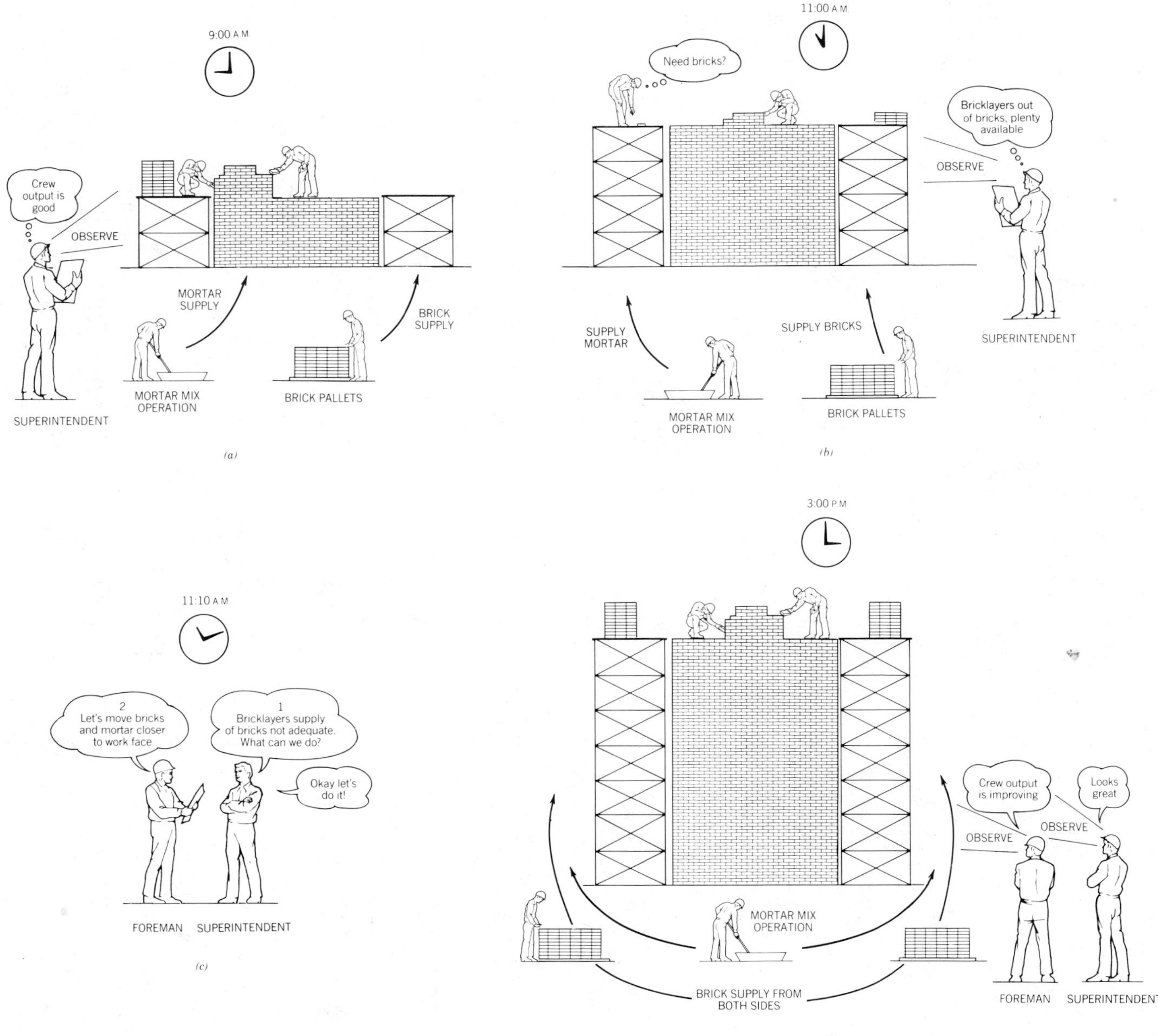

Figure 12.6 Walk-the-job troubleshooting. (*a*) Crew performance: crew members working. (*b*) Bricklayers' progress slows: supply of materials is not adequate. (*c*) Superintendent and foreman discuss and resolve problem. (*d*) Superintendent and foreman are satisfied with progress.

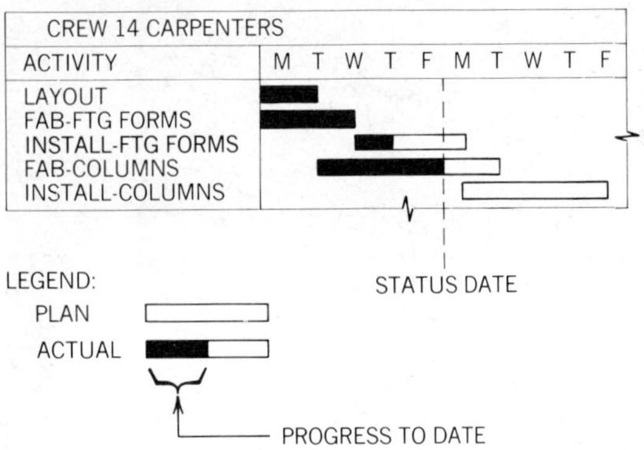

(a)

PROJECT: ______

FOREMAN: ______ WEEK ENDING: ______

CREW	LOCATION	DAY	ACTIVITY DELAYED	CAUSE	HOURS LOST
14	PROCESS AREA X	FEB 4	CONCRETE POUR	PUMP BROKE	24

(b)

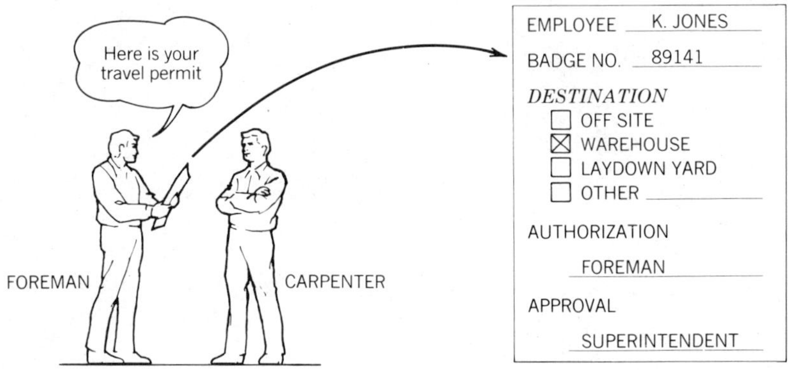

(c)

Figure 12.7 Project field monitoring information produced at the superintendent and foreman level. (*a*) Status of planned weekly work activities. (*b*) Activity delay report. (*c*) Field travel authorization.

Field observations and assessments become more specialized and penetrating when direct field observations and general purpose problem detection techniques, such as foremen delay reports and crew questionnaires, are focused on the detection of areas where preplanning has been poor, and where work is being performed inefficiently. This approach may be useful for activities with large crew sizes or those requiring several different crews and longer activity durations, and for highly technical work wherein the assistant superintendent and foreman can not readily assimilate the scope of an activity. In this instance, visual observations must be supported by field work method study techniques such as planning studies and time-lapse photography. The objective of these techniques is to aid front-line supervision in uncovering areas where better job assignments, planning, material handling, delivery, and material and craft coordination are needed. Figure 12.8 illustrates an example of field planning studies in which the observer notes the time spent by various crew members engaged in direct productive work (e.g., fabrication and erection), in delays of all kinds (e.g., personal delays such as taking a break), or in absences for whatever cause. By making repeated observations during different time periods, and identifying whether a crew member is working, delayed, or absent, the observer can quickly determine the percent time the crew spends working directly on the productive side of the activity, and on supportive work. These observations may suggest that, given a better management plan, the proportion of supportive work could be reduced. The results of the planning study

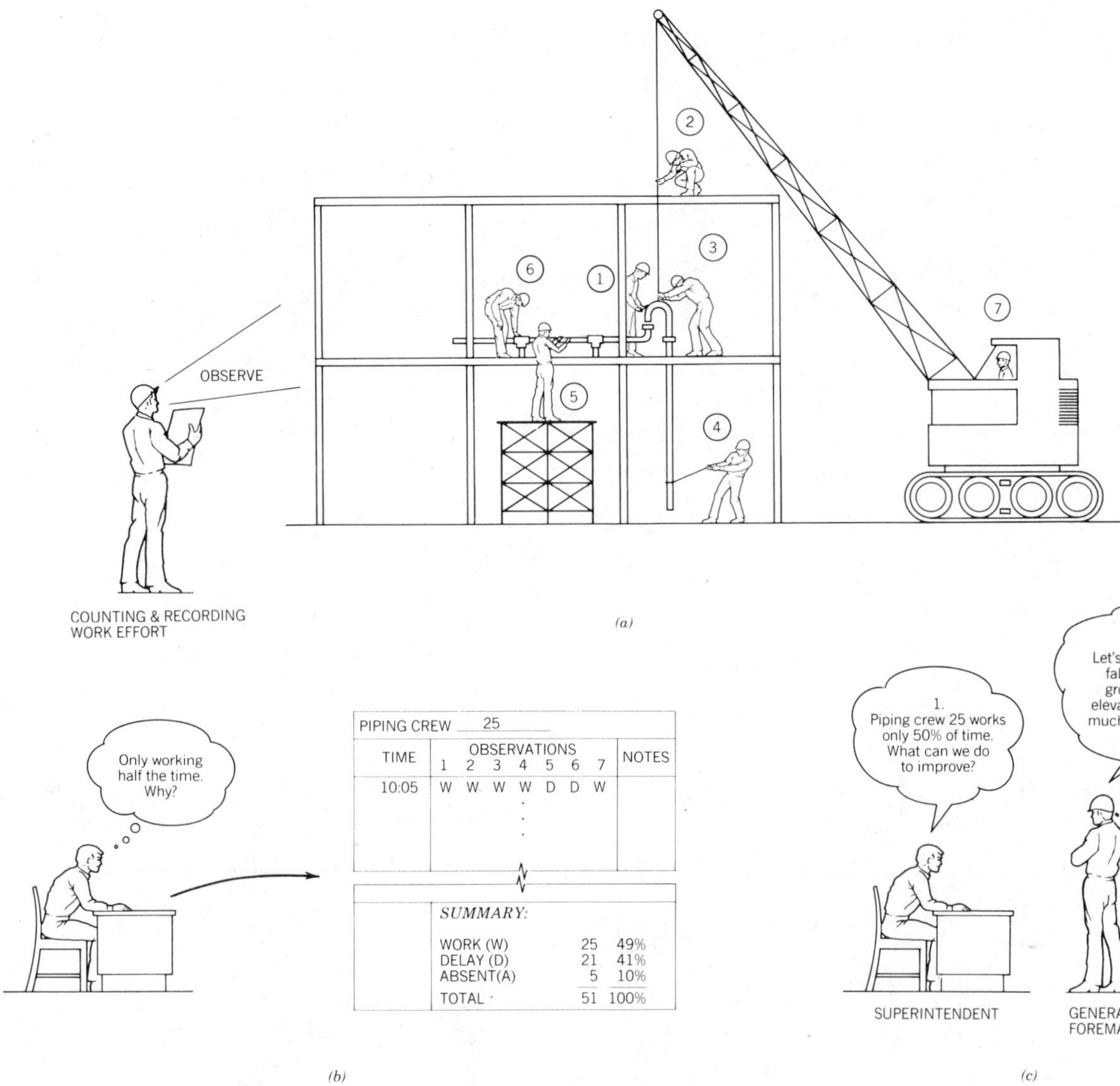

Figure 12.8 Field planning study. (*a*) Observe work effort. (*b*) Analyze data. (*c*) Discuss results and suggest improvements.

can be used to determine, for example:

1. The effectiveness of a particular preplan.
2. The verification of estimated production rates.
3. How to eliminate or expedite bottlenecks.
4. Method improvement.
5. The effectiveness of a methods study technique.

An essential part of the field observation and assessment of work activity is the recording of data for Project Status, Reporting, and Analysis. At all levels in the construction industry, a number of management functions related to field data documentation must be performed formally, either because of their importance to the contracting process, or due to the legal requirements arising out of the employment of labor. These requirements lead to the common use of field documentation of activities in progress as the grass-roots method of quantifying work progress and activity status in relation to the expenditure of labor and equipment hours and the physical placing of permanent facility materials. This approach is illustrated schematically in Figure 12.9, which depicts the four most basic areas

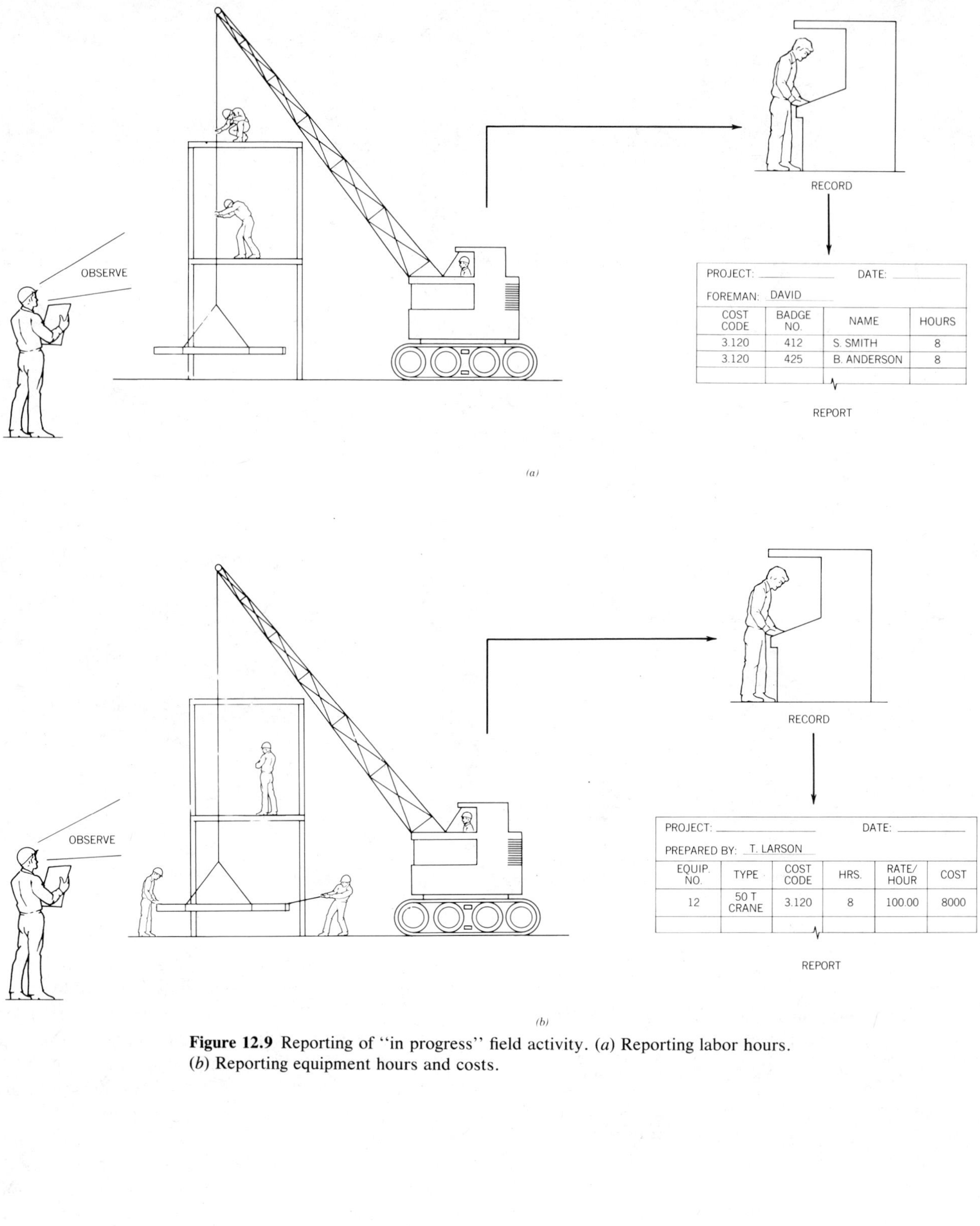

Figure 12.9 Reporting of "in progress" field activity. (*a*) Reporting labor hours. (*b*) Reporting equipment hours and costs.

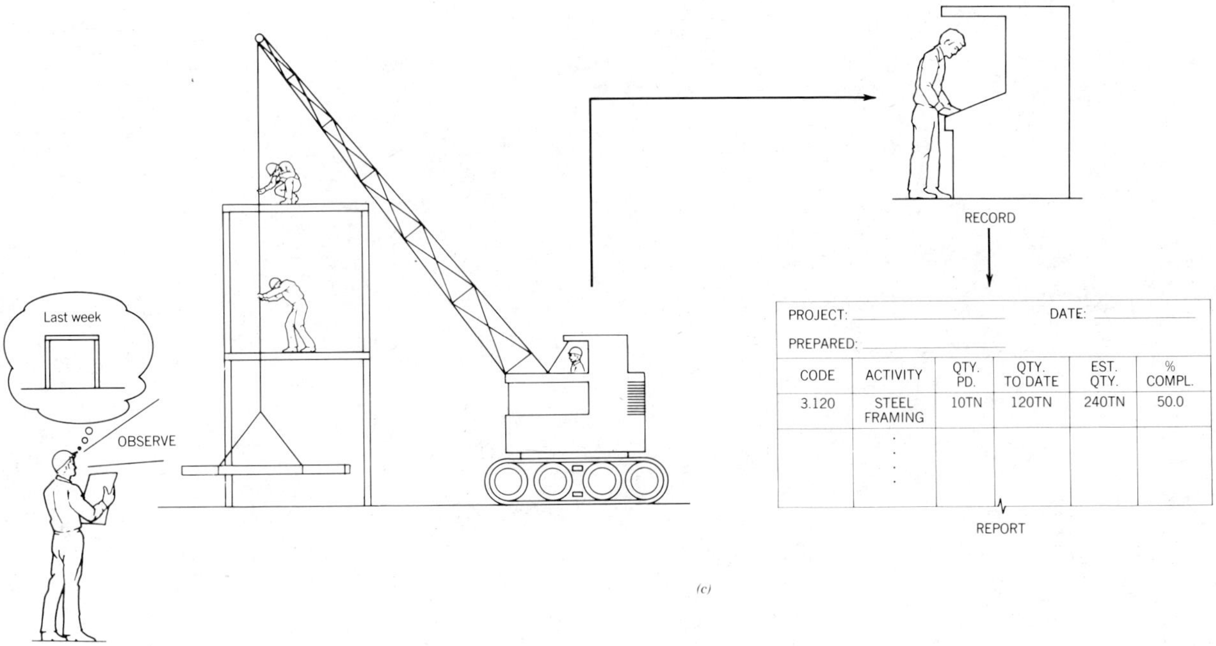

(c)

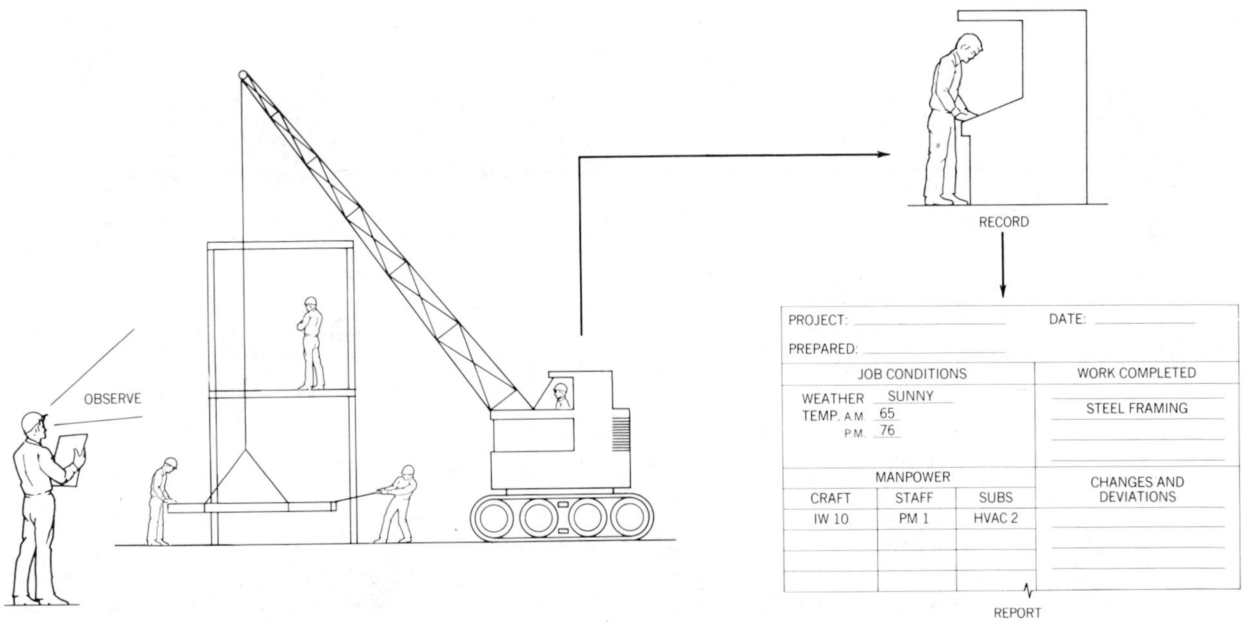

(d)

Figure 12.9 (*Continued*) (*c*) Reporting materials in place. (*d*) Reporting construction activity and status.

Figure 12.10 The extreme aspects of data gathering and processing systems. (*a*) Too little detail. (*b*) Garbage in–garbage out. (*c*) Too much detail.

where field work activity is documented. The information, level of detail, and format of these four reporting areas fluctuates widely in construction practice, and thus must be tailored by a contractor to fit its unique business practices and philosophies.*

A concept that is extremely important to the design and selection of field recording techniques and documents is that of balancing the accuracy with which field data needs to be documented with the requirements of its intended use. The significance of this concept can easily be grasped by analyzing the schematic depiction presented in Figure 12.10. While this figure represents the extremes in field data documentation and its importance to project status and progress reporting, it quite strongly emphasizes the need for the contractor to evaluate reporting requirements in terms of levels of detail, management's use of desired reports, the type of information provided by the reports, and the necessity that field personnel involved in data accumulation and reporting understand the importance of these efforts in the management process.

*Sections 12.3 through 12.6 will present typical documents related to these four areas, which are currently found in practice at the small, medium-sized, and large firm levels.

PROJECT: ______ DATE: ______

PREPARED BY: ______

CLASSIFICATION	CODE	TOTAL PAYROLL	PRESENT	ABSENCE	TERMINATED
STAFF					
MANAGER	12	1	1		
SUPERINTENDENT	10	4	3	1	
CRAFT					
CARPENTERS	08	20	17	3	1
STAFF		192	185	7	0
CRAFT		1621	1516	105	10
TOTAL		1813	1701	112	10

(a)

CONTROL ZONE	WORK PACKAGE	MAJOR CODE	DETAIL CODES	RESOURCE
(1)	(2)	(3)	(4)	(5)

(1) CRACKING UNIT
(2) STEAM SYSTEM
(3) PIPING
(4) ABOVEGROUND CARBON STEEL
(5) LABOR

(b)

LINE NO.	DRAWING NO.	SYSTEM CODE	QTY. LF	ERECTION		BOLT-UP		WELD		X-RAY		UNITS COMPL	% COMP	TOTAL UNITS
				%	UNITS	%	UNITS	%	UNITS	%	UNITS			
22-124-LF	50025	22AX	100	100	45		15	25	20		10	50	55	90
TOTAL		22AX	5210	28	5425	20	2050	21	3076	0	1055	2575	22	11606

(c)

Figure 12.11 Broad but detailed focused data-gathering techniques. (*a*) Detailed manpower report. (*b*) Detailed cost coding system. (*c*) Labor unit progress reporting system.

The requirements of field data documentation can range from the minimal data required to support the construction business, to the vast data-gathering system required to support highly specialized project reporting techniques. The use of a broad but detail-focused data gathering effort as essential input for a reliable project information system is most noticeable in large and complex project environments. This practice is often associated with rigorous techniques to support productivity, trending, construction method analysis, and the like. These techniques are frequently refinements of techniques previously introduced, such as the detailed manpower reports and control area methodologies shown in Figures 12.11*a*, and 12.11*b*, respectively, or in terms of a percent progress system in which the calculations are not directly based on quantities installed, as demonstrated in Figure 12.11*c*. In addition to these techniques, other scientific management approaches emerge including, but not limited to, methods study and work sampling techniques. Each of the techniques discussed can form the basis for specialized management reports, such as absenteeism and turnover reports, that can assist in the prediction of future manpower and productivity levels.

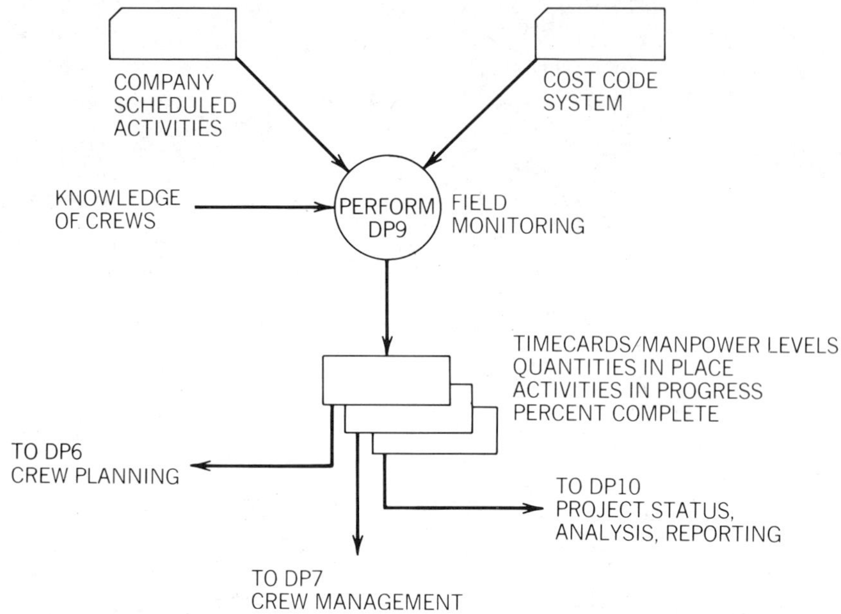

Figure 12.12 Field Monitoring Management Information System concept.

At the work face itself, the superintendent/foreman must be constantly observing and assessing the performance of individual crew members, while keeping close control of performance quality and of the safety aspects of the work environment. Field Monitoring at this level is an intense technical assessment effort that is focused intimately on labor performance. It constitutes the informal aspect of the Field Monitoring Management Information System, and consumes the greatest amount of time and effort in this work face aspect of the decision process.

In terms of the formal information flow process of Field Monitoring, the primary goals of this decision area are to collect and keep track of labor data and determine actual resource usage, activity status, and progress, all in relation to the crew activity/work face effort that is occurring over time at many different locations on a project site. The manner in which this is accomplished is determined in part by the project control philosophy, and therefore is highly correlated to project cost code systems and project scheduling approaches. The information flow and documentation of data produced through Field Monitoring provides a live database for Project Status, Analysis, and Reporting, as well as current status information for Crew Planning and Crew Management. This Management Information System concept is delineated in Figure 12.12 for the Field Monitoring decision process.

12.3 THE PROJECT TEAM COMPONENT APPROACH

In its most basic form, the observation and monitoring of construction rests with the tradesman and laborer. These workers are at the cutting edge of the construction effort and have the immediate responsibility of constantly observing the status, location, and quality of work that has been done in their immediate work station. They must be aware of how their own and other in-progress action is being performed, and know where they must position themselves in order to perform the next immediate work action. These tradesmen must necessarily focus on, or be aware of, the quality and quantity aspects of their own productive effort, as well as the work attitudes and skill levels of those around them. They will, for example, note the levelness of masonry work, the cleanness of struck joints, the defects in the surface finish of concrete, and the cooperative assistance of others as part of their ongoing responsibility as tradesmen. They will use these observations as indications of when they can move on to the next work station of the work face during the normal progression of a day's work.

Similarly, the foreman, whether acting as a working or nonworking foreman, has the key responsibility of monitoring and supervising the field construction effort and the quality of work performed. The foreman must constantly monitor the work of each member of the crew, in order to sense worker attitudes and response to directions and instances of spontaneous cooperation between crew members. He or she must also be prepared to evaluate the quality of work, and to issue corrective commands if the performance and quality of work are unsatisfactory. The technical observation and monitoring of the constructive effort is complemented and directed by the job superintendent if the project is sufficiently large as to warrant such a field manager, or by the occasional visit of the head office project manager.

The technical monitoring of field construction and crew activity, together with the continuous observation and evaluation of worker skills and attitudes, is considered the "nuts and bolts" of construction management. This field practice focus consumes most of the attention of field personnel.

Management demands on the foreman to monitor field work (which often detracts from the direct involvement in work), and to assess and report on construction activity (with its consequential demands for paperwork), are generally viewed by foremen as necessary but distracting chores. They are usually given as little attention as is required to keep head office management satisfied and maintain their own full autonomy and freedom of action in the field.

FIELD INSPECTION (M8) is the critical function in Field Monitoring, and must be performed continuously on a daily basis over the project's life. It demands from the superintendent/foreman quick and accurate evaluations of crew performance, construction methods, quality of work, activity progress, and project morale. This information enables the superintendent/foreman to act quickly to resolve problems or improve crew performance, and to make daily management changes in crew size and mix, construction methods, and material flow requirements, as well as to initiate action to solve observed work task assignment problems, personal grudge issues, human relation conflicts, and jurisdictional issues. The results of the visual evaluation of crew and activity status must be referenced against activity target completion dates and anticipated crew production, as set forth in CREW PLANNING REQUIREMENTS (P7) and CREW SCHEDULING (S7).

In addition to traditional daily field practice evaluations, Field Monitoring at the small firm level has both a labor use focus and a project status focus. Project status monitoring focuses on field data reporting and daily performance and progress analysis, while the focus of labor use is on labor man-hours expended. In its most direct form, field monitoring is achieved by the daily FIELD INSPECTION of crew action, from which the superintendent and/or foreman visually assesses both daily labor man-hour expenditures and quantity output.

The recording of data from these visual observation follows a predetermined basis. For instance, LABOR HOURS (R7) are recorded daily but summarized weekly, while QUANTITIES IN PLACE (R6) assessments of production are usually only tabulated weekly. The MONITORING OF WORK PROGRESS (M6) and RESOURCE USAGE (M7) provides the raw data for reporting to the head office.

Daily Time Sheets (P_S-DOC 5) are used to record labor hours.* Two examples of daily time sheets are shown in Figure 12.13. They typically include the employee's name, a description of the work tasks performed by the employee, and the hours expended on each work task. (The rate and amount are normally calculated in the head office.)

Apart from daily reporting of labor hours for each employee by the foremen, the job superintendent reports site activity on the Daily Construction Report (P_S-DOC 6). This document, exemplified by Figure 12.14, provides the project manager with a daily report of construction activity and includes such specifics as weather conditions, manpower totals per craft and per the total project, basic crew sizes, and the work being performed by each crew. This basic data is essential to the evaluation of work progress, and enables the project manager to keep abreast of daily construction activity without the necessity of visiting the site each day.

WORK PROGRESS monitoring is the responsibility of both the project manager and the job superintendent. (Foremen may also assume responsibility for monitoring work progress.) The project manager utilizes this function to stay informed of overall construction progress. The project manager makes periodic site assessment visits and, by *WALKING THE JOB* (P_S 18)† and making normal site inspections, helps evaluate progress of the ac-

*It should be emphasized that these time sheets are only representative of methods of time keeping. The actual document will vary from firm to firm.

†Small firm policies relevant to the Field Monitoring decision process are collected in Table 12.3.

Anderson & Woodhead
Building Contractors

PROJECT: ______ PROJECT NO.: ______

EMPLOYEE: JOHN SMITH

BADGE NO: 127 WEEK ENDING: ______

WORK DESCRIPTION CLASSIFICATION	MON.	TUES.	WED.	THURS.	FRI.	SAT.	SUN.	TOTAL HOURS	
								REG.	OT
WALL FOOTING FORMS— B. 100	8	8	8					24	
GRADE BEAMS— B. 110				8	8			16	
TOTAL	8	8	8	8	8			40	

(a)

Anderson & Woodhead
Building Contractors

EMPLOYEE: ALBERT JONES PROJECT NO.: ______

CLASSIFICATION: OPERATOR

BADGE NO: 103 WEEK ENDING: ______

MACHINE: BACK HOE #5

WORK ACTIVITY	JOB NO.	MON.	TUES.	WED.	THURS.	FRI.	SAT.	SUN.	TOTAL HOURS	
									REG.	OT
BACKFILL AROUND WALL FTGS A.130	515	8	8						16	
EXCAVATE GRADE BEAMS A. 120	515			8	8	8			24	
TOTALS		8	8	8	8	8			40	

(b)

Figure 12.13 Time Sheets (P_S-DOC 5). (*a*) Foundation operation. (*b*) Equipment operation.

tivities under construction. This is valuable input for project status reporting. The project manager, depending on specific company policy, evaluates the actual quantities placed during the desired reporting period on a basic policy document; the Quantity Distribution Report (P_S-DOC 7), a sample of which is shown in Figure 12.15. Quantities in place will be recorded on this document. Under the particular column, entries are then made for the quantity output achieved during the reporting period (usually one week) for each work task. The job superintendent will identify a work task by its cost code, using the Standard Company Cost Code (P_S-DOC 1), and will record the quantity of progress for that task. Quantity reporting is normally performed on a weekly basis to be consistent with payroll reporting. It is a requirement for defining activity progress effectively, especially when the contractor's payments from the client are fixed to a progress payment schedule.

Anderson & Woodhead
Building Contractors

PROJECT: ______________ DATE: ______________

LOCATION: ______________

ARCHITECTS: ______________ TEMPERATURE: ______________

SUPERINTENDENT: ______________ WEATHER: ______________

WORK CLASSIFICATIONS	FOREMEN	JOURNEYMEN	LABORERS	SUBCONT.	REMARKS
1. CLEAR AND GRUB					ITEM #6—BACK HOE OPERATOR
2. DEMOLITION					AND 2 LABORERS BACKFILLING
3. UNDERPINNING					WALL FOOTINGS
4. ROCK EXCAVATION					
5. COMMON EXCAVATION					ITEM #8—FORMING AND
6. BACKFILL		1	2		RE-STEEL WORK ON WALL
7. CONCRETE					FOOTINGS DWG #4510
8. CONCRETE FORMS	1	6	3		

Figure 12.14 Daily Construction Report (P_S-DOC 6).

The Quantity Distribution Report, Time Sheets, and Daily Construction Report, are generated from the field monitoring functions, and are forwarded through the job superintendent to the project manager for head office data processing.

Head office input flows into Field Monitoring through the FIELD INSPECTION function. This information flows from SUMMARY PROJECT REPORTING (M1), and consists of direct feedback from the project manager on activity status in terms of actual labor costs and progress. This kind of activity status information is prevalent only if a problem exists which requires an immediate field management response. In this case, the job superintendent will be informed if, for instance, an activity is showing a significant labor cost overrun

TABLE 12.3 POLICIES OF THE SMALL FIRM

Policy		Description
P_S 3	*USE COST CODES*	Systematic classification of work categories segregated by unique cost codes is used to identify work tasks and collect costs.
P_S13	*PAST EXPERIENCE*	Field planning, scheduling and management of manpower is based on a freewheeling ad hoc approach, with heavy reliance on past experience.
P_S17	*PERSONAL CONTACT*	Field labor management is not formally considered beyond maintaining good on-site relations via personal contact and recognition coupled with hiring and firing practices to weed out undesirable labor.
P_S 18	*WALK JOB*	In the field, activity progress and crew performance is monitored by visual inspection, and crew management decisions are based on subjective evaluation of observed progress and productivity versus planned progress and productivity.
P_S 19	*INSPECT PROGRESS*	Quantities in place versus total estimated quantities is an indicator of activity progress and evaluated by visual inspection or quantity survey.
P_S 20	*FIELD LABOR HOURS*	Labor hours are monitored and recorded by the management personnel closest to the workface.
P_S 21	*ACTIVITY STATUS*	Activity progress and labor productivity are not normally considered beyond completed activity volume, labor unit costs, and total labor dollars expended to date relative to activity-scheduled target completion dates, estimated labor unit costs, and total estimated labor cost.

Anderson & Woodhead Building Contractors				
PROJECT:		JOB NO.:		
ARCHITECT:		PREPARED:		
LOCATION:		WEEK END:		
WORK DESCRIPTION CLASSIFICATION	QTY. UNITS	QUANTITY LAST WEEK	QUANTITY THIS WEEK	QUANTITY TO DATE
DRAWING #4510 CONCRETE				
WALL FTGS—FORM B.100	SF	778	662	1450
WALL FTGS—CONC B.100	CY	0	32	32

Figure 12.15 Quantity Distribution Report (P_S-DOC 7).

or is behind schedule, and where corrective action can be implemented in a timely manner. The respective job superintendent or foreman must then, through FIELD INSPECTION, ferret out the cause of the problem, solve it, and take the necessary corrective steps through the Crew Management decision process. As mentioned previously, Field Monitoring has two basic focuses, namely the determination of project status, and of labor use. Field labor use data reporting is concerned with documenting labor hour expenditures, while project status data reporting focuses on quantities-in-place. These information flows provide the input for generating the company's payroll and cost reports, and for statusing the construction schedule in the head office. Daily performance and progress analysis is the critical problem of Field Monitoring, because it is essential for effective crew management in the field. It requires the job superintendent/foreman to make accurate and quick judgments concerning crew performance and activity progress, in order to identify and solve field problems.

The Field Monitoring decision process network is shown in Figure 12.16. This network can be summarized by its documented management information flows, as illustrated in Figure 12.17. The data input into Field Monitoring reflects the contractor's reliance on the knowledge of crew activity through visual observations combined with a comprehensive understanding of the current construction schedule and company cost code system. The execution of Field Monitoring (see Figure 12.16) results in field data output with actual man-hours, quantities, and status entered on the appropriate documents. The key document in the data collection effort of Field Monitoring is the Time Sheet, because it provides man-hour data for both payroll and cost control analysis, when transmitted to the Project Status, Analysis, and Reporting decision process.

12.4 THE TRAVELING PROJECT MANAGER

The Traveling Project Manager approach is characterized by an increased sensitivity to the merits of effective manpower management, which in turn focuses on the importance of better project scheduling and cost control. Specifically, on all projects, but particularly on larger ones, field supervisors are made responsible for collecting, processing, and reporting project information. This responsibility and management interest is reflected in the breadth and detail of information collected. This information, when properly reported, analyzed, and interpreted, can positively affect crew management by providing an objective basis for management action. Field Monitoring occurs daily and weekly, and penetrates all levels of project management in the field. It serves as the basis for making effective Crew Management decisions in the field, and provides documented input for reporting costs and evaluating scheduled progress.

While many manpower management decisions are based on project status reports derived from project data, it must be recognized that they are also made on a daily basis without reference to specific reports. These decisions are made in response to problems detected during the daily FIELD INSPECTION (M8) of crew performance and progress analysis, and are the responsibility of the field managers closest to the work (i.e., the job superintendents and craft foremen). These field supervisors rely heavily on past experience and common sense when making and implementing management decisions. It is possible that the proj-

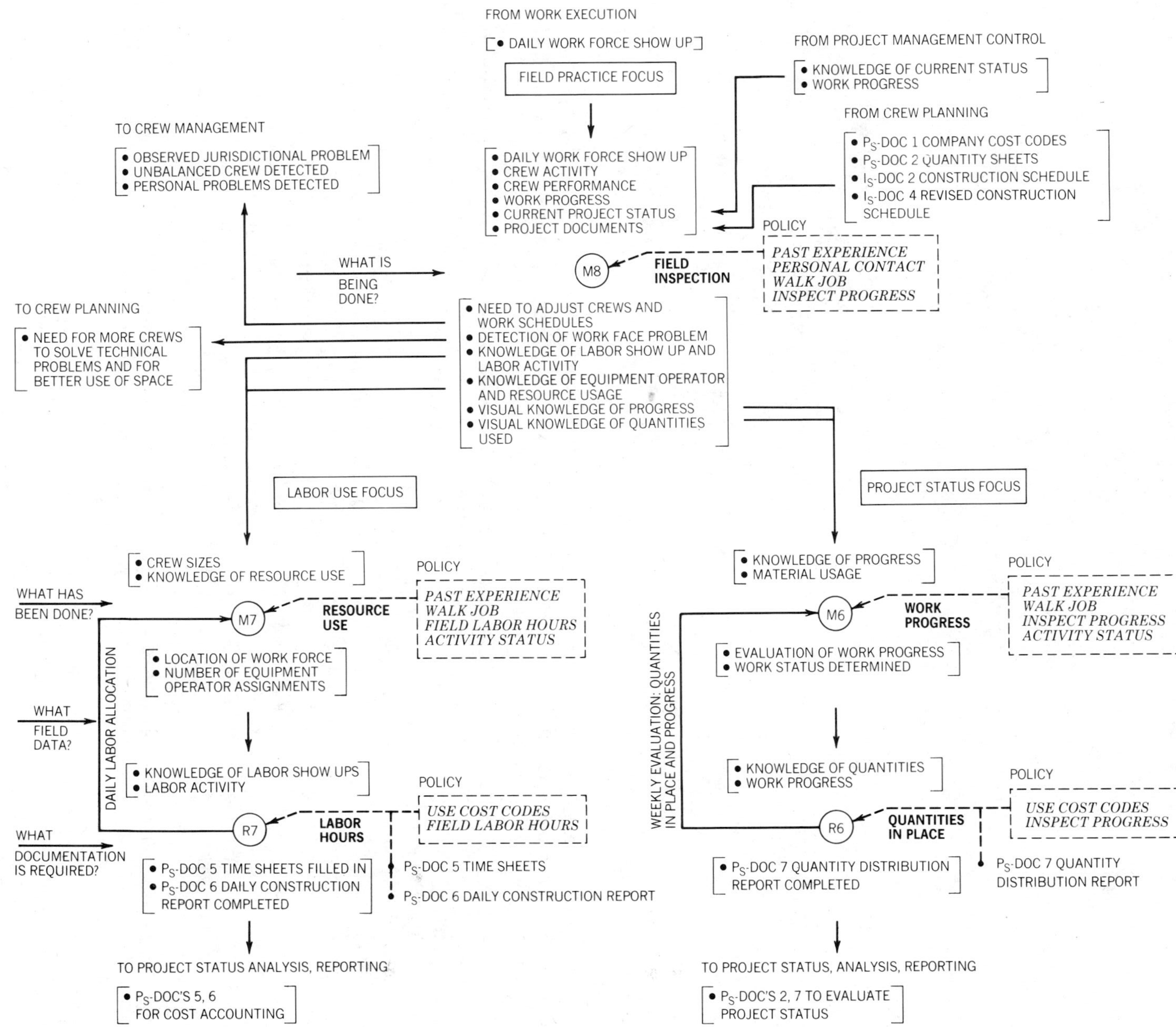

Figure 12.16 Field Monitoring decision process (Project Team Component approach).

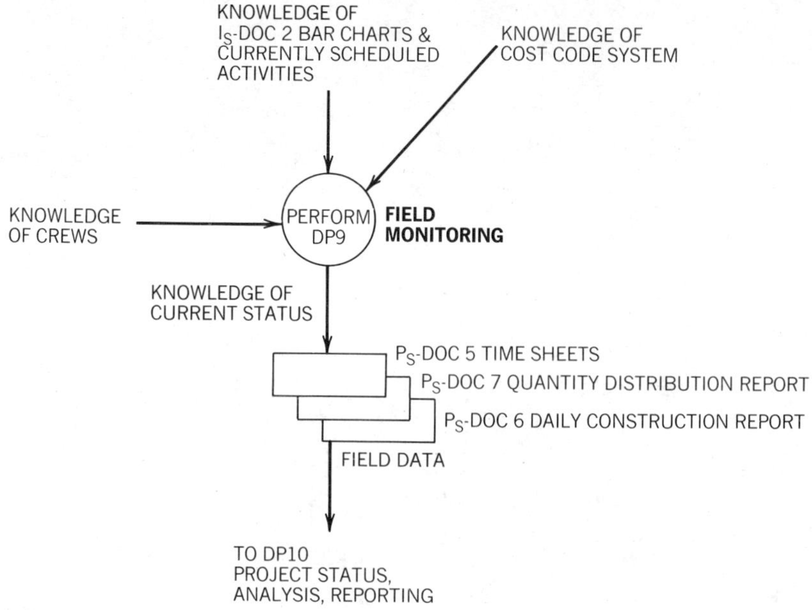

Figure 12.17 Field Monitoring Management Information System (Project Team Component approach).

ect manager will enter the daily on-site decision process if a critical problem develops. Regardless of the situation, however, the project manager is continuously apprised of significant daily field management decisions through continued interfacing with the key field superintendent and foreman.

The FIELD INSPECTION focus on an activity is determined in part by the crew selection technique used to design the crew size and mix, and the criticality of the activity. Noncritical activities for which crew design is based mainly on the historical crew size approach receive a normal FIELD INSPECTION emphasis. Crews for highly critical activities are monitored closely by the responsible craft general foreman and project superintendent. When either the trial-and-error or maximum crew size planning techniques are used, activity progress is often monitored daily, because finding the most economic crew size is dependent upon achieving optimum crew production. For example, when the trial and error approach is based on the premise that crew production (i.e., *ACTIVITY PRODUCTIVITY* (P_M 23)) is monitored regularly and the crew size and mix is modified as needed to achieve the best mix of craft for the activity in terms of each crew selection variable, close monitoring is a prerequisite. The superintendent/foremen visually evaluate the impact of actual site conditions, such as accessibility to the work face, in order to anticipate problems. If the work space is congested due to an unexpected manning overload, the crew size might be subject to revisions when planned productivity is not being achieved because of this congestion (see the *MANPOWER PLANNING* policy). Provided that both progress and crew performance are acceptable, the superintendent/foreman will not normally change the crew's profile. Activity progression is the main monitoring focus of the crew buildup approach. The superintendent and foreman must remain cognizant of crew production relative to the volume of work available, in order to properly increase crew size to maintain the level of progress consistent with the scheduled activity duration. The project manager (or field engineer on larger projects) often assists the superintendent and foreman by providing information concerning the percent complete for the activity, and the total quantities placed to date. The maximum crew size model necessitates frequent monitoring of crew production to achieve the optimum production for the given activity conditions. This means that the crew size and mix are manipulated frequently until optimum production is obtained. The responsible field supervisor must be cognizant of daily output for the activity. Optimum productivity is achieved when daily crew production reaches a constant level, allowing for slight variations due to weather, absence of a crew member, and/or interruptions of material

Anderson & Woodhead
General Contractors

PROJECT: ______ JOB NO.: ______

LOCATION: ______

PREPARED BY: ______ EMPLOYEE NO.: ______ WEEK END: ______

EMPLOYEE	CRAFT	NO.	COST CODE 0311.	0222.								TOTAL HOURS
J. SMITH	CARPENTER	75	8									8
R. ANDERSON	CARPENTER	123	8									8
F. JONES	LABORER	85	8									8
B. DAVIS	FOREMAN	53	4	4								8

Figure 12.18 Time Sheet (P_M-DOC 8).

flow. The information flow from FIELD INSPECTION enters the Crew Planning and Crew Management decision processes at various stages, depending upon the various possible outcomes of field inspections.

Field data reporting involves the monitoring and recording of actual labor hours expended and of physical progress. Monitoring labor man-hour expenditure begins with monitoring RESOURCE USE (M7). RESOURCE USE is the primary responsibility of the job superintendent and respective craft foremen. As project size and complexity increases the flow of information, a timekeeper might be employed solely to collect, record, and distribute man-hours and, in this capacity, assume responsibilities previously performed by the superintendent and foreman. Each day, *LABOR HOURS* (P_M 21) must be recorded and forwarded to Project Status, Reporting, and Analysis for payroll distribution and the analysis of labor costs to date. DOCUMENTATION OF LABOR HOURS (R7) is performed by the field supervisors described above. The policies guiding both of these functions involve the use of Time Sheets (P_M-DOC 8), as shown in Figure 12.18. Again, *DOCUMENTING ACTIVITY STATUS* (P_M 22) and *ACTIVITY PRODUCTIVITY* (P_M 23)* are the critical policies influencing RESOURCE USE and LABOR HOUR analysis, because each policy stresses the importance of realistic man-hour expenditure, evaluation, and documentation. The Time Sheet quite clearly identifies all man-hours expended by the corresponding cost code, which provides the basis for comparing actual costs against the estimate.

The documentation of daily construction activity and the correlation of that activity with LABOR HOURS is the responsibility of the job superintendent,† and is normally executed by visual observation of all field activity during FIELD INSPECTION and RESOURCE USE analysis. The job superintendent records daily project activity on the Daily Construction Report (P_M-DOC 9). A typical example of this document is shown in Figure 12.19. It includes job condition notation, contractor's labor force by craft, subcontractor's labor force if applicable, and descriptions of the work performed. The Daily Construction Report provides the project manager with basic highlights of daily activity that cannot be obtained through site visits. It is also a place for documenting problems or unusual site conditions which might impact project costs and schedules. Finally, the Daily Construction Report identifies the magnitude of the labor force for each day on the project site. All documented reports from LABOR HOURS are forwarded to the head office for processing and reporting of project status.

On smaller projects, monitoring physical progress is normally the responsibility of the project manager and superintendent. As project complexity and size expand, however, a field engineer will often be assigned responsibility for MONITORING WORK PROGRESS (M6). The project team member assigned this function visually observes *ACTIVITY PROGRESS* (P_M 20) by *WALKING THE SITE* (P_M 19), and analyzing the status of each activity in terms of the physical work completed as QUANTITIES IN PLACE (R6) since the last reporting

*The medium-sized firm policies relevant to the Field Monitoring decision process are collected in Table 12.4.

†On larger projects, the job superintendent would probably be assisted by the field engineer or timekeeper.

TABLE 12.4 POLICIES OF THE MEDIUM-SIZED FIRM

Policy		Description
P_M 4	*COST CODES*	Systematic classification of work categories segregated by unique cost codes is used to define work tasks, collect costs, and provide a means of comparing actual costs with estimated costs.
P_M 13	*MANPOWER PLANNING*	Short-range planning of manpower at the head office pre-job level, coupled with field labor manipulation, is critical to project control.
P_M 14	*PAST EXPERIENCE*	Field planning, scheduling, allocating and management of manpower is based on a freewheeling approach, past experience, and close interaction between the project manager and field management.
P_M 18	*PERSONAL CONTACT*	Manpower management begins with head office scheduling and allocation of manpower with field input, but primary emphasis remains on-site good relations via personal contact and recognition.
P_M 19	*WALK JOB*	In the field, daily activity progress and crew performance is monitored by visual inspection, and crew management decisions are based on subjective evaluation of observed progress and productivity versus planned progress and productivity.
P_M 20	*INSPECT PROGRESS*	Actual quantities in place versus estimated quantities is an indicator of physical progress and is evaluated by visual inspection and/or quantity survey.
P_M 21	*FIELD LABOR HOURS*	Labor hours are monitored by personnel closest to the workface and are recorded either by them or by personnel whose primary function is to record labor hours.
P_M 22	*DOCUMENTED ACTIVITY STATUS*	Labor costs, quantities in place, and activity status must be recorded and documented by a system yielding reliable information and viable reports for management control.
P_M 23	*ACTIVITY PRODUCTIVITY*	Labor productivity is an indicator of crew performance and is measured by activity labor dollar expenditures and quantity volume in place.
P_M 29	*PROJECT TEAM SPIRIT*	Active participation of all field managers in the management process promotes goodwill and a positive management environment.

period. This analysis is translated into a record of the actual quantities in place either by a direct take-off from the working drawings as determined by physical progress, or by estimating the percent progress of the total estimated quantity in place. Estimating the percent complete is the quickest means of evaluating progress, but it is also the least accurate. It is the best suited approach when either management time is limited or activity complexity simplifies assessing its percent complete. If a field engineer is responsible for this function, the quantity take-off approach is more suitable, because he or she has the expertise and time needed to undertake the analysis.

The Quantity Distribution Report (P_M-DOC 10) illustrated in Figure 12.20 delineates actual quantity output for each work task by their respective *COST CODES* (P_M 4). Cost codes are inserted for each item or work task to identify them, and to ensure that all quantities can be related to estimated quantities. Consequently, the Standard Project Cost Code (P_M-DOC 3) must be available to all responsible field personnel. The quantity report serves to document physical progress in terms of QUANTITIES IN PLACE for scheduling purposes, and furnishes information for calculating both actual labor unit costs and production rates in terms of labor man-hour per quantity unit of measure.

The Field Monitoring decision process network is shown in Figure 12.21. This decision process can be summarized in terms of document flow handling, as illustrated in Figure 12.22. At the medium-sized firm level the document flow for Field Monitoring would be identical to that for the small firm. The medium-sized contractor utilizes Company Cost Codes to a greater extent, however, and exercises greater care when meshing crew members with specific activities through the cost code system. When the medium-sized firm is using a computerized cost control system, the structure of the Quantity Distribution Report and the Time Sheets would be designed for direct data input into the computer (i.e., by key punching). This manipulation of field data (man-hours and quantities) would be incorporated into the Project Status, Analysis, and Reporting decision process.

Anderson & Woodhead
General Contractors

PROJECT: ____________ JOB NO.: ____________

LOCATION: ____________ DATE: ____________

WEATHER: ____________ TEMPERATURE: ____________

CRAFT	FOREMEN	CRAFT	TOTAL	SUBCONTRACTORS CRAFT	FOREMEN	CRAFT	TOTAL	MATERIALS RECEIVED DESCRIPTION QUANTITY
CARPENTERS	1	3	4					REINFORCE CONCRETE PIPE
LABORERS	1	4	5					10″ 500 L.F.
OPERATORS		2	2					
PIPEFITTERS	1	7	8					
STAFF				SUBCONTRACTORS STAFF				
FIELD SUPERINTENDENT			1					
FIELD ENGINEER			1					

WORK ACCOMPLISHED:

1. FORM FOOTINGS—BAY D-G
2. INSTALL SEWER PIPE

APPROVALS:

FIELD SUPERINTENDENT ____________ PROJECT ENGINEER ____________ OWNER'S REPRESENTATIVE ____________

Figure 12.19 Daily Construction Report (P_M-DOC 9).

Anderson & Woodhead
General Contractors

PROJECT: ____________ JOB NO.: ____________

LOCATION: ____________ DATE: ____________

CLIENT: ____________ PREPARED BY: ____________

COST CODE	WORK CATEGORY	UNIT MEASURE	TOTALS LAST PERIOD	THIS PERIOD	TO DATE
0311.0	FORMWORK—FTGS	SF	402	1850	2252

Figure 12.20 Quantity Distribution Report (P_M-DOC 10).

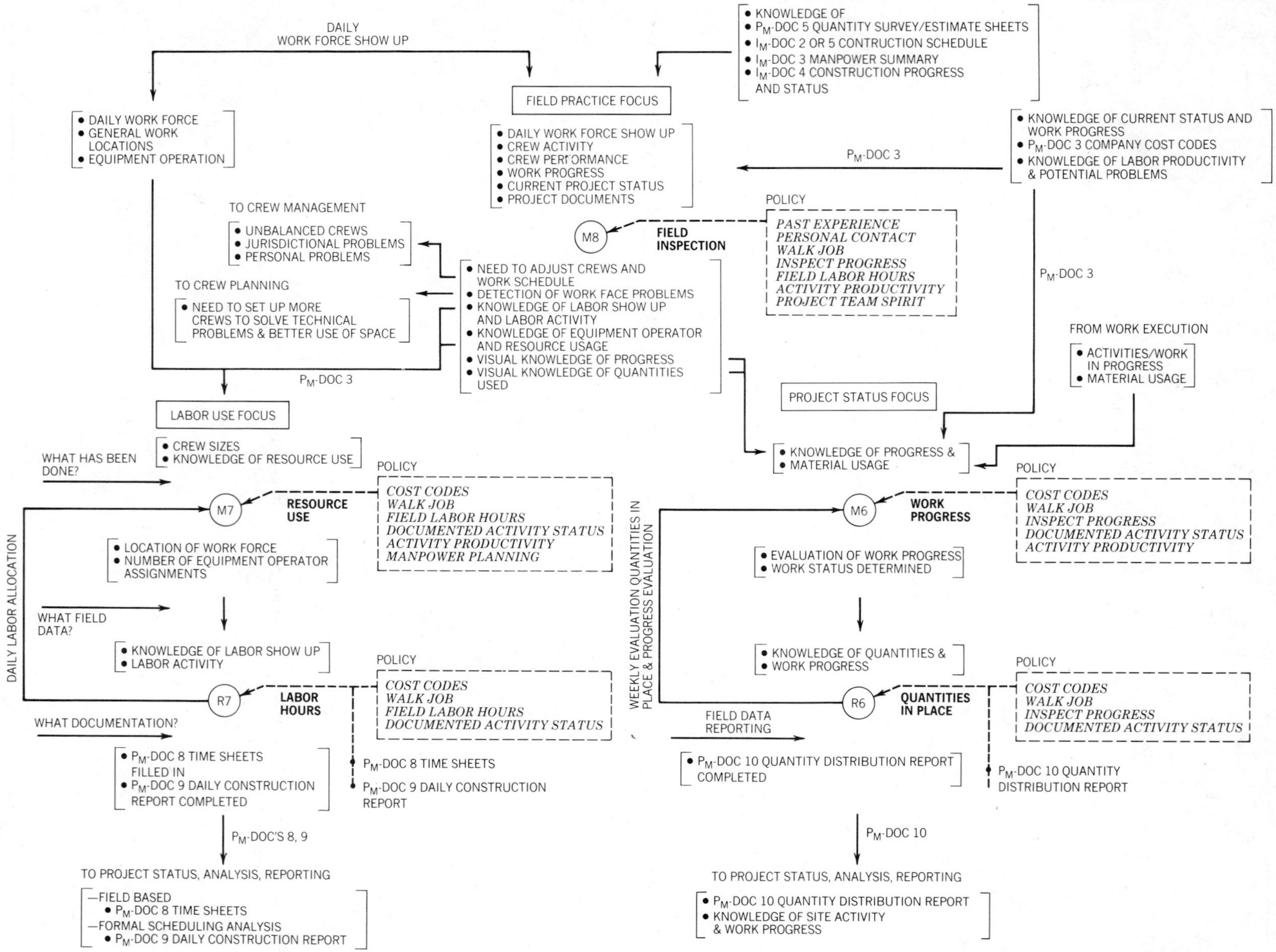

Figure 12.21 Field Monitoring decision process (Traveling Project Manager approach).

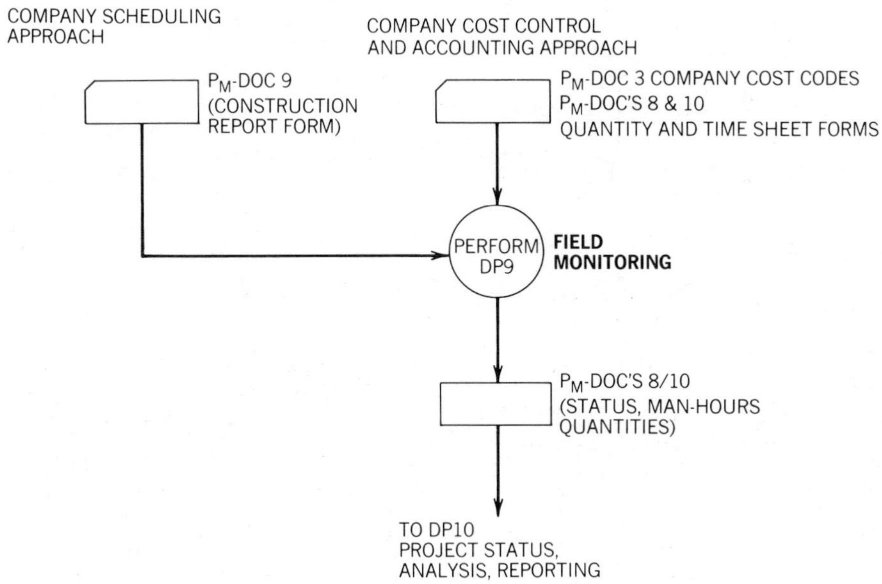

Figure 12.22 Field Monitoring Management Information System (Traveling Project Manager approach).

12.5 FIELD-BASED PROJECT MANAGEMENT

The primary management objective of the Field-Based Project Management approach is the transfer of management responsibility and authority from the head office to the field location. This changing management focus and emphasis is due in part to the technical complexity of projects in the large firm environment, the size and often remote location of these projects, and the management capacity and expertise required by the client, who is often located near the project. As a result of the shift of management responsibility from the head office to the field, an increasing management burden is imposed on field decision processes such as Field Monitoring, which normally is the sole domain of the superintendent and foreman.

While the decision process structure and problem focus of Field Monitoring remains essentially the same, Field-Based Project Management policies emphasize the need for more reliable and integrated field reporting techniques consistent with the Field-Based Project Management concept. Thus, the documented information system for Field Monitoring at the large firm level, while providing the basic information of quantities, man-hours, construction reports, and so on, changes in structure to meet the demands imposed by top management for effective and useful Project Status, Analysis, and Reporting information. Thus, the volume of data collected is related to control area practices through the project cost code system. This results in an increasing level of detailed reporting on different items such as activity status and percent complete by control zones. Finally, because of both the increased emphasis on field documentation, and the increase in project size and complexity, the Field Monitoring effort is rarely the sole responsibility of either the superintendent or the foremen. Large projects almost always demand timekeepers for collecting and reporting manpower and labor man-hours; field engineers and scheduling engineers for evaluating and reporting quantities, percent physical progress, and activity status; and cost engineers for ensuring that pertinent data are collected and assigned to the appropriate work package activity cost code.

The integration of field cost and scheduling control begins to emerge through the Field Monitoring Management Information System (MIS). This integration is important because it brings order to the collection of large volumes of data. The data structure produced by the MIS, as shown in Figure 12.23, is impacted by several key policies.

DOCUMENTED ACTIVITY STATUS (P_L 27)* defines the primary ingredients in effective cost and scheduling control, and specifies that these ingredients be documented by an integrated system yielding information useful for project management control. This approach, coupled with a *COST TO COMPLETE* (P_L 29) focus in which man-hours,

*See Table 12.5, Policies of the Large Firm.

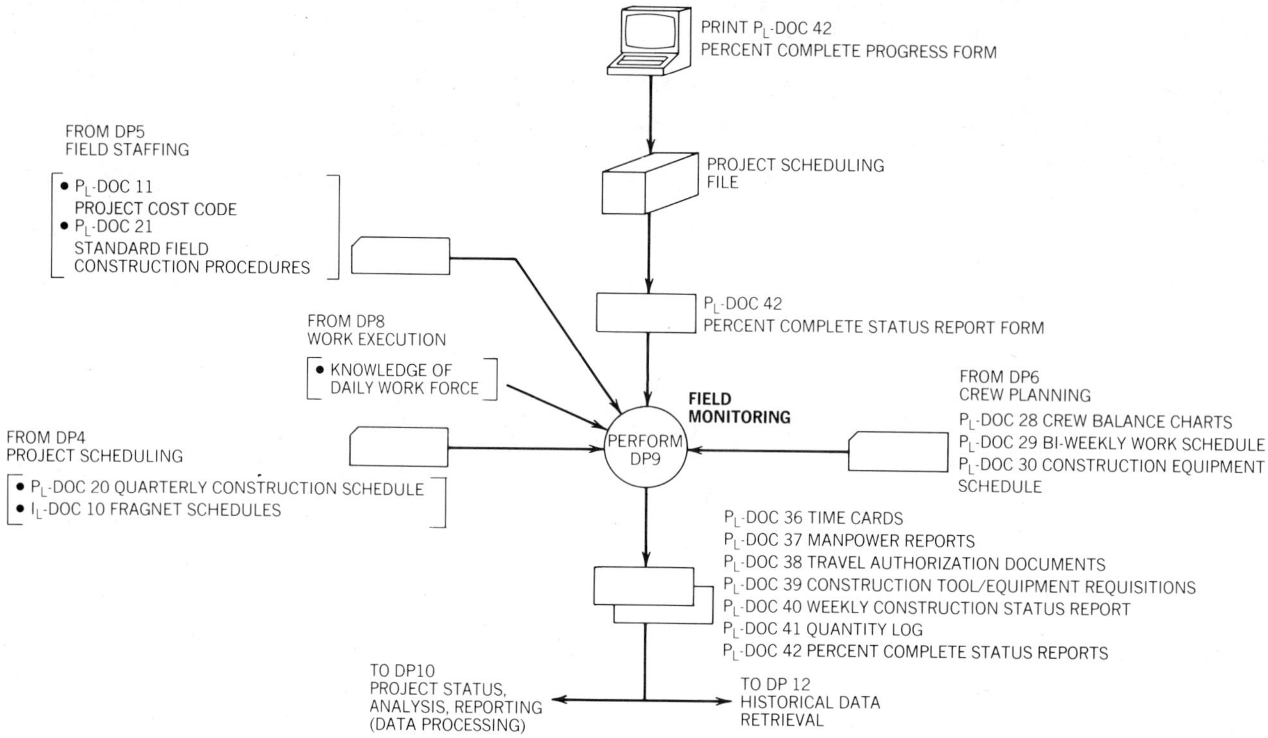

Figure 12.23 Field Monitoring Management Information System (Field-Based Project Management approach).

manpower, and labor costs are forecast, depends on quality documented field data. The accurate gathering and recording of field data is stressed through *PROJECT INSPECTIONS* (P_L 25) and the monitoring of *FIELD LABOR HOURS* (P_L 26). If either labor man-hours or percent progress are incorrectly recorded and then later processed, a distorted picture of the percent progress and productivity achieved and/or the labor man-hours expended for an activity will result. This could paint a misleading picture of the current status of an activity. Finally, since *CREW PRODUCTIVITY* (P_L 28) evaluations are based on man-hour expenditures in relation to physical production achieved, the man-hours and quantities in place must be accurately assessed and properly documented to provide a valid indicator of labor productivity.

Field Monitoring has a *labor use* focus, a *field practice* focus, and a *project status* focus. These three focuses of Field Monitoring at the large firm level are identical to those of the small and medium-sized firm in terms of the relevant functions performed. When comparing the Field Monitoring structures and information flows of the large firm with those of small and medium-sized firms, the emphasis on *field practice* is no longer the dominant field manpower management concern. Consequently, management emphasis toward field documentation increases due to top management's demands for reliable data as input for *overall* project management control through the Management Information System structure.

The increasing emphasis on field documentation first emerges during the *labor use* phase of Field Monitoring. As depicted in Figure 12.24, the evaluation of RESOURCE USE (M7) is based on an analysis of current activities in progress, equipment usage, and the daily work force level. The superintendent and foremen work closely with the field engineer and timekeeper to ensure that the proper data is gleaned, by observing daily the RESOURCE USE content of each work package activity.

Foremen are most commonly involved with the recording of LABOR HOURS (R7) because these personnel are the closest to the work face and have the most intimate knowledge of the work performed by each of their crew members. The foremen must therefore evaluate daily man-hour expenditures for each crew member and, often together with the timekeeper, must assign the crew member's time to

TABLE 12.5 POLICIES OF THE LARGE FIRM

Policy		Description
P_L 6	*COST CODES*	A systematic classification of project work categories segregated by unique physical components and by cost codes is the basis for structuring the project scope into work tasks, categorizing estimated costs, reporting and controlling costs and manpower, and developing a historical database.
P_L 10	*NETWORK-BAR CHART ANALYSIS*	Broad network-focused milestone planning is the best approach for identifying and sequencing major project components, while a combination of critical path method and bar chart model are best suited for project planning and scheduling.
P_L 11	*SCHEDULING FOCUS*	Scheduling definitions and constraints are: 1. A critical activity is one that, if delayed, will extend the completion date of the project. 2. Activity durations must be correlated to the estimated work content. 3. All constraints must be defined and incorporated into the sequence logic. 4. Manpower impact is an integral aspect of project scheduling and scheduling control. 5. The schedule must reflect optimum use of all resources. 6. Progress is based on the physical completion of work activities.
P_L 23	*LABOR RELATIONS*	A full-time labor relations expert is often required due to the complexity and manpower demands of the project.
P_L 24	*WALK JOB*	In the field, daily activity progress and crew performance are monitored by visual inspection, and crew management decisions are made on the basis of a subjective evaluation of observed progress and productivity versus planned progress and productivity.
P_L 25	*INSPECT PROGRESS*	Actual quantities in place versus estimated and/or forecasted quantities is an indicator of physical progress, and is evaluated by physical quantity take-off. Quantities must be reported by cost code for percent complete evaluation. Activity status must be evaluated in relation to start and completion dates and percent complete for each activity.
P_L 26	*FIELD LABOR HOURS*	Labor hours are monitored by field personnel closest to the workface, and recorded and processed by a staff member whose primary function is to ensure that the labor hours are allocated to the appropriate cost code.
P_L 27	*DOCUMENTED ACTIVITY STATUS*	Labor costs, manpower, man-hours, and quantities are primary ingredients to cost control and must be documented by a system yielding reliable information and viable reports for management control. Activity status must be documented in terms of start and completion dates and percent complete data.
P_L 28	*CREW PRODUCTIVITY*	Actual unit man-hour rate versus estimated man-hour rate is a measure of crew performance and a basic indicator of productivity.
P_L 29	*COST TO COMPLETE*	Projecting total project labor costs and man-hour requirements is requisite to identify problems and evaluate alternative courses of action at both the activity level and the project level.
P_L 32	*ACTIVITY ANALYSIS*	Activity analysis is geared to exception reporting mechanisms or field management desire to analyze construction methods or crew productivity in order to improve crew performance.
P_L 33	*MANAGEMENT BY EXCEPTION*	Manpower management by exception is acceptable provided that the information system is geared to exception reporting to alert management to existing or potential problems.
P_L 34	*DATA PROCESSING*	Maximum use of electronic data processing is often the most economical means of processing data into useable information.
P_L 35	*PROJECT TEAM SPIRIT*	Active participation of all field managers in the management process promotes goodwill and a positive management environment.

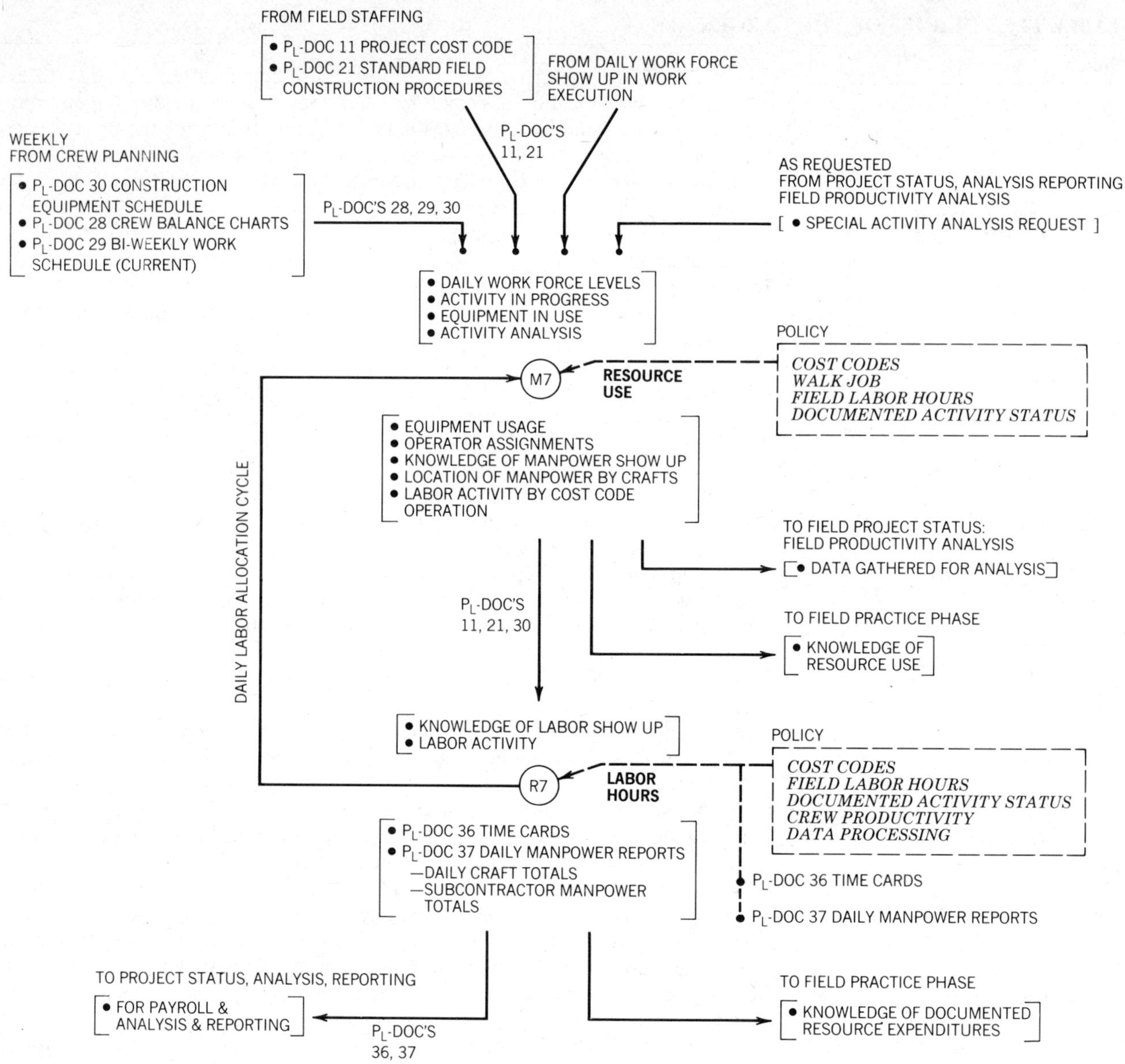

Figure 12.24 Field Monitoring decision process: labor use phase (Field-Based Project Management approach).

an activity description. The foreman together with the timekeeper records this information on the Time Card (P_L-DOC 36). The Time Card is processed daily because of the size of the work force and the volume of craft man-hours processed. As exemplified in Figure 12.25, the work performed by the employee is defined by a brief description for which the cost engineer (or cost clerk) assigns a cost code by referencing the Project Cost Code (P_L-DOC 11). In order to verify cost codes in relation to work performed, the cost engineer must continuously *WALK THE JOB* (P_L 24). This ensures that the description of work on the Time Card will accurately reflect the work being performed. The timekeeper is responsible for tabulating the site manpower levels and categorizing manpower by craft. He or she prepares this segment of the Construction Manpower Report (P_L-DOC 37) which focuses specifically on daily field manpower levels. This report, Figure 12.26, lists the number of employees by craft on the project site each day. It also identifies the subcontractors on the site and their

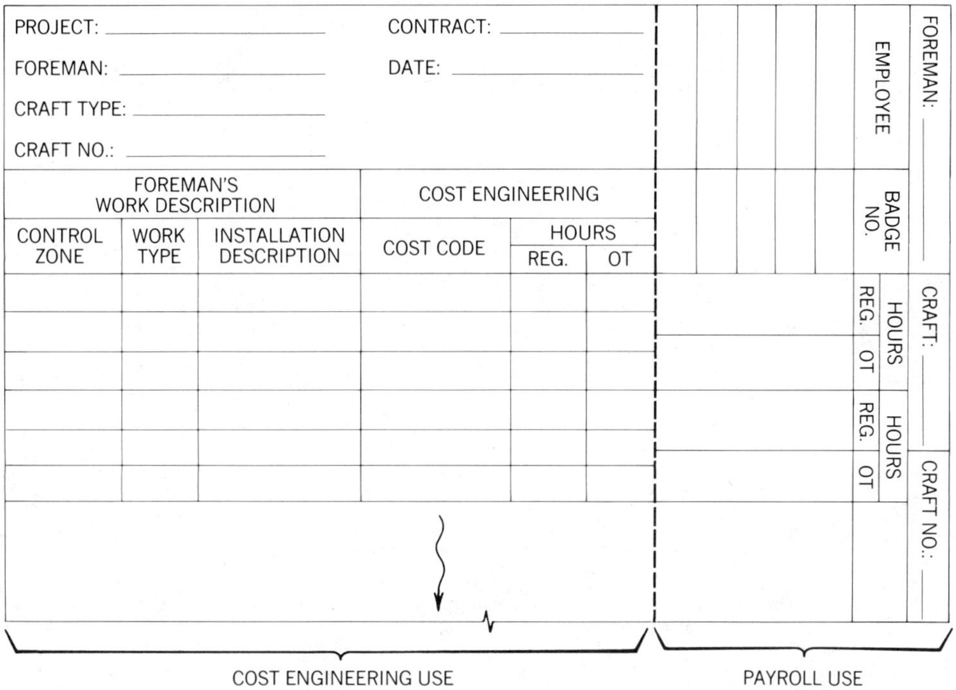

Figure 12.25 Time Cards (P_L-DOC 36).

manpower levels, and gives a brief description of the job conditions for the day (i.e., weather, strikes, etc.) and other related daily site characteristics.

In contrast to the medium-sized firm, the *field practice* focus of Field Monitoring at the large firm level has both a field observation and field documentation level as depicted in Figure 12.27. In this environment FIELD AUTHORITY DOCKETS (R8) become an integrable part of the project's Management Information System, and produce documents not necessarily required by the medium-sized firm. Due to the enlarged size of the project labor force and the geographical separation commonly found between facility areas and material warehousing locations, a systematic method of keeping track of employee movement, requisition-

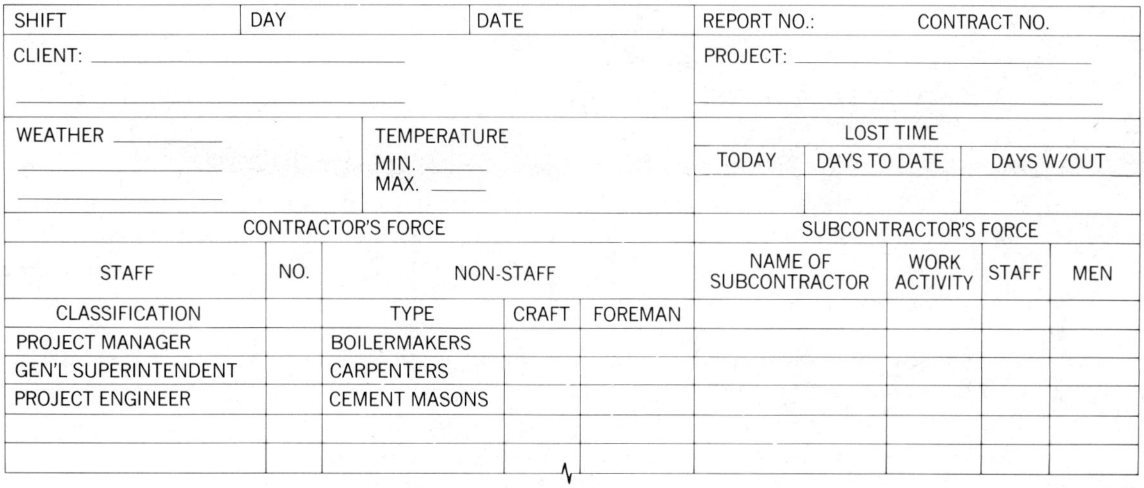

Figure 12.26 Daily Manpower Reports (P_L-DOC 37).

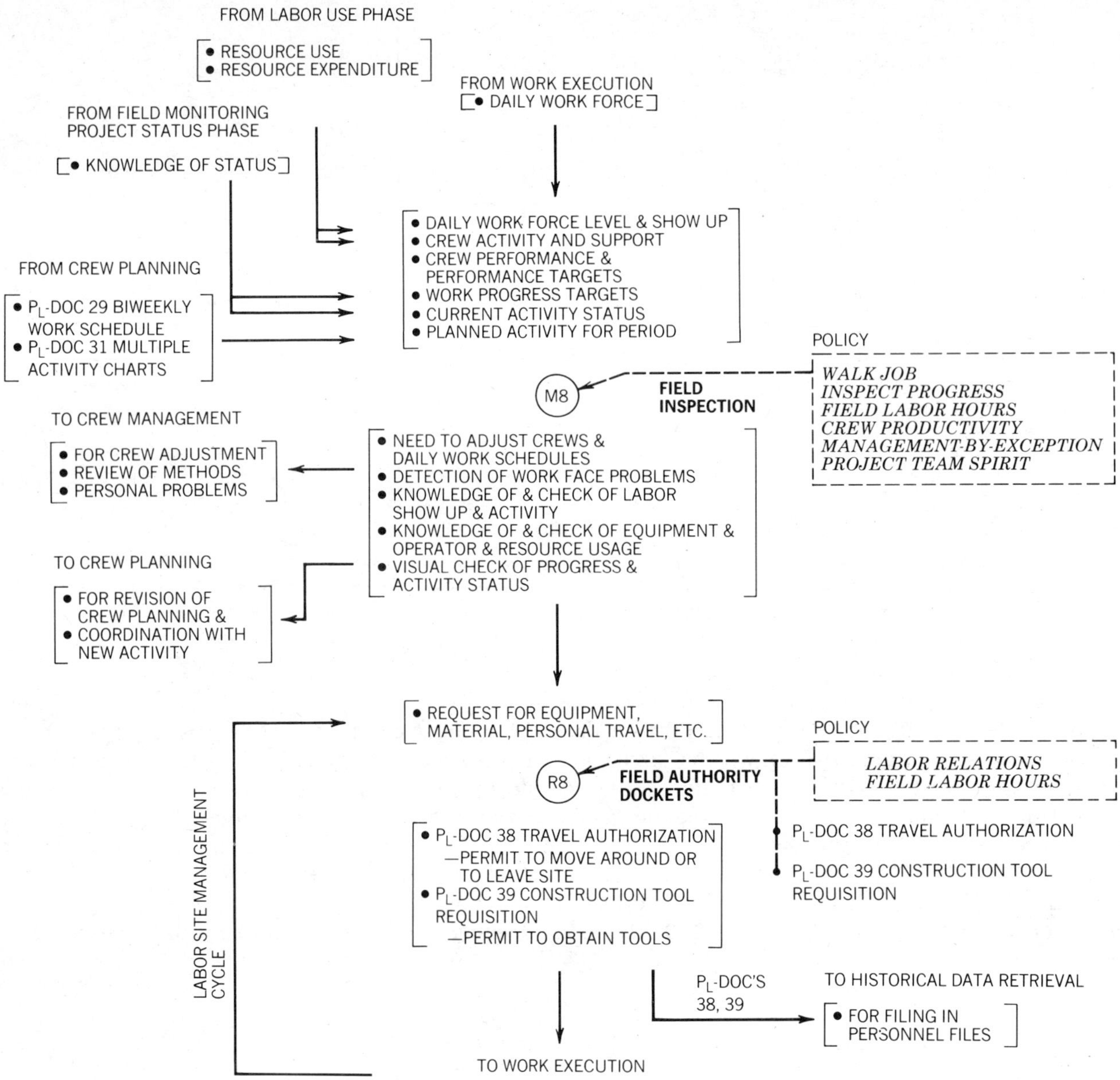

Figure 12.27 Field Monitoring decision process: field practice phase (Field-Based Project Management approach).

ing material, and locating construction tools and equipment emerges. This system may include several types of FIELD AUTHORITY DOCKETS that focus on different management problems. One type of field authority document is that which aids in the control of the movement of craft personnel by identifying the location of each employee. A Travel Authorization Docket (P_L-DOC 38) is demonstrated in Figure 12.28, and serves as a record of employees' movement around the job site and, potentially, off the site. Another example is the documents used to requisition materials and equipment from warehouses or stock piles in lay-down areas. An example of this document is the Material and Equipment Requisition (P_L-DOC 39) shown in Figure 12.29. It is used by craft personnel for procuring material and/or small tools and equipment for their work operations. The timing of material requisitions must be planned by the superintendent and general foreman/foremen as part of Crew Planning and the Execution of Work.

Crew performance, progress, and activity status

AW
Engineering & Construction Corp.

CLIENT: ______ CONTRACT: ______

PROJECT: ______

TRAVEL AUTHORIZATION

DATE	BADGE NO.	EMPLOYEE NAME	PERMIT TO TRAVEL				
			LOCATION		TIME		SUPERVISOR
			CURRENT	EXPECTED	IN	OUT	

Figure 12.28 Travel Authorization (P_L-DOC 38).

analysis are critical aspects of the FIELD INSPECTION function. These practical construction problems remain the focus of intimate daily supervision of crew activity, crew production, and progress. This supervision is carried out by visual inspection (*WALKING THE JOB*) to detect problems related to crew sizes and mixes, personal conflicts, work face problems, and so on. On larger projects, the responsibility of each superintendent and his or her foremen is limited to specific areas or components of the project such as civil or piping work. Despite this limited responsibility focus on the part of front-line supervisors, however, an awareness of the data collected and documented by other project team personnel is still necessary. Thus, as shown in Figure 12.27, the front-line supervisors gain knowledge of activity status and progress and resource expenditures through WORK PROGRESS and RESOURCE USE data that flows into the FIELD INSPECTION function. This permits them to check the various data documented as a result of recording QUANTITIES IN PLACE (R6) and LABOR HOURS (R7).

The *project status* phase of Field Monitoring, as depicted in Figure 12.30, is concerned with the weekly evaluation and documentation of status and progress at the work task activity level. The field superintendent is responsible for reporting WORK PROGRESS as compared to planned progress and status per the Bi-Weekly Work Schedule (P_L-DOC 29), and for preparing a Weekly Activity Status Report (P_L-DOC 40) covering the activity for a particular area of the project. This report contains brief statements of the work completed, in progress, and begun during the current week. It identifies problems such as material shortages, and other information needed for further progress. An example of a Weekly Activity Status Report (P_L-DOC 40) is illustrated in Figure 12.31. This report is forwarded to the field project manager for incorporation into

AW
Engineering & Construction Corp.

CLIENT: ______ CONTRACT: ______

PROJECT: ______

TOOL REQUISITION

COST CODE	SERIAL NO.	TOOL DESCRIPTION	QUANTITY	DATE	
				OUT	IN

ISSUED TO: ______	DATE: ______
BADGE NO.: ______	REQUISITION NO.: ______

Figure 12.29 Construction Tool Requisition (P_L-DOC 39).

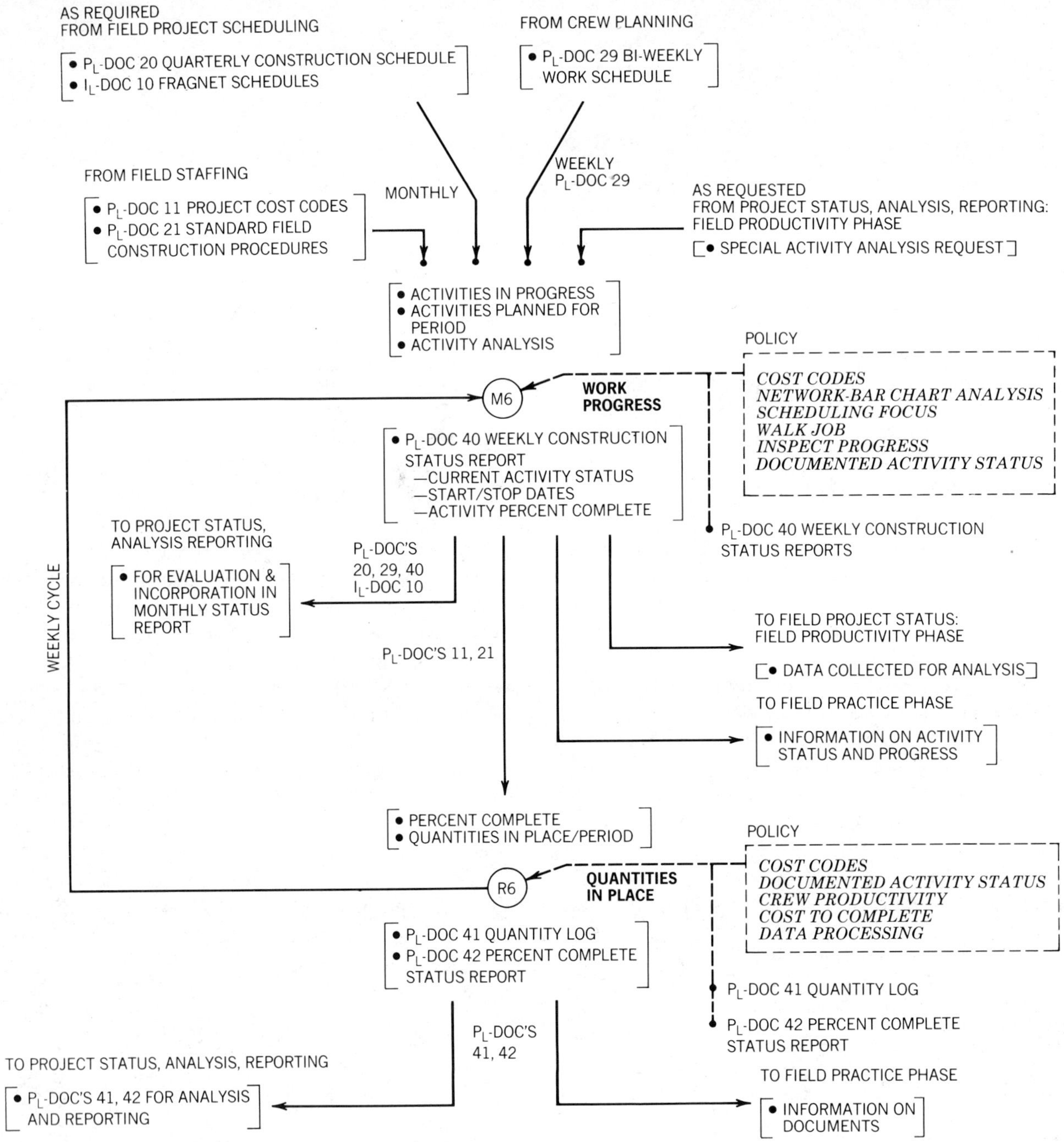

Figure 12.30 Field Monitoring decision process: project status phase (Field-Based Project Management approach).

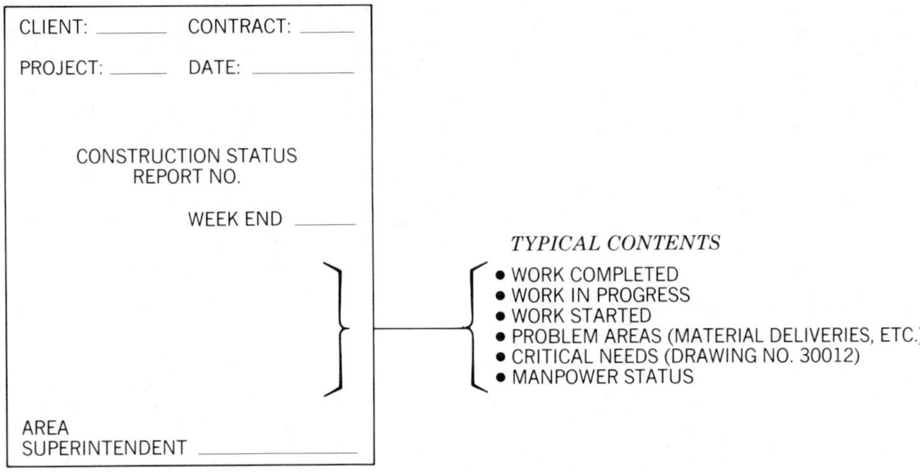

Figure 12.31 Weekly Construction Status Report (P_L-DOC 40).

project level reports and resolution of specifically identified problems.

The continuous exchange of field data between front-line superintendents/general foremen and other project team personnel increases their awareness of activity status, crew performance, and work progress. When coupled with visual field WORK PROGRESS inspections, this data exchange leads to the field documentation of QUANTITIES IN PLACE. A scheduling engineer or field engineer determines physical progress by evaluating the quantity of output for each activity for the period. They often refer to marked-up construction drawings (with completed items colored) to determine the progress achieved since the previous period. The most accurate method is to do a direct take-off from the drawing of the actual quantities placed during the period and tabulate the quantity of material placed for the work activity. However, progress evaluation for some types of work activities can be easily determined by visual observation. Quantities are recorded by the field engineer on the Quantity Log (P_L-DOC 41). A typical example appears in Figure 12.32. Each work task on the Quantity Log is identified by its unique cost code and relevant description using the Project Cost Code (P_L-DOC 9) as a guide. An assessment of a percent complete as evaluated during WORK PROGRESS inspection is recorded on a Percent Complete Status Report (P_L-DOC 42) form (see Figure 12.33). This form is normally printed by the computer with each activity identified by cost code and a corresponding space provided for entering the percent complete for the activity. Both the Quantity Log and the Percent

AW
Engineering & Construction Corp.

CLIENT: ______ CONTRACT: ______

PROJECT: ______ WEEK ENDING: ______

QUANTITY LOG

DATE	WORK DESCRIPTION	COST CODE	UNIT MEASURE	QUANTITY							
				M	T	W	T	F	S	S	TOTAL

Figure 12.32 Quantity Log (P_L-DOC 41).

AW
Engineering & Construction Corp.

CLIENT: ____________ CONTRACT: ____________

PROJECT: ____________ WEEK END: ____________

PERCENT COMPLETE

COST CODE			ACTIVITY DESCRIPTION	% COMPLETE	REMARKS
LOC.	CODE	RES.			

Figure 12.33 Percent Complete Status Report (P_L-DOC 42).

Complete Status Report forms provide valuable input for overall Project Status, Analysis, and Reporting.

12.6 TOTAL FIELD AUTONOMY

The Total Field Autonomy management approach is often a response to heavy client and company demands for efficient and effective project management. It also arises from a company's recognition that this management approach meets the challenge of a construction environment where project size, complexity, and/or location often demand the flexibility to adapt to the many unique problems associated with large projects. In this environment, management pressure is brought to bear on Field Monitoring in that it must satisfy the need to systematically collect information in sufficient detail for input to an integrated Management Information System. The structure of this MIS is established during the project definition phases of the project (i.e., control base plans, schedules and estimates). Furthermore, the control of work orders that comprise work packages imparts additional pressures on Field Monitoring, thereby requiring increasing levels of detailed documented information during the performance of this decision process. Finally, Field Monitoring has a more specific management focus at the crew level with regard to the issuing of weekly work schedules that result in detailed work activity status and progress reports such as Foreman Delay Reports.

At the Total Field Autonomy level, the MIS is similar in structure to that of the Field-Based Project Management approach. However, control area concepts incorporating project work breakdown structures produce a broader spectrum of information during Field Monitoring. In addition, there exists an increasing emphasis on specialized field observations and analyses to support activity analysis techniques. This approach necessitates another set of independent documents for data collection that lead to more penetrating observations of crew activity, construction methods, and the like, through work sampling or field study techniques.*

The integration of field data through the Management Information System becomes a critical facet of field cost and scheduling control. Thus, the data structure of the Management Information System that characterizes Field Monitoring involves a very large volume of document inputs and outputs, as illustrated in Figure 12.34.

The basic problem focus of Field Monitoring relative to the project status, labor use, and field practice phases, is similar to that of the Field-Based Project Management approach. At the Field-Based Project Management level, field documentation increases significantly in its importance to overall project management, while field observations only increase slightly against this same measure of importance. The gap between field observation and documentation closes, however, in terms of their relative importance to the overall project management process, because of the increasing influence of such management techniques as work sampling. These techniques, while requiring more documentation, are successful only when reliable data is collected through accurate and consistent field observation.

The Field Monitoring decision process is summarized in terms of the management functions, policies, and relevant documents for each phase, as shown in Figures 12.35, 12.38, and 12.40, respectively. When contrasting this decision process with that of Field-Based Project Management (see Section 12.5) a natural increase in the number of documents is apparent. However, because of the flexibility of Total Field Autonomy management, not all those documents reflected in Figures 12.35, 12.38, and 12.40 may be required. Their use will naturally depend on project size, complexity, specific project management control philosophies, and the demands of the client. For instance, the Attendance Analysis

*See Field Activity Analysis Documents (P_T-DOC 59) as discussed and illustrated in Section 13.6.

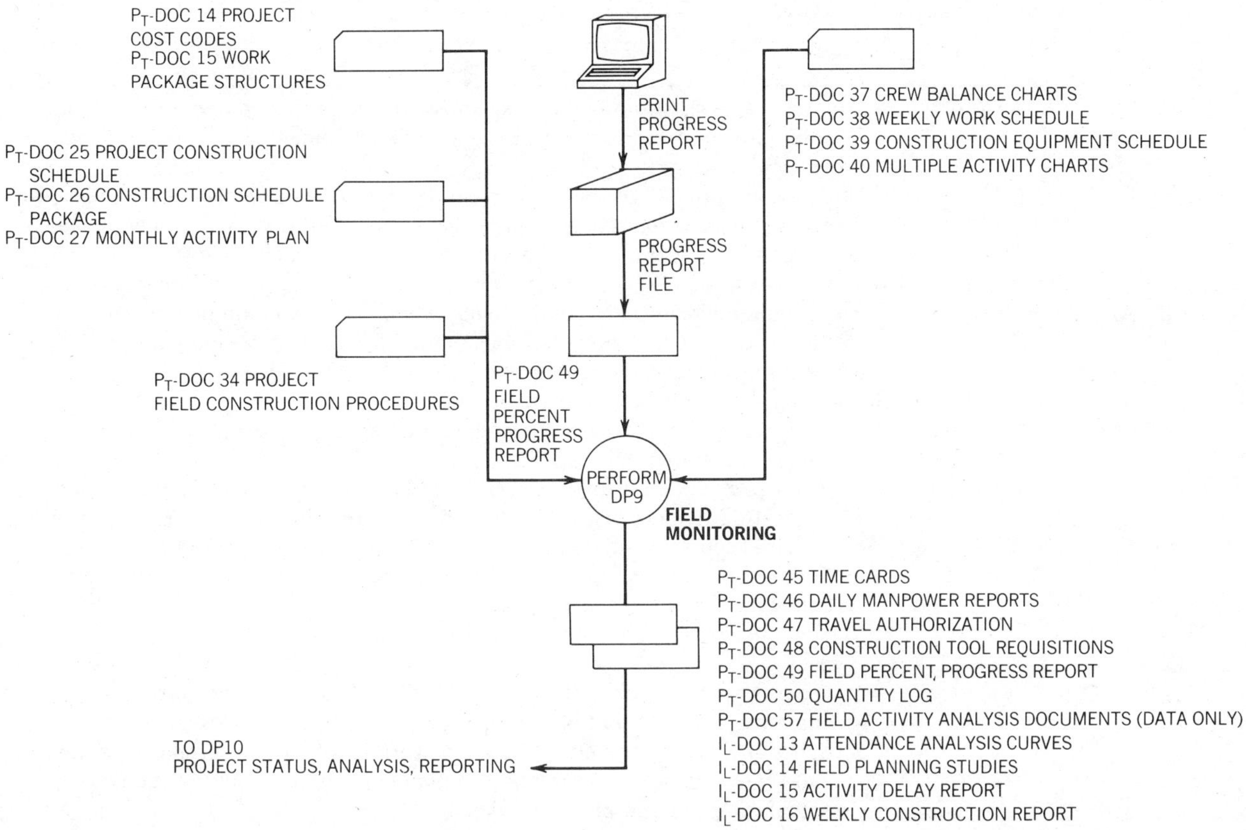

Figure 12.34 Field Monitoring Management Information System (Total Field Autonomy approach).

Report (I_T-DOC 13) may not be necessary if absenteeism is not a significant problem.

The success of field observations and documentation of data for the project status, labor use, and field practice areas requires both an increasing project team involvement and expertise, and therefore a *PROJECT TEAM SPIRIT* (P_T 37)* that promotes total integration of the project team. While the superintendent/foreman assume their technical roles in these areas of Field Monitoring, they are supported by numerous project team personnel such as cost/scheduling engineers, method engineers, field engineers, timekeepers, and so on. Consequently, the management functions related to field observations, WORK PROGRESS (M6), RESOURCE USE (M7), and FIELD INSPECTION (M8) are integrated closely with those functions of field documentation, QUANTITIES IN PLACE (R6), LABOR HOURS (R7), and FIELD AUTHORITY DOCKETS (R8) through project team interaction. This integration strengthens the decision making that evolves from the solution of problems raised through project status, labor use, and field practice analysis.

The labor use problem area, as described in Figure 12.35, is concerned with the proper allocation of *FIELD LABOR HOURS* (P_T 27) to the appropriate *COST CODE* (P_T 9), in relation to work packages as defined by the project *WORK BREAKDOWN* (P_T 8) structure. *DATA PROCESSING* (P_T 36) applications are important in the accurate distribution of labor man-hours and must be optimized in so far as they help limit the choice of cost codes used to describe the craftsman's time charged to a work category. Thus, the Project Cost Code System (P_L-DOC 13 and P_L-DOC 14) must maximize computer capabilities when distributing man-hours to cost codes, work orders, work packages, and so on, in

*See Table 12.6, Policies of the Autonomous Project Team.

TABLE 12.6 POLICIES OF THE AUTONOMOUS PROJECT TEAM

Policy		Description
P_T 8	*WORK BREAKDOWN*	A work breakdown structure representative of the project scope and control philosophy is formulated to integrate costs and schedules with work packages for project management.
P_T 9	*COST CODES*	Cost codes are applied to the work breakdown structure and serve as a basis for categorizing estimated costs, reporting and controlling costs and manpower, developing a historical database, and integrating the schedule with the estimate.
P_T 13	*NETWORK-BAR CHART ANALYSIS*	Broad network focused milestone planning is the best approach for identifying and sequencing major project components, while a combination of the critical path method and bar chart model are best suited for project planning and scheduling at the total project view level as well as at the daily level.
P_T 14	*SCHEDULING FOCUS*	Scheduling definitions and constraints are: 1. A critical work package is one that, if delayed, will extend the completion date of the project. 2. Work package durations must be correlated to the estimated work content. 3. All restraints must be defined and incorporated into the sequence logic (i.e., material delivery, subcontract coordination, drawing receipt, etc.). 4. Manpower impact is an integral aspect of project scheduling and scheduling control. 5. Schedules must reflect optimum use of all resources relative to the work package. 6. Progress is based on a realistic evaluation of the labor unit effort required to complete a component of work. 7. Schedules are portrayed to the daily work level and are prepared, analyzed, and reviewed weekly in relation to planned progress and productivity versus actual progress and productivities.
P_T 23	*LABOR RELATIONS*	When union labor is utilized, pre-job union conferences and personal contact with the unions are an integral part of successful project management. Regardless of the construction environment, a full-time labor relations expert is required because of the complexity and manpower demands of the project.
P_T 24	*PERSONAL CONTACT*	Crew management must be coupled with personal contact and good relations between field management personnel and construction workers if scientific management techniques are to be applied successfully.
P_T 25	*WALK JOB*	In the field, daily activity progress and crew performance is monitored by visual inspection, and crew management decisions are made on the basis of a subjective evaluation of observed progress and productivity versus planned progress and productivity. These decision problem areas often incorporate a number of decision makers.
P_T 26	*PROGRESS REPORTING*	A systematic method of evaluating physical progress must be based on measuring units of work for any component work tasks, and must have the capability of rolling progress up from this detailed component level to the total project level.
P_T 27	*FIELD LABOR HOURS*	Labor hours are monitored by field persons closest to the workface and are recorded and processed by a person whose primary function is to ensure that the labor hours are allocated to the appropriate cost code. Their proper allocation is essential to analyzing productivity and forecasting man-hours to complete at any level of detail.
P_T 28	*DOCUMENTED ACTIVITY STATUS*	Labor costs, manpower, man-hours, and quantities are primary ingredients to cost control and must be documented by a system yielding reliable information and viable reports for management control. These ingredients accrue as a basis for forecasting to complete at any level of detail.
P_T 29	*CREW MANAGEMENT*	Field crew management is based on past experience coupled with scientific management techniques.
P_T 30	*CREW PERFORMANCE*	Crew performance is a function of labor productivity, job conditions, and management effectiveness, and must focus on a tight but realistic performance indicator.

TABLE 12.6 ***(Continued)***

Policy		Description
P_T 31	*PRODUCTIVITY*	Productivity is an essential concept and must be rationally defined for office and field utilization. An awareness of the components that affect productivity and their relation to man-hours, activity durations, and progress is essential, also, all these must integrate and be manageable.
P_T 34	*SCIENTIFIC MANAGEMENT*	Techniques of scientific management, such as work sampling, time lapse, activity analysis, multiple activity charts, and prediction models, can contribute to improving crew performance.
P_T 35	*MANAGEMENT BY EXCEPTION*	Manpower management by exception is acceptable provided that the information system is geared to exception reporting to alert management to existing problems or potential problems which are emerging within a sufficient time frame to allow a management response.
P_T 36	*DATA PROCESSING*	A computer-based information system is the most feasible means of data processing for management control.
P_T 37	*PROJECT TEAM SPIRIT*	Active participation of all project team members in the management process promotes goodwill and a positive management environment. It ensures the success of scientific management techniques and the total integration of the project team.

order to minimize the actual number of cost codes on Time Cards (P_T-DOC 45).* The time card system often includes a daily tabulation or time card summary that can be reviewed by the responsible field supervisor for improper cost codes, absence of cost codes, or as a means of checking the crew members working each day.

When observing RESOURCE USE and recording LABOR HOURS for payroll purposes, the timekeepers tabulate the actual manpower on-site each day. These Daily Manpower Reports (P_T-DOC 46) can be prepared by control area and/or for the total project, depending on project size and control philosophy. If the Daily Manpower Report is given

*The Cost Code system is often structured so that a matrix of potential work category identifiers are contained in the computer and can be called to sort data in a number of ways, such as into work order, work packages, and so on, without the need to physically assign cost codes on the Time Card to identify these types of categories.

M7 RESOURCE USE

Policies	Documents
WORK BREAKDOWN	
COST CODES	
WALK JOB	
FIELD LABOR HOURS	
DOCUMENTED ACTIVITY STATUS	
CREW PERFORMANCE	
PRODUCTIVITY	

R7 LABOR HOURS

Policies	Documents
WORK BREAKDOWN	P_T-DOC 45 Time Cards
COST CODES	P_T-DOC 46 Daily Manpower Reports
FIELD LABOR HOURS	I_T-DOC 13 Attendance Analysis Curves
DOCUMENTED ACTIVITY STATUS	
CREW PERFORMANCE	
PRODUCTIVITY	
DATA PROCESSING	

Figure 12.35 Field Monitoring decision process: labor use phase (Total Field Autonomy approach).

MPM PROJECT — AW CORP.

DAILY MANPOWER REPORT

CONSTRUCTION MANAGER: ________ CONTROL ZONE: ________ DATE: ________

CLASSIFICATION	CODE	TOTAL PAYROLL	AT WORK	ABSENT	TERM.	GEN'L FOREMEN	JOURNEYMEN	RATIO
STAFF								
CRAFT								

Figure 12.36 Daily Manpower Reports (P_T-DOC 46).

an appropriate focus, such as that illustrated in Figure 12.36, it can be a valuable tool for analyzing craft absenteeism and turnover rates. When this data is graphically displayed, it can provide information for forecasting manpower levels by craft and predicting future productivity targets. Typical Attendance Analysis Curves (I_T-DOC 13) that depict actual and experienced absenteeism rates are shown in Figure 12.37.

When *WALKING THE JOB* (P_T 25) and visually gathering and documenting data, the superintendent/foreman and other personnel must be cognizant of the factors that influence *CREW PERFORMANCE* (P_T 30). This policy identifies a fundamental aspect of crew performance which is often overlooked, namely, management effectiveness. The normally recognized components of crew performance are job conditions (e.g., weather) and labor productivity. If the field manager, even at the foreman level, is not planning the work properly, however, conditions can develop where this ineffective planning causes crew performance to falter. For example, the crew might be overmanned, or material and equipment might not be properly supplied to the crew. The end result is poor performance. That is, more man-hours are expended than were budgeted, or time is spent idly because of lack of equipment or material which eventually leads to problems in achieving progress targets. When reviewing crew performance therefore, the supervising manager must also look at the planning approach used by the personnel in charge of the crew.

The field practice focus of Field Monitoring remains the domain of the superintendent and foreman (see Figure 12.38). However, due to the introduction of *SCIENTIFIC MANAGEMENT* (P_T 34) techniques, such as work sampling or field planning studies, these front-line supervisors are often supported by methods engineering or industrial engineering experts. Thus, while *CREW MANAGEMENT* (P_T 29) is based on past experience and sound construction savvy, it is also coupled with scientific management techniques. These techniques require both *PERSONAL CONTACT* (P_T 24) with the construction worker and a management atmosphere centered around *PROJECT TEAM SPIRIT*.

The superintendent, when *WALKING THE JOB* in a FIELD INSPECTION, may request that a field planning study (see Figure 12.8) be initiated on a critical work package. This request often results from direct observations in which an assessment of crew performance and activity progress indicates that work package progress is not proceeding in a satisfactory manner. The Field Planning Study technique is considered an internal document when used and is referred to as I_L-DOC 14.

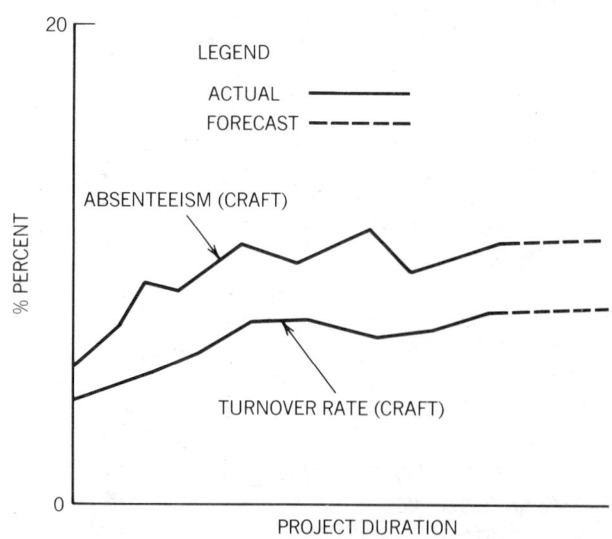

Figure 12.37 Attendance Analysis Curves (I_T-DOC 13).

M8 FIELD INSPECTION

Policies	Documents
LABOR RELATIONS	I_T-DOC 14 Field Planning Studies
PERSONAL CONTACT	I_T-DOC 15 Activity Delay Reports
WALK JOB	
PROGRESS REPORTING	
FIELD LABOR HOURS	
CREW MANAGEMENT	
CREW PERFORMANCE	
PRODUCTIVITY	
SCIENTIFIC MANAGEMENT	
MANAGEMENT-BY-EXCEPTION	
PROJECT TEAM SPIRIT	

R8 FIELD AUTHORITY DOCKETS

Policies	Documents
LABOR RELATIONS	P_T-DOC 47 Travel Authorization
FIELD LABOR HOURS	P_T-DOC 48 Construction Tool Requisitions
CREW PERFORMANCE	
PRODUCTIVITY	

Figure 12.38 Field Monitoring decision process: field practice phase (Total Field Autonomy approach).

When performing FIELD INSPECTIONS, the superintendent/foreman also relies on information from WORK PROGRESS and RESOURCE USE inspections, and documented QUANTITIES IN PLACE and field LABOR HOURS expended. An awareness of this data can assist these front-line field personnel in identifying work face layout problems, new crew requirements, and a need to adjust crews, not to mention material shortage and jurisdictional problems. *MANAGEMENT BY EXCEPTION* (P_T 35) is most often used as the best approach to solving these recurring types of field practice problems.

As part of the field practice area, managing the movement of labor, materials, and construction equipment is a continuous problem and is handled in much the same way as demonstrated in Figures 12.28 and 12.29 in Section 12.5. In addition, the assistant superintendent/foreman often maintains an Activity Delay Report (I_L-DOC 15), sometimes called the Foreman Delay Report, which identifies specific delays from factors considered to be outside of his or her normal span of control. These might include drawing design changes or construction equipment failures. Problems related to rework are often documented as well, but a certain percentage of rework is considered part of the normal construction process and should be allowed for in the budget of man-hours available. On the other hand, excessive rework should be documented, especially when it relates to an increasing number of design changes. Excessive rework may also indicate specific problems with the quality of the craftsmen involved, or possibly the improper planning of crew support activities such as the supply of the correct materials for a given activity. A typical example of an Activity Delay Report (I_L-DOC 15) is shown in Figure 12.39. This report will assist the field team in identifying problems, especially recurring ones, and will alert upper level field management to the need for corrective action in the event that problems cannot be solved by front-line supervision.

The project status area of Field Monitoring (see Figure 12.40) approaches the problems of WORK PROGRESS evaluation with a different focus than does the Field-Based Project Management approach. *PROGRESS REPORTING* (P_T 26), together with a *SCHEDULING FOCUS* (P_T 14), expresses physical progress (i.e., percent complete) based on a system of labor unit measures (see Figure 12.11*c*) which reflect the complexity of a work task as weighted against similar work tasks. For example, placing straight-run pipe on a pipe rack near ground level consumes less labor effort per linear foot of pipe than would installing the same linear foot measure of configurated pipe attached to a large piece of equipment such as a process vessel fifty feet high. Thus, quantities are not necessarily the primary ingredient in physical progress reporting. However, quantities may be helpful when

M^PM
PROJECT

AW CORP.

ACTIVITY DELAY REPORT

CONTROL ZONE: ______________

FOREMAN: ______________ WEEK ENDING: ______________

SEQ NO.	DELAY DESCRIPTION	CAUSE	M-HRS LOST	EQUIP. HRS LOST	REFERENCES
110	REPAIR WELD	FAILED X-RAY	4	3	DRAWING 30012 LINE NO. 24-LF-12

Figure 12.39 Activity Delay Report (I_T-DOC 15).

M6 WORK PROGRESS

Policies

WORK BREAKDOWN
COST CODES
NETWORK-BAR CHART ANALYSIS
SCHEDULING FOCUS
WALK JOB
PROGRESS REPORTING
DOCUMENTED ACTIVITY STATUS
CREW PERFORMANCE
PRODUCTIVITY
DATA PROCESSING

Documents

P_T-DOC 49 Field Percent Progress Report
I_T-DOC 16 Weekly Construction Report

R6 QUANTITIES-IN-PLACE

Policies

WORK BREAKDOWN
COST CODES
DOCUMENTED ACTIVITY STATUS
CREW PERFORMANCE
DATA PROCESSING

Documents

P_T-DOC 50 Quantity Log

Figure 12.40 Field Monitoring decision process: project status phase (Total Field Autonomy approach).

M^PM
PROJECT

AW CORP.

PIPE PROGRESS REPORT STATUS AS OF: ______________

CONTROL ZONE	LINE NO.	SIZE	CODE I.D.	QUANTITY L.F.	ERECTION		VALVES		WELD		X-RAY		HYDRO		UNITS COMPL	%	TOTAL UNITS
					%	UNITS	%	UNITS	%	UNITS	%	UNITS	%	UNITS			
H100	12-124-LA	4"	22 SA	100	100	45		15	50	20		5		5	55	61	90

CODE I.D. WORK PACKAGE (SYSTEM, WORK ORDER, ETC.)

Figure 12.41 Field Percent Progress Report (P_T-DOC 49).

TO: CONSTRUCTION MANAGER
GROUP X

DATE: ____________

FROM: GEN'L SUPERINTENDENT
ZONE A B & C

SUBJECT: CONSTRUCTION STATUS

ACTIVITIES
COMPLETE
IN PROGRESS

MANPOWER REQUIREMENTS

MATERIAL/DRAWINGS REQUIRED

CRITICAL NEEDS

Figure 12.42 Weekly Construction Report (I_T-DOC 16).

evaluating the percent of labor units completed for a particular work task. This assessment of progress units earned may be performed by visual observation or by analyzing marked-up drawings and assessing labor unit progress by construction operations such as welding, bolting-up valves, and fittings. A typical Field Percent Progress Report (P_T-DOC 49) based on labor units is depicted in Figure 12.41.*

The flexibility to summarize the data in the Field Percent Progress Report in a number of ways is extremely important, as this system provides data for scheduling control (both for weekly schedule analysis and total project analysis), progress and productivity analysis at many different levels, methods studies, and forecasting revised progress and manpower targets. Thus, *WORK BREAKDOWN* and *COST CODE* policies are extremely influential in the structure and format of WORK PROGRESS documents. The Project Cost Code and Work Package Structure are important documents at this level. *DATA PROCESSING* is also essential to process the multi-level sorting and summaries of the Field Percent Progress Report as requested by the various project team members for their use in field management. While this progress reporting system does not rely directly on quantities, QUANTITIES IN PLACE are recorded on documents similar to that shown in Figure 12.32, only related to work orders and work packages because quantities are primary ingredients required for forecasting labor man-hours (for example as indicated by *DOCUMENTED ACTIVITY STATUS* (P_T 28)). The progress reporting cycle can be performed weekly on an informal basis, or bi-weekly/monthly to support formal project level reporting requirements.

The analysis of physical progress, when combined with FIELD INSPECTION information and LABOR HOURS, provides the data required for evaluating the work accomplished on the Weekly Work Schedule (P_T-DOC 38). This analysis is frequently performed in weekly project team meetings wherein specific problems can be assessed and either resolved immediately or moved towards a resolution through detailed analysis and action after the meeting. This *SCHEDULING FOCUS* approach to weekly work schedule analysis gives an up-to-date view of current progress and productivity that can support present plans and/or forecasts. The information gleaned from the Weekly Work Schedule analysis provides input primarily for Crew Planning and Crew Management, but begins to highlight trends that will be further reviewed in Project Status, Analysis, and Reporting. It also serves as the information vehicle from which the general superintendent can write a Weekly Construction Report (I_T-DOC 16), highlighting activity status problems (material shortages, manpower problems), critical management decisions, and so on. A typical report is presented in Figure 12.42. Thus, many project team members are involved with these aspects of the project status area of Field Monitoring. Finally, weekly schedule analysis provides the area/craft superintendents and cost and scheduling engineers with the input necessary to status the Monthly Work Activity Plan (P_T-DOC 27) for the formal monthly construction progress report, as well as updating this document to reflect next month's work load.

*Labor units are normally labor man-hours and can be either site-or company standard-based man-hours.

CHAPTER 13
Project Status, Analysis, and Reporting

Project Status, Analysis, and Reporting is the decision phase in which the contractor melds together all the documented project status information, and processes this information into data that are pertinent to project management requirements. Documented input from status reporting sources and feedback based on visual observations are processed by the Management Information System, and formulated into project summary reports that form the basic framework from which management decisions can be made by key project managers.

The nature and demands of the construction environment significantly impact the Project Status, Analysis, and Reporting requirements, as a broad range of approaches to addressing this problem area exist in practice. In some contracting environments, cost accounting techniques which gather actual cost data and display these data against the current cost estimate provide ample information for management control. The introduction of a simplified trending program through cost-to-complete analyses provides management with pertinent information for decision making. Integrated cost systems emerge in other construction environments. These systems have the objective of identifying a broad range of potentially critical problems associated with labor performance and rates of progress that require attention and possible method analysis. This latter focus may result from the analysis of superintendent/foremen reports, and often stresses the need for decision making through direct interface with the field managers and supervisors.

The variety of construction environments also produces a broad range of problem solving approaches to scheduling analysis and reporting. In some environments, the schedule is reviewed in terms of critical subcontractor interfaces and status is displayed on the schedule in relation to planned activities. In other construction environments, the schedule provides both activity status with projections of new completion dates, and progress and manpower curves. Furthermore, there is an interest in updating schedules to determine their impact on available resources and future requirements. Finally, fully integrated scheduling control emerges when revised work packages and new scope determinations *are integrated with* current status, revised predictions of achievable productivities, labor availability, and all other remaining factors which influence the schedule. These factors include drawing issues and material deliveries.

Current project data can be documented in relation to the project schedule, work volume in place, percent complete, labor productivity and manpower profiles, and dollar expenditures incurred to date. This information becomes significant and useful to management when it is compared to the project construction plan and budget for the current period and for the total to date. Comparisons of actual versus planned status and progress yield information from which the contractor can determine the actual overall performance to date against that implied in the original plan. Specifically, the contractor is interested in determining whether the project is ahead of or behind schedule, overrunning or underrunning its budgets, and whether it is above or below anticipated productivity levels. Such analyses will assist the contractor in the selection of future courses of action, with the view of completing the project on time and within budget, and minimizing any departures from these critical goals. Finally, Project Status, Analysis, and Reporting efforts will often include a continuous program of activity analysis which focuses on the performance of critical or cost-sensitive construction methods, the evaluation of labor productivities and equipment utilization, and the use of cost comparison analyses to improve the direction of the project.

Project Status, Analysis, and Reporting is a decision area which is built around, feeds, and utilizes the contractor's Management Information System to produce standard project reports. This formal process has been conveniently contrasted with the preparation, on request, of specialized reports for key managers and top level company management, a focus which has been reserved as a function of Project Management Control.

13.1 THE BASIC DECISION PROCESS

During the life of a project, the contractor must periodically enumerate the resources used, review the effectiveness with which those resources are made available and consumed during project execution, and relate them to project needs and current status. This analysis of resource use and project status will enable the contractor to begin to evaluate whether or not the project is making or losing money, or is ahead of or behind the project sched-

TABLE 13.1 SCOPE OF PROJECT STATUS, ANALYSIS, AND REPORTING PROBLEM AREA

Major Decision Problem	Component Problem Statements
45. How are status and progress defined:	How complex is project statusing? How accurate must it be? What labor units are required? How is project status to be portrayed? How is project schedule status defined? How is current status related to progress, productivity, and milestone achievement? What range of cost accounting is required?
46. What is labor productivity:	How will labor productivity be defined? What labor productivity analysis is required? How will labor productivity be documented? What are significance and causes of productivity changes?
47. What are current status and progress:	What project milestones have been achieved? What work activities are complete? What man-hours and labor dollars have been expended? What are current manpower and craft levels? What is percent progress to date? What are actual crew and craft wage rates? What are actual labor unit rates? What is actual productivity?
48. What are planned status and progress:	What project milestones should be complete? What work activities should be complete? What man-hours and labor dollars should be expended? What should manpower and craft levels be? What are predicted productivity, crew, and craft wage rates? What is planned percent progress? What are estimated labor unit rates?
49. What deviations can be identified:	What critical milestones are behind schedule? What work activities are ahead of or behind schedule? How does expenditure of man-hours and labor dollars compare to plan? Are crew and craft wage and labor unit rates over or under current budget? How do percent progress, manpower and craft levels, and productivity compare to plan? What changes in scope have occurred? Do field reports suggest deviations in activity status and labor performance?
50. What are implications of deviations:	Will future milestones be met? Will man-hour, labor dollar, and manpower overruns or underruns occur? Can progress required for completion be achieved? Do we need more manpower? What change orders or trends should be initiated? Should activity analysis be performed? Should management consider motivational improvement measures?
51. What is impact of work scope changes:	Is schedule revision necessary? What is thc impact of changes and trends on manpower, schedules, and budgets? What is the impact of revised schedule on progress curves, manpower profiles, project milestones, crew planning, and crew schedules? Should work package scope be revised? What is the threshold limit for trending analysis?

TABLE 13.2*a* GENERAL PROCEDURE FOR PROJECT STATUS, ANALYSIS, AND REPORTING (SCHEDULING STATUS PHASE)

Procedural Step	Management Duties and Tasks
1.	Identify construction status for activities currently in progress.
2.	Evaluate current status of each scheduled activity in relation to planned status.
3.	Determine the impact of work scope changes, trends, indicators of current progress rates, current predictions of material delivered, drawing issue dates, and subcontract completion dates required to complete effort on the current construction plan and schedule.
4.	Add and delete construction activities based on work scope changes and current knowledge of scope.
5.	Revise construction sequence logic based on decisions relating to changed priorities, changes in productivity, and/or anticipated sequences of field effort.
6.	Revise the schedule of project activities remaining to be completed and revise manpower and/or progress curves.

TABLE 13.2*b* GENERAL PROCEDURE FOR PROJECT STATUS, ANALYSIS, AND REPORTING (COST, PRODUCTIVITY, AND SCOPE ANALYSIS AND REPORTING PHASE)

Procedural Step	Management Duties and Tasks
1.	Determine actual labor man-hour and labor dollar expenditures for current period and for totals to date.
2.	Evaluate man-hour and labor dollar expenditures in relation to work volume in place in current period and to date.
3.	Assess rate of work output in relation to progress achieved by crafts, work package, cost code, and crew dissections in current period and to date.
4.	Compare and document actual productivities against planned productivities in current period and to date.
5.	When necessary, initiate controlled analysis of specific construction methods, crew performances, material dumps, supply routes to work location, and equipment usage.
6.	Evaluate man-hour, labor cost, and schedule impact of client-authorized changes in scope. Implement change in scope if approved by client.
7.	Evaluate man-hour, labor cost, and schedule impact of project labor related trends, and implement trend results.
8.	Reevaluate work package scope and update work package content based on current construction status, productivities, impact of change orders and trends, and site conditions.
9.	Inform field managers/supervisors of revised work package scope and content.

ule. When the contractor compares current status against planned status, variations from the plan such as failure to achieve planned productivities and the like may be revealed, and may be shown to have influenced the direction of the project. The Management Information System is operating properly when the contractor is presented with project information that enables problems to be detected in sufficient time for appropriate corrective action to be taken. This requirement for the detection of potentially critical management problems is the charge and focus of Project Status, Analysis, and Reporting.

This decision area necessarily has an overall project performance focus, and consequently the manpower and labor components considered here represent only a portion of the scope and focus of the decision area. Thus, areas of significant management concern such as the status of material deliveries and the costs of permanent plant material and equipment are not considered here in detail but only, where necessary, in terms of their potential impact on the manpower and labor aspects of the project. However, certain general project-focused functions must be performed that, while not specifically manpower related, bear directly on the manpower content of a project.

In terms of project manpower management, Project Status, Analysis, and Reporting is concerned with:

1. The determination of the current project status in relation to the project schedule, work in place, labor man-hours, and labor dollars expended.
2. The comparison of actual status and progress versus planned status and progress.
3. Project trending and the implication of forecasted project positions.

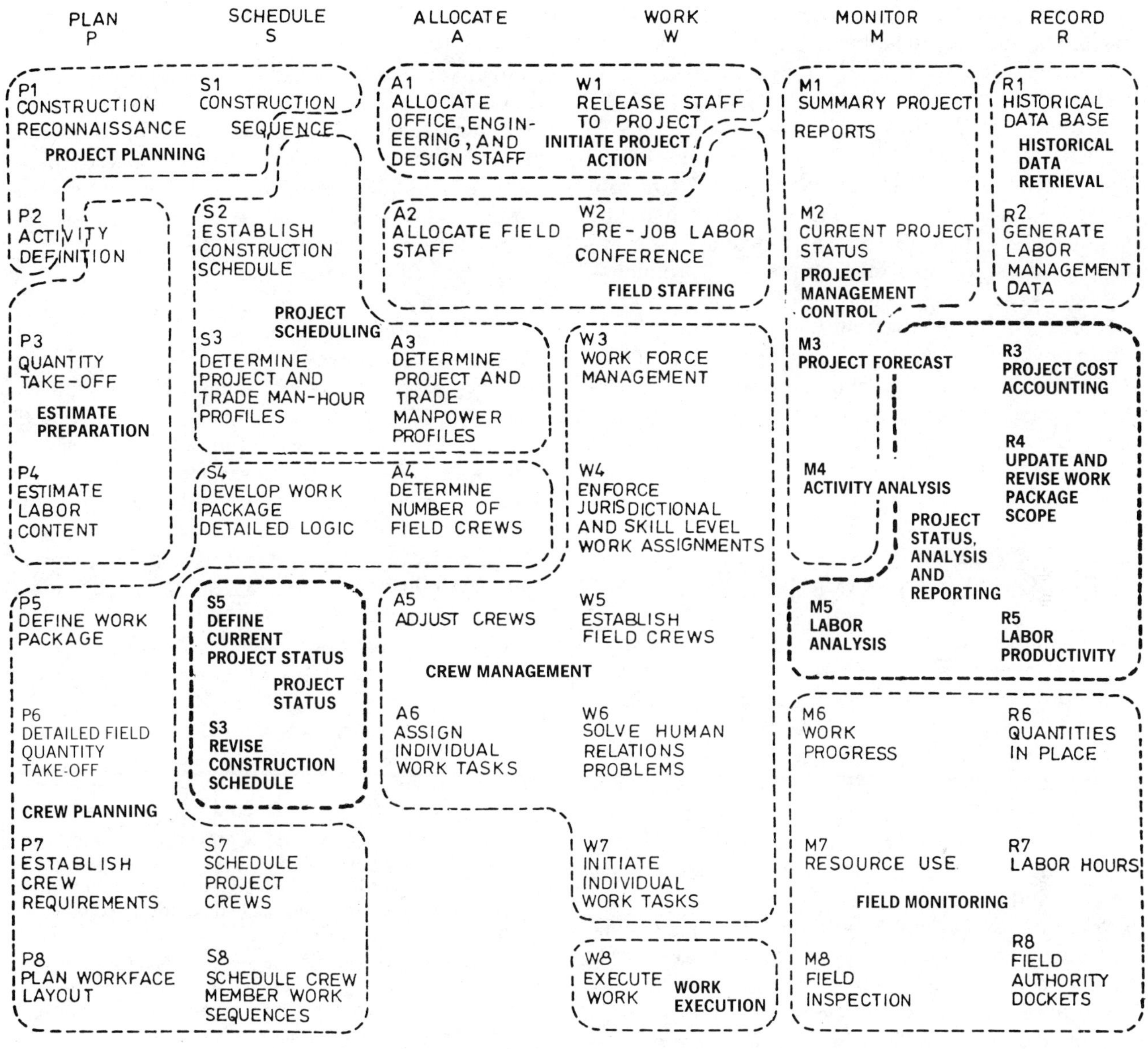

Figure 13.1 Functional plot of the Project Status, Analysis, and Reporting decision process.

4. The analysis of construction methods in relation to labor productivity.
5. The preparation of project reports.

When decision makers address the manpower management problems associated with Project Status, Analysis, and Reporting, they must focus on the component problem statements described in Table 13.1. The management action to solve the major problem areas described in Table 13.1 can be formulated into detailed management duties and tasks, as shown in Table 13.2. The general procedure prescribed in Table 13.2 is split into two decision phases. Table 13.2*a* relates specifically to scheduling status, while Table 13.2*b* focuses to a greater extent on the cost, productivity, and scope analysis and reporting phase. These two phases can be integrated to form the entire project status picture from which key project-oriented management decisions are made.

The performance of the detailed management duties and tasks indicated in Tables 13.2*a* and 13.2*b* are the essence of Project Status, Analysis, and Reporting. These duties and tasks form the basis for identifying the manpower management functions related to this decision area, as depicted in the functional plot in Figure 13.1. The functional statements for each management function in Figure 13.1 are provided in Figure 13.2. The DEFINITION OF CURRENT PROJECT STATUS (S5) and REVISING THE CONSTRUCTION SCHEDULE (S6) correspond to the duties and tasks of the scheduling status phase (Table 13.1*a*). The remaining six management functions correlate to the cost, productivity, and scope analysis and reporting phase shown in Table 13.1*b*. Together, these two decision phases are equivalent to the problem scope definition described in Table 13.1. Finally, a schematic illustration of Project Status, Analysis, and Reporting interfaces with other related decision processes is shown in Figure 13.3.

13.2 FUNCTIONAL PERFORMANCE TECHNIQUES

The management importance of and magnitude of effort expended by the contractor on the Project Status, Analysis, and Reporting area can generally be related to project size and complexity in much the same manner that field documentation of data is related to project size and complexity (see Figure 12.4). Thus, a natural correlation exists between the level of documented data produced and the level of project status and progress reporting. While the collection and documentation of data increases with project size and complexity, the fruits of a substantial documentation effort are not always realized. This is because time constraints and other factors often limit the analysis of the reports that are produced. This can be overcome, to some extent, by a top management commitment to the effective utilization of the reports and information generated. Consequently, management must have the flexibility to determine the types of project information and reports it desires as well as the management application of these reports. The focus of the project information system could range from a system that simply alerts management to overall project status and costs to one that culminates in management information produced on a multi-level-user-integrated access basis. This latter mode can effectively eliminate the possibility of a report overload, at any decision level, thus eliminating many improperly focused and/or superfluous documents.

The importance of Project Status, Analysis, and Reporting is also related to circumstances imposed by influences outside the contractor's control. For example, there are legal requirements concerning the employment of labor that result in payroll reports and other related data. Contractual arrangements between the client/owner and the contractor often precipitate formal reporting of material deliveries, work activity, subcontractor progress, and so on.

Project Status, Analysis, and Reporting, as shown in Table 13.2, has two distinct but interrelated decision phases, namely a scheduling status phase and a cost, productivity, scope analysis, and reporting phase. Both decision phases are built on a number of basic concepts that lead in practice to the use of a very broad range of functional performance techniques. This range of management techniques naturally results when the management requirements for control of projects of various types, sizes, and complexities, are combined with the contractor's flexibility to design and implement Management Information Systems that reflect a management philosophy and personal management style.

The scheduling status phase is concerned with

S5 DEFINE CURRENT PROJECT STATUS

- To establish current project status in terms of relevant indicators based upon past progress, definitive estimates, trends, and predictions of required "to-complete" effort.
- Requires detailed knowledge of project scope and definition, definitive estimates, work package definition details, current productivities, subcontractor and material delivery status, construction schedule and field inspection, and release of drawings.
- Considers accuracy of estimate, field measures of progress and productivity, choice of status indicators, milestones, and relation of planned project status to known current situation and impact of current decisions, productivities, and trends on project completion.

S6 REVISE CONSTRUCTION SCHEDULE

- Establishes new plan and schedule of project activities to project completion on basis of current project status, conditions, and decisions.
- Requires knowledge of project scope and activities, current project status, and decisions relating to changed priorities and sequences of field effort.
- Considers revised estimates, work package definitions, current work rates and productivities, material procurement problems, subcontract problems (if prime contractor), and revised delivery and drawing issue dates.

(*a*)

Figure 13.2 Functional statement of the Project Status, Analysis, and Reporting decision process. (*a*) Scheduling status phase.

M3 PROJECT FORECAST

- The assessment of changes in scope and/or trends that cause a deviation from the budget estimate and/or plan for accomplishing the work.
- Requires detailed knowledge of the current project scope and definition; budget estimate; project plan and schedule; current progress and performance-to-date in relation to budgeted quantities, man-hours, labor costs, and productivity; and a new prediction of total quantities, man-hours, and labor costs at project completion based on current progress, performance to date, and the remaining work to be completed.
- Considers the nature and cause of changes in scope and trends, the impact on project completion and costs, the impact on field man-hours and manpower, and the demand on engineering, design, and field management effort.

R3 PROJECT COST ACCCOUNTING

- Accumulates and records the actual costs and actual man-hours incurred over the life of a project.
- Requires a detailed knowledge of labor cost accounting methods, including time card applications, project cost code systems, craft wage structures, benefits, standard deductions, and payroll procedures, as well as an understanding of budget/revised estimates and project team and craft makeup.
- Considers labor man-hours expended by craft; project team man-hours expended by classification; basic wage rates, including wage differentials, overtime, burdens, and benefits; and costs to date.

M4 ACTIVITY ANALYSIS

- The controlled analysis of a construction method in relation to crew size, crew production, crew location to the workface, material and equipment location, and material flow routes.
- Requires knowledge of work package scope and details, an understanding of sophisticated field management techniques, intimate definition of construction operations for each work task, and knowledge of method improvement reporting.
- Considers the construction sequence of each detailed work task, the effective use of all resources, crew design, the needless waste of time, delays, specific location of men in relation to material supply routes, equipment location, congestion of the work space, and the production output of the crew.

R4 UPDATE AND REVISE WORK PACKAGE SCOPE

- Updates and revises the full scope and work details of project work segments based upon current field progress, status, crew productivities, and job conditions, and serves as a basis for the rebriefing and rescheduling of field construction managers/supervisors and crews.
- Requires an intimate knowledge of the current work package scope relative to physical progress, scheduling status, man-hours expended, and man-hours to complete, latest scope changes and trends, actual field conditions, and construction methods.
- Considers the current work package scope versus existing conditions and status as a means of revising work task and work assignments by trade; revising detailed logic for work package; reassessing field estimates of quantities and labor hours; reevaluating effective use of resources; and redefining procurement, equipment locations, and material flows to the workface.

M5 LABOR ANALYSIS

- Assessment of labor productivity by craft, work package, crew, and/or cost code.
- Requires a knowledge of project cost codes, estimated labor content correlated against actual experience, a predetermined measure and understanding of productivity, physical progress, and work package content.
- Considers the correlation between estimated labor data including quantities, labor unit costs, unit man-hours, and labor hours, and actual quantities, labor unit costs, unit man-hours, and labor man-hours in relation to a target performance indicator; the impact of job conditions, crew design, performance and motivation, and the effectiveness of field management.

R5 LABOR PRODUCTIVITY

- Documents labor productivity by craft, work package, crew, and/or cost code.
- Requires a knowledge of the project cost codes, estimated labor data, actual experience, performance indicators, physical progress, crafts, crew profiles, work package contents, and the pertinent recording documents.
- Considers the correlation between estimated data, including quantities, labor unit costs, unit man-hours, and labor man-hours, and actual quantities, labor unit costs, unit man-hours, and labor man-hours in relation to target performance indicators; the impact of job conditions, crew performance and motivation, and effect of field management on achieved productivity.

(*b*)

Figure 13.2 (*Continued*) (*b*) Cost, productivity, and scope analysis and reporting phase.

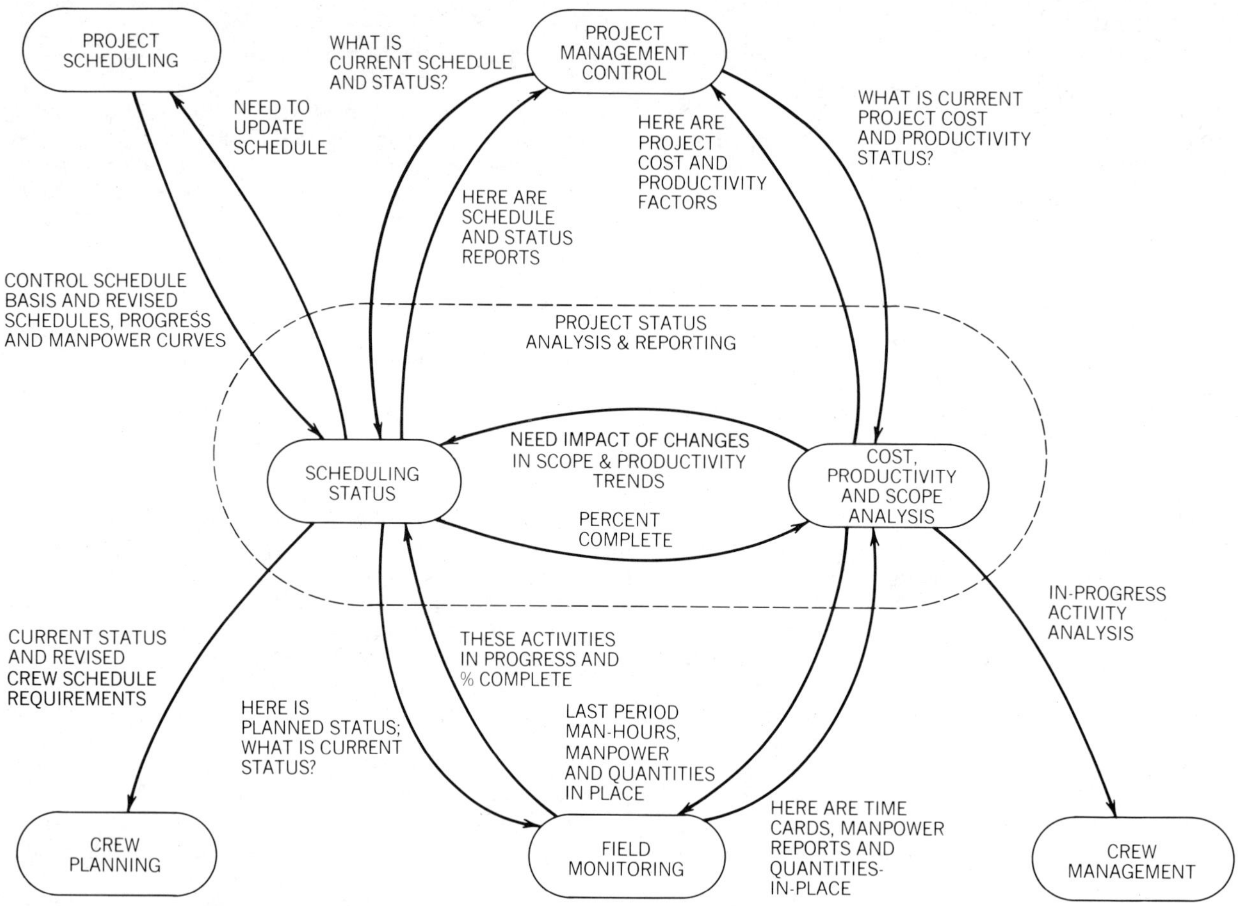

Figure 13.3 Schematic illustration of Project Status, Analysis, and Reporting decision process interfaces with other manpower management decision processes.

the general updating and revising of project schedules and progress and manpower curves. Thus, three dominant requirements emerge in the performance of the scheduling status function. First, a continuous knowledge of current project status is essential for the planning and effective use of field resources. Second, decision makers must analyze current commitments, manpower levels, and productivities in order to predict schedule requirements over the next planning horizon. Third, it is essential to establish a schedule that matches and is compatible with available resources and predicted productivities, even if this means the schedule must be revised. Consideration of these requirements and alerting management control attention to schedule problems is thus identified as a sound basis for schedule and cost control.

The scheduling techniques that implement the concepts described above are based on techniques presented in Project Scheduling,* and include activity status and progress documentation using tabular schedules, bar charts, CPM networks, monthly activity lists, progress curves, and manpower profiles. These techniques also incorporate methods of displaying revised scheduling and manpower data. Figures 13.4 through 13.9 depict a range of techniques applicable to various project types, sizes, and complexities. The correct application of each technique could aid management in the identification of potential problems. Specifically, comparisons of actual activity start dates with planned start dates (Figures 13.4 and 13.5), of actual progress versus planned progress (Figures 13.4, 13.5, and 13.6), and of actual versus planned manpower levels (Figures 13.8 and 13.9), indicate which areas of the project are behind or ahead of plan. When supplemented with

*See Section 7.2 and Figures 7.4 through 7.12.

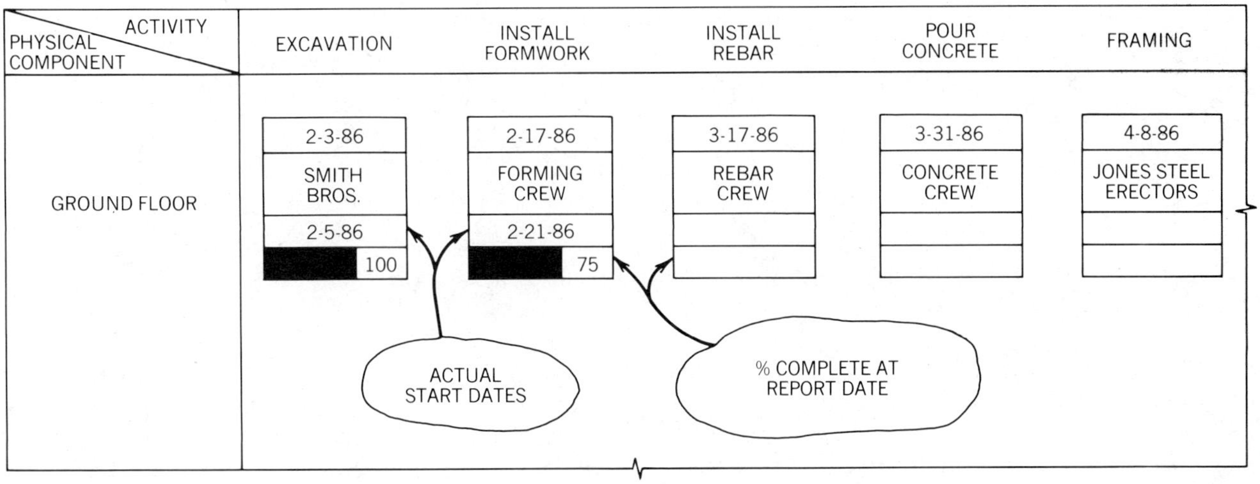

Figure 13.4 Tabular scheduling status technique.

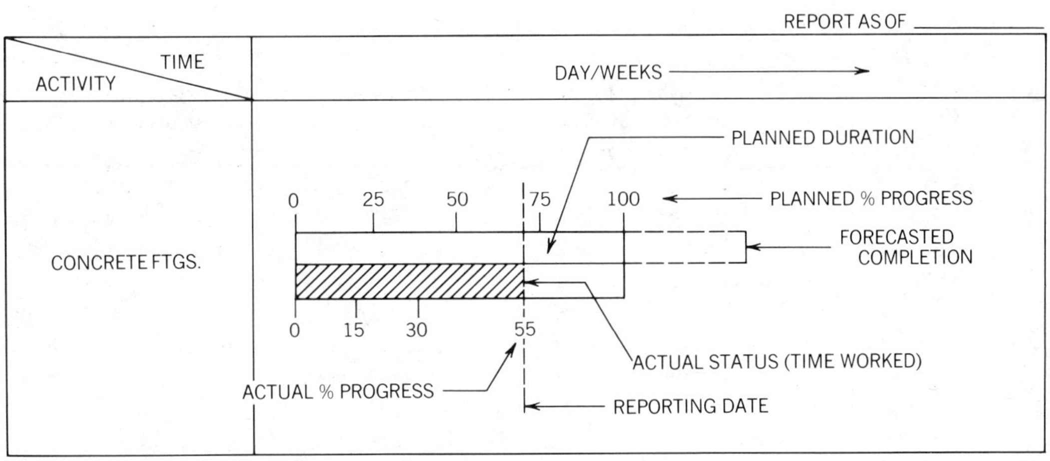

Figure 13.5 Bar chart status technique.

ACTIVITY IDENTIFIER	ACTIVITY DESCRIPTION	COST CODE	FLOAT		SCHEDULED START DATE		SCHEDULED FINISH DATE	% COMPL.	ESTIMATED OR ACTUAL FINISH DATE	CRITICAL
			FREE	TOTAL	EST.	ACT.				

Figure 13.6 Activity status list based on time-scaled network.

ACTIVITIES < 10 DAYS TOTAL FLOAT REPORT DATE AS OF ________

ACTIVITY IDENTIFIER	ACTIVITY DESCRIPTION	COST CODE	DAYS BEFORE LATEST START DATE	CRITICAL DAYS LOST	CRITICAL MAN/MACH. DAYS LOST	REASONS FOR DELAY	CORRECTIVE ACTION

Figure 13.7 Critical activities list (exception reporting).

critical activities lists (see Figure 13.7) only those activities that are in serious trouble will be highlighted for top management review on an exception basis. In large project environments these techniques often produce documents. In small project environments, however, the key decision makers know the schedule's critical activities (i.e., those activities that control the rate of progress of the schedule), and therefore carefully scrutinize their actual status and progress at the close of each reporting period. The input data for project scheduling status and analysis flows directly from Field Monitoring for certain items (e.g., activity start and completion dates and percent progress), as well as from the other decision phases of Project Status, Analysis, and Reporting (e.g., total percent progress, man-hours expended, etc.).

Project Status, Analysis, and Reporting incorporates the analysis and reporting of scope changes, labor costs, progress, and productivity. These four areas comprise the second decision phase (see Table 13.2*b*) of this decision process, and like scheduling status, are built on a broad range of concepts which produce an extensive range of related functional performance techniques.

Progress analysis and reporting, while essentially evaluated and documented during Field Monitoring for specific work activities and/or work packages, has an overall project focus. A broad project-focused concept related to progress reporting is that physical percent progress should be compatible with the project control philosophy (e.g., major cost categories, geographical areas, etc.), and that it *can be summarized* at a total project level. Several specific concepts relevant to progress analysis and reporting are: (1) physical percent completed to date, an essential indicator of the rate at which work is being accomplished; (2) physical percent yet to be completed, an essential indicator of the rate at which future work must be completed; and (3) physical percent progress, man-hours and labor productivity, which are closely interrelated.

Because of the importance of resource use, schedule, and cost interrelationships between percent progress and productivity, it is important, for both cost and schedule control purposes, to recognize the concepts associated with the formulation of productivity performance correlation techniques. At a gross level, overall performance can be evaluated in terms of labor cost overruns and underruns. In some construction environments, however, it is essential to evaluate labor performance by control areas, work packages, cost codes, work activity, and the like if useful information is to be generated for project management decision makers. Finally, it is necessary to define and document labor productivity for budget comparisons, schedule impacts, and labor forecasts.

Several techniques are presented in Figure 13.10 which illustrate a range of productivity and progress

REPORT DATE AS OF ________

ACTIVITY IDENTIFIER	ACTIVITY DESCRIPTION	COST CODE	% COMPL.	START		COMPLETE		MANPOWER		PROGRESS UNITS	
				PLAN	ACT.	PLAN	ACT.	PLAN	ACT.	PLAN	ACT.

Figure 13.8 Status of monthly activity plan.

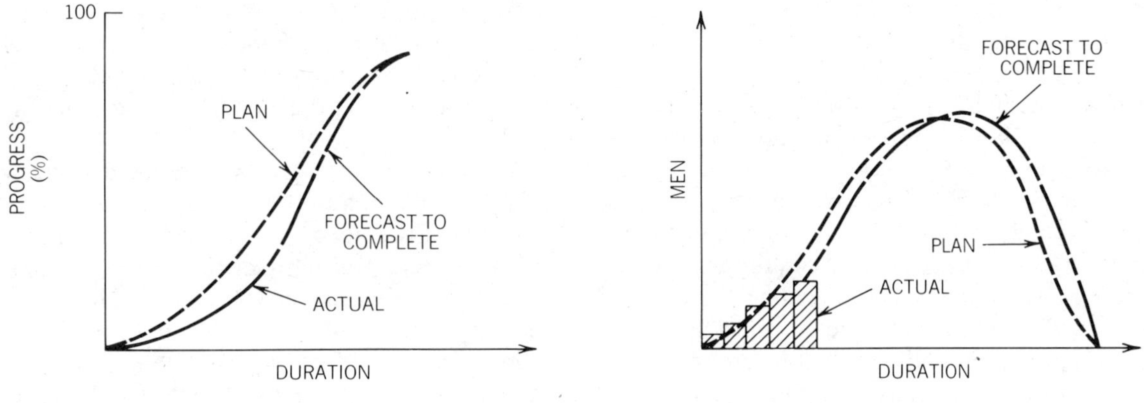

Figure 13.9 Updated and revised progress and manpower curves. (*a*) Progress. (*b*) Manpower.

COST CODE	WORK ITEM	% COMPL.	UNITS	ESTIMATE			PERIOD			TOTAL TO DATE		
				QTY.	COST	UNIT COST	QTY.	COST	UNIT COST	QTY.	COST	UNIT COST

(*a*)

CATEGORY	BUDGET MAN-HOURS	FORECAST MAN-HOURS	EXPENDED		% PROGRESS				PERFORMANCE	
			PD. M-HRS	TO DATE M-HRS	PD.		TO DATE		PD	TO DATE
					ACT.	PLAN	ACT.	PLAN		
• MAJOR COST CATEGORY • CONTROL AREA • WORK PACKAGE										

(*b*)

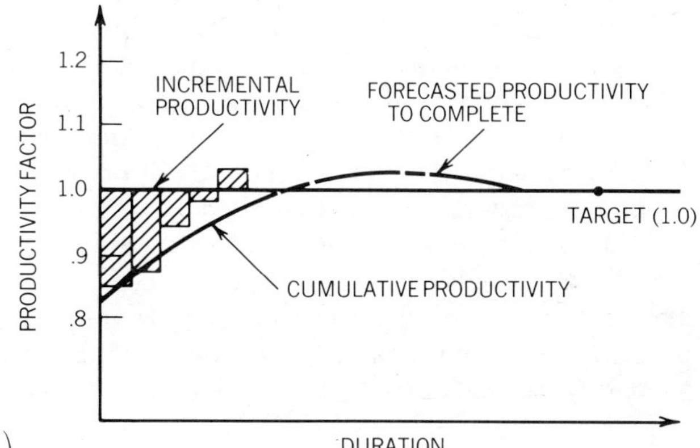

(*c*)

Figure 13.10 Progress and productivity analysis techniques. (*a*) Labor unit cost comparison analysis. (*b*) Progress and productivity

$$\left(\text{productivity} = \frac{\text{\% complete} \times \text{budget man-hours (earned value)}}{\text{expended man-hours}}\right)$$

(*c*) Productivity curves.

analysis and reporting approaches. For instance, Figure 13.10*a* provides a comparison by cost code and activity of the estimated labor unit costs versus labor cost expended for the period and for the total to date. When these comparisons are analyzed in relation to the percent complete for an activity, the decision maker can determine which activities are in trouble. An alternate approach, shown in Figure 13.10*b*, is to evaluate productivity in terms of a performance indicator as assessed against the percent complete and man-hours budgeted and expended. This technique is normally applied to broader project categories such as control areas or major cost categories, and the data can be graphically represented as displayed in Figure 13.10*c*. Comparison against planned data can reveal and identify problem areas that may require further investigation. It should be noted that total project progress can be calculated by weighting the percent progress for each constituent project item against its respective man-hour budget. From this weighting, the productivity for the total project can be assessed. The techniques presented in Figure 13.10 represent a small sampling of the techniques available in this area. Other techniques would include, but not be limited to, comparing actual labor costs versus estimated labor costs, quantities installed to date versus planned quantities to be installed, and actual man-hours per quantity unit versus budget man-hours per quantity unit. Another is to develop performance targets by estimating site man-hours and dividing by standard labor estimate man-hours. These techniques will be discussed in greater detail in Sections 13.3 through 13.6.

Management reports that tend to identify possible areas of concern, or which indicate the effects of as yet undetected problems are often of great assistance to top management, especially if they focus on project activities not normally covered by overall reporting techniques. Thus, the analysis of Activity Delay Reports (often called Foreman Delay Reports) may alert management to problems that are occurring (for example, a shortage of materials which is delaying crews), before the normal cost and productivity analyses indicate the magnitude of the effects caused by such problems. In these cases, management can be alerted to and can begin to pinpoint the type and location of problems that may need special analysis or immediate corrective decisions.

Activity analysis of construction operations is another fruitful area where analysis and reporting can assist field management in meeting schedules, labor costs, and productivity targets. Two concepts are considered as the underlying basis for activity analysis.

First, a proper method and field management analysis can pinpoint the magnitude of problems and reveal the need for corrective action and solutions. Second, for many critical cost- and schedule-sensitive activities, continuous method and field management analysis are essential for the achievement of projected labor cost savings and the meeting of project productivity and schedule requirements. The techniques most commonly associated with these concepts range from a "hunch" with a trial-and-error methods improvement approach, as schematically represented in Figure 13.11, to specific scientific management approaches, such as work sampling and method analysis, as illustrated in Figure 13.12. Figure 13.12 illustrates only one method for scientific management analysis of a work operation. Other known techniques in this area include time-lapse, crew balance charts, field ratings, and statistical analysis techniques. In addition, field labor unit cost or man-hour-per-unit studies can be conducted by work classifications or work areas, using diagnostic observational analysis. These studies, together with documented statements of corrective measures required for improvement, are currently found in construction practice.

While progress, productivity, and activity analysis focus on specific aspects of manpower management, scope analysis and project forecasting is a more general area of Project Status, Analysis, and Reporting. It is concerned with the identification of deviations from the planned schedule and budget, whether they be changes in the scope of work or trends that impact the project's labor content and schedule. The most fundamental requirement at the grass-roots level of construction is to keep track of anticipated final labor costs in relation to budget costs. This can be done using a technique similar to that shown in Figure 13.13*a*. This approach becomes more powerful when management philosophy dictates that it is essential to identify and quickly evaluate the impact of client changes in scope, design development, and/or performance trends on costs and schedules. Figure 13.13*b* illustrates a forecasting technique that is based on predicting productivities and wage rates. This technique can be utilized at any stage in the project

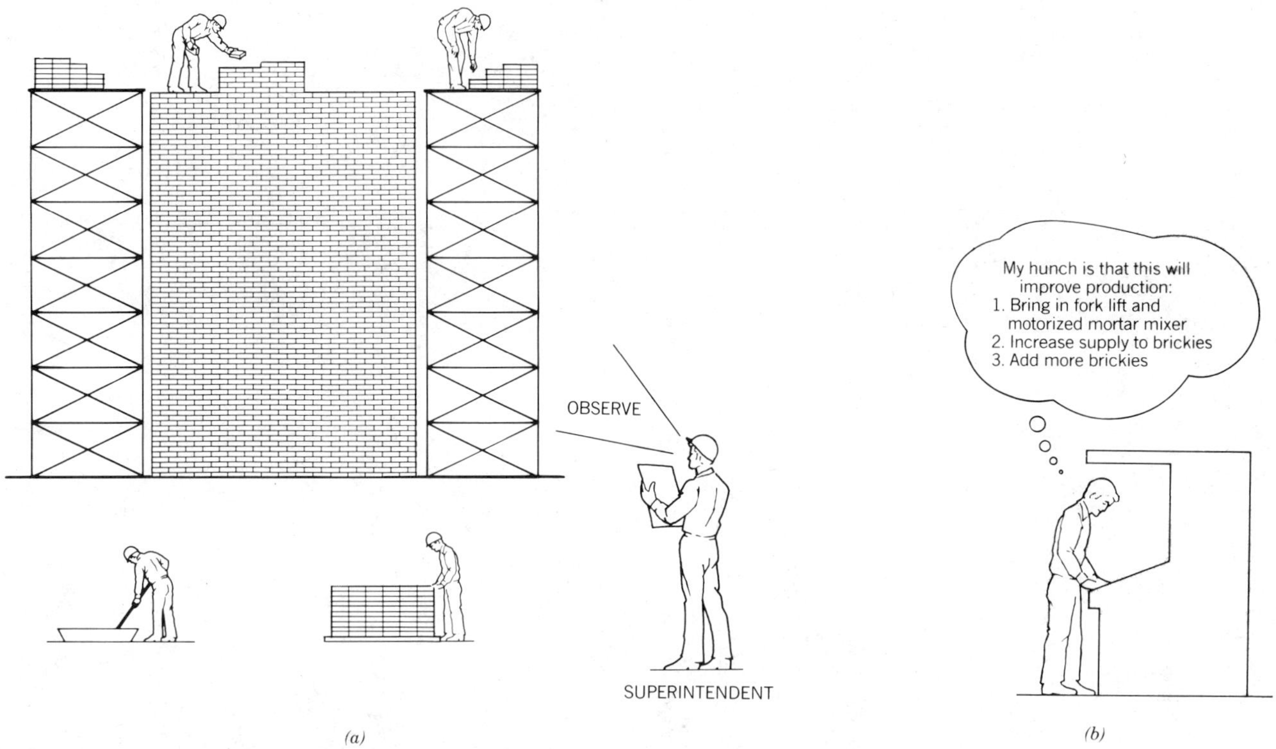

Figure 13.11 Hunch with trial-and-error activity analysis. (*a*) Observe and analyze work operation. (*b*) Change construction operation based on hunch.

when performance trends are significant enough to warrant revising the estimated labor man-hours and labor costs. Overruns can be detected early and appropriate corrective or compensating action can be taken by management. The more effective techniques of scope analysis and project forecasting include cost-to-complete assessments based on percent complete and *estimated* cost data (Figure 13.13*a*), cost-to-complete assessments based on percent complete and revised labor costs, productivities, wage rates, and so on (Figure 13.13*b*), and formalized cost engineering procedures such as a change order/trending program (Figure 13.13*c*). The techniques illustrated in Figure 13.13 must always incorporate a knowledge of the quantities required to complete the work.

Both the productivity analysis and project forecasting phases are dependent upon actual expenditures of man-hours and labor dollars. This requirement creates the need to establish and reconcile actual man-hours and labor costs to date, dissected whenever appropriate into work tasks, cost codes, work packages, and/or control areas. In addition, project accounting and financing must consider management's need to project future labor dollar expenditures over time in order to adequately assess future project payroll requirements. Several fundamental techniques are found in practice that address these functional requirements. These include standard bookkeeping practices as examplified in Figure 13.14*a*, labor cost dissection by cost categories, cost coding methods, and professional cost engineering practices wherein labor dollar expenditure curves are developed, as shown in Figure 13.14*b*. Cost accounting methodologies would also cover cost and financial issues related to progress payments, periodic billings, and short range cash flow forecasting (e.g., bi-weekly or monthly).

Project Status, Analysis, and Reporting is the focal point of the information flow through the Management Information System of a project. In this role, the MIS must focus on cost and schedule control issues to satisfy the need to feed live actual field data into the control database for comparative analysis of project cost and schedule performance

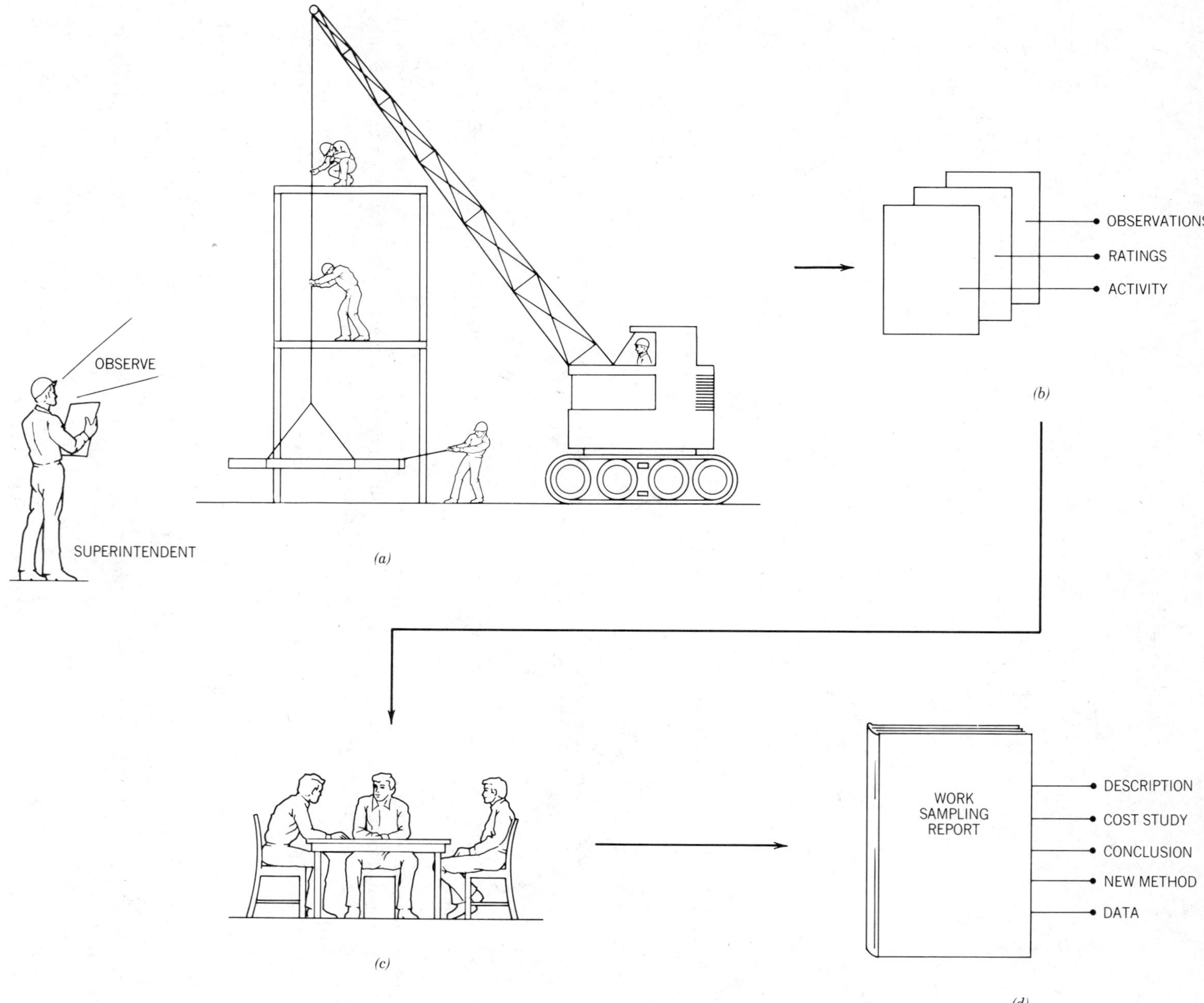

Figure 13.12 Activity analysis based on work sampling. (*a*) Observations and recording. (*b*) Documents. (*c*) Detailed analysis of operation with suggested improvements. (*d*) Report.

COST CODE	WORK ITEM	PHYSICAL COMPL.	EXPENDED TO DATE	ESTIMATE TO COMPLETE	UPDATED ESTIMATE	CURRENT ESTIMATE	NET	
							GAIN	LOSS
310.0	CONCRETE	%	$\$_1$	$\$_2$	$\$_1 + \$_2$	$\$_3$	$\$_3 - (\$_1 + \$_2)$	

(a)

[MAN-HOUR FORECAST] = [ESTIMATED MAN-HOURS REQUIRED TO ACHIEVE 1% PROGRESS] × [REMAINING % TO COMPLETE] = [MAN-HOURS TO COMPLETE] + [EXPENDED MAN-HOURS TO DATE] = [TOTAL FORECASTED MAN-HOURS]

[LABOR COST FORECAST] = [FORECASTED AVERAGE WAGE RATE TO COMPLETE] × [MAN-HOURS TO COMPLETE] = [TOTAL LABOR COST TO COMPLETE] + [LABOR COSTS EXPENDED TO DATE] = [TOTAL FORECASTED LABOR COST]

(b)

(c)

Figure 13.13 Project forecasting techniques. (*a*) Forecast based on percent progress and labor cost to complete. (*b*) Forecast based on analysis of predicted productivity and wage rates to complete in relation to remaining progress to complete. (*c*) Change order/trend procedure.

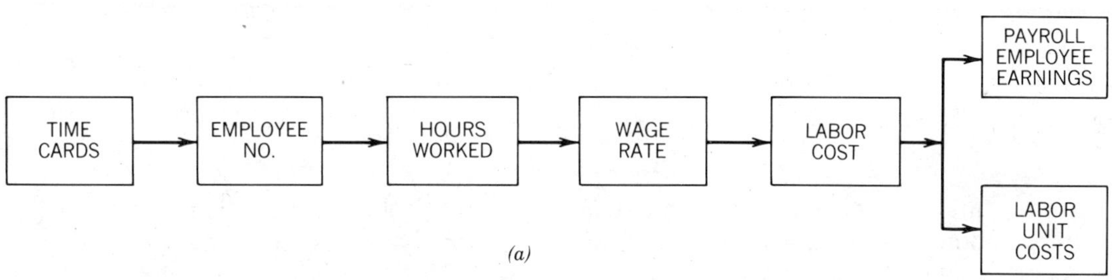

(a)

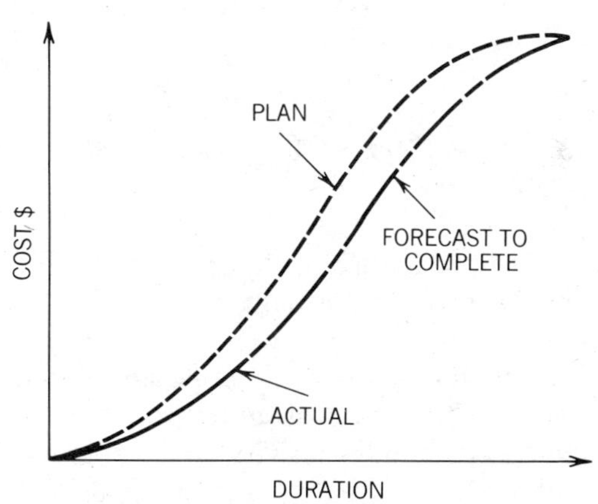

(b)

Figure 13.14 Accounting and cost engineering techniques. (*a*) Basic bookkeeping technique for labor cost expenditures. (*b*) Typical cost engineering technique for labor cost expenditure forecast.

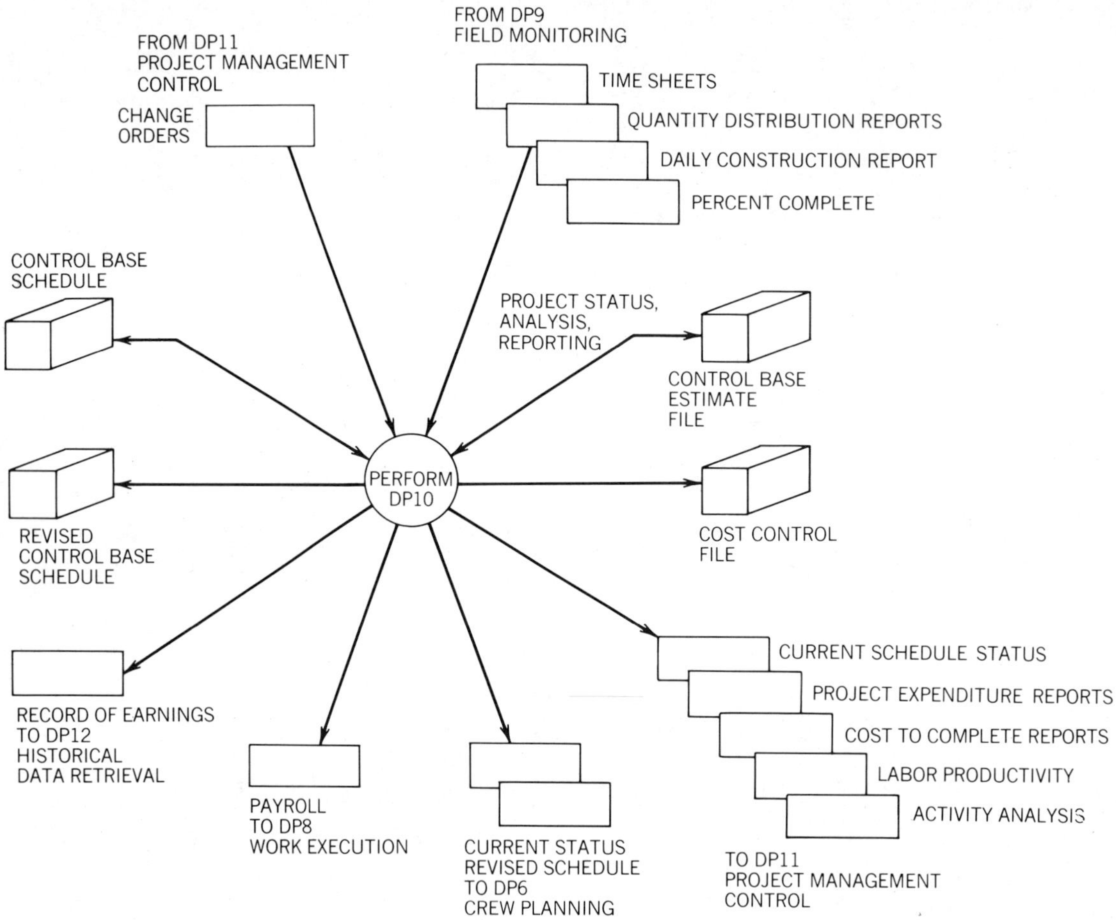

Figure 13.15 Project Status, Analysis, and Reporting Management Information System concept.

against current plans and forecasts. Thus, the Management Information System must process data, generate reports, assess differences, and subsequently distribute that information to other decision processes for management action.*

The Management Information System for the project is initially established with data derived from the control base estimate and schedule. This data is input into the contractor's project cost and scheduling system, and thus generates the active project cost and scheduling control files. The format of these files is determined to a large extent by the contractor's control philosophy. They take on characteristics of the contractor's cost code system and other management approaches. When live data is received from Field Monitoring, cost and scheduling control files are updated, reports generated, and critical problem areas identified through comparative analysis of actual data versus plan/forecast data. This MIS concept is schematically illustrated in Figure 13.15 for the Project Status, Analysis, and Reporting decision process.

13.3 THE PROJECT TEAM COMPONENT APPROACH

Due to the nature of the Project Team Component approach, the informal evaluation and field reporting aspects of Project Status, Analysis, and Reporting are very limited in scope and detail. They are confined to visual observations and quick assessments of simple activity and progress measures, and are carried out by the foreman/superintendent while monitoring field performance. This assessment is reinforced by the occasional management

*See, for example, Chapter 16.

TABLE 13.3 POLICIES OF THE SMALL FIRM

Policy		Description
P_S 3	*USE COST CODES*	Systematic classification of work categories segregated by unique cost codes is used to identify work tasks and collect costs.
P_S 6	*USE BAR CHARTS*	The bar chart planning model is the approach best suited for project scheduling.
P_S 7	*SCHEDULING FOCUS*	Scheduling definitions and constraints are: (1) critical activity is any activity that opens up work areas for other crews, and the delay of which causes work volume bottlenecks. (2) Activity durations and sequence logic incorporate the impact of seasonal weather conditions. (3) Activity scheduling is based on the earliest possible start dates. (4) Manpower impact is neglected.
P_S 19	*INSPECT PROGRESS*	Quantities in place versus total estimated quantities is an indicator of activity progress and is evaluated by visual inspection or quantity survey.
P_S 20	*FIELD LABOR HOURS*	Labor hours are monitored and recorded by management personnel closest to the workface.
P_S 21	*ACTIVITY STATUS*	Activity progress and labor productivity are not normally considered beyond completed activity volume, labor unit costs, and total labor dollars expended to date relative to activity scheduled target completion dates, estimated labor unit costs, and total estimated labor cost.
P_S 23	*MANUAL PAYROLL*	Payroll dollars are generated and recorded manually on documents showing standard deductions.

by exception visit of the project manager to the field site.

Project Status, Analysis, and Reporting is initiated when project progress data, in the form of quantities and labor man-hours, is collected in the field by the foreman and job superintendent, and is transmitted to the head office for data processing. The Daily Construction Report (P_S-DOC 6) is completed and returned to the head office on a daily basis, while the Time Sheet (P_S-DOC 5) and Quantity Distribution Report (P_S-DOC 7) are returned weekly, prior to each payroll period. Time Sheets enter the decision process through LABOR ANALYSIS (M5), while Quantity Distribution Reports enter at both the COST ACCOUNTING (R3) and DEFINITION OF CURRENT PROJECT STATUS (S5) functions. The type of data processing used is designed to be compatible with the level of detail considered essential for effective project management in the small firm's project environment. Consequently, *MANUAL DATA PROCESSING* (P_S 23)* is often used as input is periodically received from the field.†

At the small firm level the project manager and the estimator/bookkeeper are responsible for data processing. The project manager is primarily interested in actual quantities in place because these data are useful for defining project status and revising the construction schedule. He or she also assesses construction progress, reviews the labor hours expended by the field crews in order to maintain an awareness of where the man-hours are being spent, evaluates detailed labor cost reports, and makes the project management decisions. A review of the Daily Construction Report supplements the project manager's awareness of site activity and progress, which may be useful when analyzing current labor costs to date. Finally, processed labor cost information, a revised construction schedule (if required), and periodic site visits are all combined to enable the project manager to determine current project status and to prepare summary labor cost and other project reports that may be desired. The project manager will disseminate project status information to field supervisors by exception, that is, only when obvious problems appear. It then becomes the field's problem to initiate corrective action to forestall any further complications. Otherwise, the project manager assumes that the project is running smoothly.

The estimator/bookkeeper is responsible for project cost accounting, of which the primary focus is developing payroll dollars and labor costs to date. Time Sheets are reviewed for accuracy as a LABOR ANALYSIS task, and the *CORRECT USE*

*See Table 13.3 for the small firm policies relevant to Project Status, Analysis, and Reporting.

†Microcomputers provide a cost-effective means of data processing for the small contractor and are slowly replacing manual systems.

Anderson & Woodhead
Building Contractors
WEEKLY PAYROLL SUMMARY

PROJECT: ______ JOB NO.: ______
LOCATION: ______ SHEET NO. ______
PERIOD FROM ______ TO ______

EMPLOYEE NAME	EMPLOYEE NUMBER	CRAFT	MON	TUE	WED	THURS	FRI	SAT	SUN	TOTAL HOURS	RATE	TOTAL WAGES	DEDUCTIONS FED TAX	FICA	SDI			TOTAL	NET EARNINGS	CHECK NO.
JOHN SMITH	127	CARPENTER	8	8	8	8	8			40	14.50	588.00	88.20	39.44	58.80			186.44	401.56	4251
WILLIAM ANDERSON	130	IRONWORKER	8	8		8	8			32	14.75	472.00	70.80	32.96	56.64			160.40	311.60	4252
TOTALS																				

Figure 13.16 Weekly Payroll Summary (P_S-DOC 8).

OF COST CODES (P_S 3) assigned by the field is verified by the estimator/bookkeeper. This is executed weekly as time sheets are received.

Payroll dollars are generated manually and are recorded on two payroll policy documents. These are the Weekly Payroll (P_S-DOC 8) and Record of Employment and Earnings (P_S-DOC 9), as shown in Figures 13.16 and 13.17, respectively. These source documents are also input to COST ACCOUNTING. The man-hours expended per employee* are extracted from each Time Sheet and transferred to the Weekly Payroll Sheet. This document must reflect normal government legislated deductions for

*This includes overhead personnel. Their time is charged to an overhead account, however.

Anderson & Woodhead
Building Contractors

EMPLOYEE NAME JOHN SMITH S.S. NO. 536-48-4832 CRAFT CARPENTER
ADDRESS 127 NORTH DRIVE SEATTLE, WASHINGTON 98115 HOME PHONE NO. (206) 968-9721
EMPLOYEE NO. 127 HOURLY RATE 14.50 SINGLE ___ MARRIED X OTHER ___
STARTING DATE MARCH 6, 1980 TERMINATION DATE ______ EXEMPTIONS 2

PAYROLL PERIOD ENDING	TOTAL HOURS	HOURLY RATE	WEEKLY AMOUNT	TOTAL EARNINGS YEAR TO DATE	FEDERAL INC. TAX WKLY	TOTAL	F.I.C.A. WKLY	TOTAL	OTHER INSUR.		NET PAID	CHECK NO.
4-18-86	40	14.50	588.00	9996.00	88.20	1449.70	39.44	670.48	58.80		401.56	4251

Anderson & Woodhead
Building Contractors
SPECIMEN
CHECK NO. 4251
APRIL 18 1986

PAY TO THE ORDER John Smith $ 401.56
**FOUR HUNDRED ONE DOL 56 CENTS DOLLARS

UNION BANK 12-17/126

S.D. Anderson/A.W. Woodhead

Figure 13.17 Record of Employment and Earnings (P_S-DOC 9).

Anderson & Woodhead
Building Contractors

PROJECT: ____________ JOB NO.: ____________

LOCATION: ____________ SHEET NO. ____________

PREPARED BY: ____________ DATE: ____________

WEEK END: ____________

WORK CLASSIFICATION	COST CODE	CRAFT	MON	TUES	WED	THUR	FRI	SAT	SUN	TOTAL HOURS
WALL FTGS-FORM	B. 100	CARPENTERS	16	16	16	16	16			80
		LABORERS	8	8	8	8	8			40
TOTALS FOR WEEK										

(a)

Anderson & Woodhead
Building Contractors

PROJECT: ____________ JOB NO.: ____________

LOCATION: ____________ SHEET NO. ____________

PREPARED BY: ____________ DATE: ____________

WEEK ENDING: ____________

WORK CLASSIFICATION	COST CODE	CRAFT	TOTAL HOURS	HOURLY RATE	TOTAL	COST THIS PD.	QTY. PERIOD	UNIT COST PD.	COST TO DATE	QUANTITY TO DATE	UNIT COST
WALL FTGS.-FORMS	B.100	CARPENTER	80	14.50	1160.00						
		LABORER	40	12.50	500.00	1660.00	662	2.51	3944.00	1450	2.72
WALL FTGS.-CONC.	B.100	CARPENTER	8	14.50	116.00						
		LABORER	24	12.50	300.00						
		CEMENT MASON	8	13.75	110.00	526.00	32	16.43	999.96	52	19.23

(b)

Figure 13.18 Labor Distribution Report (P_S-DOC 10). (*a*) Man-hour distribution by work classification. (*b*) Labor unit cost analysis.

each employee, such as FICA and withholding taxes. At the end of the week, the man-hours for each employee are summarized, the appropriate rate is entered, and the various payroll calculations are made. The bookkeeper must also segregate those portions of the employee's hourly wage which are payable to the relevant union as union benefits.* The bookkeeper must rely on each union's current agreement for the applicable benefit patterns.

After the payroll distribution is completed, the bookkeeper can complete the Labor Distribution Report (P_S-DOC 10), as described in Figure 13.18, by inserting the wage rate for each craft, and multiplying the wage rate by the total craft man-hours to determine total labor dollar expenditures for the pe-

*The wage rate must either be the base wage rate or the base wage plus all benefits. Most contractors use the base wage rate. Regardless of which wage rate is used, it must be applied consistently to all labor documents. If the base wage rate is used, union benefits must be recorded and accumulated on separate accounting documents.

Anderson & Woodhead
Building Contractors

PROJECT: ______________ JOB NO.: ______________

LOCATION: ______________ SHEET NO.: ______________

CLASS	B. 100 WALL FTGS.-CONC		B. 100 WALL FTGS.-FORMS						TOTAL CONCRETE	
WEEK ENDING	COST $		COST $		COST $		COST $		COST $	
	WEEK	TO DATE	WEEK	TO DATE	WEEK	TO DATE	WEEK	TO DATE	WEEK	TO DATE
	2284.00	2284.00	473.96	473.96					2757.96	2757.96
	1660.00	3944.00	526.00	999.96					2186.00	4943.96

Figure 13.19 Labor Cost Summary (P_S-DOC 11).

riod. Because this document is activity oriented, labor costs are accumulated for specific work tasks by cost code. The total direct labor cost divided by the quantity of work placed, equals the labor unit cost for the period. The total direct labor cost and quantity placed per period is added to the total cost to date and total quantity in place from the previous period, generating a new total labor cost to date and total quantity in place. The new total cost to date is divided by the new total quantity in place to provide the average labor unit cost achieved for the work task from the inception of that task. The total project labor dollars (summary of the total cost column on the Labor Distribution Report) for the period must equal the total gross payroll dollars (before deductions) for that period.*

In order to compare the estimate with actual labor costs, the work tasks on the Labor Distribution Report are categorized by the activities defined in the estimate. A Labor Cost Summary (P_S-DOC 11) is used to compile this information. As shown in Figure 13.19, labor dollars for each work task are transferred from the Labor Distribution Report to their respective activities on the Labor Cost Summary. The totals for the activities are generated and transferred to the comparable activities on a Project Cost Recapitulation (P_S-DOC 12). This document, shown in Figure 13.20, depicts a comparison between estimated total costs and actual total costs to date.† If the project is complete, the columns would actually contain the final costs for the project for that general work classification. Any differences are the net gains or losses on the estimate. In terms of final costs, the Project Cost Recapitulation portrays the activities in which profits and losses occurred. If the project is incomplete, the differences in terms of labor dollars will not reflect actual profit and loss for those activities, but rather the labor dollars required

*Many small contractors' labor unit prices reflect unit costs which also exclude standard deductions. In this case, the total project labor dollars for a period must equal the total net payroll for that period. Payroll deductions would appear elsewhere within the accounting records for the project.

†The labor Cost Summary and Project Cost Recapitulation reports could be computerized on a microcomputer using a spread sheet program. The authors would urge small contractors to pursue this approach (see bibliography).

Anderson & Woodhead
Building Contractors

PROJECT COST RECAPITULATION

PROJECT: ______________ JOB NO.: ______________

ARCHITECT: ______________ SHEET NO.: ______________

LOCATION: ______________ DATE: ______________

PREPARED BY: ______________ WEEK ENDING: ______________

WORK CLASSIFICATION	CODE	BUDGET ESTIMATE				APPROVED CHANGES				CURRENT ESTIMATE				ACTUAL TO DATE			
		MAT'L	LABOR	S/C	TOTAL	MAT'L	LABOR	S/C	TOTAL	MAT'L	LABOR	S/C	TOTAL	MAT'L	LABOR	S/C	TOTAL
SITE WORK	A	1250	8700	0	9950	0	0	0	0	1250	8700	0	9950	1310	7590	0	8900
EXCAVATION AND BACKFILL	A	10100	12220		22320	0	0	0	0	10100	12220	0	22320	9125	10130	0	19255
CONCRETE	B	125352	69024		194376	0	0	0	0	125352	69024		194376	9213	4944		14157

Figure 13.20 Project Cost Recapitulation (P_S-DOC 12).

to complete, as compared with the total estimate. A comparison of estimated and actual labor unit costs, combined with an analysis of the activities' construction status (percent complete), is necessary to accurately determine profit or loss. In this case, the profit or loss could only be considered a forecast. (This type of analysis is executed as part of CURRENT PROJECT STATUSING (M2).)

The construction statusing effort is performed simultaneously with cost control, and is the responsibility of the project manager. It is initiated as progress input is received from the field. The specific documents from which progress data are extracted are the Daily Construction Reports and the Quantity Distribution Reports. Information from these documents is coupled with input gleaned from periodic site visits to yield the overall construction status. Quantitatively, activity progress is measured in terms of total quantities completed to date, and is evaluated visually or by quantity survey in accordance with *PROGRESS INSPECTIONS* (P_S 19). This measure of progress is subsequently portrayed on the now statused bar chart Construction Schedule (I_S-DOC 2), as shown in Figure 13.21*. The construction schedule must be reviewed and revised, if necessary, in order to meet target completion dates. Finally, PROJECT COST ACCOUNTING establishes the value of work progress and prepares the Progress Payment Voucher (P_S-DOC 13), Figure 13.22, for presentation to the client.

Construction statusing involves DEFINING CURRENT PROJECT STATUS (S5) and REVISING THE CONSTRUCTION SCHEDULE (S6). THE DEFINITION OF CURRENT PROJECT STATUS is achieved by first quantifying activity progress, and then portraying this progress on the construction progress schedule. The Quantity Distribution Report furnishes the project manager with total quantities completed to date for all active work tasks. Each work task is matched to its respective activity on the original Construction Schedule (I_S-DOC 2), and the total quantities completed for an activity becomes the sum of its component work tasks. Dividing the total quantity completed to date by the total estimated quantity (from P_S-DOC 3), is a method used to estimate the percent complete for each activity. This percent must be confirmed by referencing project drawings, evaluating visually the activity progress in the field, and reviewing the Daily Construction Report. This check is essential because the percent complete could be misleading if the quantities from the estimate were incorrect. When the contractor is the prime contractor on the job, the percent complete for subcontract work is furnished by the subcontractor and should be verified by either the project manager or the job superintendent. The actual percent complete is shown on the statused Construction Schedule by adding a bar reflecting activity progress. This percent can be represented either incrementally or in total. Typical examples of statused and revised Construction Progress Schedules (I_S-DOCs 3 and 4), are illustrated in Figures 13.21*a* and 13.21*b*, respectively.

In order to fully define current project status, the project manager must analyze the status of each activity on the Construction Schedule and make a determination as to where delays in activity progress will either jeopardize the completion date, or possibly restrain a subcontractor (if the contractor is a prime contractor) from commencing or finishing the work. The percent complete versus the planned progress for the reporting period is the most common measure of activity status. When this relationship is delineated on the bar chart schedule, the project manager can quickly determine those activities that are behind or ahead of schedule and planned progress. In order to accurately evaluate the project's overall status, the project manager must be cognizant of the activities which are critical in terms of scheduled target completion dates. If a critical activity is significantly behind schedule it might prevent subsequent activities from commencing in a timely manner, consequently jeopardizing the rate at which these activities can be completed. Conversely, noncritical activities have a minimal impact on scheduled completion. Only if they could ultimately influence the project schedule would the project manager show concern. All critical activities are closely scrutinized when actual progress lapses behind planned progress.

In the final analysis, if progress is suitable and appears to be on schedule, then the CURRENT PROJECT STATUS (M2) evaluation rests with Project Management Control. In addition, when the contractor's payments are linked to a progress schedule, actual progress is forwarded to cost accounting in order that payment can be requested for the work completed for the period.

*Bar chart scheduling shown here could also be microcomputerized using spread sheet programs combined with schedule graphic applications for easy updating (see bibliography).

Anderson & Woodhead
Building Contractors

PROJECT: ______________________ JOB NO.: ______________________

ARCHITECT: ______________________ STATUS DATE: 3-28-86

LOCATION: ______________________ DATE: 4-2-86

PREPARED BY: SDA

CONSTRUCTION SCHEDULE

WORK CLASSIFICATIONS	STATUS P/A	STATUS START DATES	STATUS COMPLETE DATES	1986 FEB	MAR	APR	MAY	JUN	JUL	AUG	SEP	PROGRESS % COMPLETE 50
SITE WORK	P/A	2-10/2-10	2-28/2-28									
EXCAVATION	P/A	2-17/2-17	3-19/3-21									
BACKFILL	P/A	3-12/3-12	4-14/									
CONCRETE FOOTINGS	P/A	2-26/2-26	3-19/3-26									
SLABS	P/A	3-12/3-17	4-2/									
COLUMNS	P/A	3-17/3-21	4-20/									

LEGEND:

P = PLAN DATES
A = ACTUAL DATES

SCHEDULED ACTIVITY DURATION

ACTUAL STATUS

ACTIVITY DURATION EXTENSION

REMARKS:

(a)

Anderson & Woodhead
Building Contractors

PROJECT: ______________________ JOB NO.: ______________________

ARCHITECT: ______________________ STATUS DATE: 3-28-86

LOCATION: ______________________ DATE: 4-2-86

PREPARED BY: SDA

CONSTRUCTION SCHEDULE

WORK CLASSIFICATIONS	STATUS P/A	STATUS START DATES	STATUS COMPLETE DATES	1986 FEB	MAR	APR	MAY	JUN	JUL	AUG	SEP	PROGRESS % COMPLETE 50
SITE WORK	P/A	2-10/2-10	2-28/2-28									
EXCAVATION	P/A	2-17/2-17	3-19/3-21									
BACKFILL	P/A	3-12/3-12	4-14/									
CONCRETE FOOTINGS	P/A	2-26/2-26	3-19/3-26									
SLABS	P/A	3-12/3-17	4-2/									
COLUMNS	P/A	3-17/3-21	4-20/									

LEGEND:

P = PLAN DATES
A = ACTUAL DATES

SCHEDULED ACTIVITY DURATION

ACTUAL STATUS

ACTIVITY DURATION EXTENSION

REMARKS:
1. SLABS AND COLUMNS WILL FINISH TWO TO THREE WEEKS LATE
2. BACKFILL SLIPS BECAUSE SLAB FINISH DATE SLIPPED

(b)

Figure 13.21 Scheduling status documents. (*a*) Construction Progress and Status (I_S-DOC 3). (*b*) Revised Construction Schedule (I_S-DOC 4).

Anderson & Woodhead
Building Contractors

PROGRESS PAYMENT VOUCHER

PROJECT: ______________ JOB NO.: ______________

LOCATION: ______________ SHEET NO.: ______________

PERIOD FROM ________ TO ________ DATE: ______________

LABOR COST					MATERIAL & EQUIPMENT COST			
DATE	CRAFT	HOURS	RATE	DOLLARS	QUANTITY	DESCRIPTION	UNIT COST	TOTAL COST

SUMMARY: COST $

TOTAL BILLED (LESS RETENTION) THIS BILLING .... ________

TOTAL BILLED (LESS RETENTION) LAST BILLING .... ________

TOTAL BILLED (LESS RETENTION) TO DATE .... ________

CLIENT OWNER ARCHITECT APPROVAL ______________ ANDERSON & WOODHEAD ______________

Figure 13.22 Progress Payment Voucher (P_S-DOC 13).

When an activity is significantly behind schedule and jeopardizes project completion, a potential change in the construction sequence may be warranted, provided it is feasible in terms of practical construction logic. In this instance, REVISING THE CONSTRUCTION SCHEDULE might be a necessity. The economics of changes in the construction sequence must also be evaluated, as well as any impact on subcontract work. When a change in the construction logic will not compensate for lost time, the project manager can constrict activity durations to ensure compatibility with target completion dates. In this instance, it is only necessary to revise the durations of those activities which are deemed critical to meeting project completion dates. A final option is to leave the schedule as planned and attempt to compensate for lost time through improving productivity, increasing crew manpower, or improving construction methods. The final scheduling decision is made by the project manager. The mechanics of revising the construction schedule are the same as those used in Project Scheduling, and are governed by the *USE OF BAR CHARTS* (P_S 6) and *SCHEDULING FOCUS* (P_S 7) policies. If the schedule is revised, a bar chart similar to the original Construction Schedule would be generated. This document is referred to here as a Revised Construction Schedule (I_S-DOC 4), and is illustrated in Figure 13.21*b*.

Cost and payroll analysis and reporting, together with construction statusing, are the two labor management problems that the head office must evaluate as part of Project Status, Analysis, and Reporting. Both of these problems are described on the decision network in Figure 13.23 as each relates to the functions performed to solve them.

A document flow depiction of the Project Status, Analysis, and Reporting decision process is provided in Figure 13.24. The basic input into this decision process would initially include data from the project budget estimate file, including the relevant documents describing the contractor's cost control and cost monitoring approaches. Estimated labor data from the budget estimate is then merged with the documents used to record actual costs and project payroll. This merging of data is normally a manual transfer of labor estimate data to the appropriate documents used for cost control such as the Project Cost Recapitulation (P_S-DOC 12) shown in Figure 13.20. On this document, budgeted dollars are entered in the estimate columns for ready comparison with actual costs as these costs are col-

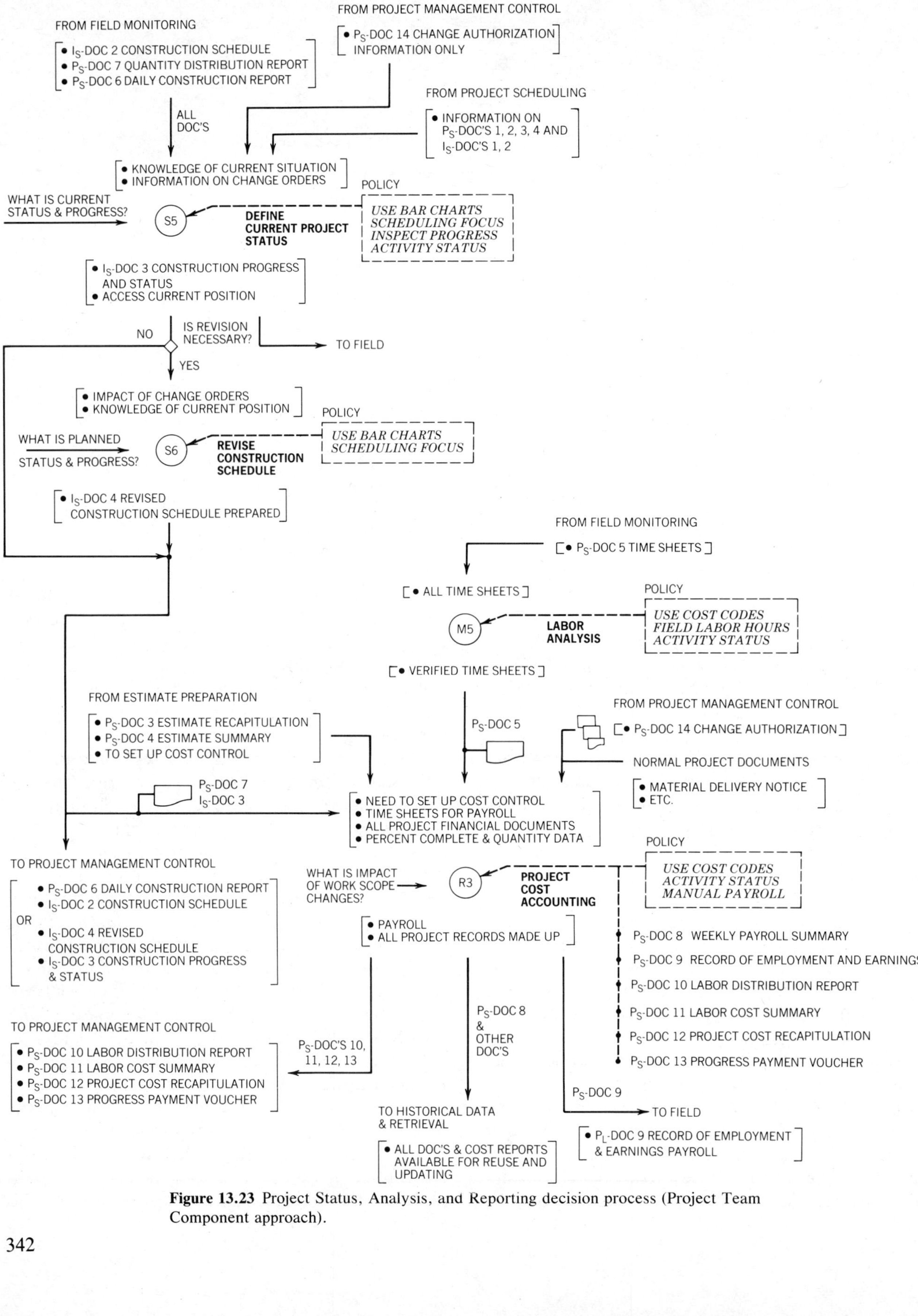

Figure 13.23 Project Status, Analysis, and Reporting decision process (Project Team Component approach).

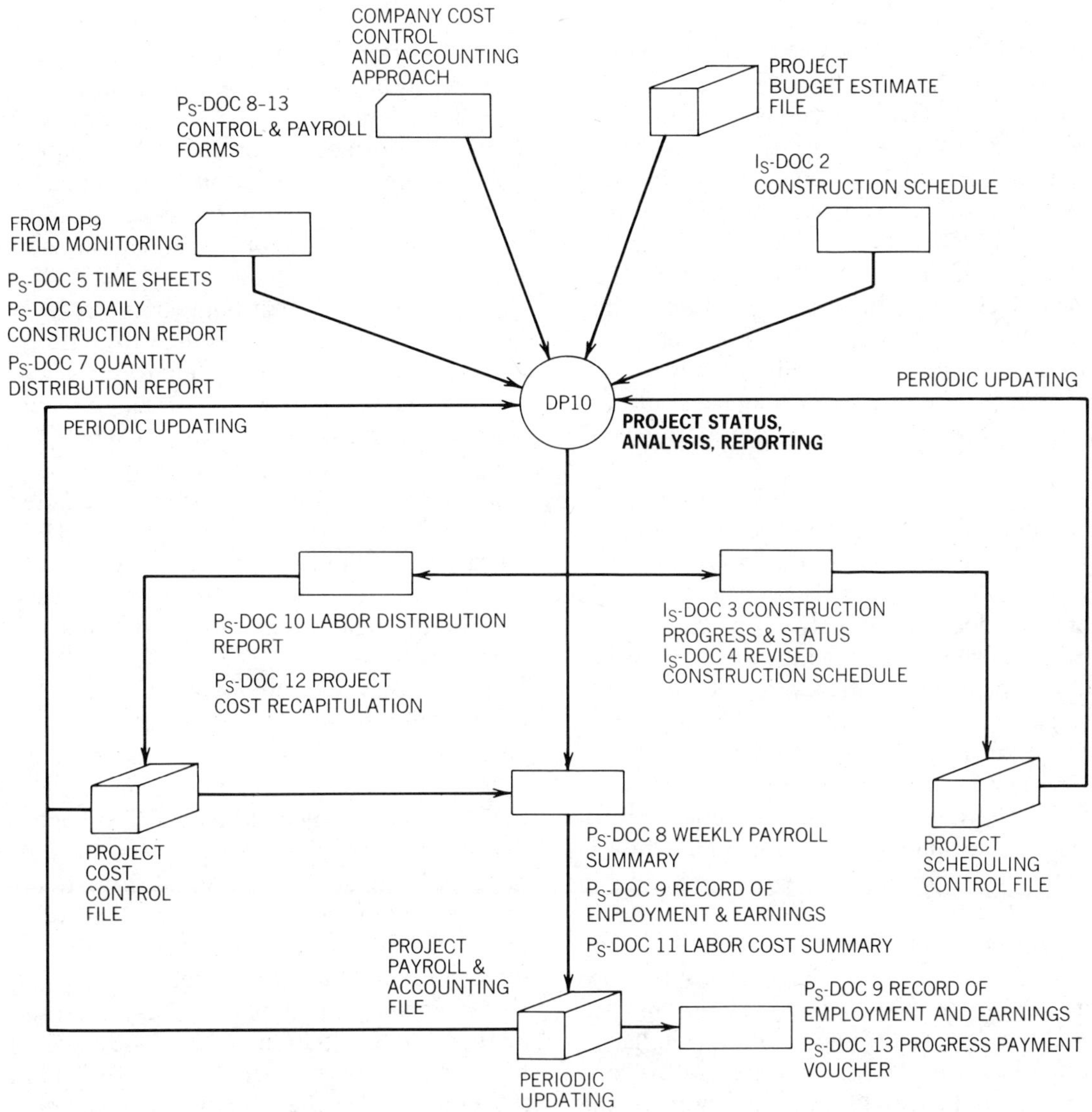

Figure 13.24 Project Status, Analysis, and Reporting Management Information System (Project Team Component approach).

lected. This compilation of budget cost data and actual cost data is the essence of the small firm's Management Information System.

Once field construction commences, actual field data enters from Field Monitoring. Then, when the Project Status decision process is executed (see Figure 13.23), the project cost control, payroll, and accounting data are updated. These payroll and accounting documents contain employee checks, records of earnings, and so on, and payment vouchers for billing clients. In addition to the cost aspects required for Project Management Control, the project schedule is statused and updated, thereby creating a project scheduling control file. All documents generated in the Project Status, Analysis, and Reporting information system, as shown in Figure 13.24, are normally produced manually by the small firm contractor with the exception of those who are now using microcomputer applications.

13.4 THE TRAVELING PROJECT MANAGER

In the medium-sized firm environment, project size and complexity often warrant the increasing involvement of the Traveling Project Manager in the planning, monitoring, and management decisions concerning field construction activities. This involvement subsequently leads to a requirement for more detailed and specially focused reports for measuring labor unit costs. These serve as a vehicle for evaluating labor performance. Thus, the medium-sized firm begins to document LABOR PRODUCTIVITY (R5) formally, and to perform limited ACTIVITY ANALYSIS (M4) and PROJECT FORECASTING (M3). In order to establish and continuously update labor unit costs, the elements of a systematic and integrated field cost control system become a feature of project management, and are reflected in the Management Information System.

Because project sizes vary in this environment, the contractor begins to consider the level of effort required for cost control as applied to a particular project. The contractor must also consider the type and number of documents that will be needed for cost control, in order to achieve the desired level of integration of cost information required for effective management of the project. The firm may choose between a computerized cost control system or a manual system or a combination of both.

When the contractor uses manual cost control and accounting techniques, the document flow is similar to the small firm's approach, as indicated in Figure 13.24. The type and structure of these documents, however, would be different in that they would reflect a more integrated cost control approach. If a computerized cost and accounting system is used the document flow would be similar to that depicted in Figure 13.25. This cost control system is best described in two phases, namely the initial establishment of the cost control file for a project (Figure 13.25*a*), and the active use of the project cost and scheduling control system (Figure 13.25*b*).

In Figure 13.25*a*, the Project Budget Estimate File, which is derived from Estimate Preparation (see Figure 6.27), is accessed for budget labor cost estimate data. When coupled with the documents that form the labor cost control and accounting system, this data enables the project cost control, payroll, and accounting files to be established through a program command when a computerized system is used. These files are then available to accept actual cost data and payroll information, and become inputs into the active project control approach as shown in Figure 13.25*b*. Other data input also enters Project Status, Analysis, and Reporting, including field-generated man-hours, quantities, construction status, the construction schedule, and change authorizations to the scope of work. The live field data, coupled with budget estimate and forecast information, provides the contractor with cost-to-complete projections. When these cost projections are reviewed in terms of informal productivity analysis (based on man-hour expenditures per quantity output and current scheduled status), a documented labor cost comparison results.

Project Status, Analysis, and Reporting operates as both a computerized system for project cost control and payroll purposes, and as a manual system for project scheduling control as illustrated in Figure 13.25*b*.* This distinction between the computer system and the manual system is denoted by the dashed lines incorporated into Figure 13.25*b*.†

At the medium-sized firm level, Project Status, Analysis, and Reporting operates in two modes, namely the field evaluation mode, and the project status mode.

The field evaluation mode relies heavily on input from Field Monitoring. As a result, a close interface between the field monitoring and field evaluation functions emerges. This interfacing is significant to the Traveling Project Manager approach, and imparts a totally different emphasis on the decision process from that of the small firm. It provides a management atmosphere for problem solving at the field level, where a number of personnel can interact to make timely decisions.

Two management problem areas are addressed in the field evaluation mode. They are, first, the need to focus on the performance of field labor and the concern for the effective coordination of the labor effort towards the achievement of scheduled milestones. Field labor performance is related to crew performance. Crew performance analysis looks at crew production for an activity in terms of

*Accounting functions are both computerized and manual. For instance, actual costs are collected and stored in a computer file, while billing and expenditure projections would normally be prepared manually. It should also be recognized that many medium-sized contractors maintain computerized scheduling control capabilities. The use of microcomputers has substantially increased the level of computer applications in the medium-sized firm.

†The detailed discussion to follow in this section will focus on documents that appear to be generated for a manual system. However, the data contained on those documents, as well as the document structure or format, is normally computerized to accommodate an information system such as that illustrated in Figure 13.25.

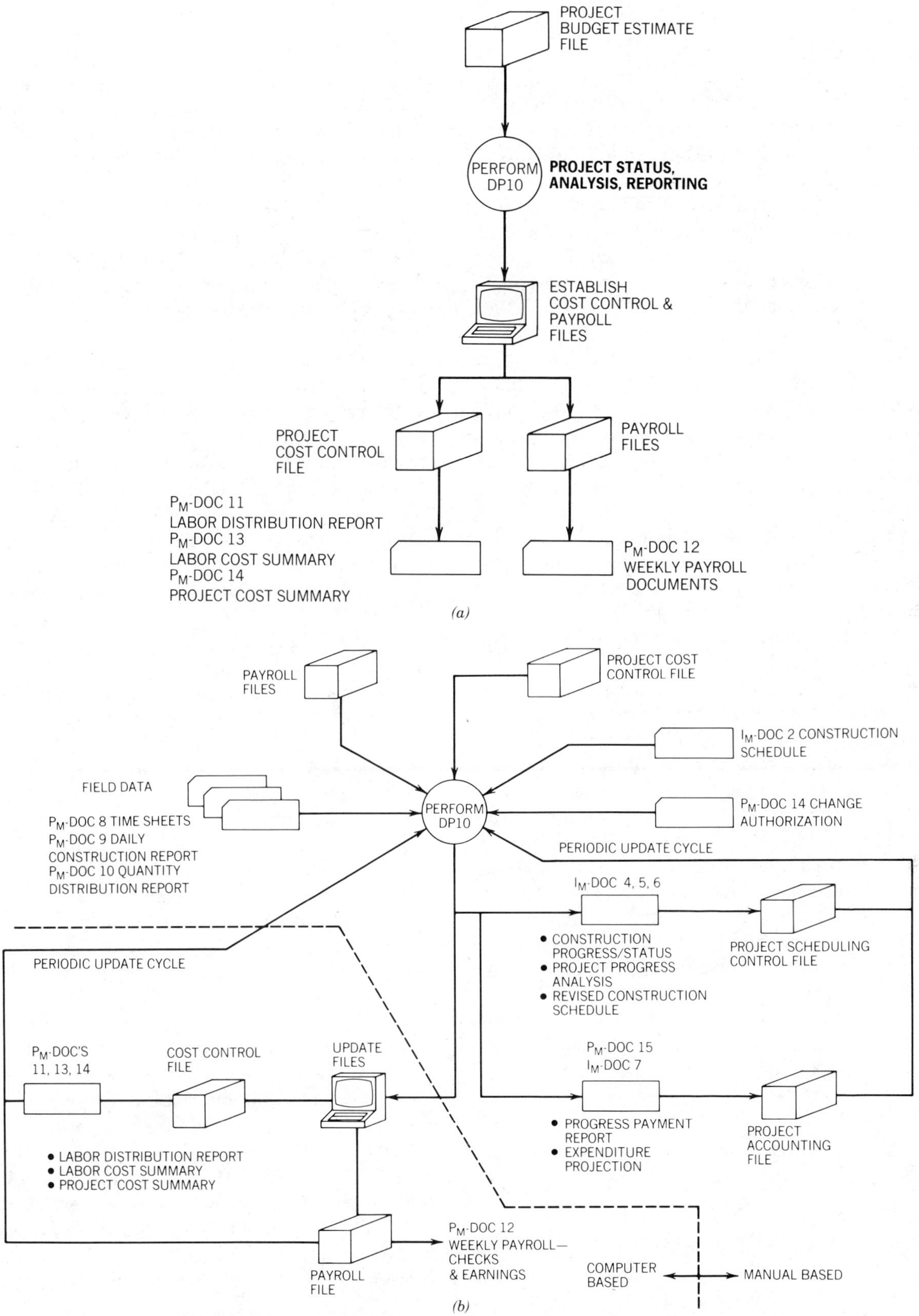

Figure 13.25 Project Status, Analysis, and Reporting Management Information System (Traveling Project Manager approach). (*a*) Establish initial project control base file. (*b*) Active project control system.

field man-hours expended to place a given quantity of material. This requires both documented evidence (labor hours per unit) coupled with visual observation of crew performance, and discussions between the project manager and the respective field superintendents/foremen. Critical problems which are discovered are analyzed and solved through on-site activity analysis. Second, the scheduling focus problem attempts to evaluate physical progress as a means of determining current activity status, and subsequently, to detect potential schedule delays or conflicts. Both of these aspects of the field evaluation mode of project status and analysis are linked directly to Field Monitoring, and reflect the professional interest of field managers to control and solve field problems before they get out of hand. Consequently, the field evaluation mode is performed *informally*.

The project status mode is mainly head office oriented and derives much of its information from field evaluation and reporting. This mode is comparable to the entire small firm's Project Status, Analysis, and Reporting decision process (see Figure 13.23). It differs from that approach in that the determination of project status can be characterized by a *formal* scheduling effort to establish the current project status, a labor productivity analysis to support project forecasts that will help spot labor cost deviations from the estimate, and activity analysis on an exception basis to solve critical site problems. Through the head office and field linking functions performed by the Traveling Project Manager, this mode of Project Status, Analysis, and Reporting provides the management atmosphere for problem solving at the field level, in which head office and field personnel can interact in order to make timely decisions.

Accordingly, it is convenient to describe the Project Status, Analysis, and Reporting decision process in terms of four distinct, but interacting, decision phases:

1. Field Labor Performance (see Figure 13.26)
2. Informal Field Scheduling (see Figure 13.28)
3. Formal Scheduling (see Figure 13.29)
4. Labor Cost Analysis (see Figure 13.32)

The field evaluation mode, which is comprised of Field Labor Performance and Informal Field Scheduling analysis, is the combined responsibility of the project manager, the job superintendent, the foremen, and the field engineering staff. Several of its functions are performed routinely by these personnel as part of the continuous management process. These include LABOR ANALYSIS (M5) and EVALUATING LABOR PRODUCTIVITY (R5), DEFINING THE CURRENT PROJECT STATUS (S5), ACTIVITY ANALYSIS (M4), and REVISING THE CONSTRUCTION SCHEDULE (R6) which are performed on an exception basis as critical problems are encountered.

The field-based labor performance analysis decision network is shown in Figure 13.26. On most projects LABOR ANALYSIS is the project manager's responsibility, except when a field engineer is assigned to the project team and assumes partial responsibility for this function. LABOR ANALYSIS initiates the LABOR PRODUCTIVITY analysis of crew performance. The information required for this function consists mainly of recorded labor hours, and assessments and comments based on visual observations. Because of the emphasis on *DOCUMENTING STATUS* (P_M 22)* and analysis of *ACTIVITY PRODUCTIVITY* (P_M 23), the project manager must fully review and analyze the labor hours expended per work task. To accomplish this, the Labor Distribution Report (P_M-DOC 11), as shown in Figure 13.27, is introduced. This document is utilized as a means of collecting the total hours expended per work task by the various cost codes.† The project manager or field engineer will supervise a timekeeper who extracts from the Time Sheets (P_M-DOC 9) the hours charged to each cost code. The hours are recorded under the appropriate cost code, and the total hours assigned to that code reflect the labor hour effort for the period. When these hours are added to the total hours to date, the total hours expended for the work task are represented by the cost code. This report provides the project manager with a basic knowledge of the crew effort, in terms of man-hours, for each active work task, and when this information is combined with physical quantities placed for each work task, a determination of crew productivity can be made. While LABOR PRODUCTIVITY is not formally documented, the knowledge of it, gleaned from analyzing the hours spent to date versus actual physical

*The relevant policies of the medium-sized firm are presented in Table 13.4.

†The labor wage rates, labor dollars, unit costs to date, and so on, are entered by cost accounting, and the various calculations are made during the COST ACCOUNTING function in the head office.

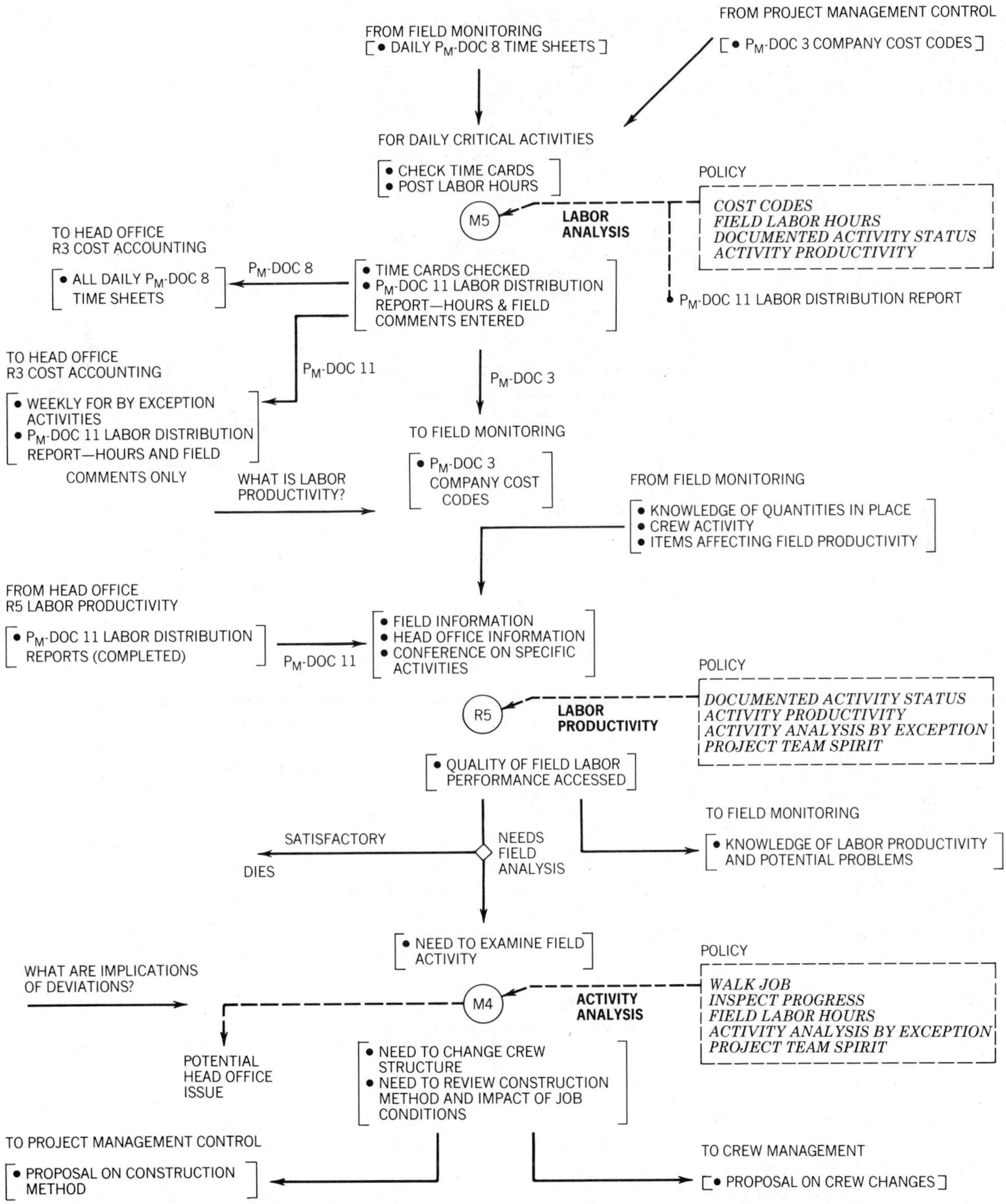

Figure 13.26 Project Status, Analysis, and Reporting decision process: field-based labor performance analysis (Traveling Project Manager approach).

TABLE 13.4 POLICIES OF THE MEDIUM-SIZED FIRM

Policy		Description
P_M 4	*COST CODES*	A systematic classification of work categories segregated by unique cost codes is used to define work tasks, collect costs, and provide a means of comparing actual costs with estimated costs.
P_M 7	*BAR CHARTS*	Bar chart model and limited application of the critical path method are best suited for project scheduling.
P_M 8	*SCHEDULING FOCUS*	Scheduling definitions and constraints are: (1) Critical activities are highly sensitive profit and loss activities and/or activities controlling the rate at which ensuing work can commence. (2) Activity durations and sequence logic incorporate the impact of seasonal weather conditions and restrictive work practices. (3) Manpower impact is considered on a conceptual basis. (4) Activity scheduling reflects the best sequencing of activities possible for effective utilization of all resources available.
P_M 19	*WALK JOB*	In the field, daily activity progress and crew performance is monitored by visual inspection, and crew management decisions are based on a subjective evaluation of observed progress and productivity versus planned progress and productivity.
P_M 20	*INSPECT PROGRESS*	Actual quantities in place versus estimated quantities is an indicator of physical progress and is evaluated by visual inspection and/or quantity survey.
P_M 21	*FIELD LABOR HOURS*	Labor hours are monitored by personnel closest to the work face, and recorded either by them or by personnel whose primary function is to record labor hours.
P_M 22	*DOCUMENTED ACTIVITY STATUS*	Labor costs, quantities in place, and activity status must be recorded and documented by a system yielding reliable information and viable reports for management control.
P_M 23	*ACTIVITY PRODUCTIVITY*	Labor productivity is an indicator of crew performance, and is measured by activity labor dollar and man-hour expenditures in relation to quantity volume in place.
P_M 24	*ACTIVITY ANALYSIS BY EXCEPTION*	Activity analysis is implemented when critical activities could potentially delay the schedule and/or potential labor dollar overruns are emerging.
P_M 25	*COST TO COMPLETE*	Projecting total project costs becomes a viable method of ferreting out potential overruns and evaluating profitability at both the activity level and total project level.
P_M 26	*MANAGEMENT BY EXCEPTION*	Manpower management by exception as problems arise is an acceptable decision approach. However, these management decisions are only made when the impact will result in a significant cost/scheduling savings.
P_M 27	*MIX PAYROLL SYSTEMS*	Payroll dollars are generated and recorded either manually or electronically on documents showing standard deductions.
P_M 28	*PERCENT COMPLETE STATUS*	Construction progress reporting must provide reliable information of activity status and total project status to be an effective tool in manpower management and control.
P_M 29	*PROJECT TEAM SPIRIT*	Active participation of all field managers in the management process promotes goodwill and a positive management environment.

progress in terms of quantities in place, does provide a basic indicator of crew performance. For example, the Labor Distribution Report in Figure 13.27 provides the man-hours per unit (i.e., production rate) by cost code for an activity.* The production rate is developed by dividing the actual hours expended by the quantities placed for the specific reporting period, and also cumulatively the total hours and quantities placed to date. These production rates may alert the project manager to potential problem areas by flagging activities which have an unusually high man-hour expenditure and a minimal amount of material-in-place.

Because the production rate is expressed weekly

*This provides excellent information for estimating purposes, especially if labor unit prices are developed based on craft production rates.

Anderson & Woodhead
General Contractors

PROJECT: ____________ JOB NO.: ____________

LOCATION: ____________ WEEK END: ____________

MAN-HOUR DISTRIBUTION

COST CODE	WORK CATEGORY DESCRIPTION	MON	TUES	WED	THURS	FRI	SAT	SUN	TOTAL HOURS	TOTAL DOLLARS
0311.	FORMWORK—FTGS	24	24	32	40	40			160	2580

(a)

Anderson & Woodhead
General Contractors

PROJECT: ____________ JOB NO.: ____________

LOCATION: ____________ WEEK END: ____________

UNIT COST ANALYSIS

COST CODE	WORK CATEGORY	QUANTITIES IN PLACE			LABOR COST			UNIT COST		
		LAST WK.	THIS WK.	TOTAL	LAST WK.	THIS WK.	TOTAL	LAST WK.	THIS WK.	TOTAL
0311.	FORMWORK—FTGS	402	1850	2252	516	2580	3096	1.28	1.39	1.27

(b)

Anderson & Woodhead
General Contractors

PROJECT: ____________ JOB NO.: ____________

LOCATION: ____________ WEEK END: ____________

UNIT MAN-HOUR ANALYSIS

COST CODE	WORK CATEGORY	LABOR HOURS			LABOR HOURS/QTY.			REMARKS
		LAST WK.	THIS WK.	TOTAL	LAST WK.	THIS WK.	TOTAL	
0311.	FORMWORK—FTGS	32	160	192	.080	.086	.085	FORMING FTGS—DRAWING—10210

(c)

Figure 13.27 Labor Distribution Report (P_M-DOC 11). (*a*) Man-hour tally. (*b*) Unit cost analysis. (*c*) Unit man-hour analysis.

and as a total to date, significant variations from week to week can be easily spotted. These types of variations may be an early indication of a crew performance problem. Furthermore, if these production rates are referenced against estimated production rates developed during planning of the crew size (if available), the project manager and key field superintendents can evaluate productivity in terms of the ratio of the actual production rate to the estimated production rate. This ratio is analyzed in a fashion similar to comparing actual and estimated labor unit costs. The actual production rate is most often analyzed on the basis of past experience, in which case the project manager and/or field superintendent will know if the crew is experiencing satisfactory productivity. This analysis alone is not sufficient, however, because these decision makers must also be cognizant of the relative activity progress to date. For example, if an activity is significantly complete (e.g., 80–90%), then further analysis of forecasted production rates may not be warranted. At this point it is too late to take corrective action. However, when changes can be made that will potentially improve future production rates at a time when only minimal progress has been achieved on an activity, then further analysis of crew performance, construction methods, and quantities to complete can be fruitful. Consequently, the project manager must decide at this point whether or not to initiate ACTIVITY ANALYSIS. The cost or schedule criticality of the activity in question would influence the decision as indicated by the policy, *ACTIVITY ANALYSIS BY EXCEPTION* (P_M 24).

Field ACTIVITY ANALYSIS, as illustrated in Figure 13.26, is a quick means of analyzing the construction method in light of existing job conditions, in order to identify probable causes of poor production. The project manager is normally assisted in this by the key field superintendent and possibly the foreman, who provide possible reasons for poor performance based on their field inspection of the crew situation. A quick analysis of the causes of poor crew performance, normally conducted in brainstorming sessions between the project manager, the field superintendent, and foremen, may provide the basis for a management decision. For example, the crew may be unbalanced because there are more men assigned to the crew than the available work can keep busy. Consequently, manhours are being wasted. The problem may be alleviated by transferring a man out of the crew or increasing the volume of work available. In the former instance, a quick decision can be made and acted on, while the latter case would probably require more analysis of material flow routes or construction equipment use. Thus, such a decision from ACTIVITY ANALYSIS can travel in two directions: (1) to Crew Management for the ADJUSTING OF CREWS (A5); or (2) to Project Management Control in the head office, with the implication that top management involvement may become necessary depending on the nature of the problem.

A second aspect of field evaluation focuses on another assessment of the effectiveness of labor performance, namely the achievement of construction schedule milestones. Informal scheduling analyses provide the field with quick basic indicators of progress relative to physical percent complete. This information ensures a better analysis of crew performance and also helps the project manager spot potential scheduling problems requiring immediate field response. Schedule changes are then enacted in the field and the impact is quickly made known to all those involved with Crew Planning and Crew Management.

The Traveling Project Manager approach itself provides the environment for informal scheduling. Field superintendents and key foremen are well aware of management's interest in achieving high production and in meeting construction goals. They are under management's spotlight, and thus are keen to demonstrate that they can perform. Consequently, they are very conscious of the critical activities in their *SCHEDULING FOCUS* (P_M 8). Thus as indicated in the Informal Field Scheduling Analysis decision network of Figure 13.28, these field managers are intimately aware of site activity and work progress, and react to head office evaluations of LABOR PRODUCTIVITY by carefully noting changes in crew unit production rates as given by the Labor Distribution Report. Thus, using this knowledge plus their assessment of current activity status, it is a simple matter for the field managers to project activity completion dates and compare these dates with those in the Construction Schedule.

If this informal scheduling analysis indicates a potential delay, then the superintendent, foreman, and the project manager act to resolve it by manipulating crew sizes to expedite activity progress. If field analysis indicates that this will not be effective, then either an ACTIVITY ANALYSIS becomes necessary, or the head office is advised that a CONSTRUCTION SCHEDULE REVISION evaluation is required, which will indicate schedule changes to be made to the Construction Schedule in the formal scheduling phase of Project Status, Analysis, and Reporting.

The head office project status mode of Project Status, Analysis, and Reporting has data processing as its main focus. Project status determination has a multidimensional focus because several functions are executed simultaneously. This is essential for adequate direction of the flow of this information, which must be assimilated into total project status reports. Project status determination is the responsibility of the project manager, superintendent, field engineer, and cost accountant, and consists of five functions, namely: (1) CURRENT PROJECT STATUS DEFINITION; (2) REVISING THE CONSTRUCTION SCHEDULE; (3) PROJECT COST ACCOUNTING; (4) LABOR PRODUCTIVITY REPORTING; and (5) PROJECT FORE-

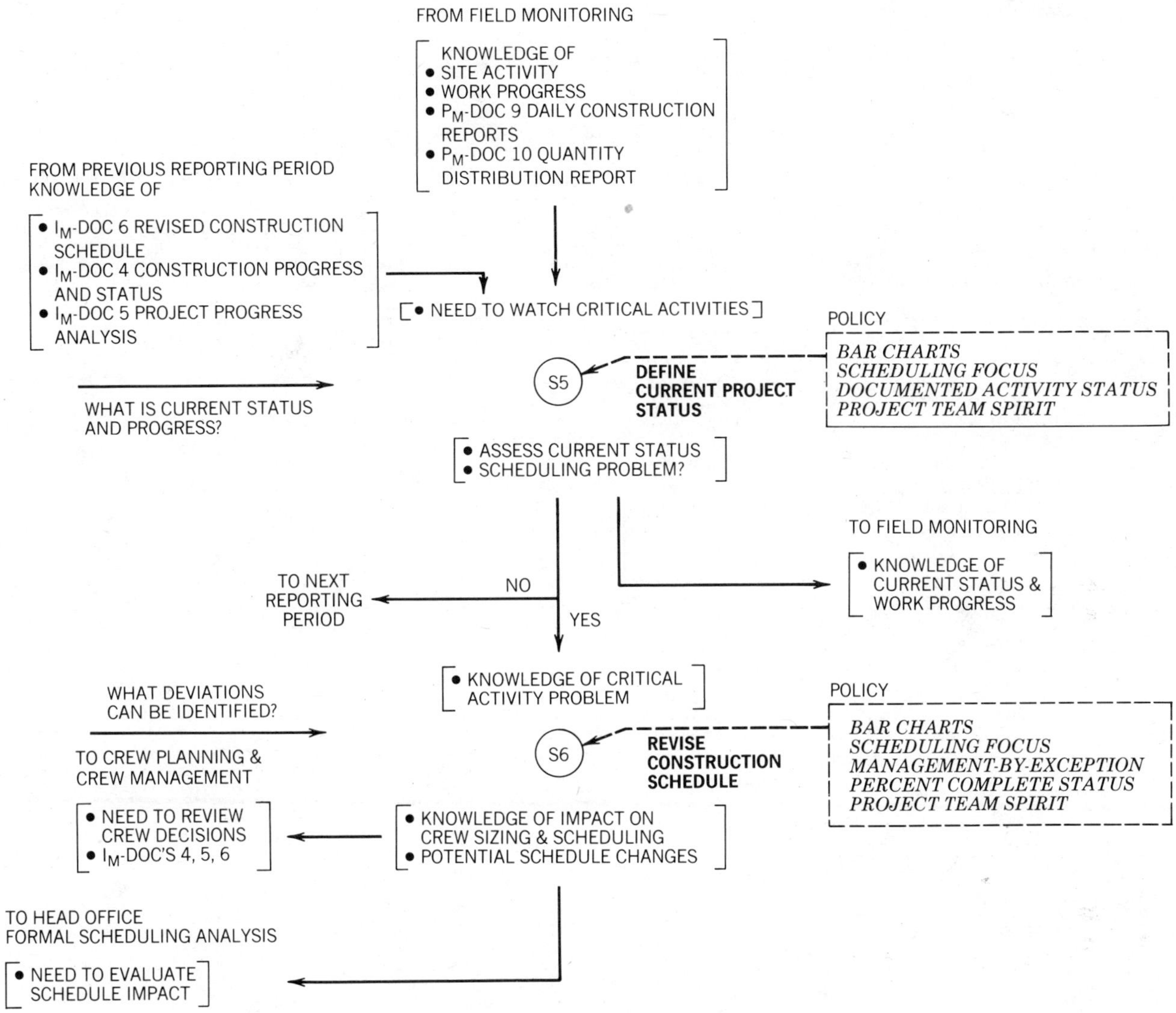

Figure 13.28 Project Status, Analysis, and Reporting decision process: informal field scheduling analysis (Traveling Project Manager approach).

CASTING. These functions have a reporting focus, in terms of formal scheduling analysis and labor cost analysis.

The decision network for the Formal Scheduling Analysis phase is illustrated in Figure 13.29, wherein the DEFINITION OF CURRENT PROJECT STATUS initiates the decision phase. This function's problem focus is toward evaluating physical progress and activity status for those activities depicted on the current construction schedule (Construction Schedule (I_M-DOC 2)) or the Revised Construction Schedule (I_M-DOC 5).

At the medium-sized firm level, the reporting techniques, as governed by *DOCUMENTED ACTIVITY STATUS* and *PERCENT COMPLETE STATUS* (P_M 28), begin to integrate the schedule with the cost function through a formalized progress report that can be used to directly tie the percent complete to a payment schedule as a means for billing the client. In order to achieve accurate reporting results and maintain schedule control, the Traveling Project Manager rarely assumes total responsibility for this function, and therefore works closely with other field personnel such as field superintendents and, particularly on larger projects, a field engineer.

The basic approach to defining activity status is first to evaluate the percent progress for each activ-

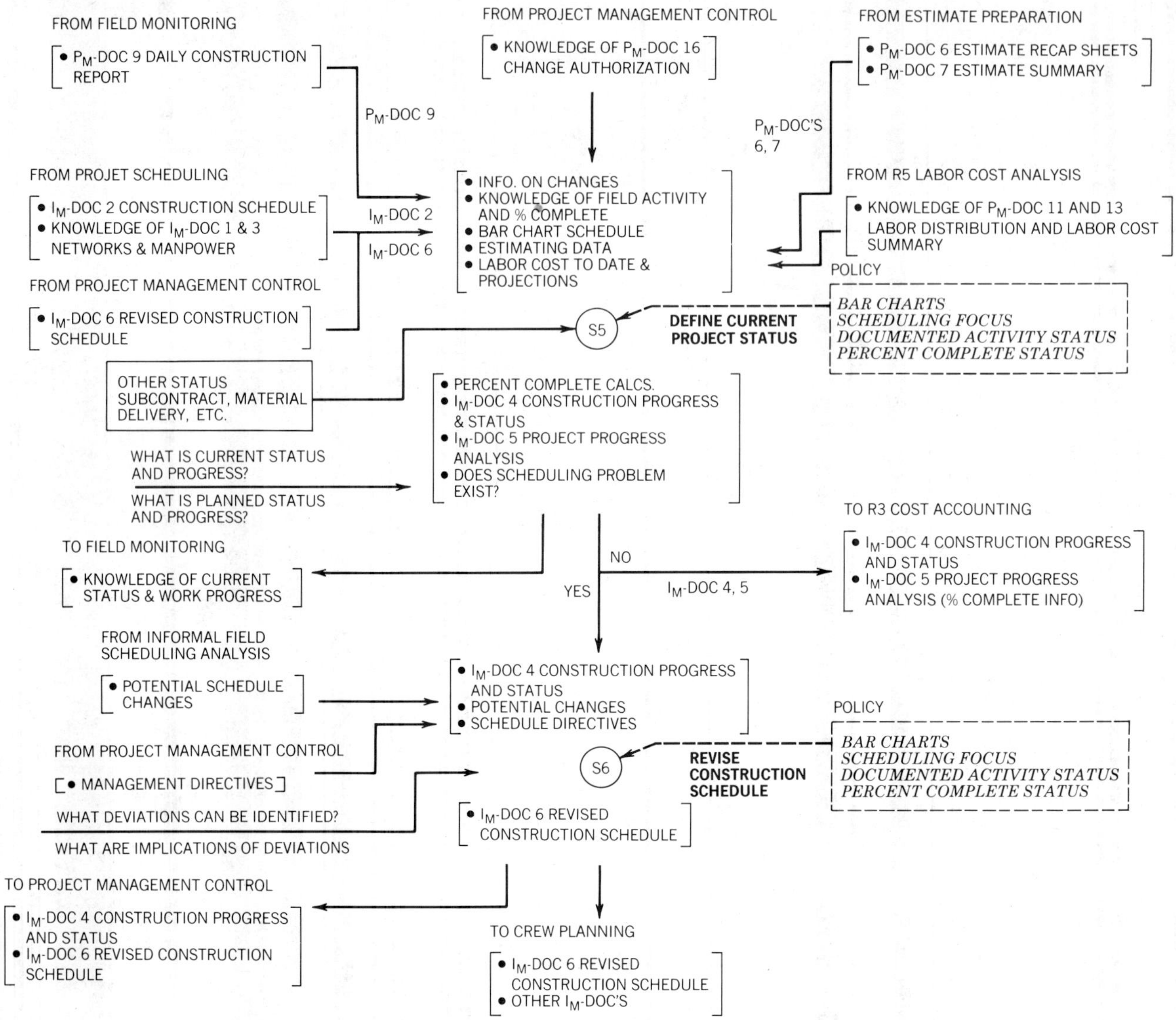

Figure 13.29 Project Status, Analysis, and Reporting decision process: formal scheduling analysis (Traveling Project Manager approach).

ity under construction as described on the Construction Schedule. The main sources of progress information are field observations and the quantity outputs per activity as recorded on the Quantity Distribution Report. As directed by *INSPECTING PROGRESS* (P_M 20), the quantities to date from this document divided by the total estimated quantity from either the Project Estimate (P_M-DOC 5), or field estimates of quantities, provide a basic indication of the percent complete for an activity. Because actual quantities to date are developed by analyzing project drawings and by physical observation of progress (i.e., counting or measuring field output), the decision maker will know whether the calculation of progress is accurate. For example, if fifty percent of the total estimated quantity is in place, but more than 50 percent of the estimated quantity is still required to complete the activity, then the activity is probably not fifty percent complete. In this case, the estimated quantity may be low, or else a significant waste of material has occurred. The true percent complete is either reevaluated using forecasted total quantities or is determined from visual observation combined with knowledge of the total effort required.

While in the process of evaluating quantities

Anderson & Woodhead
General Contractors

PROJECT: __________

ARCHITECT: __________

LOCATION: __________

CONSTRUCTION SCHEDULE

WORK CATEGORY	UNIT MEASURE	QUANTITY	LABOR COST	% PROGRESS
MOVE IN	1	LOT	15000	100.0
ROUGH GRADE SITE	1	LOT	25500	100.0
SURVEY—EXC.	1	LOT	10000	85.0
EXCAV. FTG BAY A-C	CY	220	550	100.0
EXCAV. FTG. BAY D-G	CY	220	550	100.0
EXCAV.—SEWER	CY	4000	3920	85.0
FORM—FTGS BAY A-C	SF	1365	1665	30.0
FORM—FTGS BAY D-G	SF	1365	1665	10.0
INSTALL SEWER	LF	1200	2664	0.0

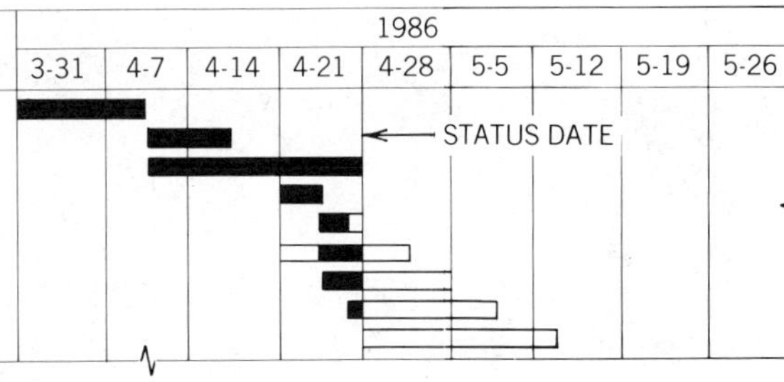

LEGEND:

PLANNED DURATION

ACTUAL STATUS

PROJECTED COMPLETION

% PROGRESS = QUANTITY IN PLACE / ESTIMATED QUANTITY

REMARKS:

1. SEWER EXCAVATION SLIGHTLY BEHIND SCHEDULE—NO MAJOR PROBLEM

(a)

Anderson & Woodhead
General Contractors

PROJECT: __________ JOB NO. __________

LOCATION: __________ DATE: __________

COST CODE	WORK CATEGORY	LABOR COST	% COMPL.	MAT'L COST	% COMPL.	SUBCONTRACT COST	% COMPL	PROJECT TOTALS ESTIMATE	COMPLETE	%
0200	SITE WORK	79,320	55.0	52,880	28.6	0	—	132,200	58,770	44.4
0300	CONCRETE	276,045	1.1	512,625	2.0	0	—	788,200	13,289	1.7
	TOTAL PROJECT							1,335,300	103,291	.8

(b)

Figure 13.30 Construction progress and status. (*a*) Construction Progress and Status (I_M-DOC 4), (*b*) Project Progress Analysis (I_M-DOC 5).

placed to date, the project manager, field superintendents, and/or field engineer can evaluate activity status in terms of start and completion dates for each activity scheduled during the monthly reporting period. According to *PERCENT COMPLETE STATUSING*, the manner in which activity status and progress is depicted is important in controlling the project. Consequently, the Construction Schedule, when statused, must clearly delineate the project position at the level of detail shown in the Construction Progress and Status (I_M-DOC 4) document illustrated in Figure 13.30*a*. The method of depicting Construction Progress and Status changes significantly at this level, because more detailed information is required for project control. As illustrated in Figure 13.30, the Construction Progress and Status report not only shows activity status by coloring the activity bars, but also identifies activity progress (percent complete) in relation to both estimated quantities and labor costs. This aids the proj-

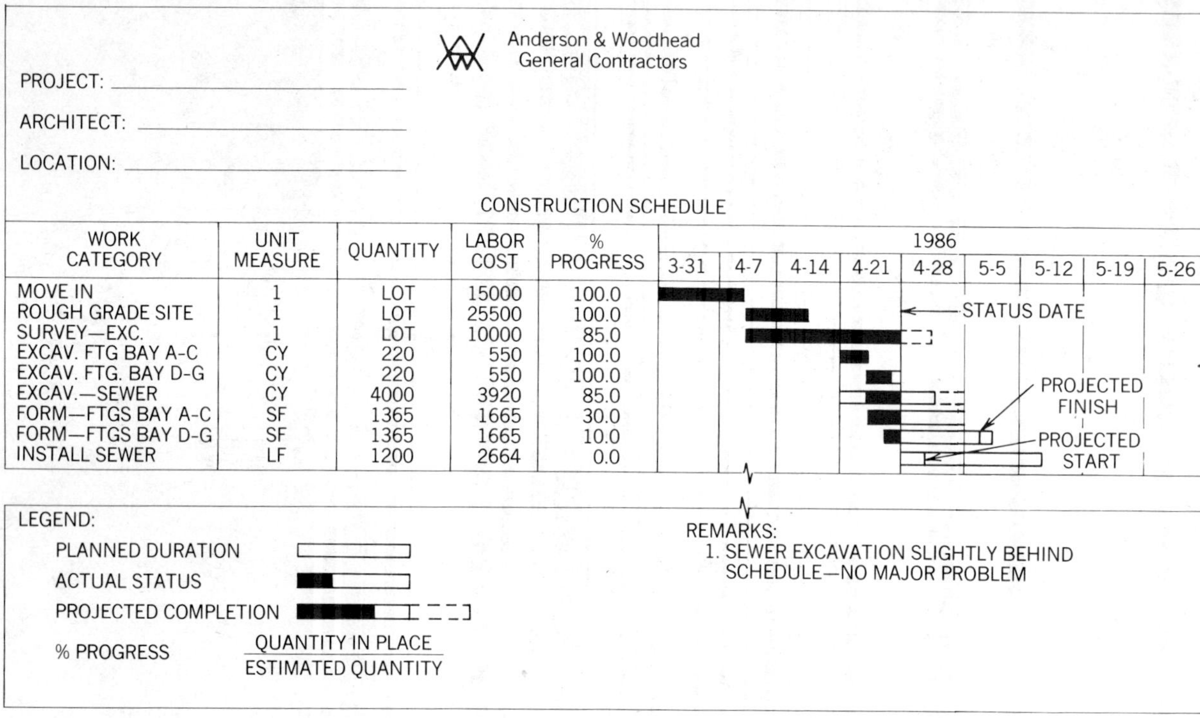

Anderson & Woodhead
General Contractors

PROJECT: ______

ARCHITECT: ______

LOCATION: ______

CONSTRUCTION SCHEDULE

WORK CATEGORY	UNIT MEASURE	QUANTITY	LABOR COST	% PROGRESS
MOVE IN	1	LOT	15000	100.0
ROUGH GRADE SITE	1	LOT	25500	100.0
SURVEY—EXC.	1	LOT	10000	85.0
EXCAV. FTG BAY A-C	CY	220	550	100.0
EXCAV. FTG. BAY D-G	CY	220	550	100.0
EXCAV.—SEWER	CY	4000	3920	85.0
FORM—FTGS BAY A-C	SF	1365	1665	30.0
FORM—FTGS BAY D-G	SF	1365	1665	10.0
INSTALL SEWER	LF	1200	2664	0.0

LEGEND:

PLANNED DURATION

ACTUAL STATUS

PROJECTED COMPLETION

% PROGRESS = QUANTITY IN PLACE / ESTIMATED QUANTITY

REMARKS:
1. SEWER EXCAVATION SLIGHTLY BEHIND SCHEDULE—NO MAJOR PROBLEM

Figure 13.31 Revised Construction Schedule (I_M-DOC 6).

ect manager by focusing on cost-critical activities and correlating them to the schedule. Finally, overall project progress can be determined by evaluating the percent complete for material and subcontract costs in addition to labor costs, and summarizing the completed costs portions of these components by major work category corresponding to cost codes. An example of a typical progress compilation is shown in Figure 13.30*b*, Project Progress Analysis (I_M-DOC 5). The Project Progress Analysis becomes the basis for calculating progress payments.

The DEFINITION OF CURRENT PROJECT STATUS often reveals scheduling problems that could affect the construction schedule. Consequently, as shown in Figure 13.29, a decision must be made relative to REVISING THE CONSTRUCTION SCHEDULE. These types of decisions are normally made on a *MANAGEMENT BY EXCEPTION* (P_M 27) basis, wherein the project manager, superintendent, and field engineer determine the scheduling action to be instituted. When a Revised Construction Schedule (I_M-DOC 6) is generated, as shown in Figure 13.31, activity durations and locations in time are adjusted to solve potential schedule conflicts, such as subcontractor interface problems and/or critical material shortages. Alternatively, schedule changes often reflect the application of significantly different construction methods, such as buying or renting construction equipment in order to save labor costs on a construction operation and insure its timely completion, especially if it is schedule critical as per the *SCHEDULING FOCUS* policy. This kind of change is normally a top level management decision when recommended by the project manager.

Once a Revised Construction Schedule is developed, the information on it must be disseminated to the appropriate decision processes such as Crew Planning and Crew Management.

A second aspect of the project status mode of Project Status, Analysis, and Reporting is the labor cost analysis phase, as depicted in Figure 13.32. This decision phase is the focal point of the project information system (see Figure 13.25), and in it the majority of data is processed. Consequently, a significant number of new documents with an increasing level of integration between them emerge. In addition, labor forecasts and productivity analyses become more formalized through the performance of the LABOR PRODUCTIVITY and PROJECT FORECASTING functions. These func-

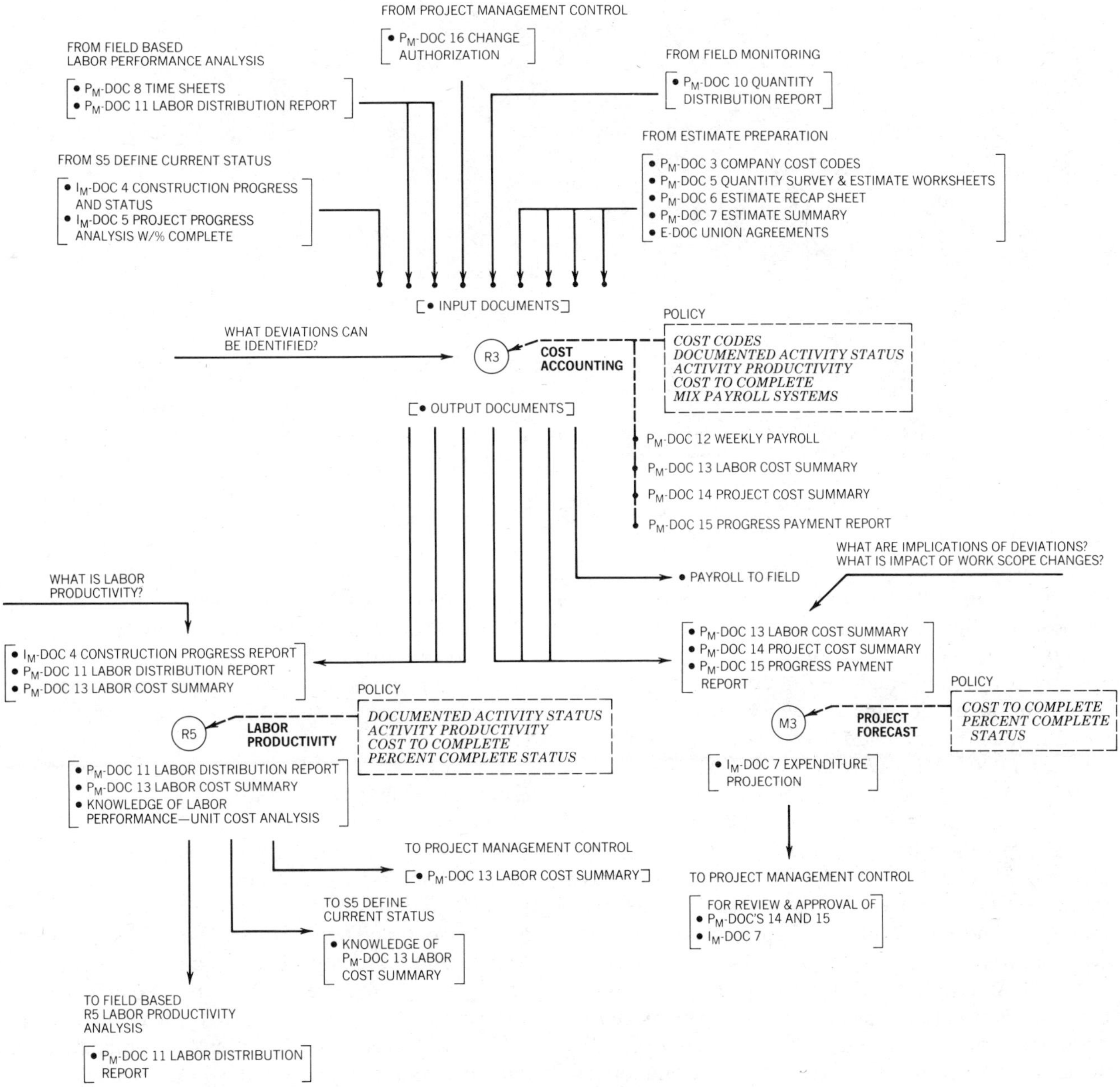

Figure 13.32 Project Status, Analysis, and Reporting decision process: labor cost analysis (Traveling Project Manager approach).

tions are the combined responsibility of the project manager and cost accountant. Both of these functions receive input data from PROJECT COST ACCOUNTING.

PROJECT COST ACCOUNTING generates labor cost data in various reporting formats, from which project cost status is analyzed and determined. The information required for cost accounting includes documents from other decision process areas, as well as policy documents governing the method of processing data and the reporting format. For example, documents from Estimate Preparation are critical input documents for the initial establishment of the cost control base (see Figure 13.25). The Quantity Survey and Estimate Work Sheets (P_M-DOC 5) are an example of such documents. Once the project is under construction, documents containing live data from the field are re-

Anderson & Woodhead
General Contractors

PROJECT: ______ JOB NO.: ______

LOCATION: ______ WEEK ENDING: ______

PREPARED BY: ______

COST CODE	WORK ITEM DESCRIPTION	UNIT MEASURE	CURRENT BUDGET			EXPENDED THIS WK.			EXPENDED TO DATE			CURRENT STATUS		PROJECTED STATUS	
			QTY.	LABOR UNIT COST	LABOR COST	QTY.	LABOR UNIT COST	LABOR COST	QTY.	LABOR UNIT COST	LABOR COST	UNDER	OVER	UNDER	OVER
0311.0	FORMWORK—FOOTINGS	SF	2730	1.22	3330	1850	1.39	2580	2250	1.37	3096		352		424
0300.0	CONCRETE—TOTAL				2760.45			2580			3096		352		424
												1.		2.	

CALCULATION
1. CURRENT STATUS = (CURRENT BUDGET QUANTITY − EXPENDED TO DATE QUANTITY) × CURRENT BUDGET LABOR UNIT COST + EXPENDED TO DATE LABOR COST − CURRENT BUDGET LABOR COST

2. PROJECT STATUS = SAME AS CALCULATION BUT REPLACE CURRENT BUDGET LABOR UNIT COST WITH PROJECTED LABOR UNIT COST TO COMPLETE SUCH AS EXPENDED LABOR UNIT COST TO DATE

Figure 13.33 Labor Cost Summary (P_M-DOC 13).

quired for payroll, accounting, and cost control purposes. The key input documents are Time Sheets, the Quantity Distribution Report, the Labor Distribution Report, and the Project Progress Analysis. The policy documents, which specify the reporting mechanisms, are a reflection of several policies. *DOCUMENTED ACTIVITY STATUS* and *COST TO COMPLETE* (P_M 25) strive for reports that not only record costs to date, but also flag potential problem areas by projecting or forecasting final costs. *ACTIVITY PRODUCTIVITY* attempts to define labor productivity and, therefore, directly influences the nature of the reports generated for cost control.

COST ACCOUNTING is performed on a weekly basis in relation to the project payroll. The primary documents are the Weekly Payroll (P_M-DOC 12) documents including Weekly Payroll Summaries and Records of Employment and Earnings. These documents would be similar in content to those described for the small firm (see Figures 13.16 and 13.17). In conjunction with the preparation of weekly payroll documents, the cost accountant, with the assistance of a timekeeper, often calculates the labor dollars assigned to each cost code as determined by the labor hours charged to each cost code. The cost accountant relies on man-hour data from the Time Sheets, and on the total distribution of man-hours per cost code from the Labor Distribution Report. When the weekly quantities per cost code are extracted from the Quantity Distribution Report (P_M-DOC 10), the cost accountant completes the pertinent information on the Labor Distribution Report (shown in Figure 13.27), including labor costs, labor unit costs, and labor hours per unit of measure for the current week and total to date.

The data contained on the Labor Distribution Report becomes the input for the Labor Cost Summary (P_M-DOC 13). As illustrated in Figure 13.33, the Labor Cost Summary provides a comparison of actual labor cost expenditures for the period, and the total to date, in relation to the current estimate for each detailed cost code. These cost data are summarized by major cost category, such as concrete. Another feature of the Labor Cost Summary is the *COST TO COMPLETE* forecast of current status. This forecast is based on the estimated labor unit cost and the projected status, using a forecasted unit cost such as the current expended labor unit cost to date.

The labor cost data, summarized by major cost category including the current projection, is carried forward to the Project Cost Summary (P_M-DOC 14). This report summarizes all project costs and incorporates approved changes to the contractor's scope of work. A typical example of the Project Cost Summary (P_M-DOC 14) is shown in Figure 13.34.

The Project Cost Summary provides all the cost data necessary for requesting payments from the client. Under the lump-sum contract, payments are tied to progress or to the estimated work completed

PROJECT: ______________________

LOCATION: ______________________

COST CODE	WORK ITEM DESCRIPTION	ORIGINAL BUDGET				APPROVED CHANGES			
		LABOR	MAT'L	SUBCON.	TOTAL	LABOR	MAT'L	SUBCON.	TOTAL
0200	SITE WORK	79,320	52,880	0	132,200	0	0	0	0
0300	CONCRETE	276,045	512,655	0	788,700	0	0	0	0
0400	MASONRY	0		1,410,000	1,410,000	3250	7580	0	10,830

Anderson & Woodhead
General Contractors

JOB NO.: ______________________

DATE: ______________________

WEEK END: ______________________

PROJECT COST SUMMARY

CURRENT BUDGET				EXPENDED TO DATE				CURRENT PROJECTION				OVER/
LABOR	MAT'L	SUBCON.	TOTAL	LABOR	MAT'L	SUBCON.	TOTAL	LABOR	MAT'L	SUBCON.	TOTAL	(UNDER)
79,320	52,880	0	132,200	35,250	23,520	0	58,770	80,120	51,320	0	131,440	<1760>
276,045	512,655	0	788,700	3096	10,193	0	13,289	276,469	512,655	0	789,124	424
3250	10,830	1,410,000	1,424,080	0	0	0	0	3250	10,830	1,410,000	1,424,080	0

Figure 13.34 Project Cost Summary (P_M-DOC 14).

to date. The percent complete for each major category of work is input from the formal schedule analysis phase through the Project Progress Analysis document. When the estimated percent complete is multiplied by the total estimated budget cost for a work category, the result represents the billable costs to date, that is the completed cost to date. The payment due from the client is determined by subtracting from the total completed cost to date through the current period the previous period's total completed cost to date. Finally, the cost to complete in terms of payments to the client is the difference between the estimated cost and the completed costs to date. This information is contained on the Progress Payment Report (P_M-DOC 15) as shown in Figure 13.35.*

*When the project is under a cost plus contract, the client is billed on a periodic predetermined basis for the actual costs incurred to date derived from the Project Cost Summary. Under unit pricing contracts, the amount billed is the actual quantity placed for the period times the contracted unit price. Each contract type is subject to a percent retention not billed to the client until the project is completed. Detailed back up will accompany each billing to the client.

The cost data generated from PROJECT COST ACCOUNTING provides information for several functions, including PROJECT FORECASTING and the evaluation of LABOR PRODUCTIVITY. At the medium-sized firm level, PROJECT FORECASTING (M3) is ideally the responsibility of the cost accountant, with the project manager involved in analyzing potential deviations from the estimate. LABOR PRODUCTIVITY is primarily the project manager's responsibility at the head office level. Because these functions follow COST ACCOUNTING, more labor cost data is available to the project manager to analyze the labor status of an activity. The basic policies guiding the performance of LABOR PRODUCTIVITY and PROJECT FORECASTING focus on *DOCUMENTING ACTIVITY STATUS* and *ACTIVITY PRODUCTIVITY*, and identifying *COSTS TO COMPLETE*. These policies provide the primary impetus for the cost control system as it relates to labor.

The data input to LABOR PRODUCTIVITY is a combination of documents from COST ACCOUNTING, coupled with knowledge of activity progress from the formal scheduling analysis phase

Anderson & Woodhead
General Contractors

PROJECT: ______ JOB NO.: ______

LOCATION: ______ PERIOD ENDING: ______

PREPARED BY: ______

COST CODE	WORK CATEGORY	CURRENT BUDGET	COST TO DATE	COST TO COMPLETE	PERCENT COMPLETE
0200.	SITE WORK	132,200	58,770	73,430	44.4
0300.	CONCRETE	788,700	13,289	775,411	1.7
0400.	MASONRY	1,424,080	0	1,424,080	

TOTAL BUDGET	13,353,300	103,291	13,250,009	0.8
WORK TO DATE	103,291			
MAT'L PURCHASE	15,231			
TOTAL EXPENDED	118,522			
LESS 10% RETENTION	11,852			
TOTAL	106,670			
LESS PREVIOUS BILLING	25,721			
PAYMENT DUE	80,949			

Figure 13.35 Progress Payment Report (P_M-DOC 15).

(see Figure 13.32). The principal documents are the Labor Distribution Report and Labor Cost Summary. These two documents can quickly alert the project manager to potential cost overruns for any given cost code category. When significant cost overruns are apparent on a cost-critical activity, the project manager will normally attempt to analyze crew performance in order to ferret out the causes of the overrun. According to *ACTIVITY PRODUCTIVITY*, the basic measures of productivity are quantity volume in place in relation to labor dollar and man-hour expenditures. These variables determine both the labor unit cost to date and labor man-hours per unit, which, when compared to the estimates, become a rough measure of productivity. For example, if the actual unit cost divided by the estimated unit cost is equal to 1.0, then the project manager assumes productivity is approximately 100% of that estimated. Alternatively, a number greater than 1.0 would indicate that productivity is less than anticipated, while a number less than 1.0 would imply that productivity is better than anticipated. This comparison is rarely documented on a labor report, rather it is retained as useful information, because it is developed on a limited basis for cost- and schedule-critical activities. The value of this type of analysis is dependent upon the quality of the estimate and the actual quantities recorded in the field. If the estimated unit cost or quantities in place are incorrect, a distorted picture of productivity might result. As a consequence, the project manager must be careful when analyzing labor unit costs or labor man-hours per unit for relative productivity indicators, especially when productivity appears low. By comparing this analysis with either visual observations or inputs from field evaluations, the project manager can determine the accuracy of the productivity indicator. When crew productivity is below acceptable limits, a more detailed analysis will be made, provided the activity is a cost-critical work category. Work categories other than cost-critical ones are often neglected because their impact on overall project costs is minor.

The project manager often works closely with the cost accountant when finalizing PROJECT FORECASTS. For example, based on the LABOR PRODUCTIVITY analysis, the project manager may recommend that labor cost projections be adjusted for certain work categories. These types of adjustments will be incorporated into the Labor Cost Summary and Project Cost Summary, in order to provide top management with the best possible picture of project cost status. Significant deviations can then to presented to top management on a *MANAGEMENT BY EXCEPTION* (P_M 26) basis for major decision making during the Project Management Control decision process.

Cash flow curves are often generated during PROJECT FORECASTING to assist the company in evaluating their short-term cash needs. A typical Expenditure Projection (I_M-DOC 7) is shown in Figure 13.36, and is represented as an internal document because it is only generated when deemed necessary. The S-curve is an extremely flexible tool

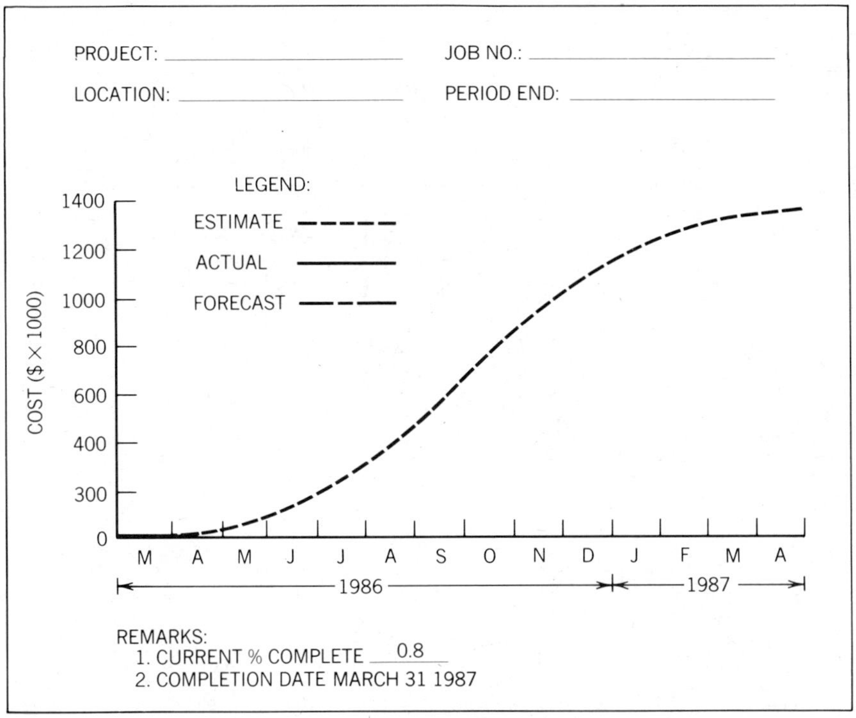

Figure 13.36 Expenditure Projection (I_M-DOC 7).

for analyzing costs to date and projected costs to complete. The information is more useful for cash flow distribution over the life of the project, and is used by the cost accountant to project cash needs on a short-term basis.

The project information generated in the Labor Cost Analysis phase, Figure 13.32, is disseminated to the field for further LABOR PRODUCTIVITY analysis, and to Formal Scheduling for incorporation into the current scheduling analysis. All other documents are forwarded to Project Management Control for top management review and action where required.

13.5 FIELD-BASED PROJECT MANAGEMENT

The heavy field commitment of the Field-Based Project Management approach is expressed in a number of ways in regard to Project Status, Analysis, and Reporting. First, all the manpower management functions relevant to the decision process are formally performed in the field. The head office supports field project status and progress reporting with summary level reporting, head office initiated scope revisions, and cost and scheduling assistance. This approach is significantly different from that of the medium-sized firm, wherein only informal field scheduling analysis and by-exception field labor performance analysis is performed. At the medium-sized firm level, all formal scheduling, productivity analysis, cost accounting, and scope change analysis decision making is head office oriented. Secondly, to complement the increasing management emphasis on field-based Project Scheduling and Crew Planning, Project Status, Analysis, and Reporting must provide more timely status, progress, and labor performance information related specifically to work packages. Finally, the analysis and reporting aspects of cost and scheduling have a broader focus and generate a greater volume of data to assist field decision makers in identifying potential problems, determining the cause of these problems, and taking the appropriate remedial action to correct or resolve these problems through Project Management Control. This leads to an increasing degree of integration in the project information system.

The Management Information System is fully computerized at the large firm level, and is concerned with the effective manipulation of often large volumes of data via control area concepts and cost code techniques. Furthermore, the information system for such a project has a multi-level report

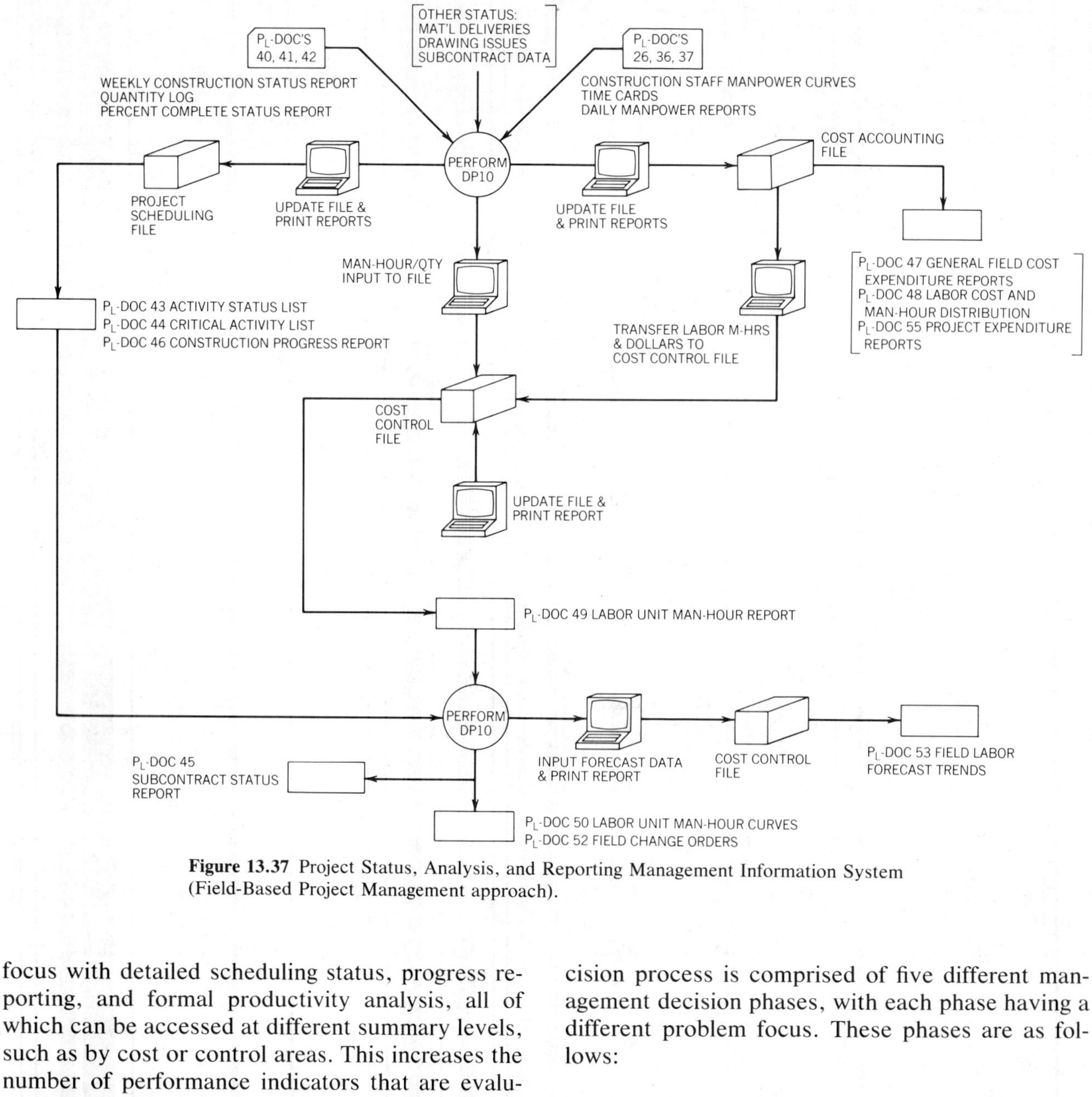

Figure 13.37 Project Status, Analysis, and Reporting Management Information System (Field-Based Project Management approach).

focus with detailed scheduling status, progress reporting, and formal productivity analysis, all of which can be accessed at different summary levels, such as by cost or control areas. This increases the number of performance indicators that are evaluated and considered in determining a project's present and future position. The project information system must guide the flow of documents and information through the decision process because data from one document is essential for other documents within the decision process. Thus, the Management Information System for Project Status, Analysis, and Reporting, as illustrated in Figure 13.37, takes live field data from Field Monitoring, and processes the data for use in generating the required cost and scheduling reports.

The Project Status, Analysis, and Reporting decision process is comprised of five different management decision phases, with each phase having a different problem focus. These phases are as follows:

1. *Field Scheduling* DEFINES CURRENT PROJECT STATUS (S5) and REVISES THE CONSTRUCTION SCHEDULE (S6) when major schedule revisions are dictated by field assessments. (The decision network is illustrated in Figure 13.38.)
2. *Field Cost Reporting* provides PROJECT COST ACCOUNTING (R3) expenditure data for incorporation into all relevant project reports. (The decision network is illustrated in Figure 13.43.)

3. *Field Productivity Analysis* is concerned with LABOR ANALYSIS (M5) and the documentation of LABOR PRODUCTIVITY (R5), from which special ACTIVITY ANALYSES (M4) are initiated. (The decision network is illustrated in Figure 13.45.)
4. *Field Scope Analysis and Reporting* focuses on PROJECT FORECASTS (M3) that document performance trends and field changes in scope. These forecasts become the basis for UPDATING AND REVISING WORK PACKAGE SCOPE (R4). Current work package scope information provides data vital for continuous Crew Planning, Crew Management, and Field Scheduling of work activities. (The decision network is illustrated in Figure 13.49.)
5. *The Head Office phase* is concerned with reporting all project cost expenditure data, including field cost data, and with the identification and preparation of head office initiated changes in scope and performance trends that influence field construction. (The decision network is illustrated in Figure 13.54.)

As reflected in these decision processes, an integrated system emerges in terms of a continuous flow of information between the various phases to support the performance of each manpower management function. In addition, information and documents are frequently passed to other decision processes to provide data for planning, scheduling, and controlling the work effort. Thus, the information system becomes problem oriented at the Field-Based Project Management level.

Project Status, Analysis, and Reporting involves many decision makers in both the head office and the field. The scheduling engineer is responsible for field scheduling status and analysis, and is always assisted by the superintendents when revising construction schedules. Crew productivity analysis is a cost engineering function involving the recording of productivity. The cost engineer works closely with the superintendents and scheduling engineers, however, when actually analyzing specific productivity problems. When activity analysis is requested a number of field decision makers may be involved. The craft superintendents are primarily responsible for any activity analysis, but the general superintendent, project field engineers, and even the field project manager or resident manager will often participate in problem solving depending on the criticality of the problem.

Both cost and scheduling engineers are involved with the analysis of project trends and scope revisions as they impact work packages. These decision makers rely on feedback from all field personnel in the performance of the scope analysis phase. Finally, payroll analysis and cost reporting is the responsibility of the field cost accountant. Head office support is provided by head office located cost and scheduling engineers who analyze and prepare changes in scope and process current budget/forecast cost data for total project cost reports. Head office cost accounting personnel are responsible for collecting and summarizing all project cost expenditure data for incorporation into project summary cost reports.

The *Field Scheduling* phase, Figure 13.38, focuses predominantly on the analysis and reporting of construction activity and progress as the means of DEFINING CURRENT PROJECT STATUS. In comparison to the medium-sized firm, there emerges at the large firm level a clearer distinction between statusing and reporting percent progress for both activities and the project as a whole. Statusing focuses on time in relation to start and finish dates of construction activities as reflected on both the Construction Schedule (P_L-DOC 19), and the short-range or Quarterly Construction Schedule (P_L-DOC 20). Reporting progress relates to physical placement of material or equipment, and is referenced in terms of percent complete for each activity. Progress reporting must also express a total project percent complete, and percent complete for component facilities. While activity status and percent progress data is collected and reviewed weekly, project-oriented status reports are normally produced on a monthly cycle. Informal scheduling analysis is performed on a weekly basis as a means for early detection of problems, and information is often passed either verbally from the scheduling engineer to the respective field managers and superintendents, or via the Weekly Construction Status Report (P_L-DOC 40).

The responsibility of DEFINING CURRENT PROJECT STATUS rests primarily with scheduling. The scheduling engineer relies on input from field superintendents and field engineers and must work closely with these personnel to accurately re-

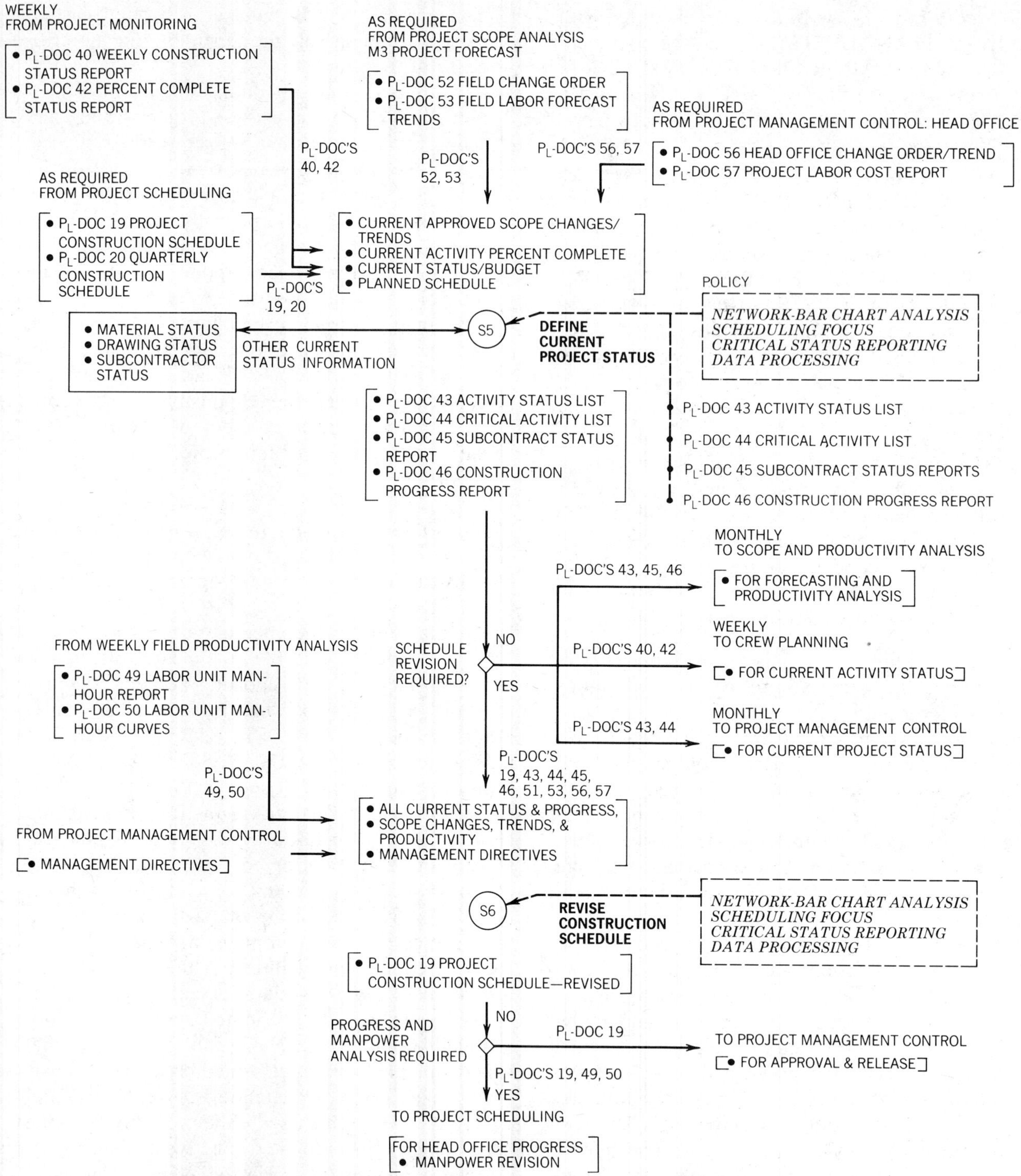

Figure 13.38 Project Status, Analysis, and Reporting decision process: field scheduling phase (Field-Based Project Management approach).

TABLE 13.5 POLICIES OF THE LARGE FIRM

Policy		Description
P_L 6	*COST CODES*	A systematic classification of project work categories segregated by unique physical components and by cost codes is the basis for structuring the project scope into work tasks, categorizing estimated costs, reporting and controlling costs and manpower, and developing a historical database.
P_L 10	*NETWORK-BAR CHART ANALYSIS*	Broad network focused milestone planning is the best approach for identifying and sequencing major project components, while a combination of the critical path method and bar chart model are best suited for project planning and scheduling.
P_L 11	*SCHEDULING FOCUS*	Scheduling definitions and constraints are: 1. A critical activity is one that, if delayed, will extend the completion date of the project. 2. Activity durations must be correlated to the estimated work content. 3. All constraints must be defined and incorporated into the sequence logic. 4. Manpower impact is an integral aspect of project scheduling and scheduling control. 5. The schedule must reflect optimum use of all resources. 6. Progress is based on the physical completion of work activities.
P_L 15	*MANPOWER PLANNING*	Long-range planning of manpower by craft and in total is a prerequisite to supporting the field effort, and must be performed in the planning and scheduling stages of the project.
P_L 27	*DOCUMENTED ACTIVITY STATUS*	Labor costs, manpower, man-hours, and quantities are primary ingredients of cost control and must be documented by a system yielding reliable information and viable reports for management control. Activity status must be documented in terms of start and completion dates and percent complete data.
P_L 28	*CREW PRODUCTIVITY*	Actual unit man-hour rates versus estimated man-hour rates is a measure of crew performance and a basic indicator of productivity.
P_L 29	*COST TO COMPLETE*	Projecting total project costs and man-hour requirements is requisite to identify problems and evaluate alternative courses of action at both the activity level and the project level.
P_L 30	*TRENDING*	A mechanism for documenting and analyzing deviations of cost and schedule consequence is an integral facet of the information system.
P_L 31	*CRITICAL STATUS REPORTING*	The construction progress reporting system must provide a comprehensive report of activity status, highlighted by a mechanism to alert management to activities at and approaching critical status.
P_L 32	*ACTIVITY ANALYSIS*	Activity analysis is geared to exception reporting mechanisms or field management desire to analyze construction methods or crew productivity in order to improve crew performance.
P_L 33	*MANAGEMENT BY EXCEPTION*	Manpower management by exception is acceptable provided that the information system is geared to exception reporting to alert management to existing problems or potential problems which are emerging.
P_L 34	*DATA PROCESSING*	Maximum use of electronic data processing is often the most economical means of processing data into useable information.
P_L 35	*PROJECT TEAM SPIRIT*	Active participation of all field managers in the management process promotes goodwill and a positive management environment.

AW
Engineering & Construction Corp.

CLIENT: ______ CONTRACT: ______

PROJECT: ______ START DATE: ______

CONTROL ZONE DESCRIPTION: ______ STATUS DATE: ______

ACTIVITY STATUS LIST

CONTROL ZONE	I NODE	J NODE	COST CODE	SPECIAL IDENT.	% COMPL.	ACTIVITY DESCRIPTION	DURATION EST. REM.	EARLY START	ACT. START	EST. FINISH	TOTAL FLOAT	REM FLOAT

↑ SORT BY SYSTEM, ETC.

← ALL ACTIVITIES CAN APPEAR ON THIS REPORT →

Figure 13.39 Activity Status List (P_L-DOC 43).

port activity status and progress. The documents most frequently referenced by the scheduling engineer are the Weekly Work Status Report, Percent Complete Forms (P_L-DOC 42), Current Schedules, and all current change orders and labor trend forecasts. The scheduling engineer must also have a working knowledge of the previous period's percent progress and status reports. Other relevant information not gleaned from these documents is knowledge of site conditions, visual assessments of activity status, and visual verification of physical progress. Thus, the scheduling engineer must have hands-on exposure to field conditions in order to keep abreast of actual site activity.

Field activity statusing is predominately influenced by the need for *CRITICAL STATUS REPORTING* (P_L 31),* and requires the scheduling engineer to assess the status of each activity that is either under construction or scheduled to start within the time frame of the next period schedule. For those activities currently under construction, the scheduling engineer must decide whether the planned completion date remains valid. If it does not, the scheduling engineer, in conjunction with the appropriate craft or assistant superintendent, may forecast a new completion date for the activity. A similar analysis is applied to activities which are scheduled to begin in the current planning period. The current planned start date for the activity may have to be rescheduled if labor or material is not available. All current activity status is depicted on an ACTIVITY STATUS LIST (P_L-DOC 43), as depicted in Figure 13.39. When utilizing computer *DATA PROCESSING* techniques, the Activity Status List can be sorted to focus only on those activities in the most current planning period as dictated by the Quarterly Construction Schedule.

During general scheduling maintenance and updating for current status, *CRITICAL STATUS REPORTING* is used to identify those activities which are critical and/or approaching critical status. This is essential on large projects because sorting on critical activities limits management's attention to the activities most likely to impact project completion. In this manner, manpower management problems concerned with critical activities can be handled quickly through Crew Planning, Crew Management, and schedule revisions. A typical Critical Activities List (P_L-DOC 44), is illustrated in Figure 13.40. In this example, critical activities are those having less than ten days of total float.

While statusing activities for which the contractor has direct control, the scheduling engineer must also analyze the current status of those subcontractors currently working on the project, as well as of those subcontractors that will commence work during the next planning horizon. This information is documented on a Subcontract Status Report (P_L-DOC 45), as illustrated in Figure 13.41, and is typically provided through a project subcontract administration group. This report provides the contractor with subcontractor percent progress and activity status as against the planned scheduled data (see Figure 13.41*a*).†

Documenting and reporting construction progress in terms of percent complete is directed by

*See Table 13.5, Policies of the Large Firm.

†The Cost Report of the Subcontract Status Report shown in Figure 13.41*b* is completed during the Field Cost Reporting Phase (Figure 13.43).

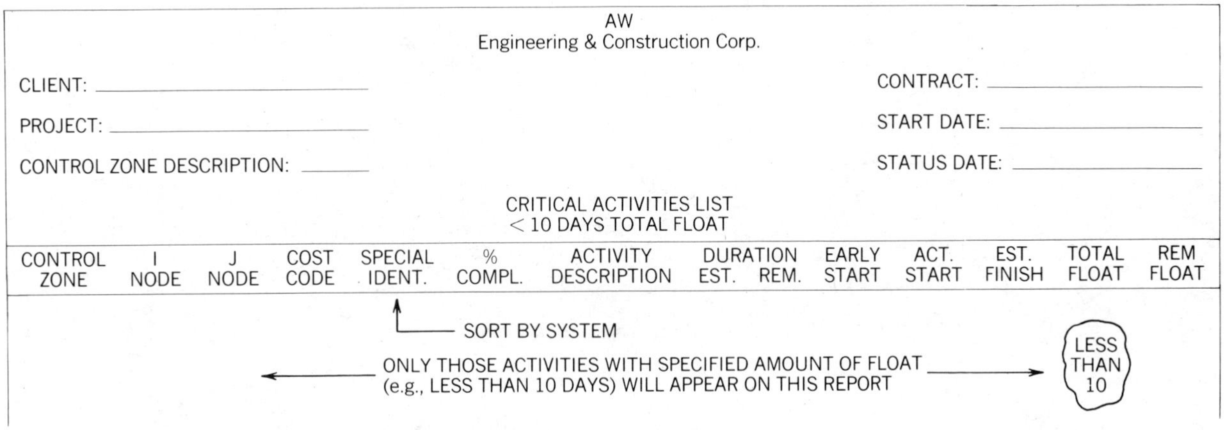

Figure 13.40 Critical Activity List (P_L-DOC 44).

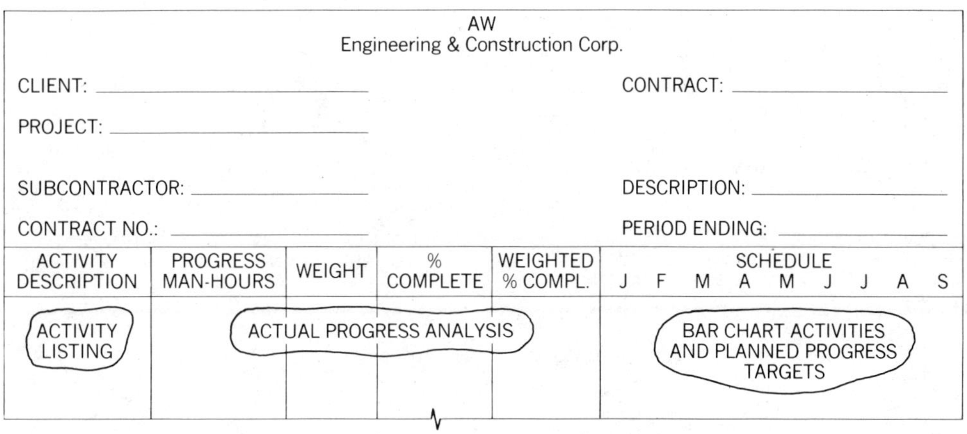

(a)

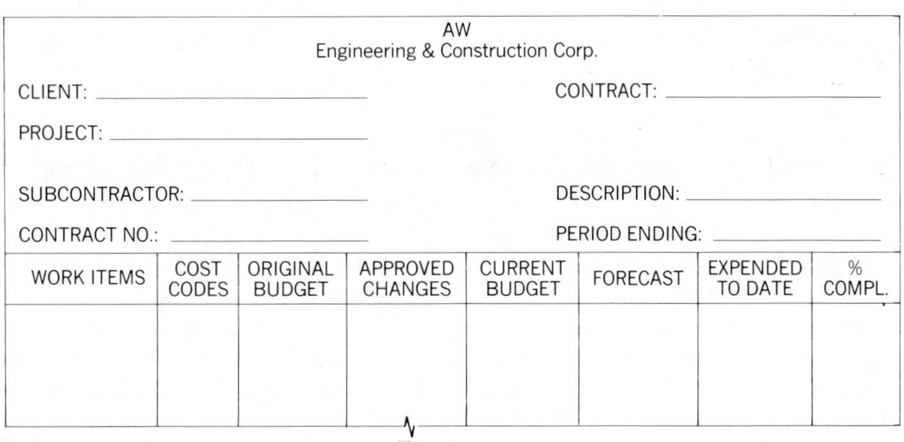

(b)

Figure 13.41 Subcontract Status Reports (P_L-DOC 45). *(a)* Progress Report. *(b)* Cost Report.

AW
Engineering & Construction Corp.

CLIENT: ______ CONTRACT: ______

PROJECT: ______ START DATE: ______

CONTROL ZONE DESCRIPTION: ______ STATUS DATE: ______

PROGRESS REPORT

CONTROL ZONE	I NODE	J NODE	COST CODE	SPECIAL IDENT.	ACTIVITY DESCRIPTION	BUDGET MAN-HOURS	% COMPL.	WEIGHTED % COMPL.
2050	←				DETAILED ACTIVITIES →	1000	50.0	5.0(1)
	SYSTEM ETC. SORT				TOTAL (FOR CONTROL ZONE)	10000		25.0(2)

NOTES:
(1) $1000 \div 10000 \times 50.0 = 5.0$
(2) SUM OF ALL DETAILED ACTIVITIES = 25.0%

(a)

AW
Engineering & Construction Corp.

CLIENT: ______ CONTRACT: ______

PROJECT: ______ START DATE: ______

TOTAL PROJECT SUMMARY: ______ STATUS DATE: ______

PROGRESS REPORT

CONTROL ZONE	COST CODE	DESCRIPTION	BUDGET MAN-HRS	% COMPL.	WEIGHTED % COMPL.
2050			10000	25.0	0.25(1)
SORT BY EITHER CONTROL ZONE OR MAJOR COST CATEGORY		TOTAL PROJECT	1000000		10.1(2)

NOTES:
(1) $10000 \div 1000000 \times 25.0 = .25$
(2) SUM OF ALL CONTROL ZONES EQUALS 10.1%

(b)

Figure 13.42 Construction Progress Report (P_L-DOC 46). (*a*) Detailed progress analysis. (*b*) Summary progress analysis.

SCHEDULING FOCUS (P_L 11). The percent complete for each activity is provided through Field Monitoring and documented on Percent Complete Forms. The percent complete for an activity is then multiplied by the weighted budget man-hours for the activity. This process is repeated for each activity for which progress has been recorded. The sum of the weighted percent complete for all activities is equal to the total percent complete for the project. This information is contained in the Construction Progress Report (P_L-DOC 46) shown in Figure 13.42. Figure 13.42*a* depicts the detailed progress analysis, while Figure 13.42*b* illustrates summary level progress analysis. This multi-level approach permits decision makers to evaluate progress to the extent necessary to identify and solve problems.

Once CURRENT PROJECT STATUS IS DEFINED, field management must decide whether a major schedule revision is required (see Figure 13.38). This decision is influenced by the current project status and progress. For example, if the project is essentially on schedule, only minor schedule revisions would be required in the form of periodic maintenance of the Project Construction Schedule. A period might consist of three to six months. If the project is seriously behind schedule, however, a major schedule revision may be initiated. Thus, both current activity status and physical progress influence THE REVISION OF THE CONSTRUCTION SCHEDULE. Schedule revisions are the responsibility of the scheduling engineer, with directions provided by the superintendents. The main documents the scheduling engineer references are the Project Construction Schedule and/or Quarterly Construction Schedule, the Activity Status Reports, the Subcontract Status Reports, the Construction Progress Report, and current productivity data, drawing status, and material deliveries. The scheduling engineer also incorporates any major management directives into schedule revisions.

Minor scheduling revisions are required on a periodic basis, at which time adjustments are made to activity start and finish dates. These adjustments normally will not influence the completion date of the project, but they can impact field production scheduling and crew planning. For instance, if the actual start of an activity is delayed beyond the planning horizon of the current Quarterly Schedule, this delay must then be considered during the next planning period. Delays in other more current activities could have an immediate impact on Crew Planning in terms of ESTABLISHING CREW REQUIREMENTS (P7), PLANNING THE WORK FACE LAYOUT (P8), and SCHEDULING THE PROJECT CREWS (S7).

While analyzing activity start and finish dates and developing revised activity durations, it is important to consider all current changes in scope and performance trends. Even when an activity slips behind schedule, it may not disrupt the overall schedule if it has a significant amount of float. However, if the float is not significant and a delay could endanger the target completion, the activity duration may be reevaluated. In this case logic changes may be required. In this situation, the scheduling engineer and the superintendent would together analyze the content of the activity to evaluate the feasibility of changing its duration or logical sequence. In addition, changes in the total estimated man-hours for an activity can influence its construction duration. If the projected man-hours are markedly higher than the budget, the construction duration of many activities might be impacted, and should be reevaluated to check whether or not existing completion dates remain feasible or if additional crew manpower is required.

Major scheduling revisions are often required to ensure that target completion dates are achieved for the project. These revisions are often the result of critical problems which arise from major material or equipment delays, strikes, labor shortages, and the like. The construction sequence logic is reevaluated in terms of current conditions and possible alternative courses of action, such as shift work or an extended work week. The basic policies of *NETWORK-BAR CHART ANALYSIS* and *SCHEDULING FOCUS* guide THE REVISION OF THE CONSTRUCTION SCHEDULE. The mechanics of performing major schedule revisions are similar to those described for Project Scheduling (see Section 7.5). These types of major revisions are reviewed by project management, both in the field and at the head office, prior to releasing a Revised Construction Schedule (P_L-DOC 19),* to the client or to the field for implementation. If a major schedule revision will significantly impact current progress and manpower targets revised progress and manpower curves will be developed to match the Revised Construction Schedule and specified completion dates. This schedule revision would be performed as part of Project Scheduling in a manner similar to that described in Section 7.5 (see Figures 7.21 through 7.30).

The Field Cost Reporting Phase, Figure 13.43, operates on a daily basis with most labor cost reports produced weekly by PROJECT COST ACCOUNTING. In terms of manpower management, the Time Cards (P_L-DOC 36), and the Field Manpower Reports (P_L-DOC 37), are the key field labor documents. Field staff man-hours are also input into PROJECT COST ACCOUNTING. In addition to these basic manpower-related documents and information flows, PROJECT COST ACCOUNTING collects and processes other field expenditure data, such as field material purchases, construction equipment rental costs, and subcontract costs.

PROJECT COST ACCOUNTING in the field is the cost accountant's functional responsibility. At the large firm level, the cost accountant has a staff of cost clerks and payroll administrators in addition

*The revised Construction Schedule would have the same format and structure as that in Figure 7.27.

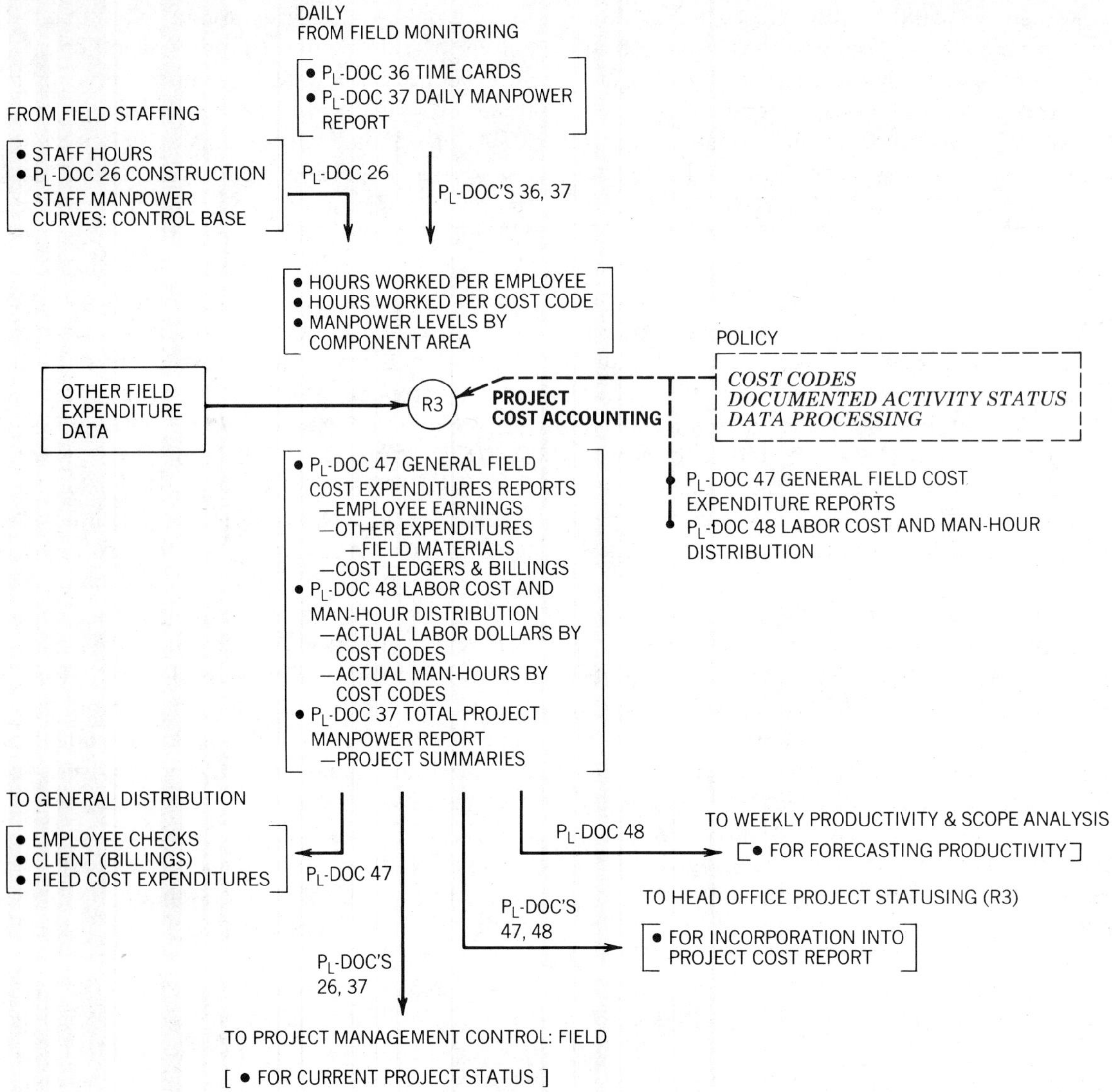

Figure 13.43 Project Status, Analysis, and Reporting decision process: field cost reporting (Field-Based Project Management approach).

to timekeeper's, who are also functionally responsible to the cost accountant. The actual number of these personnel varies with project size, location, and the level of support provided by the head office. For instance, the head office often prepares total project cost reports that incorporate the field cost reports.

Field cost reporting produces a variety of cost expenditure reports which are identified as General Field Cost Expenditure Reports (P_L-DOC 47). These include such items as employee earning records, field material expenditure reports, other cost ledgers, and so on. These types of reports would contain data similar to that presented in Figures 13.16 and 13.17. A key manpower-focused report generated through PROJECT COST ACCOUNTING is the Labor Cost and Man-hour Distribution (P_L-DOC 48). This report reflects labor man-hours expended by detailed cost code within each control component segment. The labor man-hour report also expresses labor dollars and average wage rates by cost code. An illustration of a typical such document appears in Figure 13.44*a*. The data contained here are useful for cost trending, productivity analysis, and reporting current project status, as indicated in Figure 13.43.

AW
Engineering & Construction Corp.

CLIENT: ______ CONTRACT: ______

PROJECT: ______ DATE: ______

CONTROL ZONE DESCRIPTION WEEK END: ______

COST CODE	WORK CATEGORY DESCRIPTION	CRAFT CODE	PERIOD		TO DATE		WAGE RATE	
			MAN-HRS.	COST ($)	MAN-HRS.	COST ($)	PERIOD	TO DATE
141101	CONCRETE COLS.							
	CARPENTERS	06	100	1600	512	8300	16.00	16.21
	LABORERS	16	200	2700	942	12651	13.50	13.43
	SUBTOTAL		300	4300	1454	20951	14.33	14.41

(a)

AW
Engineering & Construction Corp.

CLIENT: ______ CONTRACT: ______

PROJECT: ______ DATE: ______

SORT DESCRIPTION WEEK END: ______

COST CODE	CRAFT DESCRIPTION	CRAFT CODE	PERIOD		TO DATE		WAGE RATE		% WEIGHT	
			MAN-HRS.	COST ($)	MAN-HRS.	COST ($)	PERIOD	TO DATE	PERIOD	TO DATE
	CARPENTERS	06	712	11499	8120	130164	16.15	16.03	15.1	14.8
	LABORERS	16	1514	20061	16250	217425	13.25	13.38	29.8	30.1
	⋮									

COST CODE/CONTROL ZONE, ETC.

(b)

Figure 13.44 Labor Cost and Man-hour Distribution (P_L-DOC 48). (*a*) Report by detailed cost code. (*b*) Report by craft for major sort.

Because *DATA PROCESSING* techniques are common at the large firm level, variations of this report can be generated by sorting the information differently. For example, if each time card has a craft code assigned to it, and this code is recognized for processing, the expended labor man-hours can be collected by craft employed on the project. This craft distribution in Figure 13.44*b* is extremely useful for comparing the actual man-hours by craft required to complete the project as well as analyzing actual craft mixes versus planned. Finally, the total project manpower daily head count is summarized from the Daily Manpower Reports (P_L-DOC 37) for each control area. In summary form, the total project manpower reports become an effective tool for keeping field management apprised of the number of craft and staff personnel on the project, including subcontractors. These reports also keep management aware of the rate of absenteeism (see Figure 13.29 for an example of the manpower report format and its content).

The field productivity analysis phase, Figure 13.45, is another important area of Project Status Analysis, and Reporting. This phase focuses both on LABOR ANALYSIS and recording of LABOR PRODUCTIVITY. The cost engineer performs an analysis of both the labor man-hour expenditures relative to planned man-hour expenditures, and the labor unit man-hour rates versus current budget labor unit man-hour rates. Through the cost control system, the cost engineer accesses actual labor man-hours expended from the Labor Cost and Man-hour Distribution Report, and extracts actual quantities placed from the Quantity Log Report (P_L-

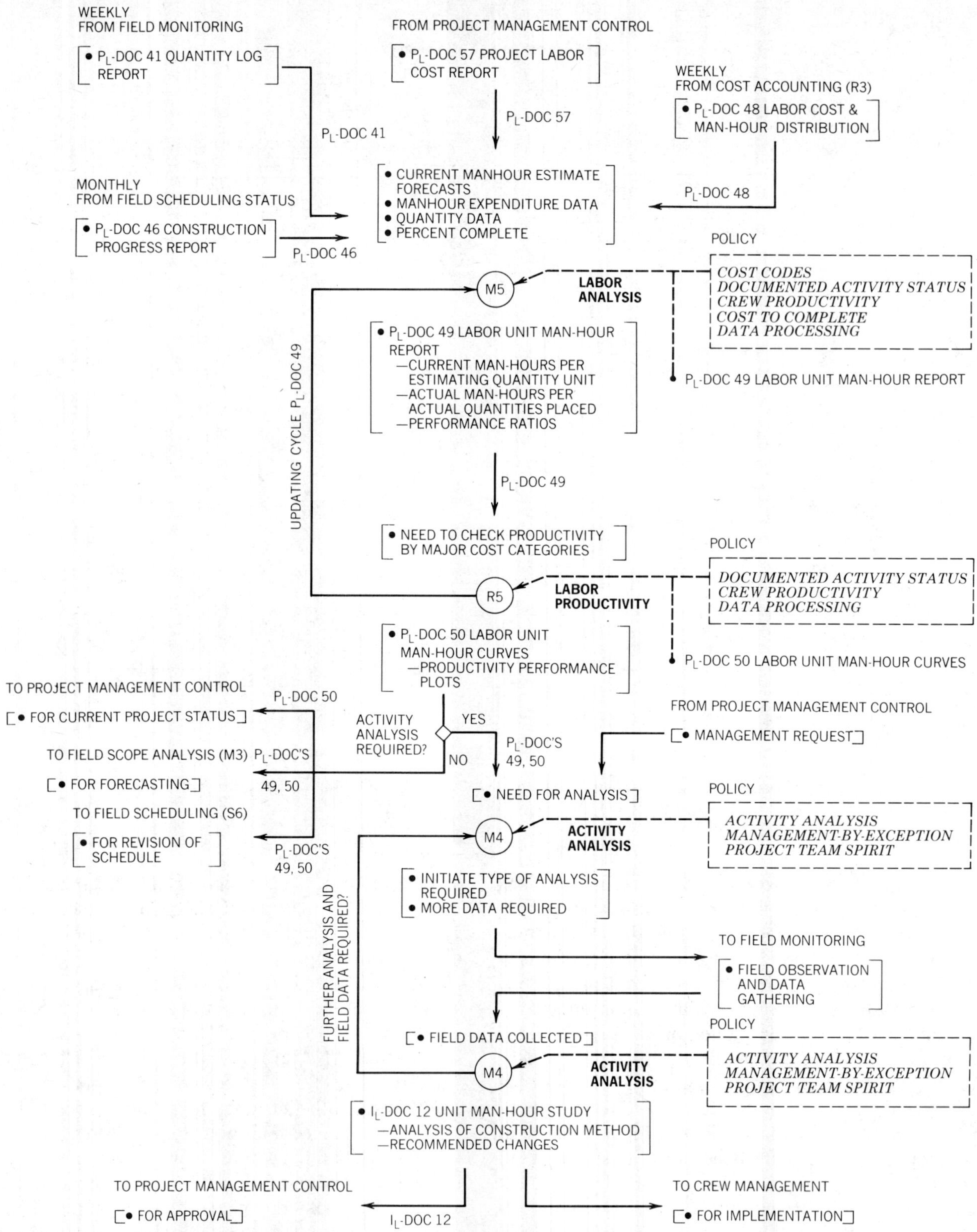

Figure 13.45 Project Status, Analysis, and Reporting decision process: field productivity analysis and reporting (Field-Based Project Management approach).

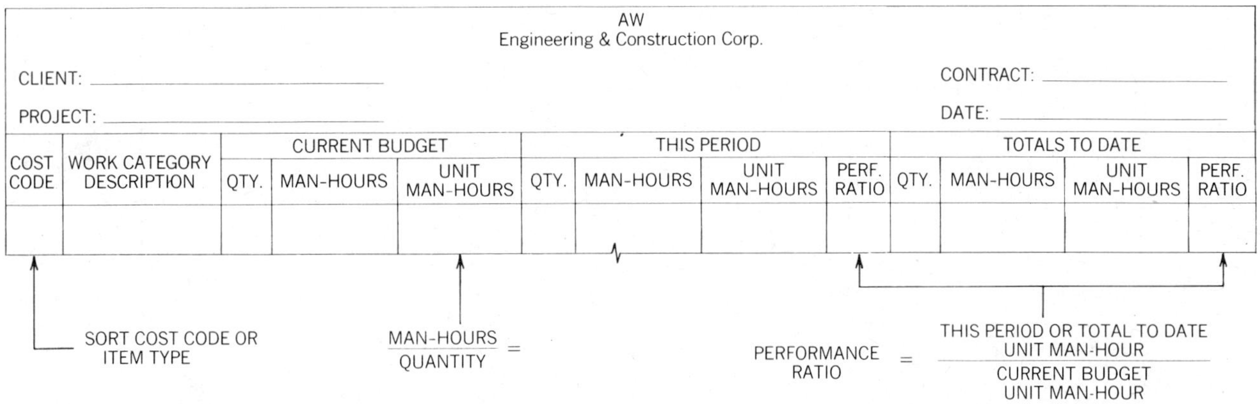

Figure 13.46 Labor Unit Man-hour Report (P_L-DOC 49).

DOC 41). Current man-hour budget data is available from the Project Cost Report (P_L-DOC 59). Once the relevant data is input to the current cost file, *DATA PROCESSING* occurs and the Labor Unit Man-hour Report (P_L-DOC 49) is generated. As shown in Figure 13.46, the Labor Unit Man-hour Report reflects a cost code level of detail. Using sorting capabilities, however, the report can isolate specified physical components or summarize data by major component categories having similar units of measure, for instance, structural concrete.

According to the *CREW PRODUCTIVITY* (P_L 28) policy, when actual labor unit man-hour rates are compared with current budget labor unit man-hour rates, a means of measuring labor performance is provided and becomes a basic indicator of labor productivity. The comparison is made by dividing the actual labor unit man-hours by the current budget labor unit man-hours, based on the assumption that the target labor unit man-hour rate is 1.0. This performance ratio is expressed as Labor Unit Manhour Curves (P_L-DOC 50) in Figure 13.47.

LABOR PRODUCTIVITY is documented for summary level project management information, as the Labor Unit Man-hour Report contains too much detail for key field managers. Graphic plots can be used to quickly display labor performance in terms of labor unit man-hour comparisons. Several curves can be generated by the cost engineer which graphically depict performance trends. For example, the actual and current budget labor unit man-hour rates can be plotted as a function of time. These plots focus on major categories of work, and reflect both incremental data (i.e., period data), and cumulative data (i.e., total to date). These curves, as illustrated in Figure 13.47, provide the key field managers with a summary of labor performance by major control area or cost category, and can help them identify problem areas which might require further dissection and analysis.

Both the Labor Unit Man-hour Report and the Labor Unit Man-hour Curves are useful tools for identifying problems, and may indicate that ACTIVITY ANALYSIS should be performed for specific areas, as shown in Figure 13.45. ACTIVITY ANALYSIS may be performed informally by the responsible superintendents through visual observations and group discussion analysis. However, a Unit Man-hour Study (I_L-DOC 12) might be requested on major work activities to assist in the analysis of the construction operation. The Unit Man-hour Study (I_L-DOC 12) presented in Figure 13.48 provides an example of typical data collected through Field Monitoring and the actual activity analysis and report results. The appropriate recommendation is implemented through Crew Management, provided management approval is not required. When the recommended plan of action requires the purchase of major construction equipment or institutes a major deviation in the construction approach, then the resident field manager will enter the decision process prior to implementation of the plan.

The Labor Unit Man-hour Report and the Labor Unit Man-hour Curves are essential documents for the PROJECT FORECASTING and FIELD SCHEDULING functions, and the Crew Planning and Crew Management decision processes. The information contained on these documents allows the various field decision makers to focus on specific problem areas when updating schedules, planning new crew activity, and/or manipulating and adjusting crews to achieve higher levels of performance.

The field scope analysis and reporting phase,

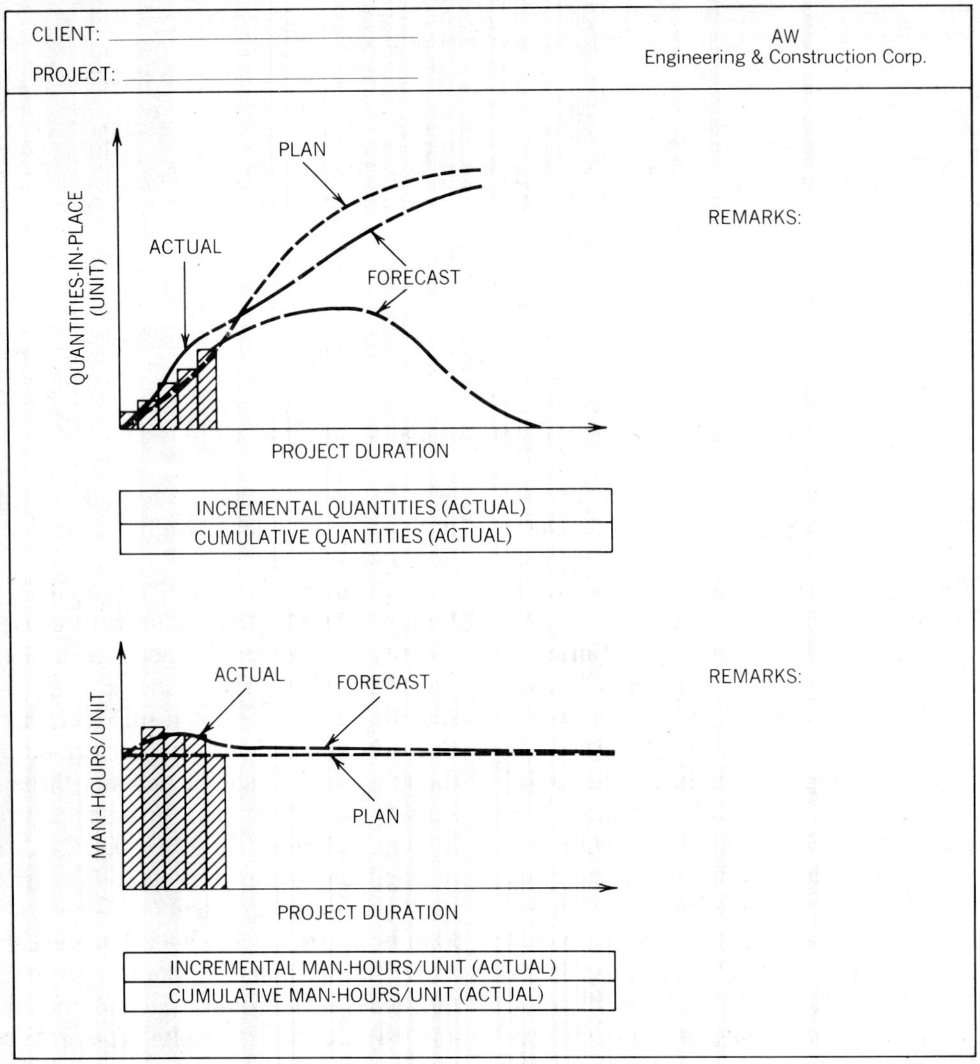

Figure 13.47 Labor Unit Man-hour Curves (P_L-DOC 50).

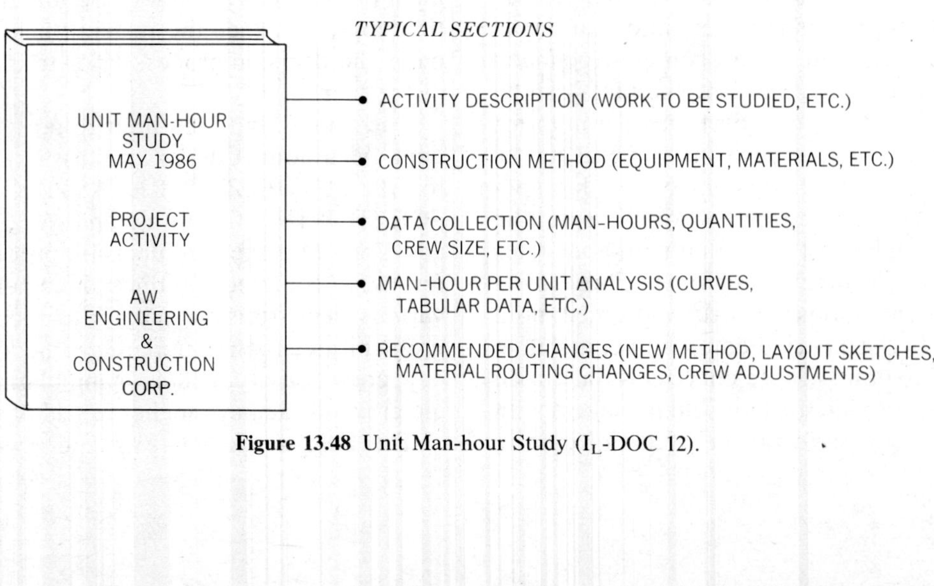

Figure 13.48 Unit Man-hour Study (I_L-DOC 12).

Figure 13.49, is a significant problem area that impacts manpower management both directly and indirectly. The significance of the field scope analysis and reporting phase often increases with the size and nature of a project. On larger design–construct projects, field scope analysis and reporting provides a method of tracking and analyzing potential changes whenever better scope definition becomes available, and also helps establish actual performance trends from the periodic monitoring of field work execution. This decision phase is traditionally considered a cost engineering function incorporating fully documented cost trending and change order procedures.

Cost trending and change order analysis problems are documented through PROJECT FORECASTING. This function normally involves many project team personnel, including cost engineers, superintendents, field engineers, and field project management. The cost engineer has primary responsibility for this function, because of the documents generated related to PROJECT FORECASTING. The cost engineer relies heavily on input from the other decision makers to successfully identify and suggest solutions to cost trending and change order related problems. The general and craft superintendents, field engineers, and the project manager (often in conjunction with the client) rely on cost trends as a basis for making decisions regarding scope changes, major schedule changes, and project performance trends.

PROJECT FORECASTING is influenced most noticeably by the need for *TRENDING* (P_L 30). This policy requirement states that a mechanism must exist within the Management Information System for documenting and analyzing deviations from costs and schedule. *TRENDING* is supported by three other policies, namely *DOCUMENTING ACTIVITY STATUS* (P_L 27), *CREW PRODUCTIVITY* (P_L 28), and *COST TO COMPLETE* (P_L 29). Of these three policies, *COST TO COMPLETE* emphasizes projecting total labor costs and man-hours as a means of identifying trends and evaluating alternative courses of action. These policies are implemented through a formalized Standard Change Order and Trend Procedure (P_L-DOC 51). Figure 13.50 provides an illustrative example of the contents of a typical change order and trend procedure.

Field changes in scope are impacted by PROJECT FORECASTING because any such changes must be incorporated into the current budget to provide an accurate reporting base. These types of scope changes can be identified by any project team member, and once sufficient scope is available a change order estimate is prepared by the field. An example of a Summary Level Field Change Order (P_L-DOC 52), is shown in Figure 13.51. Change order estimates are approved by the resident field manager, and are forwarded to the head office for final approval and issue to the client by the project manager.

The main source of information for PROJECT FORECASTING of labor performance trends is data derived from reports which delineate actual emerging trends, such as actual versus estimated unit man-hour rates. The Labor Unit Man-hour Report and the Labor Unit Man-hour Curves are excellent reports for ascertaining trends relative to man-hours, quantities, and rates of production. The Labor Unit Man-hour Report compares both estimated and actual quantities, man-hours, and unit man-hour rates for both the current period and the total to date.

Updating of work component man-hours must also reflect the impact of current progress. This requires the cost engineer to extract the actual man-hours expended to date for the particular work component from the Labor Cost and Man-hour Distribution Report, and the percent progress from the Construction Progress Report. These data are combined with the current budget man-hours for the work task to generate Field Labor Forecast Trends (P_L-DOC 53). The Field Labor Forecast Trend report in Figure 13.52 projects the man-hour delta between the current budget man-hours and the total forecasted man-hours. This calculation is based on an analysis of man-hours, quantities, and unit man-hour rates. The cost engineer will discuss labor performance trends of this nature with the superintendents to determine whether the projected man-hours are acceptable for completing the work. The superintendent may accept the current labor unit man-hour rate, or may reevaluate it based on his or her knowledge of the work area. Consequently, the man-hours projected to complete twenty percent of the remaining work might be adjusted, either higher or lower, to reflect the best forecast of man-hours required to complete. The agreed-upon total project man-hours are entered as the current forecast for the work component. This process is completed for all work components, and corrective action can be initiated for those work components with projected man-hour overruns that might significantly affect the total project. The labor

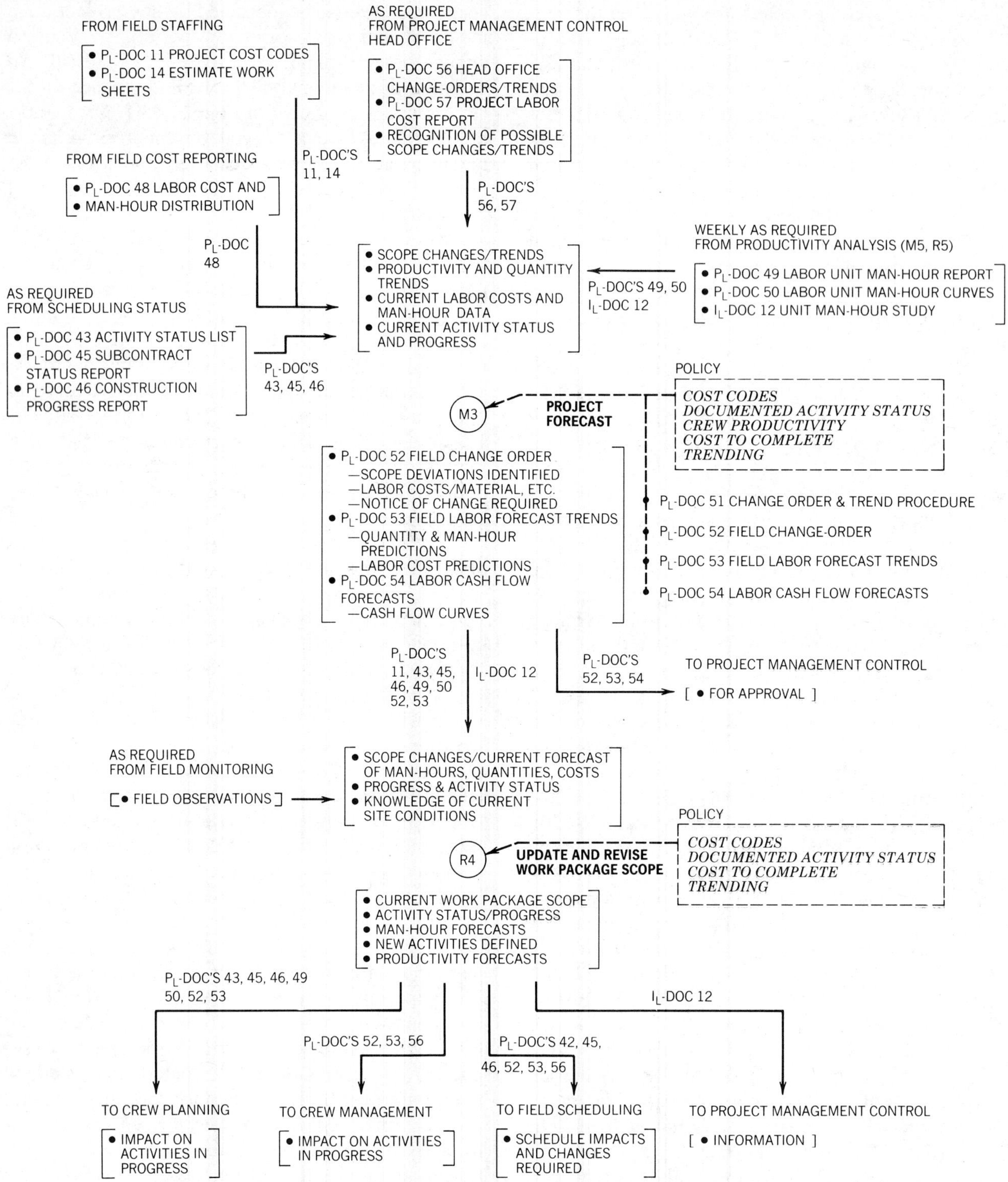

Figure 13.49 Project Status, Analysis, and Reporting decision process: field scope analysis and reporting (Field-Based Project Management approach).

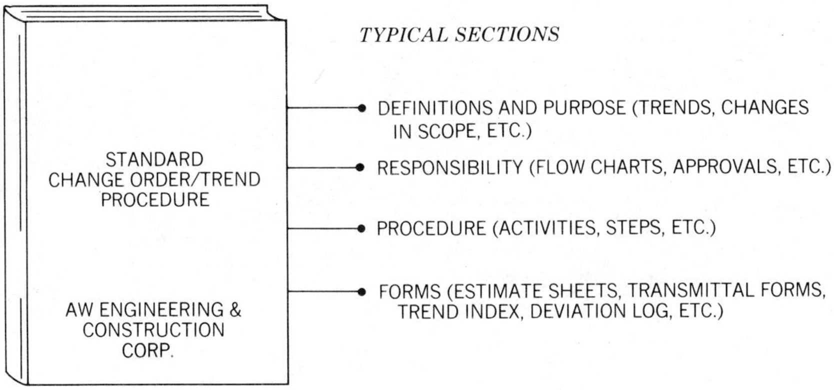

Figure 13.50 Standard Change Order and Trend Procedure (P_L-DOC 51).

man-hours are then converted to labor dollars by forecasting the work component crew wage rate to complete, and extending this wage rate times the man-hours to complete. The total labor dollars to complete plus the labor dollars expended to date (from the Labor Cost and Man-hour Distribution Report), provide the project labor dollar cost forecast for a work component. Figure 13.52 shows the labor cost data of the Field Labor Forecast.

The Field Labor Forecast provides a means of continuously monitoring the projected total man-hours against the latest budget estimate. When it appears that the man-hours are running higher than the current budget or forecast for a work component, the superintendent would then be alerted to that component as a potential problem area. The superintendent can begin to monitor problem areas closely over several reporting periods. Through

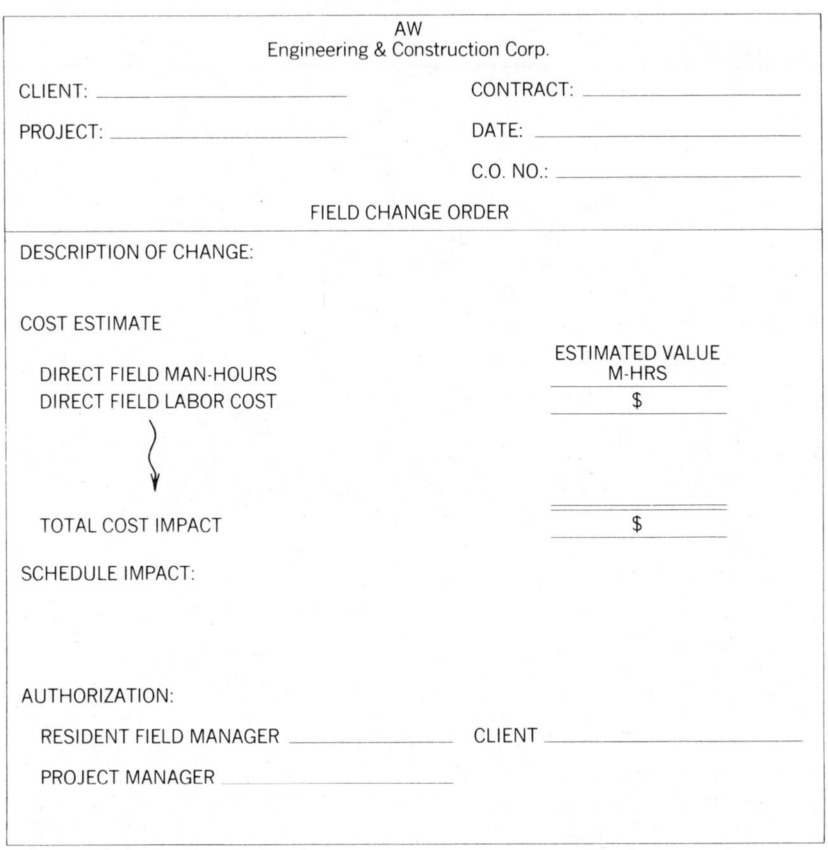

AW
Engineering & Construction Corp.

CLIENT: ______ CONTRACT: ______

PROJECT: ______ DATE: ______

C.O. NO.: ______

FIELD CHANGE ORDER

DESCRIPTION OF CHANGE:

COST ESTIMATE

	ESTIMATED VALUE
DIRECT FIELD MAN-HOURS	M-HRS
DIRECT FIELD LABOR COST	$
TOTAL COST IMPACT	$

SCHEDULE IMPACT:

AUTHORIZATION:

RESIDENT FIELD MANAGER ______ CLIENT ______

PROJECT MANAGER ______

Figure 13.51 Field Change Order (P_L-DOC 52).

AW
Engineering & Construction Corp.

CLIENT: ____________

PROJECT: ____________

CONTROL ZONE DESCRIPTION

FIELD LABOR FORECAST

COST CODE	WORK CATEGORY DESCRIPTION	EXPENDED					TO COMPLETE				
		QUANTITY	MAN-HRS.	MAN-HRS./UNIT	WAGE RATE	COST	QUANTITY	MAN-HRS.	MAN-HRS./UNIT	WAGE RATE	COST
		DATA FROM P_L-DOC 49			DATA FROM: P_L-DOC 48		BEST ESTIMATE TO COMPLETE BASED ON CURRENT INFO/STATUS				

CONTRACT: ____________

DATE: ____________

TOTAL FORECAST					BUDGET		
QUANTITY	MAN-HRS.	MAN-HRS./UNIT	WAGE RATE	COST	MAN-HRS.	COST	OVER (UNDER)
EXPENDED + TO COMPLETE					CURRENT BUDGET		BUDGET LESS FORECAST

Figure 13.52 Field Labor Forecast Trends (P_L-DOC 53).

FIELD INSPECTION (M8) he may be able to detect the problem, and, via Crew Management, adjust the crews or change the construction method accordingly.

A final requirement of the PROJECT FORECASTING function is that it produce cash flow data for project management and for the client. These data provide a means of displaying, over time, the anticipated expenditure of a project component, for instance, direct field labor. A typical direct field Labor Cash Flow Forecast (P_L-DOC 54) is shown in Figure 13.53. This cash flow curve reflects actual expenditures to date against a base plan and the forecasted expenditure to complete. These curves must be based on the current labor forecast schedule, and the latest man-hour expenditures to be realistic and beneficial to project management, and to the client, when providing funds for payroll.

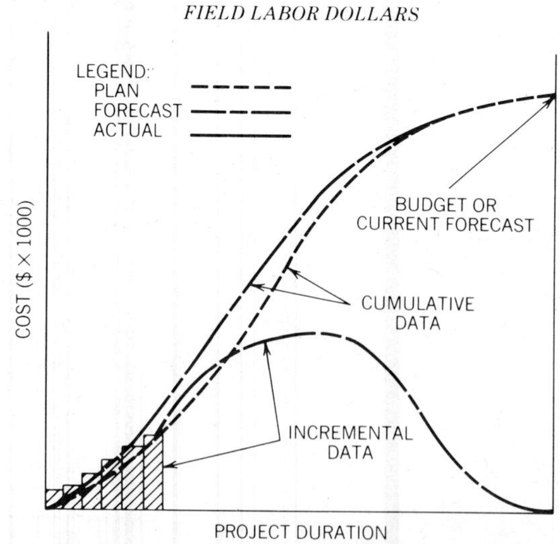

Figure 13.53 Labor Cash Flow Forecasts (P_L-DOC 54).

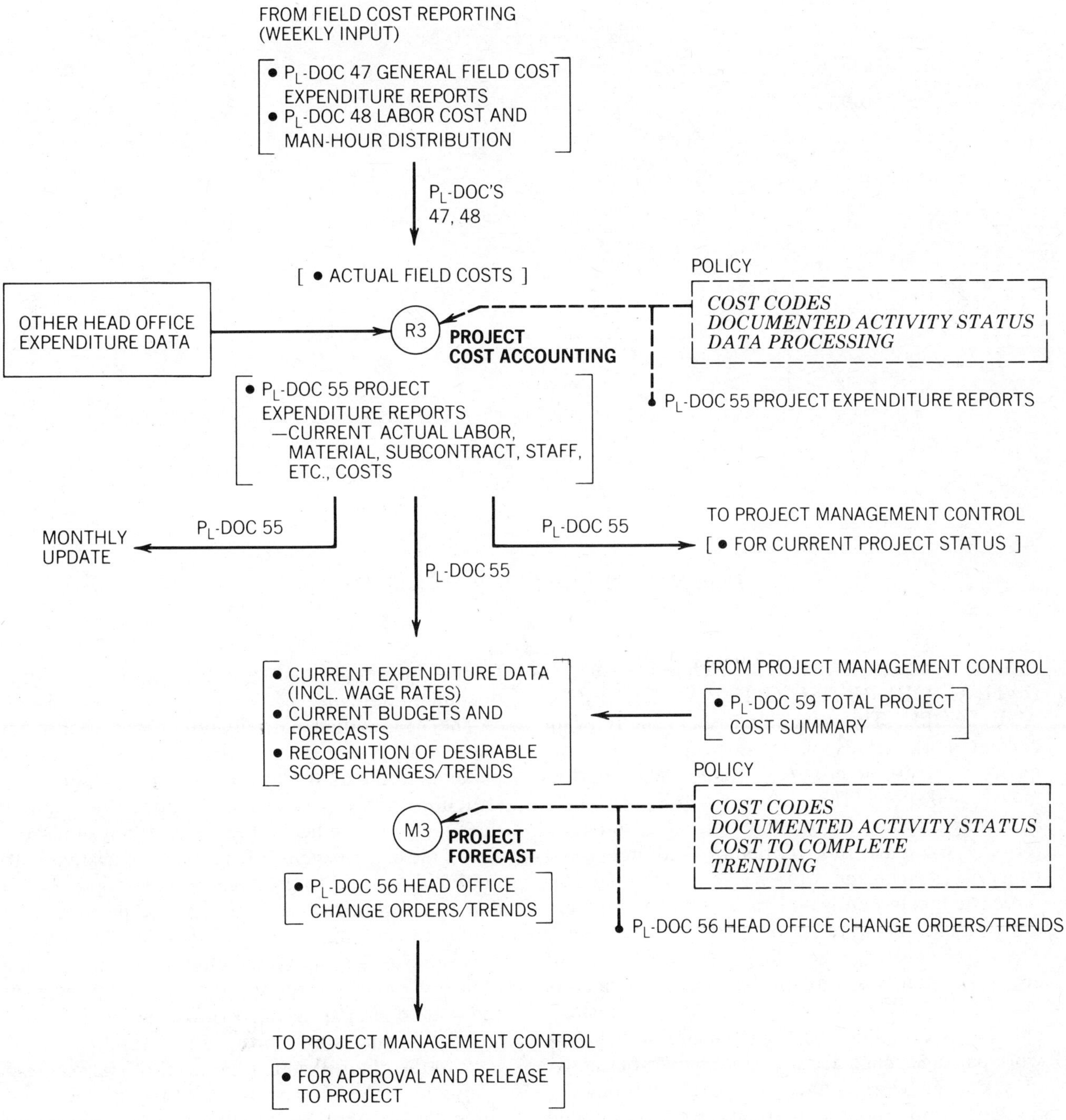

Figure 13.54 Project Status, Analysis, and Reporting decision process: head office phase (Field-Based Project Management approach).

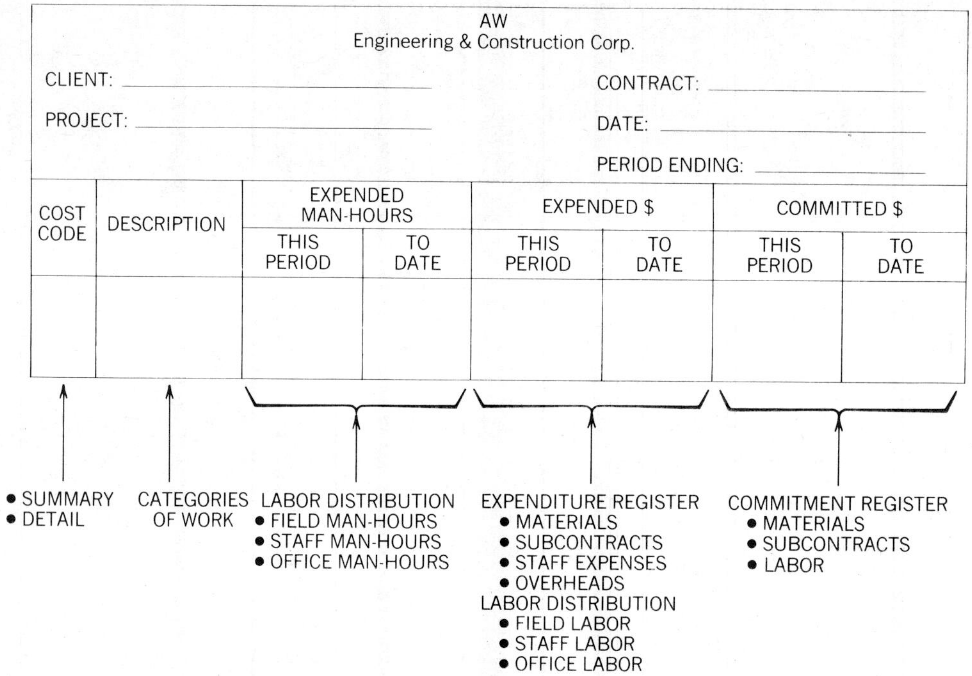

Figure 13.55 Project Expenditure Reports (P_L-DOC 55).

Work package details are considered when UPDATING AND REVISING THE WORK PACKAGE SCOPE. This function gathers all relevant current work details of active project work segments in terms of progress, status, labor performance, and scope. Thus, as shown in Figure 13.49, a variety of documents flow into this function, forming the basis for rebriefing field managers/supervisors on current work package details such as forecasted man-hours and quantities, and potentially results in the rescheduling of project crews. The cost engineer, scheduling engineers, and general superintendents are the key decision makers involved in this function. These decision makers closely review all documents to ensure that current work package scope details such as new work activities, current work activity man-hour and quantity forecasts, and productivity trends, are disseminated to the respective decision processes for incorporation into the construction management effort. For example, UPDATING AND REVISING THE WORK PACKAGE SCOPE may lead to schedule changes, further field inspection of a specific work activity with subsequent crew management changes, and/or changes to crew planning activities scheduled for the next period. This function is performed informally and rarely generates additional documentation.

The head office participation phase of Project Status, Analysis, and Reporting varies widely depending on the nature of the project. The pure construction project, for example, normally requires the support of the head office only during construction. This management role remains consistent with the Field-Based Project Management approach, in that the majority of all functions are performed in the field. The support provided by the head office in this case consists of updating and summarizing project expenditure reports for the total project, in addition to preparing head office initiated change orders. Furthermore, most formal reports are issued to the client through the head office. Alternatively, in a design–construct situation, the head office has an increased involvement because of the additional engineering and design effort associated with these projects. The head office phase of Project Status, Analysis, and Reporting is graphically displayed in Figure 13.54 for the Field-Based Project Management approach.

Head office PROJECT COST ACCOUNTING serves as the focal point for summarizing all project expenditure data, including pertinent field expenditures. The Project Expenditure Report (P_L-DOC 55), includes current expenditure data for field construction, as well as head office engineering and design costs. This report is generally updated monthly, and its information is required for current project status analysis during Project Management Control. A typical Project Expenditure Report is depicted in Figure 13.55.

Head office PROJECT FORECASTING (M3) is a particularly critical function for the design–construct firm, as it is the primary vehicle for cost control early in the project as the design develops. As indicated by the *TRENDING* policy, PROJECT FORECASTING enables any deviations from the current project control scope basis to be documented so as to reflect the impact of both design development and client-requested change orders, provided this procedure is performed in a timely manner through head office change orders/trending.

13.6 TOTAL FIELD AUTONOMY

Project Status, Analysis, and Reporting at the Total Field Autonomy level is the focal point for the integration of cost and schedule control base data with actual field data for control zones and the work packages within each control zone. This integration results from both management and client pressure to meet the need for more professional statusing, analysis, and reporting. This need reflects the increasing application of work breakdown structures and work orders containing a far greater level of detail. Consequently, more professional groups emerge to cope with the demands created by the increasing levels of information to be analyzed for problems. This growth, in turn, confers rising demands on the Management Information System in that project information must have a broader focus and multi-level points of access in order to permit the selective analysis of data when problems are uncovered. Finally, the range and complexity of data analysis that occurs during Project Status, Analysis, and Reporting leads to a management need for possibly many more formal reports than does the Field-Based Project Management approach.

At the Total Field Autonomy level, the achievement of progress and productivity targets is accomplished in part through the implementation of activity analysis techniques. These techniques, when coupled with integrated cost and scheduling systems that provide indicators of the success or failure of activity analysis methods, frequently lead to productivity improvement programs. These *SCIENTIFIC MANAGEMENT* (P_T 34)* philosophies emerge in this project management environment.

In Total Field Autonomy management, demands for continuous predictions of cost, schedule, and productivity changes resulting from scope revisions or other factors, are viewed as important mechanisms for making major management decisions. The contractor accomplishes this through the development and implementation of special project procedures that in effect force the client to make major decisions when trends (or deviations) from control base costs, productivity, and schedules are identified, quantified, and presented for decision-making action.

The Management Information System at the Total Field Autonomy level is to some degree an extension of the MIS described for the Field-Based Project Management approach (see Figure 13.37). The Total Field Autonomy MIS, however, completely integrates the cost and scheduling control system through an increasing sensitivity to work package content as defined by the project work breakdown structure. Furthermore, the MIS incorporates formal field activity analysis and project performance improvement programs that also support the cost and scheduling control systems of the project. In this way, Project Status, Analysis, and Reporting provides access to additional information from which management decisions can be formulated. The Management Information System for this decision process is described in Figure 13.56 for the Total Field Autonomy approach.

Project Status, Analysis, and Reporting is summarized in terms of the relevant manpower management functions, policies, and documents, as shown in Figures 13.57, 13.58, 13.59, 13.63, and 13.67. These figures correspond to the field schedule phase, field cost reporting phase, field productivity analysis and reporting phase, field scope analysis and reporting phase, and the head office support phase, respectively.†

*See Table 13.6, Policies of the Autonomous Project Team.

†While all functions are often relocated to the field, for the design–construct project, even in the Total Field Autonomy approach, some head office support functions are still best performed in the head office.

TABLE 13.6 POLICIES OF THE AUTONOMOUS PROJECT TEAM

Policy		Description
P_T 8	*WORK BREAKDOWN*	A work breakdown structure representative of the project scope and control philosophy is formulated to integrate costs and schedules with work packages for project management.
P_T 9	*COST CODES*	Cost codes are applied to the work breakdown structure and serve as a basis for categorizing estimated costs, reporting and controlling costs and manpower, developing a historical database, and integrating the schedule with the estimate.
P_T 13	*NETWORK-BAR CHART ANALYSIS*	Broad network focused milestone planning is the best approach for identifying and sequencing major project components, while a combination of the critical path method and bar chart model are best suited for project planning and scheduling at the total project view level, as well as at the daily level.
P_T 14	*SCHEDULING FOCUS*	Scheduling definitions and constraints are: 1. A critical work package is one that, if delayed, will extend the completion date of the project. 2. Work package durations must be correlated to the estimated work content. 3. All restraints must be defined and incorporated into the sequence logic (i.e., material delivery, subcontract coordination, drawing receipt, etc.). 4. Manpower impact is an integral aspect of project scheduling and scheduling control. 5. Schedules must reflect optimum use of all resources relative to the work package. 6. Progress is based on a realistic evaluation of the labor unit effort required to complete a component of work. 7. Schedules are portrayed to the daily work level and are prepared, analyzed, and reviewed weekly in relation to planned progress and productivity versus actual progress and productivities.
P_T 23	*LABOR RELATIONS*	When union labor is utilized, pre-job union conferences and personal contact with the unions are an integral part of successful project management. Regardless of the construction environment, a full-time labor relations expert is required because of the complexity and manpower demands of the project.
P_T 24	*PERSONAL CONTACT*	Crew management must be coupled with personal contact and good relations between field management personnel and construction workers if scientific management techniques are to be applied successfully.
P_T 25	*WALK JOB*	In the field, daily activity progress and crew performance is monitored by visúal inspection, and crew management decisions are made on the basis of subjective evaluation of observed progress and productivity versus planned progress and productivity. This decision process often incorporates a number of decision makers.
P_T 26	*PROGRESS REPORTING*	A systematic method of evaluating physical progress must be based on measuring units of work for any component work tasks, with the capability of rolling progress up from this detailed component level to the total project level.
P_T 27	*FIELD LABOR HOURS*	Labor hours are monitored by field persons closest to the work face, and recorded and processed by a person whose primary function is to ensure that the labor hours are allocated to the appropriate cost code. Their proper allocation is essential to analyzing productivity and forecasting man-hours to complete at any level of detail.
P_T 28	*DOCUMENTED ACTIVITY STATUS*	Labor costs, manpower, man-hours, and quantities are primary ingredients to cost control, and must be documented by a system yielding reliable information and viable reports for management control. These ingredients accrue as a basis for forecasting to complete at any level of detail.
P_T 29	*CREW MANAGEMENT*	Field crew management is based on past experience coupled with scientific management techniques.
P_T 30	*CREW PERFORMANCE*	Crew performance is a function of labor productivity, job conditions, and management effectiveness and must focus on a tight but realistic performance indicator.

TABLE 13.6 ***(Continued)***

Policy	Description
P_T 31 *PRODUCTIVITY*	Productivity is an essential concept and must be rationally defined for office and field utilization. An awareness of the components that affect productivity and their relation to man-hours, activity durations, and progress is essential, as is that all these integrate and are manageable.
P_T 32 *TRENDING*	A mechanism for documenting and analyzing deviations of cost and schedule consequence is an integral facet of the information system, and is essential for acquiring client-authorized scope changes, adjusting budgets and schedules, and portraying the most probable man-hours and costs of the project. It also provides a means of evaluating contractor performance on a total project basis.
P_T 33 *CRITICAL STATUS REPORTING*	A construction status reporting system must provide a comprehensive report of activity status, highlighted by a mechanism to alert management of critical activity status and activities approaching critical status within a planning horizon that allows sufficient time to react.
P_T 34 *SCIENTIFIC MANAGEMENT*	Techniques of scientific management, such as work sampling, time lapse, activity analysis, multiple activity charts, and prediction models, can contribute to improving crew performance.
P_T 35 *MANAGEMENT BY EXCEPTION*	Manpower management by exception is acceptable provided that the information system is geared to exception reporting to alert management to existing problems or potential problems which are emerging within a sufficient time frame to allow a management response.
P_T 36 *DATA PROCESSING*	A computer-based information system is the most feasible means of data processing for management control.
P_T 37 *PROJECT TEAM SPIRIT*	Active participation of all project team members in the management process promotes goodwill and a positive management environment. It ensures the success of scientific management techniques and the total integration of the project team.

The field scheduling phase (Figure 13.57), and the field cost reporting phase (Figure 13.58), generally have the same problem focus at the Total Field Autonomy level as they do under Field-Based Project Management (Section 13.5). Consequently, the types of documents generated through these two phases of Project Status, Analysis, and Reporting are the same with the exception that *WORK BREAKDOWN* (P_T 8) and *COST CODING* (P_T 9) techniques tend to influence the structure of such reports as the Activity Status List (P_T-DOC 51), Critical Activity List (P_T-DOC 52) and the Labor Cost and Man-hour Expenditure Reports (P_T-DOC 55) under the Total Field Autonomy approach.* These reports focus more keenly on statusing by work packages and by the work orders that define each work package. Thus, *DATA PROCESSING* (P_T 36) is the most feasible and quickest means of sorting project schedules into work packages for status and progress analysis when actual data is collected and integrated into the schedule. By using *CRITICAL STATUS REPORTING* (P_T 33) concepts, only those work packages that are critical and possibly in trouble are analyzed in relation to their schedule impact on the respective control zone. This *MANAGEMENT BY EXCEPTION* (P_T 35) approach of identifying critical problems is a method of focusing top management decision-making attention only to those potential problem areas likely to have significant ramifications on the overall construction schedule.

The field productivity analysis and reporting phase (Figure 13.59), reflects a changing focus toward a performance report that integrates percent progress and productivity with man-hour and manpower data. Thus, the performance analysis incorporates such data as the latest budgets and forecasted man-hours, expended man-hours, manpower, and percent progress. When these data are portrayed in relation to control base planned data, meaningful comparisons can be made, uncovering potential problem areas. If a productivity indicator is incorporated into the report, the likelihood of ex-

*These reports would be similar in structure to those shown in Figures 13.39, 13.40, and 13.44.

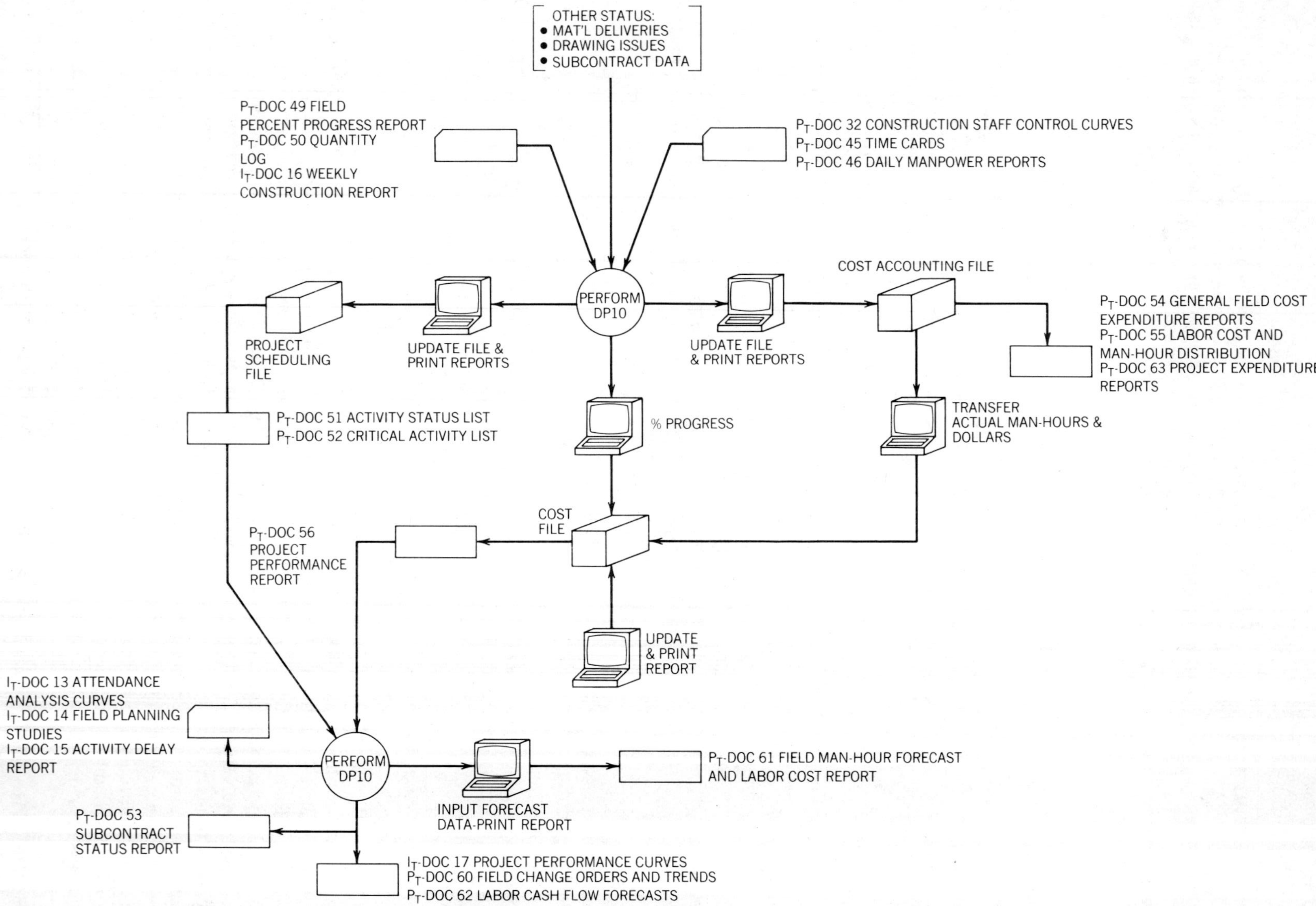

Figure 13.56 Project Status, Analysis, and Reporting Management Information System (Total Field Autonomy approach).

S5 DEFINE CURRENT PROJECT STATUS

Policies	Documents	
WORK BREAKDOWN	P_T-DOC 51	ACTIVITY STATUS LIST
NETWORK-BAR CHART ANALYSIS	P_T-DOC 52	CRITICAL ACTIVITY LIST
SCHEDULING FOCUS	P_T-DOC 53	SUBCONTRACT STATUS REPORT
DOCUMENTED ACTIVITY STATUS		
CRITICAL STATUS REPORTING		
DATA PROCESSING		

S6 REVISE CONSTRUCTION SCHEDULE

Policies	Documents
WORK BREAKDOWN	
NETWORK-BAR CHART ANALYSIS	
SCHEDULING FOCUS	
CRITICAL STATUS REPORTING	
DATA PROCESSING	

Figure 13.57 Project Status, Analysis, and Reporting decision process: field scheduling phase (Total Field Autonomy approach).

posing potential problem areas related to *CREW PERFORMANCE* (P_T 30) is increased. *PRODUCTIVITY* (P_T 31) must be defined so that it can be quantified and be meaningful to management. This can be done by developing a tight but realistic performance indicator. A common approach is to determine the earned budget man-hours (percent progress multiplied by the budget), and to compare this figure to the actual hours expended. If fewer hours are expended than earned, crew performance will be above 100 percent. Another method is to evaluate the man-hours required to perform an activity based on site conditions, construction methods, and so on, and compare these hours with standard estimating man-hours. A ratio of the two might be 1.2 or 0.8, depending on the factors affecting the activity. This ratio becomes the performance target. Finally, an awareness of those components that impact productivity is essential, and it must be recognized that these components are manageable. Both *CREW PERFORMANCE* and *PRODUCTIVITY* are important policies, and each represents a significant deviation in management from the other project team structures. A typical Project Performance Report (P_T-DOC 56) is illustrated in Figure 13.60. This report can be generated at various levels through *DATA PROCESSING* techniques, depending on project need and the desired level of management interest. For example, top-level project management would only be concerned with project level and possibly with area-level summaries, while construction managers would be interested in progress and performance reports related only to their specific areas of responsibility, including work package progress when problems are evident. The Project Performance Report is supplemented with graphic displays of progress, productivity, and manpower, as shown in Figure 13.61 on the Project Performance Curves (I_T-DOC 17). These types of curves are useful for comparing plans and current forecasts with actual experience, and portraying the path by which the project will be

R3 PROJECT COST ACCOUNTING

Policies	Documents	
COST CODES	P_T-DOC 54	GENERAL FIELD COST EXPENDITURE REPORTS
FIELD LABOR HOURS		
DOCUMENTED ACTIVITY STATUS	P_T-DOC 55	LABOR COST AND MAN-HOUR DISTRIBUTION
DATA PROCESSING		

Figure 13.58 Project Status, Analysis, and Reporting decision process: field cost reporting: (Total Field Autonomy approach).

M5 LABOR ANALYSIS

Policies

WORK BREAKDOWN
COST CODES
PROGRESS REPORTING
CREW PERFORMANCE
PRODUCTIVITY
TRENDING
DATA PROCESSING

Documents

P_T-DOC 56 PROJECT PERFORMANCE REPORT

R5 LABOR PRODUCTIVITY

Policies

CREW PERFORMANCE
PRODUCTIVITY
DATA PROCESSING

Documents

I_T-DOC 17 PROJECT PERFORMANCE CURVES

M4 ACTIVITY ANALYSIS

Policies

PERSONAL CONTACT
WALK JOB
CREW MANAGEMENT
CREW PERFORMANCE
PRODUCTIVITY
SCIENTIFIC MANAGEMENT
MANAGEMENT-BY-EXCEPTION
DATA PROCESSING
PROJECT TEAM SPIRIT

Documents

P_T-DOC 57 FIELD ACTIVITY ANALYSIS DOCUMENTS

Figure 13.59 Project Status, Analysis, and Reporting decision process: field productivity analysis and reporting (Total Field Autonomy approach).

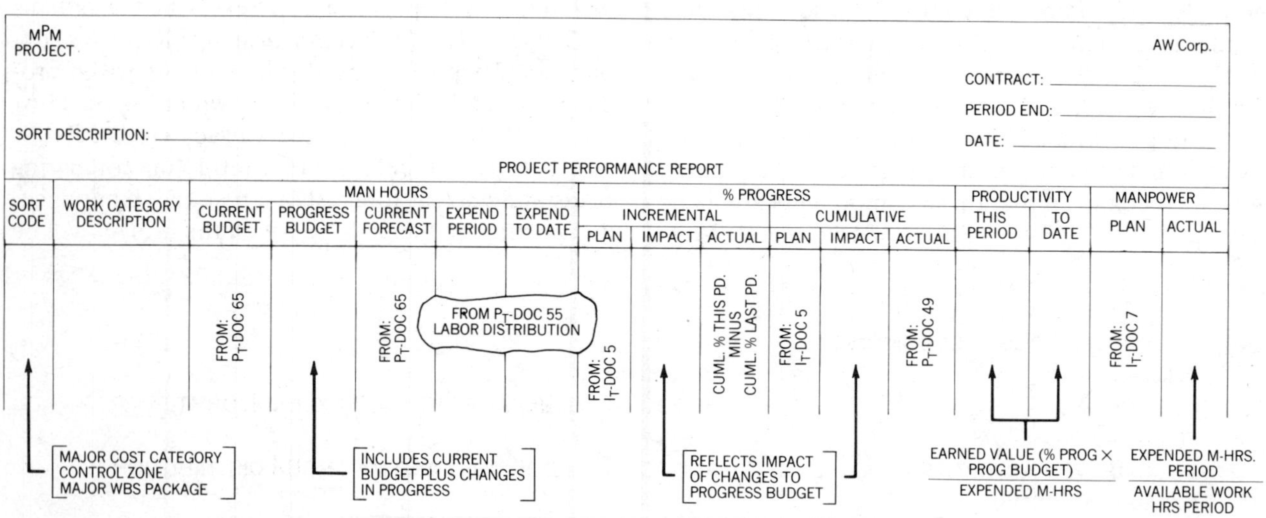

Figure 13.60 Project Performance Report (P_T-DOC 56).

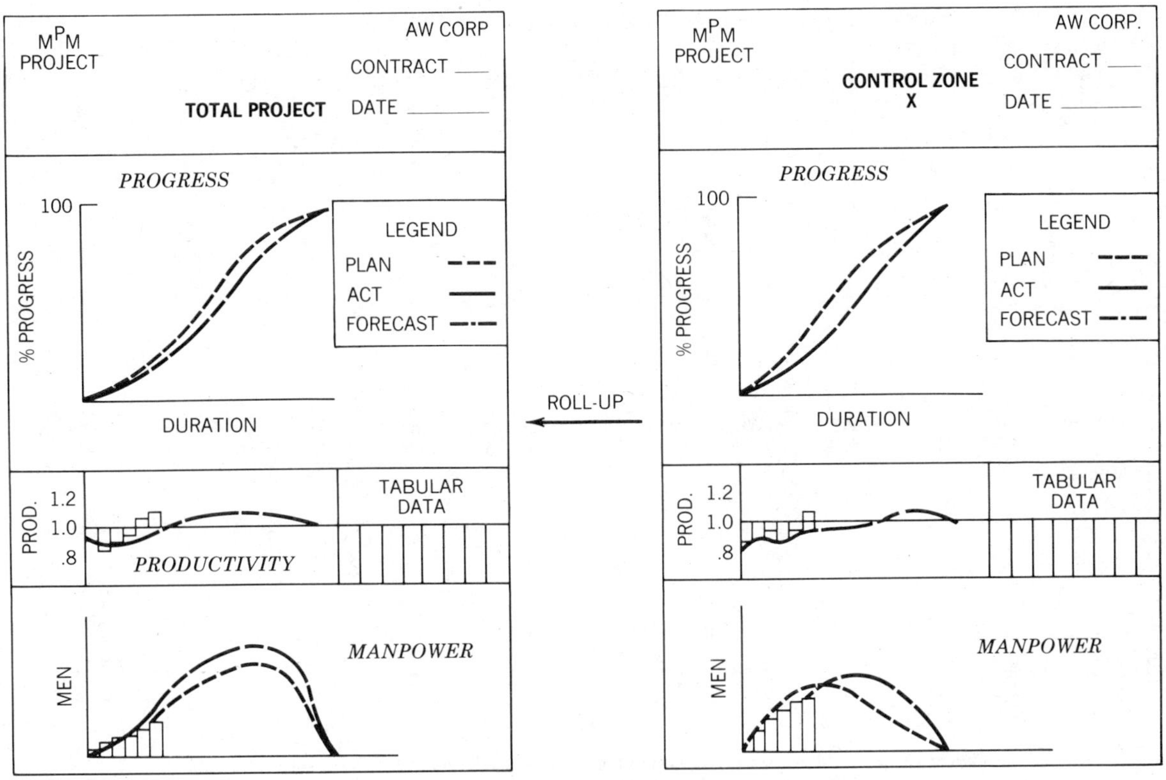

Figure 13.61 Project Performance Curves (I_T-DOC 17).

completed through revised forecasts of progress, manpower, and productivity.

Project Performance Reports and Project Performance Curves depict the project's history and current position, and forecasts to completion as determined at a given status date. At the Total Field Autonomy level, *SCIENTIFIC MANAGEMENT* techniques are often implemented to improve current productivity problems and improve progress and future productivity, in order to ensure that schedule completion dates are met within allowable budgets.

A considerable number of method study and activity analysis techniques exist. These are employed on large construction sites, either individually or collectively, in an integrated and concentrated attack on site problems. Thus, a simple work face layout and process focus on material handling flows and work station locations can often alert field management to ways of eliminating unnecessary work and repeated materials handling. The magnitude of delays identified in the Activity Delay Report, as well as a quantitative indication of the type of crew activity (or inactivity) in a construction process, can be established using work sampling techniques. On the basis of an adequate set of random observations of field activity, quantitative assessments of the quality of foreman management capabilities and of the preplanning process (together with the identification of delay causes), can be readily brought to management attention. More specifically focused techniques such as time-lapse photography can be an effective way of documenting delays and convincing top management, foremen, and crew members of the need for better work sequences. Thus, a comprehensive method study and activity analysis report can be created from the integration of individually focused studies, as indicated on the Field Activity Analysis Documents (P_T-DOC 57) (see Figure 13.62).

The field scope analysis and reporting phase (Figure 13.63) of the Project Status, Analysis, and Reporting decision process reflects a move toward greater documentation of changes in scope and project trends. Consequently, the Standard Change Order and Trend Procedure (P_T-DOC 58) is

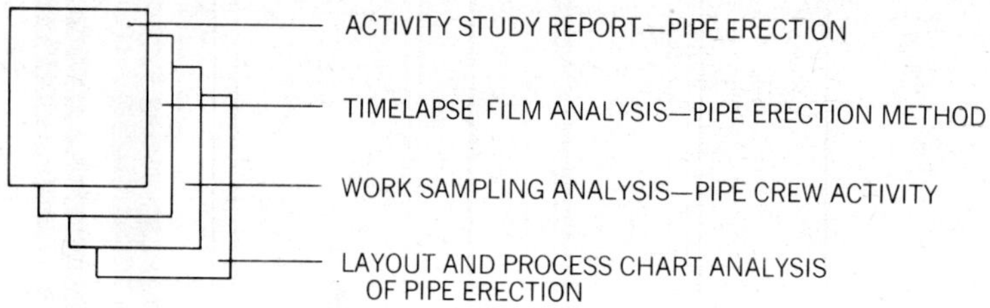

(a)

CREW	BASIC JOB				SUPPORTING ACTIVITY				DELAYS			
	THIS WEEK	LAST WEEK	PREVIOUS WEEK	3-WEEK AVERAGE	THIS WEEK	LAST WEEK	PREVIOUS WEEK	3-WEEK AVERAGE	THIS WEEK	LAST WEEK	PREVIOUS WEEK	3-WEEK AVERAGE
CARPENTER		←			HOURS PER CATEGORY					→		

1. 2. 3. 4. ↓ INDICATIVE QUESTIONS

(b)

Figure 13.62 Field Activity Analysis Documents (P_T-DOC 57). (*a*) Activity analysis techniques. (*b*) Detailed analysis of crew activity.

M3 PROJECT FORECAST

Policies

WORK BREAKDOWN
COST CODES
PROGRESS REPORTING
FIELD LABOR HOURS
DOCUMENTED ACTIVITY STATUS
PRODUCTIVITY
TRENDING
MANAGEMENT-BY-EXCEPTION
DATA PROCESSING

Documents

P_T-DOC 58 STANDARD CHANGE ORDER AND TREND PROCEDURE
P_T-DOC 59 PROJECT CHANGE ORDER AND TREND PROCEDURE
P_T-DOC 60 FIELD CHANGE ORDERS AND TRENDS
P_T-DOC 61 FIELD MAN-HOUR FORECAST AND LABOR COST REPORT
P_T-DOC 62 LABOR CASH FLOW FORECAST

R4 UPDATE AND REVISE WORK PACKAGE SCOPE

Policies

WORK BREAKDOWN
COST CODES
DOCUMENTED ACTIVITY STATUS
TRENDING
DATA PROCESSING

Documents

Figure 13.63 Project Status, Analysis, and Reporting decision process: field scope analysis and reporting (Total Field Autonomy approach).

TYPICAL SECTIONS

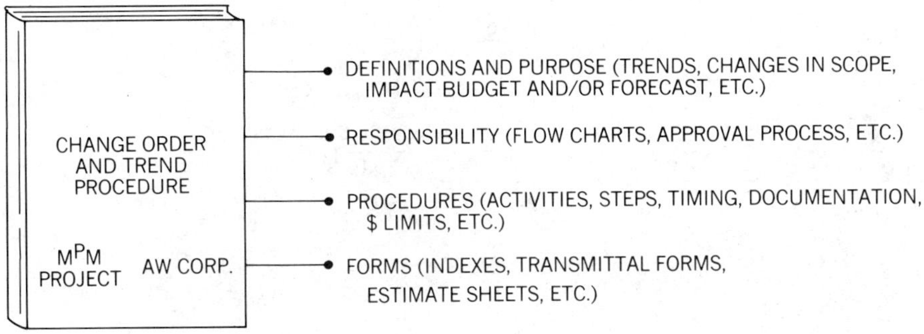

Figure 13.64 Project Change Order and Trend Procedure (P_T-DOC 59).

modified to incorporate client requirements. They also incorporate methods of ensuring that the client approves changes in scope, and that the client is constantly kept aware of project trends resulting from productivity changes. This Project Change Order and Trend Procedure (P_T-DOC 59) is based on *TRENDING* (P_T 32) combined with *MANAGEMENT BY EXCEPTION* policy approaches. The development of a Project Change Order and Trend Procedure (P_T-DOC 59) is depicted schematically in Figure 13.64. This procedure leads to defined methods for documenting changes in scope and project trend forecasts. The Field Change Order and Trend (P_T-DOC 60) is described in Figure 13.65, and contains lines for project management approval and client approval in terms of change orders and client notification of trends. Field change orders would be estimated based on quantities, unit man-

MPM
PROJECT AW Corp.

LOCATION: ______ CONTRACT: ______

DATE: ______

CHANGE ORDER NO.: ______

TYPE OF DEVIATION: CHANGE ORDER ☐ TREND ☐ TREND NO.: ______

DESCRIPTION OF CHANGE ORDER OR TREND:

ORDER OF MAGNITUDE COST ESTIMATE: ______

APPROVAL TO PROCEED: PROJECT MANAGER ______ CLIENT ______

DETAILED COST ESTIMATE:	BUDGET	FORECAST
DIRECT FIELD MAN-HOURS	______	______
DIRECT FIELD LABOR COST	______	______
TOTAL COST IMPACT	______	______

SCHEDULE IMPACT:

FINAL AUTHORIZATION TO PROCEED:

PROJECT MANAGER ______ CLIENT ______

Figure 13.65 Field Change Orders and Trends (P_T-DOC 60).

M^PM
PROJECT

CONTROL ZONE DESCRIPTION ________________

FIELD LABOR MAN-HOUR AND LABOR COST FORECAST

COST CODE	WORK CATEGORY DESCRIPTION	EXPENDED					TO COMPLETE	
		% COMPL.	MAN-HRS.	MAN-HRS./ % PROG.	WAGE RATE	COST	% TO COMPL.	MAN-HRS./ 1% PROG.
		FROM P_T-DOC 49 OR P_T-DOC 56	FROM P_T-DOC 55		FROM: P_T-DOC 55	FROM: P_T-DOC 55	100%—ACT. % COMPL. TO DATE	ESTIMATED BASED ON CURRRENT PERFORMANCE

MAN-HRS./% PROG.: $\frac{\text{EXPENDED M-HRS.}}{\text{\% COMPL.}}$

MAN-HRS./1% PROG.: % TO COMPL. × M-HRS./1% PROG. TO COMPLETE

AW Corp.

CONTRACT: ________________

DATE: ________________

			TOTAL FORECAST			BUDGET				
MAN-HRS.	WAGE RATE	COST	MAN-HRS.	WAGE RATE	COST	MAN-HRS.	WAGE RATE	COST	OVER (UNDER)	% OVER (UNDER)
	ESTIMATE BASE ON CREW MIXES	M-HRS. × WAGE RATE	EXPENDED + TO COMPLETE			P_L-DOC 65			BUDGET VS. FORECAST	

Figure 13.66 Field Man-hour Forecast and Labor Cost Report (P_T-DOC 61).

hours, and the like, while labor performance trends, for example, though far more subjective, are based on either quantities or other known factors found when developing a forecast. An example of one typical approach to this problem is the Field Man-hour Forecast and Labor Cost Report (P_T-DOC 61), as shown in Figure 13.66. This type of approach incorporates current data such as percent complete, expended man-hours and labor dollars, and the like with projections of anticipated future productivities and wage rates. These labor forecasts are supported by an analysis of quantities to complete and man-hours per unit expenditure.

The Field Change Orders and Trends and Field Man-hour Forecasts and Labor Cost Reports assist in UPDATING AND REVISING WORK PACKAGE SCOPE (R4) details. These documents, when produced in a timely manner, aid management decision making by providing a mechanism for approving changes in scope and possibly applying *SCIENTIFIC MANAGEMENT* techniques, in order to improve performance to ensure that plans and/or forecasts are achieved. Furthermore, maintaining a current scope of work eliminates rework and enhances the quality of the Crew Planning and Crew Management decision processes.

R3 PROJECT COST ACCOUNTING	
Policies	Documents
COST CODES	P_T-DOC 63 PROJECT EXPENDITURE REPORTS
DOCUMENTED ACTIVITY STATUS	
DATA PROCESSING	

M3 PROJECT FORECAST	
Policies	Documents
WORK BREAKDOWN	P_T-DOC 64 HEAD OFFICE CHANGE ORDERS AND TRENDS
COST CODES	
DOCUMENTED ACTIVITY STATUS	
TRENDING	
MANAGEMENT-BY-EXCEPTION	
DATA PROCESSING	

Figure 13.67 Project Status, Analysis, and Reporting: head office support phase (Total Field Autonomy approach).

Field Project Status, Analysis, and Reporting is always supported by the head office to some extent, especially for design–construct projects. Consequently, the head office will provide the field with support when processing Project Expenditure Reports (P_T-DOC 63) and Head Office Change Orders and Trends (P_T-DOC 64). Once approved, these documents will be forwarded to the field for incorporation of the data into the total project progress reports generated during the Project Management Control phase (see Figure 14.23). A summary of the management functions, policies, and relevant documents is shown on Figure 13.67 for the head office support phase of Project Status, Analysis, and Reporting.

CHAPTER 14

Project Management Control

Project Management Control is probably the most critical decision area for the project manager, and for the contractor's top-level management group. It is also the decision area where the effectiveness of the contractor's Management Information System is revealed to both decision makers and clients. The Project Management Control area will also demonstrate top-level management competency in relation to directing, handling, and initiating corrective action for problem solving. This is especially relevant in terms of how project management handles and resolves project manpower issues, because by their very nature these management actions and attitudes impact local communities and society in general.

In the small project environment, Project Management Control essentially focuses on the evaluative monitoring of work progress, and the status and impact of progressive accumulations of actual costs versus estimated costs. It is often based on personal observations and discussions with the key field managers. Because the scope and size of small projects are such that the top manager can visit the site, contact all the personnel involved, and readily grasp project status, decisions are frequently made without significant supporting documentation, and generally only made when the project appears to be in trouble. In construction environments where projects are large and complex, integrated cost/scheduling systems that are also sensitive to labor component performance must be devised to produce reports that can *aid* management in the process of detecting potential problems and controlling the direction of the project. In these respects, control procedures may need to be frequently reviewed and revised by project management to respond to the changing demands of the project and/or client. Thus, the contractor's Management Information Systems (MIS) should serve management. The effectiveness of an MIS is often demonstrated by its ability to identify problems, and by whether its documents are pertinent and are actually being used by management and the project team in general.

In certain management environments, the proper design and implementation of the Management Information System becomes a critical problem area of the Project Management Control decision process. This is because the MIS design, structure, and ease of access defines project status and cost indicators, establishes the level and frequency of project reports, and determines basic project procedures and policies. In addressing these issues the contractor's management group must consider how the capability of the MIS for a project will affect the manner in which decisions are made by the project team at all levels within each decision process area.

Project Management Control is the key decision process, and is the focal point for policy and authorization management decisions that have major project implications. These decisions are frequently made from an analysis of standard project reports combined with information from key field managers. Alternatively, top level management decisions are often made based on data and recommendations contained in specially prepared management reports focusing on a critical problem area. Hence, the efficient performance of Project

Management Control necessitates that the contractor be able to quickly identify problems and respond effectively to them in a manner that stays within budgets and schedules. Thus, the decision area is the interface point for relaying critical project information to the client/owner, including such items as major cost and/or schedule deviations, especially those affected by changes in labor productivity, and actions taken to minimize the impact of these deviations.

Finally, this problem area is one in which the contractor's professionalism and ability to handle and perform with large labor forces is readily apparent to the client/owner. The contractor's performance in this critical management area is vital to the firms' survival. It is the true testing ground which often makes or breaks the contractor, because it is the final job cost that the client/owner remembers after the project is completed. It is at this stage that an evaluation of the contractor is made in terms of simple criteria, such as: Was the project completed on time, within budget, and with acceptable levels of workmanship?

14.1 THE BASIC DECISION PROCESS

During the life of the construction project, managers and supervisors are continuously making decisions which impact the direction of the project. Many of these decisions are made routinely as demanded by the execution of a decision process. Any one of these routine decisions will not by itself significantly impact the direction of the project. The cumulative effect of a collection of routine decisions, however, may have a considerable impact on the project. Consequently, these decisions must be related to overall company management goals.

Top-level management and key field managers must depend on the project's Management Information System, especially on the timeliness, frequency, and accuracy of cost and scheduling reports, to reveal the success of routine decision making. Project management success may be measured at any time by evaluating the current status of the project as to whether it is within current budgets and planned schedules. However, management success can also be related to the overcoming of problems, the attainment and improvement of predicted work productivities and work progress, and so on.

Project Management Control has a natural overall project focus, because it not only reacts to manpower related problems, but also to problems in other management areas such as material procurement and material logistics around the project site. These other management areas have an impact on the manpower area, of course, but are not considered in detail in this text. Thus, in terms of project manpower management, Project Management Control is concerned with the formulation and implementation of manpower related policies and directives through the project's Management Information System, the analysis of current project status in relation to its impact on manpower, the formulation of corrective and remedial management actions and their implementation at the field level, the assessment of the effectiveness of the management effort of the project team, and the evaluation of whether or not manpower management control procedures are workable.

Before making a major project decision, management must fully understand the project's current position, and must review the effect of previous decisions that, as they affect the project environment and resources, have led to the current situation. This will often identify those factors which have contributed to the emergence of problems, and will provide a basis for selecting courses of action. Many project construction managers rely heavily on their key field managers and supervisors for direct input to problem solving. Other managers rely on specially prepared reports which address specific problem areas and which also recommend alternate decision possibilities. Once decisions are made, top managers must act decisively to implement them and to monitor their effects on project performance and on the resolution of problems encountered over the life of a project.

When project management effort is operating properly, top project managers should be constantly aware of activities in critical areas that impact the project and its status. Thus, for example, sudden changes in manpower levels or productivity factors in key crafts would require urgent consideration. Normally, however, information is filtered to top management through the management effort of the project team in conjunction with the implementation of relevant performance and control procedures. Thus, it is often left to top management's perception to identify problems or take initiatives to exploit favorable conditions. To ensure that top project managers can rely on the effectiveness of their project team efforts, they must be cognizant of project team performance in relation to the utiliza-

TABLE 14.1 SCOPE OF PROJECT MANAGEMENT CONTROL PROBLEM AREA

Major Decision Problem	Component Problem Statements
52. What is project Management Information System:	How are project status and cost indicators defined? What level of detail and frequency of reporting is required? How can this information be related to manpower performance? What are project operating procedures and policies? What standard documents are required? What lines of communication and document distribution is required? When should project procedures be implemented? How should procedures be implemented?
53. What is current position:	Where do we stand? Do we have sufficient information? Can we get more? What are the implications of the present status?
54. What should be done:	How can we proceed? What special management initiatives can be developed to address manpower and productivity issues? Who will implement our decisions? How should we operate? What checks should we implement?
55. What are procedures for implementing decisions:	How will decision information be disseminated to appropriate managers? Who will ensure that decisions are properly enacted?
56. How effective is project Management Information System:	Who is using management information system? Is it being used properly? Are problems being identified in a timely manner? Is management reacting to problems? Are effective decisions being made? Should project team structures be changed? Should staff changes be made? Should control procedures be revised? What reports should be added, deleted, or modified?

tion of reports in decision-making situations. Top management can then evaluate whether documents and reports are being used or are helpful in problem identification, and also whether project team personnel are reacting to problems and making effective decisions. This project team performance analysis may reveal work overload situations and possibly lead to changes in the project team structure or control procedures. In the case of poor performance, it can result ultimately in the releasing from the project of key managers and their replacement with new personnel.

When a contracting company begins to address the major problem areas of Project Management Control, it must be prepared to focus management attention on a range of detailed and interacting problem statements such as those contained in Table 14.1. The attention and detail in which these problems are addressed and solved by project managers depends on the contracting environment, the project size and complexity, and to some extent, the requirements of the client/owner.

The general procedure for the Project Management Control area, as it relates to solving the major problem areas in Table 14.1, is shown in Table 14.2. The procedure is described in terms of a number of management activities defined as management duties and tasks. They emerge in a sequence that follows the order of the major decision problems of

TABLE 14.2 GENERAL PROCEDURE FOR PROJECT MANAGEMENT CONTROL

Procedural Step	Management Duties and Tasks
1.	Design and implement project Management Information System.
2.	Review current project status position and summary project reports, especially in relation to labor performance indicators.
3.	Review project forecasts and implications.
4.	Identify productivity problems and areas of potential cost and schedule savings, and initiate relevant productivity improvement and activity analysis programs.
5.	Identify specific manpower management control problem areas and initiate corrective action.
6.	Monitor project team performance and propose changes in either project team staffing, project procedures, and/or project team structure.
7.	Revise information gathering and reporting procedures and techniques.

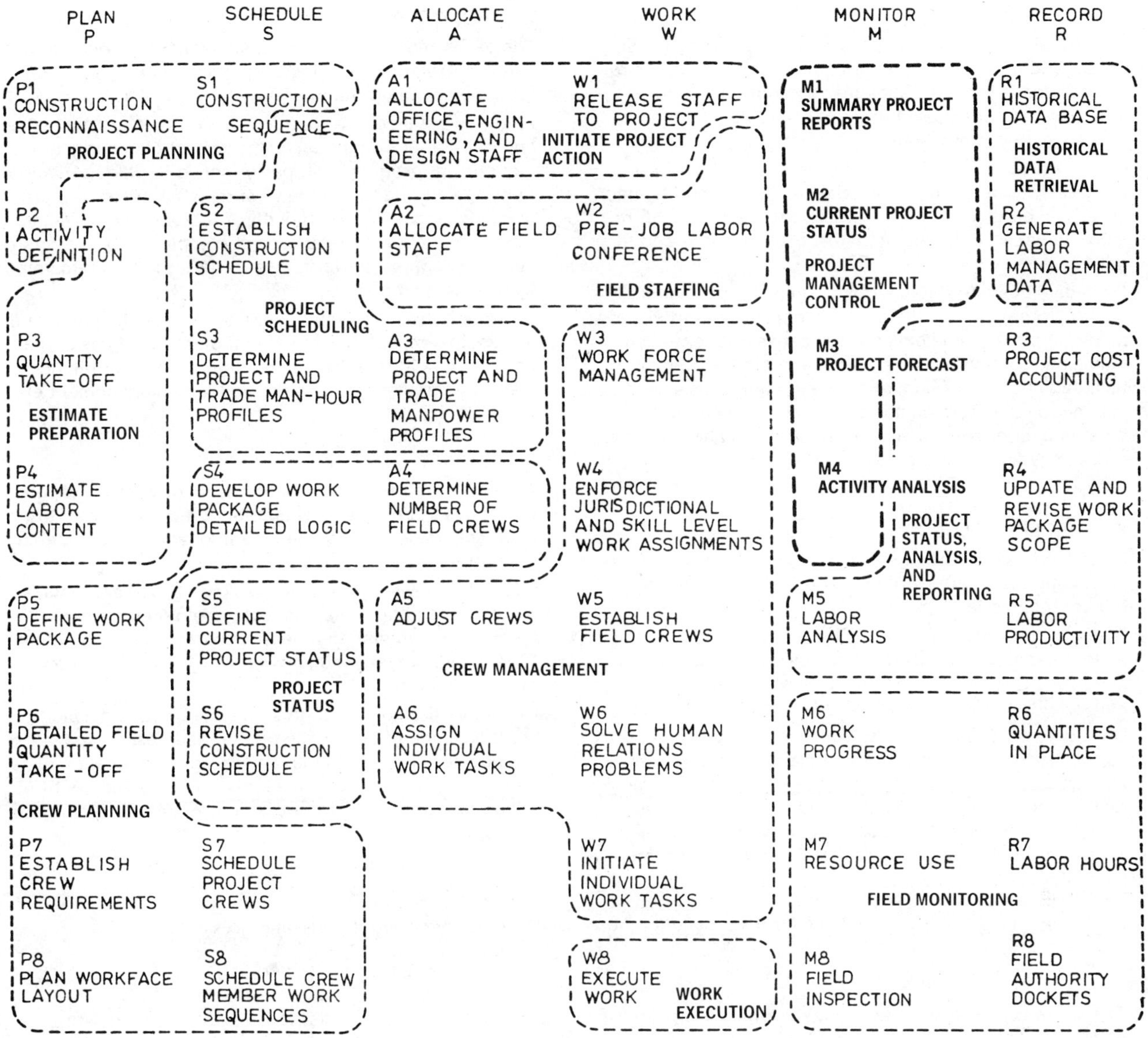

Figure 14.1 Functional plot of the Project Management Control decision process.

Table 14.1. The action of performing these duties and tasks comprises the decision process, and is the basis for identifying the relevant basic manpower management functions included in the functional plot for the Project Management Control decision process in Figure 14.1. The specific management functions in Figure 14.1 are given a full scope description, including statements of factors relevant to function performance as shown in Figure 14.2. These function statements correspond generally to the list of component problem statements and management duties and tasks in Tables 14.1 and 14.2, respectively. Finally, the interface relationships between Project Management Control and other relevant decision processes is illustrated in Figure 14.3.

14.2 FUNCTIONAL PERFORMANCE TECHNIQUES

Project managers must be continuously aware of a project's position, alert to emerging problems, ca-

M1 SUMMARY PROJECT REPORTS

- The assessment of project performance for management review and management authorization of changes in project policies, direction, estimates, schedules, and construction methods, based on current project status, performance, and predictions to complete.
- Requires knowledge of current project status, physical progress, variations from the latest estimate and planned schedule, predictions of final costs and man-hours in relation to each other, and the latest project scope and details.
- Considers costs to date; percent complete; projected cost to complete; man-hours to date; projected total man-hours to complete; variances between budget estimates and forecasts; overall labor and craft productivities; engineering, design, and construction status; work volume remaining to complete; trends and change orders; impact of current authorized changes and decisions on all aspects of the project; and client response to changes.

M2 CURRENT PROJECT STATUS

- Establishes indicators of current project status and progress compatible with work package and cost/scheduling control techniques.
- Requires knowledge of scheduling status, percent complete, revised construction schedules, current labor costs and productivities, and the latest work package scope and details in relation to predicted indicators of status.
- Considers predicted progress versus actual progress, predicted man-hour expenditure and manpower forecasts against actual man-hour expenditures and manning, actual and scheduled activity start and finish dates, revised schedules and estimates, and the impact of current decisions, productivities, change orders, and trends on project completion.

M3 PROJECT FORECAST

- The assessment of changes in scope and/or trends that cause a deviation from the budget estimate and/or plan for accomplishing the work.
- Requires detailed knowledge of the current project scope and definition; budget estimate; project plan and schedule; current progress and performance-to-date in relation to budgeted quantities, man-hours, labor costs, and productivity; and a new prediction of total quantities, man-hours, and labor costs at project completion based on current progress, performance to date, and the remaining work to be completed.
- Considers the nature and cause of changes in scope and trends, the impact on project completion and costs, the impact on field man-hours and manpower, and the demand on engineering, design, and field management effort.

M4 ACTIVITY ANALYSIS

- The controlled analysis of a construction method in relation to crew size, crew production, crew location to the workface, material and equipment location, and material flow routes.
- Requires knowledge of work package scope and details, an understanding of sophisticated field management techniques, intimate definition of construction operations for each work task, and knowledge of method improvement reporting.
- Considers the construction sequence of each detailed work task, the effective use of all resources, crew design, the needless waste of time, delays, specific location of men in relation to material supply routes, equipment location, congestion of the work space, and the production output of the crew.

Figure 14.2 Functional statement of the Project Management Control decision process.

pable of making timely decisions on new courses of action, and able to ensure that their decisions are implemented. In addition, project managers must meld the project team itself into a motivated and coordinated work group, while being prepared to modify the project team's structure, size, and method of operation to meet the changing demands over the various life phases of the project. To some extent the timeliness, quality, and effectiveness of these decisions depends upon the scope and quality of the project monitoring process, and the ability and speed with which the project's (or company's) information system can respond to management requests for project-focused and/or specialized reports.

Project data becomes information whenever the raw data is collated, processed, and/or focused to a specific problem. Consequently, a project Management Information System (MIS) must be capable of processing raw data relative to the specific needs of the project managers themselves. Therefore, the scope and design of a project MIS must reflect the basic need of the project management team to be regularly informed on project status and progress. It must also be responsive to management's need to have, on request, special one time problem-focused management reports. Finally, the MIS is often designed to meet specific requests or objectives peculiar to the client/owner.

When project management becomes aware that a problem exists, either through the identification of cost and/or schedule overruns which have been documented by the Project Status, Analysis, and Reporting decision process or identified by Field

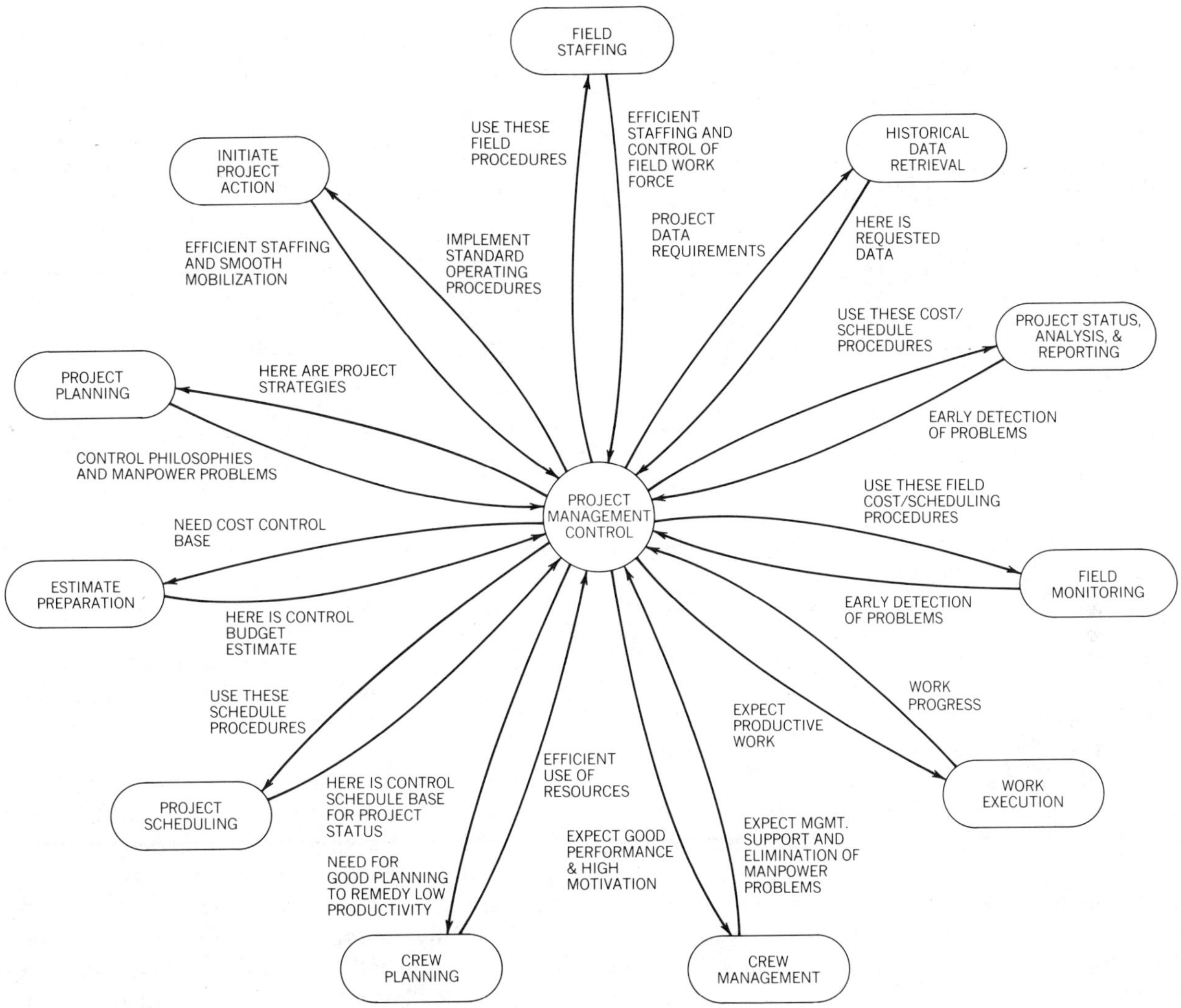

Figure 14.3 Schematic illustration of Project Management Control decision process interfaces with other manpower management decision processes.

Monitoring, it must decide on a course of action that may be investigative, remedial, or compensatory in nature. Which course of action is selected will depend upon the actual or suspected magnitude of the impact it will produce, the urgency for resolution of the particular problem, or what is known about the problem.

If little is known, management may set up a *problem solving team* composed of those personnel best qualified professionally, technically, and in terms of experience, to deal with the problem and to make a priority listing of proposed remedial courses of action for management approval. Alternatively, a *quality circle* type of effort composed of a spectrum of personnel covering all organizational levels of the project team may be set up in the hope that the pooled knowledge and experience of the group will allow them to investigate, brainstorm, and evaluate the nature and cause(s) of the problem. Based on the results of these efforts, management may decide to initiate a technically focused methods study to identify equipment, process, or procedural inadequacies, and to formulate improved technical solutions.

If the problem lies in failures or sluggishness in in-house head office or field procedures, management may wish to review and change formal procedures and initiate a *quality improvement program* that may involve the special training and motivation of key line staff.

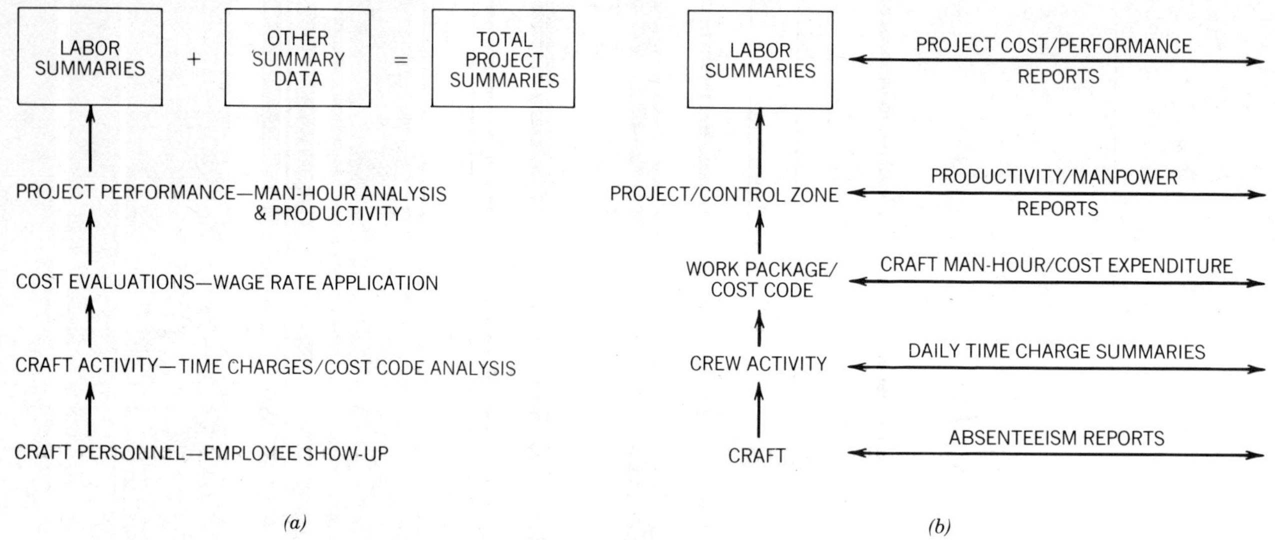

Figure 14.4 Project Management Information System structure. (*a*) Roll-up information flow. (*b*) Multilevel management report access.

If the problem is field based, management may initiate a *foreman delay survey* to tap the grass-roots knowledge of foremen in identifying causes of inefficiencies or drops in field productivities. In the event that foremen/journeymen skill levels are deficient, management may initiate *on the job training programs*. If work force attitudes and motivation are low, for whatever causes, management may wish to initiate *productivity improvement programs* or get more involved in *worker motivation activities*. These are steps to help get the work force on managements side in an effort to begin to develop a cooperative team approach to resolving the problem. In these latter cases, management must specifically address the problem of raising the tone, quality, and effort it expends on human resource management.

Generally speaking, project management monitors progress and status through regular project summary reports. Thus, the Management Information System must have the capability to produce such reports in a timely manner. A common approach is to establish an MIS which accepts raw data at the grass-roots level on labor, material, equipment, and so on, and which then processes the data and rolls it up for the production of suitable project reports that can be used by the project team for the evaluation of the project's position. Figure 14.4*a* is a simplified schematic illustration for this basic structure. In this case, craft employee show-up data leads to project labor related summaries, which can be formatted together with other summary data to provide total project summary reports as aids to project management in their decision-making processes. Similarly, it may be essential for management to access such a roll-up MIS at any desired level as a means of problem detection before relevant decisions can be made. Figure 14.4*b* indicates another simplified MIS structure which can be accessed for the production of special reports at any hierarchical stage. Thus, as shown, craft show-up data can be integrated, focused, and analyzed for absenteeism reports. Also consequential individual crew activity can be monitored for changes in daily time charges. Furthermore cost code evaluations can be made of craft manhours and cost expenditure variations. Finally project control areas/cost codes can be tapped for productivity and manpower reports. Collectively, these all lead to summary labor reports and can reflect, for example, the impacts of craft absenteeism on the labor component at various reporting levels.

Project management generally requires project reports that focus on incremental changes in status and progress, and which highlight the final project position and consequences based upon a range of predictions assuming currently perceived trends are maintained. Many project cost reports have a layout similar to that of Figures 14.5*a* and 14.5*b*. If the project MIS is built around a data structure which incorporates a cost code, control area, or work package scheme, then various dissection type re-

COST CODE	WORK ITEM	COST LAST PERIOD	COST THIS PERIOD	COST TO DATE	CURRENT BUDGET	PROJECTED COST	DEVIATION

(a)

COST CODE	WORK ITEM	CONTROL BUDGET ESTIMATE	APPROVED CHANGES	PENDING CHANGES	CURRENT APPROVED BUDGET	PROJECTED COST	DEVIATION

(b)

Figure 14.5 Typical project summary reports. (*a*) Typical summary report. (*b*) Typical project budget summary report.

ports similar to those described in Figures 14.4*b* and 14.5 provide management with the capability of probing for problem areas, or of producing in-depth project summary reports on the lines of the one outlined in Figure 14.6.

In summary, the concepts illustrated in Figures 14.4 through 14.6 describe the manner in which the Management Information System either directly produces summary level reports, or produces the relevant information to prepare specialized reports. The MIS must have selective reporting capability in order to enable decision makers to identify key problem areas and make timely decisions when corrective action is required. Figure 14.7 highlights this aspect of the Project Management Control MIS in schematic format. As shown in this figure, the key input ingredients of the MIS are the control base estimate and schedule, live data from Project Status, Analysis, and Reporting, and changes in scope authorized by the client. The processing of the data results in project level reports and other information that is disseminated to the appropriate decision processes for further action.

The actual size and structure of the project team generally reflects project needs, work loads, and project control goals. It is often necessary, however, to check for the functional overloading of key members of the project team, since this drastically affects both the time required to perform other than top priority functions and the haste or lack of care with which some decisions are made. Simple measures such as the turnaround time of documents or the use of reminder notes of action lists can often improve the efficiency of the performance of priority functions. It is essential that clear lines of responsibility, channels of communication, and information flows be established throughout the project team, because these features significantly affect the performance of Project Management Control. In this respect the simple line concept of the organizational structure shown in Figure 14.8*a* is typical of many successful field organizations. In these cases, the project manager is the principal decision maker who interacts directly with a head office manager, subcontractors' representatives, and/or the client. On larger projects there is a need to have an expanded professional staff in the field who maintain internal lines of professional communication to

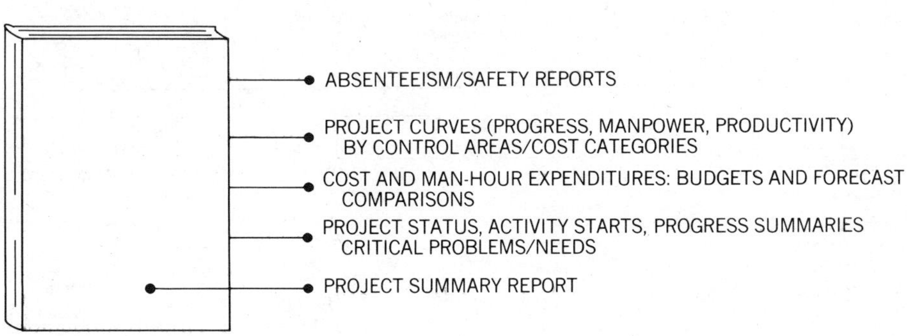

Figure 14.6 Typical structure of project reports.

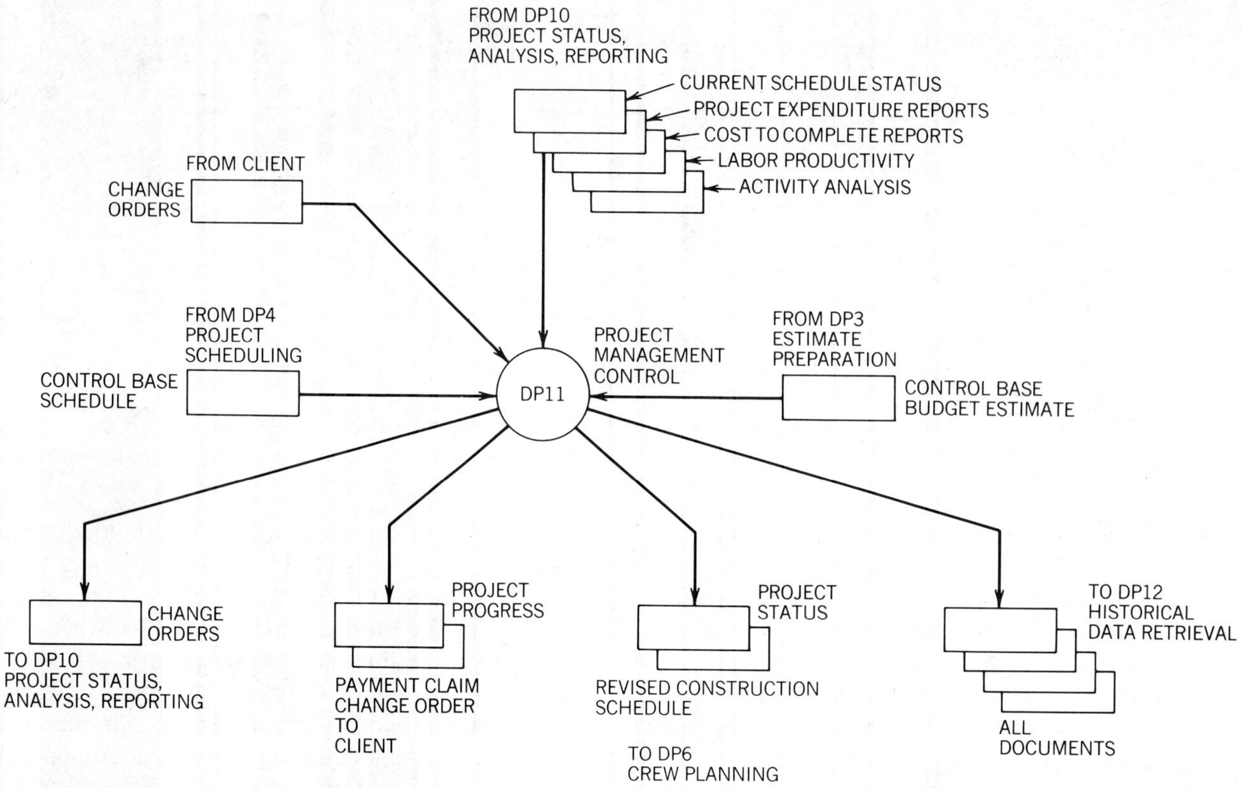

Figure 14.7 Project Management Control Management Information System concept.

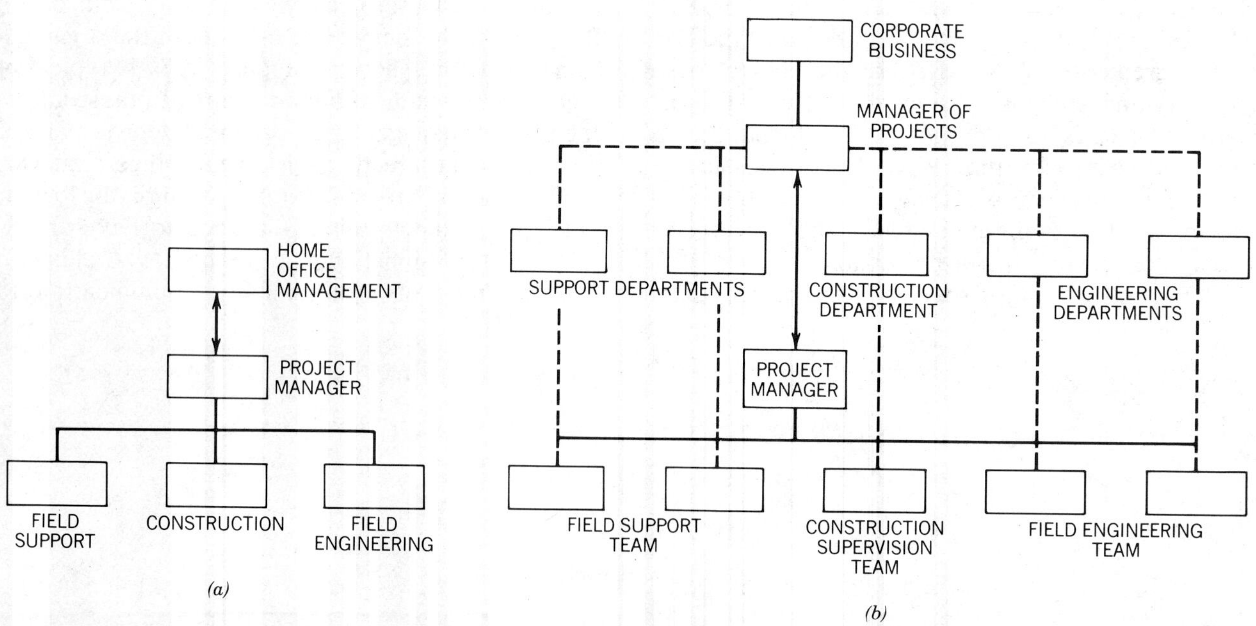

Figure 14.8 Project Management Control concepts. (*a*) Line concept. (*b*) Line and staff concept.

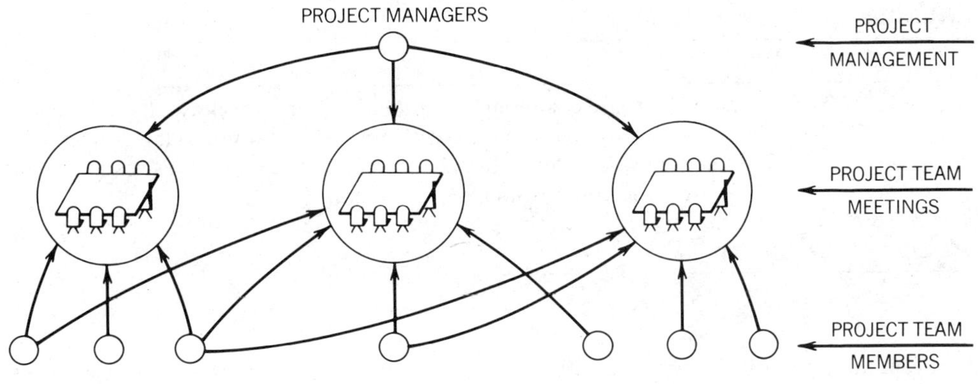

Figure 14.9 Project team management by interacting meetings.

their parent head office departments. In this way, staff assistance can be given to the field project manager with a similar effort, when required, at the head office level to the manager of projects (see Figure 14.8*b*). Thus, the field project manager can call upon significant assistance in the decision-making process. In some cases the field project team is in total control of the project from its initiation, and the Project Management Control process is performed via a series of interacting committee meetings so that the project manager need rarely to make decisions unaided. Thus, regular project conferences become the vehicle for project team consideration in the resolution of problems and effective decision making. This organizational and dynamic decision-making concept is illustrated in Figure 14.9.

On very large projects it may become necessary to formalize the Project Management Control decision process to the extent of producing project execution and project procedure manuals complete with detailed work statements, lists of duties, reporting lines, and document distribution lists for each type of management report. In such an environment, project complexity and client requirements may demand that the number of committees, their composition, functions, and frequency of meeting be described. This would necessitate the integration of field supervision staff into the project management effort and decision process, by initiating regular superintendents' and foremen's meetings and management staff meetings at all levels. Thus, toolbox meetings, weekly progress discussions, safety talks, and crew performance problem analysis sessions may lead to regular or crisis top management/crew luncheon meetings in which points of view are exchanged or grievances aired and solved. A quality circle program is another technique used to solve these types of problems. This works by promoting the continuous exchange of viewpoints between personnel closest to the problem areas.

The Project Management Control decision process is a decision area where significant pressures are brought to bear on the project team members, especially project managers. These personnel generally operate under the close and continuous scrutiny of top-level company management and client/owner management. In the small project environment, project managers often carry the additional burden of performing company business functions. While the project managers on larger projects are not normally concerned with business-related functions, these projects are often so complex that managing them effectively carries with it sufficient burdens and pressures. Consequently, in Project Management Control, decision making is constantly influenced by pressures often beyond the control of the project manager. These pressures therefore influence the manner in which the company operates in the Project Management Control area.

14.3 THE PROJECT TEAM COMPONENT APPROACH

In the small firm the principal managers are required to perform a great variety of management functions, some of which are project oriented and others of which are business oriented. This management function spread is characteristic of the small firm's organizational structure and mode of operation. Thus, Project Management Control (as distinct

TABLE 14.3 POLICIES OF THE SMALL FIRM

Policy	Description
P_S 21 *ACTIVITY STATUS*	Activity progress and labor productivity are not normally considered beyond completed activity volume, labor unit costs, and total labor dollars expended to date relative to activity scheduled target completion dates, estimated labor unit costs, and total estimated labor cost.
P_S 22 *MANAGEMENT BY EXCEPTION*	Labor management by exception as problems arise is the customary decision approach.

from the technical management of construction operations) is often a *head office activity*, because of the pressure of work and the ever present problem of function overload. In addition, the project manager is frequently involved with a number of projects, each of which require a time commitment.

The management pressures that stem from function overloads impact the Project Management Control decision process in a number of ways. First, decisions are frequently made on the run, based upon visual observations and discussions with field superintendents/foremen, and without significant reliance on cost and schedule data from the field. This approach is practical, because the overall project status of small projects can easily be comprehended by the project manager as it relates to previous experience and "gut" feelings. Second, an analysis of current costs and schedule activities is limited to those critical activities that are overrunning budget estimates or time predictions. Other activities that deviate from desired cost and schedule plans are assumed to balance out with compensating values of overruns or underruns. Finally, the project manager often focuses decision making on the key subcontractor interfaces of the project, as compared with the activities directly managed by the contractor, to ensure a smooth construction schedule that meets the desired completion date.

As mentioned in Chapter 13 the Bar Chart Construction Schedule (I_S-DOC 2), the Construction Progress and Status report (I_S-DOC 3), and the Revised Bar Chart Construction Schedule (I_S-DOC 4), when applicable, contain scheduling data inputs for CURRENT PROJECT STATUSING (M2). The Labor Distribution Report (P_S-DOC 10) and the Project Cost Recapitulation (P_S-DOC 12) provide the necessary cost input information for the performance of this function. Together, these documents aid the project manager in the evaluation of overall project status. In terms of manpower management, the labor cost documents portray total activity labor costs to date, relative to estimated labor costs. When these activity costs are compared with actual construction progress for an activity, the project manager can evaluate profit/loss potential for the activity by *ACTIVITY STATUSING* (P_S 21).* For example, if an activity is eighty-five percent complete and only fifty percent of the labor dollars are expended, the project manager might expect a potential profit for the activity. Conversely, if the activity is only fifty percent complete and eighty-five percent of the labor dollars are expended, a potential loss may be expected. In the latter case, the project manager may further attempt to analyze this activity's status in order to determine the cause of the potential cost overrun. Quick action taken by the project manager to resolve any problems that are identified may forestall further cost overruns on the activity. If the activity is progressing according to schedule, however, the project manager may neglect the potential cost overrun as being an estimate miscalculation, and expect it will be balanced by cost underruns for other activities. The criticality of the activity will also influence the management attention afforded to it, and therefore decisions relevant to critical activities are made on a *MANAGEMENT BY EXCEPTION* (P_S 22) basis. If the activity represents an insignificant percent of the total project cost, then the project manager normally will not be concerned with a cost overrun. This very pragmatic approach is often applied, especially when underruns are more prevalent than overruns.

The project manager relies heavily on a comparison of labor unit costs when tracking a problem. Problems can be tracked through labor unit costs to the work task level using the Labor Distribution Report (Figure 13.18) and the Estimate Recapitulation Sheets (P_S-DOC 3) (Figure 6.16). The Labor Distribution Report furnishes actual labor unit costs

*See Table 14.3, Policies of the Small Firm.

for each work task, and the comparable estimated labor unit costs for the same work tasks are extracted from the Estimate Recapitulation Sheets. Comparing these labor unit costs can reveal significant information. For instance, if the actual unit cost and estimated unit cost are similar, and actual labor cost expenditures in dollars are running higher than the estimate when compared to the physical percent complete, then the estimated quantity is probably low. The project manager can check material requisitions and other sources to determine the causes for the lower estimated quantities. Unless significantly high negative deviations from the budget are uncovered in labor unit cost comparisons, the project manager rarely pursues this analysis any further.

The labor cost documents portray one facet of CURRENT PROJECT STATUS, that is, the financial outlook of the project. Construction status is displayed on the bar chart Construction Schedule, the Construction Progress and Status report and the revised bar chart Construction Schedule (if applicable). These documents in themselves describe this facet of CURRENT PROJECT STATUS. When they are combined with the labor and project cost documents, they provide basic indicators of CURRENT PROJECT STATUS and, in essence, become the summary project reports for SUMMARY PROJECT REPORTING (M1). At this decision point, any new or revised management directives are disseminated to the field. For instance, significant negative cost variations, which appear on the labor cost reports, are brought to the attention of the field superintendent. Action is then taken to resolve the problem in order to forestall or minimize future losses. The field is also notified of all revisions in the schedule in order to ensure that these revisions are implemented immediately, especially when a subcontractor is affected by a shift in an activity's start date and/or finish date. Because the budget is normally fixed via the lump-sum competitive bid, the client is not necessarily as concerned with costs as with construction progress. Hence, scheduling information is continually forwarded to the client. The client and contractor interact on cost questions when potential changes in scope are identified by the contractor and/or requested by the client. In most instances, the contractor is reimbursed for these types of changes after a Change Authorization (P_S-DOC 14) document is prepared, as illustrated in Figure 14.10, and the client has approved the change and authorized the work to proceed. Frequently, minor changes requested by the client are incorporated into the construction effort without additional cost to the client. This helps provide for good working relations between the contractor and client, and may lead to additional work in the future. All changes in scope must be forwarded to the field for incorporation into the construction planning effort.

The key human management effort in the resolution of problems at the Project Team Component level can be direct, simple, and effective. Since the work force is small and there are at most three overall levels from the grass-roots work force to the company owner it is commonplace for the owner/project manager to visit the site on a *MANAGEMENT BY EXCEPTION* basis and talk personally with each worker on a first-name basis. In this way everyone has the opportunity to contribute to, and thereby get involved in, the identification and resolution of a problem. In most cases the job superintendent and/or the foreman will be able to cope with and resolve issues in the same manner. Thus, except for pathological cases, management has little trouble in establishing close rapport with the work force.

The Project Management Control decision process described in Figure 14.11 reflects informal document flows that focus primarily on comparative analysis of cost data to identify budget overruns. When coupled with periodic site visits and knowledge of schedule status and progress, this comparative analysis produces sufficient information for the project manager to make decisions relevant to potential and existing problem areas. These decisions become management directives for the project that are implemented through other decision processes such as Crew Planning, Crew Management, and Project Status, Analysis, and Reporting.

14.4 THE TRAVELING PROJECT MANAGER

Project Management Control for the medium-sized firm begins as project status is measured, portrayed on various documents, and informally analyzed, and as problem areas are uncovered. The main focus of Project Management Control is to make policy and job directive decisions which provide the impetus for solving problems. These decisions are often based on the activity analysis of critical problems, and from summary project status evaluations. Thus, as shown in Figure 14.12, Project Manage-

ANDERSON AND WOODHEAD
BUILDING CONTRACTORS

PROJECT: ________ JOB NO.: ________

LOCATION: ________ DATE: ________

CHANGE ORDER ________

ADD $ ________

DEDUCT $ ________

SUBMIT FOR APPROVAL SUM OF ________

DESCRIPTION OF CHANGE:

DETAILS OF COST:

SCHEDULE DELAY ________ DAYS

ANDERSON AND WOODHEAD

by ________

(a)

ANDERSON AND WOODHEAD
BUILDING CONTRACTORS

PROJECT: ________ JOB NO.: ________

LOCATION: ________ DATE: ________

SHEET NO.: ________

CHANGE ORDER ________ ESTIMATE

WORK	QTY.	LABOR	MAT'L	SUBCONTRACTS	TOTAL
TOTAL					

APPROVED BY: ________

(b)

ANDERSON AND WOODHEAD
BUILDING CONTRACTORS

PROJECT: ________ JOB NO.: ________

LOCATION: ________ DATE: ________

CHANGE ORDER AUTHORIZATION

In accordance with the terms of our contract...
you are hereby authorized to proceed:

DESCRIPTION OF WORK

ADDED AMOUNT $ ________

DEDUCT AMOUNT $ ________

ANDERSON AND WOODHEAD

by ________

(c)

Figure 14.10 Change Authorization (P_S-DOC 14). *(a)* Summary. *(b)* Estimate detail. *(c)* Authorization to proceed.

ment Control focuses on the performance of four functional areas, namely: CURRENT PROJECT STATUSING (M2), PROJECT FORECASTING (M3), ACTIVITY ANALYSIS (M4), and SUMMARY PROJECT STATUSING (M1).

The Traveling Project Manager typically operates under slightly different management pressures in the medium-sized firm environment than he or she would in the small firm. The project manager is often free from the function overload caused by performing business management functions that are characteristically his or her responsibility in the small firm. However, the resultant freedom from function overload provides for more direct project management involvement and responsibility, and the project manager is expected to perform in-depth analyses of labor cost reports and current schedule status. This type of pressure is brought to bear by the top-level company executives to whom the Traveling Project Manager reports during the Project Management Control decision process. The pressure is increased when the Traveling Project Manager's realm of responsibility encompasses a number of projects, each of which is typically larger

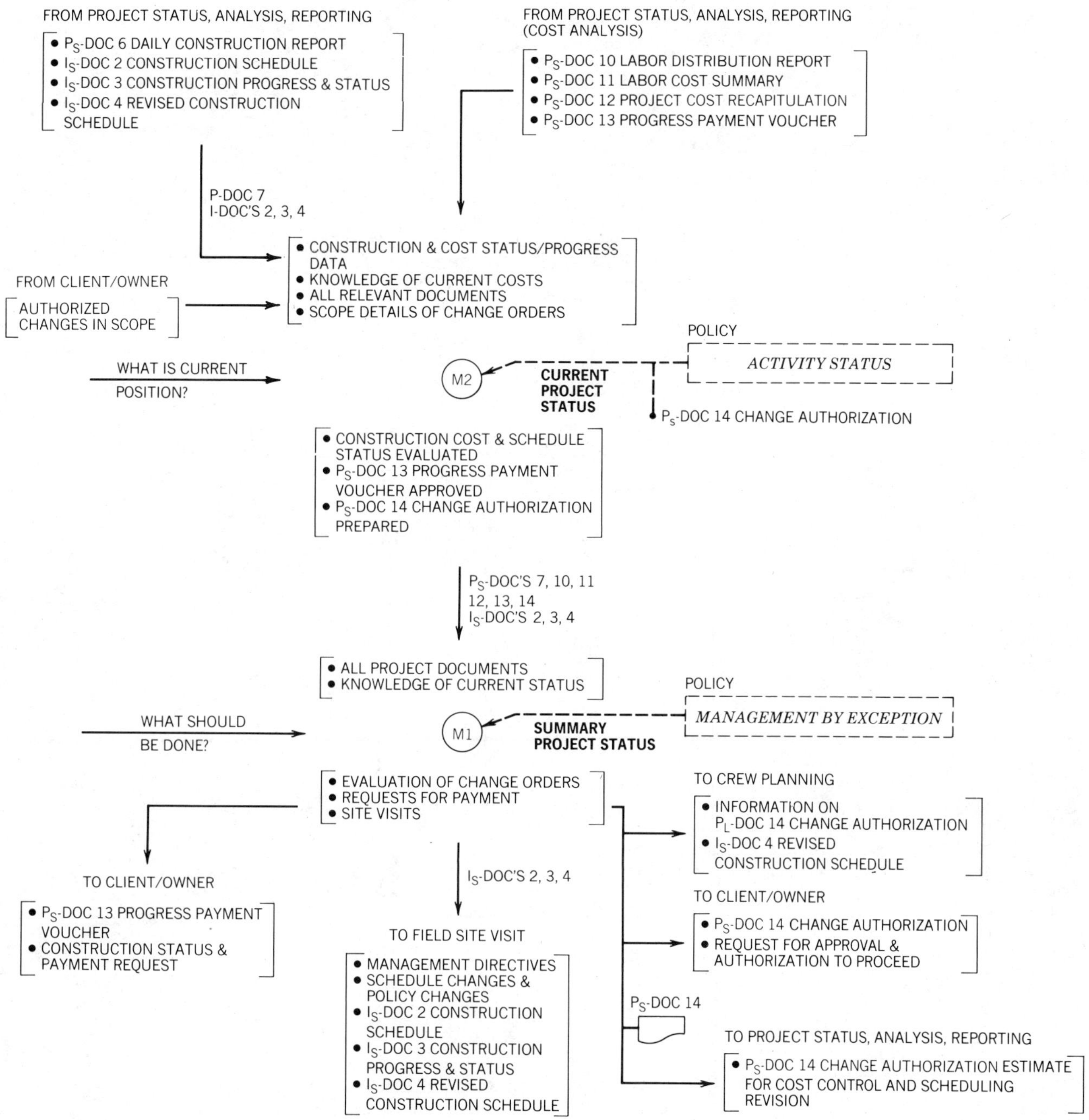

Figure 14.11 Project Management Control decision process (Project Team Component approach).

and more complex than the projects constructed by the small firm. Consequently, the project manager interacts frequently with the key field managers (i.e., superintendents, foremen, field engineers) to the extent that he or she is continuously informed of project cost and schedule status. This management approach enables the project manager to react to problems in a timely manner based on a combination of visual observation and knowledge of the project, supported by a more detailed analysis of documented labor cost and schedule data.

PROJECT FORECASTING is a management

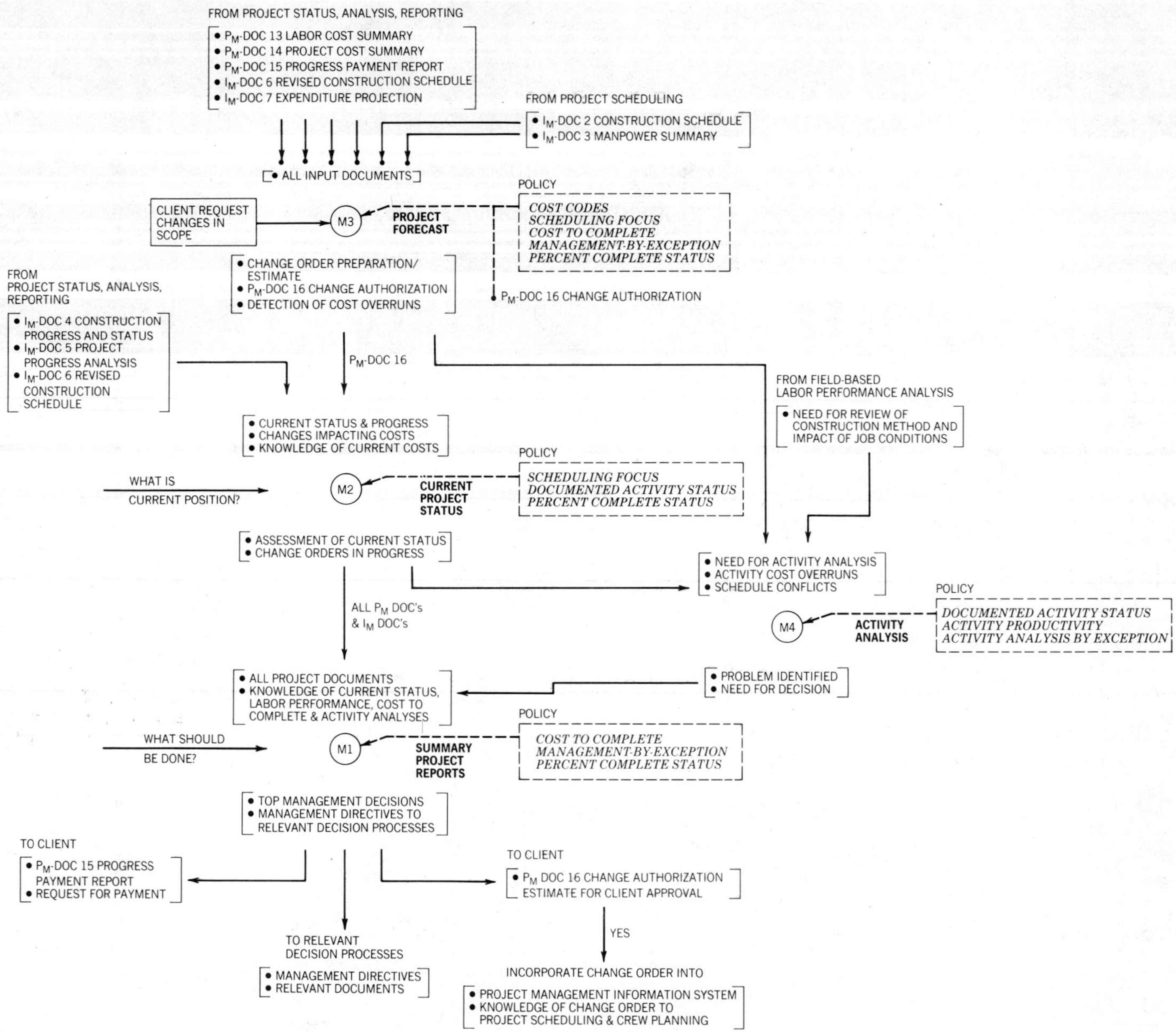

Figure 14.12 Project Management Control decision process (Traveling Project Manager approach).

TABLE 14.4 POLICIES OF THE MEDIUM-SIZED FIRM

Policy		Description
P_M 4	*COST CODES*	Systematic classification of work categories segregated by unique cost codes is used to define work tasks, collect costs, and provide a means of comparing actual costs with estimated costs.
P_M 8	*SCHEDULING FOCUS*	Scheduling definitions and constraints are: (1) Critical activities are highly sensitive profit and loss activities and/or activities controlling the rate at which ensuing work can commence. (2) Activity durations and sequence logic incorporate the impact of seasonal weather conditions and restrictive work practices. (3) Manpower impact is considered on a conceptual basis. (4) Activity scheduling reflects the best sequencing of activities possible for effective utilization of all resources available.
P_M 22	*DOCUMENTED ACTIVITY STATUS*	Labor costs, quantities in place, and activity status must be recorded and documented by a system yielding reliable information and viable reports for management control.
P_M 23	*ACTIVITY PRODUCTIVITY*	Labor productivity is an indicator of crew performance, and is measured by activity labor dollar expenditures and quantity volume in place.
P_M 24	*ACTIVITY ANALYSIS BY EXCEPTION*	Activity analysis is implemented when critical activities could potentially delay the schedule and/or potential labor dollar overruns are emerging.
P_M 25	*COST TO COMPLETE*	Projecting total project costs becomes a viable method of ferreting out potential overruns and evaluating profitability at the activity level and total project level.
P_M 26	*MANAGEMENT BY EXCEPTION*	Manpower management by exception as problems arise is an acceptable decision approach. However, management decisions are made when the impact will result in a significant cost/scheduling savings.
P_M 28	*PERCENT COMPLETE STATUS*	Construction progress reporting must provide reliable information of activity status and total project status to be an effective tool in manpower management and control.

function performed during Project Management Control by the medium-sized firm, but not by the small firm. The information input for this function includes the Labor Cost Summary (P_M-DOC 13), the Project Cost Summary (P_M-DOC 14), the Project Progress Analysis (I_M-DOC 6), the Project Progress Payment Report (P_M-DOC 15), and the Expenditure Curve (I_M-DOC 7), if generated. The Labor Cost Summary and Project Cost Summary are the reports best suited for evaluating costs per activity, because these documents report costs to the detailed work task level by cost code. The projected profit or loss shown on these documents reflects labor cost trends from the budget estimate as required by a *COST TO COMPLETE* (P_M 25)* philosophy. Cost trends which are unfavorable are closely analyzed and/or monitored by the project manager if the trend is significant and the activity is considered critical by *SCHEDULING FOCUS* (P_M 8) criteria. If trends are favorable, however, no further action is taken by the project manager. This is typical of the *MANAGEMENT BY EXCEPTION* (P_M 26) approach.

* See Table 14.4, Policies of the Medium-sized Firm.

The Project Progress Analysis and Project Progress Payment reports are useful documents for analyzing the cost status of an activity in total. By comparing the estimated costs based on percent complete from these documents against the actual costs incurred from the Project Cost Summary for each activity, the project manager can glean possible insights into the actual overruns emerging at any point in time. This may flag deviations which will prompt further analysis of the activity's cost expenditures. In terms of the labor component, the project manager would review the work tasks comprising the activity on the Labor Cost Summary to determine if labor costs are overrunning. Again, significant negative deviations could initiate an informal activity analysis.

As part of PROJECT FORECASTING the client and contractor interact on cost questions, especially where potential changes in scope are identified by the contractor and/or the client/owner. The contractor will prepare a Change Authorization

(P_M-DOC 16) document for submittal to the client/owner when a significant change is identified. The format and content of this document would be similar to that illustrated in Figure 14.10 for the small firm. Any changes in the scope of work described by the Change Authorization would be incorporated into CURRENT PROJECT STATUSING.

CURRENT PROJECT STATUSING is a project-oriented rather than an activity-oriented function. The information required to perform this function are those documents which portray construction progress, primarily the Construction Progress Report (I_M-DOC 4), the Project Progress Analysis, and the Construction Schedule (I_M-DOC 2) or Revised Construction Schedule (if the original Construction Schedule has been changed). Labor cost reports are available through PROJECT FORECASTING for information. CURRENT PROJECT STATUSING, however, mainly focuses on construction schedule status. It is a function which assesses and melds together the vital information of the Construction Schedule and the Construction Progress Report. Therefore, in accordance with the need for *DOCUMENTED ACTIVITY STATUS* (P_M 22) and *PERCENT COMPLETE STATUS* (P_M 28) information, the two documents mentioned portray the status to date, and, based on current job conditions, the activities the project manager must complete in order to finish on time. When specific, significant changes are required in the construction schedule, these recommended changes are passed on to top-level management for final approval and release to the field.

ACTIVITY ANALYSIS is performed on a limited basis and is normally executed only when severe cost or scheduling problems are uncovered, as governed by *ACTIVITY ANALYSIS BY EXCEPTION* (P_M 24) criteria. The principal focus of ACTIVITY ANALYSIS is the construction method and crew structure. The project manager, superintendent, and (at times) craft foremen analyze the existing construction method in light of actual job conditions in order to evaluate alternative methods which might improve crew productivity and reduce costs or shorten activity durations. Because ACTIVITY ANALYSIS focuses heavily on construction methods, the analysis is geared toward optimum use of construction equipment and minimum use of manpower. Hence, the project manager and superintendent will often sketch a layout of the activity components to analyze equipment position, material flow, and crew position to achieve the best mix of these variables for optimum crew productivity. The cost impact of the selected method is subsequently determined to evaluate whether the method will in fact produce a labor dollar savings or prevent further labor cost overruns on an activity. The project manager will rely on input from PROJECT FORECASTING and, in particular, the labor cost projections and possibly an S-curve, to ensure that a realistic savings is possible. Finally, the project manager will check to ensure that the problem solution is compatible with the current construction schedule.

The final function of Project Management Control is SUMMARY PROJECT REPORTING. The information used in the performance of this function comprises the various reports previously identified which depict overall construction status and the result of any special activity analysis. The documents and activity analysis results are reviewed by the executive manager in conjunction with the Traveling Project Manager. All significant deviations are discussed and recommended solutions are introduced by the project manager. The executive manager, in conjunction with the project manager, must decide the various courses of action when major policy shifts are necessary. For instance, at this level, the executive manager would decide on overtime options or a major shift from a labor-intensive to an equipment-intensive construction method, especially when either purchase or rental of equipment would result. Finally, the executive manager must agree on schedule changes and approved progress reports to be forwarded to the client. Major changes in construction approach or schedule are enacted in the field through Crew Planning and Crew Management, as directed by the project manager. If further schedule revisions are necessary, then the schedule is revised as previously described in the Project Status, Analysis, Reporting decision process under REVISING THE CONSTRUCTION SCHEDULE (S6). Client-approved and authorized changes in scope would be forwarded to the field for incorporation into the field construction plan at the Crew Planning level.

Project Management Control is concerned both with solving specific activity problems, and with rolling together all cost and scheduling reports in order to provide a total picture of a project's status. The analysis of the status of the project provides management with the basis for issuing project directives to the field including policy changes and their cost and scheduling implications. The decision network for the Project Management Control decision process, as shown in Figure 14.12, emphasizes total summary project evaluation and activity analysis.

The normal role of the project manager provides many opportunities for personal contact, conversation, and interaction with most members of the field work force. Consequently, even though the work force here is larger than that of the small firm, the frequency of the project manager's site visits, and his or her increased involvement in manpower management, gives most workers the opportunity of interacting with top head office management, and of raising issues in which they feel involved. Thus, project management has many opportunities to initiate and build a positive human resource management environment.

In the Traveling Project Manager environment, data flow remains informal and has a similar focus to that of the Project Team Component approach of the small firm. At the medium-sized firm level, however, the data analysis can be characterized by a more selective problem identification process due to the advent of documented cost forecasting. These forecasts provide an increased number of indicators for sensing problem areas, and therefore require an intensified effort by the project manager when scanning and interpreting processed data for problem identification. This effort is enhanced through knowledge gained from informal field productivity analysis information and frequent site interaction. The project manager can isolate critical problems requiring top-level management decisions for which management directives are made and dispersed to the relevant decision processes. Other minor problem areas are handled by the project manager, often without the need to consult with top company management.

14.5 FIELD-BASED PROJECT MANAGEMENT

Project Management Control for the Field-Based Project Management approach has both a field focus that effectively supports construction with timely decision making, and a head office focus for overall project management and total project summary level reporting. In addition, Project Management Control has a phase that is totally dedicated to the design and implementation of the project's Management Information System. In this way, an integrated Management Information System can be established that meets the specific needs of the project.

The primary goal of the Project Management MIS is to provide the capability of multi-level reporting with selective data entry for the generation of specially focused reports. In order to achieve this goal, the information system must be properly designed and effectively implemented in the early stages of the project. As schematically illustrated in Figure 14.13*a*, the initial establishment of a computer-based cost control file occurs once budget estimate data are available. In addition, project procedures that define operating practices of key decision processes and their relevant Management Information System structures are defined and released to the project.

Once the project is active, the details of the cost control file are updated through Project Status, Analysis, and Reporting, and, when combined with the pertinent scheduling documents, produce updated project level cost reports. These reports are analyzed and selective data from them are extracted. When these data are incorporated with project schedule status and progress, they combine to produce comprehensive documents that summarize construction and total project progress. This facet of the MIS is illustrated in Figure 14.13*b*.

The Project Management Control decision area becomes increasingly significant to the project and construction management process at the large firm level, because it is the focal point for the summarization of all relevant project status, progress, and cost reports. While various management reports yield specific data for controlling a given aspect or component of a project, it is not until these reports are summarized into total project status and progress that current indicators of the project's position can be established for top-level project management review. Consequently, an increasing level of documentation emerges at the large firm level in relation to the growth of cost and scheduling engineering. Management decision making therefore becomes much more project oriented (e.g., the decision to import rather than train welders), whereas the Traveling Project Manager of the medium-sized firm is likely to make decisions that influence only specific critical work activities (e.g., purchase of construction equipment to solve a construction methods problem for a critical activity).

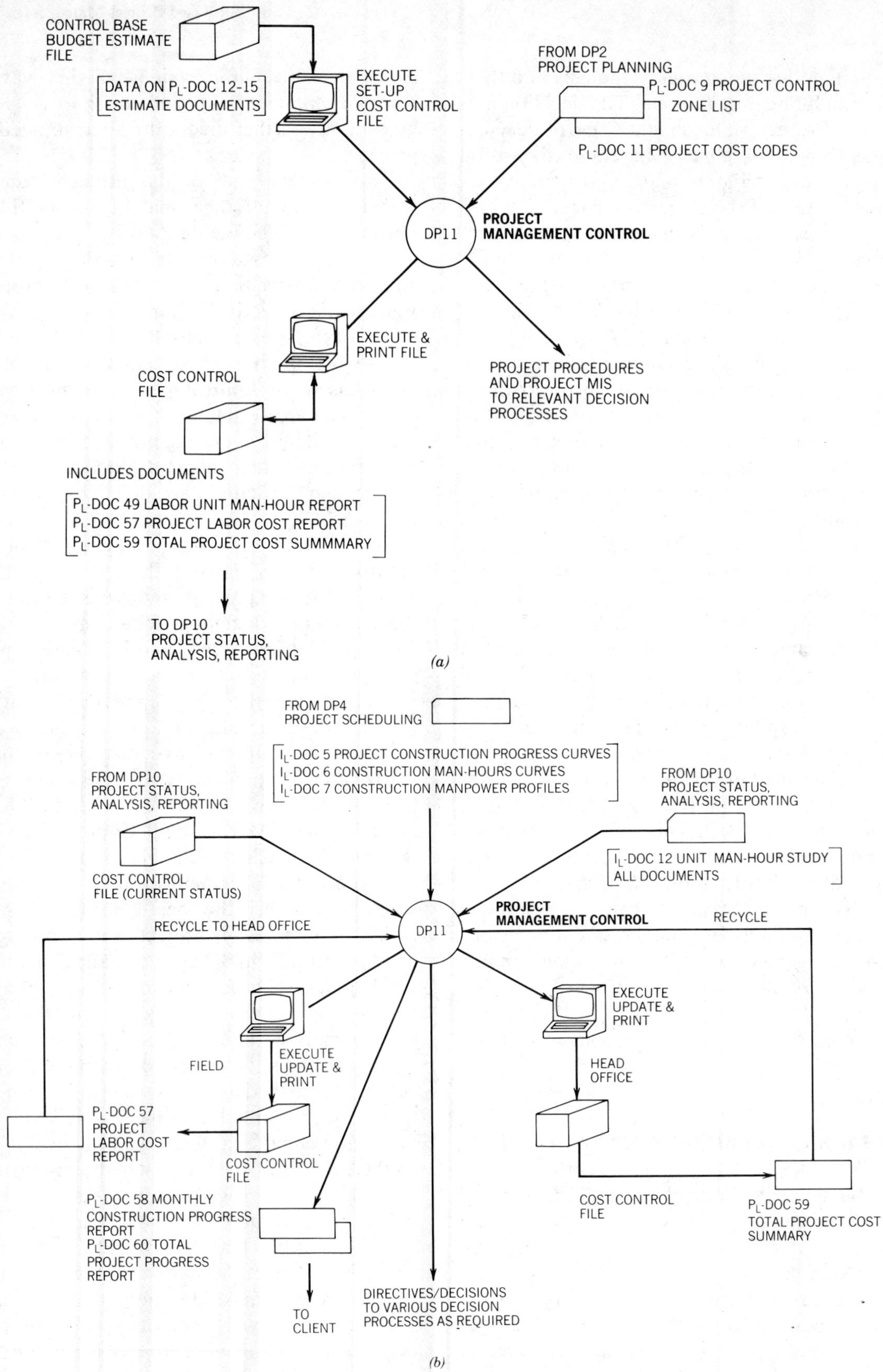

Figure 14.13 Project Management Control Management Information System (Field-Based Project Management approach). (*a*) Initial set-up. (*b*) Fully operating.

In the large firm environment, project management tends to focus their expertise and attention on a single project. The burden of performing business functions and/or managing several projects at one time is often eliminated. Because of project size, location, and technical complexity, however, project managers involved with large projects still perform Project Management Control functions under intense management pressure. These pressures arise from having to manage projects with significantly increasing levels of project information, an increased number of project team personnel, and complex technical problems. In addition, project managers are frequently confronted with tight schedules and problems associated with the overlapping of the engineering/design effort with field construction. Furthermore, top-level company management closely monitors the success and/or failure of project managers over the life of the project, and may at times change the project management personnel when the project is not proceeding successfully. Finally, because many projects in this environment are performed on a cost reimbursable basis, the client's involvement in the project (especially in the design–construct project) increases the management pressure brought to bear on project management to ensure that costs are maintained within approved budgets.

The Project Management Control decision process can be broken down into three decision phases, namely:

1. The design and implementation of the project's Management Information System (see Figure 14.14 for this decision network).
2. Fully operating field Project Management Control (see Figure 14.15).
3. Fully operating head office Project Management Control (see Figure 14.18).

The Design and Implementation of the Project's Management Information System (MIS), as described in Figure 14.14, focuses on those problems related to the determination of project indicators for evaluating the project's current position. This would include decisions regarding the level of detail to be contained in reports, the frequency of report generation, and the procedures for guiding project execution. These decisions are made as part of performing the CURRENT PROJECT STATUS (M2) function, as reflected in Figure 14.14.

In addition to designing the Management Information System, CURRENT PROJECT STATUS initially establishes the cost control system by loading the current budget estimate details, at either the Factored Estimate (P_L-DOC 11) level or the Control Budget Estimate (P_L-DOCs 14 and 15) level, into the project cost control system. The performance of the CURRENT PROJECT STATUS function is impacted by many policy directives that are concerned with the design of the project's information system.* Consequently, many decisions are made by project management to determine which Standard Project Operating Procedures (P_L-DOC 1) (see Figure 4.15) are most applicable to the control of the project, as well as which procedures best meet the requirements of the client/owner.

Once the project's Management Information System is designed and procedures selected, the MIS is implemented through the performance of SUMMARY PROJECT REPORTING (M1). This function determines the specific policy and internal documents to be used in the management of a project, and establishes the cost control system, the lines of information flow, and the distribution of documents. As shown in Figure 4.14, procedural data in the form of project directives are released to various project decision processes for actual implementation.

The design and implementation of the project Management Information System is performed in the head office, commencing early in the project's life. This decision phase of Project Management Control is frequently repeated until project procedures are established and are working effectively on the actual project, and the cost control system is operative with control budget data. Project Management, especially the project manager and other key personnel such as lead project cost engineers, project engineers, general superintendents, and the like, are involved with this decision phase.

The fully operating field phase of Project Management Control, Figure 14.15, focuses principally on the monthly summarization of CURRENT PROJECT STATUS and the development of key project reports that reflect current construction progress and status. Based on the project's current position, management directives are made that guide field management when executing the construction effort. In addition, field Project Management Control involves the approval of significant

*See Table 14.5, Policies of the Large Firm.

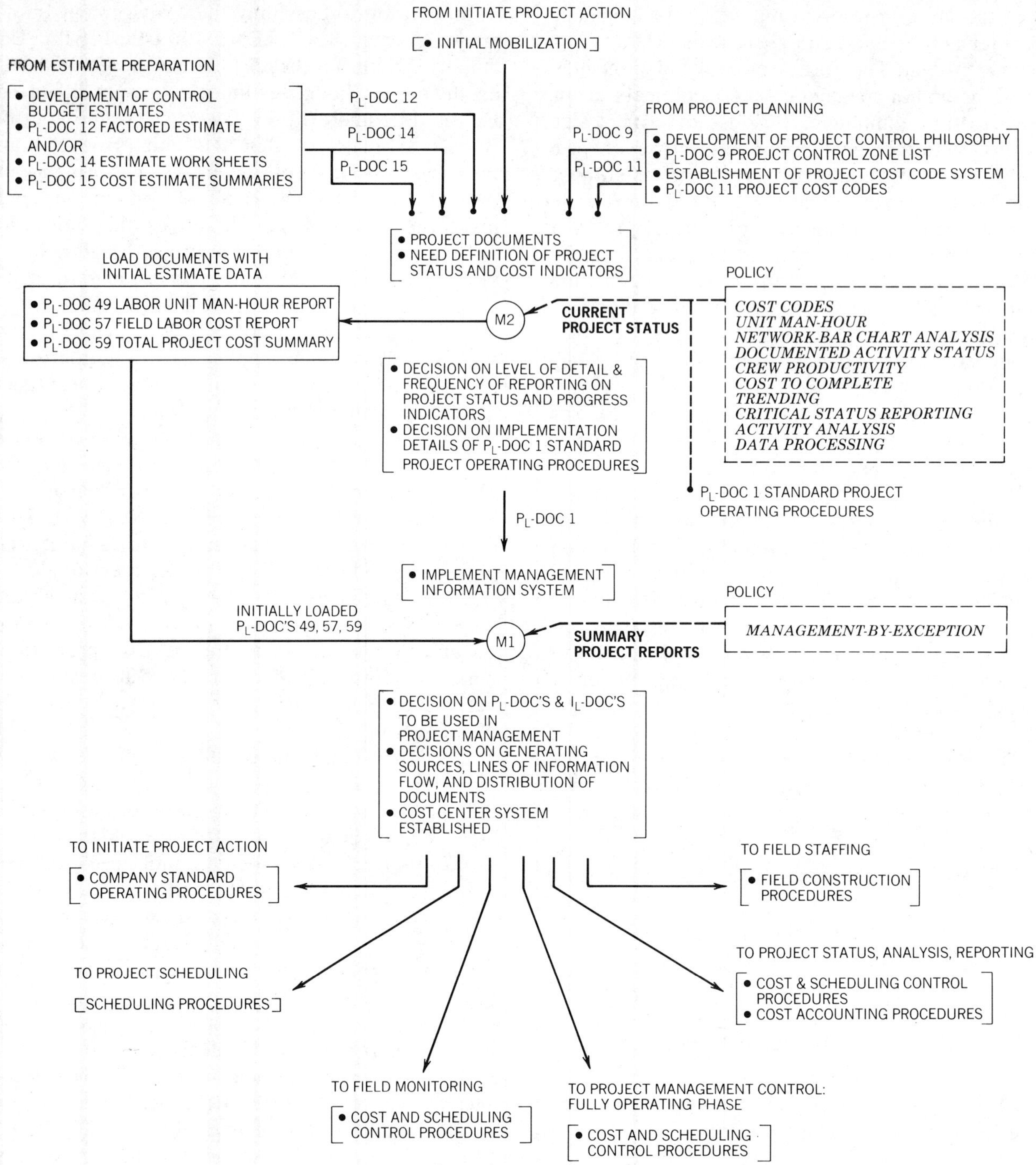

Figure 14.14 Project Management Control decision process: design and implementation of Management Information System (Field-Based Project Management approach).

TABLE 14.5 POLICIES OF THE LARGE FIRM

Policy		Description
P_L 27	*DOCUMENTED ACTIVITY STATUS*	Labor costs, manpower, man-hours, and quantities are primary ingredients to cost control and must be documented by a system yielding reliable information and viable reports for management control. Activity status must be documented in terms of start and completion dates and percent complete data.
P_L 28	*CREW PRODUCTIVITY*	Actual unit man-hour rate versus estimated man-hour rate is a measure of crew performance and a basic indicator of productivity.
P_L 29	*COST TO COMPLETE*	Projecting total project costs and man-hour requirements is requisite to identify problems and evaluate alternative courses of action at both the activity level and the project level.
P_L 30	*TRENDING*	A mechanism for documenting and analyzing deviations of cost and schedule consequence is an integral facet of the information system.
P_L 31	*CRITICAL STATUS REPORTING*	The construction progress reporting system must provide a comprehensive report of activity status highlighted by a mechanism to alert management to activities at and approaching critical status.
P_L 32	*ACTIVITY ANALYSIS*	Activity analysis is geared to exception reporting mechanisms or field management desire to analyze construction methods or crew productivity to improve crew peformance.
P_L 33	*MANAGEMENT BY EXCEPTION*	Manpower management by exception is acceptable provided that the information system is geared to exception reporting to alert management to existing problems or potential problems which are emerging.
P_L 34	*DATA PROCESSING*	Maximum use of electronic data processing is often the most economical means of processing data into useable information.
P_L 35	*PROJECT TEAM SPIRIT*	Active participation of all field managers in the management process promotes goodwill and a positive management environment.

construction methods improvement techniques, change orders, and performance trends. While Figure 14.15 focuses primarily on a monthly reporting cycle, the Project Management Control effort is a continuous decision process wherein critical problems of a total project nature are resolved.

The decision makers predominantly involved in field Project Management Control are cost and scheduling engineers, craft superintendents, general superintendents, project engineers, and the field project manager. The cost and scheduling engineers are responsible for operating the cost control system and thereby for producing summary level field project cost reports. The remaining project team members involved in field Project Management Control participate in evaluating construction progress and status, recommending solutions to specific problem areas, and approving specialized methods techniques, field change orders, and performance trends.

ACTIVITY ANALYSIS (M4) at the Project Management Control level is mainly concerned with securing management approval for a significant change in construction methodology. This request for approval is based on a detailed study of a specific problem, for which the analysis, conclusions, and recommendations are contained in a documented study such as a Unit Man-hour Study (I_L-DOC 12). Because ACTIVITY ANALYSIS decisions are governed by *MANAGEMENT BY EXCEPTION* (P_L 33), top field management is only involved with activity analysis reports that have a potentially significant impact on the project as a whole. Once a recommended construction method is approved, it is implemented in the field, and often incorporated into the monthly project progress report.

PROJECT FORECASTING (M3) has the same focus as ACTIVITY ANALYSIS because it is concerned with top field management approval of field change orders and performance trends. In the case of performance trends, Field Labor Forecasts (P_L-DOC 53) are incorporated into the project control system as they are approved. Field Change Orders (P_L-DOC 52), on the other hand, are approved by field management and forwarded to the head office for final approval and release to the client. Field Labor Forecasts will also be issued to the client from the head office by the manager of projects.

CURRENT PROJECT STATUS is established

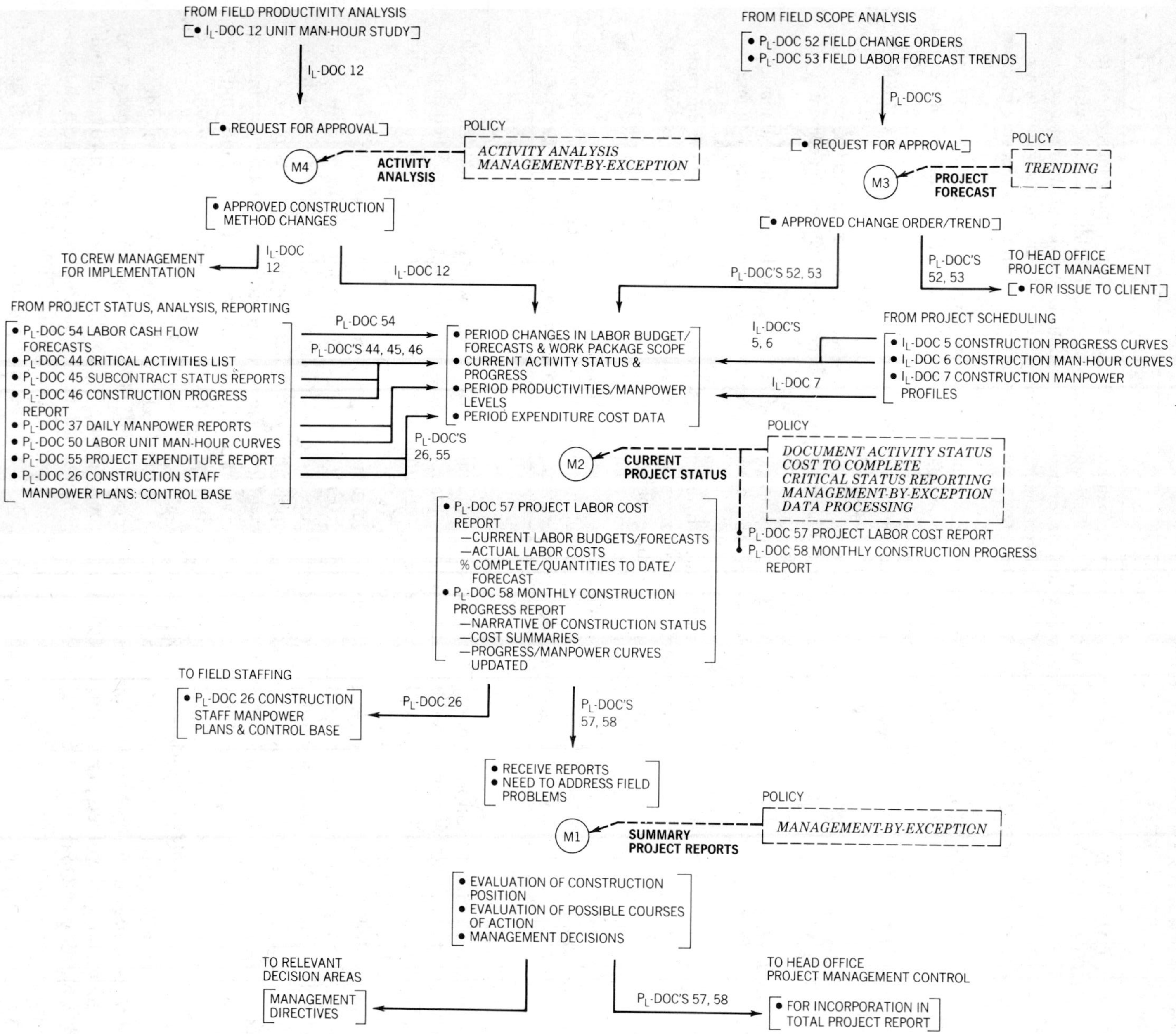

Figure 14.15 Project Management Control decision process: field-based phase (Field-Based Project Management approach).

AW
Engineering & Construction Corp.

CLIENT: ______________ CONTRACT: ______________
PROJECT: ______________ DATE: ______________
WEEK ENDING: ______________

LABOR COST REPORT

COST CODE	WORK CATEGORY DESCRIPTION	UNIT	CURRENT BUDGET			THIS PERIOD			TOTAL TO DATE			FORECAST	VARIANCE
			QTY. / RATE	MAN-HRS. / COST	UNIT MAN-HRS. / UNIT COST	QTY. / % COMP.	MAN-HRS. / COST	UNIT MAN-HRS. / UNIT COST	QTY. / % COMP.	MAN-HRS. / COST	UNIT MAN-HRS. / UNIT COST	MAN-HRS. / COST	MAN-HRS. / COST
DETAILED COST CODING	ACTIVITY DETAILED DESCRIPTIONS												

Figure 14.16 Project Labor Cost Report (P_L-DOC 57).

for field construction by combining pertinent project status and progress information into a single report. Consequently, as shown in Figure 14.15, many documents are required to perform this function. In addition to construction status and progress, current cost data in terms of budgets, forecasts, and expenditures are summarized into one field cost report. Together, the Project Labor Cost Report (P_L-DOC 57) and the Monthly Construction Progress Report (P_L-DOC 58) are generated by the project team, normally through the combined effort of cost and scheduling engineering and the key field construction managers. The most significant policies which influence the structure and content of these reports are *CRITICAL STATUS REPORTING* (P_L 31) and *MANAGEMENT BY EXCEPTION*. Each policy emphasizes the need to alert management to both existing problems and potential problems of a critical nature. Thus, these reports must identify problems by highlighting significant deviations from current plans and/or forecasts. This helps project management focus directly on the cause of problems and subsequently present timely alternate solutions for critical problem areas. Typical Field Labor Cost (P_L-DOC 57) and Monthly Construction Progress Reports (P_L-DOC 58) illustrate a means of identifying specific problem areas through comparative data analysis and the assembly of all project progress and status information into a single summary volume, as shown in Figures 14.16 and 14.17, respectively.

These documents provide field management with the information it needs to properly evaluate the current construction position, address field problems, and select possible courses of action to solve problems that are issued as management directives to the field. These decision problem areas are the main output of SUMMARY PROJECT REPORTING. As shown in Figure 14.15, management directives are released to the field for immediate implementation, while the Labor Cost Report and the Monthly Construction Progress Report are forwarded to the head office for incorporation into a total project report, and for top-level management approval. While these two documents are produced on a monthly cycle, critical problems emerging from the daily and weekly execution of field construction are resolved through the Project Management Control decision process as they arise. Consequently, in the dynamic project management process, management decisions and corresponding directives are continuously flowing to the relevant

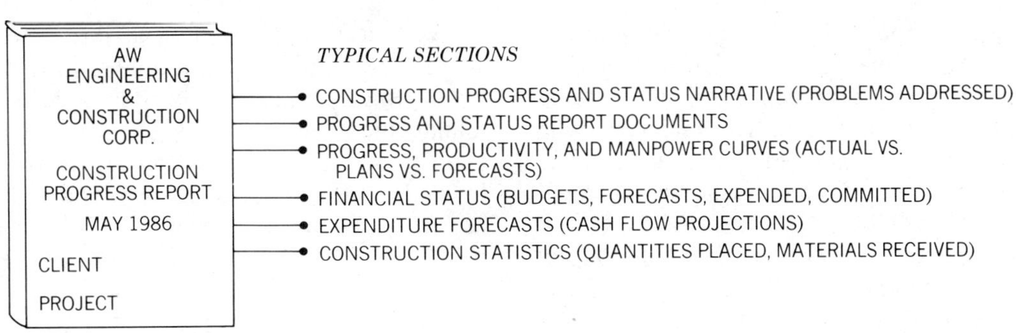

Figure 14.17 Monthly Construction Progress Report, (P_L-DOC 58).

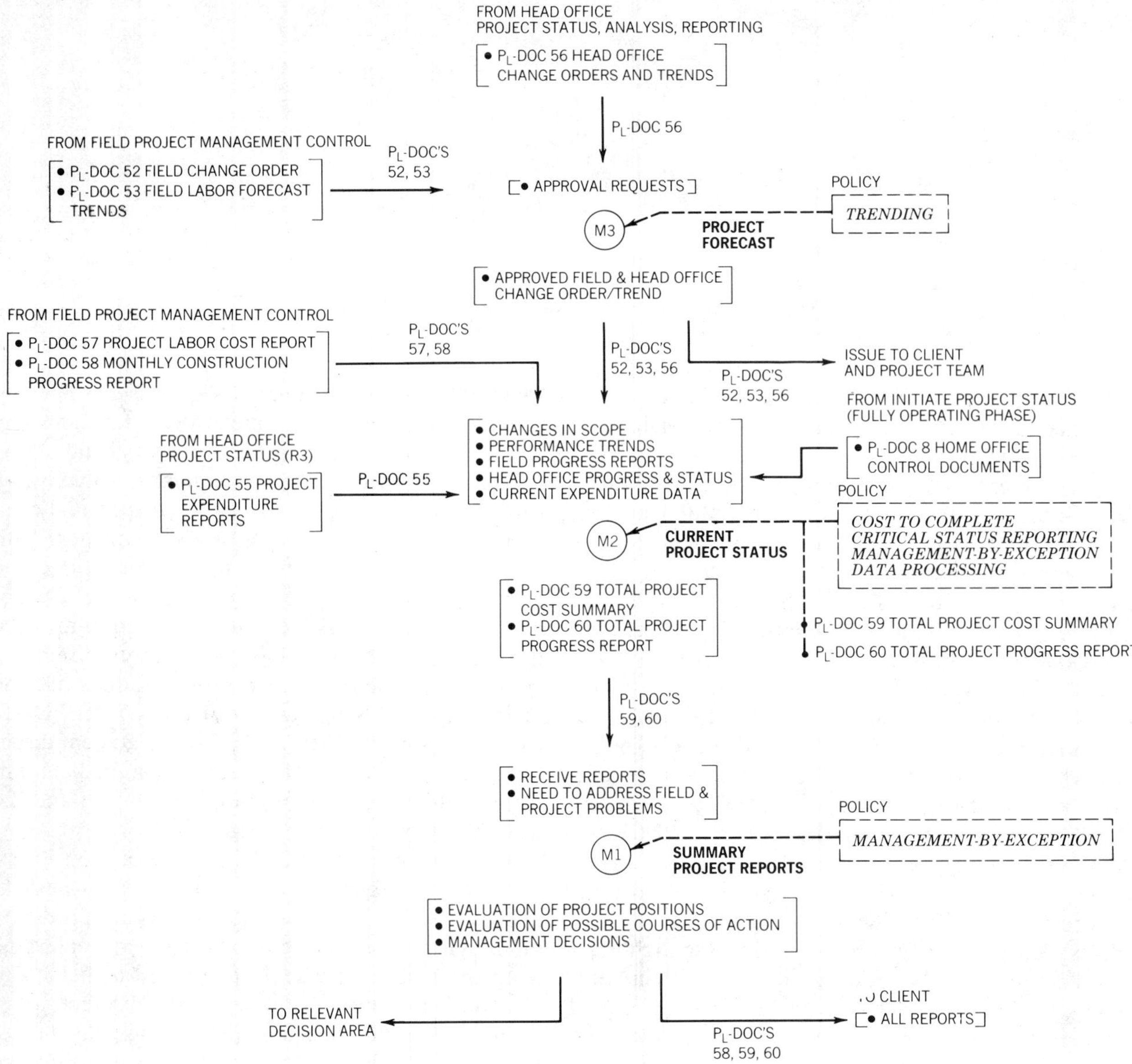

Figure 14.18 Project Management Control decision process: head office phase (Field-Based Project Management approach).

decision processes for implementation. The success of the Project Management Control effort is dependent upon the active involvement of all project team members, including the field project manager, who must receive timely project information to make correct management decisions, in the management process. Thus, *PROJECT TEAM SPIRIT* (P_L 35) becomes an important policy impacting field Project Management Control.

The Project Management Control focus in the head office is on summary level reporting of total project status and progress, as illustrated in Figure 14.18. The reports must be able to provide both top-level project management and executive company management with a quick synopsis of the current project position. Critical problems requiring top-level decision making are resolved and directives are provided to all concerned. This includes both problems relating to construction, as well as those of engineering and design, when the project is in this phase.

The three manpower management functions

AW
Engineering & Construction Corp.

CLIENT: ____________ CONTRACT: ____________

PROJECT: ____________ DATE: ____________

PERIOD ENDING: ____________

PROJECT COST SUMMARY

COST CODE	WORK CATEGORY DESCRIPTION	EXPENDED		COMMITTED		CURRENT BUDGET	CURRENT FORECAST	OVER/UNDER
		CURRENT PERIOD	TOTAL TO DATE	CURRENT PERIOD	TOTAL TO DATE			
SUMMARY LEVEL CODING	SUMMARY LEVEL DATA FOR LABOR, MATERIALS, SUBCONTRACTS, ETC.							

Figure 14.19 Total Project Cost Summary (P_L-DOC 59).

which comprise head office Project Management Control are PROJECT FORECASTING, CURRENT PROJECT STATUSING, and PROJECT SUMMARY REPORTING. These functions require the participation of the project manager of the entire project team organization, the head office project engineer, cost/scheduling engineers, and other key office construction personnel such as the head office based construction manager. The two basic documents which are generated through these functions are the Total Project Cost Summary (P_L-DOC 59) and the Total Project Progress Report (P_L-DOC 60). The project management report combines summary level status and progress data from construction and engineering with the current cost and scheduling position of the project. This report is issued to the client and to top company management on a monthly basis. Typical data contained in these two documents is depicted in Figures 14.19 and 14.20, for the Total Project Cost Summary and the Total Project Progress Report, respectively.

The head office Project Management Control phase also focuses on the final approval and issuing to the client of all project change orders and trends (P_L-DOCs 52, 53, and 56). Trends are incorporated into the control system when approved by project management and noted by the client. In comparison, change orders normally require client approval before management can fully authorize the project team to commence work on the change order scope. Once the client approves the change order, its cost and scheduling impact is incorporated into the project information system.

The key policies that influence the head office Project Management Control phase include *TRENDING, CRITICAL STATUS REPORTING,* and *MANAGEMENT BY EXCEPTION.* These policies focus on the identification of problems, and the subsequent timely alerting of management to these problems, or to potential problems which are emerging. The need to address critical problem areas leads to evaluations of possible courses of action to solve such problems, and management decisions as to the appropriate action to be implemented. At the head office level, the types of manpower management problem solutions considered are, for example, whether to extend the work week or use overtime to meet project schedules, or

Figure 14.20 Total Project Progress Report (P_L-DOC 60).

whether or not to import specialized crafts to alleviate manpower shortages. These decisions have a project focus and therefore normally incorporate top-level management personnel.

The nature and size of large projects and of the demands these projects make on company procedures and efforts (especially in the design phase and in the procurement, supply, and documentation aspects of field construction and crew planning), provide the environment in which inefficiencies, delays, rework, and so on tend to emerge. Thus, opportunities exist for management to use quality circle and problem solving team approaches to problem resolution. In this way, project management can mobilize key line and staff personnel in a way that also enhances human resource management activities and can help build project team spirit.

14.6 TOTAL FIELD AUTONOMY

At the Total Field Autonomy level, the Project Management Control decision process begins with the design and implementation of the project's Management Information System (MIS). This decision-making phase, while similar in overall focus to that of the Field-Based Project Management approach, produces a specific client-influenced and project-oriented MIS and supporting project execution procedures. The Management Information System structure and project procedures are adapted from Standard Project Operating Procedures (P_T-DOC 1) (see Figure 4.15), modified to support specific or unique client requirements and/or the contractor's need for management approaches that would be unique to the project. In addition, the effective implementation of a project designed MIS must include the authority to apply certain of its features to subcontractors under the managing contractors responsibility. This authority can only come by way of the full support and approval of the client.

In general, the Management Information System at this level has similar characteristics to that of the Field-Based Project Management approach. For instance, it is based on the selective data entry concept because in this management environment the MIS must be responsive to requests for information at any level of detail. Furthermore, the MIS has a multi-level reporting focus which is enhanced through roll-up techniques using work breakdown structures combined with the application of full cost and schedule integration. In this way, the project team can run checks and reports at many different levels to evaluate and solve problems through committee type decision-making approaches. Thus, only those problems having a project-wide consequence reach the project management level for decision making, while most other problems are solved at the point where they occur. The Project Management Control MIS is illustrated in Figure 14.21.

The management pressure brought to bear on project managers and the project team in general in the Total Field Autonomy environment is much the same as that associated with the Field-Based Project Management approach. However, two additional management pressures increase as a result of practicing in the Total Field Autonomy mode. First, due to increasing project team size and the requisite committee problem-solving approach, the project team decision-making effort becomes significantly harder to control. This is further compounded by intensified client involvement in the decision-making process. Second, the work breakdown structure, combined with work packages, produces a high volume of information to process and evaluate. This in turn increases the work load at all levels of project team involvement. If careful control is not exercised by project managers, the large volumes of data can overwhelm project team personnel to the extent that the proper and timely analysis of this data becomes extremely difficult to achieve. These management pressures are minimized through a continuous evaluation of the Management Information System and the documents it produces, in relation to their effective use in identifying problems in the management of the project.

The design and implementation of the Management Information System of Project Management Control is presented in terms of the manpower management functions, policies, and relevant documents required to perform it, as shown in Figure 14.22. The policies that influence the performance of Project Management Control address three major problem areas. These problem areas include the development of a project-oriented MIS that reflects specific client requirements, the increasing level of documents and analysis required for committee type project team decision making, and the management handling of major manpower, productivity, and motivation problems. This last leads to productivity improvement programs. For example, *PROJECT TEAM CONTROL* (P_T 2), *PLANNED ROLES*

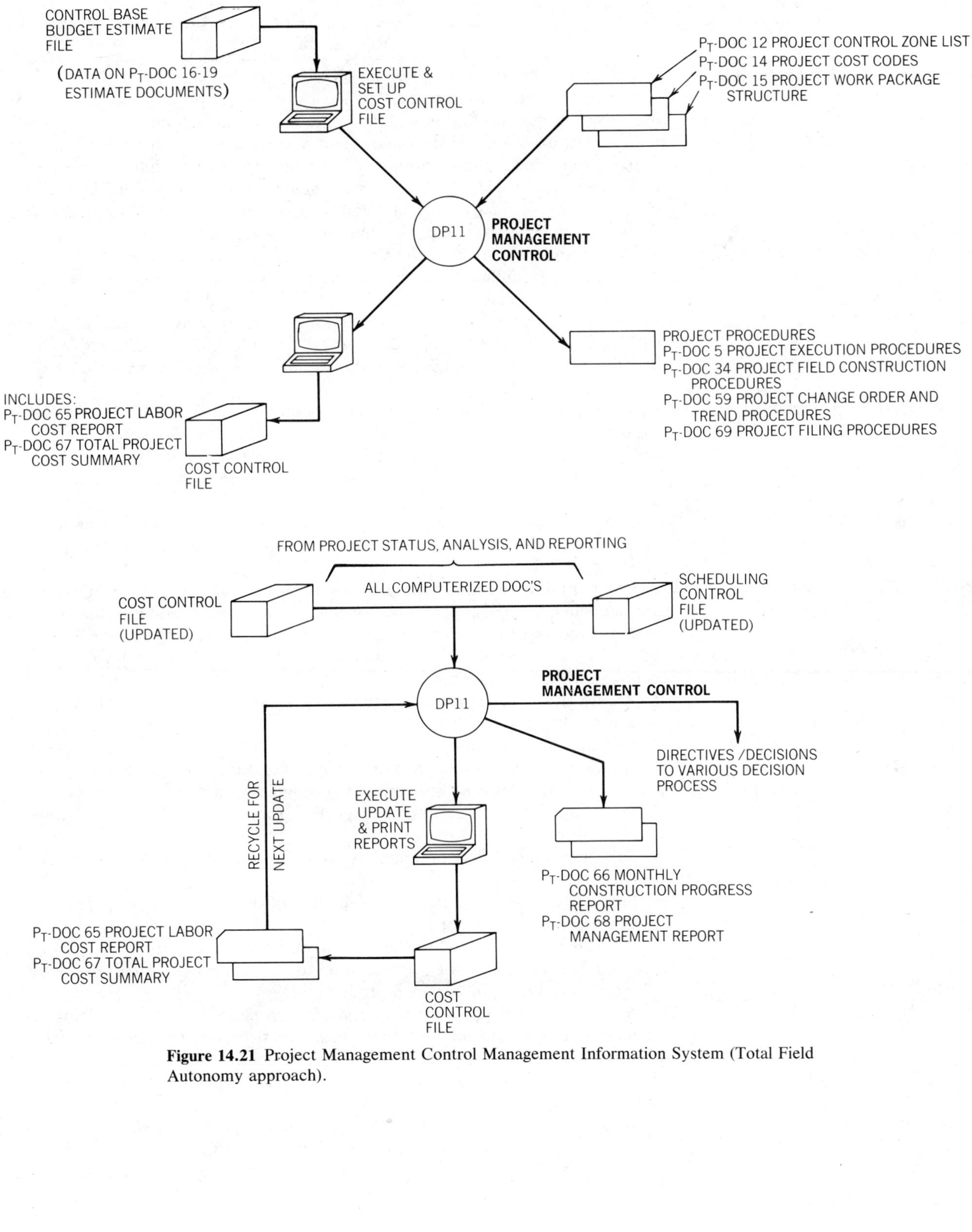

Figure 14.21 Project Management Control Management Information System (Total Field Autonomy approach).

TABLE 14.6 POLICIES OF THE AUTONOMOUS PROJECT TEAM

Policy	Description
P_T 2 *PROJECT TEAM CONTROL*	Key project team assignments, overall project team designs, and specifically designed project execution procedures, as developed for project management, are reviewed in consultation with and approved by the client.
P_T 4 *PLANNED ROLES*	The project team is a fully integrated group of managers combining their functional responsibilities through the project team structure to optimize planning techniques and provide the level of management control demanded by the client, company management, and agreed-upon project execution procedures derived from standard company operating procedures.
P_T 8 *WORK BREAKDOWN*	A work breakdown structure representative of the project scope and control philosophy is formulated to integrate costs and schedules with work packages for project management.
P_T 9 *COST CODES*	Cost codes are applied to the work breakdown structure and serve as a basis for categorizing estimated costs, reporting and controlling costs and manpower, developing a historical database, and integrating the schedule with the estimate.
P_T 10 *UNIT MAN-HOUR*	A tight but realistic evaluation of the labor content for a work task is essential for successful project planning and management control. Estimates must therefore be based on work task man-hour content in relation to work volume, construction methods, local productivities, labor availability, and site conditions.
P_T 11 *HISTORICAL DATA*	Historical labor data are basic to a realistic determination of the labor content for a work task or major project components and preliminary scheduling development.
P_T 13 *NETWORK-BAR CHART ANALYSIS*	Broad network focused milestone planning is the best approach for identifying and sequencing major project components, while a combination of the critical path method and bar chart model are best suited for project planning and scheduling at the total project view level, as well as at the daily level.
P_T 14 *SCHEDULING FOCUS*	Scheduling definitions and constraints are: 1. A critical work package is one that, if delayed, will extend the completion date of the project. 2. Work package durations must be correlated to the estimated work content. 3. All restraints must be defined and incorporated into the sequence logic (i.e., material delivery, subcontract coordination, drawing receipt, etc.). 4. Manpower impact is an integral aspect of project scheduling and scheduling control. 5. Schedules must reflect optimum use of all resources relative to the work package. 6. Progress is based on a realistic evaluation of the labor unit effort required to complete a component of work. 7. Schedules are portrayed to the daily work level and are prepared, analyzed, and reviewed weekly in relation to planned progress and productivity versus actual progress and productivities.
P_T 19 *WORK ORDER*	The work order is a key tool for field construction planning and crew sizing because it focuses on detailed estimates based on field measurements, describes construction methods, defines material and equipment mobilization, evaluates the most economical approach, and considers schedule impact, all in relation to work site conditions and the work package.
P_T 26 *PROGRESS REPORTING*	A systematic method of evaluating physical progress must be based on measuring units of work for any component work tasks, and must have the capability of rolling progress up from this detailed component level to the total project level.

Policy	Title	Description
P_T 28	*DOCUMENTED ACTIVITY STATUS*	Labor costs, manpower, man-hours, and quantities are primary ingredients to cost control and must be documented by a system yielding reliable information and viable reports for management control. These ingredients accrue as a basis for forecasting to complete at any level of detail.
P_T 30	*CREW PERFORMANCE*	Crew performance is a function of labor productivity, job conditions, and management effectiveness, and must focus on a tight but realistic performance indicator.
P_T 31	*PRODUCTIVITY*	Productivity is an essential concept and must be rationally defined for office and field utilization. An awareness of the components that affect productivity and their relation to man-hours, activity durations, and progress is essential, as is that all these integrate and are manageable.
P_T 32	*TRENDING*	A mechanism for documenting and analyzing deviations of cost and schedule consequence is an integral facet of the information system and is essential for acquiring client-authorized scope changes, adjusting budgets and schedules, and portraying the most probable man-hours and costs of the project. It also provides a means of evaluating contractor performance on a total project basis.
P_T 33	*CRITICAL STATUS REPORTING*	A construction status reporting system must provide a comprehensive report of activity status, highlighted by a mechanism to alert management to activities at and approaching critical status within a planning horizon that allows sufficient time to react.
P_T 34	*SCIENTIFIC MANAGEMENT*	Techniques of scientific management, such as work sampling, time lapse, activity analysis, multiple activity charts, and prediction models, can contribute to improving crew performance.
P_T 35	*MANAGEMENT BY EXCEPTION*	Manpower management by exception is acceptable provided that the information system is geared to exception reporting to alert management to existing problems or potential problems which are emerging within a sufficient time frame to allow a management response.
P_T 36	*DATA PROCESSING*	A computer-based information system is the most feasible means of data processing for management control.
P_T 37	*PROJECT TEAM SPIRIT*	Active participation of all project team members in the management process promotes goodwill and a positive management environment. It ensures the success of scientific management techniques and the total integration of the project team.

(P_T 4), *WORK BREAKDOWN* (P_T 8), and *COST CODES* (P_T 9),* influence the structure of the MIS in relation to client requirements. *WORK BREAKDOWN, COST CODES, WORK ORDER* (P_T 19), and *PROGRESS REPORTING* (P_T 26) are examples of policies that lead to increasing levels of documentation and subsequent problem analysis. Finally, *CREW PERFORMANCE* (P_T 30), *PRODUCTIVITY* (P_T 31), and *SCIENTIFIC MANAGEMENT* (P_T 34) can result in productivity improvement programs.

While design and implementation of the project Management Information System is heavily focused on the early stages of the project's life cycle, it is a phase of Project Management Control that is continuously under review to ensure that the project MIS is functioning properly. Thus, this phase also has a heavy field focus as the total management effort is shifted from the head office to the field and project field management control procedures are established and implemented. In this phase, all major decisions to determine the manner in which the project will be executed are made by the project team with client concurrence. These decisions are reflected in Project Execution Procedures (P_T-DOC 5) and Project Field Construction Procedures (P_T-DOC 34), as illustrated in Figure 4.24 of Section 4.6 and Figure 8.16 of Section 8.6, respectively.

*See Table 14.6, Policies of the Autonomous Project Team.

M2 CURRENT PROJECT STATUS

Policies	Documents	
PROJECT TEAM CONTROL	P_T-DOC 1	STANDARD PROJECT OPERATING PROCEDURES
WORK BREAKDOWN	P_T-DOC 5	PROJECT EXECUTION PROCEDURES
COST CODES	P_T-DOC 33	STANDARD FIELD CONSTRUCTION PROCEDURES
UNIT MAN-HOURS	P_T-DOC 34	PROJECT FIELD CONSTRUCTION PROCEDURES
HISTORICAL DATA	P_T-DOC 58	STANDARD CHANGE ORDER AND TREND PROCEDURE
NETWORK-BAR CHART ANALYSIS	P_T-DOC 59	PROJECT CHANGE ORDER AND TREND PROCEDURE
SCHEDULING FOCUS	P_T-DOC 69	PROJECT FILING PROCEDURE
WORK ORDER		
PROGRESS REPORTING		
DOCUMENTED ACTIVITY STATUS		
CREW PERFORMANCE		
PRODUCTIVITY		
TRENDING		
CRITICAL STATUS REPORTING		
SCIENTIFIC MANAGEMENT		
DATA PROCESSING		

M1 SUMMARY PROJECT REPORTS

Policies	Documents
PLANNING ROLES	
MANAGEMENT-BY-EXCEPTION	

Figure 14.22 Project Management Control decision process: design and implementation of Management Information System phase (Total Field Autonomy approach).

As project procedures are developed and implemented and the project moves forward, the fully operating phase of Project Management Control begins. The manpower management functions, policies, and relevant documents required to perform this phase are described in Figure 14.23. Unlike the Field-Based Project Management approach, fully operating Project Management Control at the Total Field Autonomy level does not reflect distinct field and head office focused phases. This is the direct result of the capability of operating an autonomous project team with full project responsibility in the field. Thus, while the fully operating phase of Project Management Control originates in the head office, it can eventually become a completely field located decision process.

Once operational, the Management Information System must process a high volume of data. This is especially due to the impact of the work breakdown structure on project execution through work packages and work orders. This problem changes the nature of the Project Labor Cost Report (P_T-DOC 65) to reflect the impact of the multi-level access feature, as depicted in Figure 14.24*a*. At the lowest level of detail, work package cost data is evaluated by those project team members who deal with the day-by-day problems of Crew Planning and Scheduling, Crew Management, and Field Monitoring. Alternatively, the construction manager and/or the area general superintendent would be interested in control zone data by major cost category. Finally, top management's interests lie with total project cost status by major control groupings. Problems identified through analysis at these various levels of cost data comparisons can then be solved by managers closest to those problems. However, top level managers can also access any level of cost detail to determine the origin of a problem, especially if it is emerging as a project-wide problem such as substantial cost overruns due to significant variations in actual craft mixes from budget or planned craft mixes.

If the Project Labor Cost Report is properly structured, it is possible to produce an alternate report option that integrates cost and schedule data (see Figure 14.24*b*). In this case, considerable *DATA PROCESSING* is required to achieve such an integration of cost and schedule by work pack-

M4 ACTIVITY ANALYSIS		
Policies	Documents	
CREW PERFORMANCE		
PRODUCTIVITY		
SCIENTIFIC MANAGEMENT		
MANAGEMENT-BY-EXCEPTION		
M3 PROJECT FORECAST		
Policies	Documents	
TRENDING		
MANAGEMENT-BY-EXCEPTION		
M2 CURRENT PROJECT STATUS		
Policies	Documents	
DOCUMENTED ACTIVITY STATUS	P_T-DOC 65	PROJECT LABOR COST REPORTS
TRENDING	P_T-DOC 66	MONTHLY CONSTRUCTION PROGRESS REPORT
CRITICAL STATUS REPORTING	P_T-DOC 67	TOTAL PROJECT COST SUMMARY
MANAGEMENT-BY-EXCEPTION	P_T-DOC 68	PROJECT MANAGEMENT REPORT
DATA PROCESSING		
PROJECT TEAM SPIRIT		
M1 SUMMARY PROJECT REPORTS		
Policies	Documents	
TRENDING		
CRITICAL STATUS REPORTING		
MANAGEMENT-BY-EXCEPTION		

Figure 14.23 Project Management Control decision process: fully operating phase (Total Field Autonomy approach).

age within the work breakdown structure for the project.

An additional feature of the Project Labor Cost Report is its unique identification of all change orders or trends that have impacted the control base budget.* This feature, often requested by the client, provides for a track record of all changes to the Control Budget Estimate (P_T-DOC 19) and keeps the client continuously apprised of the project's total cost, both in terms of the current approved budget and the current forecasted indicated total cost. This approach is depicted in Figure 14.24*c* on the Project Labor Cost Report, and is based on a *TRENDING* (P_T 32) policy.

Cost data from the Labor Cost Report is combined with construction progress and status to produce a Monthly Construction Progress Report (P_T-DOC 66). This report, while similar in format and structure to the Monthly Progress Report, (P_L-DOC 58) of the large firm, normally contains selective additional information pertaining to factors that are influencing labor performance. For example, as shown in Figure 14.25, the status of a key work sampling program is reported with specific results identified (15% welder rejection rate) and recommended solutions to problem areas stated as action items in the next reporting period (starting welder training program). These types of evaluations further strengthen the importance of the Monthly Construction Progress Report as a tool to expose the effects of critical management decisions on construction progress and status.

A similar type of presentation is developed for the project in its entirety through the Total Project Cost Summary (P_T-DOC 67) and the Project Management Report (P_T-DOC 68). Since these reports are primarily directed at high-level executive management (including the client), they expose extremely critical project-wide problems, and the appropriate solutions to those problems, as required.

* See P_T-DOCs 60, 61, and 64 for examples of change orders and trends.

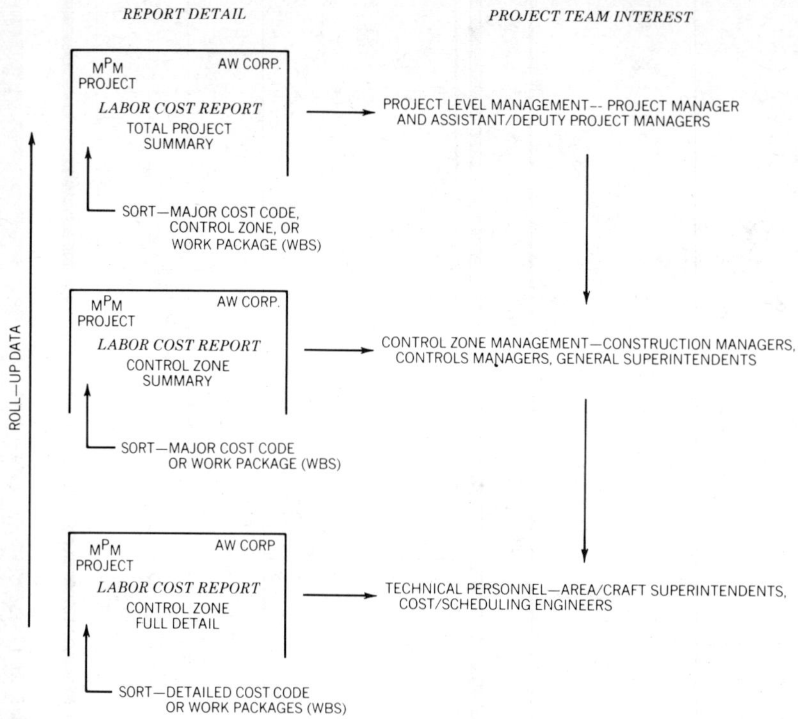

(a)

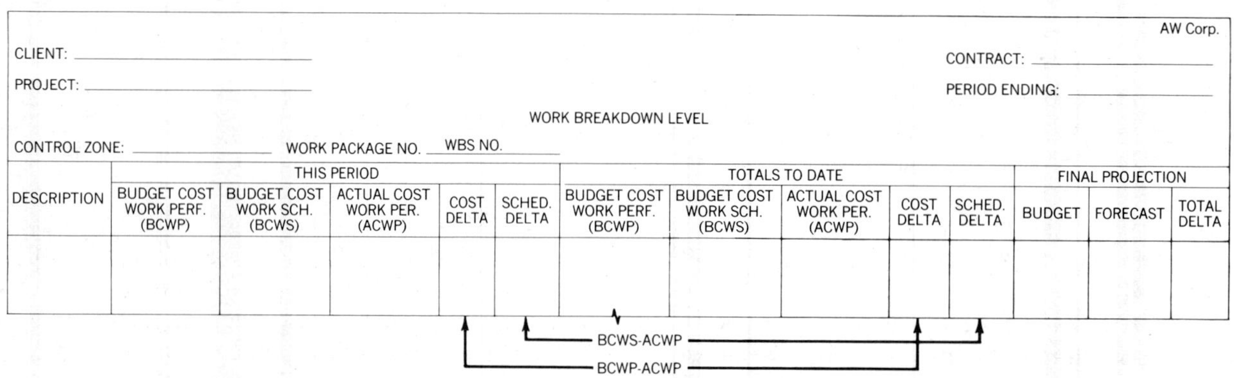

(b)

AW Corp.

CLIENT: ____________ CONTRACT: ____________

PROJECT: ____________ DATE: ____________

PERIOD ENDING: ____________

PROJECT LABOR COST REPORT
DETAILED COST CODE

COST CODE	WORK CATEGORY DESCRIPTION	MAN—HOURS			WAGE RATES			COST		
		EXPENDED TO DATE	CURRENT BUDGET	CURRENT FORECAST	EXPENDED TO DATE	CURRENT BUDGET	CURRENT FORECAST	EXPENDED TO DATE	CURRENT BUDGET	CURRENT FORECAST
	WORK ITEM		10000	10000		15.00	15.00		150000	150000
	CHANGE ORDER X		1000	1000		15.00	15.00		15000	15000
	TREND Y			500			16.00			8000
	SUBTOTAL	2000	11000	11500	14.00	15.00	15.04	28000	165000	173000

COST CODE SORT { COST CODE OR WORK PACKAGE OR SYSTEM

(c)

Figure 14.24 Project Labor Cost Report (P_T-DOC 65). (*a*) Multilevel access of labor cost data. (*b*) Report sort by work breakdown structure integrating cost and schedule. (*c*) Report sort showing impact of change orders and trends.

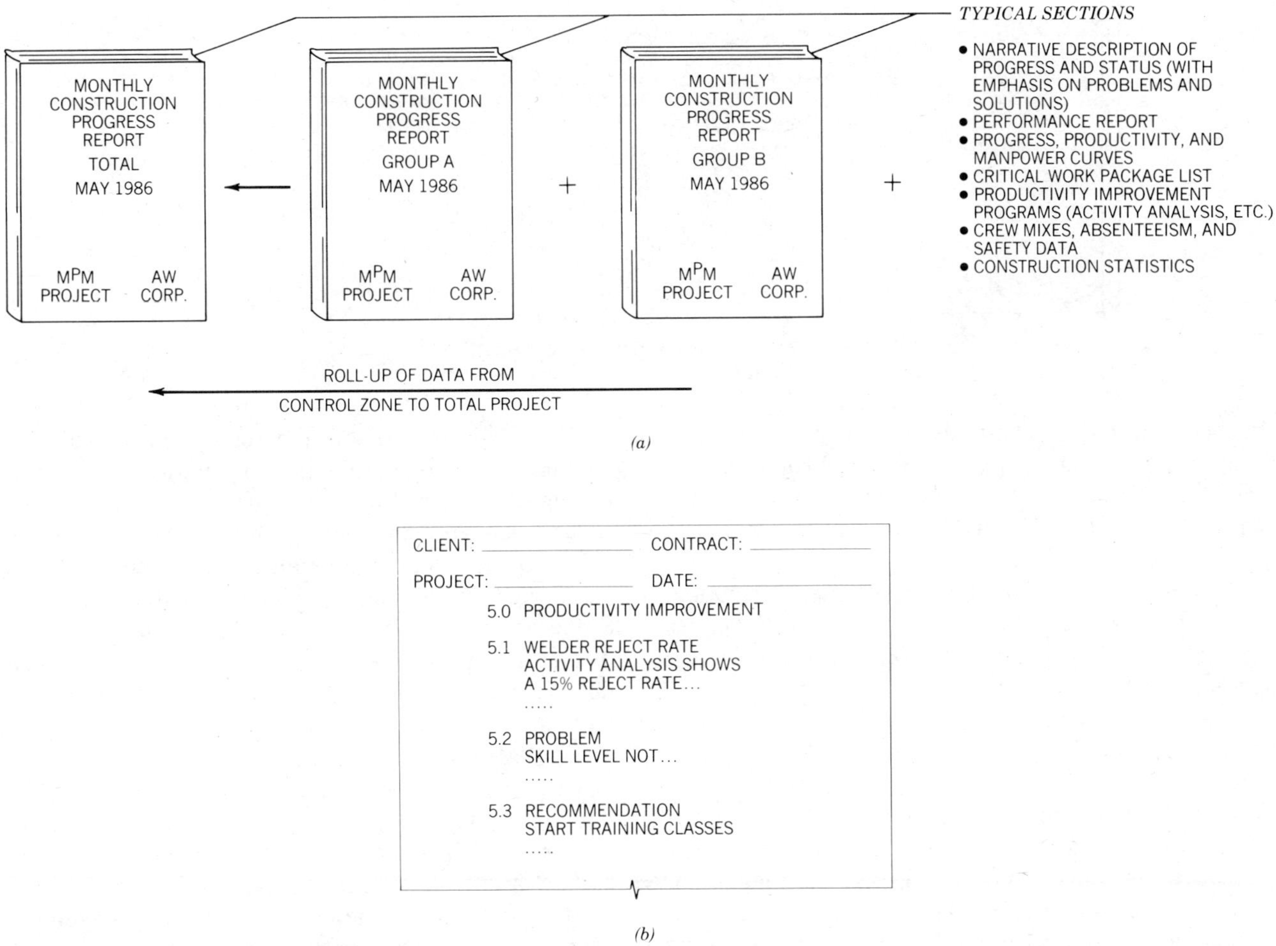

Figure 14.25 Monthly Construction Progress Report (P_T-DOC 66). (*a*) Report scope and contents. (*b*) Typical section with problem identification and recommended solution.

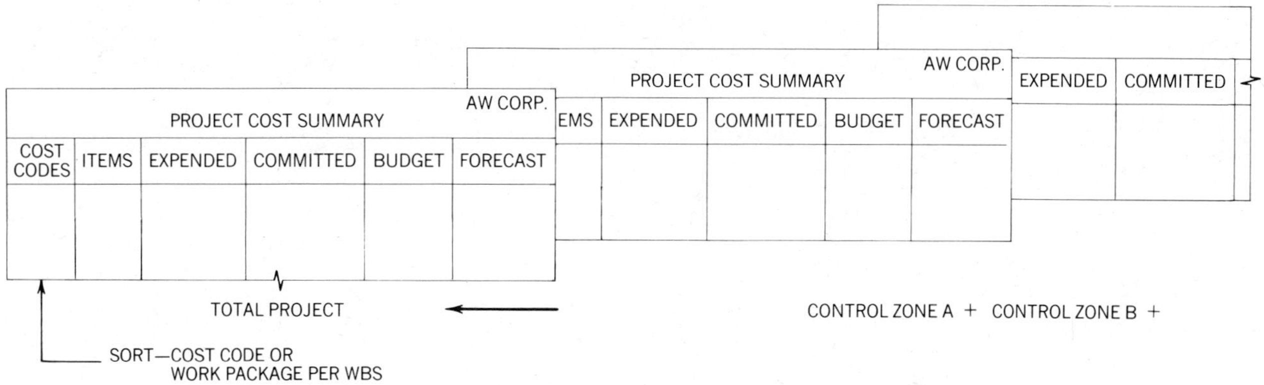

Figure 14.26 Total Project Cost Summary (P_T-DOC 67).

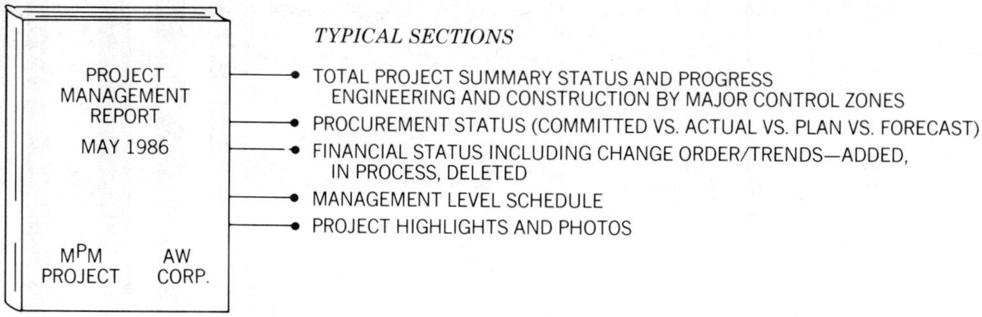

Figure 14.27 Project Management Report (P_T-DOC 68).

In this way, top level client, company, and project management is kept apprised of the project's cost and schedule status. Typical examples of these two reports are illustrated in Figures 14.26 and 14.27, respectively.

Critical problems related to labor performance, as identified in the Monthly Construction Progress and Project Management Reports, may lead to decisions regarding the initiation of major project productivity improvement or pay incentive programs if these types of programs are not currently operating. The decision to commence such programs *must* have a clear and decisive commitment and support of top-level management as well as from the project management staff. Thus, the Project Management Control decision process is the decision area responsible for producing the key decisions regarding these productivity improvement programs.

The resolution of human resource management problems on large projects is often difficult, requiring extensive management efforts and specialized techniques and programs. Thus, it is common to find project management initiating extensive work sampling efforts and specialized method studies. Since field work forces may be large, management often implements special programs such as foreman delay reporting and crew questionnaires in order to gain an understanding of worker dissatisfactions and/or attitudinal problems. This is an initial step towards effective resolution of work force problems. Because the Total Field Autonomy approach is characterized by client involvement in project management, the client participates in decisions regarding newly designed productivity improvement programs, and frequently participates in the implementation and operation of these programs.

CHAPTER 15
Historical Data Retrieval

Historical Data Retrieval is concerned with the acquisition, analysis and maintenance of reliable, up to date, historical information that relates to construction projects typically performed by the contractor. This is the decision area where the contractor must determine the types of historical data needed, the procedure for gathering these data, the method of storing and analyzing these data, and the documentation methodology which will permit the data to be released to the potential user. Historical Data Retrieval has a project focus when servicing the needs of the ongoing project, and a company focus where historical data is required for estimating, scheduling, and other related project definition functions.

In the construction industry the compilation and analysis of project information for historical reference purposes ranges from simply storing actual cost data and other relevant project information in its final form at project completion, to the full scale analysis, compilation, and seeking out of data for future use. The latter focus would involve statistical methods of data analysis, the development of a formalized database (often computerized), and the special documentation and referencing of critical information for ready use by decision makers. Some examples of this are standard unit man-hour tables or cost curves for capacity type estimating.

The Historical Data Retrieval decision process addresses a number of critical management problems which focus on the need to standardize active project information, to distill project information from previous projects, and to make such information easily available to those who need it. This requires establishing, updating, and reevaluating the basis, structure, and quality of historical information used for estimating and scheduling future projects. The resolution of these problems often bears directly on the quality of project information generated from the historical reference base and, therefore, impacts the contractor's rate of success in obtaining new projects.

Retrieving historical data is often viewed in the construction industry as a necessary evil rather than a decision area that can provide meaningful information to the contractor in the operation of the business. Considerable thought needs to be given to the form in which data is kept, however. This data structure problem is very critical, especially when a computerized database structure is being designed. The specific form of the database will affect its ability to rapidly generate processed data reports for management decisions in the future. Consequently, future needs must be considered before a database structure is chosen and implemented by the contractor. For example, a critical management problem occurs when the contractor is involved in a construction contract dispute or litigation. In this instance, the methodology with which the contractor retrieves and documents information for a project under construction may prove critical to the favorable resolution of the construction dispute. Consequently, the data retrieval problem area for an active project must focus on the documentation

TABLE 15.1 SCOPE OF HISTORICAL DATA RETRIEVAL PROBLEM AREA

Major Decision Problem	Component Problem Statements
57. What data should be kept:	How much data? What range of data? How useful and accurate are these data? Who will want these data? When are we likely to want them again?
58. How will data be kept:	Should all raw data be kept just as it comes? Should the data first be processed? What database structure should we accept?
59. How will source data be gathered:	Where is data source? How can these sources be tapped? What data gathering procedures should be implemented? Who should gather data? When should data be picked up?
60. How will data be analyzed:	Should historical data be tabulated? What historical tabular data would be useful? When should statistical data analysis be used? What data best fits statistical analysis? Who will perform this analysis?
61. How will data be released:	Must we keep original data? How will processed data be documented for use? Can master records be released? Will microfilm be used? What form is necessary or relevant?

of change orders/trends and analysis of critical cost and schedule impacts, changes in productivity, and the like.

15.1 THE BASIC DECISION PROCESS

The Historical Data Retrieval decision process has three basic segments. First, it is necessary to establish project (and company) requirements for data and, once this has been done, to design a suitable database that is economical in size while meeting these requirements. Second, once the database system has been implemented it is necessary for the duration of an active project to initiate and maintain a suitable data gathering and storage system for the project. Concern must also be given to the subsequent retainage, condensing, or dumping of final project data to suite long-range company requirements. Finally, it is necessary to initiate and develop suitable data analysis studies for both active projects and other projects to either enhance the understanding of project or company decision makers faced with specific project problems, or to use as specialized input to other management decision processes. These three basic segments are addressed by the major decision problems and component problem statements of the problem area, as shown in Table 15.1.

Any project database must address the legal requirements to maintain employment, payroll, and tax records for each employee, and also, in the case of accidents and safety violations, to produce whatever documentation is required. It is also prudent to keep labor and manpower records that may impact anticipated project problems such as cost overruns, disputes, and changes in scope. Finally, it is essential that the company maintain records that enhance the speed and accuracy with which labor and manpower questions on future projects can be evaluated.

TABLE 15.2 GENERAL PROCEDURE FOR HISTORICAL DATA RETRIEVAL

Procedural Step	Management Duties and Tasks
1.	Evaluate the types and range of labor data to be stored as historical reference information.
2.	Determine the method of storing historical labor data information and keeping database current.
3.	Assess methods of analyzing historical labor data.
4.	Gather source data.
5.	Compile, analyze, and document historical labor data.
6.	Develop means of accessing processed historical labor data.

The general procedure for the Historical Data Retrieval process in terms of sequential management duties and tasks, is presented in Table 15.2. The functional plot of the manpower management functions that address these management duties and tasks is shown in Figure 15.1. The corresponding functional statement of Historical Data Retrieval is shown in Figure 15.2, the details of which should be compared to the scope statement for the problem area in Table 15.1. The nature of the relationships

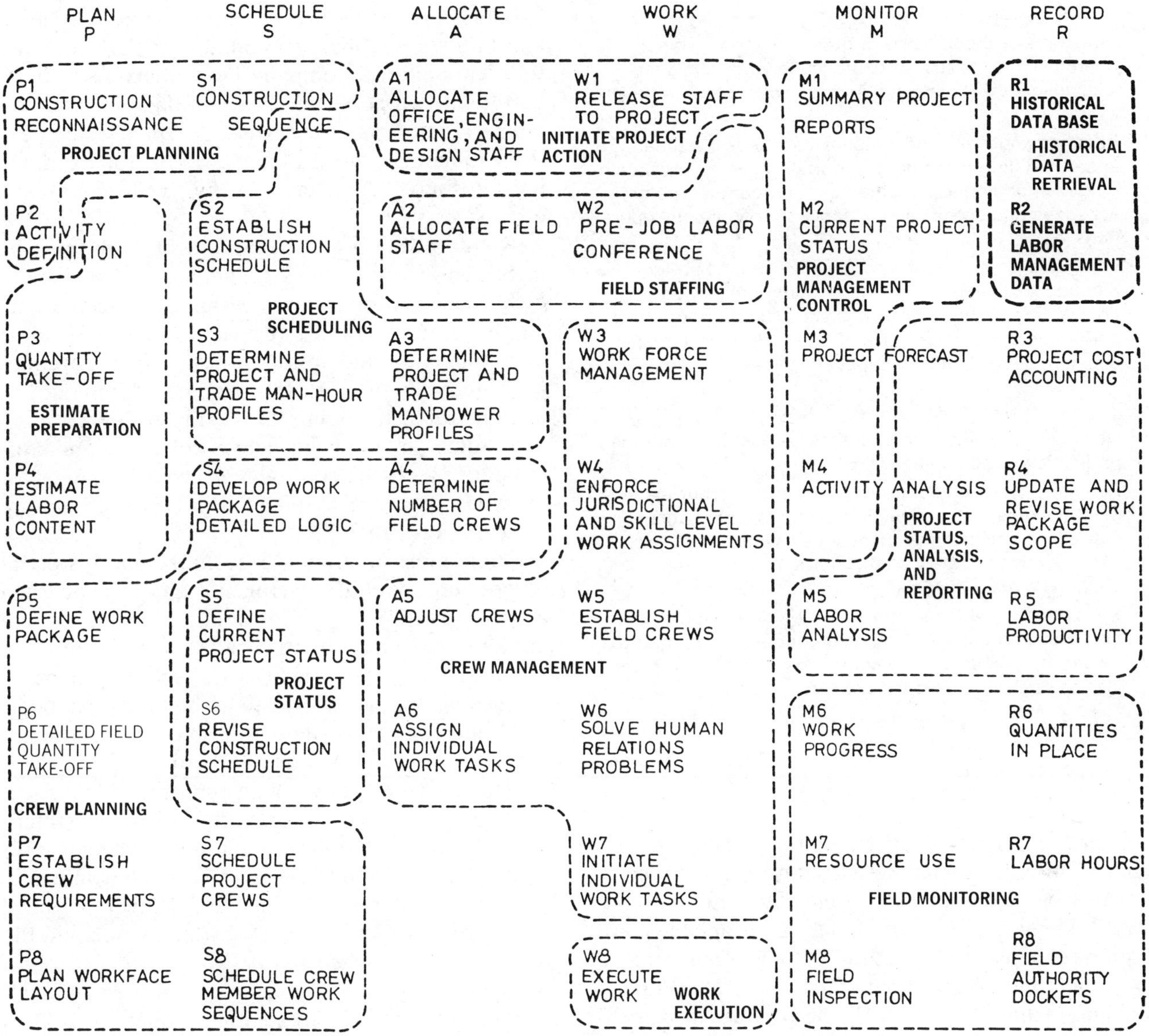

Figure 15.1 Functional plot of the Historical Data Retrieval decision process.

between the Historical Data Retrieval decision process and those other manpower management decision processes with which it may interact over the life of the project are shown in Figure 15.3.

15.2 FUNCTIONAL PERFORMANCE TECHNIQUES

In the construction industry, Historical Data Retrieval can have both a project and a company management focus. The extent to which the project-focused Historical Data Retrieval expands beyond the systematic compilation and maintenance of an up-to-date project database to include data analysis is dependent upon the company's success at retrieving from the client those costs necessarily incurred in project data analysis. Because clients are very reluctant to absorb these types of costs, most data analysis is charged to company overhead. Thus, the extent to which it is performed is correlated closely with available capital resources and top management's desire to improve, for example, their estimating, scheduling, and crew planning, and capabilities for future work. A natural problem which emerges in this decision framework occurs

R1 HISTORICAL DATA BASE

- Normalizes a broad range of project data to a standard reference base for use in conjunction with estimating, scheduling, crew planning, and project control and management.
- Requires knowledge of company cost codes, estimating and scheduling techniques, reporting systems, crew planning methodologies, and the basic structure of the project scope and specifications common to company operations.
- Considers labor data and documents from all projects, the nature of each project, its location, unique job conditions, relative labor efficiencies and productivities, crew performance indicators, size of the work force, its composition, and scheduling impact changes.

R2 GENERATE LABOR MANAGEMENT DATA

- Generates specific labor data that are useful at both the project level and company level.
- Requires knowledge of project cost codes, labor management documents, construction methods, union agreements and jurisdictional dispute decisions, characteristics of productivity, project craft profiles, and statistical methods of data analysis.
- Considers actual labor man-hours, unit man-hours, labor unit costs, progress curves, man-hour curves, manpower curves, project scope, job conditions, and schedule delays as each correlates to basic project performance.

Figure 15.2 Functional statement of the Historical Data Retrieval decision process.

when the project database (often client-oriented) has a different focus and structure from the company's desired database. While standard cost coding systems aid in overcoming this problem, there often remains a significant effort when normalizing a project database for data analysis to fit company requirements. In any case, the performance technique concepts presented herein are equally applicable to both the project- and company-focused Historical Data Retrieval decision area. The fundamental basis of Historical Data Retrieval is that historical records and data are useful. The issues relevant to this basic position focus on determining the range of data to be kept and the means of storing these data. These basic considerations apply to both project and company perspectives on data storage and manipulation, and result in historical information being collected and made readily available to project team personnel and/or company personnel. In particular, some obvious data collection and storage facilities are a project filing room, a company filing room, and libraries related to departments such as cost estimating and control, and data handling processes involving microfilm equipment and film storage files. In addition to these broader focused facilities, company individuals often maintain their own personal library of data normally related to their area of expertise. A typical example is the estimator, who probably would, for example, have resource data on labor unit costs or typical crew sizes.

Data collection and storage techniques must be integrated with process techniques to systematically locate historical data within the selected data storage framework. In respect to specific project data collection and storage, the use of filing codes for categories of work is a common technique, as demonstrated in Figure 15.4*a*. In terms of a company focus, techniques such as collecting and storing data by client, contract, or project number, or by type of project as reflected in Figure 15.4*b*, are typical approaches.

While most contractors recognize that historical records and data are useful, not all are concerned with the possibility that unsupported historical records and data may be misleading. For example, a file of historical labor unit costs would perhaps be misleading without reference to the project number, type, and completion date, and so on, from which it is derived. This additional information aids the decision maker when selecting an appropriate labor unit cost for the project the contractor might be estimating. A more penetrating technique related to preventing storage and retrieval of misleading information is described in Figure 15.5. Several typical items are listed in Figure 15.5 which would aid the decision maker when using historical data from a particular project. For example, consider the decision maker who is interested in productivity curves for a given location. After reviewing the historical files, he or she selects a project that was recently completed in the location of interest. The decision maker can then extract from the project data files the actual experience productivity curves for the project. Before applying this information to another active project, however, the decision maker should examine any unusual circumstances impacting productivity, such as a major acceleration to complete the project ahead of schedule. It would also behoove the decision maker to comprehend the relationship between the original estimated site productivity as predicted against the standard labor estimating base, and the actual field experience. This information must be included in the historical data file, as shown in Figure 15.5, to prevent the

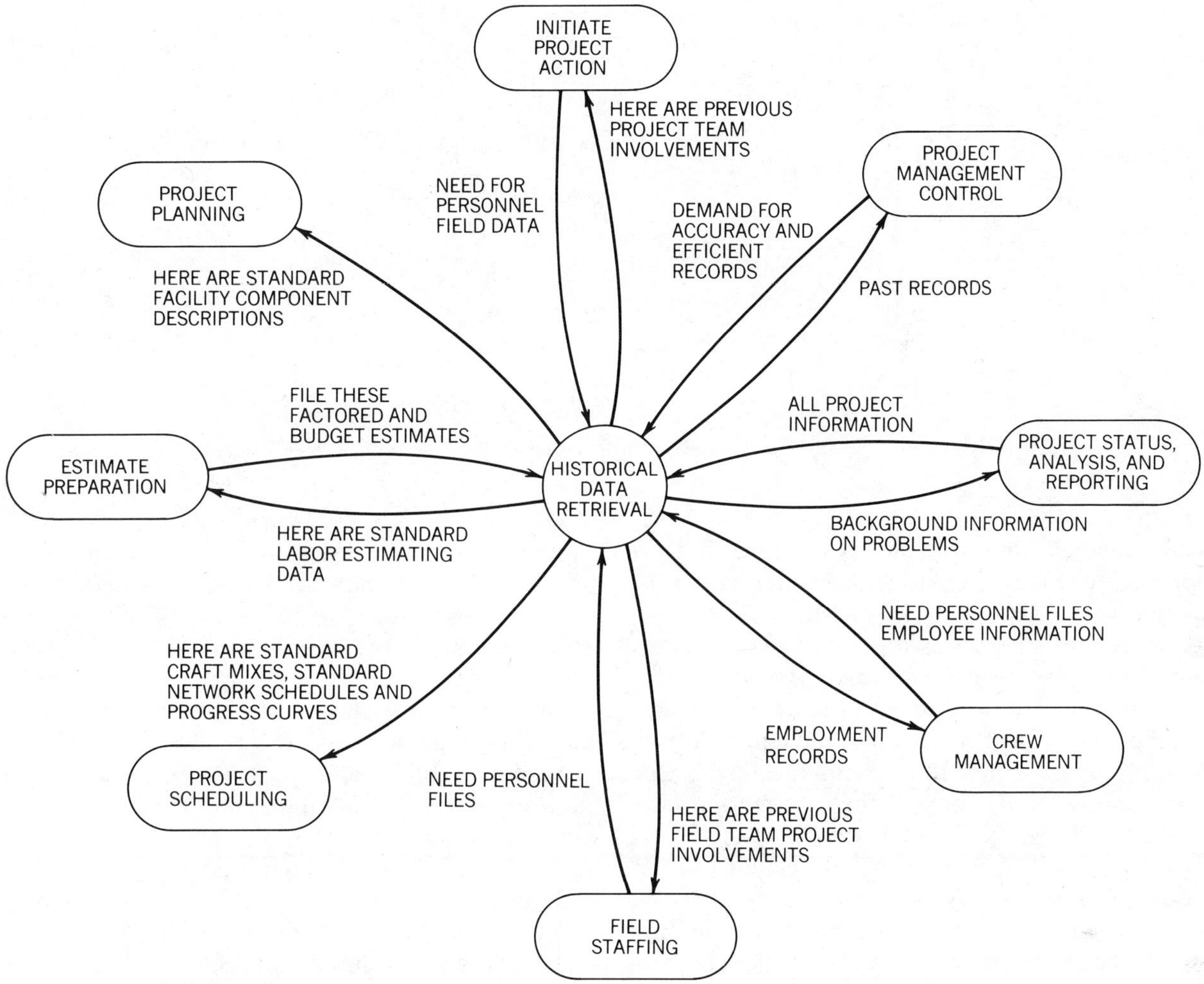

Figure 15.3 Schematic illustration of Historical Data Retrieval decision process interfaces with other manpower management decision processes.

CONTRACT 652108 WOODSON CHEMICALS PLANT	
FILING CODE	CATEGORY
200	ESTIMATES
300	SCHEDULES
400	COST REPORTS
500	CHANGE ORDERS
⋮	⋮

(a)

PROJECT TYPE	PROJECT NO.	CLIENT
OFFICE	421	JONES & ASSOCIATES
"	428	ALBERT & ALBERT
"	436	F.P. BARCLAY & SONS
⋮	⋮	⋮
WAREHOUSE	440	MACK INDUSTRIES
"	442	COMSTOCK COMPANY
"	460	A & B RENTALS
⋮	⋮	⋮

(b)

Figure 15.4 Historical data filing techniques. (*a*) Project-focused filing system by filing code and category. (*b*) Company-focused filing system by project type, number, and client.

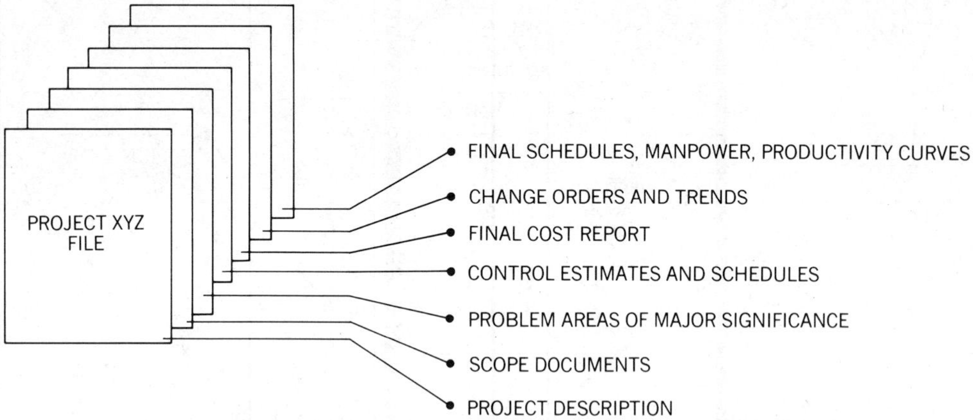

Figure 15.5 Typical project data file for background and historical reference.

decision maker from drawing conclusions that are inappropriate to the situation, and therefore misusing the historical productivity curves.

At the client's request a formal final project cost report might be generated which would contain similar data to that described in Figure 15.5. However, the final project cost report would probably be structured to meet the requirements specified by the client. Consequently, not all the information contained in it would be useful to the company in the format presented. A typical final project cost report is shown in Figure 15.6.

The gathering of project historical data in support of the concepts illustrated by Figures 15.4 through 15.6 is often accomplished via procedures either initiated specifically for an active project, or determined as standard company operating policy for all projects. The gathering of a broad range of various types of data, much of which is based on differing project conditions, raises the concept that the normalization of data to a common reference base is essential for comparative use. This approach mainly has a company focus, but it can also bear on the methods of historical data retrieval chosen by a specific project team for any given project. In the construction industry, a range of normalization techniques emerges, including cost indices, standard progress curves, standard manpower profiles and craft mixes, statistical analyses, and standard network modules, just to mention a few that are found in practice. Thus, the company standard cost code system, for example, is an effective means of collecting and systematically storing data for future use. This technique is illustrated in Figure 15.7*a* and 15.7*b*, wherein labor cost data are compiled by cost code as extracted from a range of different projects. The cost code technique provides a common reference base for the company as a whole, whereby the process of locating similar historical data over a range of projects becomes a less complicated effort. Another technique for normalizing data is to develop man-hour per unit charts by major work categories such as concrete, as illustrated in Figure 15.7*c*. Finally, cost curves can be developed through statistical analysis of standardized project data, as reflected in Figures 15.7*d* and 15.7*e*. These

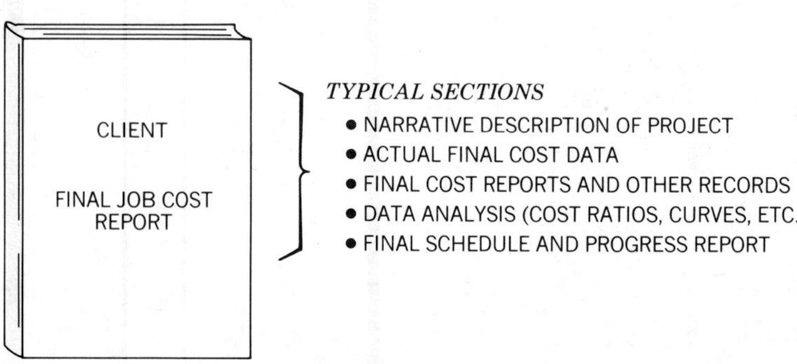

Figure 15.6 Typical final job cost report for a project.

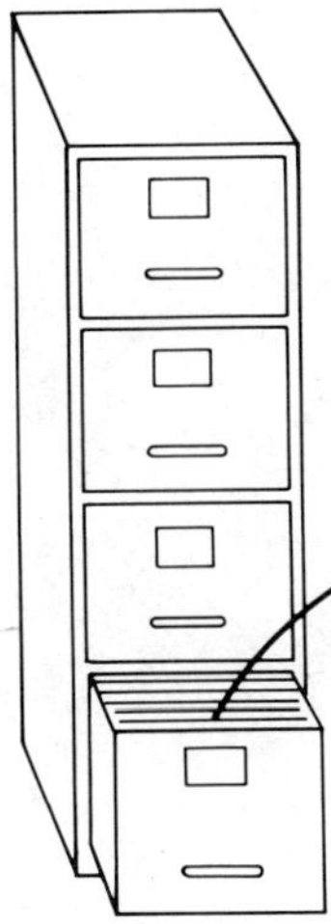

FILE CODE: 400					
SUBJECT: CONCRETE					
COST CODE	WORK CATEGORY	JOB NO.	TYPE	DATE FINISHED	LABOR UNIT COST
210.1	FOOTING FORMS	436	OFFICE	NOV 85	.66 SF
210.1	FOOTING FORMS	442	WAREHOUSE	DEC 86	.75 SF
210.1	FOOTING FORMS	435	WAREHOUSE	JUL 86	.78 SF
210.1	FOOTING FORMS	475	BANK	DEC 86	.82 SF

(a)

COST CODE: 322.0								
WORK CLASS: CONCRETE: COLUMNS								
		ESTIMATE			ACTUAL			
PROJECTS	NO.	QUANTITY	LABOR UNIT COST	LABOR COST	QUANTITY	LABOR UNIT COST	LABOR COST	ACTUAL MAN-HOURS PER QTY. UNIT
OFFICE BUILDING BANK								

(b)

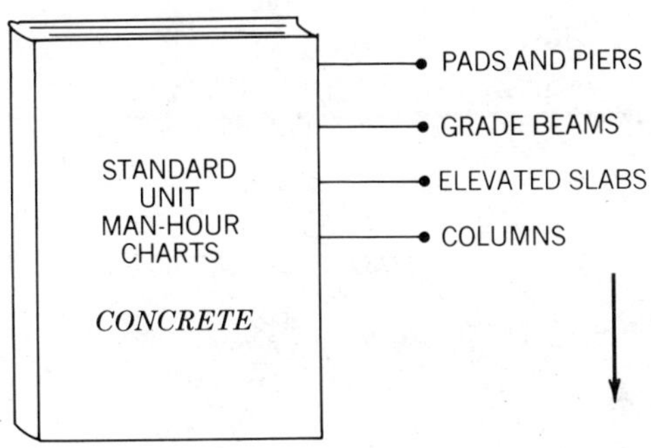

(c)

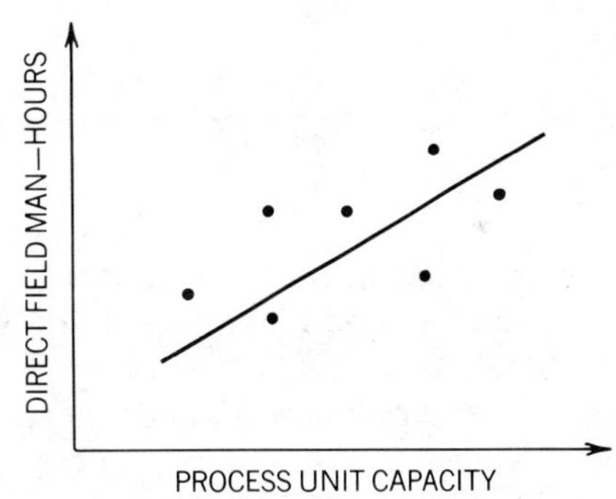

(d)

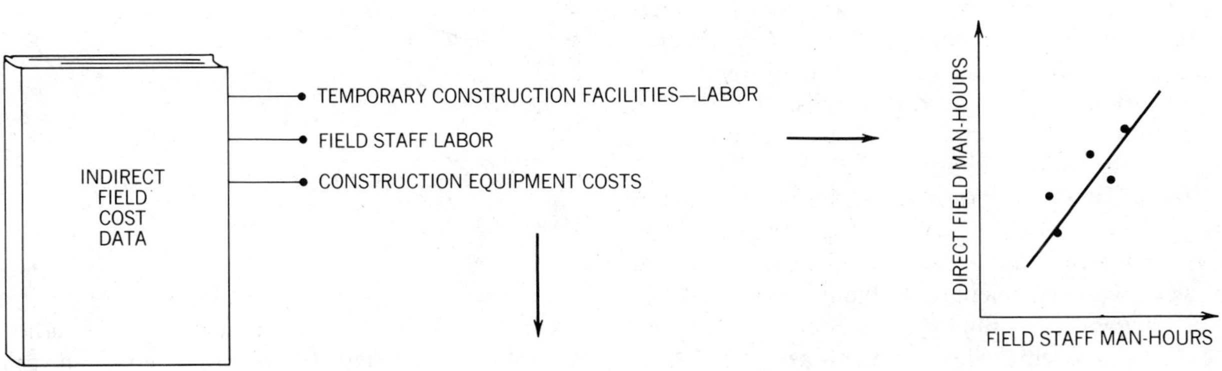

(e)

Figure 15.7 Historical data for estimating. (*a*) Labor unit cost data. (*b*) Labor cost comparisons. (*c*) Man-hour per unit chart. (*d*) Cost Curve. (*e*) Indirect field cost curves.

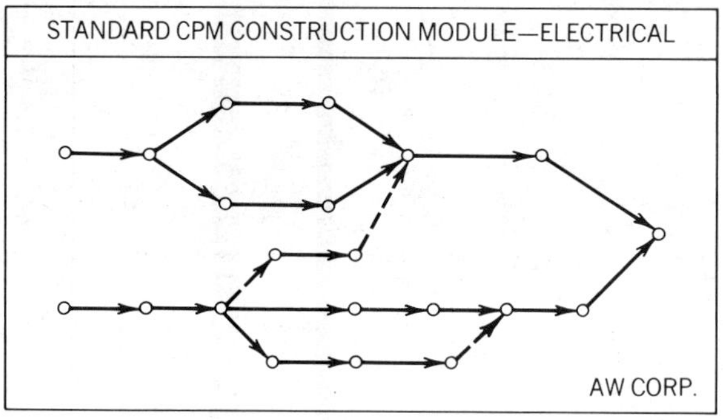

(a)

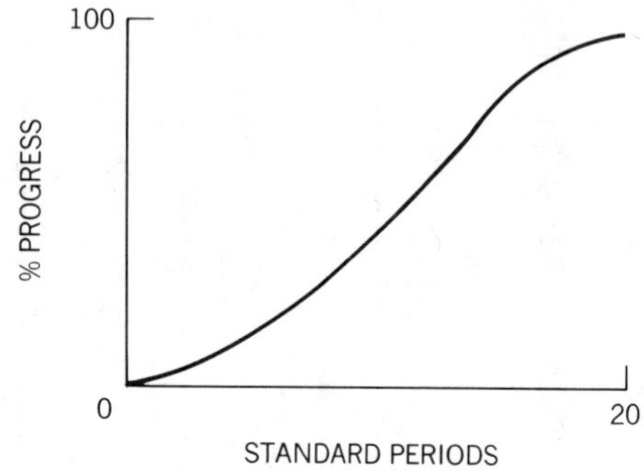

(b)

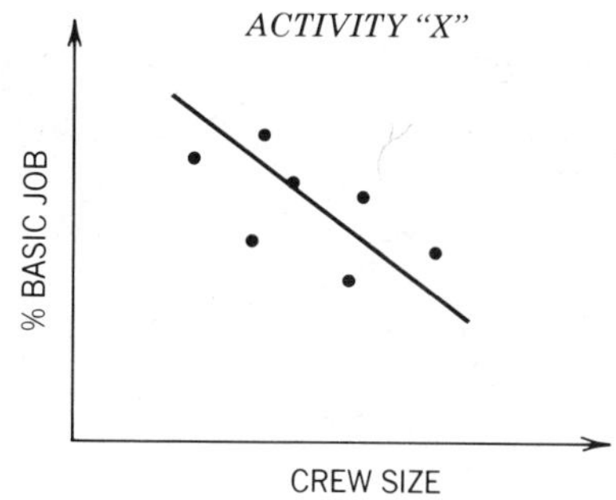

(c)

Figure 15.8 Historical data for planning, scheduling, and crew sizing. (*a*) Standard CPM logic diagram for work category. (*b*) Standard progress curve. (*c*) Crew size analysis.

types of curves are especially useful for order of magnitude estimating when limited scope information is available.

Approaches for normalizing data can also be applied to data from other decision processes such as Project Scheduling and Crew Planning. For example, standard construction network modules and standard construction progress curves can be developed for project scheduling as shown in Figures 15.8*a* and 15.8*b*, respectively. The analysis of crew size in relation to the percent of basic job (that is productive working time) can be used in crew sizing, as illustrated in Figure 15.8*c*.

Historical Data Retrieval can be helpful in the project management of an ongoing project. In this environment, historical data study has the potential for alerting management to problems. This aspect of data retrieval is normally associated with special areas of management interest such as obtaining or verifying standard labor unit man-hour rates. Historical data studies are prepared using such techniques as curves, comparisons, regression analyses, and other statistical analysis methodologies. Figure 15.9 provides an example where man-hours are plotted against the quantity installed for a unique work task. The man-hours expended and quantity installed could be based on weekly totals. The slope of the "best fit" curve would be the ac-

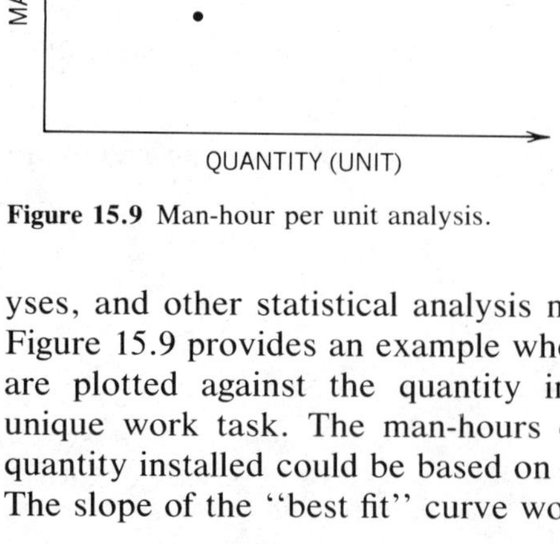

Figure 15.9 Man-hour per unit analysis.

tual site man-hour per unit for the work task. When this information is compared to the standard man-hour per unit for the same work task, an indicator of the efficiency of the site-based man-hour per unit in relation to the standard man-hour per unit is established. If the actual efficiency is different from the efficiency established for the estimate, then a potential deviation is identified and a project-wide trend may be required. Further investigations would be required prior to a final decision, employing work sampling or activity analysis techniques to verify the results.

The database concept in which information can be stored, sought, related to, and copy accessed in a logical systematic manner, is essential when large volumes of related data are involved, and/or when special file sets of data need to be set up and manipulated. In order for this to be possible some common field portion of the informational record on any item must be compatible with the relevant key file portion(s) used to establish the database structure. The structure of the database is dependent upon the number and size of the key file components involved and their demands, or upon the format of both input-stored and output-data handled, and how independent data files can be related to the main data structure. In this way, a variety of data structures emerge in practice.

The most basic database is the fully fixed word form in which there are a fixed number and size of fields. In this database the input structure is fixed, like that shown in Figure 15.7*a*. A second type of database is the fully identified single key field structure wherein a single key field(s) is defined and the remaining fields are available for any data. These remaining data fields can be structured to indicate the type of data that follows, such as labor, materials, and the like. For example, a cost code may be the single key field identifier with all other data defined by the user. A typical example of this is illustrated in Figure 15.7*b*, where the cost code reflects concrete column construction and the remaining data structure reflects types of construction and, in this example, labor related unit costs and man-hours per unit of measure. A third database structure is the relational database in which two or more distinct databases are linked together by a *user-defined common key field*. A typical example of the relational database concept is presented in Figure 15.10. This approach allows the user to formulate the database in a free format, with as many interlinked data sets as desired. It is an especially powerful tool for building a project database suited to a specific set of project requirements.* Finally, data may also be structured to meet regulatory requirements, such as those for quality control of welding. In terms of manpower, this data structure would correspond to documenting welder certification, number of welds completed, location of welds, welder rejection rate, and so on. This is an enormous problem on some large projects with a considerable piping construction content.

The primary objective of Historical Data Retrieval is to satisfy the need for a structured database geared to feeding the contractor's Management Information System. This database can range from a simple manual filing system for each project to a comprehensive computer-based data management system. Potentially, Historical Data Retrieval also has a multi-level focus wherein, on the one hand, data is retrieved and stored for company use, while on the other hand, data is retrieved and stored by a project and used throughout the project's life cycle. These MIS concepts are illustrated in Figure 15.11. The nature and structure of the database depends on the contractor's methods of operating and is generally tied to management control concepts and cost code techniques. These databases are manually maintained but are more frequently computerized for ease of access and the ready accumulation of larger volumes of data.

15.3 THE PROJECT TEAM COMPONENT APPROACH

The data retrieval system of the small firm primarily consists of a historical data file of previous projects. The process of generating labor-oriented data simply consists of collating project labor cost reports as jobs are finished, and filing them in a systematic manner. In this capacity, the files become the firm's information system for Historical Data Retrieval, and provide the documented reference source labor cost data for preparing future estimates. The action of terminating a project in itself initiates the labor data retrieval process. When Project Cost Recapitulations (P_S-DOC 12) reflect actual total costs for each activity, this document and all supporting documents become part of the firm's historical data-

*Many software programs are now available that utilize the relational database concept for specific project application. Some of these programs are identified in the bibliography.

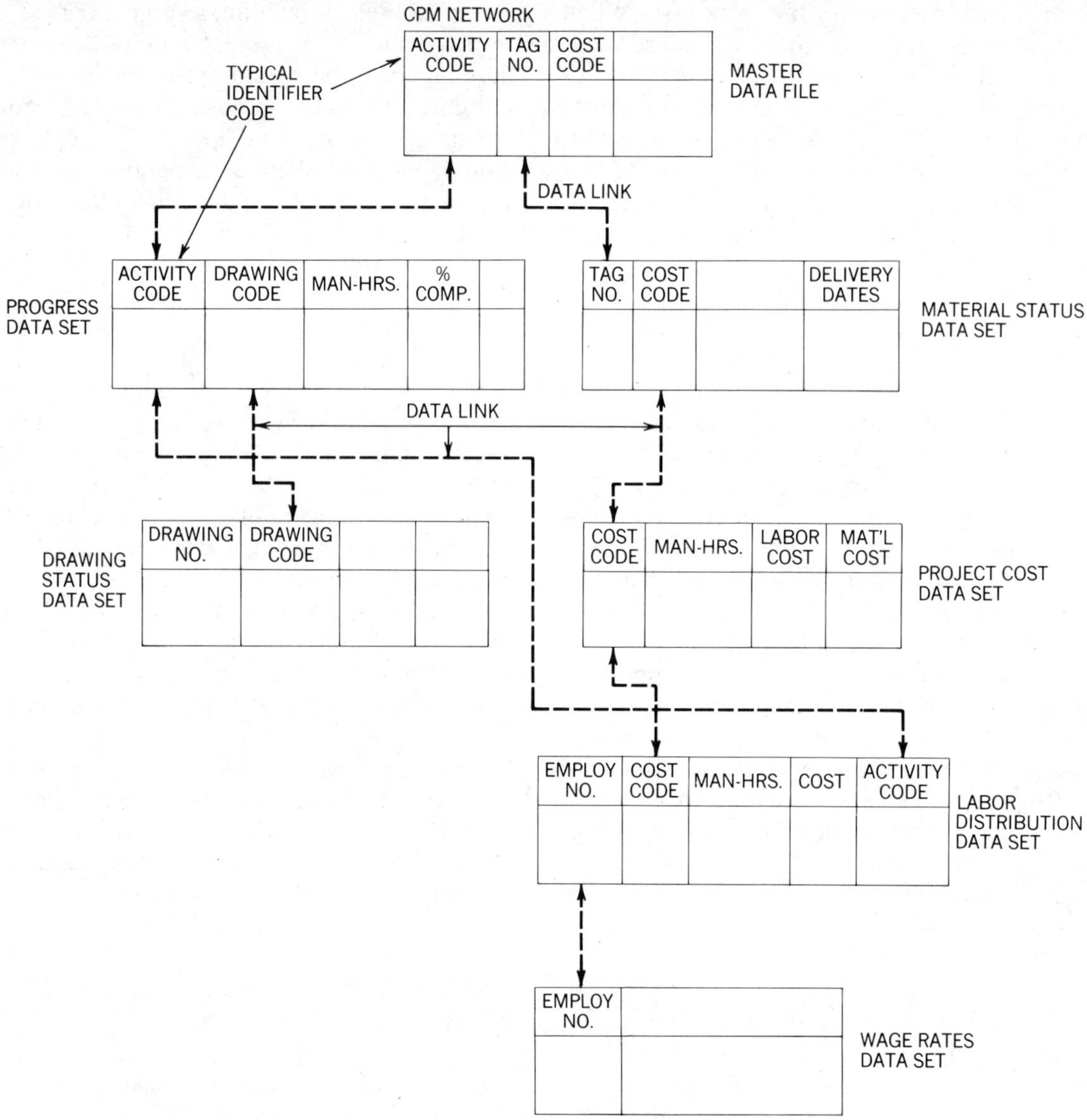

Figure 15.10 Relational database structure.

base. Hence, terminating the project GENERATES LABOR MANAGEMENT DATA (R2) which the HISTORICAL DATABASE (R1) requires to be filed in the firm's historical archives. Typical filing system techniques for a project and for the small firm as a whole are similar to the techniques illustrated in Figures 15.4*a* and 15.4*b*, respectively.

The policy directives influencing both of these functions are the requirements to *USE COST CODES* (P_S 3)* and *LABOR UNIT COSTING* (P_S 4) and the need for *ACTIVITY STATUS* (P_S 21) information. Thus, the historical database must reflect information in a format consistant with the labor unit costing policy. As a result, the small firm does not analyze labor data, but only stores it as final labor cost information for each project. This information becomes a source of input for LABOR CONTENT ESTIMATING (P4). If the labor cost reports do not reflect final project costs, these documents will flow to CURRENT PROJECT STATUSING (M2). It must be reemphasized that the estimator will also access active project files to seek

*See Table 15.3, Policies of the Small Firm.

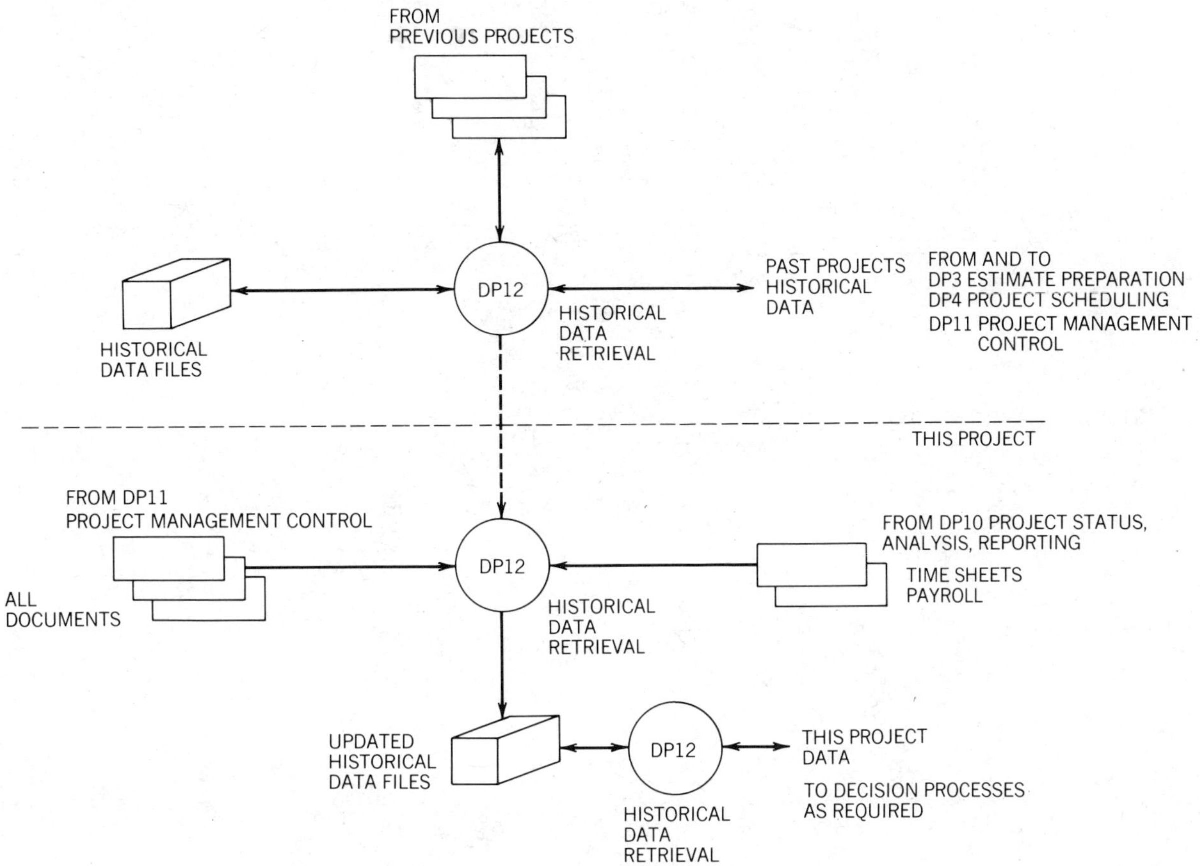

Figure 15.11 Historical Data Retrieval Management Information System concept.

TABLE 15.3 POLICIES OF THE SMALL FIRM

Policy		Description
P_S 3	*USE COST CODES*	Systematic classification of work categories segregated by unique cost codes is used to identify work tasks and collect costs.
P_S 4	*LABOR UNIT COSTS*	Labor unit costs are sufficient in project estimating provided that these costs reflect past experience from projects similar in scope and are adjusted, when necessary, to ensure compatibility with the current project scope.
P_S 21	*ACTIVITY STATUS*	Activity progress and labor productivity are not normally considered beyond completed activity volume, labor unit costs, and total labor dollars expended to date relative to activity scheduled target completion dates, estimated labor unit costs, and total estimated labor costs.

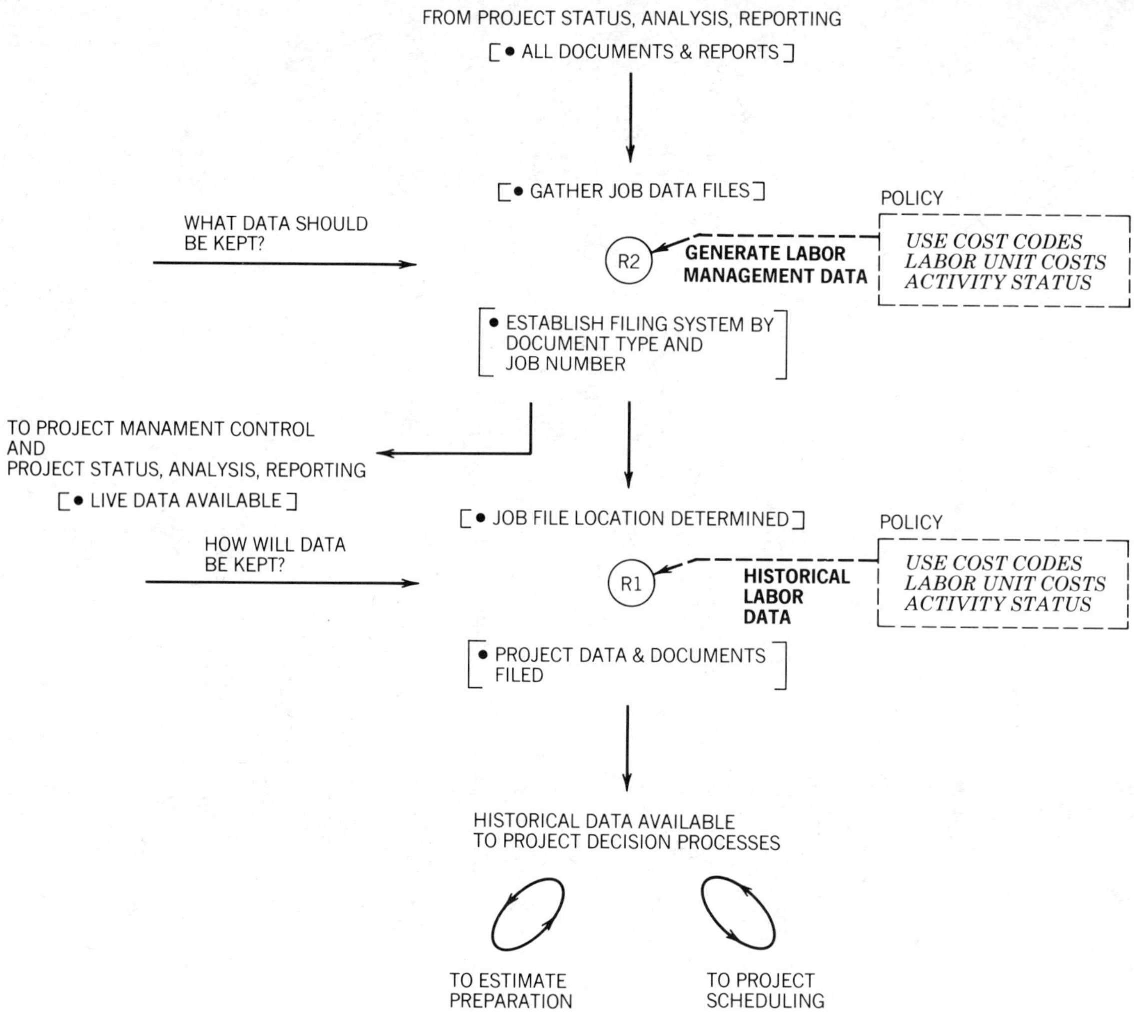

Figure 15.12 Historical Data Retrieval decision process (Project Team Component approach).

current labor unit costs when preparing an estimate for future work. The Historical Data Retrieval decision process is illustrated in Figure 15.12.

15.4 THE TRAVELING PROJECT MANAGER

A continuing aspect of Project Management Control, especially at the medium-sized firm level is the maintenance of historical labor data. The project cost accountant and the support staff generally assume responsibility for this decision phase, which requires the GENERATION OF LABOR MANAGEMENT DATA (R2) and its filing in a systematic manner in order to develop an HISTORICAL DATABASE (R1). The majority of medium-sized firms rely on their final cost reports as their historical information, and perform little analysis of the data collected. In this approach, the various labor cost reports generated over the life of the project become the input and output for these two functions. Consequently, the filing system techniques would be similar to those described for the small firm, with the exception that a manual tabulation of labor unit costs might be maintained in a similar format, as shown in Figure 15.7*a*, or with computer capabilities a more detailed labor unit cost analysis will emerge. This might be, for example, in the form of labor unit cost data by cost code that is stored in the computer. Historical information is continuously accessed by the estimator for Estimate Preparation on future work. The basic decision process

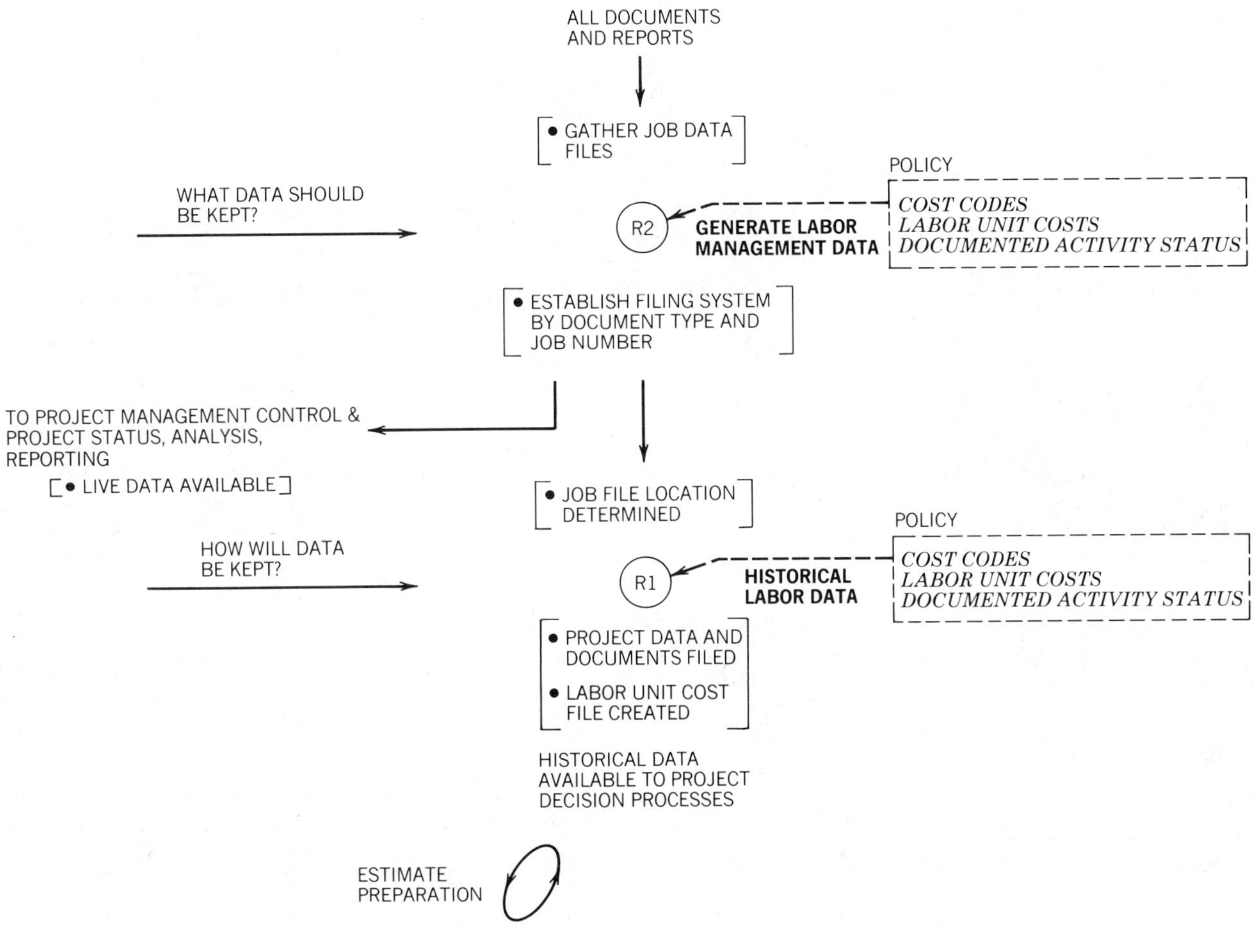

Figure 15.13 Historical Data Retrieval decision process (Traveling Project Manager approach).

for Historical Data Retrieval at this level is given in Figure 15.13. As shown in this figure *USING COST CODES* (P_M 4), *LABOR UNIT COSTS* (P_M 5), and *ACTIVITY STATUSING* (P_M 22)* are the policies that dominantly influence the manner in which Historical Data Retrieval is performed.

*See Table 15.4, Policies of the Medium-Sized Firm.

If the basic decision process is presented as document flow based on computer processing, its structure can be described as shown in Figure 15.14. The data input for Historical Data Retrieval are final project labor unit costs extracted from the computer-based cost control files on previous projects. This data is massaged to fit the structure of the database, and then input into the computer in order to

TABLE 15.4 POLICIES OF THE MEDIUM-SIZED FIRM

Policy		Description
P_M 4	*COST CODES*	Systematic classification of work categories segregated by unique cost codes is used to define work tasks, collect costs, and provide a means of comparing actual costs with estimated costs.
P_M 5	*LABOR UNIT COSTS*	A realistic value of labor dollars is sufficiently derived from labor unit costs for most tasks, provided that unit costs reflect current wages and potential production rates.
P_M 22	*DOCUMENTED ACTIVITY STATUS*	Labor costs and quantities in place must be recorded and documented by a system yielding reliable information and viable reports for management control.

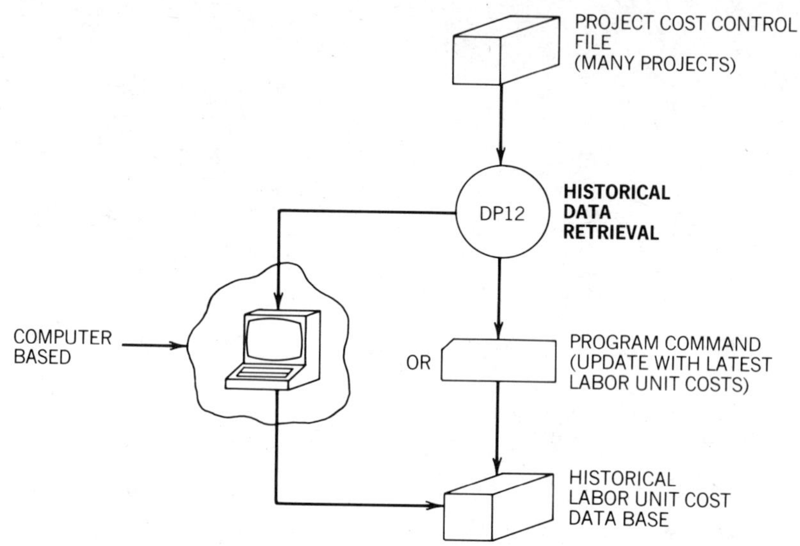

Figure 15.14 Historical Data Retrieval Management Information System (Traveling Project Manager approach).

update existing data files. The structure of the database could be similar to that illustrated in Figure 15.7*b*. In this case, the database structure is formatted in terms of cost codes reflecting unique categories of work.

15.5 FIELD-BASED PROJECT MANAGEMENT

At the large firm level, Historical Data Retrieval is complicated by a significant increase in the volume of data associated with both the active project, and the handling of historical data from an overall company perspective as projects are completed. Thus, the management of information becomes a problem beyond the simple filing of final cost data (as in small firms) and the formatting of data into a structured collection of labor unit costs (as in medium-sized firms). The Management Information System (MIS) is most often computer based with a data structure that is not only integrated through cost codes, but through the control areas concept common in the Field-Based Project Management approach. In addition, large volumes of similar work enables the large firm to build a quality database derived from an increased range of historical reference points. In this form, the MIS provides a mechanism for storing and processing data into useful forms for supporting factored estimates, detailed estimates, multi-level scheduling, progress curves, and the like. Figure 15.15 is a schematic depiction of the Historical Data Retrieval MIS.

Historical Data Retrieval encounters a significant growth development in the large firm environment. A special professional group, normally comprised of cost and scheduling engineers, is often charged with the sole function of collecting, analyzing, and reporting historical data. In terms of manpower management, the most commonly understood problem focus of Historical Data Retrieval is to systematically categorize labor data from many projects and to normalize the data, thereby establishing a labor database that can be applied to future projects. This facet of Historical Data Retrieval is company focused.

While Historical Data Retrieval is most frequently viewed in terms of its company orientation, it has first a project focus. In the project area, this decision process is mainly concerned with the collection and systematic filing of project data for reference as a database during the life of the active project. Once a project is complete, the project database is then absorbed into the overall company database.

At the large firm level, two major phases of the Historical Data Retrieval decision process emerge. The first phase is project focused, as shown in Figure 15.16*a*. The second is company oriented, Figure 15.16*b*, and reflects several significant changes in the manner in which historical labor data is analyzed and stored for future use. A typical example is the standardization of commonly referenced data (e.g., cost codes).

The project oriented phase, Figure 15.16*a*, has both a field and a head office focus and is comprised of two manpower management functions, namely

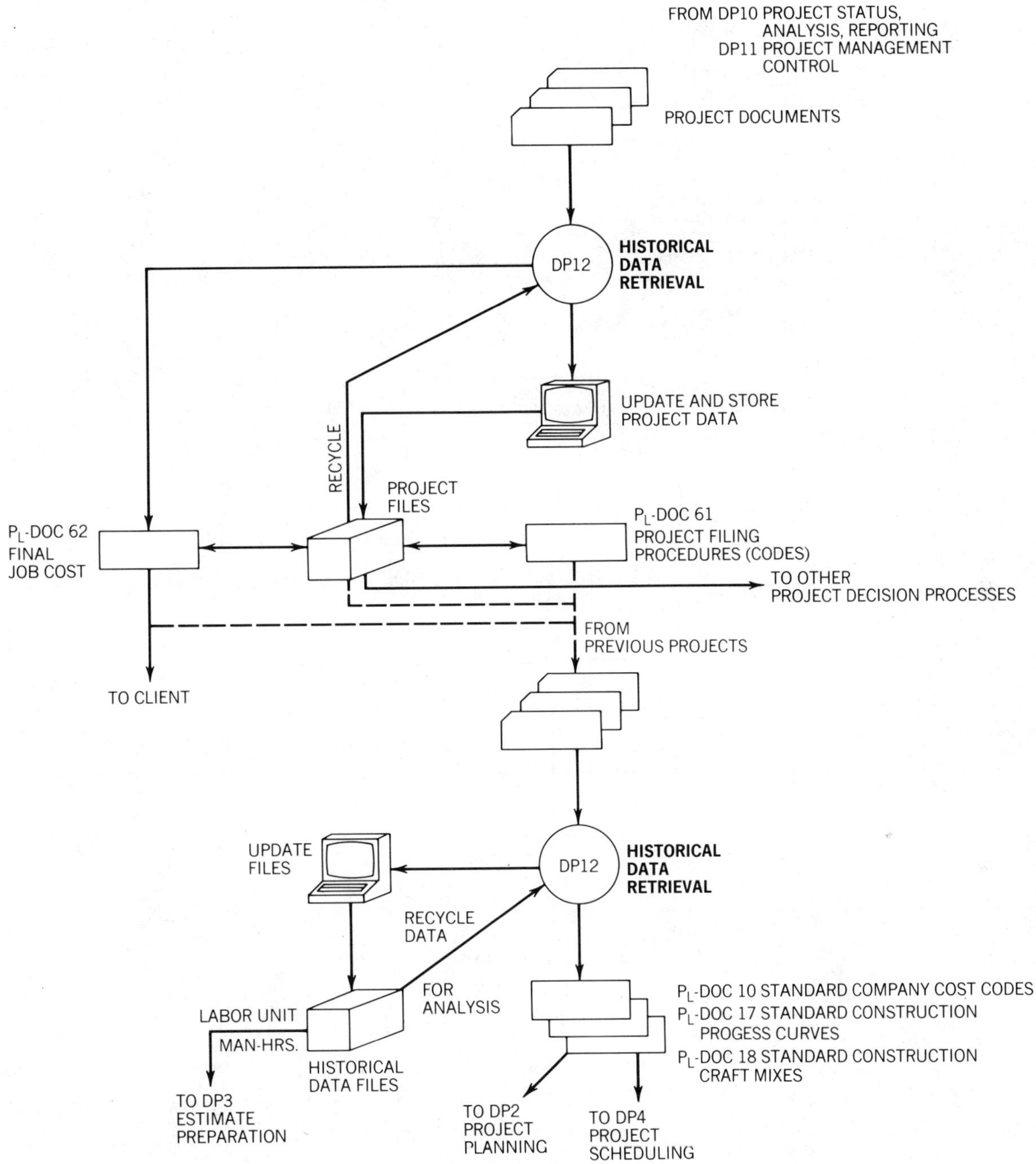

Figure 15.15 Historical Data Retrieval Management Information System (Field-Based Project Management approach).

the GENERATING OF A LABOR MANAGEMENT DATA BASE (R2), and the maintenance of an HISTORICAL DATABASE (R1). The former focuses primarily on the compilation of project labor data through a Project Filing Procedure (P_L-DOC 61). This procedure outlines a systematic method for collecting and storing active data during the project's life cycle. The nature and content of the Project Filing Procedure is illustrated schematically in Figure 15.17.

At the large firm level, much of the data input for GENERATING A LABOR DATABASE (R2) are

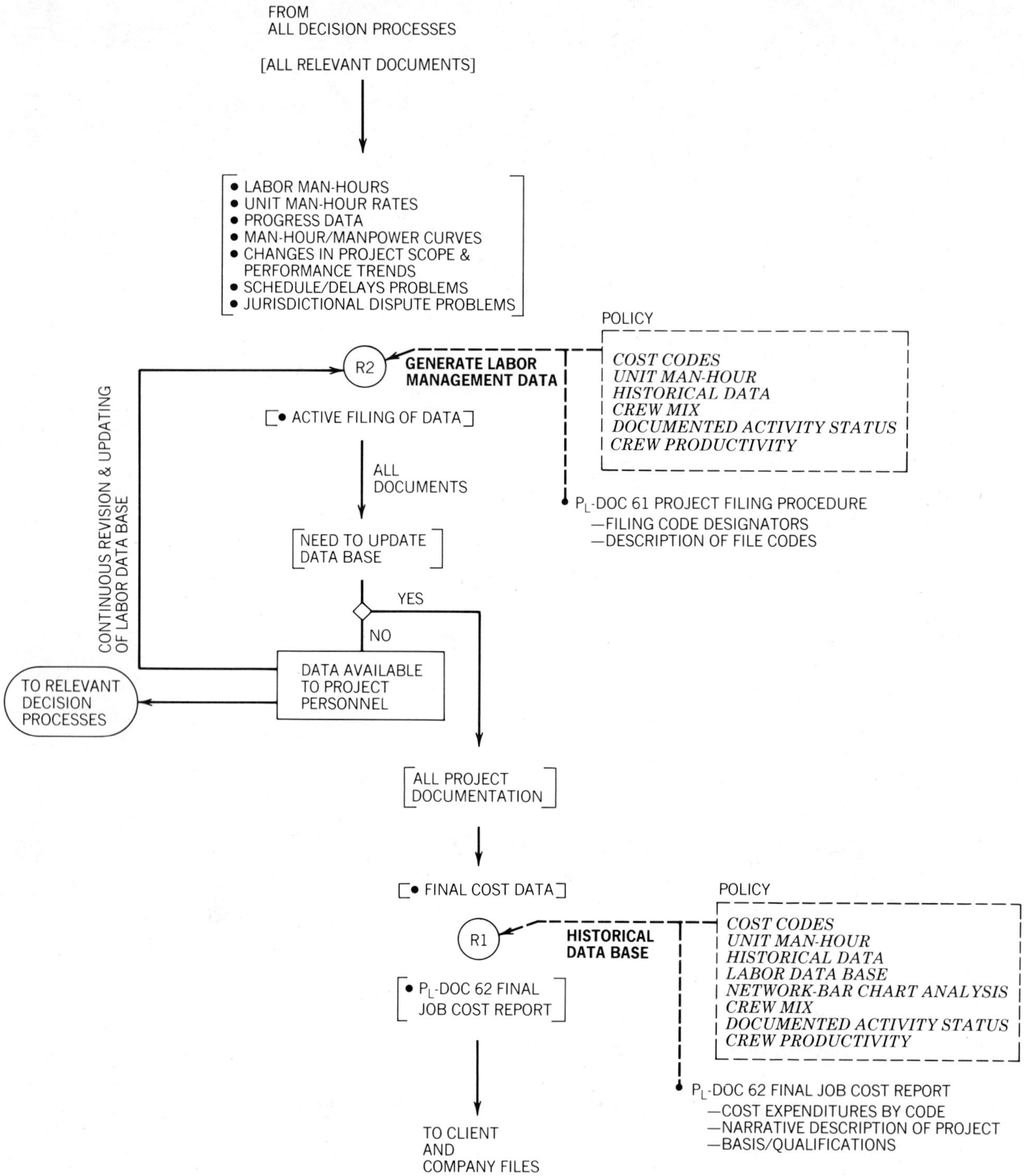

(a)

Figure 15.16 Historical Data Retrieval decision process (Field-Based Project Management approach). (*a*) Project oriented phase.

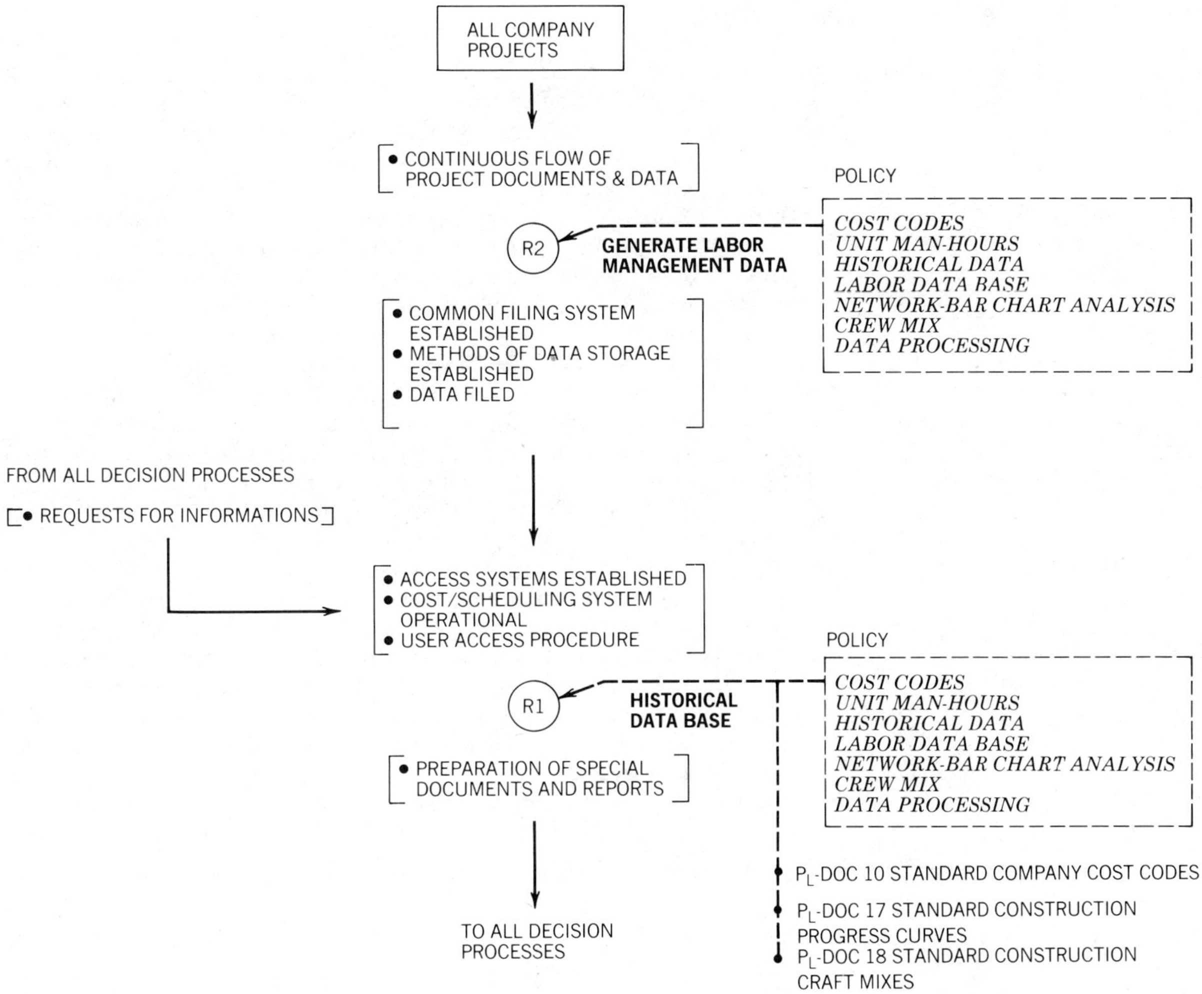

(*b*)

Figure 15.16 (*Continued*) (*b*) Company oriented phase.

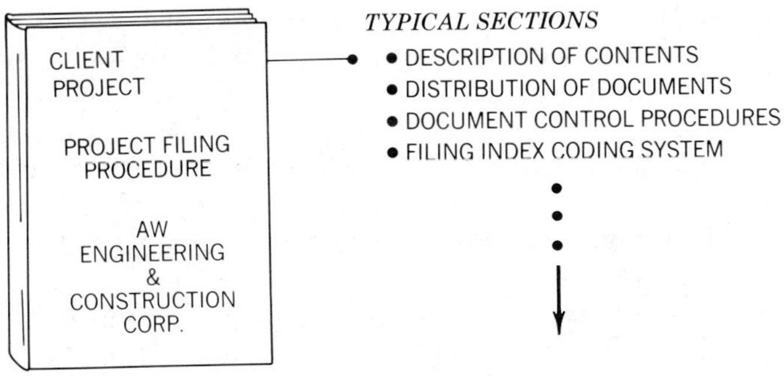

Figure 15.17 Project Filing Procedure (P_L-DOC 61).

TABLE 15.5 POLICIES OF THE LARGE FIRM

Policy		Description
P_L 6	*COST CODES*	A systematic classification of project work categories segregated by unique physical components and by cost codes is the basis for structuring the project scope into work tasks, categorizing estimated costs, reporting and controlling costs and manpower, and developing a historical database.
P_L 7	*UNIT MAN-HOUR*	Realistic evaluation of the labor content for a work task is essential for successful project planning and cost control. Therefore, labor costs must be based on work task man-hour content in relation to work volume, construction methods, local productivities, and labor availability.
P_L 8	*HISTORICAL DATA*	Historical labor data are basic to a realistic determination of the labor content for a work task as well as for major project components.
P_L 9	*LABOR DATABASE*	Historical labor data must be analyzed and documented in a meaningful way to be useful for estimating and planning manpower for future projects.
P_L 10	*NETWORK-BAR CHART ANALYSIS*	Broad network focused milestone planning is the best approach for identifying and sequencing major project components, while a combination of critical path method and bar chart model is best suited for project control level planning and scheduling.
P_L 19	*CREW MIX*	While impacted by union rules in some construction environments, the ratio of foremen to journeymen/apprentices within a crew is established to optimize both the foremen's level of supervisory control and the crew's performance.
P_L 27	*DOCUMENTED ACTIVITY STATUS*	Labor costs, manpower, man-hours, and quantities are primary ingredients to cost control and must be documented by a system yielding reliable information and viable reports for management control. Activity status must be documented in terms of start and completion dates and percent complete data.
P_L 28	*CREW PRODUCTIVITY*	Actual unit man-hour rate versus estimated man-hour rate is a measure of crew performance and a basic indicator of productivity.
P_L 34	*DATA PROCESSING*	Maximum use of electronic data processing is often the most economical means of processing data into useable information.

documents from the Project Status, Analysis, and Reporting and Project Management Control decision processes. These documents are normally filed according to a predetermined coding system as defined by the project filing procedure. Labor data may also be stored in a computer database using *COST CODE* (P_L 6)* techniques. Since new data is continuously input into the project filing system, the project files are frequently updated and revised to ensure that the most recent data is available to the project team while it is performing the other decision processes. Many different types of labor data must be retained, including *UNIT MAN-HOURS* (P_L 7), *CREW MIXES* (P_L 19), *DOCUMENTED ACTIVITY STATUS* (P_L 27), and the like.

Once the project is complete, the final cost and other relevant project data is assembled into a project HISTORICAL DATABASE through the Final Job Cost Report (P_L-DOC 62). This document normally has a predetermined format and is structured as defined by Standard Project Operating Procedures (P_L-DOC 1). It may be altered, however, to meet specific requirements established by the client/owner. A typical example of the contents of a Final job Cost Report is depicted in Figure 15.18. A variety of policies influence the type of labor data contained in this document, as shown on the decision process network in Figure 15.16*a*.

Generally speaking, minimal detailed analysis of labor data is made during the course of an active project. Labor data may be accessed for special studies and analysis at the request of top-level management, but this is the exception rather than the rule. Consequently, this phase of Historical Data Retrieval is similar to the approach of the medium-sized firm in terms of filing project data, with the exception that the increasing volume of labor data at the large firm level requires formal Project Filing Procedures and the preparation of Final Job Cost Reports.

Each project, when it is completed, contributes historical data to the company-oriented phase of

*See Table 15.5, Policies of the Large Firm.

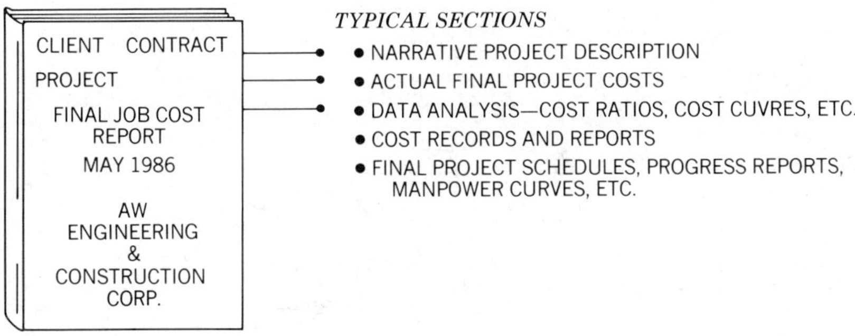

Figure 15.18 Final Job Cost Report (P_L-DOC 62).

Historical Data Retrieval, as reflected in Figure 15.16*b*. This phase of Historical Data Retrieval is exclusively a head office located decision area. The process of GENERATING A LABOR MANAGEMENT DATA BASE includes the determination of methods of data storage and specific filing procedures (see Figures 15.4 and 15.5), as well as methods for the active filing of final project data. The filing of data is normally referenced against project *COST CODES* that are based on Standard Company Cost Codes (P_L-DOC 10). Since the majority of labor data is related to the cost and scheduling functional area, cost and/or scheduling engineering personnel are primarily involved in the performance of this function. The company files become a key source for generating a company-focused HISTORICAL DATABASE. The principal focus of this function is to develop a labor database for estimating and scheduling on future projects. In terms of estimating, labor unit man-hour rates are the "guts" of the database. Thus, actual labor unit man-hours from completed projects are stored in a systematic method normally through the *COST CODE* structure as determined by the Company Standard Cost Codes. This process is often computerized because of the large volume of data to be stored. This method is practical provided that any unique project conditions associated with historical data are stripped from the data to normalize it in relation to other data of a similar type, when developing and updating the *LABOR DATABASE* (P_L 9).

Through the HISTORICAL DATABASE, many historical curves can be generated that impact both estimating and scheduling. The two most common types are the Standard Construction Progress Curves (P_L-DOC 17) and Standard Construction Craft Manpower Curves (P_L-DOC 18).* These curves are formulated by analyzing various project curves of the same type, and normalizing these curves to represent a standard which can then be applied to any project of a similar nature. Historical data in forms other than documents might include standard crews, standard crew ratios, craft efficiencies relative to reference locations and for various construction areas.

15.6 TOTAL FIELD AUTONOMY

At the Total Field Autonomy level, the Management Information System of the Historical Data Retrieval decision process is geared to producing fully normalized cost and scheduling data with a broad potential application for project use. In this environment, there emerges more standardized information such as standard labor unit man-hours, construction network modules, cost curves, and the like. When standardized data is coupled with company project control concepts including work breakdown structures, database management capabilities emerge for project execution. This provides the project-oriented facet of Historical Data Retrieval with the ability to perform data analysis for specific project problem areas, such as evaluating optimum crew sizes in relation to the productive work content of an activity. The general structure of the MIS for Historical Data Retrieval is similar to that shown in Figure 15.16 for the Field-Based Project Management approach. The only major changes include the incorporation of more standardized documents and the focus on database management for project execution. The MIS for the Total Field Autonomy approach is depicted in Figure 15.19.

The decision process structure is similar to that found in the Field-Based Project Management approach. Two phases of Historical Data Retrieval can be identified at this level. They correspond to

*See Figure 7.23 in Chapter 7, Section 7.5 for typical examples of these two documents.

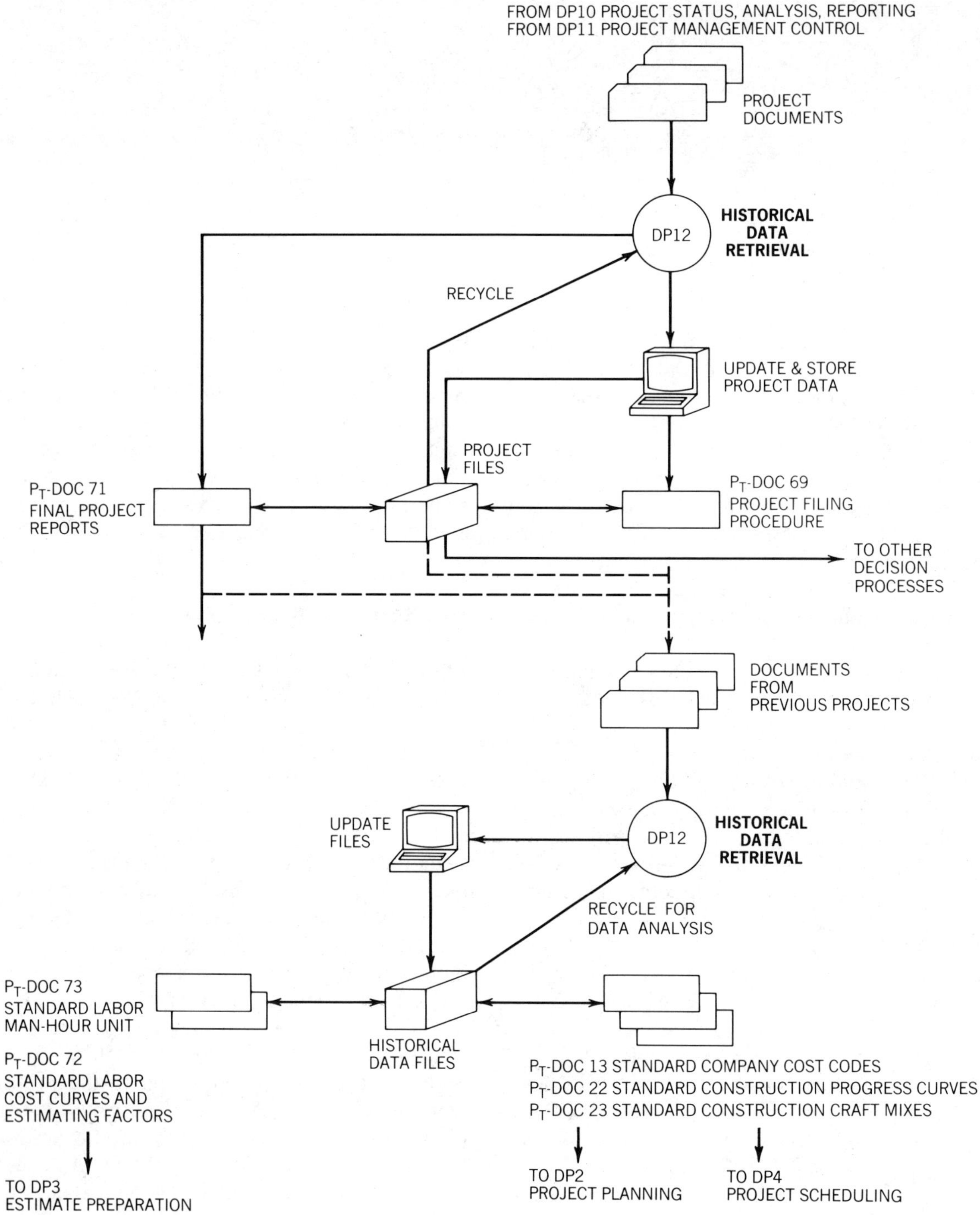

Figure 15.19 Historical Data Retrieval Management Information System (Total Field Autonomy approach).

R2 GENERATE LABOR MANAGEMENT DATA

Policies	Documents	
WORK BREAKDOWN	P_T-DOC 69	PROJECT FILING PROCEDURE
COST CODES		
UNIT MAN-HOUR		
HISTORICAL DATA		
CREW MIX		
PROGRESS REPORTING		
DOCUMENTED ACTIVITY STATUS		
SCIENTIFIC MANAGEMENT		
DATA PROCESSING		

R1 HISTORICAL DATA BASE

Policies	Documents	
WORK BREAKDOWN	P_T-DOC 70	CREW SIZE ANALYSIS REPORT
COST CODES	P_T-DOC 71	FINAL PROJECT REPORTS
UNIT MAN-HOUR		
HISTORICAL DATA		
LABOR DATA BASE		
NETWORK-BAR CHART ANALYSIS		
CREW MIX		
PROGRESS REPORTING		
DOCUMENTED ACTIVITY STATUS		
SCIENTIFIC MANAGEMENT		
DATA PROCESSING		

(*a*)

R2 GENERATE LABOR MANAGEMENT DATA

Policies	Documents
WORK BREAKDOWN	
COST CODES	
UNIT MAN-HOUR	
HISTORICAL DATA	
LABOR DATA BASE	
NETWORK-BAR CHART ANALYSIS	
CREW MIX	
PROGRESS REPORTING	
DOCUMENTED ACTIVITY STATUS	
SCIENTIFIC MANAGEMENT	
DATA PROCESSING	

R1 HISTORICAL DATA BASE

Policies	Documents	
WORK BREAKDOWN	P_T-DOC 13	STANDARD COMPANY COST CODES
COST CODES	P_T-DOC 22	STANDARD CONSTRUCTION PROGRESS CURVES
UNIT MAN-HOUR	P_T-DOC 23	STANDARD CONSTRUCTION CRAFT MIXES
HISTORICAL DATA	P_T-DOC 24	STANDARD CONSTRUCTION NETWORK MODULES
LABOR DATA BASE	P_T-DOC 72	STANDARD LABOR COST CURVES AND ESTIMATING FACTORS
NETWORK-BAR CHART ANALYSIS	P_T-DOC 73	STANDARD LABOR UNIT MAN-HOURS
CREW MIX		
PROGRESS REPORTING		
DOCUMENTED ACTIVITY STATUS		
SCIENTIFIC MANAGEMENT		
DATA PROCESSING		

(*b*)

Figure 15.20 Historical Data Retrieval decision process (Total Field Autonomy approach). (*a*) Project-oriented phase. (*b*) Company-oriented phase.

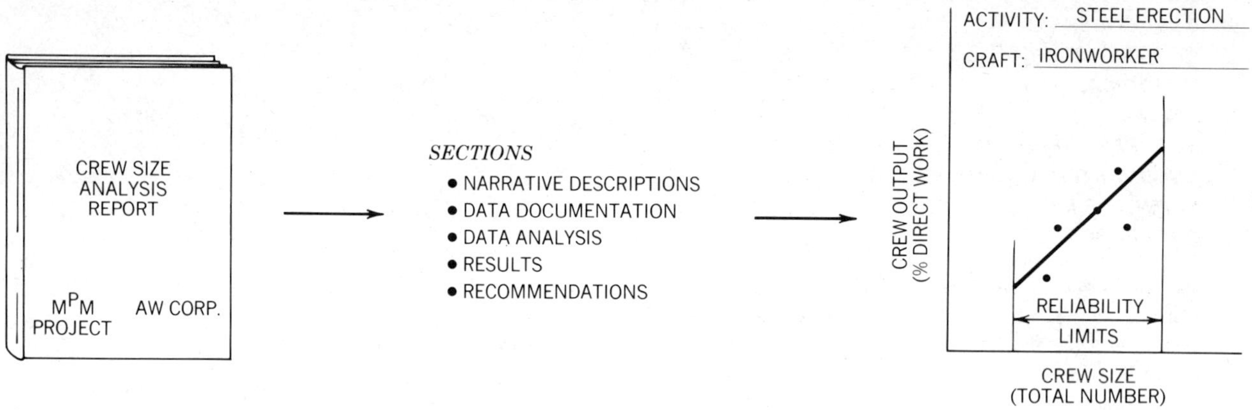

Figure 15.21 Crew Size Analysis Report (P_T-DOC 70).

TABLE 15.6 POLICIES OF THE AUTONOMOUS PROJECT TEAM

Policy		Description
P_T 8	*WORK BREAKDOWN*	A work breakdown structure representative of the project scope and control philosophy is formulated to integrate costs and schedules with work packages for project management.
P_T 9	*COST CODES*	Cost codes are applied to the work breakdown structure and serve as a basis for categorizing estimated costs, reporting and controlling costs and manpower, developing a historical database, and integrating the schedule with the estimate.
P_T 10	*UNIT MAN-HOUR*	A tight but realistic evaluation of the labor content for a work task is essential for successful project planning and management control. Therefore, estimates must be based on work task man-hour content in relation to work volume, construction methods, local productivities, labor availability, and site conditions.
P_T 11	*HISTORICAL DATA*	Historical labor data are basic to a realistic determination of the labor content for a work task or major project components, and preliminary scheduling development.
P_T 12	*LABOR DATABASE*	Historical labor data must be normalized to a standard reference location and/or base and then documented to be useful for estimating and planning manpower.
P_T 13	*NETWORK-BAR CHART ANALYSIS*	Broad network focused milestone planning is the best approach for identifying and sequencing major project components, while a combination of the critical path method and bar chart model are best suited for project planning and scheduling at the total project view level as well as at the daily level.
P_T 21	*CREW MIX*	While impacted by union rules in some construction environments, the ratio of foremen to journeymen/apprentices within a crew is established to optimize both the foreman's level of supervisory control and the crew's performance.
P_T 26	*PROGRESS REPORTING*	A systematic method of evaluating physical progress must be based on measuring units of work for any component work tasks, with the capability of rolling progress up from this detailed component level to the total project level.
P_T 28	*DOCUMENTED ACTIVITY STATUS*	Labor costs, manpower, man-hours, and quantities are primary ingredients to cost control and must be documented by a system yielding reliable information and viable reports for management control. These ingredients accrue as a basis for forecasting to complete at any level of detail.
P_T 34	*SCIENTIFIC MANAGEMENT*	Techniques of scientific management, such as work sampling, time lapse, activity analysis, multiple activity charts, and prediction models, can contribute to improving crew performance.
P_T 36	*DATA PROCESSING*	A computer-based information system is the most feasible means of data processing for management control.

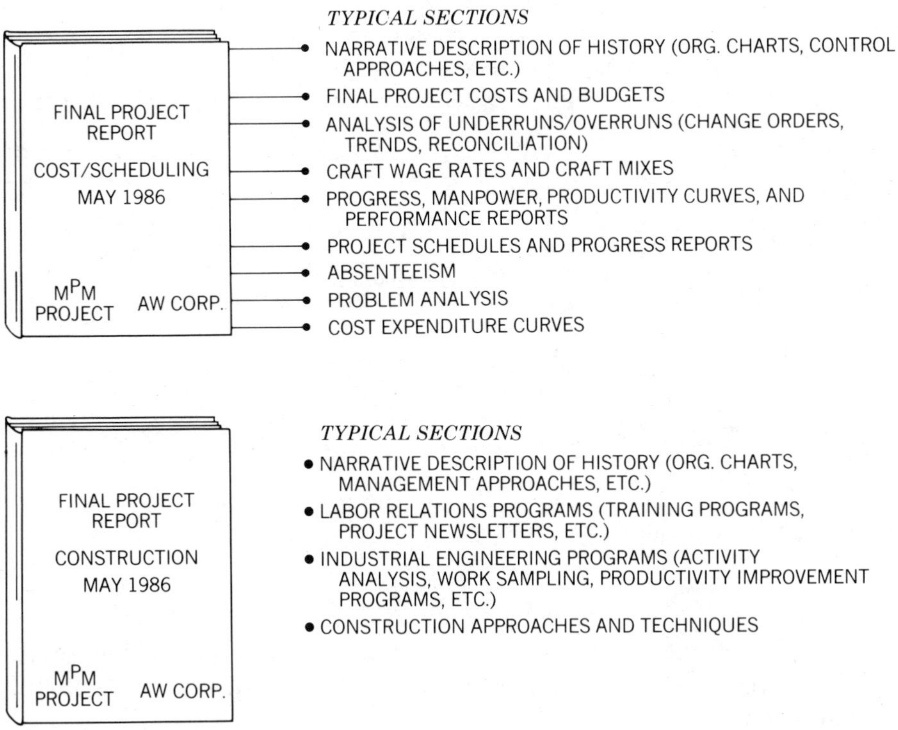

Figure 15.22 Final Project Reports (P_T-DOC 71).

the project-oriented phase in Figure 15.20*a*, and the company-oriented phase in Figure 15.20*b*. Figure 15.20 also identifies the manpower management functions performed in this decision process, and the relevant policies and documents associated with each function at both the project and company decision-making levels.

At the project level, Figure 15.20*a*, determining the structure of the project database can emerge as a significant new problem area. This occurs especially when the contractor deviates from standard control systems and devises a database structure that provides access to project data in a selective mode and format. Databased management of the selective entry nature is frequently directed at specialized reports that are requested either by the client or the project management. In these cases, the computer database for a specific area is structured so that data can be accessed at different levels of detail and printed in a format that best meets the need of the specialized report. For example, *DATA PROCESSING* (P_T 36)* techniques can be used to analyze optimum crew sizes in relation to work output for selected activities, if the database is built so that it can compile crew size and work output information in various forms and levels of detail. This might include data on productive work/delay time versus the number of crafts comprising a crew. This type of data is generally available if the contractor is implementing *SCIENTIFIC MANAGEMENT* (P_T 34) techniques. When analyzed statistically, it can produce Crew Size Analysis Reports (P_T-DOC 70), as shown in Figure 15.21.

If database management is incorporated into the project execution procedure, then a data access procedure must be developed. This problem is best solved at the inception of the project, prior to the accumulation of significant amounts of project data. In this context it is helpful if the contractor and client can agree upon those areas where database management will have a positive impact on project control. Because construction management is a dynamic process, the need for specialized reports requiring a specific type of database may only become apparent as the project proceeds. At the Total Field Autonomy level, the *DATA PROCESSING* approach must be capable of adjusting to meet these special requests by project managers. Thus, the computer-based Management Information System must be flexible to the extent that it can be quickly focused on a problem area to produce a specific problem-focused report, such as the Crew Size

*See Table 15.6, Policies of the Autonomous Project Team.

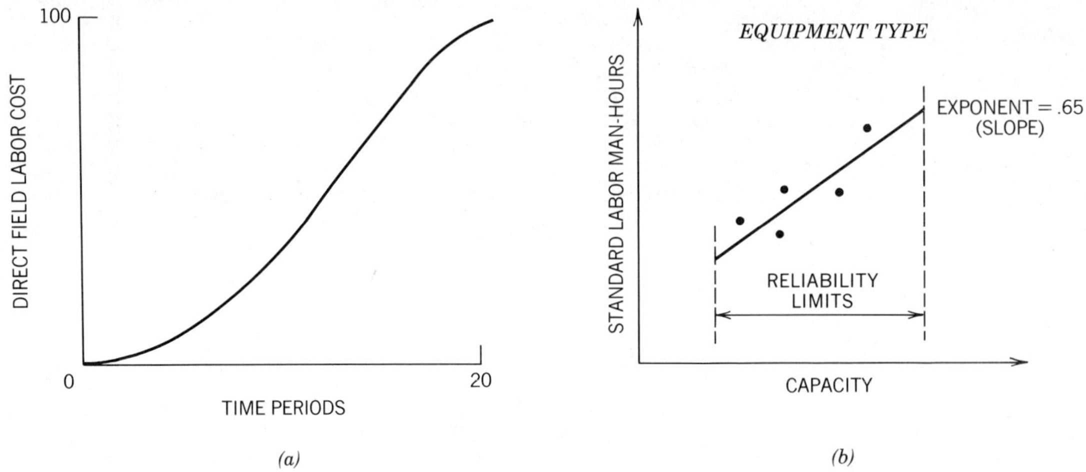

Figure 15.23 Standard Labor Cost Curves and Estimating Factors (P_T-DOC 72). (*a*) Standard labor cost curves. (*b*) Standard Labor estimating factors.

Analysis Report. Consequently, there is a critical management need for the continuous inputting, updating, and accessing of live information and data between all active decision processes. This can only be accomplished using a properly structured project database management system. Because of the increasing focus on database management and the normalization of labor and other relevant project data, the Final Project Report (P_T-DOC 71) is expanded to incorporate more than just cost data. As illustrated in Figure 15.22, the Final Project Report is frequently comprised of a number of volumes, each containing a category of data such as cost and schedule results or construction performance. The data contained in these volumes reflects more detailed information and analysis of the project's history. In this way, it becomes a meaningful source of data for updating the company database, and so becomes a part of the company-oriented phase of Historical Data Retrieval.

The company-oriented Historical Data Retrieval decision process in Figure 15.20*b* attempts to standardize a broader range of data to better support active project management and control. Consequently, the Total Field Autonomy approach often produces an increasing level of standard data. These data are normalized to typical Standard Construction Progress Curves (P_T-DOC 22) and Standard Construction Craft Mixes (P_T-DOC 23).* In addition, Standard Labor Cost Curves and Estimating Factors (P_T-DOC 72) and Standard Labor Unit Man-hours (P_T-DOC 73) are two typical documents that emerge at the Total Field Autonomy level. These types of standard documents, shown in Figure 15.23 and 15.24, respectively, are developed by the use of statistical techniques such as regression analysis, provided the data is normalized to a predetermined time frame and reference location as directed by a *LABOR DATABASE* (P_T 12) policy. This can be effectively accomplished when each project database is structured in a common format through the *WORK BREAKDOWN* (P_T 8) structure and *COST CODES* (P_T 9). Finally, company management must commit resources to developing such a standard database. This management commitment is viewed as one means of maintaining a competitive edge in the large firm environment.

PIPING ERECTION MAN-HOURS PER LINEAR FOOT				
	WALL THICKNESS			
PIPE SIZE	LIGHT WALL SCH. 5 10 20	SCH. 30 40 60	SCH. 80 100	SCH. 120 140 160
½	.161	.174	.187	.210
1	.163	.181	.209	.239
1½	.172	.198	.233	.278

INCLUDES:
TRANSPORTING PIPE AND FITTINGS...

Figure 15.24 Standard Labor Unit Man-hours (P_T-DOC 73).

*See Figure 7.23 in Chapter 7, Section 7.5.

CHAPTER 16
Project Management Overview

The previous chapters of this book present a methodological approach to project manpower management decision processes. Specifically, twelve basic decision processes are considered, each of which can be readily identified in construction practice. Each decision process has been considered in terms of the policies used, documents generated, and the flow of those documents through the constituent management functions involved in the decision process. In addition, each decision process has been formulated in terms of a formal Management Information System, together (where appropriate) with an indication of the management pressure environment and related informal information systems necessary for actual decision making. Finally, each decision process has been illustrated as it relates to small, medium, and large-sized firms in the construction industry through four different approaches to project management.

While each decision process is readily identifiable in construction practice, each in its own way contributes to the cost, schedule, and quality control of a project and to the overall project management process. Thus, these decision processes collectively expose the project Management Information System and, in the manner of their performance, reveal the management approach of the contractor to the project.

Chapter 16 looks at the different ways in which these individual decision processes can be integrated into the total project manpower management decision process. In order to accomplish this it is necessary to look at the changing requirements on a Management Information System through the different life phases of a project, together with the integrating Management Information System pressures that endure over the project's duration. In this way both the *formal* and *informal aspects* of the project manpower management process can be understood.

16.1 INTEGRATING PRESSURES FOR DECISION MAKING

The twelve basic decision processes are distributed through the project definition, field construction, and overall project management phases of construction management. Each of these phases has its own specific problems, often calling for the management efforts of different groupings of professional and skilled staff. In addition, these three phases are often separated physically and sequential in time. These naturally disjointing features, however, are countered by the need for integrating flows of information and documents, as has been indicated in each of the decision process networks. In addition, business-oriented concerns and many management policies strengthen such linkages and focus attention on the need for, and development of, integrated systems.

The project definition phase establishes contractual obligations. Prudent business acumen forces a reconciliation, if not integration, of the project planning and project estimating processes. In addition, though the eternal optimism of many contrac-

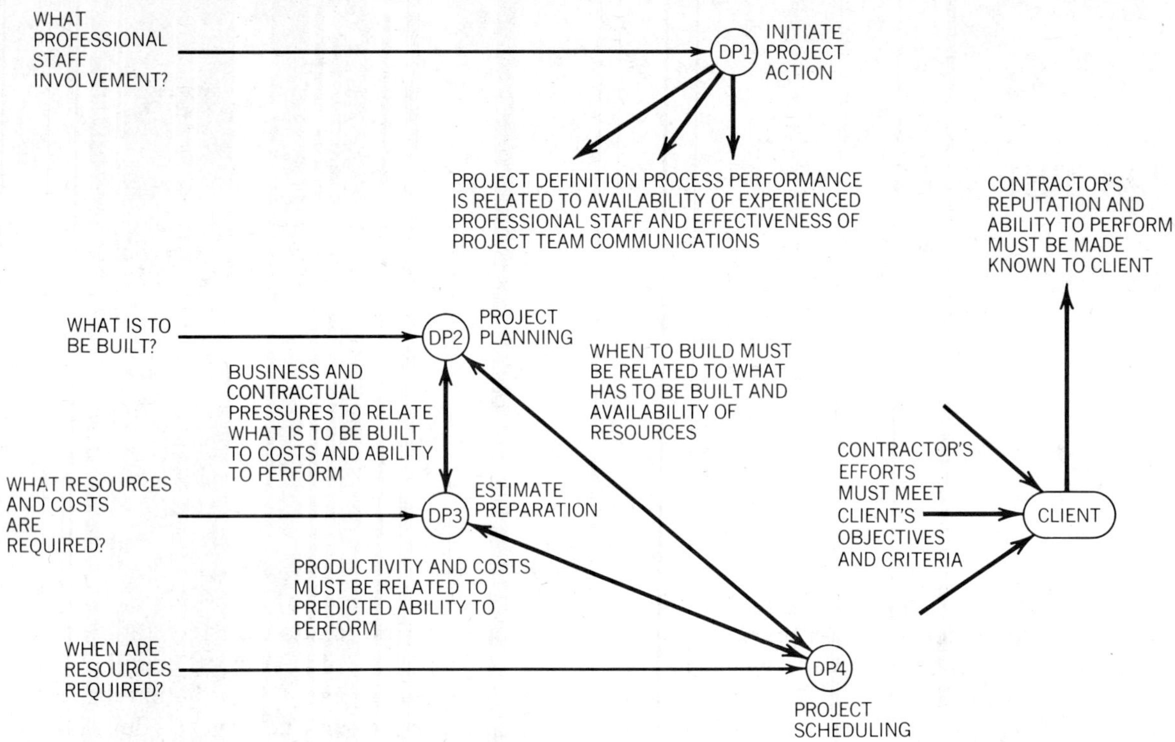

Figure 16.1 Integrating pressures on project definition phase.

tors makes this less compelling, prudent project scheduling relates promises to ability to perform. This need for integration of scheduling with project planning and estimating assumes greater significance if potentially severe cash flow commitments are involved. Finally, when key staff are in short supply and time constraints are often severe, the actual performance of the project definition phase requires the serious addressing of the problems involved in the initiation of project action effort, and even to their integration with the scheduling of this effort. A schematic illustration of these integrating pressures in the project definition phase is presented in Figure 16.1. Additionally, as indicated in this figure, the contractor is under considerable pressure as the firm's reputation and ability to perform are continually evaluated by prospective clients every time the contractor competes with others for a project. This pressure is reflected in the contractor's close interest in the project definition team, and in assessments of its performance and success.

The field construction phase is often left to the skills of the superintendent and the trade/craft foremen. Here the integration of the Field Staffing, Crew Planning, and Crew Management decision processes with the work effort is achieved by the experience and judgment of individuals. These individuals synthesize these decision processes in an *informal manner*. Project size and complexity, however, together with the growth in the professionalism and size of the project team, exert considerable management pressure on field construction personnel. In this environment, attempts are made to link the Crew Planning and Crew Management decision processes more effectively into the project definition and project management areas. This is especially true when a work package breakdown to work order documented field estimating, crew planning, and scheduling processes is introduced. Thus, the increasing integration of field procedures and documents, especially in the methods engineering and scientific management aspects of Field Monitoring and Project Status, Analysis, and Reporting, raise issues of field implementation, productivity, and management concern that link back immediately to Crew Planning and Crew Management.

The overall project management focus forces the evaluation of work progress and the effectiveness of resource use. In particular, the consideration of

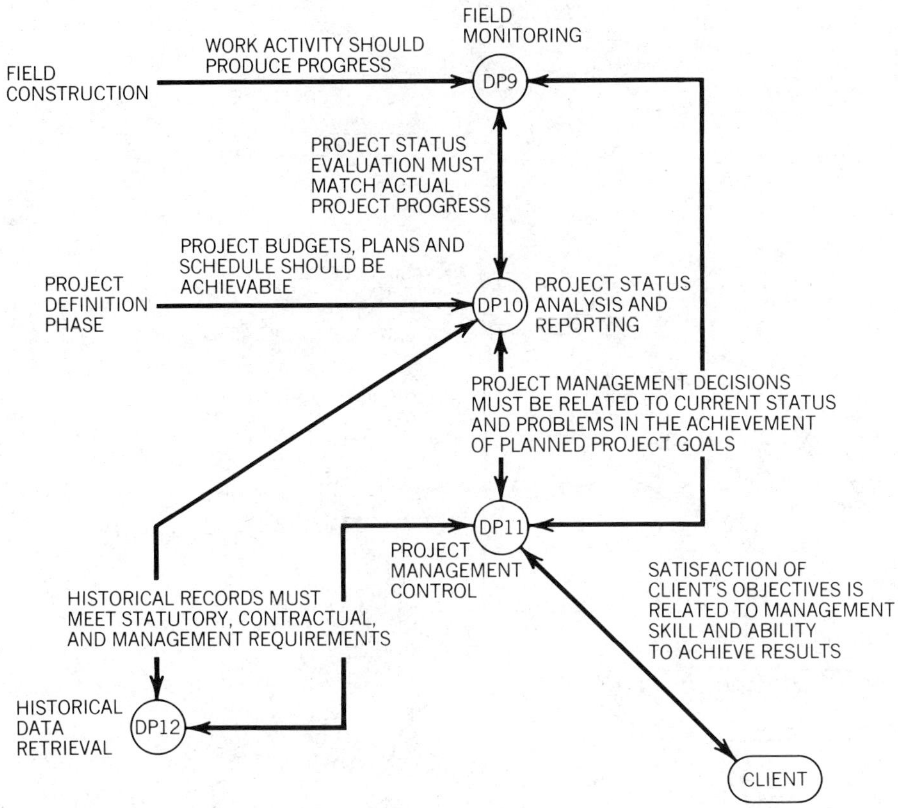

Figure 16.2 Integrating pressures on project management phase.

cash flows and other drains on financial resources, as a means of achieving profits rather than suffering losses, is a constant project management concern. These aspects of project management require that Field Monitoring of the work progress and Project Status, Analysis, and Reporting efforts be related by project management to project control budgets and schedules established in the project definition and project management decision processes. A schematic illustration of the integrating pressures operating on the project management phase is given in Figure 16.2.

Finally, Field Staffing manipulations are often viewed by some contractors as a way out, or a disposable area, for saving overhead costs if cost overruns occur due to faulty estimating or poor field achievement results. In this situation the resulting function overload of field staff can further seriously erode field management capabilities unless the contractor can feed in, by an expedient acquisition, the right troubleshooter to win the day. On large, complex projects, however, Field Staffing considerations provide the venue for the essential integration of the field construction phase with project management, and may even be a significant decision area for consideration and involvement in the project definition phase. This integration becomes complete in the Total Field Autonomy approach, when field staffing efforts are the metamorphised and continuously burgeoning form of the initial project action effort now transferred to the field.

The integrating pressures indicated above lead to the need for, and development of, compatibility of documents between decision process areas, and the emergence of integrating information system structures, whether they be formal or informal.

16.2 MANAGEMENT INFORMATION SYSTEM CONCEPTS

The document and information flow within each decision process has been highlighted in previous chapters. This has usually been done at two levels:

First, each relevant decision process network has been dissected in detail to indicate the source of documents, and the policies influencing the management functional performance which either generates the document, updates or verifies its information content, or authorizes it for further distribution

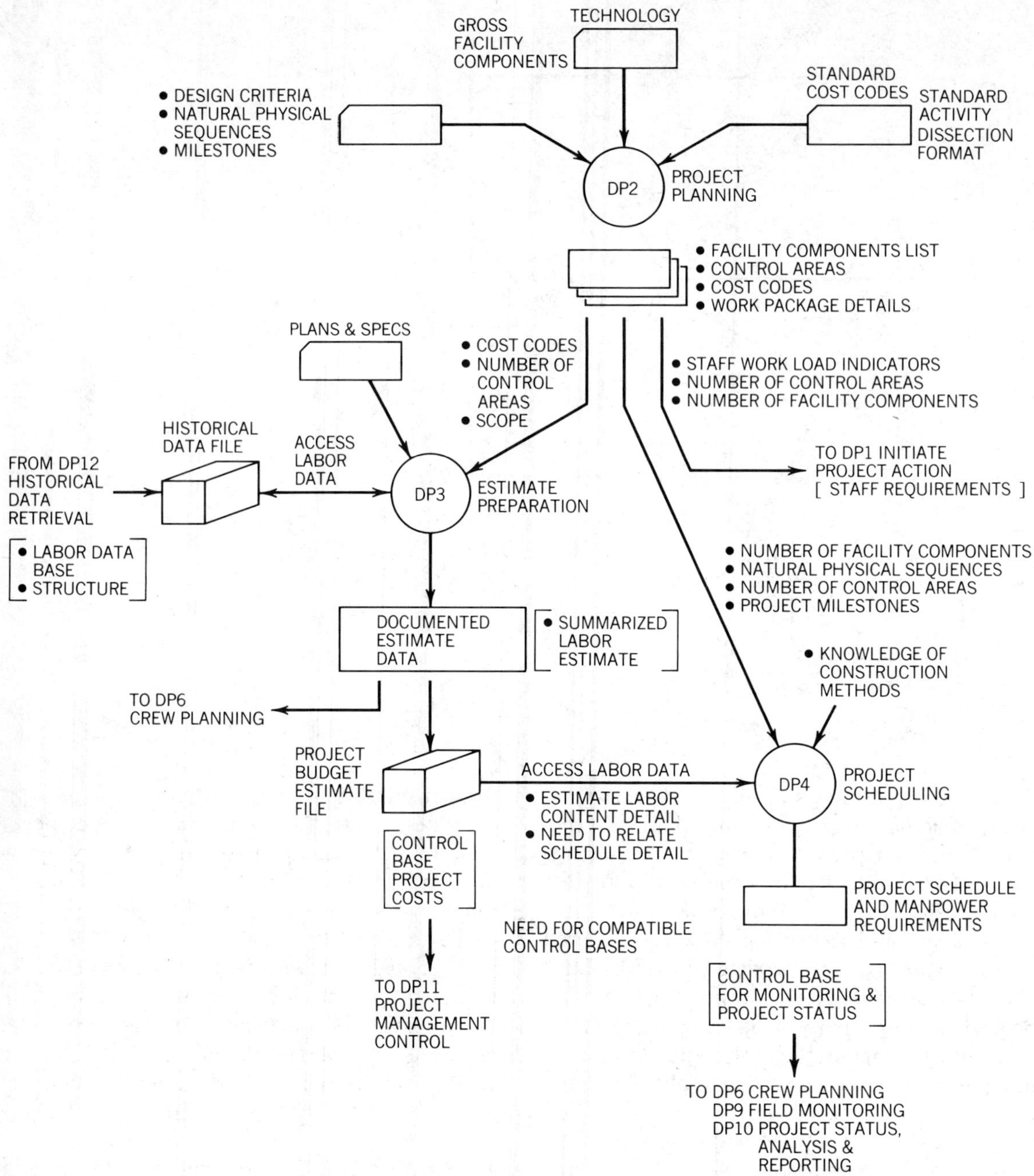

Figure 16.3 Management Information System integration concept: project definition phase.

to other management functions or decision processes. The level of detail presented in this examination of a decision process is essential if there is an interest in any of the following:

1. The technical detail involved in the decision process.
2. An examination of the management techniques that are used in different management approaches.
3. The design of the documents themselves or of the decision process itself.

Second, each relevant decision process has been described as a Management Information System in terms of the concepts associated with a broader

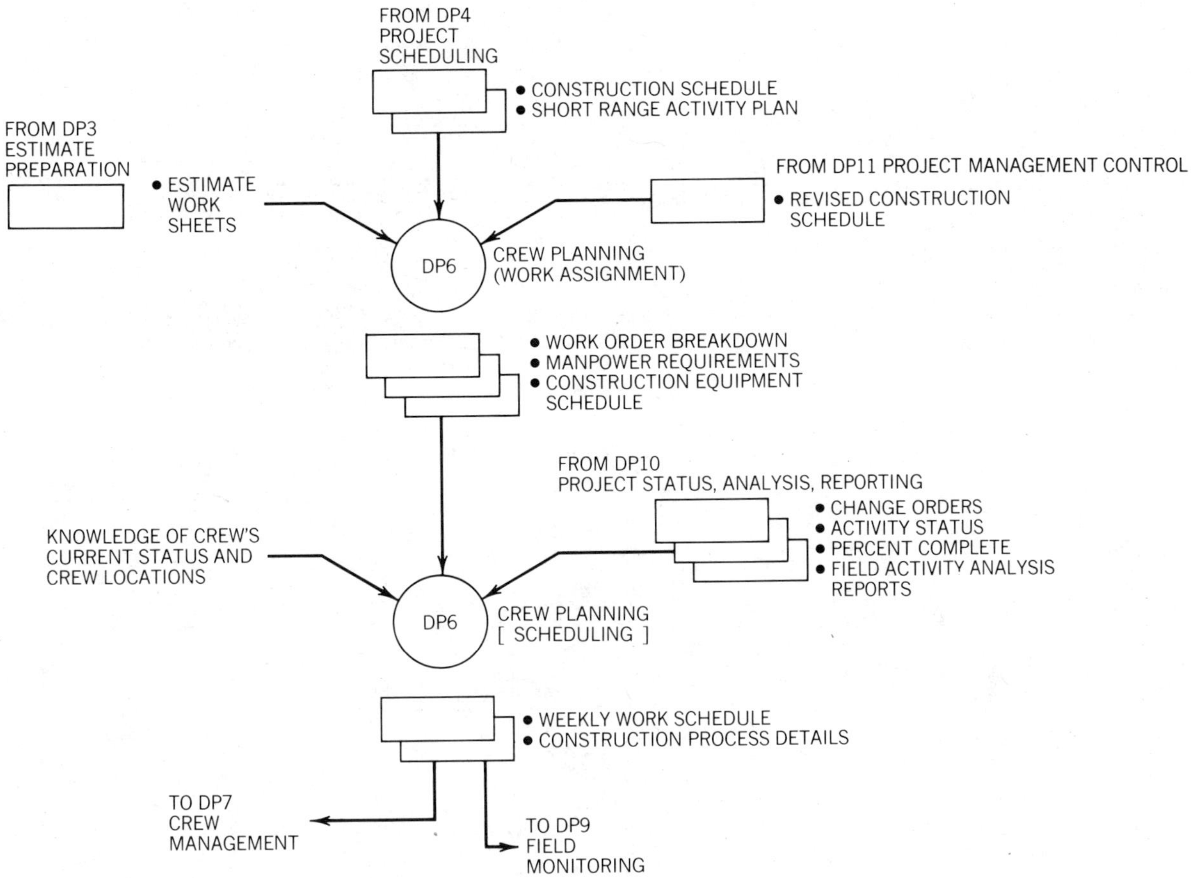

Figure 16.4 Management Information System integration concept: field construction phase.

look at document flows. This level of detail enables the reader to more readily establish the volume and nature of document and information flows that are germaine to each decision process. It thus provides the venue for an examination of the integrable aspects of the decision processes, and helps to explain, investigate the potential for, check the compatibility of, or the effectiveness of, specific Management Information Systems.

Using this broader management information system focus as a venue it is possible to schematically portray the general relationships between the decision processes. This is especially so for those decision processes in a common phase of the management of a project (i.e., project definition, field construction, or project management), or which are focused on areas of key management concern (i.e., cost control, schedule control, or totally integrated project management control). When this is done it is not surprising to find that in many cases the integrable nature of the information and document flows between the decision processes matches the management pressures for such integration indicated in Section 16.1.

At the project definition phase there is a dominant requirement to link the component decision processes, as shown in Figure 16.3, so that a coordinated approach can be taken. Thus, Project Planning (DP2) focuses on determining what has to be built and feeds that information to Estimate Preparation (DP3) and Project Scheduling (DP4). These latter two decision processes must be coordinated if the project is to be compatible with resource identification, quantification, and use as it is built.

At the field construction phase, a broad integrated MIS and document flow only emerges at the work package/work order dissection level commonly met in the Total Field Autonomy approach. Thus, as shown in Figure 16.4, Crew Planning becomes intimately linked with the project definition and project management decision processes.

The major MIS activity, however, occurs in the

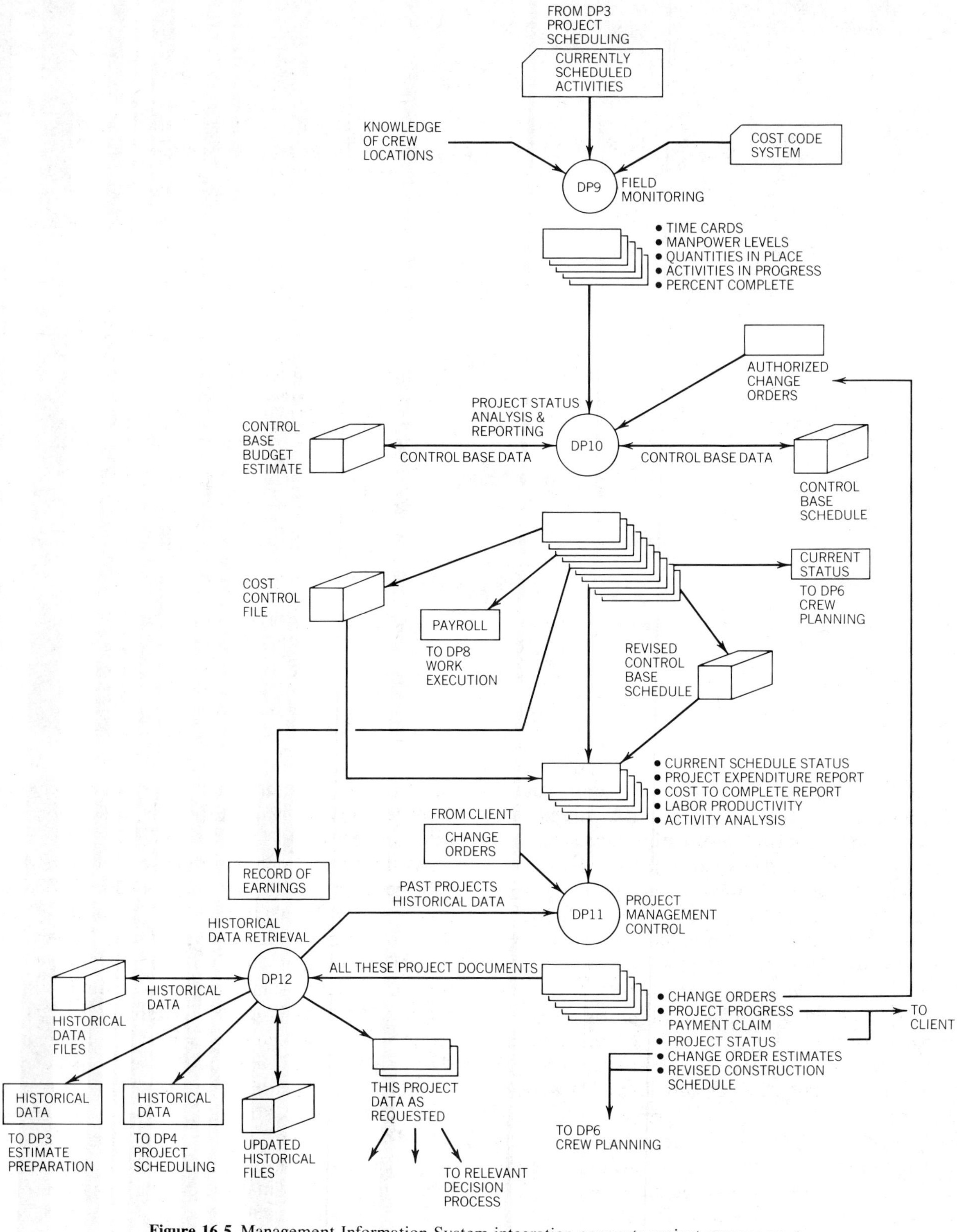

Figure 16.5 Management Information System integration concept: project management phase.

project management phase, since this decision area must be performed over the life of the project and becomes especially demanding during field construction. The nature and magnitude of general document and information flows during this phase is indicated in Figure 16.5.

A major integrating effort, however, emerges from the need of project managers to ensure that the project is being completed within cost and time, and at an acceptable quality standard. Thus, a dominant requirement of the project MIS is that cost control, schedule control, and in fact integrated cost and schedule control capabilities emerge. These aspects of the MIS will be considered in the next sections.

16.3 PROJECT COST CONTROL

A driving force for integration of basic decision processes is the control of costs, and the need to establish the cost status of a project in relation to the initial bid price or control budget (base) estimate. This sensitivity to cost control reflects the contractor's financial concern and the need to earn profits to stay in business. Thus, all contractors have, to some extent, a system that endeavors to collect all project-related costs on a continuous basis, and to periodically compile these costs into a summary form for comparison against the project cost control budget (base) estimates. The level of detail involved, and the frequency and manner in which these cost comparisons (actual versus budget estimates) are made, enables one to identify a variety of cost control systems.

The basic objectives of a cost control system are to relate current project costs (i.e., costs to date) against a preestablished, and possibly updated, financial plan (i.e., the control base budget estimate), in such a manner that accurate and timely cost assessments and cost trends, or forecasts, can be made. This enables management to investigate possible problems and take timely corrective action. In order to be accurate, all relevant cost data should be collected, and in order for timely management decisions to be possible, the cost collection system should ensure that cost information is quickly made available.

In general terms, the cost control problem impacts all three phases of the management of a construction project. The initial budget estimate control file is established in the project definition phase (usually *before* any construction effort), and its comparison with actual costs and performance to date occurs *periodically* in the project management phase, but *after* these costs have been incurred in the field construction phase. Hence, management is interested in collecting cost data at a generally finer level of detail than the summary items in the budget estimate in the hope that this will help locate cost overrun factors. In order for this detection capability to be of any use in resolving potential cost problems, the frequency of the periodic summary comparisons of costs actually incurred must be such that management has the time and the opportunity to resolve problem issues as they are detected rather than bearing their consequences.

Cost control begins in the project definition phase with the formulation of the control base cost estimate that establishes a reference point for comparing actual costs with budgeted costs. The different methods by which this initial step is accomplished can be characterized by a simple model (see Figure 16.6) that first identifies the components that have a bearing on labor cost estimation, as shown in

$C_\$$ = Labor cost of activity (e.g., \$16,900)
D = Duration of activity (e.g., days)
Q = Material content quantity (e.g., 100 cubic yards)
B = Standard man-hour content per quantity unit (e.g., 10 man-hours/cubic yard)
P = Estimated productivity relative to standard man-hours (e.g., 10% increase)
M = Estimated level of manning for work task (e.g., 10 men)
S = Scheduled work period unit (e.g., 40 hours per week based on 8 hours/day)
C_M = Percent crafts in base crew (e.g., 67% carpenters, 33% laborers)
$\$_1$ = Craft wage rates (e.g., carpenter \$17.00 and laborer \$12.00)

(*a*)

$$C_\$ = \frac{QBP \sum C_M \$_1}{MS}$$

(*b*)

Figure 16.6 Labor cost estimating model. (*a*) Labor cost components. (*b*) Labor cost estimate equation.

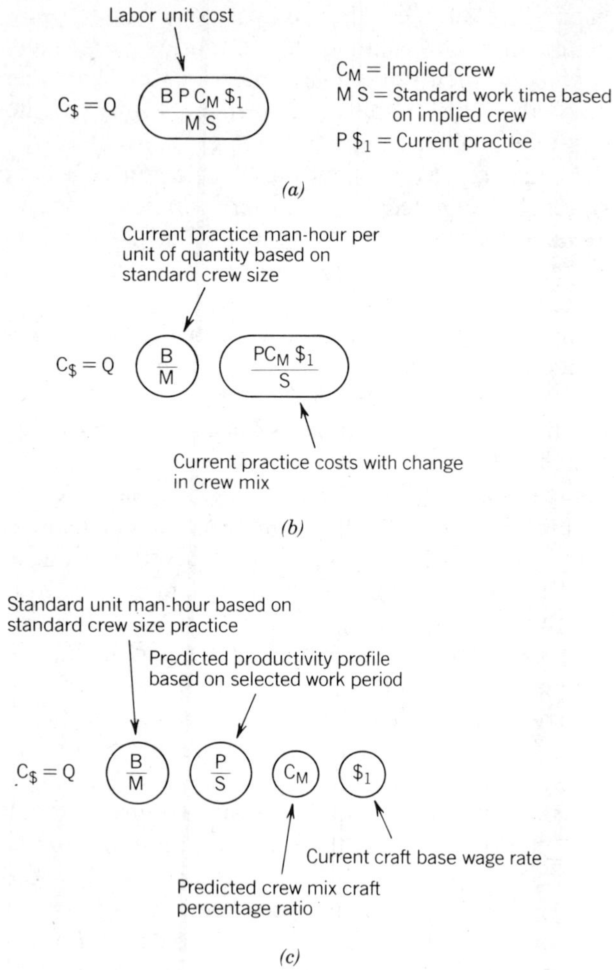

Figure 16.7 Typical labor cost estimating approaches in construction practice. (*a*) Labor unit cost approach. (*b*) Unit man-hour (standard crew) approach.(*c*) Standard unit man-hour approach.

Figure 16.6*a*. The general relationships between these components that develop during the process of establishing the estimated labor cost of an activity are depicted in Figure 16.6*b*.

The labor cost estimating model shown in Figure 16.6 can be used to describe the various types of labor estimating approaches currently found in construction practice.* Three typical examples are depicted in Figure 16.7. These correspond to the *labor unit cost* approach (small and medium-sized firms), the *unit man-hour* standard crew approach (medium-sized firm), and the *unit man-hour* approach (large firm).

When the small firm contractor (Project Team Component approach) establishes a cost control budget estimate, usually for a fixed price bid, the *labor unit cost* approach is used. In this environment, the labor unit cost reflects the aggregation of several "fixed" value variable constants established by the contractor, such as a standard crew size, forty hour work week, and productivities for the local work region known from experience. Thus, the labor unit cost equates to the product of those labor cost components, as illustrated in Figure 16.7*a*. The material content quantity for any work category is generally known and can be assumed fixed because the estimate is based on completed construction plans and specifications. If the small contractor is awarded the project, cost control then becomes a checking mechanism which simply involves comparing actual quantities and labor unit costs with those used in establishing the budget. This cost control approach is effective and practical for the small firm because actual labor costs are normally evaluated on a basis that is consistent with estimated labor unit costs.

The medium-sized contractor (Traveling Project Manager approach) normally utilizes the *labor unit cost* approach, however, under certain circumstances the contractor will analyze the crew mix (C_M) especially when operating in new work areas, unique situations or with specialized crews. In the *unit man-hour* (standard crew) approach, as depicted in Figure 16.7*b*, the labor cost estimate equation is dissected into variables evaluated from current practice man-hours per unit of quantity based on standard crew sizes (B/M), and a cost per man-hour from current practice ($PC_M\ \$_1/S$). The cost per man-hour can be adjusted by projecting a different crew mix (C_M) than the standard, and/or using current craft wage rates ($\$_1$) applicable to a specific location. Consequently, cost control not only checks and compares actual quantities and labor unit costs against budgets, but must utilize knowledge of actual man-hours per quantity and crew mixes, to forecast final costs. These forecasts aid the medium-sized contractor, especially in relation to cost-critical activities, and in identifying potential overruns.

At the large firm level, the *unit man-hour* approach is the predominant labor cost estimating tool. This tool is best applied when the labor cost estimate equation is further dissected into more fundamental and unique variables, as shown in Figure 16.7*c*. For example, the need to initiate estimating activities based on standard man-hours per unit of

*See Chapter 6, Estimate Preparation.

quantity (B/M) becomes important, because the large firm often works in diverse locations and cannot maintain a set of unit rates for every location. Consequently, this approach requires a separate special prediction of the expected site productivity in relation to a standard base. The analysis and determination of the productivity factor (P/S) becomes a significant planning decision. In addition, both the crew mix (which is otherwise normally a standard crew mix (C_M)), and craft wage rates ($\$_1$) are independently evaluated, because site locations and circumstances can vary drastically in the large firm construction environment. Finally, the material content (Q) is no longer an initially known or fixed value. This is due primarily to the nature of the projects and the facilities definition process involved in design–construct projects. Additionally, the quantities involved are generally larger than those found in the medium-sized firm project, and thus can vary over larger ranges (such as up to fifty percent higher than initial estimates) depending on the quality of the drawings at the time when the control budget is prepared.

Cost control efforts in this environment must, by necessity, focus on material content, productivity factors, crew mixes, and craft wage rates, because these components can vary significantly due to site location, work attitudes, project complexity, and so on. Thus, each component needs to be estimated separately and defined clearly for control purposes. The control budget estimate not only provides the cost control base, but it supplies man-hours for scheduling activity durations and analyses of manpower levels in relation to anticipated available resources and desired progress targets. The estimated site man-hours are the product of the material content (Q), the standard man-hour per unit quantity (B/M), and the predicted productivity (P/S).

Once the control base budget is established and after labor costs are incurred in the field construction phase, the act of comparing actual labor costs with the control base initiates the cost control system for a project. This occurs during the project management phase and is accomplished by identifying changes in the labor cost components of an activity. In order to illustrate the manner in which this may be done, the labor cost estimating model presented in Figure 16.6 can be further developed in terms of those components that impact an activities budget cost. Figure 16.8*a* defines these new cost components, while Figure 16.8*b* incorporates them into the labor cost control model. When this model is solved for the change in the activity's labor cost ($\Delta\ C_\$$), the interrelationships between the components that most dominately impact the delta change in an activity's labor cost are exposed, as illustrated in Figure 16.8*c*.

The model described in Figure 16.8*c* can be applied in the following manner. If the actual productivity, crew mix, and work hours of an activity are the same as the budget (i.e., $R_P R_M / P_M = 1$), the material content doubles (i.e., $\Delta Q = Q$), and the change in the wage rate reflects a ten percent increase (i.e., $\Delta\ \$_1 = .1\$_1$), then solving the equation for the change in the cost of the activity ($\Delta C_\$$) would reveal a twenty percent increase in the delta cost (assuming the cost of the activity is set at one (i.e. $C_\$ = 1$). Alternatively, if the change in the material content is zero (i.e., $\Delta Q = 0$) and the crew mix and work hours are the same as the budget, then when productivity drops by fifty percent and the wage rate rises by ten percent, the net change in the activity's labor cost is a twenty percent increase (assuming the cost of the activity is set at one).*

Cost control can be viewed in construction practice in terms of the manner in which the above cost equation is formulated. Thus, it is possible to identify different approaches at the small firm (PTC), medium-sized firm (TPM), and large firm (Field-Based Project Management) levels.

The structure and operation of a project cost control system at the Project Team Component level is necessarily constrained by a number of features, namely:

1. *It must be initiated, operated, and used by two or three staff only*, usually the foreman/superintendent, an office helper, and the project manager/partner.
2. The *control base estimate* is often set up in a *hasty quick effort*. This generally means that only a *gross dissection of budget items* is available in the control base.
3. Only a *simple, low volume data gathering effort* is possible in the field, and this data can only be effectively *cost coded in a limited finer detail* than that in the control base.
4. A simple *reconciliation* of budget entries with matching aggregated costs to date is sufficient for decision making by the project manager.

*The authors will leave it to the reader to work through the equation for these examples to verify the twenty percent increase.

$\Delta C_\$$ = Change in activity labor cost
ΔD = Change in activity duration
ΔQ = Change in material content quantity
R_p = Change in productivity rate
R_m = Change in manning rate
R_{cm} = Change in craft mix ratio of the crew (which is reflected in a percentage change in the composite crew wage rate)
$\Delta \$_1$ = Change in craft wage rate

(*a*)

$$C_\$ + \Delta C_\$ = (Q + \Delta Q)\frac{B}{MR_MS}PR_pC_mR_{cm}(\$ + \Delta\$_1)$$

(*b*)

$$\Delta C_\$ = C_\$\left\{\frac{R_pR_{cm}}{R_m} - 1 + \frac{R_pR_{cm}}{R_m}\left[\frac{\Delta\$_1}{\$_1} + \frac{\Delta Q}{Q} + \frac{\Delta\$_1}{S} \times \frac{\Delta Q}{Q}\right]\right\}$$

(*c*)

Figure 16.8 Labor cost control model. (*a*) Labor cost control components. (*b*) Labor cost control equation. (*c*) Labor cost control equation for delta change in labor cost.

5. Generally, if the estimator has not made a mistake, cost overruns in some items will more or less be balanced by cost underruns in other items. Also, to a limited extent the somewhat finer cost coded dissection of field activity items can be used by head office staff to trace back and detect possible cost problem areas so that corrective field action may be implemented.

At the small firm level, cost control is therefore essentially a recording effort which compares what is spent to date against budget items. This comparison is performed periodically, but mostly on completion of the item so that a simple mental analysis and summary of the (hopefully) balancing cost overruns and underruns is all that is required. This approach is very effective in terms of effort and also in terms of accuracy, because not many items are financially large and significant. It is also successful because there is very little change in productivity, crew mixes, and manning. Therefore, the project manager can focus on changes in material content and wage rates which are easily defined. However, the project manager still must maintain an intuitive feel for changes in productivity, crew mixes, and manning levels, because if the crew mix is not correct, for example, then the labor unit cost may be too high. A finer cost code dissection, together with a limited rate of field activity, provides the project manager with the time and means to intervene if a problem is uncovered.

A schematic representation of the Project Team Component project cost control system is illustrated in Figure 16.9. As indicated in that figure, typical examples of the documents involved can be found in Sections 6.3, 12.3 and 13.3 for the small firm.

As project size and complexity increases so does work volume and the number of cost-critical activities that must be closely monitored. Thus, these projects have a high rate of field activity effort and exhibit significantly increased cash flow problems that must be closely monitored if potential cost overruns emerge. Additionally, this accelerated construction rate occurs in a construction environment where past experience may not be totally relevant to the needs of the current project. Hence, it becomes necessary to implement a more rigorous cost control system that is capable of handling a wide range of construction activities in a manner than can quickly pinpoint cost overruns and provide a means of exposing the effectiveness of corrective management action.

These desirable objectives can be achieved by a more obvious use of cost codes, especially in initial cost estimates; a more professional concern for and assessment of the cost effectiveness of field construction activities; and the introduction and use of forecasting techniques that attempt to predict final costs, of an activity or for the project as a whole, *before* all these costs have been incurred. Thus, at the Traveling Project Manager level there begins to emerge an integrated cost control system incorporating cost forecasting.

Management decisions can now be based on in-

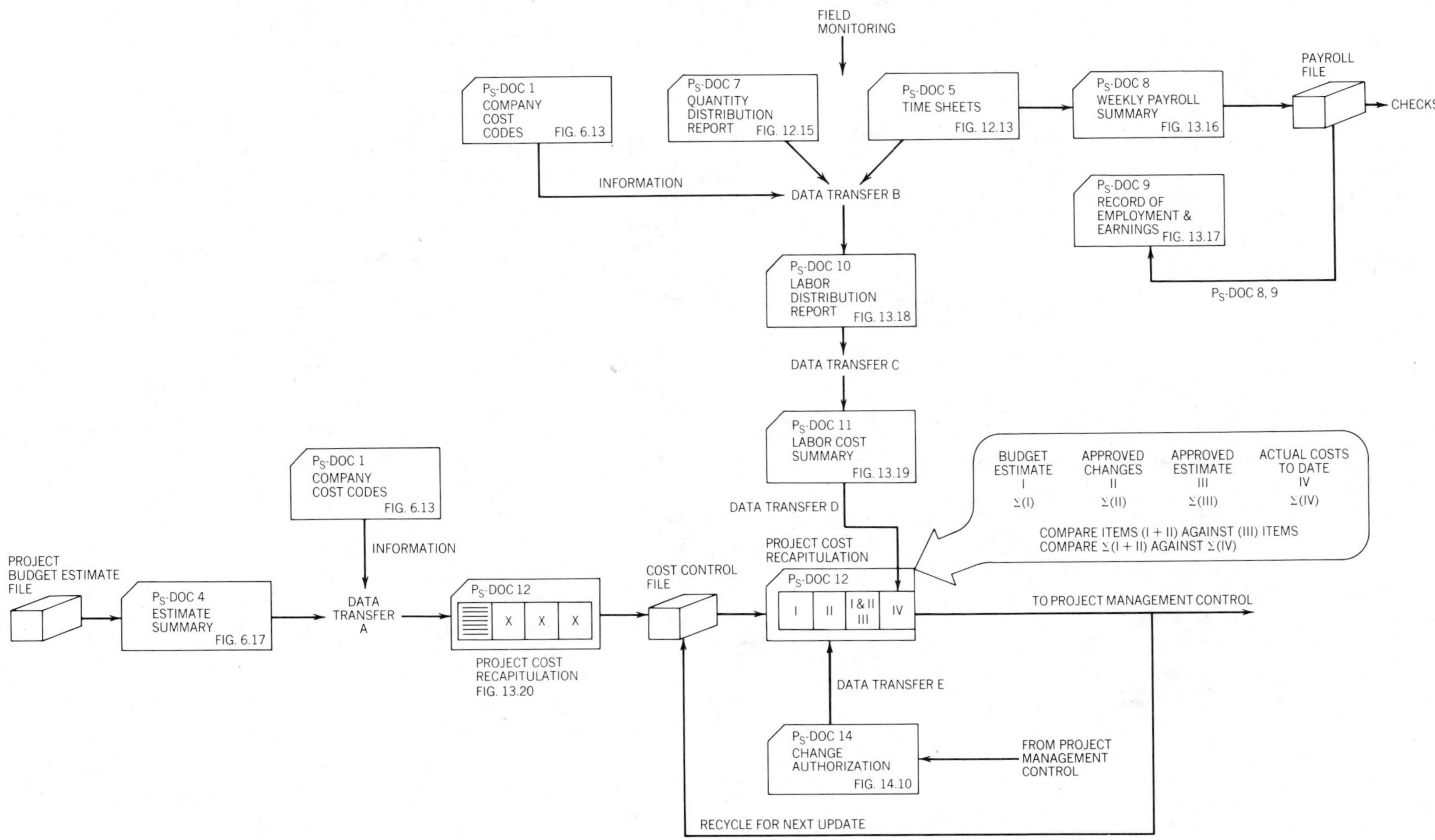

Figure 16.9 Project cost control system (Project Team Component approach).

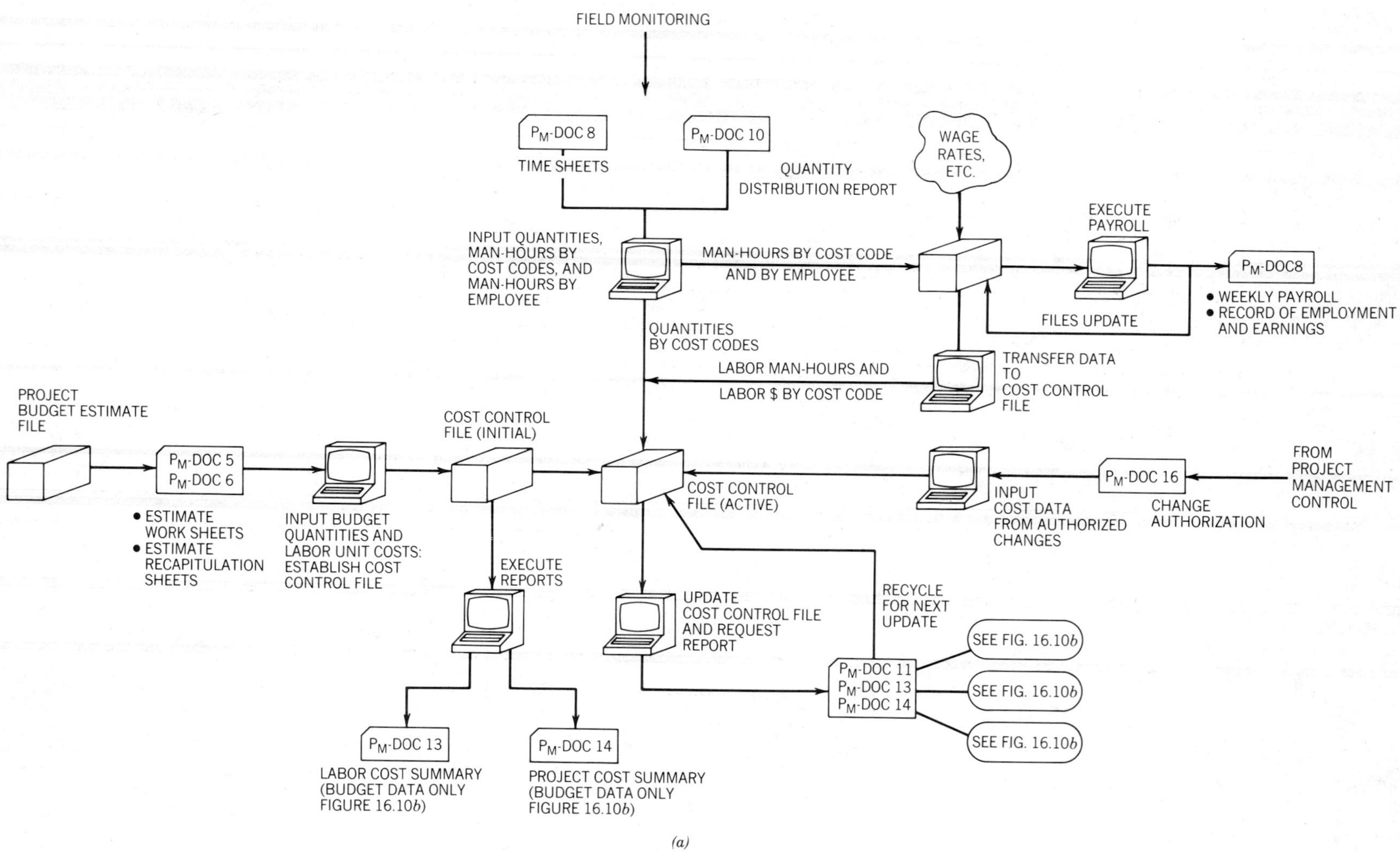

Figure 16.10 Project cost control system (Traveling Project Manager approach). (*a*) Cost control system.

P_m-DOC 11 LABOR DISTRIBUTION REPORT

COST CODE	QUANTITIES IN PLACE	LABOR HOURS	LABOR COST	LABOR UNIT COST	LABOR HOURS PER UNIT
	← ACTUALS →				

P_m-DOC 13 LABOR COST SUMMARY

COST CODE	UNIT MEASURE	CURRENT BUDGET	EXPENDED PERIOD	EXPENDED TO DATE	CURRENT STATUS OVER/UNDER	PROJECTED STATUS OVER/UNDER
		← QUANTITIES	LABOR UNIT COST	LABOR COSTS →		

P_m-DOC 14 PROJECT COST SUMMARY

COST CODE	ORIGINAL BUDGET	APPROVED CHANGES	CURRENT BUDGET	EXPENDED TO DATE	CURRENT PROJECTION	OVER/UNDER	% OVER/UNDER

(b)

Figure 16.10 (*Continued*) (*b*) Cost control documents.

formation on in-progress activities, together with the continuous monitoring by the project manager of the impact of these management decisions on field activity.

In an effort to achieve a more integrated cost control system, the change in productivity (R_P) is defined in terms of running unit costs to date and knowledge (with documentation) of actual man-hour per unit rates. This information prompts the project manager to determine the cause of perceived poor productivity by evaluating crew mixes and sizes (R_{cm} and R_m, respectively). The Traveling Project Manager can also check the change in material content (ΔQ) and wage rates ($\Delta\$_1$) as each impacts labor unit costs. By collecting and documenting these details, cost forecasts are prepared that identify the change in an activity's cost ($\Delta C_\$$) based on assessments of productivity, crew mixes, crew sizes, and changes in quantities and wage rates. This leads to increased cost control during the project management phase.

The Traveling Project Manager management approach is conducive to the effective implementation of the above mentioned cost control approach. The general structure of this cost control system is illustrated schematically in Figure 16.10*a*. The documents involved in the cost forecasting approach are indicated in Figure 16.10*b*. Thus, the Labor Distribution Report (P_M-DOC 11) reflects actual hours and costs expended, helps alert management to possible cost overruns, and aids the project manager with the analysis of the Labor Cost Summary (P_M-DOC 13). In addition, the Labor Distribution Report provides a check on the effectiveness of management action when corrective action has been initiated. The Labor Cost Summary Report first provides a measure of productivity through comparisons of budget versus experienced labor unit costs, and second provides the means of evaluating the impact of experienced productivity through forecasted projections (plus or minus) of labor costs. It also provides an indication of scheduling impact when referenced against a percentage complete evaluation (usually a mental analysis) of cost code areas. Finally, the Project Cost Summary (P_M-DOC 14) provides a measure of the impact of labor problems on total project costs. Collectively, the data and information gained from forecasts based on these three documents provides an effective cost control of projects at the medium-sized firm level. While projections are made on every cost code account, in these documents management need only focus on cost-critical activities. This management strategy enables a more effective management focus and control to be achieved. Typical examples of those documents referred to in Figure 16.10 are included in sections 6.4, 12.4, and 13.4.

At the Field-Based Project Management (large firm) level, a number of significant features force the emergence of a fully integrated cost control system. First, project complexity and uniqueness ensures that the unit cost approach to estimating is no longer feasible. Thus, budget estimates must be built around man-hours, crew mixes, quantities, wage rates, and hence, labor dollars. Consequently, a cost control system must rationally address these labor components. Second, it is no longer a simple matter to identify cost-critical activities. This is because of the magnitude of the man-hours involved in many cost areas, together with the possibility of productivity drops which may make any related activity cost critical. Consequently, there is a need to systematically look at every in-progress activity, and to develop a system for identifying, at any instant, a short list of critical activities. Thus, an effective cost control system at this level of operation must have the capability of developing critical activity lists and deviation analyses, and the sensitivity to detect cost or productivity movements in rolled-up broad categories of work. Additionally, there is a need to develop many different ways of detecting trends and deviations as they are emerging, and to incorporate techniques for forecasting the impact of these trends. As a consequence of this focus it becomes necessary to involve more people in the management and operation of the cost control system, which most often is computer based. This computerization becomes essential to handle the volume of data and the required capability of focusing on cost problems as they are detected. A compensating factor, however, is that the dollar volume of the project and the scale of work involved in many cost areas, provides time for management to detect and react to problems, and to check the impact of cost deviations and trends, through activity analysis type approaches to field operations. Finally, in order to ensure that an effective integrated cost control system is achieved, it is essential that such a system provide timely reports and interact with the management decision process.

The essential nature of the cost control system at the large firm level is that it is a rational and structured approach to the data that is to be collected, and to how this data is to be processed to provide relevant management information and document reports. These data must be effectively interrelated into the management decision process in an organization designed to act as problems arise. The efficiency of this approach is enhanced by its systematic culling of critical activity areas, and the continuous monitoring of man-hours, labor costs, and performance curves. Thus, the concept of cost-critical activities is expanded to include a focus on all forms of deviations that impact costs.

The large firm's cost control system must monitor and control the change in material content (ΔQ), because this component often becomes a large unknown that varies over the project's life. Since the quantity of material can fluctuate markedly, work package concepts evolve in Crew Planning that necessitate the use of nonstandard manning levels and crew mixes. These two variables must also be monitored and controlled as they change in relation to budget estimates. By monitoring crew mixes and manning levels (R_{CM} and R_M), the change in productivity (R_P) can be analyzed to determine if the predicted productivity used in the budget estimate is correct. In this way, revised predictions of productivity and crew mixes can be developed for cost analysis. These data also provide project scheduling with a basis for reevaluating manpower levels and progress targets. Finally, the craft wage rate may vary due to the impact of high manpower levels that could cause shortages of craft labor resulting in increased pay rates to obtain these crafts. Thus, the craft wage rate must also be monitored continuously.

Because the change in cost ($\Delta C_\$$) results from changes in quantities, crew mixes, productivity, manning levels, and wage rates (all of which are monitored for control purposes), the cost control system must incorporate a documented project change order and trending system. The purpose of change order and cost trending is to alert both project management and the client of the cost impact resulting either from changes in project scope and/or trends in material content, productivity, and the like. By identifying cost changes and documenting the magnitude of these changes, timely decisions can be made in regard to proceeding with a change in scope or, for example, implementing possible corrective action to improve productivity. Finally, through a frequent and continuous accumulation of cost changes due to changes in scope or project trends, the total project cost is updated and can be compared with current budget costs. This provides management with a macroview of the project's total cost at any time, and provides an opportunity to make major changes in the project's direction if costs trends are indicating significant overruns.

The expanded depths of cost control at the large firm Field-Based Project Management level naturally result in an increased level of documentation and computer utilization. This is reflected in the cost control system for the large firm as schematically illustrated in Figure 16.11. In Figure 16.11*a* the initial set-up of the project cost control system is shown. The format of the cost control system is based on standard company operating procedures, only taylored to match specific project requirements and cost coding structures. Once the format is established for the project, budget data from the project budget estimate file is transferred into the project cost control file. At this stage in the project, the various project cost control documents, such as Labor Unit Man-hour Reports (P_L-DOC 49) are available and will accept actual expenditures of man-hours, quantities and labor costs. Once actual data begins to accumulate as a result of construction activity, field cost control, Figure 16.11*b*, commences. The project cost control file becomes active and receives a variety of inputs including progress data (e.g., percent complete, quantities in place, etc.), and, as the project proceeds, changes in the budget and forecasted cost due to approved change orders and project trends. The project cost control file is periodically updated and the reports it generates provide the data necessary to evaluate changes in material content, productivity, crew mixes, and so on, in relation to current budget estimates. These reports must be evaluated for problem detection, and timely action must be taken by project management to resolve problems or potential problems as they are uncovered. In the Field-Based Project Management approach, total project cost control is a head office function and, as shown in Figure 16.11*c*, the project cost control file is updated with all other actual project expenditures (e.g., material costs) to produce a Total Project Cost Summary (P_L-DOC 59). These reports are then scrutinized for problem areas, and project management takes corrective action where required. Typical examples of those documents referred to in Figure 16.11 are included in sections 6.5, 12.5, 13.5, and 14.5.

The cost control system for the large firm operating in the Total Field Autonomy mode is similar in structure to that described for the Field-Based Project Management approach. Because cost control is closely integrated with project scheduling in this management approach, differences will emerge when establishing the labor control base estimate and viewing cost control in action. These differences will be discussed in Section 16.5, Cost and Scheduling Integration.

16.4 PROJECT SCHEDULE CONTROL

Project schedule control addresses the *timeliness* of field activity, and seeks to coordinate the sequence of field activity and relate it to work progress and the effective use of resources. Thus project schedule control has four major objectives, namely:

1. *Work sequence coordination:* The planning and coordination of the sequence of field activity to ensure that actual and required work progress is compatible with the achievement of project milestones and contractual completion dates. This is a focus on the contractor's ability to perform and to meet client and contractual requirements.
2. *Time schedule coordination:* The coordination of construction drawings and procurement with field activity and subcontracted work. This is a focus on ensuring the practical feasibility of the construction effort.
3. *Work progress evaluation:* To relate work progress over time in project activities to project progress status, and hence to progress payments. This is a focus on the quantification of work in place in terms of the time-integrated use of resources as a means of supporting project status evaluations and the financial control of the project.
4. *Work progress productivity:* To relate resource use against planned and achieved work progress. This is a focus on the documentation of the effective use of resources or of productivity.

Schedule control usually begins in the project definition phase for medium-sized and large firms with the Project Planning and Project Scheduling decision processes, and immediately precedes the field construction phase in the small firm environment, where it is not economical to perform this activity before knowing that the firm's bid has been successful. The control base schedule must reflect the project plan and the contractual time commit-

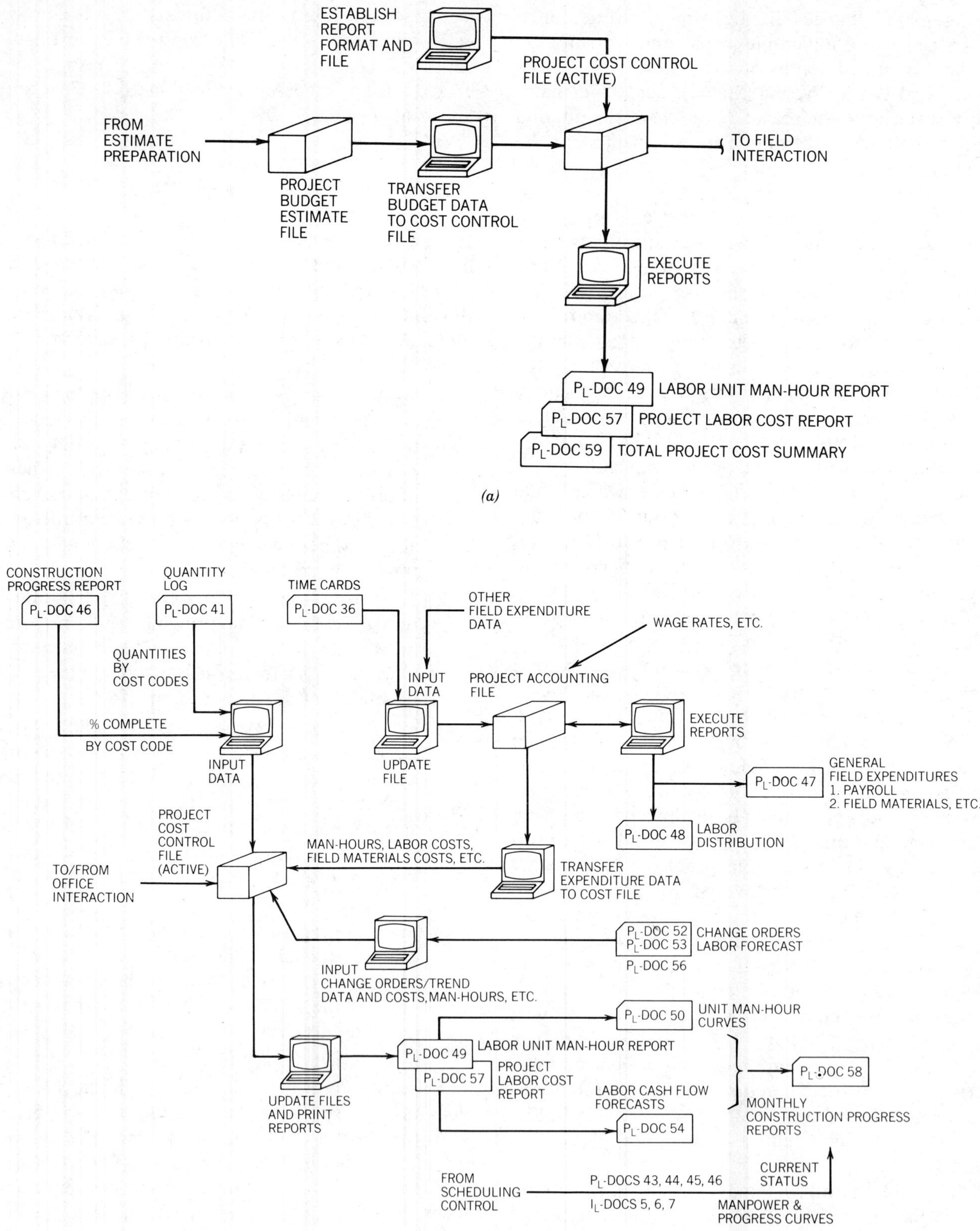

Figure 16.11 Project cost control (Field-Based Project Management approach). (*a*) Initial set-up. (*b*) Field cost control.

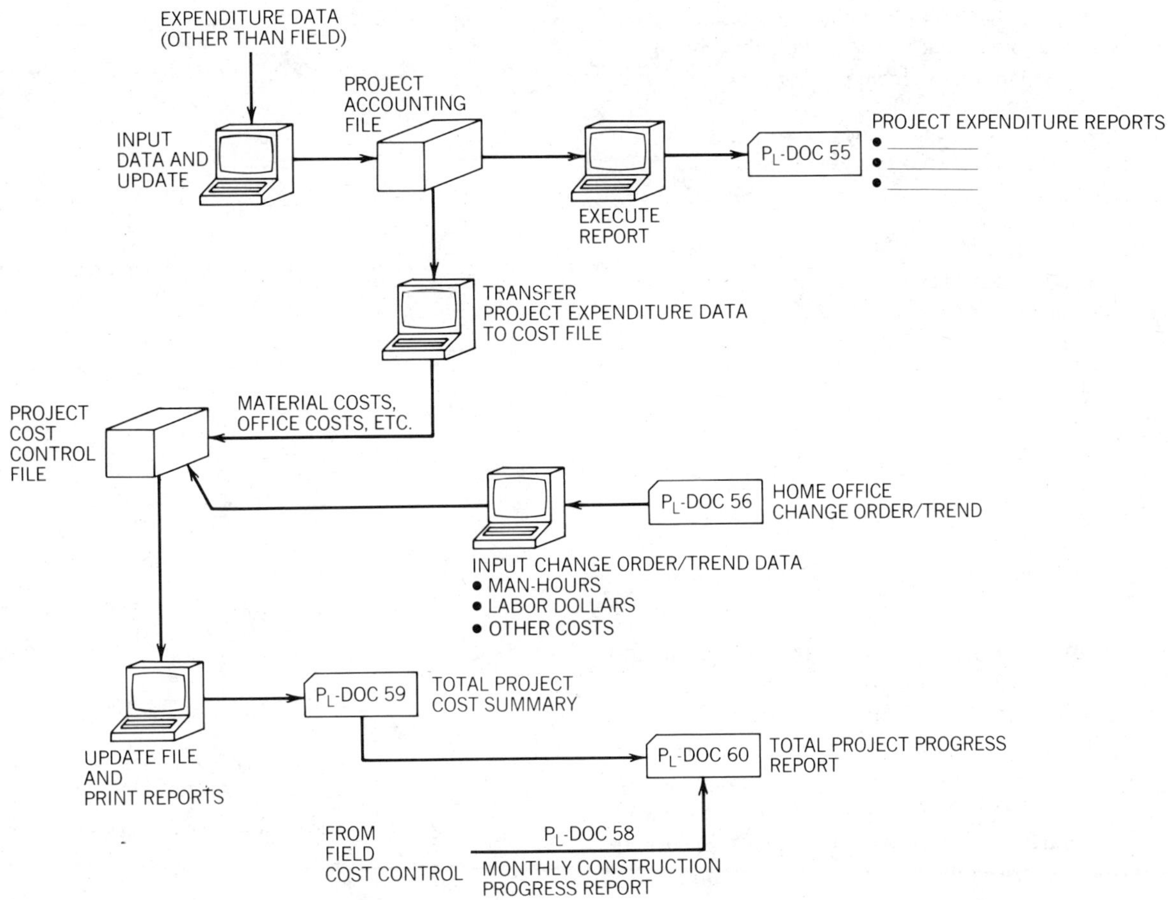

(c)

Figure 16.11 (*Continued*) (*c*) Total project cost control (office).

ments. It must also incorporate, however grossly, the logical sequence of major work activities and relate them to project milestones and schedule events of special management interest to the contractor. Usually the control base schedule is portrayed in milestone, bar chart, or network form, and the level of detail considered, and included, in it can be related to the needs and size of the project, the type of contract involved, and the management approach of the contractor.

Whatever method is employed in determining the control base schedule, consideration must be given to the time required to accomplish the work content of the activities and tasks involved. In this respect, the determination of the time required to perform and complete an individual work task is of elemental importance. A simple model for determining the scheduled duration for the completion of the work content of a work task is indicated in Figure 16.12. Figure 16.12*a* identifies the major factors that have a bearing on the work task duration, and Figure 16.12*b* depicts the general relationships between these factors. The reader will be familiar with the concept of *float* in network analysis, and no further mention will be made of it, apart from noting that its determination or inclusion in a control base schedule reflects a management strategy concerning the rate at which a project activity (or the project as a whole for that matter) will be performed. With regard to the determination of the work task duration as a function of its work content, a variety of approaches exist in practice. These different approaches reflect the nature of the scheduling problems to be solved. Two typical examples, which represent totally different approaches, will be considered here (see Figure 16.13).

The first, called the *standard crew production* approach, is illustrated in Figure 16.13*b*. This ap-

D = Activity scheduled duration (e.g., days)
Q = Material content quantity (e.g., 100 cubic yards)
B = Standard man-hour content per quantity unit (e.g., 10 man-hours/cubic yard)
P = Estimated productivity relative to standard (e.g., 10% increase due to site conditions)
M = Estimated level of manning for work task (e.g., 10-man crew)
S = Scheduled work period unit (e.g., 40 hours/week based on 8 hours/day)
F = Scheduled Float Time

(*a*)

$$D = \frac{QBP}{MS} + F$$

(*b*)

Figure 16.12 Schedule duration estimate model. (*a*) Task work content scheduled duration components. (*b*) Task work content scheduled duration model.

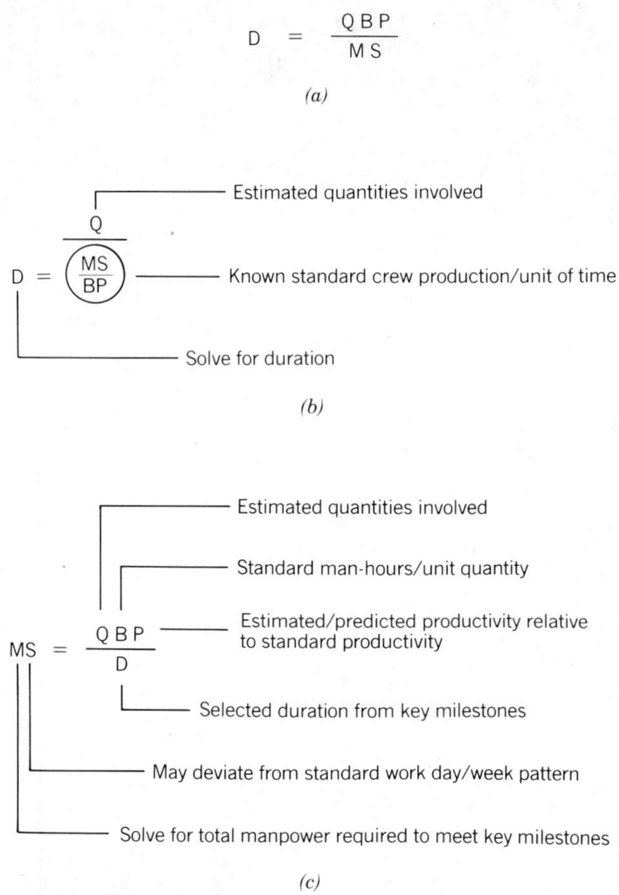

Figure 16.13 Typical control base schedule approaches. (*a*) Task work content schedule duration model. (*b*) Standard crew production approach. (*c*) Manpower prediction approach for scheduled task completion.

proach is most commonly used in the small and medium-sized firm environment, when firm and definitive estimates for familiar work activities can be made. In this approach, the required scheduled duration of a work task (D) is determined once the quantities involved (Q) are estimated by using known standard crew production rates (MS/BP). In the second or *manpower prediction* approach, the determination of the manning levels (M) required to achieve preestablished milestone periods is determined once accurate estimates are available for quantities (Q), and prediction of field productivity (P) relative to a standard man-hour per unit (B) has been made. This approach is commonly met within the large firm environment on projects where construction must proceed before, and while, definitive estimates become available, and in an environment where large work forces exist on site (see Figure 16.13*c*).

While the essential aspects of schedule control relate to the comparison of achieved schedule durations and milestones with planned schedules, a variety of factors influence field performance. The basic components that dominantly influence the achievement of planned scheduled durations (D), namely: changes in quantities (ΔQ), in productivities (R_P), and in manning levels (R_M), are shown in Figure 16.14*a*. The consequential change in scheduled duration (ΔD) can be expressed in terms of the Schedule Control Equation 16.14*b*, which when solved for the delta change in scheduled duration (ΔD), gives the first order approximation of the basic schedule control equation illustrated in Figure 16.14*c*. This schedule control equation for delta changes in task duration can be used to describe the different schedule control processes used in practice. In the Project Team Component environment of the small firm, the overall schedule control base is established after the lump-sum bid has been won and then is established in terms of a firm contract completion date. This schedule is usually portrayed in terms of a bar chart based on a gross category dissection of work items in the project to match the completion date without direct reference to the estimate. The main purpose of the schedule control bar chart is to establish subcontract interface milestone dates, and to provide a basis for progress payment evaluations. At the small firm level the activity durations associated with the major category dissection of the project are established from past experi-

ΔD = Change in scheduled duration
ΔQ = Change in quantities involved
R_p = Change in productivity rate
R_m = Change in manning rate

(*a*)

(Assuming no change in scheduled work hours, etc., other than isolated overtime)

$$D + \Delta D = \frac{(Q + \Delta Q) B P R_p}{M R_m S}$$

(*b*)

$$\Delta D = D\left(\frac{R_p}{R_m} - 1\right) + \frac{\Delta Q}{Q} \times D\left(\frac{R_p}{R_m}\right)$$

(*c*)

Figure 16.14 Schedule control model. (*a*) Schedule control components. (*b*) Schedule control equation. (*c*) Schedule control equation for delta change in task duration.

ence. Since drawings are generally complete for lump-sum contracts, there are few uncertainties in quantities and quantity changes (ΔQ) are usually small. When they do occur, these quantity changes reflect contractor estimating errors (which tend to balance each other out), or contract addendums (which if anything tend to ease project schedules). Thus, schedule control, while sensitive to the overall schedule requirement to meet contract completion dates, usually only involves the incorporation into the control base bar chart schedule of colored mark-ups based on field observations and material in place reports that establish gross percentage completion of the major category dissections. This management activity, coupled with small corrective changes to subcontractor site involvement schedules, is sufficient to effectively control project schedules. At the small firm level there is little need for any manpower focus to project scheduling.

Typical examples of the scheduling documents incorporated into the small firm's scheduling control approach can be found in Sections 7.3, 12.3, and 13.3.

At the Traveling Project Manager level of the medium-sized firm, the projects encountered are larger and more complex than those handled by the small firm. Because of this, and because more subcontractors may be involved, the control base schedule must be more carefully prepared. It is usually a logically and more finely developed bar chart with activity durations established by the project manager and superintendent, who rely mainly on past experience. Specific manpower checks are made for activity durations involving critical crafts. The control base schedule is usually not directly tied to estimate details but must grossly reflect the total quantities involved in each activity, in addition to both contract and subcontractor involvement interface milestones. It therefore begins to incorporate a finer dissection of project activities and reflects more closely the plan of field activity and the basis for detailed schedule control.

Activity durations, especially those involving the critical crafts, are determined by solving the schedule model for estimating activity durations (D). Since quantities (Q) are known from the estimate and production performance (BP) is known from experience (for fixed crews and work periods (MS)) it is fairly simple to establish durations for activities on the control base schedule. Thus this schedule begins to logically address all the components (quantities, manpower, crew size and performance productivities) that impact schedule control.

Schedule control involves the continuous monitoring of the status of critical field activities and the coloring of activities on the bar chart schedule together with continuous reassessment of completion forecast dates for critical activities, subcontractor performance and the like. Consequently, the Traveling Project Manager and field supervisors can begin to coordinate the hiring and firing of critical trades with the performance of field activities to ensure the achievement of schedule milestones and a more detailed comparison of field performance with progress payment evaluations. Thus, at this level the control base schedule is established by integrating the Project Scheduling and Crew Planning decision processes with continuous inputs into

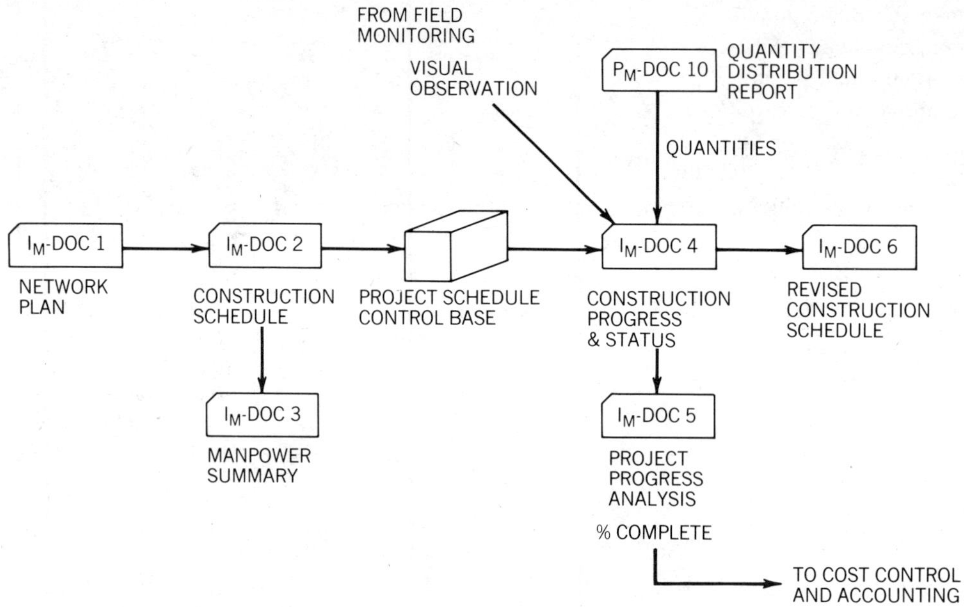

Figure 16.15 Project scheduling control system (Traveling Project Manager approach.)

the Crew Management and Project Status, Analysis, and Reporting decision processes. Also, the schedule control effort now incorporates in a continuous and focused manner Field Monitoring and work force manipulation of the Crew Management decision process. This provides project management with the up-to-date information needed to make and implement critical schedule control decisions. The general structure of the project schedule control process at the Traveling Project Manager level is illustrated in Figure 16.15. Typical examples of the scheduling control documents referred to in this figure can be found in Sections 7.4, 12.4, and 13.4 for the medium-sized firm.

At the Field-Based Project Management level, project size and the technical nature of a project, coupled with an often evolving scope definition, forces a significant change to the scheduling control practice of the large firm contractor. In the early phases of a project, the schedule is influenced by predetermined key milestone dates of major physical components that must be met to ensure overall project completion. Hence, the duration (D) of the main activities is selected based on these milestones and previous experience. The contractor must then evaluate the manpower required to achieve the selected durations. The manpower prediction model (Figure 16.13*c*) is used to determine if predicted manpower levels are achievable in relation to manpower buildup rates and the availability of crafts in the project site location. Thus, the work content (Q) and the standard man-hour rates (B), coupled with predicted productivities (P) and major component durations, are estimate inputs the contractor utilizes to solve the scheduling control model for the manpower required (M). In some cases, the contractor will be forced to vary the work periods (S) for the work force as a whole in order to enable the work to be completed within available field work force (M) constraints.

Once detailed estimates are prepared, scheduling control can begin to focus directly on fixing the base for overall construction control. At this scheduling level, manpower is still predicted based on estimated durations, work content, and activity man-hours, but in sufficient detail to establish firm project control base manpower plans, progress curves, and productivity profiles for future monitoring and control of the project. These detailed schedules are used for short-range planning and scheduling of work activity in the field once construction commences. This is accomplished by further dissecting the work planned for the next period (e.g., three months) into control schedules with work identified on a bi-weekly basis. Crew size can then be evaluated based on the actual work content (Q), the estimated man-hours (B $\times$ R_P) and the scheduled duration (D).

The process of establishing initial project schedules that evolve into detailed control level schedules is accomplished through both manual and computerized scheduling system techniques. Bar chart techniques are normally applied until sufficient estimate detail is available to produce logic networks. While bar charts are maintained for management level reporting, logic networks are developed for detailed project scheduling to ensure that major milestones can be achieved within acceptable manpower limits. This project schedule is then converted into construction control level schedules for the detailed crew planning and management of the field effort.

The scheduling control system in the Field-Based Project Management approach (large firm) is designed to provide early warning when deviations from control base plans are emerging. Special reports focus on both critical activity status and actual construction progress. When these reports are coupled with cost control information such as actual man-hours and quantities expended, they yield a viable means of evaluating the current status of the project. In addition, because control estimates are prepared prior to the completion of the design effort, variations on the work content (Q) of activities will occur. These variations are often significant. These material quantity trends, along with other changes in scope, when identified, are quantified through the change order and trending system. Their impact on scheduled activity durations and manpower requirements must be evaluated. This may result in revised schedules with newly forecasted manpower and progress curves based on updated activity durations and forecasted changes in the work content (Q). The revised schedule is reflected in detailed control level construction scheduling for crew planning and management.

When changes in activity durations (D) are reviewed and the material content is correct, then changes in productivity (R_P) and/or manpower (R_M) must be analyzed. This information is provided when actual productivity/manpower levels are compared to plans and differences identified. The impact of these differences can be evaluated in terms of, for example, increased manpower to achieve schedule progress targets, or possible schedule extensions/delays. The scheduling system must provide this information in a timely manner so that adjustments can be made by management to minimize schedule (and cost) impact, while still achieving the project's schedule goals.

The scheduling control system for Field-Based Project Management is illustrated in Figure 16.16. The scheduling system is characterized as computer based (mainframe, mini, or micro computer depending on project size). Figure 16.16*a* depicts the initial set-up of the project schedule control base. The input data consists of the control structure of the project, quantities and man-hours, network logic sequences, and project milestone dates. This data is formalized into the project control base schedule as defined by CPM listings, bar charts, progress curves, and manpower curves. The control base schedule is further manipulated to provide the lowest level of detail required for construction planning and execution in the field. This scheduling approach is shown in Figure 16.16*b*. Finally, scheduling control emerges when actual data is collected and input into the scheduling system. As described in Figure 16.16*c*, various reports are generated that identify potential schedule deviations that must be addressed by construction management before serious problems arise. The impact of these deviations are evaluated and revised schedules are generated as problems are solved. This is a continuous process over the life of the project. Typical examples of the reports referred to in Figure 16.16 are included in Sections 7.5, 9.5, 12.5, 13.5, and 14.5.

The schedule control system for the large firm operating in the Total Field Autonomy mode is similar in structure to that described for the Field-Based Project Management approach. Because scheduling control is closely integrated with project cost control in this management approach, differences will emerge when establishing the control base schedule and viewing schedule control in action. These differences will be discussed in the Section 16.5.

16.5 COST AND SCHEDULING INTEGRATION

In the management of a construction project, the contractor must confront the management problems associated with controlling both the estimated project dollars (costs), and the planned time allowed to complete the project (schedule). While each of these components (cost and schedule) has a different focus when related to project control, as demon-

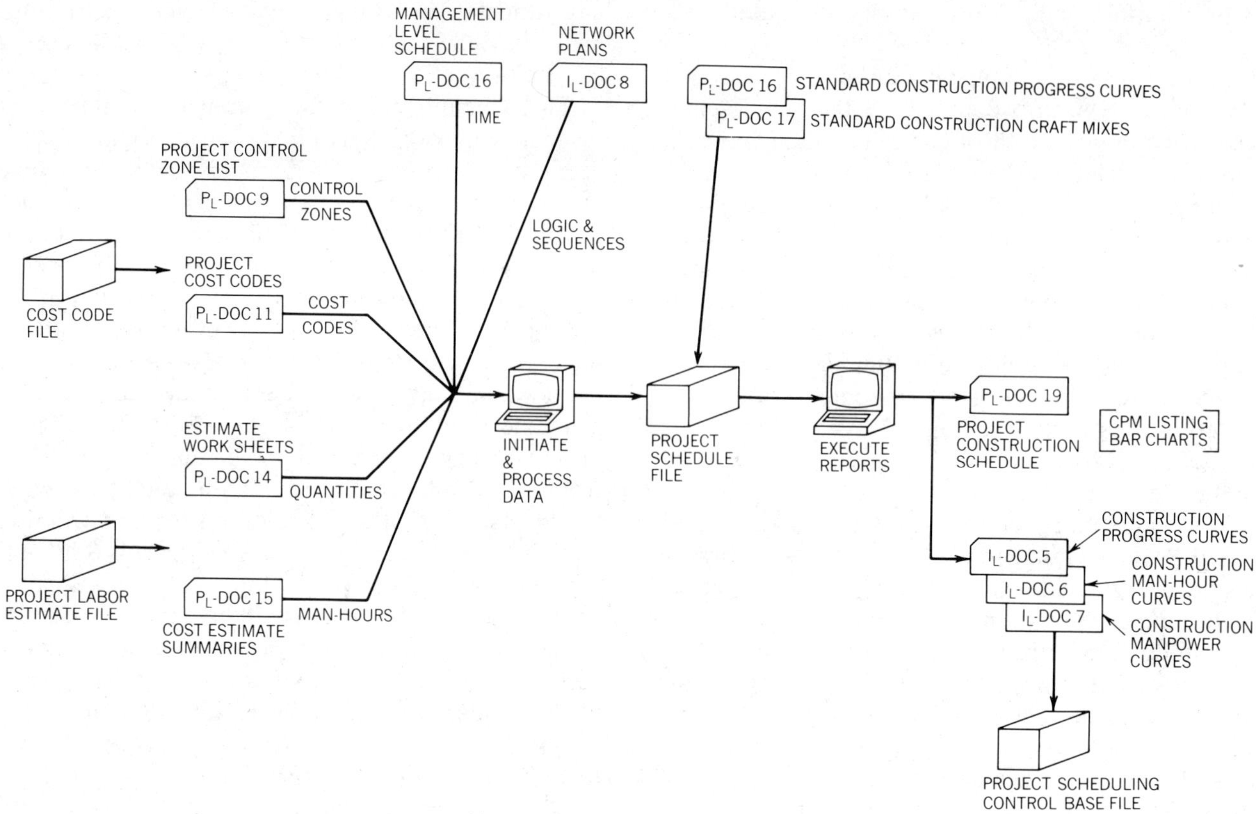

(a)

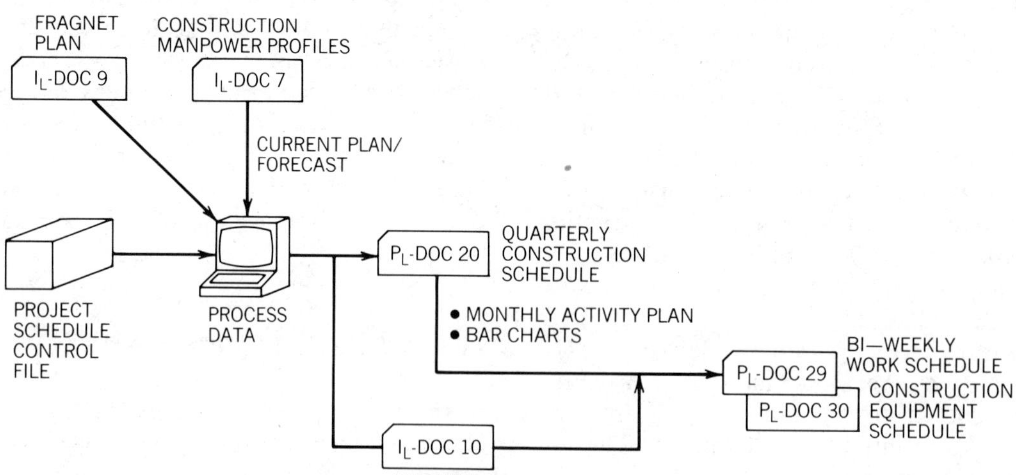

(b)

Figure 16.16 Project scheduling control system (Field-Based Project Management approach). (*a*) Establish project schedule control base. (*b*) Construction control scheduling.

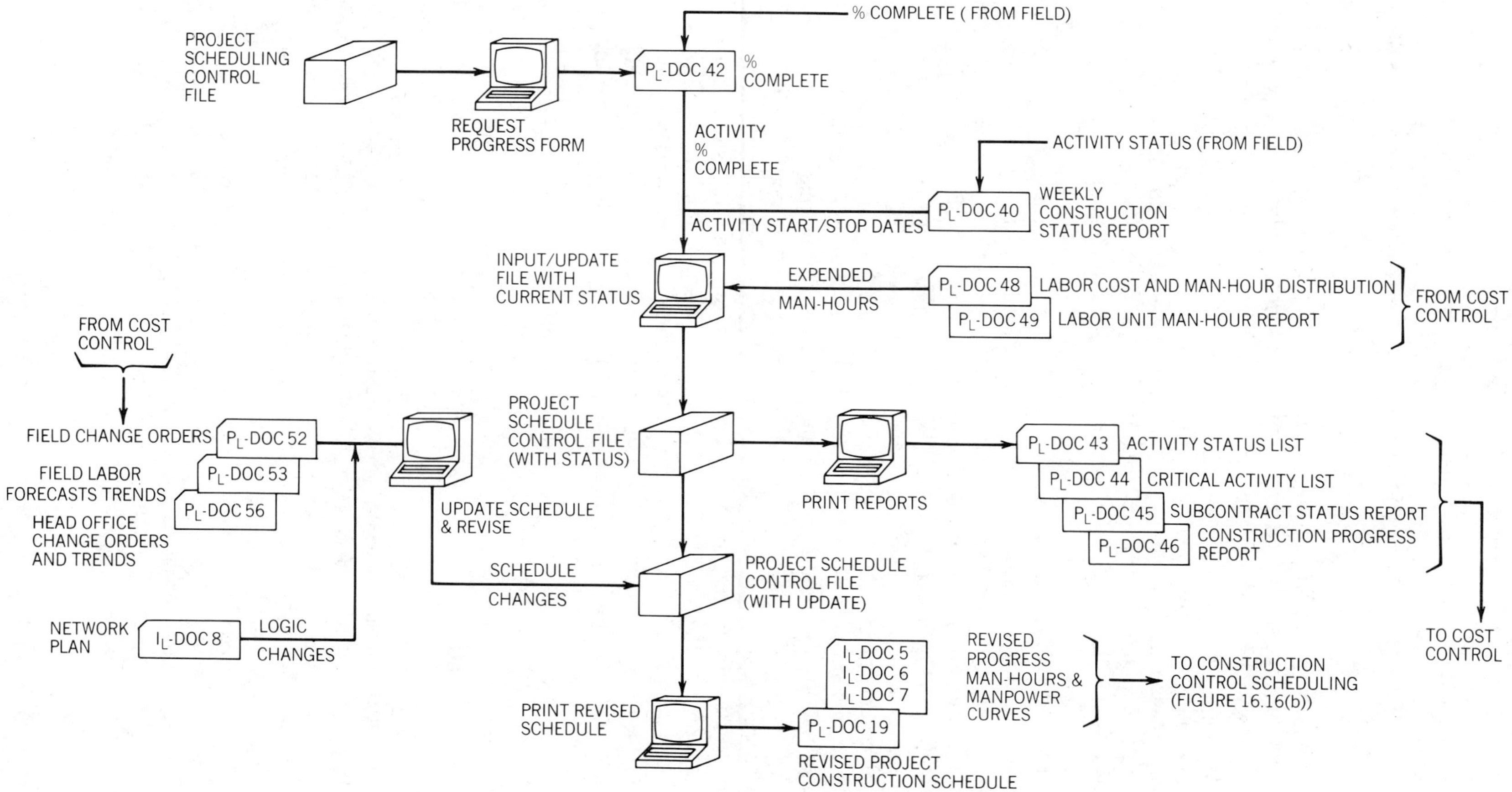

(c)

Figure 16.16 (*Continued*) (*c*) Project scheduling control.

$$(1)\ \text{Cost}\ (C_{\$}) = \frac{QBP\ \sum C_m \$_1}{MS} \quad \text{(See Figure 16.6)}$$

$$(2)\ \text{Duration}\ (D) = \frac{QBP}{MS} + F \quad \text{(See Figure 16.12)}$$

$$(3)\ \therefore \text{Cost}(C_{\$}) = (D + F) \sum C_m \$_1 \quad \text{where } D + F \text{ is expressed in man-hours}$$

Figure 16.17 Equation that integrates base cost and schedule.

strated in Sections 16.3 and 16.4, they are interrelated because the actual man-hour effort contributed to any activity over time produces the actual cost of that activity. Consequently, any success in controlling the time and effort committed to an activity, and hence its cost, depends on whether or not the budget and initial schedule for that activity are interrelated in some way either formally or informally. The manner in which this is achieved identifies the level of cost and schedule integration that is incorporated into a contractor's construction management approach. At one extreme, for example, cost and scheduling integration may be informally handled through a project manager, based on past experience. At the other extreme, however, a fully documented Management Information System is required. In the latter instance, the level of project team size increases and involvement in the project burgeons.

Cost and schedule integration can be described using the components of the Labor Cost Control Model presented in Figure 16.6, and the Scheduling Control Model described in Figure 16.12. These two equations contain common project components that influence both the change in cost and duration of an activity. These common project components are:

1. The material content quantity (Q and ΔQ).
2. The change in the productivity rate (R_P).
3. The change in the manning rate (R_M).

On the one hand, the actual duration of an activity depends on both the expended man-hours as determined by actual manning levels, and the productivity associated with installing the given work content of the activity. On the other hand, the actual labor cost of an activity is derived when the expended man-hour effort is converted to labor dollars by an actual wage rate. Therefore, the work content, schedule duration, man-hours, productivity, and costs related to an activity are interrelated. In order to control these components, the budget estimate and the initial schedule plan must be related for activities based on a commonly defined work content, an assumed productivity that produces estimated activity man-hours, and/or labor unit costs and scheduled activity durations. When translated into a model using the Labor Cost Estimate Equation (Figure 16.6*b*) and the Task Work Content Scheduled Duration Model (Figure 16.12*b*), the integration of cost ($C_{\$}$) and schedule (D) for an activity can be represented in equation form as shown in Figure 16.17.

Once a plan is defined by an approved budget and schedule, and progress and costs are incurred over time, the control process begins. Evaluations are made to determine current status against approved budgets and schedules. This status provides the basis for problem identification, especially when the project is behind schedule and over budget. The cause of these deviations must be evaluated in terms of the common components of productivity, material content, and manning rate. For instance, if productivity is deviating from the original budget estimates, then the man-hour content of an activity will be different than the estimate. This will lead to changes in scheduled duration and/or cost. Alternatively, if the estimated work content is not correct or the assumed estimate basis that defined the material content has changed, then the cost and/or schedule duration for the activity will change. The timely identification and resolution of these deviations by management is essential to successfully controlling cost and schedule, and completing a project on time and within allowable budgets.

The depth and degree of cost and scheduling integration that occurs during project execution is influenced to a large extent by project characteristics (size and complexity), the project environment (tight schedule), contractual relationships (lump-sum or cost plus), and the management approach applied to the project. Each of these influencing factors are considered in the description of the level of cost and scheduling integration that occurs in the Project Team Component approach, the Traveling Project Manager approach, the Field-Based Project Management approach, and finally the Total Field Autonomy approach, where a fully integrated cost and scheduling Management Information System approach emerges.

In the small firm environment, which facilitates the Project Team Component approach, the contractor is not likely to encounter major surprises in field labor productivity. This is because the firm has an intimate working knowledge of local work practices and uses standard crews. Furthermore, the small contractor generally works with fixed plans and specifications wherein the work content of each activity can be quantified with some assurance that these quantities will accurately reflect the materials to be purchased and installed. Apart from possible estimating errors in quantity take-offs, the labor unit cost of an activity can be estimated with some degree of reliability.

The project schedule, normally in a bar chart format, is developed based on past experience. Activity durations are formulated mentally to reflect typical start and completion sequences for each activity. While the small contractor has a working knowledge of the estimated labor cost of each activity as a frame of reference, past experience has a bigger role in establishing activity durations.

Because most small projects are paid on a lump-sum basis, the contractor's cost control emphasis is on balancing labor cost overruns against labor cost underruns. This approach provides a means of adjusting labor unit costs for future reference when preparing estimates for bids on new work. However, when significant estimating errors are uncovered that would have an adverse impact on labor costs, the contractor attempts to overcome them by cutting corners wherever possible.

Scheduling control generally revolves around simple techniques such as coloring the bar chart schedule with new forecasted activity durations. These forecasts are determined by direct communication with the field and/or subcontractors, and incorporate an activity's current progress and status. While there exists no direct formal link between schedule and costs, the contractor knows from experience which activities will cause significant cost problems if planned or forecasted activity durations and completion dates are not achieved. In this way a *loose, informal* integration of costs with schedule is achieved at the small firm level by the contractor operating in the Project Team Component mode.

At the Traveling Project Manager level in the medium-sized firm environment, a focus on cost- and schedule-critical activities begins to emerge as a basis for project control. Cost-critical activities are monitored through a continuous reading of actual labor unit costs. When these actual unit costs are compared to estimated labor unit costs, a measure of the productivity achieved over the duration of categories of work tasks, as classified by cost code, is provided. In this way, cost control has some schedule sensitivity due to the continuous tracking and analysis of labor unit costs over time. This approach is complimented by forecasting labor unit costs to complete in an attempt to quantify the magnitude of potential overruns or underruns. These forecasts to complete must be tied to the remaining time allotment for a given set of cost-critical activities, if they are to accurately reflect the impact of changing labor unit costs. Thus, the medium-sized firm contractor also becomes sensitive to the potential cost impact associated with forecasted slippages in activity durations. This is especially so when the activity is schedule critical, and influences the rate at which new activities can commence. By forecasting final project labor unit costs, the contractor can comprehend the cost implications of extending an activity's duration.

The labor unit cost forecasting technique, coupled with a realistic evaluation of scheduled progress versus actual progress, provides an effective integration of a *mixed formal and informal* cost and scheduling system. This approach frequently leads to changes in activity durations, crew sizes and mixes, and construction methods, all of which are incorporated into the bar chart schedule for schedule-critical activities. Any cost overruns that occur are borne by the contractor, due to the typically lump-sum type contracting environment in which the medium-sized contractor operates.

The Field-Based Project Management approach that emerges in the large firm environment begins to integrate cost with schedule through the man-hour content of an activity. For example, the control budget is established by using estimating techniques that focus on the man-hour involvement in a work task for an estimated work content quantity. The man-hour content is then converted into a schedule duration based on productivities and crew sizes and mixes, provided that the project cost code system is structured in a manner that facilitates a match-up between the activities in the estimate with those comprising the schedule. Thus, the cost code system gives a common reference point for both cost estimating and scheduling at the large firm level.

Project control is based on an integrated and focused information gathering system with cost coding as its common denominator. Through this system, the large contractor focuses on the man-hours expended against an activity in relation to the percent complete, the quantities in place, and the completion status of activities over time. Therefore, schedule control is linked to a man-hour base for progress reporting, and the man-hour per unit effort expended to achieve the evaluated progress in relation to the budgeted man-hours per unit and planned progress.

Cost control not only focuses on actual labor dollar expenditures in relation to the budget, but on the man-hours spent to generate these actual costs. However, this is often evaluated at a lower level of detail than that of the schedule, thus losing the effectiveness of integrating cost with schedule at all project levels. Cost deviations from budget estimates are identified and revised estimates or forecasts are made to incorporate the impact of these deviations. Several different methods of identifying performance trends emerge, for example, a unit man-hour analysis or tracking planned quantities in place versus actual quantities in place. The Field-Based Project Management approach actively promotes regular cost and schedule forecasting practices, and therefore formalized and systematic procedures for analyzing and documenting changes from the approved budget estimate and schedule emerge. In addition to forecasting the impact of performance trends, the procedure must be able to incorporate changes in scope in a timely manner so that these changes do not drastically adversely impact the project schedule. Changes in scope and/or project trends are incorporated into the cost and schedule system through the cost coding structure. While this provides a means of integrating cost and schedule, it is not necessarily the best means, because estimates and schedules are developed in different levels of detail.

At the large firm level, the contractor is frequently working in an overlapping design–construct mode. In this contracting environment, the work content quantity is constantly changing as the design develops. These changes in quantity are often significant enough to severely impact both cost and schedule. Thus, an additional burden to identify quantity changes in a timely manner is placed on the forecasting system. Project management can then respond to an increasing project growth with cost cutting measures before it is too late.

At the Total Field Autonomy level the work breakdown structure concept focuses on work packages as a basic component. These packages are used in conjunction with a cost coding system to achieve a fully integrated cost and schedule Management Information System. Each work package has a unique cost code identifier that is used to give it a budget for a defined work content. The same work package is incorporated into the schedule, and therefore it is ensured that the control base labor estimate matches the initial control base schedule for accomplishing the work. The cost estimate is based on standard man-hours per quantity unit, and, when coupled with quantities and predicted productivity factors, defines the site man-hours required for each work package. The estimated site man-hours, when combined with quantities, serve as the basis for establishing activity durations within a work package.

The key ingredient to the success of cost and schedule integration is the process of defining the project work breakdown structure and its work package details. This process *must* occur early in the planning phases of the project, and *must* involve project management, engineering, design, cost/scheduling, construction, and the client. The end result is an agreed-upon project work package structure which becomes the common reference point for all project documents, including design drawings, cost estimates, and control schedules, as well as being the common reference point for the construction and management of the project.

The work package approach generally increases the volume of information that must be gathered to effectively control the project. The project team size is therefore larger, and is often characterized by a broader range of expertise in such areas as industrial engineering for work sampling and methods engineering for introducing construction methods improvement techniques. The increasing volume of information also leads to more effective and efficient ways of processing data through computerized sorting. This requires project management to formulate project policy statements that establish specifically prescribed procedures for cost and scheduling engineering, industrial relations programs, the execution of engineering and design, progress evaluations, and work order planning techniques for crew planning and scheduling.

When project control begins and problems are uncovered, project management has available more rigorous ways of analyzing the cause of problems. This is done through special reports derived via multilevel sorting capabilities. In addition, the work

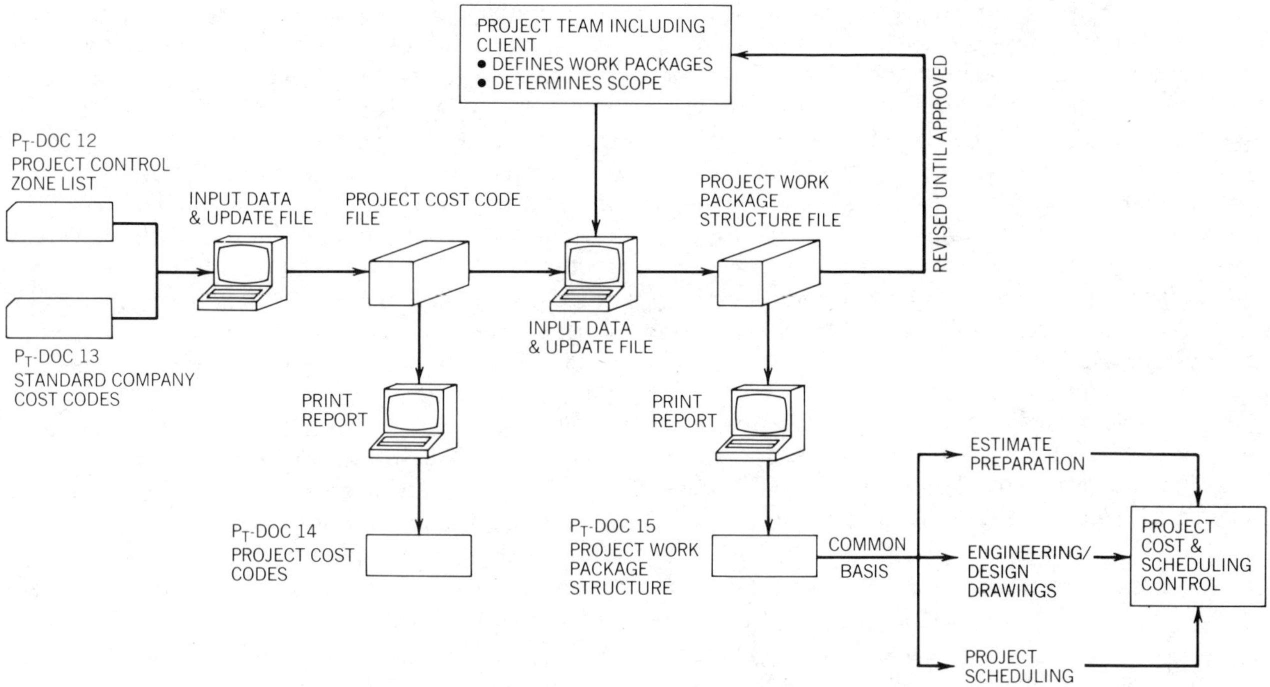

Figure 16.18 Development of project work package structure for cost and scheduling integration (Total Field Autonomy approach).

package structure allows management to evaluate the impact of selected decisions on both cost and schedule due to the common database developed for integrated work packages. For example, current productivity trends that would possibly cause an increase in man-hours, and hence potential labor cost overruns for certain areas of work, are also fed into the scheduling control system. In this way, their subsequent impact on the schedule when man-hours increase beyond budget limits can be evaluated. Based on the result of these types of analyses, a decision can be made to change the construction methods, crew sizes, and/or mixes, or possibly to change front-line supervision, in an attempt to reverse the productivity trend.

The development of a common scope statement for cost and scheduling control through a project Work Package Structure (P_T-DOC 15) is illustrated in Figure 16.18. If the front end of the system described in Figure 16.18 breaks down, then the Total Field Autonomy approach reverts back to cost coding techniques as the main basis for project control. Once the project Work Package Structure is established, however, the cost and scheduling system has many similarities to that of the Field-Based Project Management approach. The major differences would appear in areas where additional documentation would promote integration of costs with schedules. These areas include work order crew planning, detailed weekly work scheduling, labor unit oriented field progress reporting, and productivity indexing, all of which occur through work packages as defined early in the project definition phase.

16.6 THE INFORMAL COMMUNICATION SYSTEM

Man is a social being and generally seeks opportunities to interact with others. The work environment provides many opportunities for personal interaction, and therefore is potentially an inherently satisfying social environment in which to operate. In a purely social manner, close proximity or common interests provide the opportunity for personal or friendly interchanges, so that a casual sociological communication system soon emerges. Of more interest, however, is the development of an informal communication system directed to the performance and resolution of decision processes and problems. Thus, the more or less directed information flows associated with the determination of the causes of problems, the seeking of opinions based upon past experience, and the discussion of the range of possible courses of action available for problem resolution, and so on, exist in all management environments and establish the framework for an informal communication network.

This informal communication network becomes an important component of a Management Information System when it is:

1. The sole means of performing a decision process. That is to say, in cases when no formal documentation or defined decision process exists, and decision making emerges in a purely ad hoc manner.
2. Necessary to supplement the lack of detail or relevance of the formal Management Information System. That is to say, when it becomes necessary to interpret and explain data or aspects of information set down on documents.

The informal information system exists in all phases of construction management. It is the traditional mode of communication in the field environment, however, and thus is under the greatest stress to change its mode of operation, due to the increasing sophistication in the management approach. The field construction phase has extensive opportunities for informal communications. Thus, the field decision processes (i.e. Crew Planning, Crew Management, and Work Execution) are generally informal decision processes where experience, skill, and traditional rules of thumb normally prescribe the manner in which decisions are handled. Nevertheless, it is this traditional informal management area which is under the greatest pressure from the increasing sophistication of formal Management Information Systems. A consideration of this informal information system and the pressure to which it is subjected is the focus of this section.

The traditional informal information system is most evident, and is best illustrated, in the Project Team Component approach. This management approach is characterized by its small complement of staff and its minimal requirements for formal documentation. In this management environment, the informal information system is especially evident in the foreman-superintendent and superintendent-project manager interfaces. A schematic illustration of the interaction between the formal and informal information flows, and the consequential personal integration of these information flows for the various project team workers, is shown in Figure 16.19. Notice the almost totally informal nature of field information flows and field communications. The formal documentation flows at this level focus only on the essential business data needs associated with wage determination and production information related to progress payment evaluations. Notice further the totally independent nature of the roles of the job superintendent and foreman.

The progressive formalization of information flows in project management (which results from increasing project size and changing management approaches), and the consequential gradual reduction in the independence of the foreman and superintendent, is illustrated schematically in Figure 16.20. Considerable management support, and thus pressure, is brought to bear on foremen and superintendents as the project team changes in size. The inclusion of professionally trained field support staff as a result of the need to control field activity more effectively is a simple example. In this changing management environment the independent role of traditional field supervisors is progressively monitored, questioned, and finally reduced to a minimum in the Total Field Autonomy management environment. Thus, the unfettered role of the foreman/superintendent at the Project Team Component level (Figure 16.20*a*), becomes supported by field engineers and timekeepers at the Traveling Project Manager level (Figure 16.20*b*), is seriously impacted at the large firm level in the Field-Based Project Management approach (Figure 16.20*c*), and finally subverted at the Total Field Autonomy level, with the introduction of the work order approach to planning and estimating and the resulting Crew Planning and Monitoring efforts of a full complement of professional field staff (Figure 16.20*d*).

An immediate consequence of the changing foreman/superintendent role is the quality and type of person attracted to the construction field environment as considerable motivational and sub professional issues are involved. Thus, each management environment makes different professional demands on field staff and consequently usually requires different personality characteristics from these key field supervision team members. Finally, as a result of these pressures and the general loss of freedom

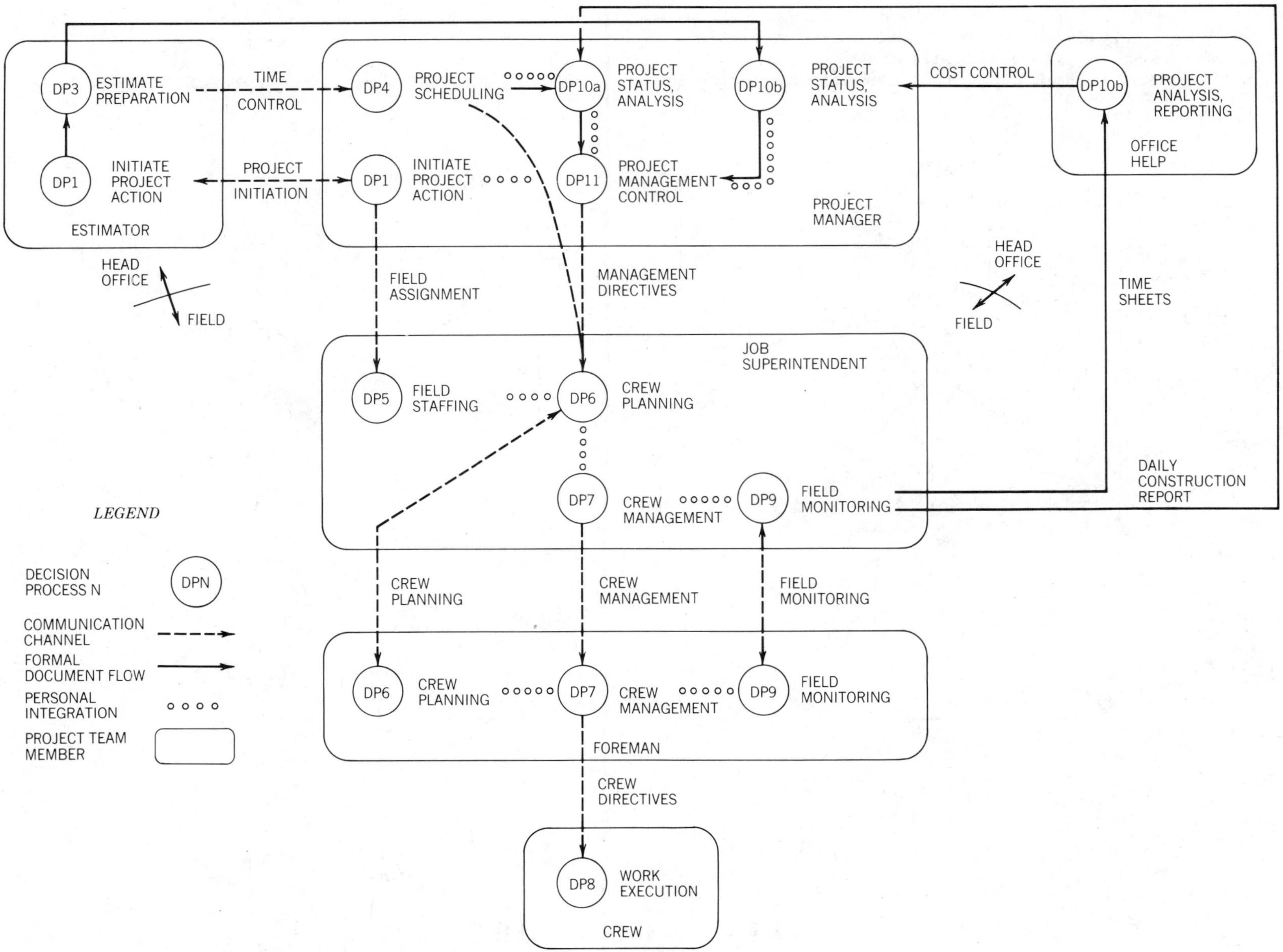

Figure 16.19 Integration of formal information system and informal communication network in Project Team Component management approach.

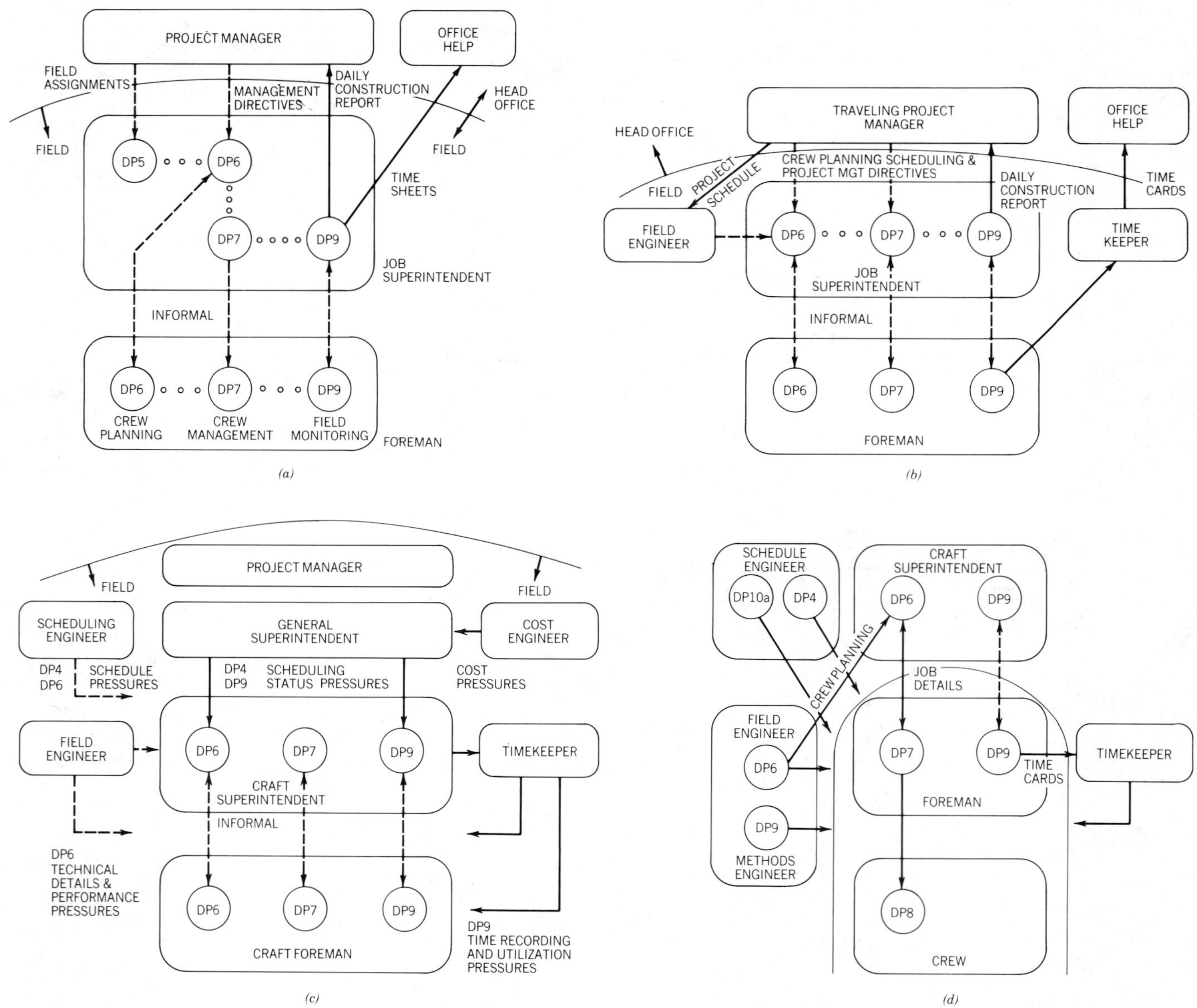

Figure 16.20 Changing superintendent and foreman management roles. (*a*) PTC approach: traditional role of job superintendent and foreman. (*b*) TPM approach: increased crew planning, scheduling, project management, technical details, and time reporting pressures on job superintendent and foreman. (*c*) Field-Based Project Management approach: pressures on all field decision processes performed by superintendents and foremen. (*d*) Total Field Autonomy approach: pressures on field crew with foreman mainly involved in crew management.

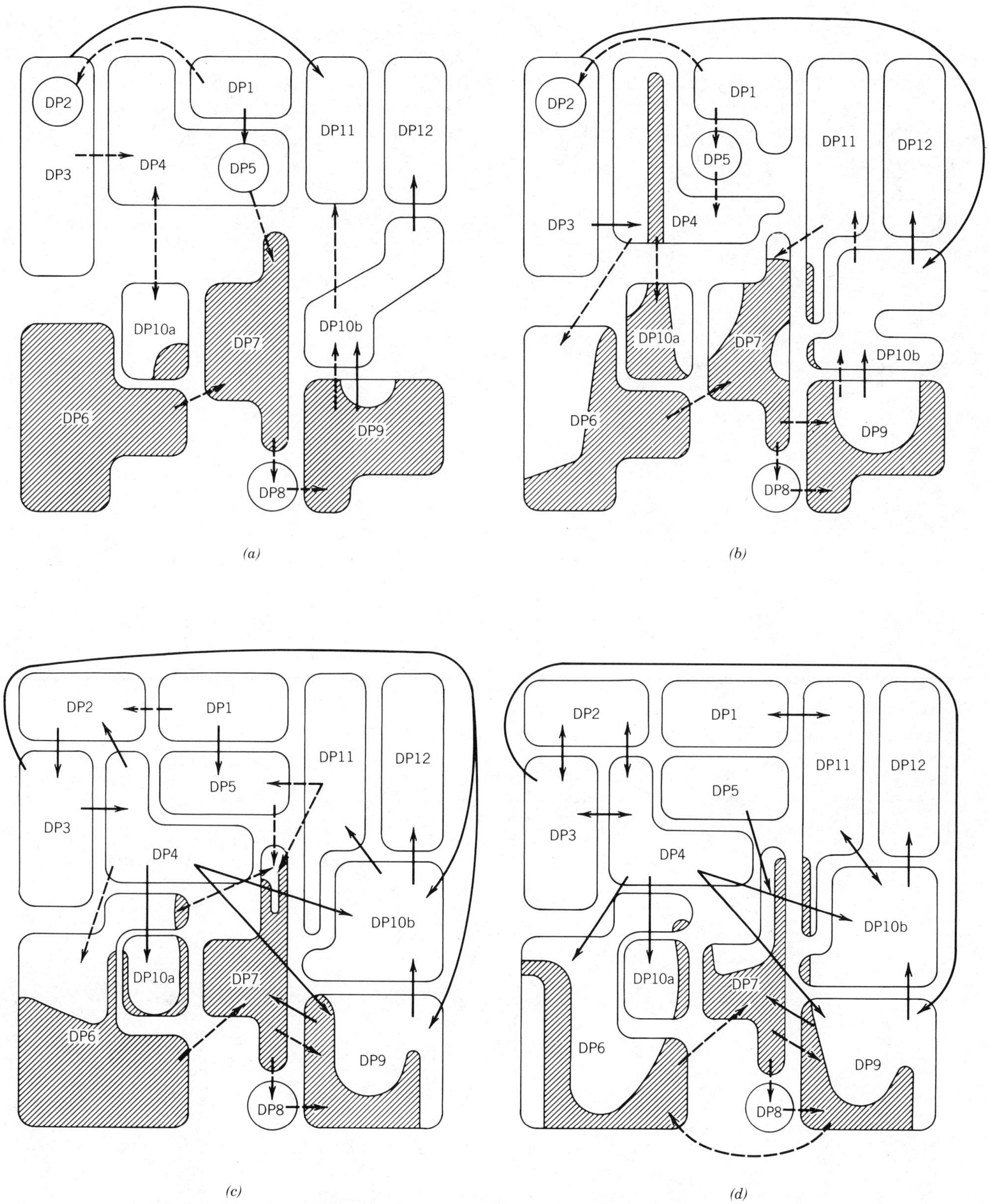

Figure 16.21 Schematics illustrating progressive growth in formal information system and growing pressure on field supervisory personnel (superintendents and foremen). (*a*) PTC approach. (*b*) TPM approach. (*c*) FBPM approach. (*d*) TFA approach.

of action in the work environment, there is the need to motivate both the foremen and crews on large projects. Otherwise these personnel may feel that their involvement in the project is both unnoticed and taken for granted.

An alternative formulation of the changing roles of foremen/superintendents in relation to their involvement in management decision processes, management practices, and in formal documentation effects, is shown in Figure 16.21. In this figure, their involvement in relation to informal information flow decision processes is shown shaded.

CHAPTER 17
The Project Management Challenge

The nature and layout of this book is unusual in that, after the introductory chapters have been read, each decision process chapter can be read independently. This unusual layout is intentional, and enables the reader to undertake an in-depth look at a single decision process from a variety of different management viewpoints. Often, each chapter also corresponds closely to a professional area such as estimating (Estimate Preparation) which is purposely not linked to other decision processes other than through suggested queries and indications of data and document source and destination flows on the decision networks. In this way the reader is encouraged to consider, or reconsider, the nature of linkages between decision processes, and the manner in which new linkages may strengthen management effort at problem resolution.

Chapter 16 presents ways in which some decision processes are integrated as a result of management, client, and problem resolution ''need'' pressures. Thus, standard project cost and schedule control practices reflect the overriding business and management interests of the contractor to be competitive and to stay in business.

This chapter introduces concepts of how other integrated linkages can be developed by management in order to address the management, and control of, emerging problem areas, or to strengthen and integrate management control in one or more functional areas of project management. Our interest in developing this flexible and lateral thinking management approach to the planned linkages of decision processes as an effective means of addressing and resolving management problems underlies the structure and layout of this book. It is our belief that this methodological approach can be applied to other management areas as a useful and practical procedural technique for the detection, description, and resolution of management problems.

17.1 DECISION PROCESS OVERVIEW

The decision process methodology developed in this book enables the scope and detail of a management area to be portrayed. In order for this approach to become effective, the scope of a management area must be established in terms of a fully representative set of component problem statements that reflect the issues, functions, and decisions that decision makers must address, handle, and resolve.

When considering the result of such an endeavor, a smaller set of major problem statements may suggest themselves to the manager, or may emerge in the activity of collecting together similarly focused component problem statements. This aggregation may reflect similar problem viewpoints, and the nature of management effort required for resolution and suitability for similar techniques when handling or processing the problem. Thus, both of these problem statement formulations present ideas on the types of questions, management action, and answers the manager must supply. In this respect, the decision maker may be called upon to give opinions off the top of his head, or to make more rational and formalized decisions based on the results of an objective analysis requiring data gathering and information system processing in a manner that can lead to the formulation of explicit procedures.

In this approach, the consideration of the component and major problem statements leads to the definition of the decision process in terms of the management functions to be performed, document and information flow structures, and procedural sequences. Finally, the manner in which particular data flows and selected functional performance techniques are built into the decision process establishes the particular way in which the now specific decision process will address and solve the management area's problems.

A number of approaches can be used in order to set up such a solution process. It is useful, for example, to first consider the adoption of well-established technique(s), and to determine which, and how many, of the problem statements can be answered by these techniques, and with what data gathering efforts. Once acceptable technique(s) have been incorporated into the decision process it is necessary to develop the relevant documents, data entry points, and information flows that establish the finer details of the decision process. In some cases, however, the adoption or retention of specific techniques for problem solving may have important consequences on the usefulness and flexibility of a decision process. The decision maker must always be alert for those decision-making formulations that lock him or her into a solution process that constrains his or her ability to effectively cope with the challenges of the decision area.

If suitable techniques are not available, the manager/decision maker must devise a means of answering problem statement questions. In this respect, it may be possible to develop useful data by the use of surrogate techniques, that provide *indicators* of (or which have some relationship to) possible or desirable manners of answering the questions. Otherwise, new techniques must be developed using new ideas or approaches. Whenever such a successful solution process emerges for a particular project management problem the opportunity exists for management to generalize it in a way that results in an advance in project management capability.

17.2 IMPROVING THE DECISION-MAKING PROCESS

The perceptive manager/supervisor is constantly striving to improve upon decision making in order to ensure a more successful management effort in resolving problems. Thus, for example, a manager may be given the responsibility to upgrade a portion of a decision process. In this case, the approach may be focused on a specific management function within an established decision process. A useful start is to examine the scope of the function, and by enlarging the details of its problem statement formulation, to better expose the nature of its performance and the way in which it can be strengthened. On the one hand, a detailed examination of functional performance may reveal that a technique is not entirely adequate for the resolution of the component problems addressed by the function. On the other hand, the technique may be effective in problem solving but further consideration and evaluation of the way the function is performed may lead to the genesis of other techniques that result in better solutions to the same problem. Consequently, the manager/supervisor could possibly change or restructure the functional performance techniques to expand the capability of the management function for problem resolution. This would naturally lead to different information flows to and from the function and would eventually impact the entire decision process.

A typical problem that management is always addressing is the timeliness with which decisions are made. If this emerges as a problem in a given decision process, the manager/supervisor would then analyze in detail specific functions relevant to the decision process. The manager/supervisor might uncover a technique that, when performed manually, requires an appreciable effort and therefore is not providing data in a timely fashion. If data processing techniques such as use of microcomputers are introduced, then the data can possibly be generated more quickly, in greater detail, and with various summary options, through data sorting. For example, manual labor cost estimating could be computerized on a microcomputer which would provide greater opportunities to analyze estimates and quickly make changes when problems in the estimates are uncovered. This capability is an extremely important feature in a decision process for decision makers working under tight bidding schedules where last minute changes are often required.

As another example, consider the case where the manner in which a decision maker usually performs a particular function, such as site reconnaissance, is found not adequate for the problems raised by a new project which is possibly larger and more remote. Management is faced with the need to develop a more capable approach. The resolution of this deficiency may require the development of new analysis techniques, the careful design of a site reconnaissance questionnaire for use by field personnel, and the preparation of procedural and briefing manuals.

Thus, the impact of such changes on the decision process will result in the need for upgraded or possibly new documents, increased data entry points, and the evaluation of changes to subsequent decision processes. Thus, management must ensure that major changes in a decision process area are thoroughly checked, and that the information and data generated supports the problem raised by the performance of those functions that comprise the decision process.

17.3 PROBLEM SOLVING

A natural tendency that results from the acceptance of a problem formulation, and which hardens with repetitive use of that formulation to resolve problems in a specific way, is to settle into a rut and to view a decision process in a fixed procedural manner. If this hypnotic attitude is to be broken then the decision maker must begin to look at a decision process in terms of how well it addresses the relevant problem area statements, and to assess the importance, or urgency, for change. This pressure for change may result from the need to consider how the decision process is connected to, and integrated with, other decision processes, or to identify desirable linkages to new management areas.

For example, a problem may result from the inadequate (or improper) linking of two decision processes that appear to be independent, but for which an effective linkage may become vital to the successful resolution of a problem. Thus, for example, the implementation of Crew Planning decisions in Crew Management situations may be a desirable way to raise field and crew productivities, or integrating a work order approach into Scheduling, Crew Planning, Field Monitoring, and Project Status activities may be a logical step to a more rational management approach to field control. In this latter case, management may wish to consider the desirability of investigating this rationalization as a function of control area definition in Project Planning.

The addressing of problems such as those indicated above may lead to the use of problem solving team techniques, and may eventually result in revised or new procedures, instructions, and documents, and new roles for the individuals and decision makers involved. The net impact may be a reevaluation of the entire approach to project management areas like material management, or the management and control of project documents.

17.4 THE MANAGEMENT CHALLENGE

Existing decision processes and current management approaches in project management are necessarily feasible, and additionally most are eminently practical. However, they may not necessarily be the best way to resolve problems that are pertinent to all areas of project management. If management is committed to achieving a competitive edge through better quality decision making, it will strive to ensure that a management environment exists in which changes can occur. The methodology presented in this book provides a new way of viewing decision-making processes that encourages this attitude first by sensitizing the reader to existing practices that may be different to those to which the reader is accustomed. The methodology, however, also provides a means of problem description that has the potential to lead the reader to question, and to suggest ways of improving, current practice.

It is our hope that through the understanding and application of the methodology developed in this book, some managers and supervisors will develop insights and a greater capacity to look critically at decision-making processes in project management. The management approach presented here relates fairly obviously to the procedures found useful by specially focused groups within project management to identify problems, and to suggest ways of solving these problems. Thus, quality circle programs and problem solving teams can often identify different ways of doing things that improve the quality of management effort, reduce cost, and enable the contractor to become more competitive.

The management challenge to the reader is to view the management environment in which a decision maker finds him or herself in a new light, and especially to review decision-making processes in ways that can open up different management approaches to both old and newly emerging problems. If this book can stimulate such endeavors, the authors will be well satisfied.

APPENDIX
Document Summaries

This appendix is a collection of document titles described in the book corresponding to the four manpower management approaches found in practice, and a reference to their location in the text. The documents, summarized, portray the information system characteristic of each management approach and indicate the way information may be integrated. The documents referenced in this book are most directly related to the labor component of a project. Thus, specific documents for material control and other areas of project management are not included.

The documents are collected by management approach so the reader can quickly review and check document sequences and formats and locate them within the text. Furthermore, the number of documents associated with a management area is a reflection of the level of management effort required to meet the demands imposed by project size, complexity, and nature. The different numbers of documents generated by each management approach, however, should not be construed as meaning that one management approach is better than another. Rather, it implies that particular management environment pressures are reflected in the nature of rigorous management effort and documented information required.

The four project manpower management approach document listings follow. In each list, the Policy Documents (P-DOC) are described first, followed by the Internal Documents (I-DOC). The document figure number and page reference are provided. In some cases, the document format and structure do not change from one management approach to the next. The document number changes but the reference to the document example corresponds to a previously identified illustration of the document.

PROJECT TEAM COMPONENTS DOCUMENTS

Policy Document	Description	Figure Number	Page Number
P_S-DOC			
1	Company Cost Codes	6.13	116
2	Quantity Sheets	6.15	117
3	Estimate Recapitulation Sheets	6.16	117
4	Estimate Summary	6.17	118
5	Time Sheets	12.13	294
6	Daily Construction Report	12.14	295
7	Quantity Distribution Report	12.15	296
8	Weekly Payroll Summary	13.16	336
9	Record of Employment and Earnings	13.17	336
10	Labor Distribution Report	13.18	337
11	Labor Cost Summary	13.19	338
12	Project Cost Recapitulation	13.20	338
13	Progress Payment Voucher	13.22	341
14	Change Authorization	14.10	402

PROJECT TEAM COMPONENTS DOCUMENTS

Internal Document	Description	Figure Number	Page Number
I_S-DOC			
1	Project Scope List	6.14	116
2	Bar Chart Construction Schedule	7.14	155
3	Construction Progress and Status	13.21*a*	340
4	Revised Construction Schedule	13.21*b*	340

TRAVELING PROJECT MANAGER DOCUMENTS

Policy Document	Description	Figure Number	Page Number
P_M-DOC			
1	Project Organization	4.11	49
2	Site Investigation Check List	6.20	121
3	Company Cost Codes	6.21	122
4	Project Scope Check List	6.22	123
5	Quantity Survey and Estimate Work Sheets	6.23	123
6	Estimate Recapitulation Sheet	6.24	124
7	Estimate Summary	6.25	125
8	Time Sheet	12.18	299
9	Daily Construction Report	12.19	301
10	Quantity Distribution Report	12.20	301
11	Labor Distribution Report	13.27	349
12	Weekly Payroll Summary and	13.16	336
	Record of Employment and Earnings	13.17	336
13	Labor Cost Summary	13.33	356
14	Project Cost Summary	13.34	357
15	Progress Payment Report	13.35	358
16	Change Authorization	14.10	402

TRAVELING PROJECT MANAGER DOCUMENTS

Internal Document	Description	Figure Number	Page Number
I_M-DOC			
1	Network Plan	7.16	157
2	Construction Schedule	7.17	158
3	Manpower Summary	7.18	159
4	Construction Progress and Status	13.30*a*	353
5	Project Progress Analysis	13.30*b*	353
6	Revised Construction Schedule	13.31	354
7	Expenditure Projection	13.36	359

FIELD-BASED PROJECT MANAGEMENT DOCUMENTS

Policy Document	Description	Figure Number	Page Number
P_L-DOC			
1	Standard Project Operating Procedures	4.15	57
2	Departmental Organization Structures	4.16	58
3	Head Office Project Organization Chart	4.17	59
4	Employee File	4.18	59
5	Head Office Man-hour Estimate	4.19*a*	60
6	Head Office Manpower Curve	4.19*b*	60
7	Discipline Organization Chart	4.19*c*	60
8	Head Office Control Documents	4.20	61
9	Project Control Zone List	5.19	92
10	Standard Company Cost Codes	5.20	93
11	Project Cost Codes	5.21	94
12	Factored Estimates	6.29	132
13	Quantity Work Sheet	6.30	133
14	Estimate Work Sheets	6.31	134
15	Cost Estimate Summaries	6.32	135
P_L-DOC			
16	Management Level Schedule	7.22	165
17	Standard Construction Progress Curve	7.23*a*	166
18	Standard Construction Craft Mixes	7.23*d*	166
19	Project Construction Schedule	7.27	169
20	Quarterly Construction Schedule	7.29	171
21	Standard Field Construction Procedures	8.9	199
22	Project Team Structure	8.10*a*	200
23	Manning Chart by Job Classification	8.10*b*	200
24	Construction Organization Chart: Control Base	8.11*a*	202
25	Construction Manning Chart: Control Base	8.11*b*	202
26	Construction Staff Manpower Curves: Control Base	8.11*c*	203
27	Manpower Requisitions	9.14	230
28	Crew Balance Chart	9.15	231
29	Bi-Weekly Work Schedule	9.18	234
30	Construction Equipment Schedule	9.19	234
31	Multiple Activity Charts	9.20	235
P_L-DOC			
32	Employment Application	10.5	249
33	Termination/Transfer Notice	10.5	249
34	Welder Certification	10.12	261
35	Brass or Clock Cards	10.7	251
36	Time Cards	12.25	307
37	Daily Manpower Reports	12.26	307
38	Travel Authorization	12.28	309
39	Construction Tool Requisition	12.29	309
40	Weekly Construction Status Report	12.31	311
P_L-DOC			
41	Quantity Log	12.32	311
42	Percent Complete Status Report	12.33	312
43	Activity Status List	13.39	364
44	Critical Activity List	13.40	365
45	Subcontract Status Reports	13.41	365
46	Construction Progress Report	13.42	366
47	General Field Cost Expenditure Reports	None	
48	Labor Cost and Man-hour Distribution	13.44	369
49	Labor Unit Man-hour Report	13.46	371
50	Labor Unit Man-hour Curves	13.47	372
P_L-DOC			
51	Standard Change Order and Trend Procedure	13.50	375
52	Field Change Order	13.51	375
53	Field Labor Forecast Trends	13.52	376
54	Labor Cash Flow Forecast	13.53	376
55	Project Expenditure Reports	13.55	378
56	Head Office Change Orders and Trends	13.51, .52	375 376
57	Project Labor Cost Report	14.16	413
58	Monthly Construction Progress Report	14.17	413
59	Total Project Cost Summary	14.19	415
60	Total Project Progress Report	14.20	415
61	Project Filing Procedure	15.17	441
62	Final Job Cost Report	15.18	443

FIELD-BASED PROJECT MANAGEMENT DOCUMENTS

Internal Document	Description	Figure Number	Page Number
I_L-DOC			
1	Facility Component Priority Lists	5.15	90
2	Site Data Reference File	5.16	91
3	Labor Reconnaissance Report	5.17	91
4	Construction Site Reconnaissance Report	5.18	92
5	Construction Progress Curves	7.23*b*	166
6	Construction Man-hour Curves	7.23*c*	166
7	Construction Manpower Profiles	7.24	166
8	Network Plans	7.26	168
9	Fragnet Plan	7.30*a*	171
10	Fragnet Schedule	7.30*b*	171
11	Activity Layout Sketch	9.16	232
12	Unit Man-hour Study	13.48	372

TOTAL FIELD AUTONOMY DOCUMENTS

Policy Document	Description	Figure Number	Page Number
P_T-DOC			
1	Standard Project Operating Procedures	4.15	57
2	Departmental Organization Structures	4.16	58
3	Project Management Level Organization Chart	4.23	66
4	Employee File	4.18	59
5	Project Execution Procedures	4.24*c*	67
6	Head Office Man-hour Estimate	4.19*a*	60
7	Head Office Manpower Curve	4.19*b*	60
8	Engineering/Design Project Organization Chart	4.25	68
9	Head Office Manning Chart	4.26	68
10	Employee Availability Report	4.27	69
11	Head Office Project Control Documents	4.28	69
12	Project Control Zone List	5.26	99
13	Standard Company Cost Codes	5.20	93
14	Project Cost Codes	5.27	100
15	Project Work Package Structure	5.28	101
P_T-DOC			
16	Factored Estimates	6.29	132
17	Quantity Work Sheet	6.30	133
18	Estimate Work Sheets	6.31	134
19	Cost Estimate Summaries	6.32	135
20	Management Estimate Summary	6.35	139
21	Management Level Schedules	7.33	176
22	Standard Construction Progress Curve	7.23*a*	166
23	Standard Construction Craft Mixes	7.23*d*	166
24	Standard Construction Network Module	7.36	179
25	Project Construction Schedule	7.37	179
26	Construction Schedule Package	7.40	181
27	Monthly Activity Plan	7.41	181
28	Project Team Structure	8.13*a*	206
29	Manning Chart by Job Classification	8.13*b*	206
30	Construction Organization Chart: Control Base	8.14*a*	207
31	Construction Manning Chart: Control Base	8.14*b*	207
32	Construction Staff Control Curves	8.15	208
33	Standard Field Construction Procedures	8.9	199
P_T-DOC			
34	Project Field Construction Procedures	8.16	208
35	Work Order	9.22	239
36	Manpower Requisitions	9.14	230
37	Crew Balance Chart	9.15	231
38	Weekly Work Schedule	9.24	240
39	Construction Equipment Schedule	9.19	234
40	Multiple Activity Charts	9.20	235
41	Employment Application	10.5	249
42	Termination/Transfer Notice	10.5	249
43	Welder Certification	10.12	261
44	Brass or Clock Cards	10.7	251
45	Time Cards	12.25	307
46	Daily Manpower Reports	12.36	316
47	Travel Authorization	12.28	309
48	Construction Tool Requisition	12.29	309
49	Field Percent Progress Report	12.41	318
50	Quantity Log	12.32	311
51	Activity Status List	13.39	364
52	Critical Activities List	13.40	365
53	Subcontract Status Report	13.41	365
54	General Field Cost Expenditure Reports	None	
55	Labor Cost and Man-hour Distribution	13.44	369
56	Project Performance Report	13.60	384
57	Field Activity Analysis Documents	13.62	386
58	Standard Change Order and Trend Procedure	13.50	375
59	Project Change Order and Trend Procedure	13.64	387
60	Field Change Orders and Trends	13.65	387
61	Field Man-hour Forecast and Labor Cost Report	13.66	388
62	Labor Cash Flow Forecast	13.53	376
63	Project Expenditure Reports	13.55	378
64	Head Office Change Orders and Trends	13.65	387
65	Project Labor Cost Report	14.24	422
66	Monthly Construction Progress Report	14.25	423
67	Total Project Cost Summary	14.26	423
68	Project Management Report	14.27	424
69	Project Filing Procedure	15.17	441
70	Crew Size Analysis Report	15.21	446
71	Final Project Reports	15.22	447
72	Standard Labor Cost Curves and Estimating Factors	15.23	448
73	Standard Labor Unit Man-hours	15.24	448

TOTAL FIELD AUTONOMY DOCUMENTS

Internal Document	Description	Figure Number	Page Number
I_T-DOC			
1	Facility Component Priority List	5.24	98
2	Site Data Reference File	5.16	91
3	Labor Reconnaissance Report	5.17	91
4	Construction Site Reconnaissance Report	5.18	92
5	Construction Progress Curve	7.34*a*	177
6	Construction Man-hour Curves	7.34*b*	177
7	Construction Manpower Curves	7.34*c*	177
8	Network Plans	7.26	168
9	Fragnet Plan	7.30*a*	171
10	Fragnet Schedule	7.30*b*	171
11	Activity Layout Sketch	9.16	232
12	Crew Location Chart	10.8	252
13	Attendance Analysis Curves	12.37	316
14	Field Planning Study	12.8	287
15	Activity Delay Report	12.39	318
16	Weekly Construction Report	12.42	319
17	Project Performance Curves	13.61	385

Bibliography

The references in this bibliography have been grouped into sections that correspond to the chapters of this book. Individual entries are not referred to in the text. They are located generally in the Bibliography according to the sections where they are considered to have the major contribution to make to the reader's knowledge. The resources listed here offer background information and/or more specific technique material. This list is by no means exhaustive, but rather is intended to provide supplementary reading for those who wish to study particular subjects such as decision-making processes, management techniques, human resource management, estimating, cost and scheduling control systems, and the like, in greater detail. Many of these materials have excellent bibliographies of their own and, therefore, provide additional sources of information for the reader. Finally, while the majority of references listed are books, there are other materials published by professional organizations that provide additional information on project manpower management issues. The publications by, for example, the American Society of Civil Engineers, Associated General Contractors, American Association of Cost Engineers, and the Construction Industry Institute are among the best available.

MANPOWER MANAGEMENT: A DECISION PROCESS APPROACH

Stuart D. Anderson and Ronald W. Woodhead, *Project Manpower Management: Management Processes in Construction Practice*, Wiley-Interscience, New York, 1981.

MANPOWER MANAGEMENT DECISION PROCESSES

Stuart D. Anderson and Ronald W. Woodhead, *Project Manpower Management: Management Processes in Construction Practice*, Wiley-Interscience, New York, 1981.

Donald S. Barrie and Boyd C. Paulson, Jr., *Professional Construction Management*, 2nd ed., McGraw-Hill, New York, 1984.

Donald S. Barrie, Editor, *Directions in Managing Construction*, Wiley-Interscience, New York, 1981.

Richard H. Clough, *Construction Project Management*, 4th ed., Wiley-Interscience, New York, 1981.

Peter F. Drucker, *Management: Tasks, Responsibilities, Practices*, Harper & Row, New York, 1973.

Edward Fisk, *Construction Project Administration*, Wiley, New York, 1978.

Stanley Goldhaber, Chandra K. Jha, and Manual C. Macero Jr., *Construction Management: Principles and Practice*, Wiley, New York, 1964.

William J. Gore, *Administrative Decision-Making: A Heuristic Model*, Wiley, New York, 1964.

Alfred P. McNulty, *Management of Small Construction Projects*, McGraw-Hill, New York, 1982.

Harry Rubey and Walter W. Milner, *Construction and Professional Management: An Introduction*, Macmillan, New York, 1966.

Leslie W. Rue and Lloyd L. Byars, *Management: Theory*

and Application, 4th ed., Irwin, Homewood, Illinois, 1986.

PORTRAYAL OF A DECISION PROCESS

"Applications of Small Computers in Construction," *Journal of Construction Engineering and Management*, Vol. III, No. 3, American Society of Civil Engineers, New York, September, 1985.

Peter F. Drucker, *Management: Tasks, Responsibilities, Practices*, Harper & Row, New York, 1973.

Edward Fisk, *Construction Project Administration*, Wiley, New York, 1978.

William J. Gore, *Administrative Decision-Making: A Heuristic Model*, Wiley, New York, 1964.

Kenneth M. Guthrie and Associates, *Managing Capital Expenditures For Construction Projects*, Craftsman Book Company of America, Solana Beach, California 1977.

Arnold Kaufmann, *The Science of Decision-Making*, World University Library, London, 1968.

Harold Kerzner, *Project Management: A Systems Approach To Planning, Scheduling and Controlling*, Van Nostrand Reinhold, New York, 1979.

Melvin W. Lifson and Edward F. Shaifer, Jr., *Decision and Risk Analysis for Construction Management*, Wiley-Interscience, New York, 1982.

Howard Raiffa, *Decision Analysis: Introductory Lectures on Choices Under Uncertainty*, Addison-Wesley, Reading, Massachusetts, 1968.

INITIATE PROJECT ACTION

Stuart D. Anderson and Ronald W. Woodhead, *Project Manpower Management: Management Processes in Construction Practice*, Wiley-Interscience, New York, 1981.

Hira N. Ahuja, *Project Management Techniques in Planning and Controlling Construction Projects*, Wiley-Interscience, New York, 1984.

Donald Barrie, Editor, *Directions in Managing Construction*, Wiley-Interscience, 1981.

Richard H. Clough, *Construction Contracting*, 4th ed., Wiley-Interscience, New York, 1981.

Edward Fisk, *Construction Project Administration*, Wiley, New York, 1981.

Kenneth M. Guthrie and Associates, *Managing Capital Expenditures for Construction Projects*, Craftsman Book Company of America, Solana Beach, California, 1977.

Stanley Goldhaber, Chandra K. Jha, and Manuel C. Macaro Jr., *Construction Management: Principles and Practice*, Wiley, New York, 1977.

Frank Harris and Ronald McCaffee, *Modern Construction Management*, Crosby Lockwood Staples, London, 1977.

Kenneth K. Humphreys, Editor, *Project and Cost Engineers' Handbook*, 2nd ed., Dekker, New York, 1984.

Harold Kerzner, *Project Management: A Systems Approach to Planning, Scheduling and Controlling*, Van Nostrand Reinhold, New York, 1979.

Ross Pilcher, *Principles of Construction Management*, McGraw-Hill, New York, 1967.

Alfred P. McNulty, *Management of Small Construction Projects*, McGraw-Hill, New York, 1982.

W. J. Stillman, *Construction Practices for Project Managers and Supervisors*, Reston Publishing Company, Reston, Virginia, 1978.

S. Peter Volpe, *Construction Management Practice*, Wiley-Interscience, New York, 1972.

Alonzo Wass, *Construction Management and Contracting*, Prentice-Hall, Englewood Cliffs, New Jersey, 1972.

Richard E. Westney, *Managing the Engineering and Construction of Small Projects*, Dekker, New York, 1985.

PROJECT PLANNING

James J. Adrian, *Quantitative Methods in Construction Management*, American Elsevier, New York, 1973.

Hira N. Ahuja, *Project Management Techniques in Planning and Controlling Construction Projects*, Wiley-Interscience, New York, 1984.

Hira N. Ahuja, *Successful Construction Cost Control*, Wiley-Interscience, New York, 1980.

Donald S. Barrie and Boyd C. Paulson Jr., *Professional Construction Management*, 2nd ed., McGraw-Hill, New York, 1984.

Donald Barrie, Ed., *Directions in Managing Construction,* Wiley-Interscience, New York, 1981.

James A. Bent, *Applied Cost and Schedule Control*, Dekker, New York, 1982.

Forest D. Clark and A. B. Lorenzoni, *Applied Cost Engineering*, 2nd ed., Dekker, New York, 1985.

Cost Control and CPM: A Manual for General Contractors, Associated General Contractors, 1968.

Harold W. Davey, *Contemporary Collective Bargaining*, Prentice-Hall, Englewood Cliffs, New Jersey, 1972.

L. L. Farkas, *Management of Technical Field Operations*, McGraw-Hill, New York, 1970.

William Haber and Harold M. Levinson, *Labor Relations and Productivity in the Building Trades*, University of Michican, Ann Arbor, 1956.

Harold Kerzner, *Project Management: A Systems Approach to Planning, Scheduling and Controlling*, Van Nostrand Reinhold, New York, 1979.

M. R. Lefkoe, "The Crisis in Construction: There is an Answer," The Bureau of National Affairs Inc., Washington, D.C., 1970.

Daniel Quinn Mills, *Industrial Relations and Manpower in Construction*, The MIT Press, Cambridge, 1972.

Harry Rubey and Walter W. Milner, *Construction and Professional Management: An Introduction*, Macmillan, New York, 1966.

MASTERFORMAT, Document MP-2-1, The Construction Specifications Institute, Alexandria, Virginia, 1983.

Richard E. Westney, *Managing the Engineering and Construction of Small Projects*, Dekker, New York, 1985.

Jack H. Willenbrock and H. Randolph Thomas, Eds, *Planning, Engineering and Construction of Electric Generation Facilities*, Wiley-Interscience, New York, 1980.

ESTIMATE PREPARATION

Building Construction Cost Data, Robert Snow Means Co., Inc., Duxbury, Massachusetts, published annually.

Building Estimator's Reference Book, 22nd ed., The Frank R. Walker Co., Chicago, 1986.

Forest D. Clark and A. B. Lorenzoni, *Applied Cost Engineering*, 2nd ed., Dekker, New York, 1985.

dBase III Plus, Ashton-Tate, Culver City, California, 1985.

"Cost Classification Index, Broadscope Edition," AGC Document No. 1111, Associated General Contractors of America, Washington, D.C., 1986.

Dodge Manual For Construction Pricing and Scheduling, McGraw-Hill Information Systems Co., New York, published annually.

Norman Foster, *Construction Estimates From Take-off To Bid*, McGraw-Hill, New York, 1972.

Kenneth M. Guthrie, *Process Plant Estimating and Control*, Craftsman Book Company of America, Solana Beach, California, 1974.

Kenneth K. Humphreys, Ed, *Project and Cost Engineers Handbook*, 2nd ed., Dekker Inc., New York, 1984.

F. C. Jelen and James H. Black, *Cost and Optimization Engineering*, 2nd ed., McGraw-Hill, New York, 1983.

M. R. Lefkoe, "The Crisis in Construction: There is an Answer," The Bureau of National Affairs Inc., Washington, D.C., 1970.

LOTUS 1-2-3, LOTUS Development Corporation, Cambridge, Massachusetts, 1983.

Gary Moselle, Ed., *National Construction Estimator*, Craftsman Book Company, Solana Beach, California, 1970.

James M. Neil, *Construction Cost Estimating for Project Control*, Prentice-Hall, Englewood Cliffs, N.J., 1982.

Robert L. Peurifoy, *Estimating Construction Costs* 3rd ed., McGraw-Hill, New York, 1975.

Practical Accounting and Cost Keeping for Contractors, 9th ed., Frank P. Walker Company, Chicago, 1982.

Rodney D. Stewart, *Cost Estimating*, Wiley-Interscience, New York 1982.

MASTERFORMAT, Document MP-2-1, The Construction Specifications Institute, Alexandria, Virginia, 1983.

S. Peter Volpe, *Construction Management Practice*, Wiley-Interscience, New York, 1972.

Richard E. Westney, *Managing the Engineering and Construction of Small Projects*, Dekker, New York, 1985.

Arthur J. Willis and Christopher J. Willis, *Practice and Procedure for the Quantity Surveyor*, Crosby Lockwood Staples, London, 1975.

PROJECT SCHEDULING

Hira N. Ahuja, *Project Management Techniques in Planning and Controlling Construction Projects*, Wiley-Interscience, New York, 1984.

James M. Antill and Ronald W. Woodhead, *Critical Path Methods in Construction Practice*, 3rd ed., Wiley, New York, 1982.

ARTEMIS, Metier Management Systems, Inc., Houston, 1986.

Donald S. Barrie and Boyd C. Paulson, Jr., *Professional Construction Management*, 2nd ed., McGraw-Hill, New York, 1984.

James A. Bent, *Applied Cost and Schedule Control*, Dekker, New York, 1982.

Richard H. Clough, *Construction Contracting*, 4th ed., Wiley-Interscience, New York, 1981.

Cost Control and CPM: A Manual for General Contractors, Associated General Contractors, Washington, D.C., 1968.

Daniel W. Halpin and Ronald W. Woodhead, *Construction Management*, Wiley, New York, 1980.

LOTUS 1-2-3, LOTUS Development Corporation, Cambridge, Massachusetts, 1983.

Harold Kerzner, *Project Management: A Systems Ap-*

proach to Planning, Scheduling and Controlling, Van Nostrand Reinhold, New York, 1979.

Joseph J. Moder and Cecil R. Phillips, *Project Management With CPM and PERT*, Van Nostrand Reinhold, New York, 1970.

James J. O'Brien and Robert G. Zilly, (eds), *Contractor's Management Handbook*, McGraw-Hill, New York, 1971.

Primavera Project Planner, Primavera Systems Inc., Bala Cynwyd, Pennsylvania, 1986.

Project 2, Project Software and Development Inc., Cambridge, Massachusetts, 1985.

Quiknet PC Project Management Software, Project Software and Development Inc., Cambridge, Massachusetts, 1984.

Schedule Graphics, Ronald L. Schultz, Distributed by Success Software, Irvine, California, 1985.

The Use of CPM in Construction, The Associated General Contractors of America, Washington, D.C., 1976.

Richard E. Westney, *Managing The Engineering and Construction of Small Projects*, Dekker, New York, 1985.

FIELD STAFFING

Stuart D. Anderson and Ronald W. Woodhead, *Project Manpower Management: Management Processes in Construction Practice*, Wiley-Interscience, New York, 1981.

Richard H. Clough, *Construction Contracting*, 4th ed., Wiley-Interscience, New York, 1981.

Dexter C. Dunphy, *Organizational Change by Choice*, McGraw-Hill, Sydney, 1981.

L. L. Farkas, *Management of Technical Field Operations*, McGraw-Hill, New York, 1970.

Edward Fisk, *Construction Project Administration*, Wiley, New York, 1978.

Stanley Goldhaber, Chandra K. Jha, and Manuel C. Macero Jr., *Construction Management: Principles and Practice*, Wiley, New York, 1977.

Harold Kerzner, *Project Management: A Systems Approach to Planning, Scheduling, and Controlling*, Van Nostrand Reinhold, New York, 1979.

Alfred P. McNulty, *Management of Small Construction Projects*, McGraw-Hill, New York, 1982.

Daniel Quinn Mills, *Industrial Relations and Manpower in Construction*, The MIT Press, Cambridge, 1972.

King Royer, *Desk Book for Construction Superintendents*, Prentice-Hall, Englewood Cliffs, New Jersey, 1967.

Alonzo Wass, *Construction Management and Contracting*, Prentice-Hall, Englewood Cliffs, New Jersey, 1972.

I. G. Weeks, *Site Foremanship*, The Construction Press Ltd., Lancaster, England, 1978.

CREW PLANNING

Philip B. Crosby, *Quality Is Free*, Mentor Books, New York, 1979.

Harvey V. Debo and Lee Diamant, *Construction Superintendent's Job Guide*, Wiley, New York, 1980.

Daniel W. Halpin and Ronald W. Woodhead, *Design of Construction and Process Operations*, Wiley, New York, 1976.

LOTUS 1-2-3, LOTUS Development Corporation, Cambridge, Massachusetts, 1983.

R. Bruce McAfee and Richard N. Feinberg, *Cases in Construction Supervision*, Reston Publishing Co., Inc., Reston, Virginia, 1977.

Alfred P. McNulty, *Management of Small Construction Projects*, McGraw-Hill, New York, 1982.

Henry W. Parker and Clarkson H. Oglesby, *Methods Improvement for Construction Managers*, McGraw-Hill, New York, 1972.

Quiknet PC Project Management Software, Project Software and Development, Inc., Cambridge, Massachusetts, 1984.

Wallace J. Richardson, *Cost Improvement, Work Sampling and Short Interval Scheduling*, Reston Publishing Company, Reston, Virginia, 1976.

King Royer, *Desk Book for Construction Superintendents*, Prentice-Hall, Englewood Cliffs, New Jersey, 1967.

I. G. Weeks, *Site Foremanship*, The Construction Press Ltd., Lancaster, England, 1978.

CREW MANAGEMENT

Robert R. Blake and Jane Srygley Mouton, *The Managerial Grid III*, Gulf Publishing Company, Houston, 1985.

John D. Borcherding, *Effective Utilization of Manpower in Construction*, National Electrical Contractors Association, Washington, D.C., 1975.

Harvey V. Debo and Leo Diamant, *Construction Superintendent's Job Guide,* Wiley, New York, 1982.

L. L. Farkas, *Management of Technical Field Operations*, McGraw-Hill, New York, 1970.

R. Bruce McAfee and Richard N. Feinberg, *Cases in Construction Supervision*, Reston Publishing Co., Reston, Virginia, 1977.

Alfred P. McNulty, *Management of Small Construction Projects*, McGraw-Hill, New York, 1982.

Daniel Quinn Mills, *Industrial Relations and Manpower*

in Construction, M.I.T. Press, Cambridge, Massachusetts, 1972.

Robert I. Patchin, *The Management and Maintenance of Quality Circles*, Dow Jones-Irwin, Homewood, Illinois, 1983.

Wallace J. Richardson, *Cost Improvement, Work Sampling and Short Interval Scheduling*, Reston Publishing Co., Reston, Virginia, 1976.

King Royer, *Desk Book for Construction Superintendents*, Prentice-Hall, Englewood Cliffs, New Jersey, 1967.

I. G. Weeks, *Site Foremanship*, The Construction Press Ltd., Lancester, England, 1978.

WORK EXECUTION

Philip Crosby, *Quality is Free*, Mentor Books, New York, 1979.

Edward Fisk, *Construction Project Administration*, Wiley, New York, 1978.

William Haberand and Harold M. Levinson, *Labor Relations and Productivity in the Building Trades*, University of Michigan, Ann Arbor, 1956.

Frederick Herzberg, "One More Time: How to Motivate Employees?" *Harvard Business Review*, January/February, 1968.

Gordon McBeath, *Productivity Through People*, Business Books, London, 1974.

Robert I. Patchin, *The Management and Maintenance of Quality Circles*, Dow Jones-Irwin, Homewood, Illinois, 1983.

King Royer, *Desk Book for Construction Superintendents*, Prentice-Hall, Englewood Cliffs, New Jersey, 1967.

Harry Rubey, and Walter W. Milner, *Construction and Professional Management: An Introduction*, Macmillan, New York, 1956.

W. J. Stillman, *Construction Practices for Project Managers and Supervisors*, Reston Publishing Company, Reston, Virginia, 1978.

Dr. Richard L. Tucker and Dr. John D. Borcherding, "Implementation and Evaluation of Productivity Improvement Programs, *Proceeding's*, Project Management Institute Joint Symposium, Boston, Massachusetts, September, 1981.

FIELD MONITORING

James J. Adrian, *Quantitative Methods in Construction Management*, American Elsevier, New York, 1973.

Hira N. Ahuja, *Successful Construction Cost Control*, Wiley-Interscience, New York, 1980.

Ralph M. Barnes, *Time Study*, Wiley, New York, 1958.

Richard H. Clough, *Construction Contracting*, 4th ed. Wiley-Interscience, New York, 1981.

Dexter C. Dunphy, *Organizational Change by Choice*, McGraw-Hill, Sydney, 1981.

Frank Harris and Ronald McCaffee, *Modern Construction Management*, Crosby Lockwood Staples, London, 1977.

Kenneth K. Humphreys, Editor, *Project and Cost Engineers Handbook*, 2nd ed., Dekker, New York, 1984.

J. A. Larkin, *Work Study: Theory and Practice*, McGraw-Hill, London, 1969.

R. Bruce McAfee and Richard N. Feinberg, *Cases in Construction Supervision*, Reston Publishing Co., Reston, Virginia, 1977.

Alfred P. McNulty, *Management of Small Construction Projects*, McGraw-Hill, New York, 1982.

James J. O'Brien, *Construction Inspection Handbook*, Van Nostrand Reinhold, New York, 1974.

Robert I. Patchin, *The Management and Maintenance of Quality Circles*, Dow Jones-Irwin, Homewood, Illinois, 1983.

Practical Accounting and Cost Keeping for Contractors, 9th ed., Frank P. Walker Company, Chicago, 1982.

Wallace R. Richardson, *Cost Improvement, Work Sampling and Short Interval Scheduling*, Reston Publishing Co., Reston, Virginia, 1976.

S. Peter Volpe, *Construction Management Practice*, Wiley-Interscience, New York, 1972.

I. E. Weeks, *Site Foremanship*, The Construction Press Ltd., Lancaster, England, 1978.

PROJECT STATUS, ANALYSIS AND REPORTING

Hira N. Ahuja, *Successful Construction Cost Control*, Wiley-Interscience, New York, 1980.

James M. Antill and Ronald W. Woodhead, *Critical Path Methods in Construction Practice,* 3rd ed., Wiley-Interscience, 1982.

ARTEMIS, Metier Management Systems, Inc., Houston, 1986.

Donald S. Barrie and Boyd C. Paulson, Jr., *Professional Construction Management*, 2nd ed., McGraw-Hill, New York, 1984.

James A. Bent, *Applied Cost and Schedule Control*, Dekker, New York, 1985.

Forrest D. Clark and A. B. Lorenzoni, *Applied Cost Engineering*, 2nd ed., Dekker, New York, 1985.

Richard H. Clough, *Construction Contracting*, 4th ed., Wiley-Interscience, New York, 1981.

Cost Control and CPM in Construction, The Associated General Contractors of America, Washington, D.C., 1968.

F. J. Drewin, *Construction Productivity*, Elsevier, New York, 1982.

Bill G. Eppes and Daniel E. Whiteman, *Cost Accounting for the Construction Firm*, Wiley-Interscience, New York, 1984.

Kenneth K. Guthrie and Associates, *Managing Capital Expenditures for Construction Projects*, Craftsman Book Company of America, Solana Beach, California, 1977.

Frank Harris and Ronald McCaffee, *Modern Construction Management*, Crosby Lockwood Staples, London, 1977.

Charles T. Horngren, *Cost Accounting: A Managerial Emphasis*, 3rd ed., Prentice-Hall, Englewood Cliffs, New Jersey, 1972.

Harold Kerzner, *Project Management: A Systems Approach to Planning, Scheduling and Controlling*, Van Nostrand Reinhold, New York, 1979.

J. A. Larkin, *Work Study: Theory and Practice*, McGraw-Hill, London, 1969.

James J. O'Brien, *Construction Inspection Handbook*, Van Nostrand Reinhold, New York, 1974.

Henry W. Parker and Clarkson H. Oglesby, *Methods Improvement For Construction Managers*, McGraw-Hill, New York, 1972.

Practical Accounting and Cost Keeping for Construction, 9th ed., Frank P. Walker Company, Chicago, 1982.

Primavera Project Planner, Primavera Systems Inc., Bala Cynwyd, Pennsylvania, 1986.

Project 2, Project Software and Development Inc., Cambridge, Massachusetts, 1984.

Wallace J. Richardson, *Cost Improvement, Work Sampling and Short Interval Scheduling*, Reston Publishing Co., Reston, Virginia, 1976.

Schedule Graphics, Ronald L. Schultz, Distributed by Success Software, Irvine, California.

S. Peter Volpe, *Construction Management Practice*, Wiley-Interscience, New York, 1972.

Richard E. Westney, *Managing the Engineering and Construction of Small Projects*, Dekker, New York, 1985.

Jack H. Willenbrock and H. Randolph Thomas, Eds, *Planning, Engineering and Construction of Electric Generation Facilities*, Wiley-Interscience, New York, 1980.

PROJECT MANAGEMENT CONTROL

Hira N. Ahuja, *Project Management: Techniques in Planning and Controlling Construction Projects*, Wiley-Interscience, New York, 1984.

Donald S. Barrie and Boyd C. Paulson, Jr., *Professional Construction Management*, 2nd ed., McGraw-Hill, New York, 1984.

James A. Bent, *Applied Cost and Schedule Control*, Dekker, New York, 1982.

John B. Bonny and Joseph P. Frein, *Handbook of Construction Management and Organization*, Van Nostrand Reinhold, New York, 1973.

Richard H. Clough, *Construction Contracting*, 4th ed., Wiley-Interscience, New York, 1981.

"Construction Management Guidelines For Use by AGC Members," Attachment to Special Contracting Methods Committee Report, The Associated General Contractors of America, Washington, D.C., February 15, 1972.

Philip Crosby, *Quality Is Free*, Mentor Books, New York, 1979.

Dexter C. Dunphy, *Organizational Change by Choice*, McGraw-Hill, Sydney, 1981.

Edward Fisk, *Construction Project Administration*, Wiley, New York, 1978.

Kenneth K. Guthrie and Associates, *Managing Capital Expenditures for Construction Projects*, Craftsman Book Company of America, Solana Beach, California, 1977.

Daniel W. Halpin and Ronald W. Woodhead, *Construction Management*, Wiley, New York, 1980.

John E. Johnston, *The Clerk of Works in the Construction Industry*, Crosby Lockwood Spales, London, 1975.

Harold Kerzner, *Project Management: A Systems Approach to Planning, Scheduling and Controlling*, Van Nostrand Reinhold, New York, 1979.

James J. O'Brien and Robert G. Zilly, (Eds.) *Contractor's Management Handbook*, McGraw-Hill, New York, 1971.

Leslie W. Rue and Lloyd L. Byars, *Management: Theory and Application*, 4th ed., Irwin, Homewood, Illinois, 1986.

W. J. Stillman, *Construction Practices for Project Managers and Supervisors*, Reston Publishing Company, Reston, Virginia, 1978.

Richard E. Westney, *Managing the Engineering and Construction of Small Projects*, Dekker, New York, 1985.

HISTORICAL DATA RETRIEVAL

"Applications of Small Computers in Construction," *Journal of Construction Engineering and Management*, Vol. III, No. 3, American Society of Civil Engineers, New York, September, 1985.

ARTEMIS, Metier Management Systems, Inc., Houston, 1986.

J. R. Britten, *Contractor's Management*, Building Research Station, Garston, Watford, 1969.

Forrest D. Clark, and A. B. Lorenzoni, *Applied Cost Engineering* 2nd ed., Dekker, New York, 1985.

dBase III Plus, Ashton-Tate, Culver City, California, 1985.

Arthur Paul Froemming, "Data Structure for a Construction Management System," University of Illinois, Urbana, Illinois, 1968.

A. Gilchrist and Kathleen Gaster, "Information Systems Relating to the Construction Industry," Building Research Station, Garston, Watford, 1969.

Kenneth M. Guthrie, *Process Plant Estimating and Control*, Craftsman Book Company of America, Solana Beach, California, 1974.

H. Murray Hohns, *Preventing and Solving Construction Contract Disputes*, Van Nostrand Reinhold, New York, 1979.

LOTUS 1-2-3, LOTUS Development Corporation, Cambridge, Massachusetts, 1983.

Henry C. Lucas, Jr. and Cyrus F. Gibson, *A Casebook for Management Information Systems*, McGraw-Hill, New York, 1976.

Practical Accounting and Cost Keeping For Construction, 9th ed., Frank P. Walker Company, Chicago, 1982.

Harry Rubey and Walter W. Milner, *Construction and Professional Management; An Introduction*, Macmillan, New York, 1956.

"A Study of Coding and Data Coordination for the Construction Industry," prepared by D. Bishop and K. Alsop for the Building Research Station," Ministry of Public Building and Works, Garston, Watford, 1969.

Michael Shave, *Data Structures*, McGraw-Hill, London, 1975.

Richard E. Westney, *Managing the Engineering and Construction of Small Projects*, Dekker, New York, 1985.

PROJECT MANAGEMENT OVERVIEW

Hira H. Ahuja, *Successful Construction Cost Control*, Wiley-Interscience, New York, 1980.

ARTEMIS, Metier Management Systems, Inc., Houston, 1986.

Donald C. Barrie and Boyd C. Paulson Jr., *Professional Construction Management*, 2nd ed., McGraw-Hill, New York, 1984.

Forrest D. Clark and A. B. Lorenzoni, *Applied Cost Engineering*, 2nd ed., Dekker, New York, 1985.

Richard Clough, *Construction Contracting*, 4th ed., Wiley-Interscience, New York, 1981.

Cost Control and CPM: A Manual for General Contractors, Associated General Contractors, Washington, D.C., 1968.

dBASE III Plus, Ashton-Tate, Culver City, California, 1985.

Stanley Goldhaber, Chandra K. Jha, and Manuel C. Macero, Jr., *Construction Management: Principles and Practice*, Wiley, New York, 1977.

Harold Kerzner, *Project Management: A Systems Approach to Planning, Scheduling and Controlling*, Van Nostrand Reinhold, New York, 1979.

Henry C. Lucas, Jr., and Cyrus F. Gibson, *A Case book for Management Information Systems*, McGraw-Hill, New York, 1976.

Edgar A. Pessemier, *New-Product Decisions: An Analytical Approach*, McGraw-Hill, 1966.

Ray Pilcher, *Appraisal and Control of Project Costs*, McGraw-Hill, 1973.

Ray Pilcher, *Principles of Construction Management*, McGraw-Hill, New York, 1967.

Linn C. Stuckenbruck, ed., *The Implementation of Project Management: The Professional's Handbook*, Addison-Wesley Publishing Co., Reading, Massachusetts, 1982.

Richard E. Westney, *Managing the Engineering and Construction of Small Projects*, Dekker, New York, 1985.

Index

This is not an ordinary index. It has the standard appearance, alphabetizing, and pagination format of a regular index but all the entries have a problem- and/or a problem-resolution focus. Most of the index is structured in terms of entries that refer to text material relevant to the causes of identification of, evaluation of, and solution approaches to a number of problems either specifically addressed, or relevant to, project manpower management. The only exceptions are those entries that locate, and scope, the twelve basic problem-solving decision processes, and their related general purpose problem-solving techniques. The unusual focus, and content, of the index has been prepared in the belief that it will encourage the user to develop innovative problem resolution decision processes, information gathering, and documentation procedures for those real life problems that beset the managers of construction projects.